OBJECTIVE FOOD SCIENCE & TECHNOLOGY

(3rd Revised & Enlarged Edition)

DEEPAK MUDGIL
Assistant Professor
Department of Dairy and Food Technology
MIDFT, Mehsana, Gujarat

&

SHEWETA BARAK MUDGIL
Assistant Professor
Department of Dairy and Food Technology
MIDFT, Mehsana, Gujarat

Published by:

SCIENTIFIC PUBLISHERS (INDIA)

5 A, New Pali Road, P.O. Box 91
Jodhpur - 342 001 INDIA
Phone : 0291-2433323
Fax : 0291-2624154
E-mail : info@scientificpub.com

Print: 2026

3rd Edition, 2019
2nd Edition, 2015
1st Edition, 2013

ISBN: 978-93-88172-86-8
eISBN: 978-93-88172-87-5

Visit the Scientific Publishers (India) website at:
http://www.scientificpub.com
Printed in India

Dedicated
to
our Family

MANSINHBHAI INSTITUTE OF DAIRY & FOOD TECHNOLOGY

FOREWORD

The urbanization & changes in life style of Indian population has given new dimension to food habits & food consumption pattern. This has generated demand of qualified professionals in Food Science & Technology. There are more entrants in the field of food science than it was in the past.

The authors of this book have tried to page objectively terminology & main aspects of Food Science & Technology which is quite handy for preparation for students in this field. This book will also serve purpose as a Hand-book & reference book for updated objective information on recent developments in the areas of Food Science & Technology.

I congratulate authors for achieving the purpose & wish them best for the future.

PRAMTHESH R PATEL
EXECUTIVE DIRCTOR
MIDFT, Mehsana

PREFACE

Food Industry in India is a sunshine industry and India has the capability to become the food basket of the world. Food technology is establishing as an emerging field as a career choice in developing as well as developed countries. In the last decade, many Indian educational organizations including university departments, institutions, colleges and deemed universities have started undergraduate, postgraduate and doctorate degree programmes related to Food Technology, Food Processing, Food Engineering and allied courses such as Dairy Technology, Post Harvest Technology, Foods and Nutrition, Agricultural and Food Process Engineering etc. Many of these institutes conduct entrance examination for admission in these courses. There are also some national level examinations conducted every year in the area of Food Science and Technology such as ASRB-NET examination, ICAR JRF-SRF examination and IIT GATE examination. Hence there was a great need of an objective type book covering all important topics of subjects related to Food Science and Technology. Hence, the first edition of this book was written to compensate the demand of students and to help them in making their career in the areas of Food Science and Technology. The first edition of Objective Food Science & Technology covered objective information of the subjects such as Food Chemistry, Food Microbiology, Food Engineering, Dairy Technology, Fruits and Vegetables Technology, Cereals Technology, Meat Fish and Poultry Processing, Food Additives, Foods and Nutrition, Bioprocess Technology, Food Packaging and Miscellaneous topics. The revised second edition of this book included four new chapters of food analysis, functional foods, emerging food processing technologies and food biochemistry.

Third revised edition of this book has included previous five years solved question papers of ICAR-NET Exam question papers. The subjective keynotes of each chapter has been revised and updated. There are around 1500 subjective keynotes in the third edition of this book. Hence the revised edition can serve as a single platform for preparation of competitive examinations in Food Science and Technology Discipline as it is having updated content of all areas of food science and technology. We wish to acknowledge authors and editors of all the books and research publications which helped us in the preparation of this book. The patience of Scientific Publishers (India) and the assistance provided by Scientific Publishers staff members are highly appreciated. Last but not least we must thank our family for their love, support and encouragement.

Deepak Mudgil
Sheweta Mudgil

PREFACE (2nd Edition)

In India, food technology is coming as an emerging field from career point of view. In the last decade, many Indian educational organizations including university departments, institutions, colleges and deemed universities have started undergraduate, postgraduate and doctorate degree programmes related to Food Technology, Food Processing, Food Engineering and allied courses such as Dairy Technology, Post Harvest Technology, Foods and Nutrition, Agricultural and Food Process Engineering etc. Many of these institutes conduct entrance examination for admission in these courses. There are also some national level examinations conducted every year in the area of Food Science and Technology such as ASRB-NET examination, ICAR JRF-SRF examination and IIT GATE examination. Hence there was a great need of an objective type book covering all important topics of subjects related to Food Science and Technology. Hence, the first edition of this book was written to compensate the demand of students and to help them in making their career in the areas of Food Science and Technology. The first edition of Objective Food Science & Technology covered objective information of the subjects such as Food Chemistry, Food Microbiology, Food Engineering, Dairy Technology, Fruits and Vegetables Technology, Cereals Technology, Meat Fish and Poultry Processing, Food Additives, Foods and Nutrition, Bioprocess Technology, Food Packaging and Miscellaneous topics. The revised second edition of this book included four new chapters of food analysis, functional foods, emerging food processing technologies and food biochemistry. Hence the revised edition can serve as a single platform for preparation of competitive examinations in Food Science and Technology Discipline as it is having updated content of all areas of food science and technology. We wish to acknowledge authors and editors of all the books and research publications which helped us in the preparation of this book. The patience of Scientific Publishers (India) and the assistance provided by Scientific Publishers staff members are highly appreciated. Last but not least we must thank our family for their love, support and encouragement.

Deepak Mudgil

Sheweta Mudgil

PREFACE (1st Edition)

In the last decade, many Indian educational organizations including university departments, institutions, colleges and deemed universities have started undergraduate, postgraduate and doctorate degree programmes related to Food Technology, Food Processing, Food Engineering and allied courses such as Dairy Technology, Post Harvest Technology, Foods and Nutrition, Agricultural and Food Process Engineering etc. Many of these institutes conduct entrance examination for admission in these courses. There are also some national level examinations conducted every year in the area of Food Science and Technology such as ASRB-NET examination, ICAR JRF-SRF examination and IIT GATE examination. Hence there is a great need of an objective type book covering all important topics of subjects related to Food Science and Technology. Hence, this book is written to compensate the demand of students and to help them in making their career in the areas of Food Science and Technology. This book includes the objective information of the subjects such as Food Chemistry, Food Microbiology, Food Engineering, Dairy Technology, Fruits and Vegetables Technology, Cereals Technology, Meat Fish and Poultry Processing, Food Additives, Foods and Nutrition, Bioprocess Technology, Food Packaging and Miscellaneous topics. Hence this book can serve as a single platform for preparation of competitive examinations in Food Science and Technology Discipline. We wish to acknowledge authors and editors of all the books and research publications which helped us in the preparation of this book. The patience of Scientific Publishers (India) and the assistance provided by Scientific Publishers staff members are highly appreciated. Also a special thanks to Prof. B.S. Khatkar and Sh. P.R. Patel for their support, inspiration and encouragement. Last but not least we must thank our family for their love, support and encouragement.

Deepak Mudgil

Sheweta Mudgil

CONTENTS

MEMORY BASED ASRB NET QUESTION PAPER

MEMORY BASED ASRB NET QUESTION PAPER

CHAPTER 1

FOOD CHEMISTRY

1.1 SUBJECTIVE KEYNOTES

1. Food chemistry, a major aspect of food science, deals with the composition and properties of food and the chemical changes it undergoes during handling, processing, and storage.
2. Carbohydrates, proteins and lipids are three major components of any food system.
3. Carbohydrates occur in plants, animals and microorganisms in various form such as sugars, starches, celluloses and other complex substances.
4. The general scientific term for polysaccharides is glycans and are of two types-homoglycan & hetroglycan
5. Carbohydrates provide 4 kcal per gram and are ideal food component for energy.
6. Carbohydrates are represented by Fischer projection formula, Haworth cyclic structure and conformational formula.
7. Carbohydrates are classified into three main groups– monosaccharides, oligosaccharides and polysaccharides.
8. Monosaccharides are categorised into aldoses (aldehyde group) and ketoses (ketone group).
9. Monosaccharides are designated as trioses (3C), tetroses (4C), pentoses (5C), hexoses (6C) etc.
10. Monosaccharide derivatives include amino sugars, deoxy sugars and glycosides.
11. Glucose is the most abundant monosaccharide with six carbon atoms. Glucose is the basic building blocks of starch and cellulose.
12. Glucose in blood is the source of energy for animals.
13. Glycogen is the storage polysaccharide in animals.
14. Glucose, fructose (fruit sugar), galactose, xylose (wood sugar) and ribose are included in monosaccharides.
15. Sucrose (table sugar), lactose (milk sugar), maltose (malt sugar), raffinose, stachyose, verbascose are included in oligosaccharides.
16. Invert sugar is produced by hydrolysis of sucrose in presence of mild acid or enzyme invertase.
17. Cellulose is the most abundant polysaccharide in biological system.

18. Starch, glycogen, cellulose, hemicelluloses, lignin, pectin, modified starches, carboxymethyl cellulose, chitin and arabinoxylans are included in polysaccharides.
19. Some of these polysaccharides which resist the action of intestinal secretion act as dietary fiber. Dietary fibers are of two types: soluble fiber and insoluble fiber.
20. Soluble dietary fibers include β-glucan, pectin, gums and mucilages.
21. Insoluble fiber includes cellulose, hemicelluloses and lignin.
22. Example of gums include xanthan gum, guar gum, locust bean gum, carrageenan, alginate etc.
23. Cereal starches generally contain 25 % amylose and 75% amylopectin.
24. Waxy starches (waxy corn starch) contain little or no amylose or 100% amylopectin.
25. Amylose contributes to the gel properties of cooked and cooled starches. Amylopectin is responsible for the thickened properties of starch preparations but it does not contribute to gel formation.
26. The temperature at which the starch granules start to swell rapidly and lose birefringence is called gelatinization temperature.
27. Main characteristics of carbohydrates are mutarotation, caramelization, and crystallization.
28. Mutarotation is change in optical rotation that occurs by epimerization.
29. Enantiomers are stereoisomers that are complete mirror images of each other, much as one's left and right hands are "the same" but opposite.
30. Configurations are the permanent geometry that results from the spatial arrangement of its bonds.
31. Conformations are the arrangement of the parts of an object.
32. Caramelization is also known as non-enzymatic browning.
33. Caramel formation occurs when sugars are subjected to heat in absence of water or heated or heated in concentrated solution leads to caramel formation.
34. Galactose, sucrose and glucose caramelize at 160°C whereas fructose at 110°C and maltose at 180°C.
35. Caramel has pungent taste and is often bitter. It is non-crystalline and is soluble ij water.
36. Crystallization is an important property of sugars which enables it to form crystals.
37. Maillard reaction is the reaction between amino group of protein and hydroxyl group of reducing sugar.
38. Sucrose is a non-reducing sugar.
39. Proteins are made up of about twenty different amino acids.
40. Amino acids in protein are joined together with peptide bonds.
41. Amino acids are grouped into polar and non-polar amino acids based on their polarity.
42. Wheat proteins such as gliadins and glutenins contain low levels of polar side chains or amino acids.
43. Animal products are utilized as protein source in non-vegetarian diets.

44. Legume products are utilized as protein source vegetarian diets.
45. Out of twenty amino acids, only nine amino acid acids are essential means these should be included in diets.
46. All amino acids (except glycine) have a chiral centre.
47. Albumin proteins are water and salt solution soluble. Globulin proteins are soluble in salt solution. Glutelin proteins are acid or alkali soluble. Prolamin proteins are alcohol soluble.
48. Iso-electric point of casein is 4.6
49. Proteins are generally divided into three groups: simple, conjugated and derived proteins.
50. Those proteins which yield only amino acids on hydrolysis are known as simple proteins such as albumin, globulin, glutelins, prolamins, scleroprotein, histones and protamines.
51. Conjugated proteins are those containing an amino acid combine with a non protein material such as lipid, nucleic acid or carbohydrate.
52. Examples of conjugated proteins are phosphoproteins, lipoproteins, nucleoproteins, glycoproteins and chromoproteins.
53. The lipids in lipoproteins are triglycerides, phospholipids, cholesterol or cholesterol derivatives.
54. Glycoproteins are the protein which contain heterosaccharides as prosthetic group such as hexosamine, galactose, mannose, fucose and sialic acid.
55. Metalloproteins are complexes of protein and heavy metals e.g. haemoglobin, myoglobin, ferritin, haemosiderin, conalbumin etc.
56. Derived proteins are obtained by obtained by chemical or enzymatic methods. Examples of derived proteins include proteoses, peptones and peptides.
57. Proteins yield 4.1 kcal energy per gram
58. Sources of proteins are meat, milk, eggs, pulses, beans, nuts etc.
59. Protein molecules exhibit four structural forms: primary, secondary, tertiary, quaternary.
60. Denaturation refers to changes in molecular structures of proteins.
61. Proteins are also involved in non-enzymatic browning i.e. Maillard reaction.
62. Important terms related to proteins are Biological Value (BV), Digestibility (D), Net Protein Utilization (NPU)
63. Functional properties of proteins include emulsifying, foaming, water absorption, binding and texture improvement.
64. Zein is a class of prolamin protein found in maize (corn).
65. Hordein is a prolamin glycoprotein, present in barley.
66. Oryzenin is a Glutelin protein found in rice grains.
67. Proteins related to legumins present in starchy endosperm of wheat are called triticins.

68. All of the prolamins of the Triticeae (wheat, barley and rye) are classify into three broad groups: sulphur - rich (S - rich), sulphur - poor (S - poor) and high molecular weight (HMW) prolamins, with several subgroups within the S - rich group.
69. Lipids are substances mostly made up of esters of fatty acids and glycerol.
70. Fats yield 9.0 kcal energy per gram
71. Fatty acids are generally classified into 3 groups: short chain, medium chain and long chain.
72. Fatty acids are of two types on the basis of presence of multiple bonds i.e. saturated and unsaturated.
73. Saturated fatty acids are those having single bonds between C-C.
74. Unsaturated fatty acids are those having double or triple bonds between C-C.
75. Examples of saturated fatty acids are butyric, caproic, caprylic, capric, lauric, myristic, palmitic, stearic, arachidic, valeric etc.
76. Examples of unsaturated fatty acids are myristoleic, palmitoleic, oleic, linoleic, linolenic, arachidonic, EPA, DHA etc.
77. Hydrogenation of fats consist addition of hydrogen at double bonds of unsaturated acyl groups.
78. Hydrogenation process is important as it converts oil into plastic fats for production of margarine.
79. Oxidative rancidity of fats refers to oxidation of fat which results in deterioration in flavor of fat.
80. Hydrolytic rancidity of fats refers to hydrolytic changes in fat resulting from enzyme activity.
81. Trans fats increases the serum cholesterol and thus contribute to heart disease.
82. Short-chain fatty acids (SCFA) are fatty acids with aliphatic tails of five or lesser carbons e.g. butyric acid.
83. Medium-chain fatty acids (MCFA) are fatty acids with aliphatic tails of 6 to 12 carbons.
84. Long-chain fatty acids (LCFA) are fatty acids with aliphatic tails of 13 to 21 carbons.
85. Acid value is the number of mg of KOH required to neutralize the Free Fatty Acids (FFA) in 1 g of fat. Acid vale (AV) expresses the number of free fatty acid (FFA) in oil.
86. Saponification is Hydrolysis of ester (Tri-acylglycerol) under alkaline condition.
87. Saponification value is number of mg of KOH required to saponify 1 g of fat. Higher SV means lower molecular weight of Tri-acylglycerol.
88. Ester value is mg of KOH required to saponify glyceride in 1 g of fat.
89. Iodine value is grams of iodine absorbed by 100 g of oil.
90. Peroxide value is number of ml of standard $Na_2S_2O_3$ required to titrate 1 g of fat or express by iodine content (%). It expresses the degree of oxidation.
91. Water is an essential constituent of food found in two forms: free and bound water.
92. On the basis of solubility in water, vitamins are of two types- water soluble (vitamin B & C), fat soluble (vitamin A, D, E & K).

93. Chemical names of vitamin A, B1, B2, B3, B6, B9, C, D2, D3, E, K are retinol, thiamin, riboflavin, niacin, pyridoxine, folic acid, ascorbic acid, ergocalciferol, cholecalciferol, tocopherol, phylloquinone, respectively.

94. Protopectin is found in immature fruits and is a high-molecular-weight methylated galacturonic acid polymer. It is insoluble in water.

95. Pectinic acid is a methylated form of galacturonic acid that is formed by enzymatic hydrolysis of protopectin as a fruit ripens. High-molecular-weight pectinic acids are known as pectins.

96. Pectic acid is a shorter-chain derivative of pectinic acid that is formed as fruit overripens.

97. The seaweed polysaccharides include the agars, alginates, and carrageenans.

98. Xanthan gum, gellan gum, dextran, and curdlan are the gums which are produced by microorganisms.

99. The water activity of a food is not similar to its moisture content. Water activity (a_w) refers to unbound and available water. It is the ratio of vapour pressure of food substrate to that of pure water.

100. Intermediate moisture foods (IMF) have water activity (a_w) values in the range of 0.65- 0.90.

1.2 MULTIPLE CHOICE QUESTIONS

1. Furanose is ring structure with
   (a) 5 member (b) 6 member
   (c) 4 member (d) 8 member

2. Pyranose is ring structure with
   (a) 4 member (b) 5 member
   (c) 6 member (d) 8 member

3. In D series, highest numbered a symmetric carbon has OH group directed towards
   (a) left (b) right
   (c) none (d) both

4. In L series, highest numbered a symmetric carbon has OH group directed towards
   (a) left (b) right
   (c) none (d) both

5. Most important monosaccharide in biological system is
   (a) fructose (b) glucose
   (c) galactose (d) xylose

6. At room temperature, an aqueous solution of glucose can exist in ________ tautomeric form

7. At room temperature, an aqueous solution of glucose can exist in ________ tautomeric form
   (a) one (b) two
   (c) three (d) four

8. Naturally occurring sugars are mostly
   (a) pentoses (b) hexoses
   (c) heptoses (d) octoses

9. Milk contain (%) total sugar about
   (a) 3.2 (b) 4.9
   (c) 5.9 (d) 7.0

10. Honey contains (%) total sugar about
    (a) 82 (b) 70
    (c) 60 (d) 72.5

11. Sweet corn contain total sugar (%) about
    (a) 12 (b) 35
    (c) 10 (d) 22

12. Potato contain total sugar (%) about
(a) 11 (b) 25
(c) 17 (d) 40

13. Apple contain total sugar (%) about
(a) 4.5 (b) 14.5
(c) 24.5 (d) 34.5

14. Grape contain total sugar (%) about
(a) 7.3 (b) 17.3
(c) 27.3 (d) 37.3

15. Carrot contain total sugar (%) about
(a) 10 (b) 20
(c) 25 (d) 30

16. Sweet potato contain total sugar (%) about
(a) 25 (b) 40
(c) 50 (d) 45

17. Maple syrup contain total sugar (%) about
(a) 45 (b) 55
(c) 65 (d) 80

18. Glycosides are stable to
(a) acid (b) enzyme
(c) alkali (d) all

19. Banana contains xylitol (mg/100g of dry weight)
(a) 11 (b) 21
(c) 31 (d) 41

20. Pears contains sorbitol (mg/100g of dry weight)
(a) 2600 (b) 3600
(c) 3600 (d) 4600

21. Raspberries contains xylitol (mg/100g of dry weight)
(a) 168 (b) 268
(c) 368 (d) 468

22. Strawberries contains xylitol (mg/100g of dry weight)
(a) 162 (b) 262
(c) 362 (d) 462

23. Peaches contains sorbitol (mg/100g of dry weight)
(a) 560 (b) 760
(c) 960 (d) 460

24. Cauliflower contains xylitol (mg/100g of dry weight)
(a) 100 (b) 200
(c) 300 (d) 400

25. White mushrooms contains mannitol (mg/100g of dry weight)
(a) 276 (b) 376
(c) 476 (d) 576

26. White mushrooms contains xylitol (mg/100g of dry weight)
(a) 186 (b) 128
(c) 286 (d) 228

27. White mushrooms contains galactitol (mg/100g of dry weight)
(a) 48 (b) 68
(c) 88 (d) 108

28. Oligosaccharides contain monosaccharide units of about
(a) one (b) 2-10
(c) 15-20 (d) >20

29. Sucrose is ———— in nature
(a) reducing (b) non-reducing
(c) both (d) none

30. Caramel is formed when sugar is heated (ºC) to
(a) 100 (b) 200
(c) 300 (d) 400

31. Which of the following is not an oligosaccharide
(a) sucrose (b) galactose
(c) lactose (d) maltose

32. Hydrolysis of sucrose results in formation of equal quantities of D-glucose and __________
(a) D-glucose (b) D-fructose
(c) galactose (d) lactose

33. Specific rotation (degree) of sucrose is
(a) +66.5 (b) –66.5
(c) +55.5 (d) –55.5

34. Specific rotation (degree) of D-glucose is
(a) +32.2 (b) –32.2
(c) +52.2 (d) –52.2

35. Staling of bread is related to
(a) gluten (b) starch
(c) albumins (d) none

36. __________ is known as milk sugar
(a) sucrose (b) lactose
(c) maltose (d) raffinose

37. Average lactose content (%) in cow milk is about
(a) 4.9 (b) 3.5
(c) 3 (d) 6

38. Average lactose content (%) in human milk is about
(a) 3 (b) 7
(c) 11 (d) 14

39. Lactose is hydrolyzed by
(a) α-galactosidase
(b) β-galactosidase
(c) α-mannase
(d) β-mannase

40. Maltose is __________ disaccharide
(a) Reducing (b) non-reducing
(c) both (d) none

41. Cellobiose is __________ disaccharide
(a) Reducing (b) non-reducing
(c) both (d) none

42. Raffinose and stachyose from legumes are responsible for
(a) Allergy (b) flatulence
(c) vomiting (d) diarrhea

43. Fructo-oligosaccharides are commercially manufactured by __________ enzyme on sucrose.
(a) α-fructofuranosidase
(b) β-fructofuranosidase
(c) α-mannase
(d) β-mannase

44. The rate of mutarotation increases __________ times for every 10°C rise in temperature
(a) 1.5-3.0 (b) 10-15
(c) 15-20 (d) 25

45. Formation of caramel can be considered non-enzymatic browning reaction in absence of
(a) C-compound
(b) N-compound
(c) glucose
(d) fructose

46. The pigment caramelan tastes
(a) sweet (b) sour
(c) bitter (d) salty

47. Melting point (°C) of caramelan is
(a) 112 (b) 120
(c) 138 (d) 175

48. Heating of caramelen results in formation of
(a) caramelan (b) caramelin
(c) glucose (d) fructose

49. Melting point (°C) of caramelen is
(a) 154 (b) 164
(c) 174 (d) 184

50. Caramelin is also known as
(a) caramelan (b) caramel
(c) humin (d) none

51. In commercial production, crystallization of sugar is an important step for
(a) neutralization
(b) purification
(c) sweetening
(d) all

52. Which of the following affects growth of sucrose crystals
(a) temperature
(b) relative velocity of crystal
(c) super saturation
(d) all

53. Which of the following affects growth of sucrose crystals
(a) Nature of impurities
(b) relative velocity of solution
(c) concentration of impurities
(d) all

54. In crystal growth of sucrose, mass transfer of sucrose molecules to surface of crystal is
(a) First order process
(b) Second order process
(c) Zero order process
(d) None

55. In crystal growth of sucrose, incorporation of molecules to the surface of crystal is
(a) First order process
(b) Second order process
(c) Zero order process
(d) None

56. Lactose can occur in __________ crystalline form
(a) One (b) two
(c) three (d) four

57. Which of the following is/are crystalline form of lactose
(a) α-hydrate (b) β-anhydrous
(c) both (d) none

58. α-hydrate form of lactose has a melting point (°C) of about
(a) 102 (b) 152
(c) 202 (d) 252

59. Melting point (°C) of β-anhydrous form of lactose is about
(a) 102 (b) 152
(c) 202 (d) 252

60. Specific rotation (degree) of α-hydrate form of lactose is
(a) +79.4 (b) -79.4
(c) +89.4 (d) -89.4

61. Specific rotation (degree) of β-anhydrous form of lactose is
(a) +79.4 (b) +35
(c) –79.4 (d) +52.5

62. Solubility (g/100ml) of α-hydrate form of lactose in water at 20 °C
(a) 8 (b) 16
(c) 32 (d) 55

63. Solubility (g/100ml) of β-anhydrous form of lactose in water at 20 °C
(a) 8 (b) 16
(c) 32 (d) 55

64. Specific gravity of α-hydrate form of lactose at 20°C
(a) 1.34 (b) 1.38
(c) 1.54 (d) 1.59

65. Specific gravity of β-anhydrous form of lactose at 20°C
(a) 1.34 (b) 1.38
(c) 1.54 (d) 1.59

66. Specific heat of α-hydrate form of lactose is ______ than β-anhydrous form
(a) higher (b) lower
(c) equal (d) none

67. Starch can be hydrolyzed by
(a) acid (b) enzyme
(c) both (d) none

68. Which of the following is not a source of starch
(a) corn (b) wheat
(c) potato (d) guar gum

69. Hydrolysis product of starch is
(a) Glucose monomer
(b) glucose dimer
(c) oligosaccharide
(d) all

70. Enzyme used for starch hydrolysis is
(a) protease
(b) amylase
(c) galactosidase
(d) cellulase

71. Endo-α-amylase attacks on
(a) α -1,4 glycosidic bond
(b) α -1,6 glycosidic bond
(c) β -1,4 glycosidic bond
(d) β -1,6 glycosidic bond

72. Isomerization of glucose to fructose is done with
(a) glucose isomerase
(b) galacto isomerase
(c) mannose isomerase
(d) none

73. Polyols are produced industrially from corresponding saccharide by
(a) Catalytic hydration
(b) catalytic hydrogenation
(c) catalytic oxidation
(d) none

74. Sorbitol is found in
(a) plums (b) apples
(c) pears (d) all

75. Sorbitol is commercially produced by catalytic hydrogenation of
(a) D-mannose (b) D-glucose
(c) D-galactose (d) all

76. Mannitol is commercially produced by catalytic hydrogenation of
(a) D-mannose (b) D-glucose
(c) D-galactose (d) all

77. Mannitol is found in
(a) mushrooms (b) celery
(c) olives (d) all

78. Disaccharide alcohols include
(a) isomalt (b) maltitol
(c) lactitol (d) all

79. Maltitol is hydrogenated form of
(a) mannose (b) maltose
(c) glucose (d) all

80. Which of the following has highest sweetness compared to sugar
(a) isomalt (b) maltitol
(c) lactitol (d) none

81. Lactitol is hydrogenated form of
(a) lactic acid (b) lactose
(c) maltose (d) glucose

82. Isomalt is also known as
(a) hydrogenated isomaltose
(b) hydrogenated palatinose
(c) both
(d) none

83. Isomalt consists of
(a) D-glucopyranosyl- α-(1-1)-D-mannitol
(b) D-glucopyranosyl- α-(1-6)-D-sorbitol
(c) both
(d) none

84. Which of the following is used in cereals quality laboratory
(a) Falling number apparatus
(b) alveogarph
(c) SKCS
(d) all

85. Starch is a polymer made up of monomer units of
(a) galactose (b) sucrose
(c) glucose (d) fructose

86. Starch is found as storage carbohydrate in plants in form of
(a) grains (b) granules
(c) crystal (d) none

87. Granular structure of starch can be analyzed using
(a) ordinary microscope
(b) polarized light microscopy
(c) X-ray diffraction
(d) all

88. Starch molecule is composed of
(a) amylase (b) amylopectin
(c) both (d) none

89. Relative sweetness of xylitol with respect to sucrose solution at 20°C is
(a) 50-60 (b) 60-70
(c) 70-80 (d) 80-100

90. Relative sweetness of sorbitol with respect to sucrose solution at 20°C is
(a) 50-60 (b) 60-70
(c) 70-80 (d) 80-100

91. Relative sweetness of mannitol with respect to sucrose solution at 20°C is
(a) 60-65 (b) 50-60
(c) 70-80 (d) 80-100

92. Relative sweetness of lactitol with respect to sucrose solution at 20°C is
(a) 30-40 (b) 40-50
(c) 50-60 (d) 60-70

93. Relative sweetness of maltitol with respect to sucrose solution at 20°C is
(a) 50-60 (b) 60-70
(c) 70-80 (d) 80-90

94. Relative sweetness of isomalt with respect to sucrose solution at 20°C is
(a) 50-60 (b) 60-70
(c) 70-80 (d) 80-100

95. The test used for detection of starch
(a) glucose test (b) sodium test
(c) iodine test (d) salt test

96. Starch granules are
(a) crystalline
(b) amylopectin
(c) partly crystalline
(d) none

97. Waxy starch contains
(a) amylose (b) amylopectin
(c) both (d) none

98. Potato starch granules have a diameter range (micron)
(a) 100-200 (b) 200-250
(c) 15-100 (d) 400

99. Wheat starch granules have a diameter range (micron)
(a) 50-100 (b) 154-165
(c) 2-35 (d) 70-86

100. The temperature at which granules swell in aqueous medium is known as __________ temperature
(a) crystallization
(b) caramelization
(c) gelatinization
(d) none

101. Gelatinization temperature also exhibit
(a) Optical birefringence disappearance
(b) loss of crystallinity
(c) swelling
(d) all

102. Starch with large granules swell at __________ temperature than those with small granules
(a) higher (b) lower
(c) same (d) 0 °C

103. Potato starch generally swells at
(a) 49-55 °C (b) 55-60 °C
(c) 64-72 °C (d) 90-95 °C

104. Amylase acts on
(a) protein (b) carbohydrate
(c) fat (d) minerals

105. Swelling temperature of starch is influenced by
(a) pH (b) heating rate
(c) salt & sugar (d) all

106. Retrogradation is related to
(a) protein (b) fat
(c) starch (d) none

107. Cereal starches contain fatty acids at levels (%) of about
(a) 0.1-0.2 (b) 0.5-0.7
(c) 1.2-1.5 (d) 1.5-1.8

108. Starch is composed of
(a) amylose (b) amylopectin
(c) both (d) none

109. Staling of bread occurs due to
(a) saccharification
(b) retrogradation
(c) caramelization
(d) browning

110. Rate of staling is mainly ________ dependent
(a) pH (b) salt
(c) temperature (d) all

111. Retrogradation is ________ at low temperature
(a) slower (b) faster
(c) intermediate (d) none

112. Bread stales quickly in ________ temperature
(a) room (b) elevated
(c) refrigerated (d) none

113. Freezing condition ________ the retrogradation
(a) retard (b) prevent
(c) speed up (d) do not affect

114. Which of the following is not a cereal starch
(a) corn (b) sorghum
(c) rice (d) cassava

115. Which of the following is not a root or tuber starch
(a) potato (b) cassava
(c) tapioca (d) millet

116. Which of the following starch paste forms clear gel on cooling
(a) corn (b) wheat
(c) rice (d) cassava

117. Which of the following starch paste forms opaque gel on cooling
(a) potato (b) tapioca
(c) cassava (d) sorghum

118. Waxy starches shows property of
(a) clear paste
(b) heavy body paste
(c) both
(d) none

119. High amylase starch require ________ temperature for gelatinization
(a) high (b) low
(c) room temperature (d) none

120. Which of the following is not a property of high amylase starch gel
(a) firm (b) opaque
(c) short bodied (d) none

121. Starches are used in food products due to their
(a) viscosity (b) gel formation
(c) cohesiveness (d) all

122. At gelatinization temperature, viscosity of starch gel is
(a) maximum (b) minimum
(c) do not change (d) none

123. Modified starches can be prepared by treatment with
(a) acid (b) enzyme
(c) oxidation (d) all

124. Modified starches can be prepared by
(a) cross bonding (b) substitution
(c) heat (d) all

125. Acid treated starches shows ________ viscosity
(a) medium (b) low
(c) high (d) zero

126. Acid conversion of starch is carried out with
(a) hydrochloric acid
(b) sulfuric acid (c) both
(d) none

127. Example of substituted starch is
(a) starch acetate
(b) starch monophosphates
(c) hydroxyl propyl starch
(d) all

128. Acetylation of starch is carried out with
(a) acetic anhydride
(b) vinyl acetate (c) both
(d) none

129. Starch monophosphates give dispersions with
(a) higher viscosity (b) better clarity
(c) better stability (d) all

130. Oxidised starch is prepared by treating starch with
(a) chlorine (b) hypochlorite
(c) iodine (d) hydrogen

131. Modified starches is related to ________ of starch
(a) oxidation (b) esterification
(c) etherification (d) all

132. Etherification of starches leads to
(a) stabilization
(b) low temperature storage
(c) both
(d) none

133. Cross linking of starches leads to
(a) thickening (b) stabilization
(c) texturizing (d) all

134. Oxidation of starches leads to
(a) stabilization
(b) adhesion gelling
(c) clarification
(d) all

135. Dextrins is used as ________ in food products
(a) coating agent (b) binder
(c) both (d) none

136. The monomer unit of glycogen is
(a) galactose (b) glucose
(c) mannose (d) fructose

137. In glycogen molecules, branching chains are linked by
(a) α 1-4 linkage (b) β 1-4 linkage
(c) α 1-6 linkage (d) none

138. Glycogen gives ________ color with iodine
(a) black (b) green
(c) reddish brown (d) purple

139. Cellulose is polymer made up of
(a) glucose (b) galactose
(c) mannose (d) all

140. Glucose molecules are linked with ____ linkage in cellulose
(a) α 1-4 (b) β 1-4
(c) α 1-6 (d) β 1-6

141. High degree of crystallinity in cellulose molecules results in
(a) high elastic modulus
(b) high tensile strength
(c) both (d) none

142. Cellulose generally have a degree of polymerization of
(a) 1000 (b) 10000
(c) 2000 (d) 20000

143. Crystalline nature of cellulose can be checked by
(a) polarizing microscope
(b) X-ray diffraction
(c) both (d) none

144. Unit cell of cellulose crystal consist of ________ cellobiose units
(a) one (b) two
(c) three (d) four

145. Which of the following is non starch polysaccharide
(a) hemicelluloses
(b) pentosans
(c) galactomannan
(d) all

146. Hemicellulose is the ________ polysaccharide
(a) water insoluble (b) non starch
(c) both (d) none

147. Pentosan refer to ________ polysaccharide
(a) water soluble (b) non starch
(c) both (d) none

148. Hemicellulose refer to ________ polysaccharide
(a) water insoluble (b) non starch
(c) both (d) none

149. Pentosan refer to ________ polysaccharide
(a) water insoluble (b) non starch
(c) both (d) none

150. Which of the following is hemicellulose
(a) xylan (b) mannan
(c) galactan (d) all

151. Xylan consists of
(a) mannose (b) xylose
(c) both (d) none

152. Mannans consist of
(a) galactose (b) mannose
(c) both (d) none

153. Hemicellulose of wheat bran consist about ____% carbohydrate
(a) 23 (b) 33
(c) 43 (d) 53

154. Which of the following is present in wheat bran
(a) L-arabinose (b) D-xylose
(c) D-galacturonic acid
(d) all

155. Hemicelluose of wheat bran has about ____________ degree of polymerization
(a) 100 (b) 300
(c) 500 (d) 800

156. Wheat endosperm contains about ________% hemicelluloses
(a) 2.4 (b) 4.5
(c) 6.7 (d) 8.2

157. Wheat endosperm hemicelluloses contains
(a) L-arabinose (b) D-xylose
(c) D-glucose (d) all

158. Water soluble pentosans occurs in wheat flour at the level of ______%
(a) 1-1.5 (b) 2-3
(c) 4-5 (d) 7-8

159. Water soluble pentosans of wheat are
(a) highly branched
(b) highly viscous
(c) gel forming
(d) all

160. Water soluble pentosans of wheat contain
(a) arabinose (b) xylose
(c) both (d) none

161. Which of the following is not a polysaccharide
(a) cellulose (b) starch
(c) glycogen (d) none

162. Which of the following is not a source of dietary fiber
(a) cellulose (b) glucose
(c) hemicelluloses (d) lignin

163. Which of the following statement is true
(a) lignin provides mechanical support
(b) lignin is present in mature plant cells
c) lignin provide resistance to microbial degradation
(d) all

164. Lignin is always associated in cell walls with
(a) cellulose (b) hemicelluloses
(c) both (d) none

165. Lignin occurs in
(a) plant cell wall (b) wood
(c) both (d) none

166. Cyclodextrin may contain ________ glucose units
(a) 6 (b) 7
(c) 8 (d) all

167. Cyclic polymers obtained by enzymatic treatment of starch is
(a) cyclodextrose (b) cyclodextrin
(c) cellulose (d) hemicellulose

168. Enzyme used in preparation of cyclodextrin is
(a) glycosyl transferase
(b) lactase
(c) amylase
(d) cellulose

169. Which of the following is the property of cyclodextrins
(a) hollow cavity
(b) hydrophobicity
(c) both
(d) none

170. Cyclodextrin is used in
(a) enzyme preparation
(b) encapsulation
(c) retrogradation
(d) all

171. Polydextrose is a randomly bonded condensation polymer of
(a) fructose (b) glucose
(c) maltose (d) starch

172. Which of the following is used in preparation of polydextrose
(a) sorbitol (b) citric acid
(c) both (d) none

173. Polydextrose can be used for
(a) bulkiness (b) no sweetness
(c) both (d) none

174. Pectic substances form ________ on hydrolysis
(a) chitin (b) pectin
(c) galactose (d) sugar

175. Pectin forms gel in the presence of
(a) calcium (b) sugar
(c) acid (d) all

176. Properties of pectin gel depends on
(a) degree of polymerization
(b) nature of side chain
(c) degree of methylation
(d) all

177. Low methoxyl pectin has ________% degree of methylation
(a) 25-50 (b) 0-20
(c) 50-80 (d) 80-90

178. High methoxyl pectin has ________% degree of methylation
(a) 0-20 (b) 25-50
(c) 50-80 (d) 80-90

179. Low methoxyl pectin forms
(a) calcium gel (b) acid gel
(c) both (d) none

180. High methoxyl pectin forms
(a) calcium gel (b) acid gel
(c) both (d) none

181. Commercially pectin require ______ pH for gel formation
(a) 1-2 (b) 2.1-2.5
(c) 3.2-3.5 (d) 5.0-5.5

182. Commercially pectin require ______ % sugar content for gel formation
(a) 40-50 (b) 55-60
(c) 65-70 (d) 80-85

183. Commercially pectin concentration used for gel formation is
(a) 0.2-1.5% (b) 1.8-2.5%
(c) 2.5-3.0% (d) 4.5%

184. Slow-set pectin has a degree of methoxylation (%)
(a) 20-30 (b) 30-50
(c) 50-70 (d) 80-90

185. Polysaccharide that give highly viscous solution at low concentration
(a) pectin (b) starch
(c) gum (d) cellulose

186. Gums are used in food industries as ________ agent
(a) gelling (b) stabilizing
(c) suspending (d) all

187. Galactomannan can be obtained from
(a) guar (b) locust bean
(c) tara (d) all

188. Alginates can be obtained from
(a) Macrocystic (b) Ascophyllum
(c) Laminaria (d) all

189. Xanthan gum is a type of
(a) seed gum
(b) seaweed gum
(c) microbial gum
(d) exudate gum

190. Carrageenan is obtained from
(a) Gracilaria (b) Eucheuma
(c) Gigartina (d) all

191. Polysaccharide with strong acid group is
(a) furcellaran (b) Carrageenan
(c) both (d) none

192. Pectin is used in making
(a) jam (b) jelly
(c) confectionery (d) all

193. CMC is used as stabilizer and thickener in
(a) ice cream (b) cake mixes
(c) meat products (d) all

194. Gum arabic is a type of
(a) seed gum
(b) microbial gum
(c) exudate gum (d) seaweed gum

195. Gum acacia is also known as
(a) guar gum
(b) gum tragacanth
(c) gum arabic (d) xanthan gum

196. Gum arabic may contain which of the following anions
(a) calcium (b) magnesium
(c) potassium (d) all

197. Gum arabic consists of
(a) L-arbinose (b) L-rhamnose
(c) D-galactose (d) all

198. Gum arabic generally has a molecular weight of about
(a) 30000 (b) 300000
(c) 800000 (d) 1000000

199. Gum arabic form coacervates with
(a) starch (b) glucose
(c) protein (d) lipid

200. Locust bean gum is also known as
(a) tamarind gum
(b) carob bean gum
(c) guar gum
(d) none

201. Commercial locust bean gum contains ______ % of galactomannan
(a) 50 (b) 60
(c) 70 (d) 88

202. Which of the following gum was used to adulterate locust bean gum
(a) xanthan (b) guar gum
(c) carrageenan (d) alginate

203. Guar gum is also known as
(a) cluster bean gum
(b) carob gum
(c) gum arabic (d) none

204. In guar gum molecule, mannose units are linked with ______ linkage
(a) β 1-4 (b) α 1-4
(c) α 1-6 (d) β 1-6

205. In guar gum molecule, galactose units are linked with mannose units by
(a) α 1-4 (b) β 1-4
(c) α 1-6 (d) β 1-6

206. Ratio of galactose to mannose in guar gum is
(a) 1:2 (b) 1:3
(c) 1:4 (d) 2:2

207. 1% aqueous solution of native guar gum show viscosity (cps) of about
(a) 1000 (b) 6000
(c) 15000 (d) 20000

208. Agar is extracted from
(a) algae (b) fungi
(c) bacteria (d) crops

209. Which of the following is true
(a) agar is soluble in boiling water
(b) agar is soluble in cold water
(c) both
(d) none

210. Agar generally contains
(a) agarose (b) agaropectin
(c) both (d) none

211. Agar is obtained from
(a) red sea weed (b) plants
(c) both (d) none

212. Alginic acid contains
(a) mannuronic acid
(b) D-glucuronic acid
(c) both (d) none

213. Most common form of alginic acid is
(a) calcium alginate
(b) sodium alginate
(c) potassium alginate
(d) none

214. Which of the following property is exhibited by algin
(a) thickening (b) suspending
(c) film forming (d) all

215. Algin is soluble in
(a) hot water (b) cold water
(c) both (d) none

216. Carrageenan is a ________ gum
(a) seed (b) seaweed
(c) exudate (d) microbial

217. Which of the following is a type of carrageenan
(a) kappa (b) lambda
(c) iota (d) all

218. Carrageenan is used as suspending agent in chocolate milk at a level (%) of
(a) 0.5 (b) 0.03
(c) 1.0 (d) 1.75

219. Which of the following is modified cellulose
(a) hydroxymethyl cellulose
(b) hydroxypropyl cellulose
(c) carboxymethyl cellulose
(d) all

220. Methyl cellulose is soluble in water at ________ degree of substitution
(a) 1.3-2.6 (b) 4.5-7.5
(c) 13-26 (d) 10-12

221. Methyl cellulose is soluble in
(a) hot water (b) cold water
(c) both (d) none

222. Crude fiber is the residue remaining after
(a) acid extraction
(b) alkali extraction
(c) both
(d) none

223. Dietary fiber are group of plant substances resistant to
(a) intestinal enzyme
(b) hormones
(c) acids
(d) alkali

224. Which of the following is not an insoluble fiber
(a) cellulose (b) hemicelluloses
(c) pectin (d) lignin

225. TDF estimation is based on
(a) enzymatic digestion
(b) acid digestion
(c) alkali digestion
(d) none

226. Crude fiber content of cellulose is ________ (g/100g dry weight)
(a) 5.5 (b) 10.5
(c) 25.5 (d) 72.5

227. TDF means
(a) Total Dietary Fiber
(b) Total Day's Fiber
(c) Total Dairy Fat
(d) none

228. Pea hulls generally contains ________ (g/100g dry weight) crude fiber
(a) 16 (b) 26
(c) 36 (d) 56

229. Pea hulls generally contains ________ (g/100g dry weight) TDF
(a) 21 (b) 31
(c) 51 (d) 71

230. Corn bran generally contains ______ (g/100g dry weight) crude fiber
(a) 9 (b) 19
(c) 39 (d) 49

231. Corn bran generally contain TDF content of about ________ %
(a) 58 (b) 68
(c) 72 (d) 88

232. Wheat bran generally contains _____ (g/100g dry weight) crude fiber
(a) 8.7 (b) 18.7
(c) 28.7 (d) 38.7

233. Wheat bran generally contains _____ (g/100g dry weight) TDF
(a) 29.8 (b) 39.8
(c) 49.8 (d) 59.8

234. Citrus pulp generally contains _____ (g/100g dry weight) crude fiber
(a) 4.4 (b) 14.4
(c) 24.4 (d) 34.4

235. Citrus pulp generally contains _____ (g/100g dry weight) TDF
(a) 4.8 (b) 14.8
(c) 24.8 (d) 34.8

236. Dried apple generally contains ______ (g/100g dry weight) TDF
(a) 10 (b) 20
(c) 30 (d) 40

237. Citrus peel contains ________ (g/100g dry weight) TDF
(a) 10-20 (b) 30-65
(c) 65-80 (d) >90

238. Potato powder generally contains ______ (g/100g dry weight) TDF
(a) 7.6 (b) 17.6
(c) 27.6 (d) 37.6

239. Soya flour generally contains _______ (g/100g dry weight) TDF
(a) 11.1 (b) 21.1
(c) 31.1 (d) 41.1

240. Tomatoes generally contains _______ (g/100g dry weight) TDF
(a) 19.6 (b) 29.6
(c) 39.6 (d) 49.6

241. Lettuce contains ______ (g/100g dry weight) TDF
(a) 16 (b) 26
(c) 36 (d) 46

242. Carrots contains ______ (g/100g dry weight) TDF
(a) 13.9 (b) 23.9
(c) 33.9 (d) 43.9

243. Mushrooms contains ______ (g/100g dry weight) TDF
(a) 19.2 (b) 29.2
(c) 39.2 (d) 49.2

244. Green peas contains ______ (g/100g dry weight) TDF
(a) 14.6 (b) 24.6
(c) 34.6 (d) 44.6

245. Apples generally contains ________ (g/100g dry weight) TDF
(a) 4.7 (b) 14.7
(c) 24.7 (d) 28.9

246. Strawberries generally contains ______ (g/100g dry weight) TDF
(a) 4.2 (b) 14.2
(c) 24.2 (d) 34.2

247. Wheat bran contain approximately ________ % protein
(a) 4.3 (b) 14.3
(c) 24.3 (d) 30.5

248. Wheat bran contain approximately ________ % lipid
(a) 5.2 (b) 15.2
(c) 25.2 (d) 35.2

249. Wheat bran contain approximately ________ % phytic acid
(a) 3.4 (b) 0.6
(c) 13.4 (d) 15.8

250. Wheat bran contain approximately ________ % lignin
(a) 3.2 (b) 10.8
(c) 14.5 (d) 19.8

251. Wheat bran contain approximately ______% ash
(a) 5.1 (b) 10.1
(c) 15.1 (d) 20.1

252. Which is not a type of fiber
(a) crude fiber
(b) acid detergent fiber
(c) neutral detergent fiber
(d) cellulose

253. Dietary fiber include which of the following
(a) structural polysaccharide
(b) structural non-polysaccharide
(c) non- structural polysaccharide
(d) all

254. Which of the following is not included in pentoses
(a) ribose (b) ribulose
(c) mannose (d) ribose

255. Which of the following is not included in hexoses
(a) arabinose (b) galactose
(c) mannose (d) fructose

256. Which of the following is not included in trioses
(a) glyceraldehydes
(b) dihydroxyacetone
(c) both
(d) none

257. Ribose is a structural element of
(a) ATP (b) nucleic acid
(c) coenzymes (d) all

258. Amino sugar includes which of the following
(a) glucosamine (b) galactosamine
(c) mannosamine (d) all

259. Sucrose is also termed as
(a) cane sugar (b) beet sugar
(c) both (d) none

260. Sucrose is obtained from
(a) sugar cane (b) sugar beet
(c) both (d) none

261. Melting point of lactose is about ______°C
(a) 103 (b) 203
(c) 353 (d) 303

262. Which of the following is source of trehalose
(a) insect haemolymph
(b) mushroom
(c) yeast (d) all

263. Melting point of trehalose is ______°C
(a) 110 (b) 210
(c) 150 (d) 310

264. Maltose give ______ molecules of glucose upon hydrolysis
(a) 2 (b) 3
(c) 4 (d) 5

265. Upon hydrolysis, lactose yields
(a) glucose (b) galactose
(c) both (d) none

266. Upon hydrolysis, sucrose yields
(a) glucose (b) fructose
(c) both (d) none

267. Amylose is a ______ chained polysaccharide
(a) straight (b) branched
(c) highly branched (d) all

268. Amylopectin is a ______ chained polysaccharide
(a) straight (b) branched
(c) both (d) none

269. Amylose contains about ______ glucose units
(a) 2500-3000 (b) 250-300
(c) 1000-1500 (d) 100

270. Amylopectin contains about ______ glucose units
(a) 100 (b) 250
(c) 1000 (d) 10000

271. Enzyme which react with starches is
(a) α- amylase (b) β-amylase
(c) both (d) none

272. α-amylase is also known as ______ amylase
(a) liquefying (b) dextrogenic
(c) both (d) none

273. α-amylase is also known as ______ amylase
(a) liquefying (b) dextrogenic
(c) saccharifying (d) none

274. Diastase is commercially extracted from
(a) wheat (b) barley
(c) rice (d) sorghum

275. Amylose content in tapioca starch is ______%
(a) 8 (b) 18
(c) 28 (d) 38

276. Amylose content in barley starch is ______%
(a) 7 (b) 11
(c) 27 (d) 47

277. Amylose content in buckwheat starch is ______%
(a) 7 (b) 11
(c) 28 (d) 0.5

278. Pectic substances include
(a) protopectin (b) pectinic acid
(c) pectic acid (d) all

279. Baker's yeast produce little or no CO2 with
(a) glucose (b) fructose
(c) maltose (d) lactose

280. Instrument used to measure optical activity is
(a) optical meter (b) polarimeter
(c) tintometer (d) all

281. Optical activity is related to
(a) specific rotation
(b) optical density
(c) surface tension (d) viscosity

282. Dextrinization starch occurs on ______ portion of bread
(a) crust (b) crumb
(c) air cells (d) none

283. Which of the following is not a pentose
(a) arabinose (b) xylose
(c) rhammose (d) ribose

284. Which of the following is not disaccharide
(a) cellobiose (b) gentiobiose
(c) isomaltose (d) panose

285. Cellobiose is a
(a) disaccharide
(b) trisaccharide
(c) tetrasaccharide
(d) polysaccharide

286. Gentiobiose is a
(a) disaccharide
(b) trisaccharide
(c) tetrasaccharide
(d) polysaccharide

287. Fucosidolactose is a
(a) disaccharide
(b) trisaccharide
(c) tetrasaccharide
(d) polysaccharide

288. Isomaltose is a
(a) disaccharide
(b) trisaccharide
(c) tetrasaccharide
(d) polysaccharide

289. Lactulose is a
(a) disaccharide
(b) trisaccharide
(c) tetrasaccharide
(d) polysaccharide

290. Stachyose is a
(a) disaccharide
(b) trisaccharide
(c) tetrasaccharide
(d) polysaccharide

291. Maltulose is a
(a) disaccharide
(b) trisaccharide
(c) tetrasaccharide
(d) polysaccharide

292. Melibiose is a
(a) disaccharide
(b) trisaccharide
(c) tetrasaccharide
(d) polysaccharide

293. Gentianose is a
(a) disaccharide
(b) trisaccharide
(c) tetrasaccharide
(d) polysaccharide

294. Neotrehalose is a
(a) disaccharide
(b) trisaccharide
(c) tetrasaccharide
(d) polysaccharide

295. Nigerose is a
(a) disaccharide
(b) trisaccharide
(c) tetrasaccharide
(d) polysaccharide

296. Isokestose is a
(a) disaccharide
(b) trisaccharide
(c) tetrasaccharide
(d) polysaccharide

297. Kestose is a
(a) disaccharide
(b) trisaccharide
(c) tetrasaccharide
(d) polysaccharide

298. Palatinose is a
(a) disaccharide
(b) trisaccharide
(c) tetrasaccharide
(d) polysaccharide

299. Rutinose is a
(a) disaccharide
(b) trisaccharide
(c) tetrasaccharide
(d) polysaccharide

300. Maltotriose is a
(a) disaccharide
(b) trisaccharide
(c) tetrasaccharide
(d) polysaccharide

301. Maltotetraose is a
(a) disaccharide
(b) trisaccharide
(c) tetrasaccharide
(d) polysaccharide

302. Trehalose is a
(a) disaccharide
(b) trisaccharide
(c) tetrasaccharide
(d) polysaccharide

303. Panose is a
(a) disaccharide
(b) trisaccharide
(c) tetrasaccharide
(d) polysaccharide

304. Raffinose is a
(a) disaccharide
(b) trisaccharide
(c) tetrasaccharide
(d) polysaccharide

305. Umbelliferose is a
(a) disaccharide
(b) trisaccharide
(c) tetrasaccharide
(d) polysaccharide

306. Sophorose is a
(a) disaccharide
(b) trisaccharide
(c) tetrasaccharide
(d) polysaccharide

307. Gum ghatti is a type of
(a) seed gum (b) exudate gum
(c) microbial gum (d) seaweed gum

308. Gum tragacanth is a type of ________ gum
(a) seed gum (b) exudate gum
(c) microbial gum (d) seaweed gum

309. Karaya gum is a type of ________ gum
(a) seed gum (b) exudate gum
(c) microbial gum (d) seaweed gum

310. Diameter of wheat starch granule is about ______ (microns)
(a) 2-38 (b) 0.5
(c) 40-50 (d) 54-58

311. Diameter of rice starch granule is about ______(microns)
(a) 3-8 (b) 12-15
(c) 17-31 (d) 35-60

312. Diameter of potato starch granule is about ______ (microns)
(a) 15-100 (b) 10
(c) 100-150 (d) 140

313. Diameter of barley starch granule is about ______ (microns)
(a) 2-5 (b) 12-15
(c) 17-31 (d) 35-60

314. Diameter of corn starch granule is about ______ (microns)
(a) 2-3 (b) 5-25
(c) 30-50 (d) 105-135

315. Gelatinization temperature (°C) of wheat starch granules ranges between
(a) 40-50 (b) 53-65
(c) 70-75 (d) 81-84

316. Crystallinity of wheat starch granules is about ________ %
(a) 16 (b) 26
(c) 36 (d) 46

317. Crystallinity of potato starch granules is about ________ %
(a) 15 (b) 25
(c) 35 (d) 38

318. Crystallinity of rice starch granules is about ________ %
(a) 18 (b) 28
(c) 38 (d) 48

319. Shape of the wheat starch granules is
(a) lenticular (b) polyhedral
(c) both (d) none

320. Shape of the corn starch granules is
(a) lenticular (b) polyhedral
(c) spherical (d) oval

321. Shape of the rice starch granules is
(a) lenticular (b) polyhedral
(c) spherical (d) oval

322. Shape of the sorghum starch granules is
(a) spherical (b) polyhedral
(c) both (d) none

323. Shape of the potato starch granules is
(a) polyhedral (b) spherical
(c) lenticular (d) elliptical

324. Shape of the rye starch granules is
(a) lenticular (b) polyhedral
(c) spherical (d) elliptical

325. Shape of the barley starch granules is
(a) lenticular (b) polyhedral
(c) spherical (d) oval

326. Which of the following is a type of resistant starch
(a) RSI (b) RS II
(c) RS I, II, III, IV (d) none

327. Inulin occurs as a reserve carbohydrate in
(a) chicory (b) rye
(c) onion (d) all

328. Cereal proteins are deficient in
(a) lysine (b) threonine
(c) both (d) none

329. Protein content in beef is ________ (g/100g)
(a) 6.5 (b) 16.5
(c) 26.5 (d) 30

330. Protein content in pork is ________ (g/100g)
(a) 5.2 (b) 10.2
(c) 20.2 (d) 30.2

331. Protein content in milk is ________ (g/100g)
(a) 1.6 (b) 2.6
(c) 3.6 (d) 7.9

332. Protein content in egg is ________ (g/100g)
(a) 2.9 (b) 12.9
(c) 22.9 (d) 32.9

333. Protein content in potato is ________ (g/100g)
(a) 2 (b) 6
(c) 8 (d) 10

334. Protein content in corn is ________ (g/100g)
(a) 2.8-3.7 (b) 10-12
(c) 14-16 (d) 16-20

335. Isoleucine content of milk is about ________ (mg/g total nitrogen)
(a) 299 (b) 399
(c) 499 (d) 599

336. Isoleucine content of wheat is about ________ (mg/g total nitrogen)
(a) 104 (b) 204
(c) 304 (d) 404

337. Isoleucine content of peas is about ________ (mg/g total nitrogen)
(a) 137 (b) 167
(c) 267 (d) 367

338. Isoleucine content of corn is about ________ (mg/g total nitrogen)
(a) 30 (b) 130
(c) 180 (d) 230

339. Leucine content of milk is about ________ (mg/g total nitrogen)
(a) 182 (b) 382
(c) 782 (d) 1282

340. Leucine content of egg is about ________ (mg/g total nitrogen)
(a) 151 (b) 251
(c) 351 (d) 551

341. Leucine content of wheat is about ________ (mg/g total nitrogen)
(a) 217 (b) 317
(c) 417 (d) 517

342. Leucine content of peas is about ________ (mg/g total nitrogen)
(a) 425 (b) 525
(c) 625 (d) 725

343. Leucine content of corn is about ________ (mg/g total nitrogen)
(a) 483 (b) 583
(c) 683 (d) 783

344. Lysine content of milk is about ________ (mg/g total nitrogen)
(a) 150 (b) 250
(c) 350 (d) 450

345. Lysine content of egg is about ________ (mg/g total nitrogen)
(a) 236 (b) 336
(c) 436 (d) 536

346. Lysine content of wheat is about ________ (mg/g total nitrogen)
(a) 179 (b) 279
(c) 379 (d) 479

347. Lysine content of peas is about ________ (mg/g total nitrogen)
(a) 170 (b) 270
(c) 370 (d) 470

348. Lysine content of corn is about ________ (mg/g total nitrogen)
(a) 67 (b) 167
(c) 267 (d) 367

349. Methionine content of milk is about ________ (mg/g total nitrogen)
(a) 56 (b) 156
(c) 256 (d) 350

350. Methionine content of egg is about ________ (mg/g total nitrogen)
(a) 70 (b) 110
(c) 210 (d) 310

351. Methionine content of wheat is about ________ (mg/g total nitrogen)
(a) 34 (b) 94
(c) 134 (d) 194

352. Methionine content of peas is about ________ (mg/g total nitrogen)
(a) 27 (b) 57
(c) 127 (d) 157

353. Methionine content of corn is about ________ (mg/g total nitrogen)
(a) 20 (b) 120
(c) 180 (d) 220

354. Phenylalanine content of milk is about ________ (mg/g total nitrogen)
(a) 134 (b) 234
(c) 334 (d) 434

355. Phenylalanine content of egg is about ________ (mg/g total nitrogen)
(a) 158 (b) 258
(c) 358 (d) 458

356. Phenylalanine content of wheat is about ________ (mg/g total nitrogen)
(a) 182 (b) 282
(c) 382 (d) 482

357. Phenylalanine content of peas is about ________ (mg/g total nitrogen)
(a) 187 (b) 287
(c) 387 (d) 487

358. Phenylalanine content of corn is about ________ (mg/g total nitrogen)
(a) 105 (b) 205
(c) 305 (d) 405

359. Threonine content of milk is about ________ (mg/g total nitrogen)
(a) 178 (b) 278
(c) 378 (d) 478

360. Threonine content of egg is about ________ (mg/g total nitrogen)
(a) 120 (b) 220
(c) 320 (d) 420

361. Threonine content of wheat is about ________ (mg/g total nitrogen)
(a) 183 (b) 283
(c) 383 (d) 483

362. Threonine content of peas is about ________ (mg/g total nitrogen)
(a) 154 (b) 254
(c) 354 (d) 454

363. Threonine content of corn is about ________ (mg/g total nitrogen)
(a) 125 (b) 225
(c) 325 (d) 425

364. Valine content of milk is about ________ (mg/g total nitrogen)
(a) 163 (b) 263
(c) 363 (d) 463

365. Valine content of egg is about ________ (mg/g total nitrogen)
(a) 128 (b) 228
(c) 328 (d) 428

366. Valine content of wheat is about ________ (mg/g total nitrogen)
(a) 176 (b) 276
(c) 376 (d) 476

367. Valine content of peas is about ________ (mg/g total nitrogen)
(a) 194 (b) 294
(c) 394 (d) 494

368. Valine content of corn is about ________ (mg/g total nitrogen)
(a) 103 (b) 203
(c) 303 (d) 403

369. Histidine content of milk is about ________ (mg/g total nitrogen)
(a) 114 (b) 214
(c) 314 (d) 414

370. Histidine content of egg is about ________ (mg/g total nitrogen)
(a) 152 (b) 252
(c) 352 (d) 452

371. Histidine content of wheat is about ________ (mg/g total nitrogen)
(a) 143 (b) 243
(c) 343 (d) 443

372. Histidine content of peas is about ________ (mg/g total nitrogen)
(a) 143 (b) 173
(c) 243 (d) 273

373. Isoleucine content of beef is about ________ (mg/g total nitrogen)
(a) 101 (b) 201
(c) 301 (d) 401

374. Leucine content of beef is about ________ (mg/g total nitrogen)
(a) 107 (b) 307
(c) 507 (d) 707

375. Lysine content of beef is about ________ (mg/g total nitrogen)
(a) 456 (b) 556
(c) 656 (d) 756

376. Methionine content of beef is about ________ (mg/g total nitrogen)
(a) 69 (b) 169
(c) 269 (d) 369

377. Phenylalanine content of beef is about ________ (mg/g total nitrogen)
(a) 175 (b) 275
(c) 350 (d) 450

378. Valine content of beef is about ________ (mg/g total nitrogen)
(a) 113 (b) 213
(c) 313 (d) 413

379. Limiting essential amino acid in wheat is
(a) leucine (b) lysine
(c) phenylalanine (d) isoleucine

380. Which of the following is not a globulin protein
(a) oryzenin
(b) lactoglobulin
(c) myosin
(d) actin

381. Oryzenin is a ________ type of protein
(a) globulin (b) glutelin
(c) prolamin (d) histone

382. Leucosin is a ________ type of protein
(a) albumin (b) globulin
(c) glutelin (d) none

383. Myosin is a ________ type of protein
(a) albumin (b) globulin
(c) glutelin (d) prolamin

384. Glycinin is a ________ type of protein
(a) albumin (b) globulin
(c) glutelin (d) prolamin

385. Glutenin is a ________ type of protein
(a) albumin (b) globulin
(c) glutelin (d) prolamin

386. Zein is a ________ type of protein
(a) albumin (b) globulin
(c) glutelin (d) prolamin

387. Gliadin is a ________ type of protein
(a) albumin (b) prolamin
(c) globulin (d) glutelin

388. Hordein is a ________ type of protein
(a) prolamin (b) albumin
(c) globulin (d) glutelin

389. Gelatin is a ________
(a) Histone (b) scleroprotein
(c) albumin (d) globulin

390. Histone proteins are rich in
(a) lysine (b) arginine
(c) both (d) none

391. Oryzenin is present in
(a) wheat (b) rice
(c) barley (d) soybean

392. Glutenin is present in
(a) wheat (b) rice
(c) sorghum (d) corn

393. Gliadin is present in
(a) wheat (b) corn
(c) rice (d) sorghum

394. Zein is present in
(a) wheat (b) corn
(c) barley (d) rice

395. Hordein is present in
(a) wheat (b) corn
(c) barley (d) soybean

396. Glycinin is present in
(a) wheat (b) rice
(c) barley (d) soybean

397. Actin is present in
(a) wheat (b) meat
(c) milk (d) rice

398. Whey protein is present in
(a) wheat (b) meat
(c) milk (d) rice

399. Casein protein is present in
(a) wheat (b) milk
(c) rice (d) corn

400. Which of the following is considered as best quality protein
(a) cereal protein (b) soya protein
(c) egg protein (d) none

401. The molecular weight of lactoglobulin is ________ Dalton
(a) 35000 (b) 64500
(c) 68300 (d) 140000

402. The molecular weight of hemoglobin is ________ Dalton
(a) 35000 (b) 64500
(c) 68300 (d) 140000

403. The molecular weight of myosin is ________ Dalton
(a) 75000 (b) 150000
(c) 325000 (d) 475000

404. Denaturation of protein does not involve
(a) change in molecular structure
(b) loss of biological quality
(c) breaking of peptide bonds
(d) change in solubility

405. Meat proteins are denatured in temperature range of ________ °C
(a) 27-35 (b) 40-45
(c) 57-75 (d) all

406. Caseinate micelles of milk are stable towards
(a) heating (b) freezing
(c) both (d) none

407. Casein coagulates on ________
(a) boiling
(b) pasteurization
(c) sterilization
(d) none

408. Heat stability of casein is due to high content of ________
(a) proline
(b) hydroxyproline
(c) both
(d) none

409. Coagulation temperature of bovine casein is about ________ °C
(a) 40-80 (b) 90-110
(c) 120-135 (d) 160-200

410. Coagulation temperature of egg albumin is about ________ °C
(a) 36 (b) 56
(c) 76 (d) 96

411. Coagulation temperature of bovine serum albumin is about ________ °C
(a) 36 (b) 56
(c) 67 (d) 96

412. Coagulation temperature of bovine milk albumin is about ________ °C
(a) 36 (b) 56
(c) 72 (d) 96

413. Coagulation temperature of bovine β-lactoglobulin is about ________ °C
(a) 50 (b) 70
(c) 90 (d) 40

414. The cysteine content of milk albumin is ______ %
(a) 3.4 (b) 6.4
(c) 14.8 (d) 18.9

415. The cysteine content of bovine serum albumin is ______ %
(a) 0.3 (b) 1.3
(c) 2.3 (d) 3.3

416. The cystine content of bovine serum albumin is ______ %
(a) 1.2 (b) 2.5
(c) 5.7 (d) 9.8

417. The cystine content of egg albumin is ______ %
(a) 0.1 (b) 0.5
(c) 1.5 (d) 2.1

418. The cysteine content of egg albumin is ______ %
(a) 1.4 (b) 2.4
(c) 3.4 (d) 4.4

419. The cysteine content of β- lactoglobulin is ______ %
(a) 1.1 (b) 2.1
(c) 3.1 (d) 4.1

420. The cystine content of β- lactoglobulin is ______ %
(a) 1.3 (b) 2.3
(c) 3.3 (d) 4.3

421. The cystine content of casein is ______ %
(a) 0.3 (b) 1.3
(c) 2.3 (d) 3.3

422. Phenylalanine content of gelatin is ______ g amino acid/100g protein
(a) 2.2 (b) 5.2
(c) 9.2 (d) 12.8

423. Isoleucine content of gelatin is ______ g amino acid/100g protein
(a) 0.7 (b) 1.7
(c) 17.4 (d) 11.2

424. Leucine content of gelatin is ______ g amino acid/100g protein
(a) 0.3 (b) 1.3
(c) 2.3 (d) 3.3

425. Lysine content of gelatin is ______ g amino acid/100g protein
(a) 1.5 (b) 2.5
(c) 4.5 (d) 11.5

426. Methionine content of gelatin is ______ g amino acid/100g protein
(a) 0.9 (b) 2.9
(c) 4.9 (d) 6.9

427. Threonine content of gelatin is ______ g amino acid/100g protein
(a) 0.2 (b) 2.2
(c) 4.5 (d) 15.6

428. Valine content of gelatin is ______ g amino acid/100g protein
(a) 0.2 (b) 1.2
(c) 2.6 (d) 4.9

429. Proline content of gelatin is ______ g amino acid/100g protein
(a) 6.4 (b) 11.4
(c) 16.4 (d) 21.4

430. Glycine content of gelatin is ______ g amino acid/100g protein
(a) 5.5 (b) 14.5
(c) 27.5 (d) 39.5

431. Alanine content of gelatin is ______ g amino acid/100g protein
(a) 4.5 (b) 11.0
(c) 21.8 (d) 35.4

432. Glycine content of serum albumin is ______ g amino acid/100g protein
(a) 0.6 (b) 1.8
(c) 2.9 (d) 4.6

433. Alanine content of serum albumin is ______ g amino acid/100g protein
(a) 1.3 (b) 3.3
(c) 6.3 (d) 9.3

434. Valine content of serum albumin is ______ g amino acid/100g protein
(a) 11.2 (b) 5.9
(c) 17.4 (d) 25.6

435. Leucine content of serum albumin is __________ g amino acid/100g protein

(a) 12.3 (b) 24.3
(c) 35.3 (d) 38.2

436. Isoleucine content of serum albumin is ________ g amino acid/100g protein

(a) 0.9 (b) 2.6
(c) 3.9 (d) 5.6

437. Threonine content of serum albumin is ________ g amino acid/100g protein

(a) 1.4 (b) 2.6
(c) 5.8 (d) 7.2

438. Methionine content of serum albumin is ________ g amino acid/100g protein

(a) 0.8 (b) 1.9
(c) 3.8 (d) 5.4

439. Phenylalanine content of serum albumin is ____________ g amino acid/100g protein

(a) 2.8 (b) 3.5
(c) 4.8 (d) 6.6

440. Proline content of serum albumin is __________ g amino acid/100g protein

(a) 3.2 (b) 4.8
(c) 7.9 (d) 10.5

441. Glutamic acid content of serum albumin is __________ g amino acid/100g protein

(a) 6.5 (b) 7.9
(c) 11.5 (d) 16.5

442. Lysine content of serum albumin is __________ g amino acid/100g protein

(a) 12.8 (b) 17.4
(c) 0.5 (d) 3.4

443. Glycine content of casein is _______ g amino acid/100g protein

(a) 0.5 (b) 1.9
(c) 2.5 (d) 3.9

444. Valine content of casein is ________ g amino acid/100g protein

(a) 1.2 (b) 3.6
(c) 6.8 (d) 8.4

445. Leucine content of casein is ______ g amino acid/100g protein

(a) 3.8 (b) 5.9
(c) 9.2 (d) 12.8

446. Isoleucine content of casein is __________ g amino acid/100g protein

(a) 5.6 (b) 15.6
(c) 25.6 (d) 19.6

447. Threonine content of casein is _____ g amino acid/100g protein

(a) 4.4 (b) 3.4
(c) 2.2 (d) 7.4

448. Methionine content of casein is __________ g amino acid/100g protein

(a) 0.8 (b) 1.8
(c) 2.8 (d) 3.8

449. Phenylalanine content of casein is __________ g amino acid/100g protein

(a) 2.3 (b) 3.3
(c) 4.3 (d) 5.3

450. Proline content of casein is __________ g amino acid/100g protein

(a) 7.5 (b) 13.5
(c) 17.5 (d) 4.5

451. Glutamic acid content of casein is __________ g amino acid/100g protein

(a) 4.5 (b) 14.4
(c) 17.5 (d) 24.5

452. Lysine content of casein is ________ g amino acid/100g protein

(a) 2.3 (b) 5.4
(c) 8.9 (d) 11.2

453. Which of the following is associated with non-enzymatic browning

(a) carbohydrate
(b) proteins
(c) both
(d) none

454. Which of the following group of protein is used in maillard reaction

(a) hydrogen (b) alkyl
(c) amino (d) all

455. Which of the following is used in maillard reaction
(a) reducing sugar
(b) non reducing sugar
(c) both
(d) none

456. Viscous nature of wheat dough is contributed by
(a) albumin (b) globulin
(c) gliadin (d) glutenin

457. Elastic nature of wheat dough is contributed by
(a) albumin (b) globulin
(c) gliadin (d) glutenin

458. Viscoelastic nature of wheat dough is due to
(a) gliadin (b) glutenin
(c) both (d) none

459. Functional properties of protein include
(a) viscosity (b) coagulation
(c) tenderization (d) all

460. Functional properties of protein include
(a) turbidity
(b) emulsification
(c) foaming
(d) all

461. Functional properties of protein include
(a) chewiness (b) cohesiveness
(c) film forming (d) all

462. Which of the following is associated with functionality of proteins in whipped toppings
(a) viscosity (b) elasticity
(c) gelation (d) foaming

463. Which of the following is associated with functionality of proteins in meat products
(a) viscosity (b) elasticity
(c) solubility (d) foaming

464. Proteins can act as surfactant for stabilization of
(a) emulsions (b) foam
(c) both (d) none

465. Amphiphilic molecule contain_______ moiety
(a) hydrophobic (b) hydrophilic
(c) both (d) none

466. Hydrophobicity of protein does not contribute in
(a) dough making
(b) heat coagulation
(c) foaming
(d) solubility

467. Surface hydrophobicity of protein contribute in
(a) emulsification
(b) fat binding
(c) both
(d) none

468. Hydrophobicity of protein does not support
(a) solubility
(b) water-holding
(c) both
(d) none

469. Charge frequency of protein does not contribute to
(a) foaming
(b) heat coagulation
(c) dough making (d) all

470. Basic type of emulsions are of
(a) two (b) three
(c) four (d) five

471. Proteins can form gels by ________ treatment
(a) acid (b) enzyme
(c) heat (d) all

472. Which of the following is a property of gel
(a) viscosity (b) elasticity
(c) plasticity (d) all

473. Which of the following is an example of protein gel
(a) yoghurt (b) tofu
(c) mayonnaise (d) all

474. Casein is precipitated at a pH of about
(a) 2.8 (b) 3.2
(c) 4.6 (d) 5.9

475. Molecular weight of β-lactoglobulin is about ________ dalton
(a) 3600 (b) 7200
(c) 36000 (d) 72000

476. Which of the following is not a milk protein
(a) casein (b) whey
(c) lactalbumin (d) none

477. Muscle fibers are enclosed in membrane called
(a) sarcomere
(b) sarcolemma
(c) film
(d) dual membrane

478. Muscle fibers also contain
(a) mitochondria (b) ribosomes
(c) lysosomes (d) all

479. Extracellular space present in muscle is ______%
(a) 18 (b) 28
(c) 38 (d) 48

480. A bands in muscles are
(a) isotropic (b) anisotropic
(c) both (d) none

481. In muscle fiber, I band is dissected by a dark line called
(a) I line (b) A line
(c) M line (d) Z line

482. In muscle fiber, A band contains a light area called
(a) A zone (b) M zone
(c) Z line (d) H zone

483. Gelatin is soluble protein derived from
(a) collagen (b) lignin
(c) cellulose (d) casein

484. Which of the following is a part of egg protein
(a) egg white (b) egg yolk
(c) both (d) none

485. Ovalbumin is a type of
(a) lipoprotein
(b) phosphoprotein
(c) scleroprotein
(d) all

486. Conalbumin has a molecular weight of about
(a) 7000 (b) 70000
(c) 11000 (d) 111000

487. Ovalbumin has a molecular weight of about
(a) 3500 (b) 4500
(c) 35000 (d) 45000

488. Ovalbumin constitute __________% of egg white protein
(a) 24 (b) 34
(c) 44 (d) 54

489. Conalbumin constitute __________% of egg white protein
(a) 3 (b) 13
(c) 23 (d) 33

490. Ovomucoid constitute __________% of egg white protein
(a) 5 (b) 11
(c) 17 (d) 27

491. Lysozyme constitute __________% of egg white protein
(a) 3.5 (b) 6.5
(c) 9.5 (d) 11.5

492. Ovomucin constitute __________% of egg white protein
(a) 1.5 (b) 4.2
(c) 5.9 (d) 7.4

493. Avidin constitute ________% of egg white protein
(a) 0.05 (b) 0.5
(c) 2.5 (d) 3.5

494. Isoelectric point of ovalbumin is about
(a) 2.6 (b) 3.6
(c) 4.6 (d) 5.2

495. Isoelectric point of conalbumin is about
(a) 3 (b) 4
(c) 5 (d) 6

496. Isoelectric point of ovomucoid is about
(a) 1.3 (b) 2.3
(c) 3.3 (d) 4.3

497. Isoelectric point of lysozyme is about
(a) 4.7 (b) 6.7
(c) 8.7 (d) 10.7

498. Isoelectric point of avidin is about
(a) 5.5 (b) 7.5
(c) 9.5 (d) 10.5

499. Cereal seed proteins are low in
(a) leucine (b) isoleucine
(c) lysine (d) all

500. Peanut protein is poor in
(a) lysine (b) tryptophan
(c) methionine (d) all

501. Wheat proteins are classified into ________ groups
(a) two (b) three
(c) four (d) five

502. Which of the following is a water soluble protein
(a) albumin (b) globulin
(c) gliadin (d) glutenin

503. Which of the following protein is soluble in salt solution
(a) albumin (b) globulin
(c) gliadin (d) glutenin

504. Which of the following is an alcohol soluble protein
(a) albumin (b) globulin
(c) gliadin (d) glutenin

505. Which of the following is a dilute acid/alkali soluble protein
(a) albumin (b) globulin
(c) gliadin (d) glutenin

506. Gliadin and glutenin are ________ protein of wheat
(a) storage
(b) gluten forming
(c) both
(d) none

507. Gluten proteins are rich in
(a) lysine
(b) methionine
(c) glutamine
(d) tryptophan

508. Gluten proteins have low amount of
(a) lysine (b) methionine
(c) tryptophan (d) all

509. Gliadin protein has a molecular weight (kDa) of about
(a) 10-20 (b) 30-75
(c) 80-100 (d) 120-135

510. Glutenin protein has a molecular weight (kDa) of about
(a) 10-20 (b) 30-55
(c) 70-200 (d) 250-325

511. Which of the following is non gluten forming protein
(a) albumin (b) globulin
(c) both (d) none

512. Wheat gliadin protein is of ________ types
(a) one (b) two
(c) three (d) four

513. Protein content of maize is about
(a) 10 (b) 20
(c) 25 (d) 30

514. Which of the following is a major type of maize protein
(a) zein (b) glutelin
(c) both (d) none

515. Which of the following is a minor type of maize protein
(a) albumin (b) zein
(c) glutelin (d) none

516. Maize protein contain low amount of ________ amino acid
(a) lysine (b) tryptophan
(c) both (d) none

517. Which of the following is a low molecular weight maize protein
(a) zein (b) glutelin
(c) both (d) none

518. Which of the following is a high molecular weight maize protein
(a) zein (b) glutelin
(c) both (d) none

519. Zein protein has molecular weight (kDa) of about
(a) 5-10 (b) 22-24
(c) 32-40 (d) 40-45

520. Zein molecules contain about _____ % α-helix
(a) 15 (b) 35
(c) 45 (d) 60

521. Zein molecules contain about _____ % β-sheet
(a) 15 (b) 35
(c) 45 (d) 60

522. Maize glutelin have higher amount of ________ amino acid than zein
(a) lysine (b) arginine
(c) histidine (d) all

523. Maize glutelins are joined together by ________ bonds
(a) disulfide (b) hydrogen
(c) glycosidic (d) glucosidic

524. Protein content of polished rice is _______ as compared to wheat
(a) lower (b) higher
(c) equal (d) little higher

525. Major protein in rice bran is
(a) albumin (b) globulin
(c) both a & b (d) glutelin

526. Milled rice is rich in
(a) glutelin (b) albumin
(c) globulin (d) none

527. Soy protein is not a good source of
(a) phenylalanine (b) isoleucine
(c) leucine (d) methionine

528. Soy protein is a good source of
(a) methionine (b) tryptophan
(c) both a & b (d) lysine

529. Soy proteins are generally classified as
(a) albumin (b) globulin
(c) both a & b (d) none

530. Proteins are amphoteric due to presence of
(a) free amino group
(b) carboxyl group
(c) both a & b
(d) none

531. Total charge on protein molecule at isoelectric point is
(a) +1 (b) -1
(c) 0 (d) +2

532. Which of the following is associated with isoelectric point of protein
(a) cation (b) anion
(c) zwitter ion (d) none

533. Which of the following cause denaturation in proteins
(a) UV light (b) high pressure
(c) ultrasonic waves
(d) all

534. Which of the following is protein estimation method
(a) Kjeldahl (b) Dumas
(c) Lowry's (d) all

535. Casein content of cow milk is about ____%
(a) 1.5-2% (b) 3-3.5%
(c) 5-6% (d) 7.5-8.5%

536. Casein content of human milk is about ______%
(a) 0.3-0.6 (b) 1-1.5
(c) 2.5-3 (d) 3-3.5

537. Which of the following is a type of milk casein
(a) α (b) β
(c) γ (d) all

538. α- casein contain ____% lysine
(a) 3.9 (b) 8.9
(c) 11.9 (d) 16.9

539. α- casein contain ____% tryptophan
(a) 1.6 (b) 2.6
(c) 3.6 (d) 4.6

540. α- casein has a total nitrogen content of ________ %
(a) 5.5 (b) 15.5
(c) 25.5 (d) 30.5

541. α- casein contain ___% histidine
(a) 1.9 (b) 2.9
(c) 3.9 (d) 4.9

542. α- casein contain ___% phosphorus
(a) 0.1 (b) 0.99
(c) 2.9 (d) 1.9

543. β-casein contain ___% phosphorus
(a) 0.22 (b) 0.33
(c) 0.66 (d) 0.99

544. β- casein has a total nitrogen content of______%
(a) 5.4 (b) 15.4
(c) 20.4 (d) 22.4

545. β- casein contain ____% histidine
(a) 1.1 (b) 2.1
(c) 3.1 (d) 4.1

546. β- casein contain ____% lysine
(a) 2.2 (b) 4.4
(c) 6.6 (d) 8.8

547. β- casein contain ____% tryptophan
(a) 0.2 (b) 0.6
(c) 1.4 (d) 1.9

548. γ- casein contain ____% phosphorus
(a) 0.02 (b) 0.04
(c) 0.06 (d) 0.08

549. α- casein contain ____% arginine
(a) 2.3 (b) 4.3
(c) 14.3 (d) 7.3

550. β- casein contain ____% arginine
(a) 0.4 (b) 0.8
(c) 2.4 (d) 3.4

551. Keratin protein is present in
(a) nails (b) hair
(c) horn (d) all

552. Which of the following is albumin protein
(a) leucosine (b) legumeline
(c) ovalbumin (d) all

553. Which of the following is globulin protein
(a) ovalbumin (b) glycinine
(c) legumine (d) all

554. Which of the following is prolamin protein
(a) gliadin (b) zein
(c) hordein (d) all

555. Which of the following is scleroprotein
(a) collagen (b) elastin
(c) keratin (d) all

556. Chromoproteins are the proteins with
(a) carbohydrate (b) lipid
(c) pigment (d) water

557. Which of the following is chromoprotein
(a) hemoglobin
(b) cytochrome
(c) flavoprotein
(d) all

558. Mucoproteins are the proteins which contain ________as prosthetic group
(a) carbohydrate (b) lipid
(c) pigment (d) water

559. Mucoproteins are also known as
(a) glycoproteins
(b) phosphoprotein
(c) chromoprotein
(d) all

560. Which of the following is not a protein
(a) elastin (b) hordein
(c) gluten (d) none

561. Lipids are not soluble in
(a) ether (b) water
(c) benzene (d) chloroform

562. Short chain fatty acids are relatively more soluble in
(a) ether (b) water
(c) benzene (d) chloroform

563. Which of the following is associated with phospholipid
(a) sphingomyelin
(b) phosphatidyl ester
(c) plasmalogens
(d) all

564. Fats are ____________ at room temperature
(a) solid (b) liquid
(c) both (d) none

565. Oils are _______ at room temperature
(a) solid (b) liquid
(c) both (d) none

566. Which of the following contains invisible fat
(a) egg (b) meat
(c) fruits (d) all

567. Which of the following contains visible fat
(a) lard (b) butter
(c) margarine (d) all

568. Which of the following do not contain invisible fat
(a) fish (b) vegetables
(c) grains (d) shortening

569. Which of the following do not contains visible fat
(a) butter (b) shortening
(c) cooking oil (d) poultry product

570. Which of the following oil is not seed oil
(a) soybean oil (b) palm oil
(c) both (d) none

571. Diacylglycerol has ________ fatty acid ester linkage
(a) 1 (b) 2
(c) 3 (d) none

572. Cod liver oil contains about ______ % saturated fat
(a) 23 (b) 43
(c) 33 (d) 53

573. Ham contains about _______ % fat
(a) 3.1 (b) 13.1
(c) 23.1 (d) 31

574. Chicken contains about _______ % fat
(a) 0.7 (b) 4
(c) 7 (d) 14

575. Hamburger contains about _______ % fat
(a) 3 (b) 13
(c) 23 (d) 30

576. Butter contains about _______ % fat
(a) 60 (b) 70
(c) 80 (d) 90

577. Cow milk contains about _______ % fat
(a) 3.5 (b) 7.5
(c) 11.5 (d) 13.5

578. Sunflower contains about ______ % fat
(a) 8 (b) 18
(c) 28 (d) 38

579. Soybean contains about ______ % fat
(a) 7 (b) 17
(c) 27 (d) 34

580. Peanut contains about ______ % fat
(a) 9 (b) 19
(c) 29 (d) 49

581. Coconut contains about ______ % fat
(a) 14 (b) 34
(c) 54 (d) 74

582. Walnut contains about ______ % fat
(a) 18 (b) 38
(c) 58 (d) 78

583. Rice contains about ______ % fat
(a) 1.4 (b) 3.4
(c) 5.4 (d) 6.4

584. Barley contains about ______ % fat
(a) 1.9 (b) 3.9
(c) 4.9 (d) 5.9

585. Wheat contains about ______ % fat
(a) 2 (b) 4
(c) 8 (d) 12

586. Oat contains about ______ % fat
(a) 3.1 (b) 4.4
(c) 6.1 (d) 7.9

587. Asparagus contains about ______ % fat
(a) 0.25 (b) 2.5
(c) 3.5 (d) 4.5

588. Which of the following is an unsaturated fatty acid
(a) butyric (b) caproic
(c) caprylic (d) none

589. Which of the following is an unsaturated fatty acid
(a) capric (b) lauric
(c) myristic (d) none

590. Which of the following is an unsaturated fatty acid
(a) palmitic (b) stearic
(c) arachidonic (d) none

591. Which of the following is a saturated fatty acid
(a) valeic (b) enanthic
(c) pelargonic (d) all

592. Butyric acid is also known as
(a) n- butanoic (b) n- hexanoic
(c) n-octanoic (d) none

593. Caproic acid is ______
(a) n-butanoic (b) n- hexanoic
(c) n-octanoic (d) none

594. Caprylic acid is ______
(a) n-butanoic (b) n- hexanoic
(c) n-octanoic (d) none

595. Capric acid is ______
(a) n-butanoic (b) n-decanoic
(c) n-octanoic (d) none

596. Lauric acid is ______
(a) n- octanoic (b) n- decanoic
(c) n- dodecanoic (d) none

597. Myristic acid is ______
(a) n- tetradecanoic (b) n-hexanoic
(c) n- dodecanoic (d) none

598. Palmitic acid is ______
(a) n- tetradecanoic
(b) n- hexadecanoic
(c) n-octanoic
(d) none

599. Stearic acid is ______
(a) n- tetradecanoic
(b) n- hexadecanoic
(c) n-octanoic
(d) none

600. Arachidic acid is ______
(a) n- eicosanoic
(b) n- hexadecanoic
(c) n-octanoic
(d) none

601. Valeric acid is ______
(a) n- pentanoic
(b) n- hexadecanoic
(c) n-octanoic (d) none

602. Number of carbon atom present in butyric acid

(a) 3 (b) 4
(c) 5 (d) 6

603. Number of carbon atom present in caproic acid

(a) 3 (b) 4
(c) 5 (d) 6

604. Number of carbon atom present in caprylic acid

(a) 5 (b) 6
(c) 7 (d) 8

605. Number of carbon atom present in lauric acid

(a) 8 (b) 10
(c) 12 (d) 16

606. Number of carbon atom present in capric acid

(a) 8 (b) 10
(c) 12 (d) 16

607. Number of carbon atom present in myristic acid

(a) 10 (b) 12
(c) 14 (d) 16

608. Number of carbon atom present in palmitic acid

(a) 10 (b) 12
(c) 14 (d) 16

609. Number of carbon atom present in stearic acid

(a) 18 (b) 20
(c) 22 (d) 24

610. Number of carbon atom present in arachidonic acid

(a) 18 (b) 20
(c) 22 (d) 24

611. Number of carbon atom present in valeric acid

(a) 20 (b) 24
(c) 5 (d) 24

612. Which of the following is beneficial fatty acid

(a) EAP (b) EPA
(c) ETP (d) ETA

613. Which of the following is not a beneficial fatty acid

(a) EPA (b) DHA
(c) both (d) none

614. Which of the following is saturated fatty acid

(a) myristoleic (b) palmitoleic
(c) petroselinic (d) none

615. Which of the following is saturated fatty acid

(a) oleic (b) vaccenic
(c) linoleic (d) none

616. Which of the following is saturated fatty acid

(a) linolenic (b) elaeostearic
(c) arachidonic (d) none

617. Which of the following is an unsaturated fatty acid

(a) EPA (b) DHA
(c) oleic (d) all

618. Myristoleic acid is also known as

(a) tetradec-9-enoic
(b) hexadec-9-enoic
(c) octadec-6-enoic
(d) octadec-9-enoic

619. Palmitoleic acid is also known as

(a) tetradec-9-enoic
(b) hexadec-9-enoic
(c) octadec-6-enoic
(d) octadec-9-enoic

620. Petroselinic acid is also known as

(a) tetradec-9-enoic
(b) hexadec-9-enoic
(c) octadec-6-enoic
(d) octadec-9-enoic

621. Oleic acid is also known as

(a) tetradec-9-enoic
(b) hexadec-9-enoic
(c) octadec-6-enoic
(d) octadec-9-enoic

622. Linoleic acid is also known as
(a) Octadeca-9:12-dienoic
(b) Octadeca-9:12:15-trienoic
(c) Hexadec-9-enoic
(d) none

623. Linolenic acid is also known as
(a) Octadeca-9:12-dienoic
(b) Octadeca-9:12:15-trienoic
(c) Hexadec-9-enoic
(d) none

624. Gadoleic acid is also known as
(a) Eicos-9-enoic acid (b) EPA
(c) DHA (d) all

625. Number of carbon atoms present in myristoleic acid
(a) 14 (b) 16
(c) 18 (d) 20

626. Number of carbon atoms present in palmitoleic acid
(a) 14 (b) 16
(c) 18 (d) 20

627. Number of carbon atoms present in petroselinic acid
(a) 14 (b) 16
(c) 18 (d) 20

628. Number of carbon atoms present in oleic acid
(a) 14 (b) 16
(c) 18 (d) 20

629. Number of carbon atoms present in vaccenic acid
(a) 14 (b) 16
(c) 18 (d) 20

630. Number of carbon atoms present in linoleic acid
(a) 18 (b) 20
(c) 22 (d) 24

631. Number of carbon atoms present in linolenic acid
(a) 16 (b) 18
(c) 20 (d) 22

632. Number of carbon atoms present in elaeostearic acid
(a) 16 (b) 18
(c) 20 (d) 22

633. Number of carbon atoms present in gadoleic acid
(a) 16 (b) 18
(c) 20 (d) 22

634. Number of carbon atoms present in arachidonic acid
(a) 20 (b) 22
(c) 24 (d) 26

635. Number of carbon atoms present in EPA
(a) 18 (b) 20
(c) 22 (d) 24

636. Number of carbon atoms present in DHA
(a) 18 (b) 20
(c) 22 (d) 24

637. Iodine value of beef fat is about
(a) 35 (b) 50
(c) 60 (d) 70

638. Which of the following is highest component fatty acid (%) in cow milk fat
(a) 16:0 (b) 16:1
(c) 18:0 (d) 18:1

639. Which of the following is lowest component fatty acid (%) in cow milk fat
(a) 16:0 (b) 16:1
(c) 18:0 (d) 18:1

640. Which of the following is highest com ponent fatty acid (%) in goat milk fat
(a) 16:0 (b) 16:1
(c) 18:0 (d) 18:1

641. Which of the following is lowest component fatty acid (%) in goat milk fat
(a) 16:0 (b) 16:1
(c) 18:0 (d) 18:1

642. Which of the following is highest component fatty acid (%) in sheep milk fat
(a) 16:0 (b) 18:1
(c) both (a) & (b) (d) 18:0

643. Which of the following is lowest component fatty acid (%) in sheep milk fat
(a) 16:0 (b) 16:1
(c) 18:0 (d) 18:1

644. Iodine value of milk fat is about
(a) 20 (b) 35
(c) 55 (d) 65

645. Total short chain fatty acid (%) in cow milk fat is about
(a) 7.6 (b) 17.6
(c) 27.6 (d) 32.6

646. Total short chain fatty acid (%) in goat milk fat is about
(a) 9.3 (b) 15.3
(c) 29.3 (d) 32.6

647. Total short chain fatty acid (%) in sheep milk fat is about
(a) 2.7 (b) 22.7
(c) 40 (d) 45.7

648. Total saturated fatty acids (%) in egg yolk
(a) 16.2 (b) 22.2
(c) 36.2 (d) 48.2

649. Total monounsaturated fatty acid (%) in egg yolk
(a) 16.2 (b) 22.2
(c) 36.2 (d) 48.2

650. Total polyunsaturated fatty acid (%) in egg yolk
(a) 14.7 (b) 24.7
(c) 34.7 (d) 49.8

651. Which of the following oil has highest 16:0 fatty acid component
(a) canola (b) cottonseed
(c) sunflower (d) palm

652. Which of the following oil has lowest 16:0 fatty acid component
(a) canola (b) cottonseed
(c) olive (d) soybean

653. Which of the following oil has highest 18:0 fatty acid component
(a) canola (b) cottonseed
(c) soybean (d) cocoa butter

654. Which of the following oil has lowest 18:0 fatty acid component
(a) olive (b) sunflower
(c) palm (d) soybean

655. Which of the following oil has highest 18:1 fatty acid component
(a) canola (b) olive
(c) soybean (d) palm

656. Which of the following oil has lowest 18:1 fatty acid component
(a) canola (b) peanut
(c) cottonseed (d) palm

657. Which of the following oil has highest 18:2 fatty acid component
(a) olive (b) canola
(c) palm (d) sunflower

658. Which of the following oil has lowest 18:2 fatty acid component
(a) cocoa butter (b) olive
(c) canola (d) cottonseed

659. 16:0 fatty acid component (%) in palm oil is
(a) 4 (b) 24
(c) 14 (d) 44

660. 16:0 fatty acid component (%) in sunflower oil is
(a) 5 (b) 15
(c) 25 (d) 35

661. 18:0 fatty acid component (%) in cocoa butter is
(a) 4 (b) 14
(c) 24 (d) 34

662. 18:1 fatty acid component (%) in olive oil is
(a) 38 (b) 58
(c) 78 (d) 28

663. 18:1 fatty acid component (%) in sunflower oil is
(a) 10 (b) 20
(c) 30 (d) 40

664. 18:2 fatty acid component (%) in sunflower oil is
(a) 37 (b) 53
(c) 69 (d) 41

665. 16:0 fatty acid component (%) in soybean oil is
(a) 4 (b) 11
(c) 22 (d) 53

666. 18:0 fatty acid component (%) in soybean oil is
(a) 4 (b) 11
(c) 22 (d) 53

667. 18:1 fatty acid component (%) in soybean oil is
(a) 4 (b) 11
(c) 22 (d) 53

668. 18:2 fatty acid component (%) in soybean oil is
(a) 4 (b) 11
(c) 22 (d) 53

669. 16:0 fatty acid component (%) in peanut oil is
(a) 3 (b) 13
(c) 38 (d) 41

670. 18:0 fatty acid component (%) in peanut oil is
(a) 3 (b) 13
(c) 38 (d) 41

671. 18:1 fatty acid component (%) in peanut oil is
(a) 3 (b) 13
(c) 38 (d) 41

672. 18:2 fatty acid component (%) in peanut oil is
(a) 3 (b) 13
(c) 38 (d) 41

673. 16:0 fatty acid component (%) in cottonseed oil is
(a) 2 (b) 18
(c) 27 (d) 51

674. 18:0 fatty acid component (%) in cottonseed oil is
(a) 2 (b) 18
(c) 27 (d) 51

675. 18:1 fatty acid component (%) in cottonseed oil is
(a) 2 (b) 18
(c) 27 (d) 51

676. 18:2 fatty acid component (%) in cottonseed oil is
(a) 2 (b) 18
(c) 27 (d) 51

677. 18:0 fatty acid component (%) in palm oil is
(a) 4 (b) 11
(c) 39 (d) 44

678. 18:1 fatty acid component (%) in palm oil is
(a) 4 (b) 11
(c) 39 (d) 44

679. 18:2 fatty acid component (%) in palm oil is
(a) 4 (b) 11
(c) 39 (d) 44

680. 12:0 fatty acid component (%) in coconut fat is
(a) 1 (b) 9
(c) 18 (d) 46.4

681. 14:0 fatty acid component (%) in coconut fat is
(a) 1 (b) 9
(c) 18 (d) 46.4

682. 16:0 fatty acid component (%) in coconut fat is
(a) 1 (b) 9
(c) 18 (d) 46.4

683. 18:0 fatty acid component (%) in coconut fat is
(a) 1 (b) 9
(c) 18 (d) 46.4

684. 18:1 fatty acid component (%) in coconut fat is
(a) 1.6 (b) 7.6
(c) 9 (d) 18

685. 18:2 fatty acid component (%) in coconut fat is
(a) 1.6 (b) 7.6
(c) 9 (d) 18

686. Crude cottonseed oil contains phospholipid content (%) of
(a) 2-3 (b) 5-6
(c) 8-10 (d) 12

687. Crude soybean oil contains phospholipid content (%) of
(a) 2 (b) 5
(c) 8 (d) 11

688. Fish muscle tissue contain phospholipid content (%) of
(a) 0.7 (b) 1.7
(c) 2.7 (d) 3.7

689. Phospholipids contains ________ portion
(a) lipophilic (b) hydrophilic
(c) both (d) none

690. Phospholipids from soybean oil are used as
(a) sweetener
(b) anticaking agent
(c) emulsifier
(d) preservative

691. Soybean phospholipids contain
(a) lecithin (b) cephalin
(c) both (d) none

692. Soybean phospholipid contain ____% lecithin
(a) 25 (b) 35
(c) 55 (d) 65

693. Soybean phospholipid contain ____% cephalin
(a) 25 (b) 35
(c) 55 (d) 65

694. Unsaponifiable fraction of fat consists of
(a) sterols
(b) terpenic alcohols
(c) aliphatic alcohols
(d) all

695. Major component of unsaponifiable fraction of most of the fat is
(a) sterol
(b) terepenic alcohols
(c) aliphatic alcohols
(d) all

696. Which of the following is not a sterol
(a) sitosterol (b) campesterol
(c) stigmasterol (d) none

697. Cholesterol acetate generally has a melting point of about _____ °C
(a) 54 (b) 74
(c) 114 (d) 126

698. Phytosterol acetate generally has a melting point range of about _____ °C
(a) 100-125 (b) 126-137
(c) 151-165 (d) 170-178

699. Sterol content of lard is ______%
(a) 0.12 (b) 1.2
(c) 1.56 (d) 0.9

700. Sterol content of milk fat is ______%
(a) 0.1 (b) 0.3
(c) 0.5 (d) 0.9

701. Sterol content of cottonseed oil is _____%
(a) 0.4 (b) 1.4
(c) 2.4 (d) 3.4

702. Sterol content of soybean oil is _____%
(a) 0.1 (b) 0.7
(c) 1.4 (d) none

703. Sterol content of corn oil is ______%
(a) 0.1 (b) 0.5
(c) 1.0 (d) 2.0

704. Sterol content of rapeseed is ______%
(a) 0.4 (b) 0.8
(c) 1.2 (d) 1.6

705. Sterol content of coconut oil is ______%
(a) 0.02 (b) 0.08
(c) 0.2 (d) 0.8

706. Sterol content of cocoa butter is ______%
(a) 0.2 (b) 0.4
(c) 0.6 (d) 0.8

707. Which of the following is a part of auto-oxidation reaction
(a) initiation (b) propagation
(c) termination (d) all

708. Which of the following is involved during heating of fats
(a) dimer formation
(b) cis-trans isomerization
(c) formation of conjugated fatty acid
(d) all

709. Commercial frying temperature ranges from______ °C
(a) 100-140 (b) 160-195
(c) 225-260 (d) 300-345

710. Lower frying temperature leads to
(a) longer frying
(b) light product color
(c) higher oil uptake
(d) all

711. Iodine value of soybean oil is about
(a) 100 (b) 130
(c) 35 (d) 58

712. Iodine value of sunflower oil is about
(a) 100 (b) 120
(c) 140 (d) 150

713. Iodine value of high oleic sunflower oil is about
(a) 50 (b) 70
(c) 90 (d) 110

714. Iodine value of corn oil is about
(a) 50 (b) 70
(c) 90 (d) 110

715. Iodine value of cottonseed oil is about
(a) 58 (b) 78
(c) 98 (d) 118

716. Iodine value of canola oil is about
(a) 50 (b) 70
(c) 90 (d) 100

717. Iodine value of peanut oil is about
(a) 76 (b) 84
(c) 92 (d) 58

718. Iodine value of lard is about
(a) 40 (b) 60
(c) 80 (d) 100

719. Iodine value of olive oil is about
(a) 88 (b) 108
(c) 118 (d) 128

720. Iodine value of palm oil is about
(a) 25 (b) 35
(c) 55 (d) 65

721. Iodine value of palmolein oil is about
(a) 38 (b) 58
(c) 78 (d) 28

722. Iodine value of palm stearin oil is about
(a) 35 (b) 55
(c) 75 (d) 85

723. Iodine value of coconut oil is about
(a) 9 (b) 19
(c) 29 (d) 39

724. Melting point of oleic acid (cis) is about____ °C
(a) 13 (b) 44
(c) 70 (d) -5

725. Melting point of elaidic acid (trans) is about_____ °C
(a) 13 (b) 44
(c) 70 (d) -5

726. Melting point of stearic acid is about____ °C
(a) 13 (b) 44
(c) 70 (d) -5

727. Melting point of butyric acid is about____ °C
(a) 13 (b) -5
(c) 28 (d) -8

728. Emulsifiers with low HLB tend to form __________ emulsion
(a) w/o (b) o/w
(c) both (d) none

729. Emulsifiers with intermediate HLB tend to form __________ emulsion
(a) w/o (b) o/w
(c) both (d) none

730. HLB scale ranges about
(a) 0-20 (b) 10-20
(c) 0-35 (d) 0-50

731. HLB value of sorbitan tristearate is about
(a) 2.1 (b) 3.8
(c) 4.7 (d) 6.7

732. HLB value of glycerol monostearate is about
(a) 2.1 (b) 3.8
(c) 4.7 (d) 6.7

733. HLB value of sorbitan monostearate is about
(a) 2.1 (b) 3.8
(c) 4.7 (d) 6.7

734. HLB value of sorbitan monopalmitate is about
(a) 2.1 (b) 3.8
(c) 4.7 (d) 6.7

735. HLB value of GMS is about
(a) 3.8 (b) 4.7
(c) 6.7 (d) 11.0

736. Which of the following is a fat replacer
(a) olestra (b) salatrin
(c) both (d) none

737. Melting point of palmitic acid is about ____ °C
(a) 38 (b) 52
(c) 63 (d) 44

738. Melting point of caprylic acid is about ____ °C
(a) 3.4 (b) 5.7
(c) 16.7 (d) 23.2

739. Melting point of capric is about __ °C
(a) 3.4 (b) 5.7
(c) 16.7 (d) 31.6

740. Melting point of lauric acid is about ____ °C
(a) 14.2 (b) 34.2
(c) 44.2 (d) 24.2

741. Which of the following affect rate of oxidation of fats and oils
(a) amount of O_2
(b) degree of unsaturation
(c) both
(d) none

742. Smoke point of virgin olive oil is__ °C
(a) 119 (b) 169
(c) 155 (d) 199

743. Smoke point of refined soybean oil is____ °C
(a) 176 (b) 196
(c) 226 (d) 256

744. Flash point of virgin olive oil is___ °C
(a) 191 (b) 222
(c) 321 (d) 421

745. Flash point of refined soybean oil is ___ °C
(a) 196 (b) 226
(c) 326 (d) 356

746. Fire point of refined soybean oil is ___ °C
(a) 256 (b) 296
(c) 326 (d) 356

747. Fire point of virgin olive oil is____ °C
(a) 321 (b) 339
(c) 361 (d) 381

748. Reichert Meissl value of sunflower oil is about
(a) 0.1 (b) 0.5
(c) 0.92 (d) 4.45

749. Reichert Meissl value of peanut oil is about
(a) 0.1 (b) 0.2
(c) 0.4 (d) 0.9

750. Reichert Meissl value of cottonseed oil is about
(a) 0.15 (b) 0.35
(c) 0.65 (d) 0.95

751. Reichert Meissl value of corn oil is about
(a) 1.3 (b) 4.3
(c) 2.3 (d) 3.3

752. Reichert Meissl value of cod-liver oil is about
(a) 0.02 (b) 0.2
(c) 0.6 (d) 0.8

753. Reichert Meissl value of almond oil is about
(a) 0.5 (b) 1.0
(c) 1.5 (d) 2.0

754. Specific gravity of coconut oil is ____
(a) 0.895 (b) 0.926
(c) 0.964 (d) 0.974

755. Which of the following is polymorphic form of triacylalcohol
(a) α (b) β
(c) β' (d) all

756. Cholesterol content in egg yolk is _____ mg/100g
(a) 240 (b) 340
(c) 1010 (d) 540

757. Cholesterol content in butter is _____ mg/100g
(a) 240 (b) 340
(c) 440 (d) 540

758. Cholesterol content in pork meat (lean) is _____ mg/100g
(a) 50 (b) 60
(c) 70 (d) 240

759. Cholesterol content in beef is ______ mg/100g
(a) 50 (b) 60
(c) 70 (d) 240

760. Which of the following includes sphingolipids
(a) ceramides
(b) sphingomyelin
(c) cerebroside
(d) all

761. Which of the following is not associated with lipid processing
(a) degumming
(b) bleaching
(c) deodorization
(d) sterilization

762. Molecular weight of water molecule is
(a) 14 (b) 18
(c) 36 (d) 3

763. Tomato has about _______% water
(a) 60 (b) 70
(c) 80 (d) 90

764. Cabbage has about _______% water
(a) 60 (b) 70
(c) 80 (d) 90

765. Beer has about _______% water
(a) 60 (b) 70
(c) 80 (d) 90

766. Orange has about _______% water
(a) 67 (b) 77
(c) 87 (d) 94

767. Apple juice has about _______% water
(a) 67 (b) 77
(c) 87 (d) 94

768. Milk has about _______% water
(a) 67 (b) 77
(c) 87 (d) 94

769. Potato has about _____% water
(a) 67 (b) 77
(c) 87 (d) 94

770. Chicken has about _______% water
(a) 60 (b) 70
(c) 80 (d) 90

771. Meat has about ______% water
(a) 65 (b) 45
(c) 85 (d) 95

772. Bread has about ______% water
(a) 25 (b) 35
(c) 45 (d) 65

773. Jam has about _____% water
(a) 18 (b) 28
(c) 38 (d) 48

774. Honey has about _______% water
(a) 20 (b) 30
(c) 40 (d) 50

775. Butter has about _______% water
(a) 10 (b) 16
(c) 24 (d) 34

776. Wheat flour has about _______% water
(a) 2 (b) 12
(c) 24 (d) 34

777. Rice has about _______% water
(a) 6 (b) 12
(c) 18 (d) 24

778. Milk powder has about _______% water
(a) 0 (b) 3
(c) 6 (d) 7.5

779. Shortening has about _______% water
(a) 0 (b) 3
(c) 10 (d) 15

780. Density of water at 20°C is about _______ g/cm3
(a) 0.9998 (b) 0.9982
(c) 0.9718 (d) 0.9583

781. Specific heat of water at 20°C is about _______ cal/g °C
(a) 1.0074 (b) 0.9988
(c) 0.9994 (d) 1.0023

782. Thermal conductivity of water at 20°C is _________ kcal/m2h °C
(a) 0.486 (b) 0.590
(c) 0.515 (d) 0.561

783. Surface tension of water at 20°C is _________ dynes/cm
(a) 72.65 (b) 75.62
(c) 69.55 (d) 66.17

784. Viscosity of water at 20°C is _______ cps
(a) 1.792 (b) 0.653
(c) 0.466 (d) 1.002

785. Refractive index of water at 20°C is _______
(a) 1.3272 (b) 1.3306
(c) 1.333 (d) 1.3338

786. Dielectric constant of water at 20°C
(a) 88 (b) 80.4
(c) 73.3 (d) 66.7

787. Vapor pressure (mm Hg) of ice at 0°C is
(a) 4.58 (b) 3.01
(c) 1.95 (d) 1.24

788. Heat of fusion of ice at 0°C is _____ cal/g
(a) 7.98 (b) 62.9
(c) 79.8 (d) 53.7

789. Heat of sublimation of ice at 0°C is _______ cal/g
(a) 662.3 (b) 666.7
(c) 672.3 (d) 677.8

790. Density of ice at 0°C is _______ g/cm3
(a) 0.917 (b) 0.9168
(c) 0.9171 (d) 0.9175

791. Specific heat of ice at 0°C is _______ cal/g °C
(a) 0.4873 (b) 0.4770
(c) 0.4647 (d) 0.4504

792. Heat capacity of ice at 0°C is _______ joule/g
(a) 1.94 (b) 2.06
(c) 3.03 (d) 2.59

793. Bound water remain unfrozen at _______ °C
(a) 0 (b) -5
(c) -10 (d) all

794. Which of the following is associated with a_w range of 0.20-0.30
(a) WMP (b) corn flakes
(c) dried vegetables (d) all

795. Which of the following is associated with a_w range of 0.30-0.40
(a) cookies (b) crackers
(c) bread crust (d) all

796. Which of the following is associated with a_w range of 0.50-0.60
(a) pasta (b) spice
(c) dried herbs (d) all

797. Which of the following is associated with a_w range of 0.60-0.65
(a) toffees (b) caramel
(c) honey (d) all

798. Which of the following is associated with a_w range of 0.65-0.75
(a) jelly (b) molasses
(c) raw cane sugar (d) all

799. Which of the following is associated with a_w range of 0.75-0.80
(a) jam (b) marmalade
(c) marshmallows (d) all

800. Which of the following is associated with a_w range of 0.80-0.87
(a) fruit juice concentrate
(b) sweetened condensed milk
(c) chocolate syrup
(d) all

801. Which of the following is associated with a_w range of 0.87- 0.91
(a) salami (b) sponge cake
(c) margarine (d) all

802. Which of the following is associated with a_w range of 0.91-0.95
(a) cheddar cheese (b) ham
(c) swiss cheese (d) all

803. Which of the following is associated with a_w range of 0.95-1.00
(a) highly perishables
(b) canned fruits
(c) vegetables (d) all

804. a_w = ERH/?
(a) 10 (b) 50
(c) 75 (d) 100

805. Co-ordination number of water molecule in ice is
(a) 2 (b) 4
(c) 6 (d) 8

806. Melting point of water is _______ °C
(a) -5 (b) 0
(c) 50 (d) 100

807. Density of water is maximum at _______ °C
(a) 2.98 (b) 3.98
(c) 4.98 (d) 5.98

808. A plot of water content of food vs _______ is known as moisture absorption isotherm
(a) (t/t0)p (b) (p/po)T
(c) (d/do)m (d) (v/vo)m

809. Water plays critical role in following process in food
(a) physical (b) chemical
(c) microbiological (d) all

810. Solute addition _______ the freezing point of water
(a) increase (b) decrease
(c) no effect (d) none

811. Solute addition _______ the boiling point of water
(a) increase (b) decrease
(c) no effect (d) none

812. Which of the following is essential nutritive element
(a) Fe (b) Cu
(c) I (d) all

813. Which of the following is essential nutritive element
(a) Co (b) Mn
(c) Zn (d) all

814. Which of the following is essential nutritive element
(a) Cr (b) Ni
(c) Si (d) all

815. Which of the following is essential nutritive element
(a) F (b) Mo
(c) Se (d) all

816. Which of the following is not an essential nutritive element
(a) Mn (b) Si
(c) Se (d) Sb

817. Which of the following is non nutritive non toxic element
(a) Al (b) B
(c) Sn (d) all

818. Which of the following is non nutritive non toxic element
(a) Hg (b) Pb
(c) As (d) all

819. Which of the following is non nutritive non toxic element
(a) Cd (b) Sb
(c) both (d) none

820. Minerals in food are determined by
(a) freeze drying (b) incineration
(c) boiling (d) none

821. Ashing temperature is ______°C
(a) 350 (b) 550
(c) 750 (d) 800

822. Metal complexes are formed by
(a) ionic bond
(b) co-ordinate covalent bond
(c) hydrogen bond (d) none

823. In chelate, ligand functions as a
(a) lewis acid (b) lewis base
(c) both (d) none

824. In chelate, metal ion acts as a
(a) lewis acid (b) lewis base
(c) both (d) none

825. Which of the following is donor group in chelate
(a) $-NH_2$ (b) $=NH$
(c) $-COOH$ (d) all

826. Sodium content of cow milk is ________ mg/100ml
(a) 50 (b) 120
(c) 145 (d) 13

827. Potassium content of cow milk is ________ mg/100ml
(a) 50 (b) 120
(c) 145 (d) 13

828. Calcium content of cow milk is ________ mg/100ml
(a) 50 (b) 120
(c) 145 (d) 12

829. Magnesium content of cow milk is ________ mg/100ml
(a) 50 (b) 120
(c) 145 (d) 13

830. Total phosphorus content of cow milk is ________ mg/100ml
(a) 95 (b) 120
(c) 10 (d) 20

831. Chloride content of cow milk is ________ mg/100ml
(a) 95 (b) 100
(c) 10 (d) 20

832. Sulfate content of cow milk is ________ mg/100ml
(a) 95 (b) 100
(c) 10 (d) 20

833. Carbonate (as CO_2) content of cow milk is ________ mg/100ml
(a) 95 (b) 100
(c) 10 (d) 120

834. Citrate (as citric acid) content of cow milk is ______ mg/100ml
(a) 75 (b) 35
(c) 175 (d) 235

835. Total calcium content of beef is ______ mg/100g
(a) 8.6 (b) 3.8
(c) 17.7 (d) 24.4

836. Total magnesium content of beef is ______ mg/100g
(a) 8.6 (b) 3.8
(c) 17.7 (d) 24.4

837. Total citrate content of beef is ______ mg/100g
(a) 8.2 (b) 3.8
(c) 17.7 (d) 24.4

838. Total inorganic phosphorus content of beef is______ mg/100g
(a) 133 (b) 173
(c) 233 (d) 168

839. Sodium content of beef is______ mg/100g
(a) 133 (b) 173
(c) 233 (d) 168

840. Potassium content of beef is______ mg/100g
(a) 133 (b) 244
(c) 174 (d) 168

841. Chloride content of beef is______ mg/100g
(a) 48 (b) 68
(c) 128 (d) 158

842. Average potassium content of wheat grain is______ %
(a) 0.02 (b) 0.04
(c) 0.08 (d) 0.4

843. Average phosphorus content of wheat grain is______ %
(a) 0.02 (b) 0.04
(c) 0.08 (d) 0.4

844. Which of the following contains highest amount of calcium
(a) apple (b) orange
(c) grapes (d) banana

845. Which of the following contains highest amount of magnesium
(a) banana (b) apple
(c) orange (d) banana

846. Which of the following contains highest amount of potassium
(a) banana (b) apple
(c) orange (d) pear

847. Which of the following contains highest amount of phosphorus
(a) apple (b) orange
(c) pear (d) pineapple

848. Which of the following contains lowest amount of calcium
(a) orange (b) cherry
(c) apple (d) grapes

849. Which of the following contains lowest amount of magnesium
(a) orange (b) cherry
(c) apple (d) pear

850. Which of the following contains lowest amount of phosphorus
(a) pineapple (b) banana
(c) grape (d) orange

851. Which of the following contains lowest amount of potassium
(a) grape (b) cherry
(c) banana (d) apple

852. Which of the following is a good source of nickel
(a) nuts (b) legumes
(c) cocoa (d) all

853. Which of the following is a poor source of nickel
(a) cashew nut (b) cocoa
(c) peanut (d) spinach

854. Which of the following is a highest source of nickel

(a) red kidney beans
(b) frozen pea
(c) spinach
(d) milk chocolate

855. Calcium content (g/kg) present human body
(a) 0.4-0.5 (b) 1.0-1.2
(c) 10-20 (d) 2.0-2.5

856. Which of the following is not a trace element
(a) Fe (b) Zn
(c) Cu (d) K

857. Which of the following is a trace element
(a) K (b) Mg
(c) P (d) Cu

858. Which of the following is not a trace element
(a) I (b) Mn
(c) Co (d) all

859. Minerals perform which of the following function
(a) electrolyte
(b) enzyme constituent
(c) building material
(d) all

860. DRI is
(a) Daily reference intake
(b) Dietary required intake
(c) Dietary reference intake
(d) Daily required intake

861. Which of the following factor influence mineral bioavailability of food
(a) solubility of mineral
(b) food ligand (c) reductant
(d) all

862. Which of the following is a mineral bioavailability enhancer
(a) ascorbic acid (b) citric acid
(c) lactic acid (d) all

863. Which of the following is a mineral bioavailability inhibitor
(a) phytic acid (b) polyphenol
(c) tannin (d) all

864. Mineral composition of plant foods is affected by
(a) plant genetics (b) soil fertility
(c) environment (d) all

865. Food function of copper include
(a) catalyst (b) color modifier
(c) Texture stability (d) all

866. Which of the following is a dough improver
(a) Cu (b) Fe
(c) I (d) Se

867. Which of the following is a dough improver
(a) bromine (b) iodine
(c) both (d) none

868. Food function of aluminium include
(a) colorant
(b) leavening acid
(c) emulsifying agent
(d) all

869. Which of the following is a fat soluble vitamin
(a) A (b) D & E
(c) K (d) all

870. Which of the following is a water soluble vitamin
(a) B (b) C
(c) both (d) none

871. Which of the following is not a fat soluble vitamin
(a) A (b) B
(c) D (d) E

872. Milk and milk products provide which of the following vitamin
(a) A (b) D
(c) both (d) none

873. Which of the following is a source of thiamin
(a) meat (b) fish
(c) poultry (d) all

874. Bread contains which of the following vitamin
(a) thiamin (b) riboflavin
(c) niacin (d) all

875. Chemical name of vitamin A is
(a) tocopherol (b) calcipherol
(c) retinol (d) none

876. Chemical name of vitamin C is
(a) tocopherol (b) calcipherol
(c) ascorbic acid (d) retinol

877. Chemical name of vitamin D is
(a) tocopherol (b) calcipherol
(c) ascorbic acid (d) retinol

878. Chemical name of vitamin E is
(a) tocopherol (b) calcipherol
(c) ascorbic acid (d) none

879. Which of the following is a component of vitamin B-complex
(a) thiamin (b) riboflavin
(c) niacin (d) all

880. Vitamin H is also known as
(a) thiamin (b) niacin
(c) biotin (d) riboflavin

881. Citrus fruits contain
(a) β-cryptoxanthin
(b) β-carotene
(c) α-carotene
(d) all

882. Vitamin A is used for the fortification of
(a) skim milk (b) margarine
(c) both (d) none

883. Vitamin A is stable to heat in absence of
(a) water (b) oxygen
(c) carbon dioxide (d) none

884. Which of the following is a form of vitamin D
(a) ergocalciferol
(b) cholecalciferol
(c) both
(d) none

885. Which of the following lacks vitamin D
(a) plant products (b) egg
(c) milk (d) fish liver oil

886. Precursor of vitamin D2 is
(a) cholecalciferol
(b) ergocalciferol
(c) ergosterol
(d) 7-dehydrocholesterol

887. Precursor of vitamin D3 is
(a) cholecalciferol
(b) ergocalciferol
(c) ergosterol
(d) 7-dehydrocholesterol

888. Vitamin D content (μg/1000g edible portion) of eggs is about
(a) 0.9 (b) 2.5
(c) 44 (d) 12

889. Vitamin D content (μg/1000g edible portion) of milk is about
(a) 0.9 (b) 2.5
(c) 44 (d) 12

890. Which of the following affect vitamin D stability in milk
(a) Pasteurization (b) boiling
(c) sterilization (d) none

891. Cottonseed oil contain which of the following tocopherol
(a) alpha (b) beta
(c) gamma (d) all

892. Cereal germ oil contain tocopherol content (mg/100g) is
(a) 0.5-1.5 (b) 10-50
(c) 55-110 (d) 150-500

893. Vitamin K1 contains _________ double bond in side chain
(a) one (b) two
(c) three (d) four

894. Which of the following is a good source of vitamin K
(a) spinach (b) cabbage
(c) peas (d) all

895. Total tocopherol content (mg/100g edible portion) of wheat is about
(a) 2.0-2.5 (b) 0.4
(c) 2.9 (d) 7-10

896. Total tocopherol content (mg/100g edible portion) of polished rice is about
(a) 2.0-2.5 (b) 0.4
(c) 2.9 (d) 7-10

897. Total tocopherol content (mg/100g edible portion) of unpolished rice is about
(a) 2.0-2.5 (b) 0.4
(c) 2.9 (d) 7-10

898. Total tocopherol content (mg/100g edible portion) of corn is about
(a) 6.5 (b) 7.5
(c) 8.5 (d) 9.5

899. Vitamin C content (mg/100g edible portion) of cauliflower is about
(a) 7 (b) 37
(c) 70 (d) 117

900. Vitamin C content (mg/100g edible portion) of cabbage is about
(a) 40 (b) 60
(c) 80 (d) 100

901. Vitamin C content (mg/100g edible portion) of spinach is about
(a) 40 (b) 60
(c) 80 (d) 100

902. Vitamin C content (mg/100g edible portion) of orange is about
(a) 50 (b) 75
(c) 100 (d) 150

903. Vitamin C content (mg/100g edible portion) of lemon is about
(a) 50 (b) 75
(c) 100 (d) 150

904. Vitamin C content (mg/100g edible portion) of pea is about
(a) 25 (b) 50
(c) 75 (d) 100

905. Vitamin C content (mg/100g edible portion) of tomato is about
(a) 10 (b) 20
(c) 40 (d) 65

906. Vitamin C content (mg/100g edible portion) of apple is about
(a) 5 (b) 25
(c) 37.5 (d) 45

907. Thiamin content (mg/100g edible portion) of almond is about
(a) 0.11 (b) 0.24
(c) 0.37 (d) 0.03

908. Thiamin content (mg/100g edible portion) of corn is about
(a) 0.11 (b) 0.24
(c) 0.37 (d) 0.03

909. Thiamin content (mg/100g edible portion) of egg is about
(a) 0.11 (b) 0.24
(c) 0.37 (d) 0.03

910. Thiamin content (mg/100g edible portion) of milk is about
(a) 0.11 (b) 0.24
(c) 0.37 (d) 0.03

911. Thiamin content (mg/100g edible portion) of pea is about
(a) 0.8 (b) 0.18
(c) 0.28 (d) 0.48

912. Thiamin content (mg/100g edible portion) of potatoes is about
(a) 0.10 (b) 0.20
(c) 0.30 (d) 0.40

913. Riboflavin content (mg/100g edible portion) of beef is about
(a) 0.16 (b) 0.05
(c) 0.30 (d) 0.29

914. Riboflavin content (mg/100g edible portion) of cabbage is about
(a) 0.16 (b) 0.05
(c) 0.30 (d) 0.19

915. Riboflavin content (mg/100g edible portion) of eggs is about
(a) 0.16 (b) 0.05
(c) 0.30 (d) 0.19

916. Riboflavin content (mg/100g edible portion) of chicken is about
(a) 0.06 (b) 0.05
(c) 0.30 (d) 0.19

917. Riboflavin content (mg/100g edible portion) of pea is about
(a) 3.26 (b) 0.29
(c) 0.20 (d) 0.04

918. Riboflavin content (mg/100g edible portion) of beef liver is about
(a) 3.26 (b) 0.29
(c) 0.20 (d) 0.04

919. Riboflavin content (mg/100g edible portion) of spinach is about
(a) 3.26 (b) 0.29
(c) 0.20 (d) 0.04

920. Riboflavin is a type of
(a) vitamin (b) protein
(c) mineral (d) lignin

921. Vitamin B6 content (μg/g) of whole wheat bread is about
(a) 1.0 (b) 4.2
(c) 0.35 (d) 1.51

922. Vitamin B6 content (μg/g) of white bread is about
(a) 1.0 (b) 4.2
(c) 0.35 (d) 1.51

923. Vitamin B6 content (μg/g) of apple juice is about
(a) 1.0 (b) 4.2
(c) 0.35 (d) 1.51

924. Vitamin B6 content (μg/g) of tomatoes is about
(a) 1.0 (b) 4.2
(c) 0.35 (d) 1.51

925. Vitamin B6 content (μg/g) of yeast is about
(a) 10 (b) 20
(c) 35 (d) 50

926. Vitamin B6 is also known as
(a) niacin
(b) pyridoxine
(c) riboflavin
(d) cyanocobalamine

927. Vitamin B2 is also known as
(a) niacin
(b) pyridoxine
(c) riboflavin
(d) cyanocobalamine

928. Vitamin B1 is also known as
(a) thiamin (b) niacin
(c) riboflavin (d) pyridoxine

929. Vitamin B12 is also known as
(a) niacin
(b) pyridoxine
(c) riboflavin
(d) cyanocobalamine

930. Niacin content (mg/100g of edible portion) of mushroom is about
(a) 0.1 (b) 4.2
(c) 2.9 (d) 17.2

931. Niacin content (mg/100g of edible portion) of milk is about
(a) 0.1 (b) 4.2
(c) 2.9 (d) 17.2

932. Niacin content (mg/100g of edible portion) of peanuts is about
(a) 0.1 (b) 4.2
(c) 2.9 (d) 17.2

933. Niacin content (mg/100g of edible portion) of pea is about
(a) 0.1 (b) 4.2
(c) 2.9 (d) 17.2

934. Niacin content (mg/100g of edible portion) of beef liver is about
(a) 2.9 (b) 4.2
(c) 13.6 (d) 17.2

935. Niacin content (mg/100g of edible portion) of potatoes is about
(a) 36.7 (b) 4.3
(c) 0.6 (d) 1.5

936. Niacin content (mg/100g of edible portion) of spinach is about
(a) 36.7 (b) 4.3
(c) 0.6 (d) 1.5

937. Niacin content (mg/100g of edible portion) of wheat is about
(a) 36.7 (b) 4.3
(c) 0.6 (d) 1.5

938. Niacin content (mg/100g of edible portion) of dry yeast is about
(a) 36.7 (b) 4.3
(c) 0.6 (d) 1.5

939. Vitamin B12 content (μg/100g) of milk is about
(a) 0.28-1.55 (b) 3.2-12.4
(c) 14-152 (d) 600-970

940. Vitamin B12 content (μg/100g) of beef liver is about
(a) 0.28-1.55 (b) 3.2-12.4
(c) 14-152 (d) 600-970

941. Vitamin B12 content (μg/100g) of shellfish is about
(a) 0.28-1.55 (b) 3.2-12.4
(c) 14-152 (d) 600-970

942. Vitamin B12 content (μg/100g) of egg yolk is about
(a) 0.28-1.55 (b) 3.2-12.4
(c) 14-152 (d) 600-970

943. Folic acid contains
(a) pterin
(b) aminobenzoic acid
(c) glutamic acid
(d) all

944. RDA value (μg) of folic acid for adults is about
(a) 100 (b) 200
(c) 300 (d) 400

945. Folate content (μg/g) of boiled beef is about
(a) 0.03 (b) 0.07
(c) 0.16 (d) 0.30

946. Folate content (μg/g) of roasted chicken is about
(a) 0.03 (b) 0.07
(c) 0.16 (d) 0.30

947. Folate content (μg/g) of fried cod is about
(a) 0.03 (b) 0.07
(c) 0.16 (d) 0.30

948. Folate content (μg/g) of boiled egg is about
(a) 0.03 (b) 0.07
(c) 0.16 (d) 0.30

949. Folate content (μg/g) of boiled cabbage is about
(a) 0.29 (b) 0.12
(c) 2.0 (d) 0.11

950. Folate content (μg/g) of lettuce is about
(a) 0.29 (b) 0.12
(c) 2.0 (d) 0.11

951. Folate content (μg/g) of boiled potato is about
(a) 0.29 (b) 0.12
(c) 2.0 (d) 0.11

952. Folate content (μg/g) of boiled spinach is about
(a) 0.29 (b) 0.12
(c) 2.0 (d) 0.11

953. Folate content (μg/g) of tomato is about
(a) 0.18 (b) 0.45
(c) 0.0028 (d) 0.38

954. Folate content (μg/g) of orange is about
(a) 0.18 (b) 0.45
(c) 0.0028 (d) 0.38

955. Folate content (µg/g) of milk is about
(a) 0.18 (b) 0.45
(c) 0.0028 (d) 0.38

956. Pantothenic acid content (µg/g) of lean beef is about
(a) 10 (b) 6.5
(c) 1.0 (d) 0.7

957. Pantothenic acid content (µg/g) of wheat is about
(a) 5 (b) 11
(c) 6.5 (d) 1.0

958. Pantothenic acid content (µg/g) of potatoes is about
(a) 5 (b) 11
(c) 6.5 (d) 1.0

959. Pantothenic acid content (µg/g) of tomatoes is about
(a) 5 (b) 11
(c) 6.5 (d) 1.0

960. Pantothenic acid content (µg/g) of orange is about
(a) 5 (b) 0.7
(c) 6.5 (d) 1.0

961. Pantothenic acid content (µg/g) of walnut is about
(a) 8 (b) 2.7
(c) 16.5 (d) 46

962. Pantothenic acid content (µg/g) of broccoli is about
(a) 8 (b) 2.7
(c) 16.5 (d) 46

963. Biotin content (µg/100g) of tomatoes is about
(a) 5.2 (b) 1.0
(c) 2.6 (d) 96

964. Biotin content (µg/100g) of wheat is about
(a) 5.2 (b) 1.0
(c) 2.6 (d) 96

965. Biotin content (µg/100g) of beef is about
(a) 5.2 (b) 1.0
(c) 2.6 (d) 96

966. Biotin content (µg/100g) of beef liver is about
(a) 5.2 (b) 1.0
(c) 2.6 (d) 96

967. Biotin content (µg/100g) of lettuce is about
(a) 16 (b) 3.1
(c) 0.6 (d) 6.9

968. Biotin content (µg/100g) of mushroom is about
(a) 16 (b) 3.1
(c) 0.6 (d) 6.9

969. Biotin content (µg/100g) of potatoes is about
(a) 16 (b) 3.1
(c) 0.6 (d) 6.9

970. Biotin content (µg/100g) of spinach is about
(a) 16 (b) 3.1
(c) 0.6 (d) 6.9

971. Biotin content (µg/100g) of apples is about
(a) 6.9 (b) 0.9
(c) 1.9 (d) 34

972. Biotin content (µg/100g) of orange is about
(a) 6.9 (b) 0.9
(c) 1.9 (d) 34

973. Biotin content (µg/100g) of peanuts is about
(a) 6.9 (b) 0.9
(c) 1.9 (d) 34

974. Which of the following system is used to describe food color
(a) Munsell (b) Hunter
(c) Lovibond (d) all

975. CIE system is used for the determination of food
(a) flavor (b) color
(c) texture (d) all

976. Numbers of primary colors used in CIE system
(a) two (b) three
(c) four (d) five

977. Which of the following is used as primary color in CIE system
(a) red (b) blue
(c) green (d) all

978. Which of the following is used in Munsell system
(a) hue (b) value
(c) chroma (d) all

979. Number of attributes used in Munsell system to describe the color
(a) two (b) three
(c) four (d) one

980. Which of the following is used in Hunter system
(a) L (b) a
(c) b (d) all

981. Number of filters used in lovibond system
(a) two (b) three
(c) four (d) five

982. Lycopene content (mg/100g) of ripened tomato is
(a) 0.11 (b) 0.84
(c) 5.43 (d) 7.85

983. Carotene content (mg/100g) of ripened tomato is
(a) 0.11 (b) 0.73
(c) 5.43 (d) 0.06

984. Xanthphyll content (mg/100g) of ripened tomato is
(a) 0.11 (b) 0.73
(c) 5.43 (d) 0.06

985. Number of basic taste are
(a) one (b) two
(c) three (d) four

986. Glutamic acid content (%) in wheat gluten
(a) 16 (b) 17.6
(c) 24.5 (d) 36

987. Glutamic acid content (%) in zein
(a) 16 (b) 17.6
(c) 24.5 (d) 36

988. Glutamic acid content (%) in egg albumin
(a) 16 (b) 17.6
(c) 24.5 (d) 36

989. Glutamic acid content (%) in casein
(a) 16 (b) 22
(c) 36 (d) 42

990. Reversible deformation is also known as
(a) flow (b) elasticity
(c) both (d) none

991. Irreversible deformation is also known as
(a) flow (b) elasticity
(c) both (d) none

992. Rheological phenomenon depends on
(a) time (b) temperature
(c) both (d) none

993. Which of the following is not a type of viscosity
(a) intrinsic (b) relative
(c) dynamic (d) none

994. Which of the following is a unit of viscosity
(a) P (b) Pa.s
(c) cP (d) all

995. Viscosity of water at 20°C is ______ cP
(a) 1.0 (b) 4.28
(c) 2.12 (d) 84

996. Viscosity of whole milk at 20°C is ______ cP
(a) 1.0 (b) 4.28
(c) 2.12 (d) 84

997. Viscosity of molasses at 20°C is __ cP
(a) 1200 (b) 42.8
(c) 212 (d) 6600

998. Which of the following is a type of non Newtonian material
(a) dilatent (b) pseudoplastic
(c) plastic (d) all

999. Which of the following is not a type of viscometer
(a) capillary (b) rotational
(c) falling ball (d) none

1000. Enzyme are ______ in nature
(a) carbohydrate (b) lipid
(c) protein (d) all

1.3 ANSWER KEY

Q.	A.	Q.	A.	Q.	A.	Q.	A.	Q.	A.	Q.	A.
1.	(a)	2.	(c)	3.	(b)	4.	(a)	5.	(b)	6.	(d)
7.	(d)	8.	(b)	9.	(b)	10.	(a)	11.	(d)	12.	(c)
13.	(b)	14.	(b)	15.	(a)	16.	(a)	17.	(c)	18.	(c)
19.	(b)	20.	(c)	21.	(b)	22.	(c)	23.	(c)	24.	(c)
25.	(c)	26.	(b)	27.	(a)	28.	(b)	29.	(b)	30.	(b)
31.	(b)	32.	(b)	33.	(a)	34.	(c)	35.	(b)	36.	(b)
37.	(a)	38.	(b)	39.	(b)	40.	(a)	41.	(a)	42.	(b)
43.	(b)	44.	(a)	45.	(b)	46.	(c)	47.	(c)	48.	(b)
49.	(a)	50.	(c)	51.	(b)	52.	(d)	53.	(d)	54.	(a)
55.	(b)	56.	(b)	57.	(c)	58.	(c)	59.	(d)	60.	(c)
61.	(b)	62.	(a)	63.	(d)	64.	(c)	65.	(d)	66.	(a)
67.	(c)	68.	(d)	69.	(d)	70.	(b)	71.	(a)	72.	(a)
73.	(b)	74.	(d)	75.	(b)	76.	(a)	77.	(d)	78.	(d)
79.	(b)	80.	(b)	81.	(b)	82.	(c)	83.	(c)	84.	(d)
85.	(c)	86.	(b)	87.	(d)	88.	(c)	89.	(d)	90.	(a)
91.	(b)	92.	(a)	93.	(d)	94.	(a)	95.	(c)	96.	(c)
97.	(b)	98.	(c)	99.	(c)	100.	(c)	101.	(d)	102.	(b)
103.	(b)	104.		105.	(d)	106.	(c)	107.	(b)	108.	(c)
109.	(b)	110.	(d)	111.	(b)	112.	(c)	113.	(b)	114.	(d)
115.	(d)	116.	(d)	117.	(d)	118.	(c)	119.	(a)	120.	(d)
121.	(d)	122.	(a)	123.	(d)	124.	(d)	125.	(b)	126.	(c)
127.	(d)	128.	(c)	129.	(d)	130.	(b)	131.	(d)	132.	(c)
133.	(d)	134.	(d)	135.	(c)	136.	(b)	137.	(c)	138.	(c)
139.	(a)	140.	(b)	141.	(c)	142.	(b)	143.	(c)	144.	(b)
145.	(d)	146.	(c)	147.	(c)	148.	(b)	149.	(b)	150.	(d)
151.	(b)	152.	(b)	153.	(c)	154.	(d)	155.	(b)	156.	(a)
157.	(d)	158.	(b)	159.	(d)	160.	(c)	161.	(d)	162.	(b)

Q. A.	Q. A.	Q. A.	Q. A.	Q. A.	Q. A.
163. (d)	164. (c)	165. (c)	166. (d)	167. (b)	168. (a)
169. (c)	170. (d)	171. (b)	172. (c)	173. (c)	174. (b)
175. (d)	176. (d)	177. (a)	178. (c)	179. (a)	180. (b)
181. (c)	182. (c)	183. (a)	184. (c)	185. (c)	186. (d)
187. (d)	188. (d)	189. (c)	190. (d)	191. (c)	192. (d)
193. (d)	194. (c)	195. (c)	196. (d)	197. (d)	198. (b)
199. (c)	200. (b)	201. (d)	202. (b)	203. (a)	204. (a)
205. (c)	206. (a)	207. (b)	208. (a)	209. (a)	210. (c)
211. (a)	212. (c)	213. (b)	214. (d)	215. (c)	216. (b)
217. (d)	218. (b)	219. (d)	220. (a)	221. (b)	222. (c)
223. (a)	224. (c)	225. (a)	226. (d)	227. (a)	228. (c)
229. (c)	230. (b)	231. (d)	232. (a)	233. (b)	234. (b)
235. (c)	236. (a)	237. (b)	238. (b)	239. (a)	240. (b)
241. (b)	242. (b)	243. (a)	244. (b)	245. (b)	246. (c)
247. (c)	248. (b)	249. (a)	250. (a)	251. (a)	252. (d)
253. (d)	254. (c)	255. (a)	256. (c)	257. (d)	258. (d)
259. (c)	260. (c)	261. (b)	262. (d)	263. (b)	264. (a)
265. (c)	266. (c)	267. (d)	268. (b)	269. (b)	270. (b)
271. (c)	272. (c)	273. (c)	274. (b)	275. (b)	276. (c)
277. (c)	278. (d)	279. (d)	280. (b)	281. (a)	282. (a)
283. (c)	284. (d)	285. (a)	286. (a)	287. (b)	288. (a)
289. (a)	290. (c)	291. (a)	292. (a)	293. (b)	294. (a)
295. (a)	296. (b)	297. (b)	298. (a)	299. (a)	300. (b)
301. (b)	302. (a)	303. (b)	304. (b)	305. (b)	306. (a)
307. (b)	308. (b)	309. (b)	310. (a)	311. (a)	312. (a)
313. (a)	314. (b)	315. (b)	316. (c)	317. (b)	318. (c)
319. (c)	320. (b)	321. (b)	322. (c)	323. (d)	324. (a)
325. (a)	326. (c)	327. (d)	328. (c)	329. (b)	330. (b)
331. (c)	332. (b)	333. (a)	334. (a)	335. (b)	336. (b)
337. (c)	338. (d)	339. (c)	340. (d)	341. (c)	342. (a)
343. (d)	344. (d)	345. (c)	346. (a)	347. (d)	348. (b)
349. (b)	350. (c)	351. (b)	352. (b)	353. (b)	354. (d)
355. (c)	356. (c)	357. (b)	358. (c)	359. (b)	360. (c)
361. (a)	362. (b)	363. (b)	364. (d)	365. (d)	366. (b)
367. (b)	368. (c)	369. (b)	370. (a)	371. (a)	372. (a)

Q.	A.	Q.	A.	Q.	A.	Q.	A.	Q.	A.	Q.	A.
373.	(c)	374.	(c)	375.	(b)	376.	(b)	377.	(b)	378.	(c)
379.	(b)	380.	(a)	381.	(b)	382.	(a)	383.	(b)	384.	(b)
385.	(c)	386.	(d)	387.	(b)	388.	(a)	389.	(b)	390.	(c)
391.	(b)	392.	(a)	393.	(a)	394.	(b)	395.	(c)	396.	(d)
397.	(b)	398.	(c)	399.	(b)	400.	(c)	401.	(a)	402.	(b)
403.	(d)	404.	(c)	405.	(c)	406.	(a)	407.	(d)	408.	(c)
409.	(d)	410.	(b)	411.	(c)	412.	(c)	413.	(b)	414.	(b)
415.	(a)	416.	(c)	417.	(b)	418.	(a)	419.	(a)	420.	(b)
421.	(a)	422.	(a)	423.	(b)	424.	(d)	425.	(c)	426.	(a)
427.	(b)	428.	(c)	429.	(c)	430.	(c)	431.	(b)	432.	(b)
433.	(c)	434.	(b)	435.	(a)	436.	(b)	437.	(c)	438.	(a)
439.	(d)	440.	(b)	441.	(d)	442.	(a)	443.	(b)	444.	(c)
445.	(c)	446.	(a)	447.	(a)	448.	(b)	449.	(d)	450.	(b)
451.	(d)	452.	(c)	453.	(c)	454.	(c)	455.	(a)	456.	(c)
457.	(d)	458.	(c)	459.	(d)	460.	(d)	461.	(d)	462.	(d)
463.	(b)	464.	(c)	465.	(c)	466.	(d)	467.	(c)	468.	(c)
469.	(d)	470.	(a)	471.	(d)	472.	(d)	473.	(d)	474.	(c)
475.	(c)	476.	(d)	477.	(b)	478.	(d)	479.	(a)	480.	(b)
481.	(d)	482.	(d)	483.	(a)	484.	(c)	485.	(b)	486.	(b)
487.	(d)	488.	(d)	489.	(b)	490.	(b)	491.	(a)	492.	(a)
493.	(a)	494.	(c)	495.	(d)	496.	(d)	497.	(d)	498.	(c)
499.	(c)	500.	(d)	501.	(c)	502.	(a)	503.	(b)	504.	(c)
505.	(d)	506.	(c)	507.	(c)	508.	(d)	509.	(b)	510.	(c)
511.	(c)	512.	(d)	513.	(a)	514.	(c)	515.	(a)	516.	(c)
517.	(a)	518.	(b)	519.	(b)	520.	(c)	521.	(a)	522.	(d)
523.	(a)	524.	(a)	525.	(c)	526.	(a)	527.	(d)	528.	(d)
529.	(b)	530.	(c)	531.	(c)	532.	(c)	533.	(d)	534.	(d)
535.	(b)	536.	(a)	537.	(d)	538.	(b)	539.	(a)	540.	(b)
541.	(b)	542.	(b)	543.	(c)	544.	(b)	545.	(c)	546.	(c)
547.	(b)	548.	(c)	549.	(b)	550.	(d)	551.	(d)	552.	(d)
553.	(d)	554.	(d)	555.	(d)	556.	(c)	557.	(d)	558.	(a)
559.	(a)	560.	(c)	561.	(b)	562.	(b)	563.	(d)	564.	(a)
565.	(b)	566.	(d)	567.	(d)	568.	(d)	569.	(d)	570.	(b)
571.	(b)	572.	(a)	573.	(d)	574.	(c)	575.	(d)	576.	(c)
577.	(a)	578.	(c)	579.	(b)	580.	(d)	581.	(b)	582.	(c)

Q. A.	Q. A.	Q. A.	Q. A.	Q. A.	Q. A.
583. (a)	584. (a)	585. (a)	586. (b)	587. (a)	588. (d)
589. (d)	590. (d)	591. (d)	592. (a)	593. (b)	594. (c)
595. (b)	596. (c)	597. (a)	598. (b)	599. (c)	600. (a)
601. (a)	602. (b)	603. (d)	604. (d)	605. (c)	606. (b)
607. (c)	608. (d)	609. (a)	610. (b)	611. (c)	612. (b)
613. (d)	614. (d)	615. (d)	616. (c)	617. (d)	618. (a)
619. (b)	620. (c)	621. (d)	622. (a)	623. (b)	624. (a)
625. (a)	626. (b)	627. (c)	628. (c)	629. (c)	630. (a)
631. (b)	632. (c)	633. (c)	634. (a)	635. (b)	636. (c)
637. (b)	638. (a)	639. (b)	640. (a)	641. (b)	642. (c)
643. (b)	644. (b)	645. (b)	646. (c)	647. (b)	648. (c)
649. (d)	650. (a)	651. (d)	652. (a)	653. (d)	654. (a)
655. (b)	656. (c)	657. (d)	658. (a)	659. (d)	660. (a)
661. (d)	662. (c)	663. (b)	664. (c)	665. (b)	666. (a)
667. (c)	668. (d)	669. (b)	670. (a)	671. (c)	672. (d)
673. (c)	674. (a)	675. (b)	676. (d)	677. (a)	678. (c)
679. (b)	680. (d)	681. (c)	682. (b)	683. (a)	684. (b)
685. (a)	686. (a)	687. (a)	688. (a)	689. (c)	690. (c)
691. (c)	692. (b)	693. (d)	694. (d)	695. (a)	696. (d)
697. (c)	698. (b)	699. (a)	700. (b)	701. (b)	702. (b)
703. (c)	704. (a)	705. (b)	706. (a)	707. (d)	708. (d)
709. (b)	710. (d)	711. (b)	712. (b)	713. (c)	714. (d)
715. (c)	716. (d)	717. (c)	718. (b)	719. (a)	720. (c)
721. (b)	722. (a)	723. (a)	724. (a)	725. (b)	726. (c)
727. (d)	728. (a)	729. (b)	730. (a)	731. (a)	732. (b)
733. (c)	734. (d)	735. (d)	736. (c)	737. (c)	738. (c)
739. (d)	740. (c)	741. (c)	742. (d)	743. (d)	744. (c)
745. (c)	746. (d)	747. (c)	748. (b)	749. (c)	750. (d)
751. (b)	752. (b)	753. (a)	754. (b)	755. (d)	756. (c)
757. (a)	758. (c)	759. (b)	760. (d)	761. (d)	762. (b)
763. (d)	764. (d)	765. (d)	766. (c)	767. (c)	768. (c)
769. (b)	770. (b)	771. (a)	772. (b)	773. (b)	774. (a)
775. (b)	776. (b)	777. (b)	778. (b)	779. (a)	780. (b)
781. (b)	782. (c)	783. (a)	784. (d)	785. (c)	786. (b)
787. (a)	788. (c)	789. (d)	790. (b)	791. (a)	792. (b)

Q. A.	Q. A.	Q. A.	Q. A.	Q. A.	Q. A.
793. (d)	794. (d)	795. (d)	796. (d)	797. (d)	798. (d)
799. (d)	800. (d)	801. (d)	802. (d)	803. (d)	804. (d)
805. (b)	806. (b)	807. (b)	808. (b)	809. (d)	810. (b)
811. (a)	812. (d)	813. (d)	814. (d)	815. (d)	816. (d)
817. (d)	818. (d)	819. (c)	820. (b)	821. (b)	822. (b)
823. (b)	824. (a)	825. (d)	826. (a)	827. (c)	828. (b)
829. (d)	830. (a)	831. (b)	832. (c)	833. (d)	834. (c)
835. (a)	836. (d)	837. (a)	838. (c)	839. (d)	840. (b)
841. (a)	842. (d)	843. (d)	844. (b)	845. (d)	846. (a)
847. (b)	848. (c)	849. (c)	850. (a)	851. (d)	852. (d)
853. (d)	854. (d)	855. (c)	856. (d)	857. (d)	858. (d)
859. (d)	860. (c)	861. (d)	862. (d)	863. (d)	864. (d)
865. (d)	866. (c)	867. (c)	868. (d)	869. (d)	870. (c)
871. (b)	872. (c)	873. (d)	874. (d)	875. (c)	876. (c)
877. (b)	878. (a)	879. (d)	880. (c)	881. (d)	882. (c)
883. (b)	884. (c)	885. (a)	886. (c)	887. (d)	888. (c)
889. (a)	890. (d)	891. (d)	892. (d)	893. (a)	894. (d)
895. (d)	896. (b)	897. (c)	898. (d)	899. (c)	900. (b)
901. (b)	902. (a)	903. (a)	904. (a)	905. (b)	906. (a)
907. (b)	908. (c)	909. (a)	910. (d)	911. (c)	912. (a)
913. (a)	914. (b)	915. (c)	916. (d)	917. (b)	918. (a)
919. (c)	920. (a)	921. (b)	922. (a)	923. (c)	924. (d)
925. (d)	926. (b)	927. (c)	928. (a)	929. (d)	930. (b)
931. (a)	932. (d)	933. (c)	934. (c)	935. (d)	936. (c)
937. (b)	938. (a)	939. (b)	940. (c)	941. (d)	942. (a)
943. (d)	944. (d)	945. (a)	946. (b)	947. (c)	948. (d)
949. (d)	950. (c)	951. (b)	952. (a)	953. (a)	954. (b)
955. (c)	956. (a)	957. (b)	958. (c)	959. (d)	960. (b)
961. (a)	962. (d)	963. (b)	964. (a)	965. (c)	966. (d)
967. (b)	968. (a)	969. (c)	970. (d)	971. (b)	972. (c)
973. (d)	974. (d)	975. (b)	976. (b)	977. (d)	978. (d)
979. (b)	980. (d)	981. (c)	982. (d)	983. (b)	984. (d)
985. (d)	986. (d)	987. (d)	988. (a)	989. (b)	990. (b)
991. (a)	992. (c)	993. (d)	994. (d)	995. (a)	996. (c)
997. (d)	998. (d)	999. (d)	1000. (c)		

CHAPTER 2

FOOD MICROBIOLOGY

2.1 SUBJECTIVE KEYNOTES

1. Antonie Philips van Leeuwenhoek was a Dutch scientist known as "Father of Microbiology". Yeast cells were firstly observed by him in 1680
2. Appertization or canning was the first attempt for preservation of food via heat processing. Nicholas Appert is known as "Father of Food Processing". Pasteurization of milk generally destroys most of the pathogenic bacteria.
3. In 1857, Louis Pasteur demonstrated that souring in milk was caused by growth of microorganisms.
4. In 1902, Schimdt-Nielsen used the term psychrophile for microorganisms that grow at 0°C.
5. Richter coined the term osmophilic to describe yeasts that grow well in an environment of high osmotic pressure in 1912.
6. First isolation of *Bacillus coagulans* from coagulated milk was carried out by B. W. Hammer in 1915.
7. Microbial food spoilage may be due to bacteria, yeast and mold. Microorganisms are most dominant cause of food deterioration. Microorganisms are also used in preservation of food products. Lactic acid bacteria is used in cheese making and fermented dairy products. Yeast serves as leavening agent in bread production.
8. Bacterial growth curve is divided into four phases: lag, log, stationery and death phase.
9. Bacteria are classified according to their shape of cells. Cocci are spherical, bacilli are elongated cylindrical, spirilla and vibrio are spiral forms.
10. W. Taylor discovered that milk as a transmitter of typhoid fever in 1857.
11. Bacterial spores are more resistant to most processing conditions than yeast and mold spores.
12. Yeasts are unicellular plants widely distributed in nature. Yeast grow well in slightly acidic medium in presence of water & sugar.
13. Most fermentative yeasts are favored by pH of 4.0-4.5 and they cannot grow in alkaline pH.
14. Molds can grow over pH range of (1.5-11.0), yeasts (1.5-9.0) and bacteria (4.4-9.8)

15. Gram staining of bacterial cells was developed by Hans Christian Gram in 1884.
16. Molds are multicellular, filamentous fungi having cottony appearance in food.
17. Optimum temperature for growth of bacteria, yeasts and moulds ranges between 16-38°C.
18. Microorganisms that can grow at low temperature (below 7°C) storage are known as psychrophiles. Microorganisms that can grow at high temperature (55-65°C) are known as thermophiles. Microorganisms that can grow at 30-40°C are known as mesophiles.
19. Refrigerator temperature do not destroy pathogenic or spoilage microbes. It slow down the growth of microbes present in food.
20. Algae are associated with production of alginates, single cell protein (SCP) & toxin.
21. Most probable number (MPN) method for direct enumeration was introduced by McCrady in 1915.
22. Bacterial spoilage can cause food infection or intoxication. Clostridium botulinum can produce toxin which causes botulism.
23. Thermal death time curve is used to explain the relationship between number of microorganisms and heat treatment time at a particular temperature.
24. Microorganisms can be destroyed or inactivated by heat or cold treatment. Heat treatment used for microbial destruction involves pasteurization and sterilization. Cold treatments of food to check microbial growth are chilling, freezing and refrigeration
25. Nitrites and nitrates prevent botulism in meat products. Sulphur dioxide prevents further degradation in fruit products. Sorbates and benzoates are antifungal used in jams, salads, cheese and pickles.
26. Most of the Tetra Pak & bottled juices are pasteurized before packing.
27. Bacteria that can transmit food borne infections in humans are Clostridium, Salmonella, Shigella, Vibrio, Streptococcus, Bacillus and Campylobacter.
28. Microorganisms in food can also be controlled by drying treatment.
29. Water activity is generally the amount of water available for microorganisms.
30. Most bacteria require water activity in range of 0.90-1.00.
31. Some yeasts and molds can grow at water activity as low as 0.65.
32. Presence of sugar and salts also inhibits the growth of microorganisms.
33. Smoke can also be used for preservation of food products.
34. Chemical preservative are used to inhibit the growth of microorganisms are: sodium benzoate, sorbic acid, propionates, formate and sulphur dioxide.
35. Microorganisms can be inactivated by irradiation techniques such as x-rays, microwaves, UV rays and ionizing radiation.
36. Irradiation is also known as cold sterilization as it can sterilize the food without raising its temperature.
37. Microorganisms are also used in food fermentation processes such as alcohol production, beer making, cheese making, yoghurt manufacture, amino acid

production, bread production etc. Baker's yeast is used for bread fermentation. Brewer's yeast is used for beer fermentation

38. *Clostridium botulinum* is not viable in acidic foods (less than 4.6) hence these can be heat processed just below 100°C.
39. In low acid foods (pH greater than 4.6), sterilization is generally carried out in steam vessels called retort at temperature ranging from 116 °C-129 °C.
40. Intermediate moisture food (IMF) generally has water activity range of 0.60-0.80 and moisture content of 15-50%.
41. Lactobacillus bulgaricus and Streptococcus thermophilus are used as culture in yoghurt manufacture.
42. Bacillus species generally causes spoilage in UHT milk.
43. Kefir can be prepared using culture of Acetobacter, Lactobacillus and Lactococcus.
44. Sufu, tempeh and Tao-si are fermented products of soybean.
45. Saccharomyces cerevisiae is used in production of scotch whisky.
46. Leuconostoc species is used in idli fermentation.
47. Glycerol, glycols and sorbitols are used as humectants in food industry.
48. Aflatoxin, citrinin and ochrotoxin are mycotoxins.
49. With respect to gram staining, bacteria are divided into two groups: gram negative and gram positive.
50. Food infection is ingestion of viable pathogenic microorganisms. It is generally caused by Salmonella, *Bacillus cereus* and *E. coli*.
51. The term fermentation is derived from the Latin verb ***"fevere"*** which means **"to boil"** since the bubbling and foaming of early fermenting beverages seemed closely similar to boiling. Fermentation is one of the oldest food preservation method throughout the world.
52. Bacteriocins are the antimicrobial peptides produced by lactic acid bacteria and other bacteria.
53. Lactic acid bacteria is a natural group of gram-positive, non-motile, non-spore forming, rod-shaped and coccus-shaped microorganism that is capable of fermenting carbohydrates to form lactic acid.
54. Yeasts are unicellular microbes found in diverse habitats. Most important yeast in terms of desirable food fermentation is *Sachharomyces & Kluyeromyces.*
55. Molds are multicellular microorganisms with fiber like structures which can be observed on spoiled bread. These are having an important role in ripening and flavor development e.g. *Penicillium*.
56. Popular fermented milk products are acidophilus milk (*Lactobacillus acidophilus*), Bulgarian milk (*Lactobacillus bulgaricus*), dahi (Lactococcus sp. Lactobacillus sp.), cultured buttermilk (*Lactobacillus acidophilus*), yoghurt (*S. thermophilus, L. bulgaricus*), Kefir (*Lactococci, Leuconostoc, Lactobacilli, Acetobacter acetii, yeast*), Koumiss (*L. bulgaricus, L. acidophilus, yeast*).

57. Popular fermented vegetable products are Sauerkraut (*Leuconostoc mesenteroids, Lactobacillus plantarum, L. brevis*), Pickles (*Lactobacillus plantarum, Pediococcus cerevisiae*), Soy sauce (*Lactobacillus* sp., *Aspergillus oryzae*).
58. Commercially used probiotics are *Lactobacillus acidophilus, Lactobacillus casei, Lactobacillus rhamnosus, Lactobacillus reuteri, Bifidobacterium bifidum, Bifidobacterium longum, Saccharomyces boulardi.*
59. The starter culture was first used by Storch and Conn (1890) for developing flavor and aroma in butter.
60. Commercial culture were placed in market by Dannish farm of Hansen, today this is known as Chr. Hansen's Lab, the world supplier of starter culture.
61. Probiotic microorganism in Yakult probiotic product is *L. casei.*
62. Fermenting microorganisms in Dhokla are LAB & yeast.
63. Fermenting microorganisms in Dosa are *Leuconostoc, Lb. fermentum, Saccharomyces.*
64. Fermenting microorganisms in Jalebi are yeasts and lactobacilli.
65. Fermenting microorganism in Nan is yeast.
66. Fermenting microorganisms in Vada are *Leuconostoc, H. anomala, Saccharomyces.*
67. Toxic infection is the ingestion of toxigenic micro-organisms and formation of toxin inside the human system.
68. Food intoxication is the ingestion of microbial toxin which is pre-formed by microorganisms.
69. Contamination of raw milk upto magnitude of 100-1000 microorganisms/ml is practically unavoidable.
70. Coliforms are facultative anaerobes with an optimum growth at temperature of 37ºC. Coliforms are indicator microorganisms as they are closely associated with presence of pathogens.
71. Microorganisms that can grow in foods at refrigerated temperatures but have temperature optima above 20 ºC are called psychotrophs.
72. Minimum water activity required by most gram-negative (0.97), most gram-positive (0.90), halophilic bacteria (0.75), respectively.
73. Minimum water activity required by most yeasts (0.88), osmophilic yeasts (0.62), respectively.
74. Minimum water activity required by most filamentous fungi (0.80), xerotolerant fungi (0.71), xerophilic fungi (0.61), respectively.
75. Two main species of yeast used in beer manufacture are *Saccharomyces cerevisiae* (top-fermenting, lager beer) & *Saccharomyces uvarum* (bottom-fementing, ale beer).
76. Flavor/aroma changes by spoilage organisms include souring, alcoholic flavor, garlic, fruity, potato like flavor.
77. Texture problems caused by spoilage organisms include sliminess, ropiness, holes, softening/rotting, curdling.
78. Color changes by spoilage organisms include red spot, browning, blackening, discoloration.

79. Intrinsic parameters of food related to food spoilage are pH, moisture, nutrient, antimicrobials, oxidation-reduction potential.
80. Spoilage bacteria require minimum water activity of 0.90-0.91 for their growth.
81. Minimum water activity requirement for growth of spoilage yeast is 0.87-0.94.
82. *Clostridium botulinum* require minimum water activity of 0.70-0.80 for their growth.
83. Minimum water activity requirement for growth of Salmonella is 0.95-0.96.
84. *S. aureus* require minimum water activity of 0.86-0.92 for their growth.
85. Eggs contain antimicrobial agents like lysozyme that weakens cell wall of Gram-positive bacteria; conalbumin that chelates metals; avidin that binds biotin ovoflavoprotein that sequesters riboflavin.
86. Hypothiocyanite is an oxidizing agent present in foods that can inhibit growth and metabolism of many gram-positive and Gram-negative bacteria.
87. Lactoferrin is an iron binding protein inhibitor that inhibits microbes.
88. Gram-negative rods include coliforms like *Escherichia, Enterobacter, Klebsiella, Proteus* etc.
89. Gram-positive sporeformers include *Bacillus* and *Clostridium.*
90. Acidophiles are the microorganisms that grow on foods with pH 4.6 or below e.g. spoiled juice, pickle, fermented sausage.
91. Spoilage of meat due to surface slime is produced by species of *Psedomonas, Acinetobacter, Moraxella, Alcaligens, Micrococcus, Bacillus, Streptococcus, Leuconostoc.*
92. Budding growth of molds causes surface stickiness of meat.
93. In freezing condition, a limited amount of mycelial growth may take place without sporulation and produce growth known as whiskers.
94. Decomposition of protein is known as putrefaction which is produced by anaerobic microbes with foul-smelling compounds such as hydrogen sulphide, mercaptans, indole, ammonia etc.
95. Mycotoxins are the secondary metabolites produced during growth of many species of molds.
96. The chief pathogens involved in food poisoning are bacteria, viruses or protozoa.
97. Some of the common bacteria causing food borne illnesses include *Bacillus cereus, Aeromonas, hydrophila, Campylobacter jejuni, Clostridium botulinum; Clostridium perfringens,* some strains of *E. coli, Salmonella paratyphi, Shigella spp, Staphylococcus aureus, Vibrio cholerae, V. parahemolyticus Listeria monocytogenes, Salmonella typhi,* and *Yersinia enterocolitica.*
98. Common viruses that can cause food borne illness are Hepatitis A virus, Norovirus and Rotavirus .
99. Amongst protozoa, common pathogens associated with food borne illness are *Entamoeba histolytica , Giardia lamblia and Toxoplasma gondii .*
100. *Campylobacter jejuni* is one of the most prominent organism causing food borne infections.

2.2 MULTIPLE CHOICE QUESTIONS

1. Who invented canning of food
(a) Louis Pasteur
(b) Nicolas Appert
(c) T. Kensett (d) E. Daggett

2. Preservation of food by canning was patented by Appert in
(a) 1710 (b) 1782
(c) 1810 (d) 1834

3. Appert was awarded ______ francs by French Ministry for his invention
(a) 6000 (b) 8000
(c) 12000 (d) 30000

4. Pasteurization of milk began in Germany in year
(a) 1860 (b) 1880
(c) 1843 (d) 1890

5. Commercial pasteurization of milk began in United States in year
(a) 1860 (b) 1880
(c) 1843 (d) 1890

6. First bacteriological study of canning was made by Russel in
(a) 1890 (b) 1886
(c) 1895 (d) 1907

7. In ______, E. Metchnikoff and coworkers isolated and named one of the yoghurt bacteria, *Lactobacillus delbrueckii* subsp. *bulgaricus*
(a) 1890 (b) 1886
(c) 1895 (d) 1907

8. In ______, sodium benzoate was given official sanction by United States as preservative in certain foods
(a) 1892 (b) 1895
(c) 1887 (d) 1908

9. In ______, the first commercial use of controlled atmospheric storage of apples was made in Europe
(a) 1908 (b) 1915
(c) 1928 (d) 1940

10. The D value concept came into general use in
(a) 1910 (b) 1920
(c) 1930 (d) 1950

11. Nisin was accorded GRAS status in US in
(a) 1958 (b) 1968
(c) 1978 (d) 1988

12. Irradiation of poultry was approved in US in
(a) 1960 (b) 1990
(c) 1970 (d) 1980

13. Leeuwenhoek was the first to observe yeast cells in
(a) 1580 (b) 1680
(c) 1780 (d) 1880

14. Pasteur showed that souring of milk was caused by micro-organisms in it in year
(a) 1757 (b) 1840
(c) 1857 (d) 1887

15. 70 S ribosome unit is composed of
(a) 30 S (b) 50 S
(c) both (a) & (b) (d) 40 S

16. 50 S subunit is composed of
(a) 23 S (b) 5 S
(c) both (a) & (b) (d) 16 S

17. 30 S subunit is composed of
(a) 23 S (b) 16 S
(c) 5 S (d) all

18. Which of the following is associated with alpha subclass of proteobacteria
(a) Acetobacter (b) Gluconobacter
(c) Zymomonas (d) all

19. Which of the following is associated with beta subclass of proteobacteria
(a) Alkaligenes
(b) Chromobacterium
(c) Telluria
(d) all

20. Which of the following is associated with gamma subclass of proteobacteria
(a) Aeromonas (b) Azomonas
(c) Flavobacterium (d) all

21. Which of the following is associated with gamma subclass of proteobacteria
(a) Psychrobacter (b) Vibrio
(c) Xanthomonas (d) all

22. Proteobacteria are ______ bacteria
(a) gram +ve (b) gram –ve
(c) both (a) & (b) (d) none

23. Which of the following is not a bacteria
(a) Acinetobacter (b) Aeromonas
(c) Alcaligenes (d) Alternania

24. Which of the following is not a bacteria
(a) Arcobacteria (b) Bacillus
(c) Brochothrix (d) Botrytis

25. Which of the following is not a bacteria
(a) Campylobacter
(b) Carnobacterium
(c) Citrobacter
(d) Byssochlamys

26. Which of the following is not a bacteria
(a) Clostridium
(b) Cladosporium
(c) Cornebacterium
(d) Enterobacter

27. Which of the following is not a bacteria
(a) Enterococcus (b) Erwinia
(c) Escherichia
(d) Colletotrichum

28. Which of the following is not a bacteria
(a) Hafnia (b) Fusarium
(c) Flavobacterium (d) Kocuria

29. Which of the following is not a bacteria
(a) Geotrichium (b) Lactococcus
(c) Lactobacillus (d) Listeria

30. Which of the following is not a bacteria
(a) Mucor (b) Moraxella
(c) Pandoraea (d) Pantoea

31. Which of the following is not a bacteria
(a) Leuconostoc (b) Micrococcus
(c) Monilia (d) Moraxella

32. Which of the following is not a bacteria
(a) Pediococcus (b) Proteus
(c) Pseudomonas (d) Penicillium

33. Which of the following is not a bacteria
(a) Rhizopus
(b) Psychrobacter
(c) Serratia (d) Salmonella

34. Which of the following is not a bacteria
(a) Serratia (b) Shewanella
(c) Trichothecium (d) Shigella

35. Which of the following is not a bacteria
(a) Sphingomonas (b) Wallemia
(c) Staphyllococcus (d) Shigella

36. Which of the following is a bacteria
(a) Yersinia (b) Shigella
(c) Serratia (d) all

37. Which of the following is a bacteria
(a) Pantoea (b) Pandoraea
(c) Hafnia (d) all

38. Which of the following is a bacteria
(a) Erwinia (b) Kocuria
(c) Pediococcus (d) all

39. Which of the following is a bacteria
(a) Monilia (b) Botrytis
(c) Proteus (d) all

40. Which of the following is a bacteria
(a) Alcaligenes
(b) Colletotrichum
(c) both (a) & (b)
(d) none

41. Which of the following is not a mold
(a) Candida (b) Alternaria
(c) Aspergillus (d) all

42. Which of the following is not a mold
(a) Cryptococcus
(b) Aureobasidium
(c) Botritis
(d) all

43. Which of the following is not a mold
(a) Byssochlamys
(b) Cladosporium
(c) Pichia
(d) all

44. Which of the following is not a mold
(a) Kluyveromyces
(b) Colletotrichum
(c) Fusarium
(d) all

45. Which of the following is not a mold
(a) Monilia (b) Geotrichum
(c) Rhodotorula (d) all

46. Which of the following is not a mold
(a) Mucor (b) Penicillium
(c) Rhizopus (d) Saccharomyces

47. Which of the following is not a mold
(a) Trichothecium (b) Wallemia
(c) Yarowia (d) all

48. Which of the following is not a mold
(a) Trichosporon (b) Xeromyces
(c) Wallemia (d) all

49. Which of the following is a mold
(a) Trichothecium (b) Clostridium
(c) Kocuria (d) Pichia

50. Which of the following is a mold
(a) Candida (b) Cryptococcus
(c) Botrytis (d) all

51. Which of the following is not a bacteria
(a) Enterococcus (b) Lactococcus
(c) Erwinia (d) Cryptococcus

52. Which of the following is a type of yeast
(a) Brettanomyces (b) Candida
(c) Cryptococcus (d) all

53. Which of the following is a type of yeast
(a) Debaryomyces
(b) Hanseniaspora
(c) Issatchenkia
(d) all

54. Which of the following is a type of yeast
(a) Kluyveromyces (b) Pichia
(c) Rhodotorula (d) all

55. Which of the following is a type of yeast
(a) Saccharomyces
(b) Schizasaccharomyces
(c) Torulaspora
(d) all

56. Which of the following is a type of yeast
(a) Trichosporon
(b) Yarowia
(c) Zygosaccharomyces
(d) all

57. Which of the following is a protozoa
(a) Cyclospora (b) Entamoeba
(c) Toxoplasma (d) all

58. Which of the following is a bacterial plant pathogen
(a) Pseudomonas (b) Xanthomonas
(c) Corynebacterium (d) all

59. Which of the following affect microbial growth
(a) pH (b) moisture
(c) oxidation-reduction potential
(d) all

60. Which of the following affect microbial growth
(a) nutrients (b) antimicrobial
(c) both (a) & (b) (d) none

61. Which of the following is an intrinsic parameter for microbial growth
(a) pH (b) moisture
(c) oxidation-reduction potential
(d) all

62. Which of the following survive over broader pH range
(a) bacteria (b) yeast
(c) mold (d) none

63. Fruits generally undergo ________ spoilage
(a) yeast (b) mold
(c) both (a) & (b) (d) bacterial

64. Approximate pH of fresh broccoli is about
(a) 5.5 (b) 5.8
(c) 6.5 (d) 7.0

65. Approximate pH of cauliflower is about
(a) 5.0 (b) 5.6
(c) 6.5 (d) 7.0

66. Approximate pH of fresh sweet corn is about
(a) 5.3 (b) 6.0
(c) 6.3 (d) 7.3

67. Approximate pH of fresh cucumber is about
(a) 3.8 (b) 4.8
(c) 5.8 (d) 6.8

68. Approximate pH of potatoes is about
(a) 3.6 (b) 4.6
(c) 5.6 (d) 6.6

69. Minimum pH value required for the growth of *Aeromonas hydrophila*
(a) 4 (b) 5
(c) 6 (d) 7

70. Minimum pH value required for the growth of *Asaia siamensis*
(a) 3 (b) 4
(c) 5 (d) 6

71. Minimum pH value required for the growth of *Alicyclobacillus acidocaldarius*
(a) 2 (b) 3
(c) 4 (d) 5

72. Minimum pH value required for the growth of *Bacillus cereus*
(a) 3.9 (b) 4.9
(c) 5.9 (d) 6.9

73. Minimum pH value required for the growth of *Botrytis cinerea*
(a) 2 (b) 3
(c) 4 (d) 5

74. Minimum pH value required for the growth of *Clostridium botulinum* Group I
(a) 3.6 (b) 4.6
(c) 5 (d) 5.6

75. Minimum pH value required for the growth of *Clostridium botulinum* Group II
(a) 3.6 (b) 4.6
(c) 5 (d) 5.6

76. Minimum pH value required for the growth of *Clostridium perifringens*
(a) 3 (b) 4
(c) 5 (d) 6

77. Minimum pH value required for the growth of *E. coli* O157:H7
(a) 3.5 (b) 4.5
(c) 5.5 (d) 6.5

78. Minimum pH value required for the growth of *Gluconobacter* spp.
(a) 3.6 (b) 4.6
(c) 5.6 (d) 5.6

79. Minimum pH value required for the growth of *Lactobacillus brevis*
(a) 2.16 (b) 3.16
(c) 4.16 (d) 5.16

80. Minimum pH value required for the growth of *Lactobacillus plantarum*
(a) 3.12 (b) 3.34
(c) 4.12 (d) 4.34

81. Minimum pH value required for the growth of *Lactobacillus lactis*
(a) 2.3 (b) 3.3
(c) 4.3 (d) 5.3

82. Minimum pH value required for the growth of *Listeria monocytogenes*
(a) 2.1 (b) 3.1
(c) 4.1 (d) 5.1

83. Minimum pH value required for the growth of *Penicillium roqueforti*
(a) 2 (b) 3
(c) 4 (d) 5

84. Minimum pH value required for the growth of *Propionibacterium cyclohexanicum*
(a) 2.2 (b) 3.2
(c) 4.2 (d) 5.2

85. Minimum pH value required for the growth of *Plesiomonas shigelloides*
(a) 2.5 (b) 3.5
(c) 4.5 (d) 5.2

86. Minimum pH value required for the growth of *Pseudomonas fragi*
(a) 2 (b) 3
(c) 4 (d) 5

87. Minimum pH value required for the growth of *Salmonella* spp.
(a) 3.05 (b) 4.05
(c) 5.05 (d) 5.45

88. Minimum pH value required for the growth of *Shewanella putrefaciens*
(a) 2.4 (b) 3.4
(c) 4.4 (d) 5.4

89. Minimum pH value required for the growth of *Shigella flexneri*
(a) 4.75-5.5 (b) 3.75-4.0
(c) 4.0-4.5 (d) 3.0-3.75

90. Minimum pH value required for the growth of *Shigella sonnei*
(a) 3.5-4.0 (b) 4.5-5.0
(c) 5.5-6.0 (d) 4.0-4.5

91. Minimum pH value required for the growth of *Staphylococcus aureus*
(a) 2 (b) 3
(c) 4 (d) 5

92. Minimum pH value required for the growth of *Vibrio parahaemolyticus*
(a) 3.8 (b) 4.8
(c) 5.8 (d) 6.8

93. Minimum pH value required for the growth of *Yersinia enterocolitica*
(a) 3.18 (b) 4.18
(c) 5.18 (d) 6.18

94. Minimum pH value required for the growth of *Zygosaccharomyces bailii*
(a) 1.8 (b) 2.8
(c) 3.8 (d) 4.8

95. Approximate pH of butter is about
(a) 4.8-5.2 (b) 5.5-5.8
(c) 5.8-6.0 (d) 6.1-6.4

96. Approximate pH of buttermilk is about
(a) 4.5 (b) 5.5
(c) 6.5 (d) 6.0

97. Approximate pH of milk cream is about
(a) 4.5 (b) 5.5
(c) 5.8 (d) 6.5

98. Approximate pH of cheddar cheese is about
(a) 3.9 (b) 4.9
(c) 5.9 (d) 6.9

99. Approximate pH of chicken is about
(a) 4.2-4.5 (b) 4.8-5.5
(c) 5.5-5.8 (d) 6.2-6.4

100. Approximate pH of ham is about
(a) 4.2-4.5 (b) 4.8-5.5
(c) 5.9-6.1 (d) 6.5-6.9

101. Approximate pH of crabs is about
(a) 4 (b) 5
(c) 6 (d) 7

102. Which of the following require higher water activity for growth
(a) bacteria (b) fungi
(c) mold (d) none

103. Most spoilage bacteria do not grow below________ water activity
(a) 0.80 (b) 0.86
(c) 0.91 (d) none

104. *Staphylococcus aureus* can grow at lowest water activity of about
(a) 0.86 (b) 0.90
(c) 0.92 (d) 0.94

105. *Clostridium botulinum* does not grow below _______ water activity
(a) 0.86 (b) 0.90
(c) 0.92 (d) 0.94

106. Approximate minimum water activity value for growth of most of spoilage yeast
(a) 0.90 (b) 0.88
(c) 0.80 (d) none

107. Approximate minimum water activity value for growth of most of spoilage molds
(a) 0.90 (b) 0.88
(c) 0.80 (d) none

108. Approximate minimum water activity value for growth of most of halophilic bacteria
(a) 0.80 (b) 0.61
(c) 0.75 (d) none

109. Approximate minimum water activity value for growth of most of xerophilic molds
(a) 0.80 (b) 0.61
(c) 0.75 (d) none

110. Approximate minimum water activity value for growth of most of osmophilic yeasts
(a) 0.80 (b) 0.61
(c) 0.75 (d) none

111. Approximate minimum water activity value for growth of most of *Pseudomonas* spp.
(a) 0.97 (b) 0.95
(c) 0.94 (d) 0.93

112. Approximate minimum water activity value for growth of *E. coli*
(a) 0.82 (b) 0.96
(c) 0.94 (d) 0.93

113. Approximate minimum water activity value for growth of *Acinetobacter* sp.
(a) 0.82 (b) 0.96
(c) 0.94 (d) 0.93

114. Approximate minimum water activity value for growth of *Bacillus subtilis*
(a) 0.82 (b) 0.97
(c) 0.95 (d) 0.93

115. Approximate minimum water activity for growth of *Enterobacter aerogenes*
(a) 0.82 (b) 0.97
(c) 0.95 (d) 0.93

116. Approximate minimum water activity for growth of *Clostridium botuli-num* type A & type B
(a) 0.97 (b) 0.96
(c) 0.95 (d) 0.94

117. Approximate minimum water activity for growth of *Clostridium botulinum* type E
(a) 0.97 (b) 0.96
(c) 0.95 (d) 0.94

118. Approximate minimum water activity for growth of *Candida utilis*
(a) 0.97 (b) 0.96
(c) 0.95 (d) 0.94

119. Approximate minimum water activity for growth of *Vibrio parahaemolyticus*
(a) 0.97 (b) 0.96
(c) 0.95 (d) 0.94

120. Approximate minimum water activity for growth of *Botrytis cinerea*
(a) 0.93 (b) 0.94
(c) 0.95 (d) 0.96

121. Approximate minimum water activity for growth of *Rhizopus stolonifer*
(a) 0.94 (b) 0.93
(c) 0.96 (d) 0.97

122. Approximate minimum water activity for growth of *Candida scottii*
(a) 0.94 (b) 0.93
(c) 0.92 (d) 0.97

123. Approximate minimum water activity for growth of *Geotrichum candidum*
(a) 0.90 (b) 0.92
(c) 0.93 (d) 0.94

124. Approximate minimum water activity for growth of *Trichothecium* sp.
(a) 0.90 (b) 0.92
(c) 0.93 (d) 0.94

125. Which of the following is a *Microaerophilic bacteria*
(a) Lactobacilli (b) Campylobacter
(c) both (a) & (b) (d) none

126. Facultative anaerobes can grow under ________ condition
(a) aerobic (b) anaerobic
(c) both (a) & (b) (d) none

127. Eh (mV) of minced meat is generally around
(a) 20 (b) 200
(c) 2000 (d) 2500

128. ________ can lower the Eh of their environment
(a) aerobes (b) anaerobes
(c) both (a) & (b) (d) none

129. Natural antimicrobial agent present in cloves
(a) allicin (b) eugenol
(c) cinnamic aldehyde
(d) all

130. Natural antimicrobial agent present in garlic
(a) allicin (b) eugenol
(c) cinnamic aldehyde
(d) all

131. Natural antimicrobial agent present in cinnamon
(a) allicin
(b) eugenol
(c) cinnamic aldehyde
(d) all

132. Natural antimicrobial agent present in mustard
(a) eugenol
(b) allicin
(c) allylisothocyanate
(d) all

133. Natural antimicrobial agent present in sage
(a) eugenol (b) thymol
(c) both (a) & (b) (d) none

134. Natural antimicrobial agent present in cow milk
(a) lactoferrin
(b) conglutenin
(c) lactoperoxidase system
(d) all

135. Hydroxy cinnamic acid derivates found in
(a) tea (b) molasses
(c) vegetables (d) all

136. Lactoferrin is an iron binding
(a) lipoprotein (b) glycoprotein
(c) both (a) & (b) (d) none

137. Which of the following contain glucosinolates
(a) cabbage (b) broccoli
(c) turnip (d) all

138. Isothiocyantes shows ________ activity
(a) antifungal (b) antibacterial
(c) both (a) & (b) (d) none

139. Lactoperoxidase system consists of
(a) lactoperoxidase
(b) thiocyanate
(c) hydrogen peroxide
(d) all

140. Bovine milk normally contain lactoperoxidase (ppm)
(a) 30 (b) 110
(c) 160 (d) 240

141. Which of the following is an intrinsic parameter for microbial growth
(a) pH (b) temperature
(c) moisture (d) O/R potential

142. ____________ is extrinsic parameters for microbial growth
(a) temperature
(b) R.H.
(c) other microorganisms
(d) all

143. With respect to temperature requirement, microorganisms are of ______ type
(a) two (b) three
(c) four (d) five

144. Organisms that grow well at or below 7°C are called
(a) mesophiles (b) thermophiles
(c) psychrophiles (d) none

145. Organisms that grow well at or above 55°C are called
(a) mesophiles (b) thermophiles
(c) psychrophiles (d) none

146. Which of the following is a psychrotroph
(a) Pseudomonas (b) Enterococcus
(c) Micrococcus (d) all

147. Which of the following is a psychrotroph
(a) Psychrobacter
(b) Corynebacterium
(c) Flavobacterium
(d) all

148. Which of the following include thermophilic bacteria
(a) Clostridium
(b) Alicyclobacillus
(c) Thermoanerobacter
(d) all

149. Which of the following mold strain can grow at refrigerated temperature
(a) Aspergillus
(b) Cladosporium
(c) Thamnidium
(d) all

150. Ozone gas should not be used in storage of high ______ food
(a) carbohydrate (b) protein
(c) lipid (d) none

151. Which of the following is produced by food borne organisms and inhibits growth of other microorganisms
(a) antibiotics
(b) bacteriocins
(c) hydrogen peroxide
(d) all

152. Stiffening of muscles upon slaughter is due to formation of
(a) actin (b) myosin
(c) actomosin (d) none

153. Upon slaughter, O/R potential ____________
(a) decreases (b) increases
(c) do not change (d) none

154. Upon slaughter, fat in animal gets
(a) liquefy (b) solidify
(c) soluble (d) none

155. Upon slaughter, temperature of animal body
(a) decreases (b) increases
(c) do not change (d) none

156. Glycolysis after slaughtering decreases the pH level approximately upto
(a) 7.4 (b) 6.9
(c) 6.2 (d) 5.6

157. Which of the following is associated with rigor mortis
(a) stiffening of muscles
(b) O/R potential reduction
(c) reduction in animal body temperature
(d) all

158. Which of the following is associated with rigor mortis
(a) pH lowering
(b) protein denaturation
(c) solidification of body fat
(d) all

159. Which of the following is frequently reported in fresh meats
(a) Acinetobacter (b) Aeromonas
(c) Enterococcus (d) all

160. Which of the following is frequently reported in fresh meats
(a) Moraxella (b) Pseudomonas
(c) Psychrobacter (d) all

161. Which of the following is not occur in fresh meats
(a) Campylobacter (b) Clostridium
(c) Micrococcus (d) Salmonella

162. Which of the following occur in fresh meats
(a) Alcaligenes
(b) Corynebacterium
(c) Escherichia
(d) all

163. Which of the following occur in fresh meats
(a) Flavobacterium (b) Lactococcus
(c) Leuconostoc (d) all

164. Which of the following occur in fresh meats
(a) Listeria
(b) Microbacterium
(c) Micrococcus
(d) all

165. Which of the following occur in fresh meats
(a) Pediococcus
(b) Staphylococcus
(c) Citrobacter
(d) all

166. Which of the following is frequently reported in fresh livers
(a) Micrococcus
(b) Carnobacterium
(c) Bacillus
(d) all

167. Which of the following occur in fresh livers
(a) Acinetobacter (b) Alcaligenes
(c) Brochothrix (d) all

168. Which of the following occur in fresh livers
(a) Corynebacterium
(b) Enterococcus
(c) Escherchia
(d) all

169. Which of the following occur in fresh livers
(a) Flavobacterium (b) Kocuria
(c) Leuconostoc (d) all

170. Which of the following occur in fresh livers
(a) Moraxella
(b) Staphylococcus
(c) Weissella
(d) all

171. Which of the following is frequently reported in poultry
(a) Acinetobacter
(b) Campylobacter
(c) Corynebacterium
(d) all

172. Which of the following is frequently reported in poultry
(a) Listeria (b) Micrococcus
(c) Pseudomonas (d) all

173. Which of the following is frequently reported in poultry
(a) Vagococcus (b) Arcobacter
(c) Kocuria (d) Pediococcus

174. Which of the following is not occur in poultry
(a) Arcobacter (b) Carnobacter
(c) Caseobacter (d) all

175. Which of the following is not occur in poultry
(a) Escherchia (b) Hafnia
(c) Lactobacillus (d) all

176. Which of the following is not occur in poultry
(a) Lactococcus (b) Leuconostoc
(c) Pediococcus (d) all

177. Which of the following mold is frequently occur in fresh and refrigerated meats
(a) Cladosporium (b) Geotrichum
(c) Mucor (d) all

178. Which of the following mold is frequently occur in fresh and refrigerated meats
(a) Rhizopus (b) Sporotrichum
(c) Thamnidium (d) all

179. Which of the following yeast is frequently occur in fresh and refrigerated meats
(a) Candida
(b) Torulopsis
(c) both (a) & (b)
(d) Saccharomyces

180. Which of the following mold occur in fresh and refrigerated meats
(a) Alternaria (b) Aspergillus
(c) Aureobasidium (d) all

181. Which of the following mold occur in fresh and refrigerated meats
(a) Eurotium (b) Fusarium
(c) Monascus (d) all

182. Which of the following mold occur in fresh and refrigerated meats
(a) Monilia (b) Neurospora
(c) Penicillium (d) all

183. Which of the following yeast occur in fresh and refrigerated meats
(a) Cryptococcus
(b) Debaryomyces
(c) Hansenula (d) all

184. Which of the following yeast occur in fresh and refrigerated meats
(a) Pichia (b) Rhodotorula
(c) Trichosporon (d) all

185. Which of the following mold occur in poultry
(a) Alternaria (b) Aspergillus
(c) Cladosporium (d) all

186. Which of the following mold occur in poultry
(a) Geotrichum (b) Mucor
(c) Rhizopus (d) all

187. Which of the following yeast frequently found in poultry
(a) Candida
(b) Debaryomyces
(c) Rhodotorula (d) all

188. Which of the following yeast occur in poultry
(a) Cryptococcus (b) Pichia
(c) Saccharomyces (d) all

189. Which of the following yeast occur in poultry
(a) Torulopsis (b) Trichosporon
(c) both (a) & (b) (d) none

190. Which of the following has higher pH
(a) MDM
(b) Conventionally processed meat
(c) both (a) & (b) (d) none

191. Which of the following produce "whiskers" on beef
(a) Thamnidium (b) Mucor
(c) Rhizopus (d) all

192. Cladosporium causes ______ in beef
(a) whiskers (b) black spot
(c) green patches (d) white spots

193. Penicillium causes______ in beef
(a) whiskers (b) black spot
(c) green patches (d) white spots

194. Sporotrichum causes ______ in beef
(a) whiskers (b) black spot
(c) green patches (d) white spots

195. Chrysosporium causes ______ in beef
(a) whiskers (b) black spot
(c) green patches (d) white spots

196. Which of the is produced during spoilage of meats at low temperature
(a) ammonia
(b) hydrogen sulfide
(c) indole
(d) all

197. ______ is a off color spoilage compound in meats
(a) ammonia
(b) hydrogen sulfide
(c) indole
(d) all

198. Lysine is converted into ______ by decarboxylase
(a) cadaverine (b) putrescine
(c) both (a) & (b) (d) none

199. Ornithine is converted into ______ by decarboxylase
(a) cadaverine (b) putrescine
(c) both (a) & (b) (d) none

200. Arginine is converted into ______ by decarboxylase
(a) cadaverine (b) putrescine
(c) both (a) & (b) (d) none

201. Which of the following is present in fresh livers
(a) glycogen (b) lactic acid
(c) ammonia (d) all

202. Which of the following is a classic meat curing agent
(a) NaCl (b) nitrite
(c) nitrate (d) all

203. Which of the following is a major ingredient for curing
(a) NaCl (b) nitrite
(c) nitrate (d) sugar

204. Which of the following is curing adjunct
(a) phosphate
(b) sodium ascorbate
(c) potassium sorbate
(d) all

205. Which of the following is curing adjunct
(a) monosodium glutamate
(b) hydrolyzed vegetables protein
(c) lactates (d) all

206. In pickle curing of meat, which is used with NaCl
(a) water (b) spices
(c) oil (d) none

207. Function performed by sugar in curing
(a) color stabilization
(b) flavoring
(c) substrate for lactic fermentation
(d) all

208. Which of the following is used in curing brine
(a) sodium tripolyphosphate
(b) sodium hexametaphosphate
(c) both (a) & (b) (d) none

209. Gram negative bacteria most frequently reported in processed meat
(a) Lactobacillus (b) Pseudomonas
(c) Clostridium (d) all

210. Which of the following bacteria occur in processed meat
(a) Aeromonas (b) Alcaligenes
(c) Bacillus (d) all

211. Which of the following bacteria occur in processed meat
(a) Bronchothrix
(b) Carnobactreium
(c) Corynebacterium
(d) all

212. Which of the following bacteria occur in processed meat
(a) Enterobacter (b) Enterococcus
(c) Hafnia (d) all

213. Which of the following bacteria occur in processed meat
(a) Kocuria (b) Kurthia
(c) Lactococcus (d) all

214. Which of the following bacteria occur in processed meat
(a) Leuconostoc (b) Listeria
(c) Microbacterium (d) all

215. Which of the following bacteria occur in processed meat
(a) Micrococcus (b) Moraxella
(c) Paenibacillus (d) all

216. Which of the following bacteria occur in processed meat
(a) Pediococcus (b) Serratia
(c) Staphylococcus (d) all

217. Which of the following bacteria occur in processed meat
(a) Vibrio (b) Yersinia
(c) Macrococcus (d) all

218. Which of the following fungi is most frequently occurred in processed meat
(a) Debaryomyces (b) Aspergillus
(c) Penicillium (d) all

219. Which of the following yeast occur in processed meat
(a) Candida
(b) Saccharomyces
(c) Trichospora
(d) all

220. Which of the following mold occur in processed meat
(a) Alternaria (b) Botrytis
(c) Cladosporium (d) all

221. Which of the following mold occur in processed meat
(a) Fusarium (b) Geotrichum
(c) Monilia (d) all

222. Which of the following mold occur in processed meat
(a) Mucor (b) Rhizopus
(c) Scopulariopsis (d) all

223. Which of the following is a type of spoilage in processed meat
(a) sliminess (b) souring
(c) greening (d) all

224. Slimy spoilage in processed meats may be due to
(a) Lactobacillus (b) Enterococcus
(c) Weissela (d) all

225. Moldiness in bacon may be due to
(a) Aspergillus (b) Alternaria
(c) Fusarium (d) all

226. Moldiness in bacon may be due to
(a) Mucor (b) Rhizopus
(c) Botrytis (d) all

227. Bacteria which causes ham sours
(a) Bacillus (b) Pseudomonas
(c) Lactobacillus (d) all

228. Bacteria which causes ham sours
(a) Proteus (b) Micrococcus
(c) Clostridium (d) all

229. Which of the following is most frequently reported in spoiled fish and sea foods
(a) Aeromonas (b) Alcaligenes
(c) Bacillus (d) Enterobacter

230. Which of the following is most frequently reported in spoiled fish and sea foods
(a) Corynebacterium
(b) Enterococcus
(c) Pseudomonas
(d) Escherichia

231. Which of the following is most frequently reported in spoiled fish and sea foods
(a) Shewanella
(b) Lactobacillus (c) Listeria
(d) Microbacterium

232. Which of the following is most frequently reported in spoiled fish and sea foods
(a) Psychrobacter (b) vibrio
(c) Moraxella (d) Weissela

233. Which of the following is most frequently reported in spoiled fish and sea foods
(a) Debaryomyces (b) Hansenula
(c) Candida (d) Pichia

234. Which of the following is most frequently reported in spoiled fish and sea foods
(a) Cryptococcus
(b) Sporobolomyces
(c) Trichosporon
(d) Listeria

235. Which of the following is most frequently reported in spoiled fish and sea foods
(a) Pichia (b) Acinetobacter
(c) Flavobacterium (d) Rhodotorula

236. Which of the following is most frequently reported in spoiled fish and sea foods
(a) Aureobasidium
(b) Aspergillus
(c) Penicillium
(d) Scopulariopsis

237. Fresh iced fish are generally spoiled by
(a) bacteria (b) yeast
(c) mold (d) none

238. Salted and dried fish are generally spoiled by
(a) bacteria (b) fungi
(c) both (a) & (b) (d) none

239. Which of the following decreases during fish spoilage
(a) trimethylamine oxide
(b) creatine
(c) taurine (d) all

240. Which of the following is produced by fish spoilers
(a) propanol (b) ethanol
(c) isopropanol (d) all

241. In detection of microbial fish spoilage, TMAO is reduced into
(a) TMV (b) TMN
(c) TMA (d) none

242. Which of the following is not found in fresh fish
(a) TMAO (b) TMA
(c) both (a) & (b) (d) none

243. Which of the following is used as fish spoilage indicator
(a) histamine (b) diamine
(c) total volatiles (d) all

244. Histamine is produced from histidine by
(a) protease (b) lipase
(c) decarboxylase (d) none

245. Which of the following produce tyramine
(a) *Carnobacterium pissicola*
(b) *Weissella viridescens*
(c) both (a) & (b) (d) none

246. Which of the following is included in total volatile compound in fish
(a) TVB (b) TVA
(c) TVN (d) all

247. Total volatile acids in fish includes
(a) acetic (b) propionic
(c) both (a) & (b) (d) none

248. Which of the following is most predominant bacteria in shrimp spoilage at 0°C
(a) Pseudomonas (b) Moraxella
(c) Proteus (d) all

249. Which of the following is most predominant bacteria in shrimp spoilage at 11°C
(a) Pseudomonas (b) Moraxella
(c) Proteus (d) all

250. Which of the following is most predominant bacteria in shrimp spoilage at 22°C
(a) Pseudomonas (b) Moraxella
(c) Proteus (d) all

251. Which of the following generally causes spoilage in oysters
(a) Serratia (b) Pseudomonas
(c) Proteus (d) all

252. Which of the following generally causes spoilage in oysters
(a) Clostridium (b) Bacillus
(c) Escherichia (d) all

253. Which of the following generally causes spoilage in oysters
(a) Enterobacter
(b) Pseudoalteromonas
(c) Shewanella
(d) all

254. Which of the following generally causes spoilage in oysters
(a) Lactobacillus
(b) Flavobacterium
(c) Micrococcus
(d) all

255. Which of the following pH indicate good microbial quality of oysters
(a) 5.2 & below (b) 5.9-6.2
(c) 5.6 (d) 5.5-5.7

256. Aerobic plate count of common vegetables is generally around _____ log_{10} cfu/g
(a) 3-4 (b) 0-2
(c) 6-7 (d) 11-12

257. Bacterial soft rot in vegetables is caused by
(a) Erwinia
(b) Pectobacterium
(c) both (a) & (b)
(d) none

258. Which of the following is associated with vascular wilt, leaf and fruit spot on tomatoes
(a) *Cornybacterium michiganenese*
(b) *Cornybacterium nebraskense*
(c) *Cornybacterium sepedonicum*
(d) *Pseudomonas corrupata*

259. Which of the following is associated with leaf spot, leaf blight and wilt of corn
(a) *Cornybacterium michiganenese*
(b) *Cornybacterium nebraskense*
(c) *Cornybacterium sepedonicum*
(d) *Pseudomonas corrupata*

260. Which of the following is associated with tuber root of white potatoes
(a) *Cornybacterium michiganenese*
(b) *Cornybacterium nebraskense*
(c) *Cornybacterium sepedonicum*
(d) *Pseudomonas corrupata*

261. Which of the following is associated with bacterial wilt of beans
(a) *Curtobacterium flaccumfaciens*
(b) *Acidovorax valerianelle*
(c) *Pseudomonas pisi*
(d) Pseudomonas lachrymans

262. Which of the following is associated with drippy gill of mushroom
(a) *Pseudomonas agarici*
(b) *Pseudomonas tolassi*
(c) both (a) & (b)
(d) none

263. Which of the following is associated with tomato pith necrosis
(a) *Pseudomonas corrupta*
(b) *Pseudomonas agarici*
(c) *Pseudomonas tolassi*
(d) none

264. Which of the following is associated with bacterial zonite spot of cabbage
(a) *Pseudomonas cichorii*
(b) *Pseudomonas corrupta*
(c) *Pseudomonas agarici*
(d) none

265. Which of the following is associated with halo blight of beans
(a) *Pseudomonas phaseolicola*
(b) *Pseudomonas syringae*
(c) *Pseudomonas pisi*
(d) none

266. Which of the following is associated with bacterial canker of stonefruit trees
(a) *Pseudomonas phaseolicola*
(b) *Pseudomonas syringae*
(c) *Pseudomonas pisi*
(d) none

267. Which of the following is a tomato disease
(a) nailhead spot
(b) bacterial spot
(c) bacterial canker
(d) all

268. Which of the following is a lima bean disease
(a) pod blight (b) seed spotting
(c) yeast spot (d) all

269. Which of the following is a pea disease
(a) pod spot (b) Anthracnose
(c) both (a) & (b) (d) none

270. Black rot in onion may be due to
(a) *Aspergillus niger*
(b) *Ceratocystis fibriata*
(c) *Penicillium digtatum*
(d) none

271. Black rot in cabbage may be due to
(a) *Aspergillus niger*
(b) *Ceratocystis fibriata*
(c) *Penicillium digtatum*
(d) none

272. Black rot in sweet potatoes may be due to
(a) *Aspergillus niger*
(b) *Ceratocystis fibriata*
(c) *Penicillium digtatum*
(d) none

273. Blue mold rot in citrus fruits may be due to
(a) *Aspergillus niger*
(b) *Ceratocystis fibriata*
(c) *Penicillium digtatum*
(d) none

274. Brown rot in peaches may be due to
(a) Monilinia
(b) *Phytophthora* spp.
(c) both (a) & (b)
(d) none

275. Brown rot in cherries may be due to
(a) Monilinia
(b) *Phytophthora* spp.
(c) both (a) & (b)
(d) none

276. Brown rot in citrus fruit may be due to
(a) Monilinia
(b) *Phytophthora* spp.
(c) both (a) & (b)
(d) none

277. Crown rot in bananas may be due to
(a) *Colletotrichium musae*
(b) *Fusarium roseum*
(c) *Verticillium theobromae*
(d) all

278. Crown rot in bananas may be due to
(a) *Colletotrichium musae*
(b) *Ceratocystis paradoxa*
(c) both (a) & (b)
(d) all

279. Dry rot in potatoes may be due to
(a) Fusarium (b) Botrytis
(c) Penicillium (d) none

280. Gray mold rot in grapes is due to
(a) Fusarium (b) Botrytis
(c) Penicillium (d) none

281. Green mold rot in citrus fruits may be due to
(a) Fusarium (b) Botrytis
(c) Penicillium (d) none

282. Lenticel rot in apples may be due to
(a) *Cryptosporiopsis malicortis*
(b) *Phyletaena vagabunda*
(c) both (a) & (b)
(d) none

283. Pineapple black rot may be due to
(a) *Ceratocytis paradoxa*
(b) *Colletotrichum coccodes*
(c) both (a) & (b)
(d) none

284. Slimy brown rot in vegetables may be due to
(a) Rhizocotonia (b) Penicillium
(c) Aspergillus (d) None

285. Smut is also known as
(a) green mold rot
(b) black mold rot
(c) brown mold rot
(d) none

286. Sour rot in tomatoes may be due to
(a) Rhizopus
(b) Phytophora
(c) Geotrichum candidum
(d) none

287. Sour rot in citrus fruits may be due to
(a) Rhizopus
(b) Phytophora
(c) Geotrichum candidum
(d) none

288. Watery soft rot in carrots may be due to
(a) Sclerotinia (b) Fusarium
(c) Ustilago (d) none

289. Wheat scab in wheat may be due to
(a) Sclerotinia (b) Fusarium
(c) Ustilago (d) none

290. Corn smut in corn may be due to
(a) Sclerotinia (b) Fusarium
(c) Ustilago (d) none

291. Rice blast in rice may be due to
(a) *Magnaporaathe grisea*
(b) *Phytophthora infestens*
(c) Cinnamomi (d) Sojae

292. Potato blight in potatoes may be due to
(a) *Magnaporaathe grisea*
(b) *Phytophthora infestens*
(c) Cinnamomi (d) Sojae

293. Blight in chest nut tree may be due to
(a) *Magnaporaathe grisea*
(b) *Phytophthora infestens*
(c) Cinnamomi (d) Sojae

294. Rice blast in rice may be due to
(a) *Magnaporaathe grisea*
(b) *Phytophthora infestens*
(c) Cinnamomi
(d) Sojae

295. Root rot in soybeans may be due to
(a) *Magnaporaathe grisea*
(b) *Phytophthora infestens*
(c) Cinnamomi
(d) Sojae

296. Gray mold rot generally occurs in
(a) onion (b) lettuce
(c) cauliflower (d) all

297. Gray mold rot generally occurs in
(a) onion (b) garlic
(c) carrots (d) all

298. Sour rot generally occurs in
(a) broccoli (b) radish
(c) turnip (d) all

299. Photophthora rot generally occurs in
(a) Watermelons (b) tomatoes
(c) egg plant (d) all

300. Anthracnose generally occurs in
(a) cucumber (b) pumpkin
(c) pepper (d) all

301. Blue mold rot generally occurs in
(a) apples (b) pears
(c) both (a) & (b) (d) none

302. Which of the following in an onion disease
(a) white rot
(b) black mold rot
(c) diplodia stain (d) all

303. Which of the following is not a fermented product
(a) ripened cheese (b) sauerkraut
(c) pickles (d) none

304. Koji is a fermented product made from
(a) wheat (b) rice
(c) soybean (d) all

305. For ale production, ________ fermentation yeast is used
(a) top (b) bottom
(c) both (a) & (b) (d) none

306. For lager beer production, ________ fermentation yeast is used
(a) top (b) bottom
(c) both (a) & (b) (d) none

307. Which of the following is a lactic acid bacteria
(a) Carnobacterium (b) Enterococcus
(c) Lactococcus (d) all

308. Which of the following is a lactic acid bacteria
(a) Lactobacillus (b) Lactosphaera
(c) Leuconostoc (d) all

309. Which of the following is a lactic acid bacteria
(a) Oenococcus
(b) Pediococcus
(c) Paralactobacillus
(d) all

310. Which of the following is a lactic acid bacteria
(a) Streptococcus
(b) Tetragenococcus
(c) Vagococcus (d) all

311. Which of the following is a lactic acid bacteria
(a) Pediococcus (b) Weissekka
(c) both (a) & (b) (d) none

312. Which of the following are related to LAB but not considered into this group
(a) Aerococcus
(b) Microbacterium
(c) Propionibacterium
(d) all

313. Which of the following Lactobacillus (LAB) is homo-fermentative
(a) *L. acetotolerans* (b) *L. brevis*
(c) *L. buchneri* (d) none

314. Which of the following Lactobacillus (LAB) is homo-fermentative
(a) *L. cellobiosus* (b) *L. acidipiscis*
(c) *L. corprophilus* (d) none

315. Which of the following Lactobacillus (LAB) is homo-fermentative
(a) *L. acidophilus* (b) *L. fermentum*
(c) *L. hilgardii* (d) all

316. Which of the following Lactobacillus (LAB) is homo-fermentative
(a) *L. sanfranciscensis*
(b) *L. trichoides*
(c) *L. alimentarius* (d) all

317. Which of the following Lactobacillus (LAB) is homo-fermentative
(a) *L. casei* (b) *L. pontis*
(c) *L. fructivorans* (d) all

318. Which of the following Lactobacillus (LAB) is homo-fermentative
(a) *L. kimchii*
(b) *L. fructivorans*
(c) *L. coryniformis* (d) all

319. Which of the following Lactobacillus (LAB) is homo-fermentative
(a) *L. curvatus*
(b) *L. nilgardii*
(c) *L. paralimentarius*
(d) all

320. Which of the following Lactobacillus (LAB) is homo-fermentative
(a) *L. pontis* (b) *L. panis*
(c) *L. melibiosus* (d) all

321. Which of the following Lactobacillus (LAB) is homo-fermentative
(a) *L. delbrueckii* (b) *L. kimchii*
(c) *L. sakei* (d) all

322. Which of the following lactic acid bacteria (LAB) is homo-fermentative
(a) *L. carnosus* (b) *L. bulgaricus*
(c) *L. panis* (d) all

323. Which of the following Lactobacillus (LAB) is homo-fermentative
(a) *L. lactis*
(b) *L. coprophilus*
(c) *L. brevis* (d) all

324. Which of the following Lactobacillus (LAB) is homo-fermentative
(a) *L. brevis* (b) *L. buchneri*
(c) *L. fuchuensis* (d) all

325. Which of the following Lactobacillus (LAB) is homo-fermentative
(a) *L. helveticus* (b) *L. sakei*
(c) *L. trichoides* (d) all

326. Which of the following Lactobacillus (LAB) is homo-fermentative
(a) *L. cellobiosus* (b) *L. jugurti*
(c) *L. hilgardii* (d) all

327. Which of the following Lactobacillus (LAB) is homo-fermentative
(a) *L. jensenii*
(b) *L. kefiranofaciens*
(c) *L. kefirgranum* (d) all

328. Which of the following Lactobacillus (LAB) is homo-fermentative
(a) *L. leichmannii* (b) *L. mindensis*
(c) *L. plantarum* (d) all

329. Which of the following Lactobacillus (LAB) is homo-fermentative
(a) *L. salivarius* (b) *L. sakei*
(c) *L. ferementum* (d) all

330. Which of the following Lactobacillus (LAB) is hetero-fermentative
(a) *L. brevis* (b) *L. buchneri*
(c) *L. cellobiosus* (d) all

331. Which of the following Lactobacillus (LAB) is hetero-fermentative
(a) *L. coprophilus* (b) *L. fermentum*
(c) *L. hilgardii* (d) all

332. Which of the following Lactobacillus (LAB) is hetero-fermentative
(a) *L. sanfranciscensis*
(b) *L. trichoides* (c) *L. pontis*
(d) all

333. Which of the following Lactobacillus (LAB) is hetero-fermentative
(a) *L. fructivorans* (b) *L. casei*
(c) *L. acidipiscis* (d) all

334. Which of the following Lactobacillus (LAB) is hetero-fermentative
(a) *L. jugurti* (b) *L. curvatus*
(c) *L. kimchii* (d) all

335. Which of the following Lactobacillus (LAB) is hetero-fermentative
(a) *L. paralimentarius*
(b) *L. bulgaricus*
(c) *L. coryniformis* (d) all

336. Which of the following Lactobacillus (LAB) is hetero-fermentative
(a) *L. panis* (b) *L. jensenii*
(c) L. mindensis (d) all

337. Which of the following Lactobacillus (LAB) is hetero-fermentative
(a) L. plantarum (b) *L. salivarius*
(c) *L. sakei* (d) all

338. Which of the following Lactobacillus (LAB) is hetero-fermentative
(a) *L. helveticus* (b) *L. carnosus*
(c) *L. fuchuensis* (d) all

339. Which of the following Lactobacillus (LAB) is hetero-fermentative
(a) *L. panis* (b) *L. sakei*
(c) *L. kimchii* (d) all

340. Which of the following Lactobacillus (LAB) is not hetero-fermentative
(a) *L. jugurti* (b) *L. pontis*
(c) *L. hilgardii* (d) all

341. Which of the following Lactococcus (LAB) is homo-fermentative
(a) *L. lactis* (b) *L. cremoris*
(c) *L. diacetylactis* (d) all

342. Which of the following Lactococcus (LAB) is homo-fermentative
(a) L. hordniae (b) L. garvieae
(c) L. plantarum (d) all

343. Which of the following LAB is homo-fermentative
(a) Lactobacillus brevis
(b) Paralactobacillus selangorensis
(c) Leuconostoc citreus
(d) Oenococcus oeni

344. Which of the following Pediococcus (LAB) is homo-fermentative
(a) P. acidilactici (b) P. claussenii
(c) P. pentosaceus (d) all

345. Which of the following Pediococcus (LAB) is homo-fermentative
(a) P. damnosus (b) P. dextrinicus
(c) P. inopinatus (d) all

346. Which of the following LAB is homo-fermentative
(a) Pediococcus parvulus
(b) Leuconostoc citreus
(c) Leuconostoc fallax
(d) Weissella minor

347. Which of the following Leuconostoc (LAB) is hetero-fermentative
(a) L. argentinum (b) L. citreus
(c) L. fallax (d) all

348. Which of the following Leuconostoc (LAB) is hetero-fermentative
(a) L. carnosum (b) L. gelidum
(c) L. inhae (d) all

349. Which of the following Leuconostoc (LAB) is hetero-fermentative
(a) *L. kimchii* (b) *L. lactis*
(c) L. mesenteroides (d) all

350. Which of the following Leuconostoc (LAB) is hetero-fermentative
(a) L. cremoris
(b) L. dextranicum
(c) *L. kimchii* (d) all

351. Which of the following Carnobactreium (LAB) is hetero-fermentative
(a) C. divergens (b) C. gallinarum
(c) C. mobile (d) all

352. Which of the following Carnobactreium (LAB) is hetero-fermentative
(a) C. piscicola (b) C. viridians
(c) both (a) & (b) (d) none

353. *Oenococcus oeni* is ____________ lactic acid bacteria
(a) homofermentative
(b) heterofermentative
(c) both (a) & (b) (d) none

354. Which of the following Weissella (LAB) is hetero-fermentative
(a) W. cibaria (b) W. confusa
(c) W. hellenica (d) all

355. Which of the following Weissella (LAB) is hetero-fermentative
(a) W. halotolerans (b) W. kandleri
(c) W. kimchii (d) all

356. Which of the following Weissella (LAB) is hetero-fermentative
(a) W. minor
(b) W. thialandensis
(c) W. paramesenteroides
(d) all

357. Which of the following Weissella (LAB) is hetero-fermentative
(a) W. viridescens (b) W. koreensis
(c) both (a) & (b) (d) none

358. Which of the following Streptococcus (LAB) is homo-fermentative
(a) S. bovis (b) S. salivarius
(c) S. thermophilus (d) all

359. *Tetragenococcus halophilus* is ________ lactic acid bacteria
(a) homofermentative
(b) heterofermentative
(c) both (a) & (b)
(d) none

360. *Tetragenococcus muriaticus* is ________ lactic acid bacteria
(a) homofermentative
(b) heterofermentative
(c) both (a) & (b)
(d) none

361. Carnobacterium mobile is ________ lactic acid bacteria
(a) homofermentative
(b) heterofermentative
(c) both (a) & (b)
(d) none

362. Vagococcus fluvialis is ________ lactic acid bacteria
(a) homofermentative
(b) heterofermentative
(c) both (a) & (b)
(d) none

363. Vagococcus salmoninarum is ________ lactic acid bacteria
(a) homofermentative
(b) heterofermentative
(c) both (a) & (b)
(d) none

364. Lactobacillus species are
(a) homofermentative
(b) heterofermentative
(c) both (a) & (b) (d) none

365. *Lactobacillus kimchii* is ________ bacteria
(a) homofermentative
(b) heterofermentative
(c) both (a) & (b)
(d) none

366. *Lactobacillus kefirgranum* is ________ bacteria
(a) homofermentative
(b) heterofermentative
(c) both (a) & (b) (d) none

367. All members of the genera of Pediococcus and Streptococcus are
(a) homofermentative
(b) heterofermentative
(c) both (a) & (b) (d) none

368. All members of the genera of Leuconostoc and Oenococcus are
(a) homofermentative
(b) heterofermentative
(c) both (a) & (b) (d) none

369. All members of the genera of Lactococcus and Vagococcus are
(a) homofermentative
(b) heterofermentative
(c) both (a) & (b) (d) none

370. All members of the genera of Weisella and Carnobacterium are
(a) homofermentative
(b) heterofermentative
(c) both (a) & (b) (d) none

371. Which of the following produce acetylaldehyde and diacetyl
(a) homolactis (b) heterolactis
(c) both (a) & (b) (d) none

372. Which of the following is a subgenera of Lactobacillus genus
(a) Betabacterium
(b) Streptobacterium
(c) Thermobacterium
(d) all

373. Which of the following genera belongs to acetic acid bacteria
(a) Acetobacter (b) Asaia
(c) Acidomonas (d) all

374. Which of the following genera belongs to acetic acid bacteria
(a) Gluconobacter
(b) Gluconoacetobacter
(c) Kozakia
(d) all

375. Mastitis in cows may be caused by
(a) Streptococcus agalactiae
(b) Streptococcus uberis
(c) *Staphylococcus aureus*
(d) all

376. Mastitis in cows may be caused by
(a) Streptococcus agalactiae
(b) Streptococcus dysgelactiae
(c) both (a) & (b)
(d) none

377. Which of the following organism is not destroyed in pasteurization of milk
(a) Clostridium tyrobutyricum
(b) Clostridium sporogenes
(c) both (a) & (b)
(d) none

378. Which of the following organism is not destroyed in pasteurization of milk
(a) Bacillus cereus
(b) Clostridium sporogenes
(c) both (a) & (b)
(d) none

379. Which of the following belongs to thermoduric gram positive bacteria
(a) Enterococcus (b) Streptococcus
(c) Microbacterium (d) all

380. Which of the following belongs to thermoduric gram positive bacteria
(a) Lactobacillus
(b) Mycobacterium
(c) Corynebacterium
(d) all

381. Generally aerobic plate count of milk from healthy cows is less than
(a) 10^3 (b) 10^5
(c) 10^6 (d) 10^7

382. Which of the following cause spoilage in UHT milk
(a) Bacillus cereus
(b) Bacillus licheniformis
(c) Bacillus badius (d) all

383. During cold storage of pasteurized milk, Bacillus wiehenstephanensis causes sweet curding due to production of
(a) amylase (b) lipase
(c) protease (d) none

384. Alcaligenes viscolactis in raw milk causes
(a) sandiness (b) ropiness
(c) curdling (d) all

385. Which of the following Bacillus sp. is most often used as probiotic
(a) Bacillus clausii
(b) Bacillus pumilus
(c) bacillus cereus (d) all

386. Culture used in yoghurt manufacture is
(a) Streptococcus thermophilus
(b) Lactobacillus bulgaricus
(c) both (a) & (b) (d) none

387. *Streptococcus thermophilus* and *Lactobacillus bulgaricus* are used in _______ ratio in yoghurt manufacture
(a) 1:2 (b) 1:1
(c) 2:3 (d) 1:4

388. Kefir can be prepared by bacteria
(a) Acetobacter (b) Lactobacillus
(c) Lactococcus (d) all

389. Which of the following yeast can be used for kefir preparation
(a) Candida
(b) Kluyveromyces
(c) Saccharomyces (d) all

390. Which of the following Lactobacillus sp. is used for kefir preparation
(a) L. kefiri
(b) L. parakefiri
(c) L. kefiranofaciens
(d) all

391. The inoculation is done at ______°C for acidophilus milk
(a) 27 (b) 30
(c) 37 (d) 42

392. Which of the following is used for Bulgarian buttermilk preparation
(a) *L. bulgaricus*
(b) *L. acidophilus*
(c) both (a) & (b)
(d) none

393. Which of the following is type of spoilage in butter
(a) putridity (b) rancidity
(c) both (a) & (b) (d) none

394. Which of the following I used in fermentation of Bioghurt
(a) *L. acidophilus*
(b) Lactococcus lactis
(c) both (a) & (b)
(d) none

395. Which of the following fungi cause spoilage in butter
(a) Cladosporium (b) Alternaria
(c) Aspergillus (d) all

396. Which of the following fungi cause spoilage in butter
(a) Mucor (b) Rhizopus
(c) Penicillium (d) all

397. Cottage cheese undergoes spoilage by
(a) bacteria (b) yeasts
(c) molds (d) all

398. The bacteria that causes slimy curd condition in cottage cheese is
(a) Alcaligenes (b) Pseudomonas
(c) Proteus (d) all

399. The bacteria that causes slimy curd condition in cottage cheese is
(a) Enterobacter (b) Acinetobacter
(c) both (a) & (b) (d) none

400. Penicillium in cottage cheese causes ______ flavor
(a) stale (b) moldy
(c) musty (d) all

401. Mucor in cottage cheese causes ______ flavor
(a) yeasty (b) stale
(c) musty (d) all

402. Which of the following is used in sauerkraut formation
(a) Leuconostoc mesenteroides
(b) Lactobacillus plantarum
(c) Leuconostoc fallax
(d) all

403. Rotted sauerkraut may be caused by
(a) bacteria (b) yeast
(c) mold (d) all

404. Which of the following is a type of microbial spoilage of sauerkraut
(a) soft kraut (b) slimy kraut
(c) rotted kraut (d) all

405. Slimy kraut in sauerkraut is caused by rapid growth of
(a) Lactobacillus cucumeris
(b) Lactobacillus plantarum
(c) both (a) & (b)
(d) none

406. Pink krut is caused by surface growth of
(a) Torula sp.
(b) *Lactobacillus* sp.
(c) *Leuconostoc* sp.
(d) all

407. Ropiness in beer is caused by
(a) Acetobacter (b) Lactobacillus
(c) Pedicoccus (d) all

408. Coffee beans are fermented by
(a) Erwinia
(b) Saccharomyces
(c) both (a) & (b)
(d) none

409. Gari is a fermented product of
(a) cassava
(b) coffee cherries
(c) olives
(d) all

410. Fermentor used in gari preparation is
(a) Erwinia
(b) Corynebacterium
(c) Rhizopus
(d) all

411. Fermentor used in Kenkey preparation is
(a) Aspergillus (b) Penicillium
(c) Lactobacilli (d) all

412. Kenkey is a fermented product of
(a) cassava (b) corn
(c) olives (d) coffee

413. Miso is a fermented product of
(a) soybean (b) moong dal
(c) cucumber (d) all

414. Fermentor used in Miso preparation is
(a) Aspergillus
(b) Zygosaccharomyces
(c) both (a) & (b)
(d) none

415. Ogi is a fermented product of
(a) corn (b) wheat
(c) taro roots (d) all

416. Fermentor used in Ogi preparation is
(a) L. plantarum
(b) *L. lactis*
(c) Zygosaccharomyces
(d) all

417. Poi is a fermented product of
(a) corn (b) wheat
(c) taro roots (d) all

418. Fermentor used in olives fermentation is
(a) *L. mesenteroides*
(b) *L. plantarum*
(c) both (a) & (b)
(d) none

419. Fermentor used in soy sauce is
(a) A. oryzae (b) A. soyae
(c) both (a) & (b) (d) none

420. Fermentor used in soy sauce is
(a) Z. rouxii (b) *L. delbrueckii*
(c) both (a) & (b) (d) none

421. Sufu is a fermented product of
(a) soybean (b) rice
(c) wheat (d) corn

422. Tempeh is a fermented product of
(a) soybean (b) rice
(c) wheat (d) corn

423. Tao-si is a fermented product of
(a) rice (b) soybean
(c) wheat (d) corn

424. Fermentor used in sufu fermentation
(a) Mucor (b) A. oryzae
(c) Rhizopus (d) none

425. Fermentor used in Tao-si fermentation
(a) Mucor (b) A. oryzae
(c) Rhizopus (d) none

426. Fermentor used in Tempeh fermentation
(a) Mucor (b) A. oryzae
(c) Rhizopus (d) none

427. Arrack is a fermented beverage obtained from
(a) rice (b) cereal wort
(c) corn (d) wheat

428. Beer is a fermented beverage obtained from
(a) rice (b) cereal wort
(c) corn (d) wheat

429. Binubaran is a fermented beverage obtained from
(a) rice (b) cereal wort
(c) corn (d) wheat

430. Bourbon whisky is a fermented beverage obtained from
(a) rice (b) cereal wort
(c) corn (d) wheat

431. Bourbon whisky is a fermented beverage obtained from
(a) rye (b) cereal wort
(c) corn (d) wheat

432. Bouza beer is a fermented beverage obtained from
(a) rye (b) cereal wort
(c) corn (d) wheat

433. Cider is a fermented beverage obtained from
(a) apple (b) cereal wort
(c) corn (d) wheat

434. Kaffir beer is a fermented beverage obtained from
(a) kaffir corn (b) kaffir apple
(c) both (a) & (b) (d) none

435. Magon is a fermented beverage obtained from
(a) rye (b) cereal wort
(c) corn (d) wheat

436. Mezcal is a fermented beverage obtained from
(a) century plant (b) money plant
(c) both (a) & (b) (d) none

437. Oo is a fermented beverage obtained from
(a) rice (b) apple
(c) corn (d) wheat

438. Sake is a fermented beverage obtained from
(a) rice (b) apple
(c) corn (d) wheat

439. Which of the following is yeast fermented beverage
(a) Arrack
(b) Binuburan
(c) Bourbon whiskey
(d) all

440. Which of the following is yeast fermented beverage
(a) Mezcal (b) Oo
(c) Pulgue (d) all

441. Which of the following is yeast fermented product
(a) Sake (b) Magon
(c) Kimchi (d) all

442. Mezcal is a fermented beverage from
(a) Japan (b) Thailand
(c) Mexico (d) all

443. Oo is a fermented beverage from
(a) Japan (b) Thailand
(c) Mexico (d) all

444. Sake is a fermented product from
(a) Japan (b) Thailand
(c) Mexico (d) all

445. Binuburan is a fermented beverage from
(a) United States (b) Philippines
(c) Egypt (d) all

446. Bourbon whisky is a fermented beverage from
(a) United States (b) Philippines
(c) Egypt (d) all

447. Bouza beer is a fermented beverage from
(a) United States (b) Philippines
(c) Egypt (d) all

448. Which of the following is a Mexican fermented beverage
(a) Pulque (b) Mezcal
(c) both (a) & (b) (d) none

449. Scotch whisky is a fermented product from
(a) wheat (b) barley
(c) pearl millet (d) all

450. Fermentor used in Scotch whisky is
(a) S. cerevisiae
(b) Acetobacter
(c) Schizosaccharomyces
(d) all

451. Thuma is a fermented product of
(a) wheat (b) millet
(c) rice (d) corn

452. Tibi is a fermented product of
(a) dried figs (b) dried grapes
(c) both (a) & (b) (d) none

453. Vodka is a fermented product of
(a) wheat (b) potato
(c) rice (d) corn

454. Wine is a fermented product of
(a) wheat (b) millet
(c) grapes (d) corn

455. Which of the following is an Indian fermented product
(a) Scotch whisky (b) Teekwass
(c) Thumba (d) Vodka

456. Which of the following is prepared from wheat flour
(a) cakes
(b) sourdough bread
(c) biscuits
(d) all

457. Fermentor in idli fermentation is
(a) Leuconostoc
(b) Saccharomyces
(c) both (a) & (b)
(d) none

458. Which of the following is fermented by Leuconostoc sp.
(a) idli
(b) sour pumpernickel
(c) both (a) & (b)
(d) none

459. Single cell protein is ________ protein
(a) algae (b) fungal
(c) bacterial (d) all

460. Which of the following algae is a source of SCP
(a) Chlorella (b) Candida
(c) Pichia (d) all

461. Which of the following is not an algae
(a) Chlorella (b) Scenedesmus
(c) Rhodotorula (d) none

462. Which of the following yeast is used for SCP
(a) Candida (b) Rhodotorula
(c) Saccharomyces (d) all

463. Which of the following yeast is used for SCP
(a) Candida utilis
(b) Candida lipolytica
(c) Candida tropicalis
(d) all

464. Which of the following is a filamentous fungi
(a) Agaricus (b) Fusarium
(c) Penicillium (d) all

465. Which of the following is a filamentous fungi
(a) Saccharomycopsis
(b) Trichosporon
(c) both (a) & (b)
(d) none

466. Which of the following support growth of *Chlorella pyrenoidosa* in SCP production
(a) CO_2 & sunlight (b) methane
(c) n-alkanes (d) all

467. Which of the following support growth of *Spirulina maxima* in SCP production
(a) CO_2 & sunlight (b) methane
(c) n-alkanes (d) all

468. Which of the following support growth of *Candida lipolytica* in SCP production
(a) CO_2 & sunlight (b) methane
(c) n-alkanes (d) all

469. Which of the following support growth of *Methylomonas* sp. in SCP production
(a) CO_2 & sunlight (b) methane
(c) n-alkanes (d) all

470. Which of the following support growth of *Trichoderma* sp. in SCP production
(a) CO_2 & sunlight (b) methane
(c) n-alkanes (d) all

471. Which of the following support growth of *Candida tropicalis* in SCP production
(a) CO_2 & Sunlight (b) methane
(c) n-alkanes (d) all

472. Which of the following support growth of *Alcaligenes eutrophus* in SCP production
(a) H_2 & CO_2 (b) gas oil
(c) methanol (d) ethanol

473. Which of the following support growth of *Methylomonas methanica* in SCP production
(a) H_2 & CO_2 (b) gas oil
(c) methanol (d) ethanol

474. Which of the following support growth of *Acinetobacter calcoaceticus* SCP production
(a) ethanol (b) gas oil
(c) both (a) & (b) (d) none

475. Which of the following support growth of *Candida utilis* in SCP production
(a) H_2 & CO_2 (b) gas oil
(c) n-alkane (d) ethanol

476. Which of the following support growth of *Cellumonas* sp. SCP production
(a) cellulose (b) starch
(c) sugar (d) none

477. Which of the following support growth of *Trichoderma viride* SCP production
(a) cellulose (b) starch
(c) sugar (d) none

478. Which of the following support growth of *Saccharomycopsis fibuligera* SCP production
(a) cellulose (b) starch
(c) sugar (d) none

479. Which of the following support growth of *Saccharomyces cerevisiae* SCP production
(a) cellulose (b) starch
(c) sugar (d) none

480. Which of the following support growth of *Candida utilis* SCP production
(a) cellulose (b) starch
(c) sugar (d) none

481. Propionic acid is active against growth of ________ in bread
(a) yeasts (b) molds
(c) both (a) & (b) (d) none

482. Benzoic acid is active against growth of ________ in margarine
(a) yeasts (b) molds
(c) both (a) & (b) (d) none

483. Sorbic acid is active against growth of ________ in cheeses
(a) yeasts (b) molds
(c) both (a) & (b) (d) none

484. Parabens are active against growth of ________ in pickles
(a) yeasts (b) molds
(c) both (a) & (b) (d) none

485. Ethylene oxide is active against growth of ________ in spices
(a) yeasts (b) molds
(c) both (a) & (b) (d) none

486. Sodium acetate is active against growth of ________ in bread
(a) yeasts (b) molds
(c) both (a) & (b) (d) none

487. Ethyl formate is active against growth of ________ in dried fruits
(a) yeasts (b) molds
(c) both (a) & (b) (d) none

488. SO_2 is active against growth of ________ in molasses
(a) yeasts (b) molds
(c) bacteria (d) all

489. Nisin is active against growth of ________ in cheese spread
(a) bacteria (b) yeasts
(c) molds (d) none

490. Caprylic acid is active against growth of ________ in cheese wraps
(a) bacteria (b) yeasts
(c) molds (d) none

491. Sodium lactate is active against growth of ________ in pre cooked meats
(a) bacteria (b) yeasts
(c) mold (d) none

492. Which of the following has antimicrobial activity
(a) lactoferrin (b) ozone
(c) sodium chlorite (d) all

493. Which of the following antibiotic is widely used in foods
(a) subtilin (b) tylosin
(c) nisin (d) all

494. Natamycin is also known as
(a) pimaricin (b) tennecetin
(c) myprozine (d) all

495. Which of the following is an antifungal agent
(a) Thiabendazole (b) Benomyl
(c) Biphenyl (d) all

496. Nisin is generally produced by
(a) Lactococcus (b) Lactobacillus
(c) Clostridium (d) all

497. Volatile compound produced by Shewanella putrefaciens in sterile fish muscles at 1-2°C
(a) dimethyl sulfide
(b) dimethyl trisulfide
(c) methyl mercaptan
(d) all

498. Volatile compound produced by Shewanella putrefaciens in sterile fish muscles at 1-2°C
(a) trimethylamine
(b) propionaldehyde
(c) both (a) & (b)
(d) none

499. Which of the following bacteria are generally more resistant to irradiation
(a) gram +ve (b) gram –ve
(c) both (a) & (b) (d) none

500. Which of the following bacteria are generally more resistant to irradiation
(a) non-spore formers
(b) spore formers
(c) both (a) & (b)
(d) none

501. Minimum water activity required for growth of *Candida utilis*
(a) 0.91 (b) 0.92
(c) 0.93 (d) 0.94

502. Minimum water activity required for growth of *Botrytis cinerea*
(a) 0.91 (b) 0.92
(c) 0.93 (d) 0.94

503. Minimum water activity required for growth of *Rhizopus stolonifer*
(a) 0.91 (b) 0.92
(c) 0.93 (d) 0.94

504. Minimum water activity required for growth of *Mucor spinosus*
(a) 0.91 (b) 0.92
(c) 0.93 (d) 0.94

505. Minimum water activity required for growth of *Candida scottii*
(a) 0.91 (b) 0.92
(c) 0.93 (d) 0.94

506. Minimum water activity required for growth of *Trichosporon pullulans*
(a) 0.91 (b) 0.92
(c) 0.93 (d) 0.94

507. Minimum water activity required for growth of *Candida zeylanoides*
(a) 0.70 (b) 0.90
(c) 0.84 (d) 0.89

508. Minimum water activity required for growth of *Saccharomycopsis vernalis*
(a) 0.70 (b) 0.90
(c) 0.84 (d) 0.89

509. Minimum water activity required for growth of *Alternaria citri*
(a) 0.70 (b) 0.90
(c) 0.84 (d) 0.89

510. Minimum water activity required for growth of *Aspergillus glaucus*
(a) 0.70 (b) 0.90
(c) 0.84 (d) 0.89

511. Minimum water activity required for growth of *Aspergillus echinulatus*
(a) 0.62 (b) 0.64
(c) 0.70 (d) 0.80

512. Minimum water activity required for growth of *Zygosaccharomyces rouxii*
(a) 0.62 (b) 0.64
(c) 0.70 (d) 0.80

513. "Alarm water" content of whole milk powder
(a) 8 (b) 10-11
(c) 13-15 (d) 15

514. "Alarm water" content of dehydrated whole eggs
(a) 8 (b) 10-11
(c) 13-15 (d) 15

515. "Alarm water" content of wheat flour
(a) 8 (b) 10-11
(c) 13-15 (d) 19

516. "Alarm water" content of fat free hydrated meat
(a) 8 (b) 10-11
(c) 12 (d) 15

517. "Alarm water" content of pulses
(a) 8 (b) 10-11
(c) 12 (d) 15

518. "Alarm water" content of starch
(a) 8 (b) 12
(c) 15 (d) 18

519. Intermediate moisture foods (IMF) have a_w
(a) 0.40-0.45 (b) 0.45-0.55
(c) 0.60-0.85 (d) 0.90

520. Intermediate moisture foods (IMF) have moisture content of
(a) 15-50% (b) 50-60%
(c) 60-75% (d) 75-855

521. Water Activity of dried fruits range from
(a) 0.60-0.90 (b) 0.60-0.75
(c) 0.73-0.83 (d) 0.96

522. Water Activity of fruit cake range from
(a) 0.60-0.90 (b) 0.60-0.75
(c) 0.73-0.83 (d) 0.96

523. Water Activity of ripened cheese is about
(a) 0.60-0.90 (b) 0.60-0.75
(c) 0.73-0.83 (d) 0.96

524. Water Activity of maple syrup is about
(a) 0.60 (b) 0.75
(c) 0.90 (d) 0.86

525. Water Activity of sweetened condensed milk is about
(a) 0.90 (b) 0.75
(c) 0.83 (d) 0.96

526. Water Activity of honey is about
(a) 0.88 (b) 0.75
(c) 0.60 (d) 0.96

527. Water Activity of dried fruits range from
(a) 0.60-0.90 (b) 0.60-0.75
(c) 0.73-0.83 (d) 0.96

528. Which of the following is not an IMF
(a) dried fruits (b) cake
(c) pastry (d) none

529. Which of the following is not an IMF
(a) sugar syrup (b) candies
(c) pastry fillings (d) none

530. Which of the following is not an IMF
(a) honey (b) jams
(c) fruit juice concentrates
(d) none

531. Which of the following is not an IMF
(a) maple syrup
(b) sweetened condensed milk
(c) both (a) & (b) (d) none

532. Which of the following is humectant
(a) glycerol (b) glycols
(c) sorbitol (d) all

533. Which of the following is negatively associated with fresh cider quality
(a) Acetobacter
(b) *Bacillus* sp.
(c) *Byssochlamys* sp. (d) all

534. Which of the following is negatively associated with bread dough quality
(a) Acetobacter
(b) *Bacillus* sp.
(c) *Byssochlamys* sp. (d) all

535. Which of the following is negatively associated with canned fruits quality
(a) Acetobacter
(b) *Bacillus* sp.
(c) *Byssochlamys* sp. (d) all

536. Which of the following is negatively associated with hard cheese quality
(a) *Clostridium* sp.
(b) flat sour spores
(c) Geotrichum
(d) all

537. Which of the following is negatively associated with canned vegetables quality
(a) *Clostridium* sp.
(b) flat sour spores
(c) Geotrichum
(d) all

538. Which of the following is negatively associated with fruit cannery sanitation
(a) Clostridium sp.
(b) flat sour spores
(c) Geotrichum (d) all

539. Which of the following is negatively associated with beer quality
(a) Lactic acid bacteria
(b) Lactococcus lactis
(c) Leuconostoc mesenteroides
(d) none

540. Which of the following is negatively associated with wine quality
(a) Lactic acid bacteria
(b) Lactococcus lactis
(c) Leuconostoc mesenteroides
(d) none

541. Which of the following is negatively associated with refrigerated raw milk quality
(a) Lactic acid bacteria
(b) Lactococcus lactis
(c) Leuconostoc mesenteroides
(d) none

542. Which of the following is negatively associated with sugar quality
(a) Lactic acid bacteria
(b) Lactococcus lactis
(c) Leuconostoc mesenteroides
(d) none

543. Which of the following is negatively associated with beer quality
(a) Pectinatus cerevisiiphilus
(b) Pseudomonas putrefaciens
(c) both (a) & (b)
(d) none

544. Which of the following is negatively associated with butter quality
(a) Pectinatus cerevisiiphilus
(b) Pseudomonas putrefaciens
(c) both (a) & (b)
(d) none

545. Which of the following is negatively associated with mayonnaise quality
(a) Bacillus
(b) Acetobacter
(c) Zygosaccharomyces
(d) none

546. Which of the following metabolite is negatively associated with vacuum packaged beef quality
(a) cadaverine (b) putrescine
(c) both (a) & (b) (d) none

547. Which of the following metabolite is negatively associated with frozen juice concentrate quality
(a) diacetyl (b) ethanol
(c) histamine (d) lactic acid

548. Which of the following metabolite is negatively associated with apple juice quality
(a) diacetyl (b) ethanol
(c) histamine (d) lactic acid

549. Which of the following metabolite is negatively associated with fishery product quality
(a) diacetyl (b) ethanol
(c) histamine (d) lactic acid

550. Which of the following metabolite is negatively associated with canned tuna quality
(a) diacetyl (b) ethanol
(c) histamine (d) lactic acid

551. Which of the following metabolite is negatively associated with canned vegetable quality
(a) diacetyl (b) ethanol
(c) histamine (d) lactic acid

552. Which of the following metabolite is negatively associated with fish quality
(a) diacetyl (b) ethanol
(c) histamine (d) trimethylamine

553. Which of the following metabolite is negatively associated with butter quality
(a) diacetyl (b) ethanol
(c) histamine
(d) volatile fatty acids

554. Which of the following metabolite is negatively associated with cream quality
(a) diacetyl (b) ethanol
(c) histamine
(d) volatile fatty acids

555. Which of the following metabolite is negatively associated with seafood's quality
(a) total volatile bases
(b) total volatile nitrogen
(c) both (a) & (b)
(d) none

556. Which of the following bacteria is included in food borne pathogens
(a) Staphylococcus
(b) Bacillus cereus
(c) Bacillus anthracis
(d) all

557. Which of the following bacteria is included in food borne pathogens
(a) Clostridium botulinum
(b) Clostridium argentinensis
(c) Clostridium perfringens
(d) all

558. Which of the following bacteria is included in food borne pathogens
(a) Listeria monocytogenes
(b) Mycobacterium avium
(c) both (a) & (b)
(d) none

559. Which of the following bacteria is included in food borne pathogens
(a) Salmonella (b) Shigella
(c) Escherichia (d) all

560. Which of the following bacteria is included in food borne pathogens
(a) Yersinia (b) Vibrio
(c) Campylobacter (d) all

561. Which of the following bacteria is included in food borne pathogens
(a) Aeromonas (b) Brucella
(c) Plesiomonas (d) all

562. Which of the following virus is considered as food borne pathogens
(a) Hepatitis A (b) Novoviruses
(c) Rotaviruses (d) all

563. Which of the following mycotoxin is considered as food borne pathogens
(a) Aflatoxins (b) Fumonisins
(c) Ochratoxins (d) all

564. Which of the following protozoa is considered as food borne pathogens
(a) Giardia (b) Toxoplasma
(c) Cyclospora (d) all

565. Which of the following protozoa is considered as food borne pathogens
(a) Sarcocystis
(b) Cryptosporium
(c) both (a) & (b)
(d) none

566. Which of the following is a mycotoxin
(a) Aflatoxin (b) Citrinin
(c) Ochratoxins (d) all

567. Which of the following is a mycotoxin
(a) Patulin
(b) Penicillic acid
(c) Sterigmatocystin
(d) all

568. Which of the following is a mycotoxin
(a) Fumonisins (b) Sambutoxin
(c) both (a) & (b) (d) none

569. Which of the following is a gram positive bacteria (genera)
(a) Aflatoxin (b) Citrinin
(c) Ochratoxins (d) all

570. Which of the following is a distilled fermentation product
(a) cider (b) brandy
(c) beer (d) all

571. Which of the following is a distilled fermentation product
(a) whiskey (b) wine
(c) toddy (d) all

572. Which of the following is a distilled fermentation product
(a) perry (b) pulque
(c) tequila (d) all

573. Which of the following is a distilled fermentation product
(a) brandy (b) sake
(c) kaffir beer (d) all

574. Which of the following is a distilled fermentation product
(a) cider (b) beer
(c) rum (d) all

575. Which of the following is a non-distilled fermentation product
(a) cider (b) beer
(c) pulque (d) all

576. Which of the following is a non-distilled fermentation product
(a) wine (b) toddy
(c) perry (d) all

577. Which of the following is a non-distilled fermentation product
(a) mead (b) sake
(c) beer (d) all

578. Which of the following is a filamentous fungi
(a) Aspergillus
(b) Aureobasidium
(c) Fusarium
(d) all

579. Which of the following is a filamentous fungi
(a) Mucor (b) Candida
(c) Cryptococcus (d) all

580. Which of the following is a filamentous fungi
(a) Geotrichum (b) Cryptococcus
(c) Neurospora (d) all

581. Which of the following is a filamentous fungi
(a) Penicillium
(b) Brettanomyces
(c) Debaromyces
(d) all

582. Which of the following is a filamentous fungi
(a) Hansenula (b) Rhizomucor
(c) Kloeckera (d) all

583. Which of the following is a filamentous fungi
(a) Kluyveromyces (b) Monascus
(c) Rhizopus (d) all

584. Which of the following is a filamentous fungi
(a) Trichoderma (b) Pichia
(c) Rhodotorula (d) all

585. Which of the following is a non-filamentous fungi
(a) Saccharomyces
(b) Saccharomycopsis
(c) Torulopsis
(d) all

586. Which of the following is a non-filamentous fungi
(a) Trichosporon (b) Rhizopus
(c) Penicillium (d) all

587. Which of the following is a non-filamentous fungi
(a) Neurospora (b) Mucor
(c) Yarowia (d) all

588. Which of the following is a non-filamentous fungi
(a) Zygosaccharomyces
(b) Fusarium
(c) Aspergillus
(d) all

589. Which of the following is a major source of α-amylase
(a) Aspergillus (b) Bacillus
(c) both (a) & (b) (d) none

590. Which of the following is a major source of β-amylase
(a) Bacillus (b) Streptomyces
(c) Rhizopus (d) all

591. Which of the following is a major source of glucoamylase
(a) Aspergillus (b) Rhizopus
(c) both (a) & (b) (d) none

592. Which of the following is a major source of glucose isomerase
(a) Arthrobacter (b) Streptomyces
(c) both (a) & (b) (d) none

593. Which of the following is a major source of pullulanase
(a) Klebsiella (b) Bacillus
(c) both (a) & (b) (d) none

594. Which of the following is a major source of invertase
(a) Kluyveromyces
(b) Saccharomyces
(c) both (a) & (b)
(d) none

595. Which of the following is a major source of glucose oxidase
(a) Penicillium (b) Bacillus
(c) Rhizopus (d) all

596. Which of the following is a major source of glucose oxidase
(a) Penicillium (b) Aspergillus
(c) both (a) & (b) (d) none

597. Which of the following is a major source of pectinase
(a) Aspergillus (b) Arthrobacter
(c) Rhizopus (d) none

598. Which of the following is a major source of β-glucanases
(a) Bacillus (b) Penicillium
(c) Trichoderma (d) all

599. Which of the following is a major source of proteinases
(a) Aspergillus (b) Bacillus
(c) Lactococcus (d) all

600. Which of the following is a major source of catalase
(a) Micrococcus (b) Klebsiella
(c) Rhizopus (d) all

601. Which of the following is a major source of catalase
(a) Micrococcus
(b) Corynebacterium
(c) Aspergillus
(d) all

602. Which of the following is a major source of lipases
(a) Aspergillus (b) Bacillus
(c) both (a) & (b) (d) none

603. Which of the following is a major source of lipases
(a) Rhizopus (b) Rhotorula
(c) both (a) & (b) (d) none

604. Which of the following is a major source of urease
(a) Lactobacillus (b) Cryptococcus
(c) both (a) & (b) (d) none

605. Which of the following is a major source of tannase
(a) Micrococcus (b) Aspergillus
(c) both (a) & (b) (d) none

606. Which of the following is a major source of β- galactosidase
(a) Aspergillus (b) Bacillus
(c) Kluyveromyces (d) all

607. Which of the following is a major source of pentosanases
(a) Cryptococcus (b) Trichosporon
(c) both (a) & (b) (d) none

608. Which of the following is a mold
(a) Saccharomyces
(b) Staphylococcus
(c) Rhizopus
(d) all

609. Which of the following is a mold
(a) Penicillium (b) Aspergillus
(c) Rhizopus (d) all

610. Which of the following is a fungus
(a) Saccharomyces cerevisiae
(b) Saccharomyces pastorianus
(c) Staphylococcus carnosus
(d) all

611. Which of the following is ale yeast
(a) Saccharomyces cerevisiae
(b) Saccharomyces pastorianus
(c) both (a) & (b)
(d) none

612. Which of the following is lager yeast
(a) Saccharomyces cerevisiae
(b) Saccharomyces pastorianus
(c) both (a) & (b)
(d) none

613. Which of the following yeast can ferment the sugar melibiose
(a) ale (b) lager
(c) both (a) & (b) (d) none

614. Which of the following yeast produce α-galactosidase
(a) lager (b) ale
(c) both (a) & (b) (d) none

615. α-galactosidasein lager yeast strain convert melibiose into
(a) glucose (b) galactose
(c) both (a) & (b) (d) none

616. Beer is resistant to infection because of
(a) low pH
(b) presence of ethanol
(c) anaerobic environment
(d) all

617. Which of the following grow slowly in beer
(a) Lactobacillus brevis
(b) Lactobacillus plantarum
(c) Lactobacillus lindner
(d) all

618. Which of the following grow quickly in beer

(a) Lactobacillus brevis
(b) Lactobacillus plantarum
(c) both (a) & (b)
(d) none

619. Which of the following causes spoilage in beer
(a) Lactobacillus (b) Pediococcus
(c) both (a) & (b) (d) none

620. Which of the following causes spoilage in beer
(a) Acetobacter
(b) Gluconobacter
(c) Escherchia (d) all

621. Which of the following causes spoilage in beer
(a) Aerobacter (b) Klebsiella
(c) Citrobacter (d) all

622. Which of the following causes spoilage in beer
(a) Obesumbacterium
(b) Zymomonas
(c) both (a) & (b) (d) none

623. Which of the following causes spoilage in beer
(a) Pectinatus (b) Megasphaera
(c) both (a) & (b) (d) none

624. Gluconobacter produce ________ flavor in beer
(a) citrus (b) bitter
(c) vinegary (d) none

625. Acetic acid bacteria produce________ in beer
(a) Vinegary flavor (b) Ropy slime
(c) both (a) & (b) (d) none

626. Aerobacter produce ________ aroma in beer
(a) cabbagy (b) vegetable
(c) eggy (d) all

627. Organism involved in kefir making
(a) Lactococcus lactis
(b) Lactococcus cremoris
(c) both (a) & (b) (d) none

628. Organism involved in kefir making
(a) Lactobacillus bulgaricus
(b) Lactobacillus helveticus
(c) Lactococcus lactis
(d) all

629. Organism involved in kefir making
(a) Lactobacillus casei
(b) Lactobacillus brevis
(c) Lactobacillus kefir
(d) all

630. Baker's yeast used in bread making is
(a) top fermenting
(b) bottom fermenting
(c) both (a) & (b)
(d) none

631. Homofermentative sourdough starter organism is
(a) Lactobacillus acidophilus
(b) Lactobacillus casei
(c) Lactobacillus plantarum
(d) all

632. Heterofermentative sourdough starter organism (*Lacto bacillus*) is
(a) *L. brevis* (b) *L. buchneri*
(c) *L. fermentum* (d) all

633. Heterofermentative sourdough starter organism (Lactobacillus) is
(a) L. lindneri
(b) *L. fructivorans*
(c) both (a) & (b)
(d) none

634. Which of the following yeast is used as sourdough starter
(a) Cadida crusei
(b) Pichia saitoi
(c) Saccharamyces cerevisiae
(d) all

635. Which of the following yeast is used as sourdough starter
(a) Torulopsis holmii
(b) Lactobacillus brevis
(c) both (a) & (b) (d) none

636. Bagel is a type of fermented ________
(a) bread (b) beer
(c) wine (d) none

637. Basi is a type of fermented ________
(a) bread (b) beer
(c) wine (d) none

638. Cachaca is a type of fermented ______
(a) bread (b) beer
(c) wine (d) spirit

639. Chicha is a type of ________ beverage
(a) alcoholic (b) non-alcoholic
(c) both (a) & (b) (d) none

640. Jerky is a type of ________ product
(a) bakery (b) meat
(c) wine (d) beer

641. Kanji is a type of ________ beverage
(a) alcoholic (b) non-alcoholic
(c) both (a) & (b) (d) none

642. Organism involved in fermentation of kanji
(a) Hansenula anomala
(b) Candida guilliermondii
(c) Candida tropicalis
(d) all

643. Organism involved in fermentation of kanji
(a) Candida tropicalis
(b) Geotrichum candidum
(c) both (a) & (b)
(d) none

644. Organism involved in fermentation of kimchi
(a) Leuconstoc mesenteroides
(b) Streptococcus fecalis
(c) Pediococcus
(d) all

645. Organism involved in fermentation of kimchi
(a) Lactobacillus plantarum
(b) Lactobacillus brevis
(c) Alcaligenes
(d) all

646. Organism involved in fermentation of kimchi
(a) Flavobacterium
(b) Pseudomonas
(c) Bacillus megaterium
(d) all

647. Mead is a type of fermented
(a) bread
(b) meat
(c) alcoholic beverage
(d) all

648. Mead is obtained by fermentation of
(a) honey (b) wheat
(c) rice (d) oats

649. Pisco is a type of fermented beverage
(a) alcoholic (b) non-alcoholic
(c) both (a) & (b) (d) none

650. Pisco is a ________ alcoholic beverage
(a) distilled (b) non-distilled
(c) both (a) & (b) (d) none

651. Which of the following help in Tempe fermentation
(a) Klebsiella (b) Rhizopus
(c) both (a) & (b) (d) none

652. Tequilla is ________ fermentaed beverage
(a) distilled (b) non-distilled
(c) both (a) & (b) (d) none

653. Which of the following is a microbial gum
(a) xanthan (b) gellan
(c) pullulan (d) all

654. Which of the following factor affect water activity requirement of micro-organisms
(a) temperature (b) oxygen
(c) pH (d) all

655. Which of the following is not a characteristic of mold
(a) Chlorophyll (b) root
(c) stem (d) all

656. Which of the following yeast can tolerate high levels of alcohol
(a) Hansenula (b) Pichia
(c) both (a) & (b) (d) none

657. Which of the following yeast can tolerate high levels of acid
(a) Candida (b) Trichosporon
(c) Hansenula (d) all

658. Which of the following is thermoduric bacteria
(a) Clostridium (b) Bacillus
(c) Micrococcus (d) all

659. Which of the following is thermoduric bacteria
(a) Streptococcus (b) Lactobacillus
(c) Microbacterium (d) all

660. Which of the following is thermoduric mold
(a) Byssochlamys (b) Aspergillus
(c) Penicillium (d) all

661. Which of the following is a halophilic bacteria
(a) Pseudomonas (b) Moraxella
(c) Flavobacterium (d) all

662. Which of the following is a psychrotrophic bacteria
(a) Pseudomonas
(b) Achromobacter
(c) Flavobacterium
(d) all

663. Which of the following bacterial genera contain psychrotrphs species
(a) Alcaligenes (b) Lactobacillus
(c) Micrococcus (d) all

664. Moderate halophilic bacteria grow at ________ % salt
(a) 0.5-3.0 (b) 3.0-15
(c) 15-30 (d) none

665. Slightly halophilic bacteria grow at ________ % salt
(a) 0.5-3.0 (b) 3.0-15
(c) 15-30 (d) none

666. Extreme halophilic bacteria grow at ________ % salt
(a) 0.5-3.0 (b) 3.0-15
(c) 15-30 (d) none

667. Which of the following cause ropiness in milk
(a) Alcaligenes viscolactis
(b) Enterobacter aerogenes
(c) Klebsiella oxytoca
(d) all

668. Which of the following genera contains gas forming bacteria
(a) Leuconostoc
(b) Lactobacillus
(c) Propionibacterium (d) all

669. Which of the following is a stable or nonperishable food
(a) Sugar (b) flour
(c) dry beans (d) all

670. Which of the following is a semiperishable food
(a) potatoes (b) apples
(c) nutmeats (d) all

671. Which of the following is a perishable food
(a) meat (b) fish
(c) poultry (d) all

672. Which of the following is a perishable food
(a) fruits (b) vegetables
(c) both (a) & (b) (d) none

673. Putrefaction is related to hydrolysis of
(a) starch (b) protein
(c) pullulan (d) all

674. Oxidative deamination of amino acids produce
(a) keto acid (b) ammonia
(c) both (a) & (b) (d) none

675. Hydrolytic deamination of amino acids produce
(a) hydroxyacid (b) ammonia
(c) both (a) & (b) (d) none

676. Reductive deamination of amino acids produce
(a) saturated acid (b) ammonia
(c) both (a) & (b) (d) none

677. Hydrolysis of proteins by microbial proteinases gives
(a) sour (b) bitter
(c) sweet (d) salty

678. Radiation inactivation of microorganisms can be achieved by
(a) radicidation (b) radurization
(c) radappertization (d) all

679. Heat inactivation of microorganisms can be achieved by
(a) pasteurization (b) sterilization
(c) both (a) & (b) (d) none

680. Microbial growth curve is composed of
(a) lag phase (b) log phase
(c) stationery phase (d) all

681. Log phase in microbial growth curve is divided into ______ portions
(a) two (b) three
(c) four (d) none

682. Which of the following phase is included in log phase of microbial growth curve
(a) +ve acceleration
(b) –ve acceleration
(c) both (a) & (b)
(d) none

683. Microbial growth curve is generally divided into______ phases
(a) two (b) three
(c) four (d) five

684. In thermal death time curve, D-value represents
(a) time
(b) temperature
(c) number of microorganisms
(d) all

685. In thermal death time curve, z-value represents
(a) time
(b) temperature
(c) number of microorganisms
(d) all

686. In thermal death time curve, F-value represents
(a) time
(b) temperature
(c) number of microorganisms
(d) all

687. Pressurized packaged foods with CO_2 does not inhibit
(a) lactic acid bacteria
(b) Bacillus coagulans
(c) Streptococcus faecalis
(d) all

688. Pressurized packaged foods with CO_2 does not inhibit
(a) aerobic bacteria
(b) molds
(c) both (a) & (b)
(d) none

689. Minimum temperature (°C) for growth of *Bacillus cereus* is about
(a) 7 (b) 4
(c) 5.2 (d) 10

690. Minimum temperature (°C) for growth of *E. coli* is about
(a) 7 (b) 4
(c) 5.2 (d) 10

691. Minimum temperature (°C) for growth of *Salmonella* is about
(a) 7 (b) 4
(c) 5.2 (d) 10

692. Minimum temperature (°C) for growth of *Staphylococcus aureus* is about
(a) 7 (b) 4
(c) 5.2 (d) 10

693. Minimum temperature (°C) for growth of *Plesiomonas shigelloides* is about
(a) 3 (b) 27
(c) 8 (d) 5

694. Minimum temperature (°C) for growth of *Lysteria monocytogenes* is about
(a) 3 (b) 27
(c) 8 (d) 5

695. Minimum temperature (°C) for growth of *Campylobacter jejuni* is about
(a) 3 (b) 27
(c) 8 (d) 5

696. Minimum temperature (°C) for growth of Vibrio parahaemolyticus is about
(a) 3 (b) 27
(c) 8 (d) 5

697. Appoximate lethal dose (kilo grays) for Salmonella is about
(a) 1.7-2.7 (b) 2.7-3.5
(c) 3.7-4.8 (d) none

698. Kilo gray unit is used for
(a) radiation (b) temperature
(c) purity (d) mass

699. Appoximate lethal dose (kilo grays) for Taphilococcus aureus is about
(a) 0.4-1.7 (b) 1.4-7.0
(c) 0.4-0.9 (d) none

700. Appoximate lethal dose (kilo grays) for Corynebacterium diptheriae is about
(a) 1.2 (b) 2.2
(c) 3.2 (d) 4.2

701. Appoximate lethal dose (kilo grays) for *E. coli* is about
(a) 0.5-0.8 (b) 0.2-0.5
(c) 1.0-2.3 (d) 2.2-3.5

702. Appoximate lethal dose (kilo grays) for Pseudomonas fluorescens is about
(a) 1.2-2.3 (b) 2.2-3.4
(c) 3.2-4.3 (d) 2.5-3.0

703. Appoximate lethal dose (kilo grays) for Enterobacter aerogenes is about
(a) 0.4-0.8 (b) 1.4-1.8
(c) 2.4-2.8 (d) 3.4-3.8

704. Appoximate lethal dose (kilo grays) for Lactobacillus spp. is about
(a) 0.23-0.38 (b) 1.2-1.3
(c) 2.2-2.8 (d) 2.8-3.4

705. Appoximate lethal dose (kilo grays) for bacterial spores of *Bacillus subtitlis* is about
(a) 1.2-1.8 (b) 3.2-3.8
(c) 6.2-7.8 (d) 12-18

706. Appoximate lethal dose (kilo grays) for bacterial spores of *Bacillus coagulans* is
(a) 2.5 (b) 5
(c) 10 (d) 25

707. Appoximate lethal dose (kilo grays) for bacterial spores Clostridium Botulinum (A)
(a) 9-12 (b) 15-18
(c) 19-37 (d) 38-52

708. Appoximate lethal dose (kilo grays) for bacterial spores Clostridium Botulinum (E)
(a) 9-12 (b) 15-18
(c) 19-37 (d) 38-52

709. Radiation dose of 20-70 kilograys in meat performs
(a) sterilization
(b) delay spoilage
(c) both (a) & (b) (d) none

710. Radiation dose of 1-10 kilograys in meat performs
(a) sterilization
(b) delay spoilage
(c) both (a) & (b)
(d) none

711. Radiation dose of ______ kilo grays inhibits sprouting in potatoes
(a) 0.05-0.15 (b) 0.25-0.50
(c) 0.5-0.7 (d) 1.0-1.5

712. Ropiness of bread may be due to
(a) *Bacillus subtilis*
(b) *Bacillus licheniformis*
(c) *Bacillus panis*
(d) all

713. Ropiness area in bread show ________ color
(a) yellow (b) brown
(c) both (a) & (b) (d) none

714. Which of the following mold causes spoilage in bread
(a) Rhizopus (b) Penicillium
(c) Aspergillus (d) all

715. *Aspergillus niger* causes ______- color in bread
(a) green
(b) purplish brown
(c) black
(d) all

716. Which of the following causes red color in bread
(a) Serratia marcescens
(b) Monilia sitophila
(c) Geotrichum aurantiacum
(d) all

717. Which of the following bacteria is generally found in sugarcane
(a) Enterobacter (b) Leuconostoc
(c) Flavobacterium (d) all

718. Which of the following bacteria is generally found in sugarcane
(a) Xanthomonas (b) Bacillus
(c) Erwinia (d) all

719. Which of the following bacteria is generally found in raw sugarcane juice
(a) Micrococcus (b) Lactobacillus
(c) Actinomyces (d) all

720. Which of the following mold is generally found in raw sugarcane juice
(a) Aspergillus
(b) Cladosporium
(c) Monilia (d) all

721. Which of the following yeast is generally found in raw sugarcane juice
(a) Saccharomyces (b) Candida
(c) Pichia (d) all

722. Which of the following bacteria is present in raw sugar
(a) Bacillus (b) Clostridium
(c) both (a) & (b) (d) none

723. Which of the following osmophillic yeast is present in raw sugar
(a) Aspergillus (b) Penicillium
(c) both (a) & (b) (d) none

724. Which of the following osmophillic mold is present in raw sugar
(a) Hansenula (b) Pichia
(c) Saccharomyces (d) all

725. Ropiness or stringiness in maple sap is due to
(a) Enterobacter (b) Leuconostoc
(c) both (a) & (b) (d) none

726. Cloudiness in maple sap is due to
(a) Pseudomonas (b) Alcaligenes
(c) Flavobacterium (d) all

727. Which of the following bacteria can occur in fresh and refrigerated meat
(a) Acinetobacter (b) Moraxella
(c) Pseudomonas (d) all

728. Which of the following bacteria can occur in fresh and refrigerated meat
(a) Aeromonas (b) Alcaligenes
(c) Micrococcus (d) all

729. Which of the following mold can occur in fresh and refrigerated meat
(a) Cladosporium (b) Geotrichum
(c) Sporotrichum (d) all

730. Which of the following mold can occur in fresh and refrigerated meat
(a) Mucor (b) Thamnidium
(c) both (a) & (b) (d) none

731. Which of the following yeast can occur in fresh and refrigerated meat
(a) Candida (b) Torulopsis
(c) both (a) & (b) (d) none

732. Which of the following yeast can occur in fresh and refrigerated meat
(a) Debaryomyces (b) Rhodotorula
(c) both (a) & (b) (d) none

733. Which of the following bacteria can occur in processed and cured meats
(a) Lactobacillus (b) Acinetobacter
(c) Bacillus (d) all

734. Which of the following bacteria can occur in processed and cured meats
(a) Micrococcus (b) Serratia
(c) Staphylococcus (d) all

735. Which of the following mold can occur in processed and cured meats
(a) Aspergillus (b) Penicillium
(c) both (a) & (b) (d) none

736. Which of the following mold can occur in processed and cured meats
(a) Rhizopus (b) Thamnidium
(c) both (a) & (b) (d) none

737. Which of the following yeast can occur in processed and cured meats
(a) Debaryomyces (b) Torula
(c) Torulopsis (d) all

738. Which of the following yeast can occur in processed and cured meats
(a) Trichosporon (b) Candida
(c) both (a) & (b) (d) none

739. Physchrotrophic bacteria occur in fish during freezing
(a) Pseudomonas
(b) Acinetobacterium
(c) Moraxella (d) all

740. Physchrotrophic bacteria occur in fish during freezing
(a) Alcaligenes
(b) Flavobacterium
(c) both (a) & (b) (d) none

741. Microbial spoilage in fish can be reduced by
(a) dry salting (b) heating
(c) freezing (d) all

742. Spoilage in chilled shrimp may be due to
(a) Acinetobacter (b) Moraxella
(c) Vibrio (d) all

743. Crabmeat at chilling temperature is deteriorated by
(a) Pseudomonas (b) Acinetobacter
(c) Moraxella (d) all

744. Which of the following is more responsible for egg spoilage at afrm level
(a) bacteria (b) molds
(c) both (a) & (b) (d) none

745. Bacterial spoilage in eggs is also known as
(a) roots (b) spots
(c) rots (d) all

746. Egg rots are generally _______ types
(a) two (b) three
(c) four (d) seven

747. Green rots in egg is chiefly caused by
(a) Pseudomonas (b) Aeromonas
(c) both (a) & (b) (d) none

748. Colorless rots in egg may by caused by
(a) Pseudomonas (b) Acinetobacter
(c) Alcaligenes (d) all

749. Black rots in egg may be caused by
(a) Proteus (b) Pseudomonas
(c) Aeromonas (d) all

750. Which of the following mold cause spoilage of eggs
(a) Penicillium (b) Cladosporium
(c) Sporotrichum (d) all

751. Which of the following mold cause spoilage of eggs
(a) Mucor (b) Thamnidium
(c) Botrytis (d) all

752. Fungal red rot in eggs is due to
(a) Sporotrichum
(b) Cladosporium
(c) both (a) & (b)
(d) none

753. Black spot of eggs is caused by
(a) Sporotrichum
(b) Cladosporium
(c) both (a) & (b)
(d) none

754. Mustiness in eggs may be caused by
(a) Achromobacter (b) Pseudomonas
(c) both (a) & (b) (d) none

755. Bacteria occur in processed poultry
(a) Pseudomonas (b) Acinetobacter
(c) Flavobacterium (d) all

756. Yeast occur in processed poultry
(a) Trichosporon (b) Torulopsis
(c) Rhodotorula (d) all

757. Mold occur in processed poultry
(a) Penicillium (b) Alternaria
(c) Aspergillus (d) all

758. Surface ropiness in milk is mainly due to
(a) Alcaligenes (b) Vibrio
(c) Enterococcus (d) all

759. Ropiness in milk may be due to
(a) Enterobacter (b) Streptococcus
(c) Lactobacillus (d) all

760. Bacterial ropiness generally develops at _____ temperature
(a) low (b) high
(c) both (a) & (b) (d) none

761. Ropiness decreases with__________ in acidity of milk
(a) decrease (b) increase
(c) do not change (d) none

762. Bacterial ropiness in milk are of _____ types
(a) two (b) three
(c) four (d) five

763. Off-flavor caused by microorganisms in milk is
(a) sour (b) bitter
(c) burnt (d) all

764. Bacteria produce sour flavor in milk is
(a) Streptococcus (b) Leuconostoc
(c) Clostridium (d) all

765. Which of the following produce bitter flavor in milk
(a) Cocci (b) Actimycetes
(c) both (a) & (b) (d) none

766. Which of the following produce burnt flavor in milk
(a) Streptococcus (b) *E. coli*
(c) Aeromonas (d) all

767. Barny flavor in milk is due to
(a) Pseudomonas sp.
(b) Enterobacter sp.
(c) Micrococcus sp. (d) all

768. Soapy flavor in milk is due to
(a) Pseudomonas (b) Enterobacter
(c) Micrococcus (d) all

769. Turnip like flavor in milk is due to
(a) Pseudomonas (b) Enterobacter
(c) Micrococcus (d) all

770. Malty flavor in milk is due to
(a) Pseudomonas (b) Enterobacter
(c) Micrococcus (d) all

771. Fruity flavor in milk is due to
(a) Pseudomonas (b) Enterobacter
(c) Micrococcus (d) all

772. Potato like flavor in milk is due to
(a) Pseudomonas (b) Enterobacter
(c) Micrococcus (d) all

773. Fishy flavor in milk is due to
(a) Pseudomonas (b) Enterobacter
(c) Aeromonas (d) all

774. Earthy or musty flavor in milk is due to
(a) Actinomycetes (b) Aeromonas
(c) Pseudomonas (d) all

775. Putrefaction in milk is due to
(a) Clostridium (b) Pseudomonas
(c) both (a) & (b) (d) none

776. Blue color in milk is due to
(a) Pseudomonas (b) Streptococcus
(c) both (a) & (b) (d) none

777. Blue color in milk is due to
(a) Actinimycetes (b) Geotrichum
(c) both (a) & (b) (d) none

778. Yellow color in milk is due to
(a) Pseudomonas synxantha
(b) Flavobacterium
(c) both (a) & (b) (d) none

779. Red color in milk may be due
(a) Serratia (b) Brevibacterium
(c) Micrococcus (d) all

780. Brown color in milk is due to
(a) Pseudomonas (b) Serratia
(c) both (a) & (b) (d) none

781. Bitterness in evaporated milk may be due to
(a) Bacillus (b) Clostridium
(c) both (a) & (b) (d) none

782. Which of the following is a type of spoilage in sweetened condensed milk
(a) gas formation (b) thickening
(c) buttons (d) all

783. Buttons in sweetened condensed milk is
(a) bacteria (b) yeast
(c) mold (d) none

784. Which of the following is a spoilage condition of food cans
(a) flipper (b) springer
(c) soft swell (d) all

785. Which of the following is a spoilage condition of food cans
(a) hard swell (b) breather
(c) both (a) & (b) (d) none

786. Microbial spoilage of canned foods are generally of __________ types
(a) two (b) three
(c) four (d) five

787. Which of the following is a type of microbial spoilage of canned foods
(a) flat sour (b) TA spoilage
(c) both (a) & (b) (d) none

788. Which of the following is a type of microbial spoilage of canned foods
(a) Putrefaction
(b) hydrogen swell
(c) both (a) & (b) (d) none

789. Flat sour spoilage chiefly occurs in ______ foods
(a) low acid (b) high acid
(c) both (a) & (b) (d) none

790. Flat sour spoilage is chiefly caused by
(a) Salmonella (b) *E. coli*
(c) *Bacillus* sp. (d) none

791. TA spoilage is chiefly caused by
(a) Clostridium (b) *E. coli*
(c) both (a) & (b) (d) none

792. Which of the following non spore-forming bacteria causes spoilage in canned foods
(a) Micrococcus (b) Lactobacillus
(c) Microbacterium (d) all

793. Which of the following mold causes spoilage in canned foods
(a) Aspergillus (b) Penicillium
(c) Citromyces (d) all

794. *Bacillus* sp. causes __________ in canned meat and fish
(a) softening (b) souring
(c) both (a) & (b) (d) putrid swell

795. *Clostridium* sp. causes ________ in canned meat and fish
(a) softening (b) souring
(c) both (a) & (b) (d) putrid swell

796. Which of the following gas is produced during spoilage of canned foods
(a) hydrogen (b) carbondioxide
(c) both (a) & (b) (d) none

797. Which of the following bacteria is most active in microbial spoilage of canned foods
(a) Thermophillic (b) mesophillic
(c) psychrophillic (d) none

798. Which of the following yeast cause spoilage in soft drinks
(a) Candida (b) Torulopsis
(c) both (a) & (b) (d) none

799. Which of the following bacteria occasionally cause spoilage in soft drinks
(a) Gluconobacter
(b) Lactobacillus
(c) Leuconostoc (d) all

800. Which of the following bacteria is used in rye bread making
(a) Lactobacillus (b) Leuconostoc
(c) Streptococcus (d) all

2.3 ANSWER KEY

Q.	A.	Q.	A.	Q.	A.	Q.	A.	Q.	A.	Q.	A.
1.	(b)	2.	(c)	3.	(c)	4.	(b)	5.	(d)	6.	(c)
7.	(c)	8.	(d)	9.	(c)	10.	(d)	11.	(d)	12.	(b)
13.	(b)	14.	(c)	15.	(c)	16.	(c)	17.	(b)	18.	(d)
19.	(d)	20.	(d)	21.	(d)	22.	(b)	23.	(d)	24.	(d)
25.	(d)	26.	(b)	27.	(d)	28.	(b)	29.	(d)	30.	(d)
31.	(c)	32.	(d)	33.	(d)	34.	(c)	35.	(b)	36.	(d)
37.	(d)	38.	(d)	39.	(c)	40.	(d)	41.	(d)	42.	(d)
43.	(c)	44.	(d)	45.	(c)	46.	(d)	47.	(c)	48.	(d)
49.	(d)	50.	(c)	51.	(d)	52.	(d)	53.	(d)	54.	(d)
55.	(d)	56.	(d)	57.	(d)	58.	(d)	59.	(d)	60.	(c)
61.	(d)	62.	(c)	63.	(c)	64.	(c)	65.	(b)	66.	(d)
67.	(d)	68.	(c)	69.	(c)	70.	(d)	71.	(d)	72.	(b)
73.	(d)	74.	(b)	75.	(c)	76.	(c)	77.	(b)	78.	(d)
79.	(b)	80.	(b)	81.	(c)	82.	(c)	83.	(b)	84.	(b)
85.	(c)	86.	(d)	87.	(b)	88.	(d)	89.	(d)	90.	(b)
91.	(c)	92.	(b)	93.	(b)	94.	(d)	95.	(d)	96.	(d)
97.	(d)	98.	(c)	99.	(d)	100.	(c)	101.	(d)	102.	(d)
103.	(c)	104.	(d)	105.	(d)	106.	(b)	107.	(c)	108.	(c)
109.	(b)	110.	(b)	111.	(d)	112.	(b)	113.	(b)	114.	(c)
115.	(c)	116.	(d)	117.	(d)	118.	(d)	119.	(d)	120.	(d)
121.	(b)	122.	(c)	123.	(d)	124.	(d)	125.	(c)	126.	(c)
127.	(b)	128.	(d)	129.	(b)	130.	(d)	131.	(c)	132.	(c)
133.	(c)	134.	(d)	135.	(d)	136.	(b)	137.	(d)	138.	(c)
139.	(d)	140.	(d)	141.	(d)	142.	(d)	143.	(b)	144.	(c)

Q.	A.	Q.	A.	Q.	A.	Q.	A.	Q.	A.	Q.	A.
145.	(b)	146.	(d)	147.	(d)	148.	(d)	149.	(d)	150.	(c)
151.	(d)	152.	(c)	153.	(d)	154.	(b)	155.	(d)	156.	(d)
157.	(d)	158.	(d)	159.	(d)	160.	(d)	161.	(d)	162.	(d)
163.	(d)	164.	(d)	165.	(d)	166.	(d)	167.	(d)	168.	(d)
169.	(d)	170.	(d)	171.	(d)	172.	(d)	173.	(d)	174.	(d)
175.	(d)	176.	(d)	177.	(d)	178.	(d)	179.	(c)	180.	(d)
181.	(d)	182.	(d)	183.	(d)	184.	(d)	185.	(d)	186.	(d)
187.	(d)	188.	(d)	189.	(c)	190.	(d)	191.	(d)	192.	(b)
193.	(c)	194.	(d)	195.	(d)	196.	(d)	197.	(d)	198.	(d)
199.	(b)	200.	(b)	201.	(d)	202.	(d)	203.	(d)	204.	(d)
205.	(d)	206.	(d)	207.	(d)	208.	(c)	209.	(b)	210.	(d)
211.	(d)	212.	(d)	213.	(d)	214.	(d)	215.	(d)	216.	(d)
217.	(d)	218.	(d)	219.	(d)	220.	(d)	221.	(d)	222.	(d)
223.	(d)	224.	(d)	225.	(d)	226.	(d)	227.	(d)	228.	(d)
229.	(d)	230.	(c)	231.	(d)	232.	(b)	233.	(c)	234.	(d)
235.	(d)	236.	(d)	237.	(d)	238.	(b)	239.	(d)	240.	(d)
241.	(c)	242.	(b)	243.	(d)	244.	(c)	245.	(c)	246.	(d)
247.	(c)	248.	(d)	249.	(b)	250.	(c)	251.	(d)	252.	(d)
253.	(d)	254.	(d)	255.	(b)	256.	(c)	257.	(c)	258.	(d)
259.	(b)	260.	(c)	261.	(d)	262.	(c)	263.	(d)	264.	(d)
265.	(d)	266.	(b)	267.	(d)	268.	(d)	269.	(c)	270.	(d)
271.	(d)	272.	(b)	273.	(c)	274.	(d)	275.	(d)	276.	(b)
277.	(d)	278.	(c)	279.	(d)	280.	(b)	281.	(c)	282.	(c)
283.	(b)	284.	(d)	285.	(b)	286.	(c)	287.	(c)	288.	(d)
289.	(b)	290.	(c)	291.	(d)	292.	(b)	293.	(c)	294.	(d)
295.	(d)	296.	(d)	297.	(d)	298.	(d)	299.	(d)	300.	(d)
301.	(c)	302.	(d)	303.	(d)	304.	(c)	305.	(d)	306.	(b)
307.	(d)	308.	(d)	309.	(d)	310.	(d)	311.	(c)	312.	(d)
313.	(d)	314.	(b)	315.	(d)	316.	(c)	317.	(d)	318.	(c)
319.	(d)	320.	(c)	321.	(d)	322.	(b)	323.	(d)	324.	(c)
325.	(d)	326.	(b)	327.	(d)	328.	(d)	329.	(d)	330.	(d)
331.	(d)	332.	(d)	333.	(d)	334.	(c)	335.	(d)	336.	(d)
337.	(c)	338.	(b)	339.	(d)	340.	(d)	341.	(d)	342.	(d)
343.	(b)	344.	(d)	345.	(d)	346.	(d)	347.	(d)	348.	(d)

Q. A.	Q. A.	Q. A.	Q. A.	Q. A.	Q. A.
349. (d)	350. (d)	351. (d)	352. (c)	353. (b)	354. (d)
355. (d)	356. (d)	357. (c)	358. (d)	359. (d)	360. (d)
361. (b)	362. (d)	363. (d)	364. (c)	365. (b)	366. (d)
367. (d)	368. (b)	369. (d)	370. (b)	371. (b)	372. (d)
373. (d)	374. (d)	375. (d)	376. (c)	377. (c)	378. (c)
379. (d)	380. (d)	381. (d)	382. (d)	383. (c)	384. (b)
385. (d)	386. (c)	387. (b)	388. (d)	389. (d)	390. (d)
391. (c)	392. (d)	393. (c)	394. (c)	395. (d)	396. (d)
397. (d)	398. (d)	399. (c)	400. (d)	401. (d)	402. (d)
403. (d)	404. (d)	405. (c)	406. (d)	407. (d)	408. (c)
409. (d)	410. (b)	411. (d)	412. (b)	413. (d)	414. (c)
415. (d)	416. (d)	417. (c)	418. (c)	419. (c)	420. (c)
421. (d)	422. (d)	423. (b)	424. (d)	425. (b)	426. (c)
427. (d)	428. (b)	429. (d)	430. (c)	431. (d)	432. (d)
433. (d)	434. (d)	435. (c)	436. (d)	437. (d)	438. (d)
439. (d)	440. (d)	441. (d)	442. (c)	443. (b)	444. (d)
445. (b)	446. (d)	447. (c)	448. (c)	449. (b)	450. (d)
451. (b)	452. (d)	453. (b)	454. (c)	455. (c)	456. (d)
457. (d)	458. (c)	459. (d)	460. (d)	461. (c)	462. (d)
463. (d)	464. (d)	465. (c)	466. (d)	467. (d)	468. (c)
469. (b)	470. (b)	471. (c)	472. (d)	473. (c)	474. (c)
475. (d)	476. (d)	477. (d)	478. (b)	479. (c)	480. (c)
481. (b)	482. (c)	483. (b)	484. (c)	485. (c)	486. (b)
487. (c)	488. (d)	489. (d)	490. (c)	491. (d)	492. (d)
493. (c)	494. (d)	495. (d)	496. (d)	497. (d)	498. (c)
499. (d)	500. (b)	501. (d)	502. (c)	503. (c)	504. (c)
505. (b)	506. (d)	507. (b)	508. (d)	509. (c)	510. (d)
511. (b)	512. (d)	513. (d)	514. (b)	515. (c)	516. (d)
517. (d)	518. (d)	519. (c)	520. (d)	521. (b)	522. (c)
523. (d)	524. (c)	525. (c)	526. (b)	527. (b)	528. (d)
529. (d)	530. (d)	531. (d)	532. (d)	533. (d)	534. (b)
535. (c)	536. (d)	537. (b)	538. (c)	539. (d)	540. (d)
541. (b)	542. (c)	543. (d)	544. (b)	545. (c)	546. (c)
547. (d)	548. (b)	549. (b)	550. (c)	551. (d)	552. (d)

Q.	A.	Q.	A.	Q.	A.	Q.	A.	Q.	A.	Q.	A.
553.	(d)	554.	(d)	555.	(c)	556.	(d)	557.	(d)	558.	(c)
559.	(d)	560.	(d)	561.	(d)	562.	(d)	563.	(d)	564.	(d)
565.	(c)	566.	(d)	567.	(d)	568.	(c)	569.	(d)	570.	(b)
571.	(d)	572.	(c)	573.	(d)	574.	(c)	575.	(d)	576.	(d)
577.	(d)	578.	(d)	579.	(d)	580.	(c)	581.	(d)	582.	(b)
583.	(c)	584.	(d)	585.	(d)	586.	(d)	587.	(c)	588.	(d)
589.	(c)	590.	(d)	591.	(c)	592.	(c)	593.	(c)	594.	(c)
595.	(d)	596.	(c)	597.	(d)	598.	(d)	599.	(d)	600.	(d)
601.	(d)	602.	(c)	603.	(c)	604.	(d)	605.	(b)	606.	(d)
607.	(c)	608.	(c)	609.	(d)	610.	(d)	611.	(d)	612.	(b)
613.	(b)	614.	(d)	615.	(c)	616.	(d)	617.	(c)	618.	(c)
619.	(c)	620.	(d)	621.	(d)	622.	(c)	623.	(c)	624.	(c)
625.	(c)	626.	(d)	627.	(c)	628.	(d)	629.	(d)	630.	(d)
631.	(d)	632.	(d)	633.	(c)	634.	(d)	635.	(d)	636.	(d)
637.	(c)	638.	(d)	639.	(d)	640.	(b)	641.	(d)	642.	(d)
643.	(c)	644.	(d)	645.	(d)	646.	(d)	647.	(c)	648.	(d)
649.	(d)	650.	(d)	651.	(c)	652.	(d)	653.	(d)	654.	(d)
655.	(d)	656.	(c)	657.	(d)	658.	(d)	659.	(d)	660.	(d)
661.	(d)	662.	(d)	663.	(d)	664.	(b)	665.	(d)	666.	(c)
667.	(d)	668.	(d)	669.	(d)	670.	(d)	671.	(d)	672.	(c)
673.	(b)	674.	(c)	675.	(c)	676.	(c)	677.	(b)	678.	(d)
679.	(c)	680.	(d)	681.	(d)	682.	(c)	683.	(c)	684.	(d)
685.	(b)	686.	(d)	687.	(d)	688.	(c)	689.	(d)	690.	(b)
691.	(c)	692.	(d)	693.	(c)	694.	(d)	695.	(b)	696.	(d)
697.	(c)	698.	(d)	699.	(b)	700.	(d)	701.	(c)	702.	(d)
703.	(b)	704.	(d)	705.	(d)	706.	(c)	707.	(c)	708.	(b)
709.	(c)	710.	(b)	711.	(d)	712.	(d)	713.	(c)	714.	(d)
715.	(d)	716.	(d)	717.	(d)	718.	(d)	719.	(d)	720.	(d)
721.	(d)	722.	(c)	723.	(c)	724.	(d)	725.	(c)	726.	(d)
727.	(d)	728.	(d)	729.	(d)	730.	(c)	731.	(c)	732.	(c)
733.	(d)	734.	(d)	735.	(c)	736.	(c)	737.	(d)	738.	(c)
739.	(d)	740.	(c)	741.	(d)	742.	(d)	743.	(d)	744.	(d)
745.	(c)	746.	(b)	747.	(d)	748.	(d)	749.	(d)	750.	(d)
751.	(d)	752.	(d)	753.	(b)	754.	(c)	755.	(d)	756.	(d)

Q.	A.	Q.	A.	Q.	A.	Q.	A.	Q.	A.	Q.	A.
757.	(d)	758.	(d)	759.	(d)	760.	(d)	761.	(b)	762.	(d)
763.	(d)	764.	(d)	765.	(c)	766.	(d)	767.	(b)	768.	(d)
769.	(d)	770.	(c)	771.	(d)	772.	(d)	773.	(c)	774.	(d)
775.	(c)	776.	(c)	777.	(c)	778.	(c)	779.	(d)	780.	(d)
781.	(c)	782.	(d)	783.	(c)	784.	(d)	785.	(c)	786.	(c)
787.	(c)	788.	(c)	789.	(d)	790.	(c)	791.	(d)	792.	(d)
793.	(d)	794.	(c)	795.	(d)	796.	(c)	797.	(d)	798.	(c)
799.	(d)	800.	(d)								

CHAPTER 3

FOOD ENGINEERING

3.1 SUBJECTIVE KEYNOTES

1. Food engineering deals with study of use of engineering principles and unit operations in food processing industry.
2. The main engineering prospects used in food industries are heat and mass transfer, fluid flow, thermodynamics, material and energy balance and chemical kinetics.
3. Rheology is the study of flow (liquids) and deformation (solids).
4. Stress, strain, storage modulus and loss modulus are associated with rheological analysis of foods.
5. Newton's law states that shear stress is directly proportional to shear rate.
6. Stress relaxation tests, creep recovery tests, small amplitude oscillatory tests are associated with rheological analysis of foods.
7. Power law, Newtonian, Bingham plastic and Herschel-Bulkley models are used to describe the rheological behavior of materials.
8. Different types of viscosity include relative viscosity, apparent viscosity, intrinsic viscosity and dynamic viscosity.
9. Glass transition is the change in physical state of amorphous materials which involves transformation of non-equilibrium solid to a viscous liquid state.
10. Differential scanning calorimetry is thermal technique used for analyzing phase and state transitions including glass transition in foods.
11. Heat transfer occurs by three modes-conduction (solids), convection (liquid) and radiation (gas).
12. Heat exchangers used in food processing for heating and cooling purpose are plate heat exchangers and tubular heat exchangers.
13. Specific heat capacity, thermal conductivity, compressibility, thermal expansion and glass transitions are important properties relevant to thermal processing of foods.
14. Heated probe method is used for measurement of thermal conductivity.
15. Enthalpy of frozen foods are generally determined using calorimetric methods.
16. Emissivity, reflectivity, absorptivity and transmissivity are important properties relevant to radiative heat transfer in foods.

17. Moisture sorption isotherm is a graphical expression of relationship between total moisture content & the corresponding water activity of a food at constant temperature.
18. Equilibrium moisture content is the moisture content remaining in the dry material when drying rate drops to 0 at specified conditions.
19. Molecular diffusion is the transfer of the mass caused by random movement of molecules.
20. Important application of mass transfer to food system includes moisture transfer, diffusion in porous foods, diffusion of solutes and diffusion of aroma compounds.
21. Extraction, distillation, gas absorption, crystallization and food packaging are associated with mass transfer in foods.
22. Electrical conductivity of foods generally exhibits a linear increase with temperature.
23. Flash 18 is related to thermal processing of canned foods.
24. D_{250} value is number of minutes at 250 °F to accomplish 1 log cycle reduction.
25. The difference in equilibrium moisture content between adsorption and desorption curves is called hysteresis. Hysteresis is generally observed in hygroscopic products.
26. All engineering quantities are expressed in fundamental dimensions. These are mass, length and time. A forth fundamental dimension is added to the system of measurement is temperature.
27. Water activity is the ratio of vapor pressure of water in food and vapor pressure of pure water.
28. At equilibrium condition, the water activity of the food is equal to relative humidity of air.
29. The process of water gain is termed adsorption and that of water loss is termed desorption.
30. The amount of water absorbed generally decreases with increase in temperature.
31. Dilatometry was one of the earlier methods used to find glass transition tempe-rature by measuring change in volumetric thermal expansion co-efficient.
32. Glass transition temperature can be determined using Dilatometry, dynamic thermal mechanical analysis (DMTA) and differential scanning calorimetry.
33. Porosity is defined as the ratio of empty space volume to total volume.
34. Shrinkage is defined as the relative volume change to the original volume.
35. Fick's law is often used to describe a moisture diffusion process.
36. The first step in sizing a spray drier is the determination of the droplet size.
37. In freeze drying, the product is frozen and the frozen water is removed by sublimation under vacuum.
38. Industrial freeze driers are of two types: tray freeze drier and tunnel freeze drier.
39. Membrane processing of foods include reverse osmosis, nano filtration, ultra filtration and microfiltration.
40. Membrane fouling restricts the large scale use of membrane technology in food industry.
41. Membrane processing is generally used in beverage industry.

42. Physical type evaporators include short tube evaporator, long tube vertical rising film evaporator, long tube vertical falling film evaporator, forced circulation evaporator, scraped surface thin film evaporator, plate evaporator and thin film spinning cone evaporator.
43. The energy input into an MVR evaporator is altered by fan speed.
44. The study of fluid behavior at rest is called fluid statics. Fluid dynamics involves the study of flow and the forces causing the flow.
45. Planck's equation can be used for determination of freezing time. Freezing times are basis design criteria for freezing substances.
46. Freezing systems are of two types: direct contact and indirect contact.
47. Types of heat exchangers are : tubular heat exchanger, triple tube heat exchanger, plate heat exchanger and scraped surface heat exchangers.
48. Flow of granular solid is affected by gravitational force, friction, cohesion and adhesion.
49. Friction factor represents the ratio of pressure drop to the viscous energy.
50. Polymorphism is the existence of number of alternative crystal structures and can be studied by Dilatometry, temperature controlled X rays diffraction, NMR and DSC techniques.
51. The manometers are the devices used for measuring the pressure by balancing the column of liquid by the same or other fluid column. The common manometric fluid is mercury. Gauge pressure is measured in manometers where as the absolute pressure is measured using mechanical gauges like Bourdon tube pressure gauge.
52. Fluid kinematics is the branch of fluid mechanics, which describes the fluid motion without considering its causes.
53. In flow pattern, stream line is a line in every point the velocity vector is tangent to it. Path line is the actual line followed by a fluid particle. Streak line is the line formed by all fluid particles that have passed through a given fixed point. Time line is a definite line formed by set of points at any instant.
54. When the fluid particles pass through a well defined path or streamline, the flow is called laminar flow ($NRe \leq 2100$). In turbulent flow ($NRe \geq 4000$), the fluid particles move randomly and change the direction in every instant. The flow is characterized by a dimensionless number Reynold's number which is a ratio of inertial force to viscous force.
55. The minimum velocity beyond which the flow becomes turbulent is critical velocity.
56. Venturi meter is a device used to measure the discharge in a pipe flow.
57. Orifice meter is a device used to measure the discharge in a pipe flow (small volume pipelines). It is basically a flat plate with a circular hole drilled in it.
58. Thermal conductivity can be measured using Fourier's equation of heat conduction. Thermal conductivity depends on the molecular arrangement of a substance. Thermal conductivity of food materials varies from 0.6-0.7 W/mK.
59. The Nusselt number describes about the relative dominance of convection over conduction. The rate of heat transfer increases as a result of increased in Nusselt number.

60. Prandtl number is the ratio of hydrodynamic layer and thermal boundary layer. The Prandtl number for gas is about 0.7 and for water it is around 10.
61. Emissivity is the ratio of the radiant energy flux of a body to that of a perfect black body whose absorptivity is unity.
62. Kirchhoff's law state that the emissivity and absorptivity of any substance are same at thermal equilibrium.
63. A body is said to be perfectly black if it absorbs all the incident radiation to it. The absorptivity of a black body is unity and hence the transmissivity and reflectivity are zero.
64. Stefan-Boltzmann law states that the total emissive power is the total radiation energy per unit area leaving a surface with temperature T over all wavelengths.
65. D-value refers to decimal reduction time and is the time required at a certain temperature to kill 90% population of the organisms.
66. Thermal death time (F) is the time required to kill a specific number of microbes at a specific temperature.
67. Z Value is the number of degrees of temperature required to reduce the D Value by one log cycle.
68. The rate of heat removal from a system is termed as refrigeration load.
69. The refrigerants are volatile liquids used in mechanical vapour compression system.
70. The basic principle of mechanical refrigeration cycle is based on cyclic thermodynamic process known as Rankine-cycle or vapour compression cycle. The cycle consists of four components- compression, condensation, expansion & evaporation.
71. Cooling load is the rate of removal of heat from a defined space in order to lower the temperature of that space to a desired level. The cooling loads of refrigeration systems are designated with a common unit called ton of refrigeration (TR).
72. One ton of refrigeration (TR) is equivalent to latent heat of fusion of one ton of ice in 24 hours. The value of one TR is 3.52 kW. So, a mechanical refrigeration system that has the capacity to absorb heat at the rate of 3.52 kW is rated as one TR of refrigeration.
73. Coefficient of performance (COP) checks the efficacy of a refrigeration cycle. It is the ratio of refrigeration effect and the work done to achieve this effect.
74. Psychrometry is the study of thermodynamic properties of air-water vapour mixture
75. Specific heat is the heat required to raise the temperature to one unit of a unit mass of air. The average specific heat of dry air at one atmospheric temperature is 1.005 kJ/kgK. The specific heat of water vapour is 1.88 kJ/kgK
76. Dry bulb temperature is the temperature sensed by an ordinary thermometer when in thermal equilibrium with air.
77. *Gibbs-Dalton law* states that the total pressure exerted by a mixture of air water vapour is same as that exerted by the individual components independently.
78. Dew point is the temperature at which the water vapour at a particular pressure condenses to water at the same rate at which it evaporates. At dew point, the relative humidity is 100%.

79. Wet bulb temperature is the temperature at which water gets evaporated and brings the air-water vapour to a state of saturation adiabatically at the same temperature. It is also known as adiabatic saturation temperature.
80. The graphical representation of the thermodynamic properties of air-water vapour mixture is known as psychrometric chart.
81. The rate of evaporation in an evaporator is determined by the amount of heat transferred to the solution by the heat exchanger. The steam gives off its latent heat to the surface of the exchanger which is in contact with the feed solution. The latent heat at atmospheric pressure is 2257 kJ/kg.
82. The void fraction in a packed bed is the fraction of the flow cross-sectional area occupied by the gas phase and represented as the ratio between volumes of voids in the bed to total volume of the bed.
83. The hydraulic radius is the ratio between the cross sectional area available for flow and wetted perimeter.
84. Solid – liquid extraction is the removal of a soluble component from a solid by contact with a liquid solvent. It is also known as leaching.
85. Osmodehydration is the extraction of water using a low water activity solution (e.g. concentrated sugar solution), accompanied by diffusion of other solutes into the solid.
86. Crystals are the solids composed of atoms, ions, or molecules which are arranged in an orderly and repetitive manner in space lattices. The interatomic distances and the angles between these planes are measured by X-ray diffraction (XRD).
87. Distillation is a method of separating mixtures based on differences in volatility of components in a boiling liquid mixture.
88. Flotation is a technique in which different types of solid particles in a liquid are separated out, the principle being that some particles will absorb water while others will not. The use of bubbles to float fine particles in a liquid is commonly known as flotation process.
89. Emulsification is the process by which two immiscible liquids or solid-liquid can be dispersed one into the other by severe mixing, stirring and beating.
90. Colloid mill is suitable for homogenization of medium and high viscosity food products.
91. The efficiency of homogenization is calculated by comparing the actual energy required to form an emulsion and the minimum theoretical energy required.
92. Extrusion is a method of forming substances by forcing them through a perforated plate or die to produce tubes, rods, or other desired shapes. Extrusion cooking the process of cooking moistened starchy and or proteinaceous food materials by combination of heat, pressure and mechanical shear.
93. Extruder is a thermodynamic device to shape by forcing material through a specially designed opening often after a previous heating of the material.
94. Ohmic heating is defined as a process where in electric current is passed through materials with the primary purpose of heating them. It is also called as is also called electrical resistance heating, Joule heating, or electro-heating.
95. Pulsed electric field (PEF) processing is a non-thermal method for food preservation that uses short bursts of electricity for microbial inactivation and causes minimal or

no detrimental effect on food quality attributes. The basic principle of the PEF technology is the application of short pulses of high electric fields with duration of micro to milliseconds and intensity in the order of 10-80 kV/cm.

96. Electroporation is the phenomenon in which a cell exposed to high voltage electric field pulses temporarily destabilizes the lipid bilayer and proteins of cell membranes.

97. A pulsed Electric Field processing system consists of a high-voltage power source, an energy storage capacitor bank, a charging current limiting resistor, a switch to discharge energy from the capacitor across the food and a treatment chamber.

98. High Pressure Processing (HPP) is a method of Food processing where food is subjected to elevated pressure 6000 atmospheres approximately (700MPa). It also called as High- hydrostatic pressure (HHP).

99. Food irradiation is treatment of foods by subjecting them to ionizing radiation in order to control foodborne pathogens, reduce microbial load and insect infestation, inhibit the germination of root crops, and extend the durable life of perishable produce.

100. Ultrasonic wave producing system contains the generator, transducer and the application system. Generator produces electrical or mechanical energy and transducer converts this energy into the sound energy at ultrasonic frequencies.

3.2 MULTIPLE CHOICE QUESTIONS

1. Which of the following is unit of length
(a) s (b) m
(c) A (d) K

2. Which of the following is unit of mass
(a) s (b) m
(c) A (d) g

3. Which of the following is unit of time
(a) s (b) m
(c) A (d) K

4. Which of the following is unit of electric current
(a) s (b) m
(c) A (d) K

5. Which of the following is unit of temperature
(a) s (b) m
(c) A (d) K

6. Which of the following is unit of amount of substance
(a) concentration (b) strength
(c) mole (d) none

7. Which of the following is not a derived unit
(a) newton (b) ohm
(c) farad (d) kelvin

8. Which of the following is not a basic unit
(a) candela (b) mole
(c) ampere (d) lumen

9. Which of the following is a basic unit
(a) electric current
(b) amount of substance
(c) luminous intensity
(d) all

10. Which of the following is a derived unit
(a) joule (b) volt
(c) weber (d) all

11. Joule is a unit of
(a) work
(b) energy
(c) both
(d) none

12. Newton is the unit of
(a) energy
(b) electric potential
(c) force
(d) electric charge

13. Watt is the unit of
(a) energy (b) force
(c) power (d) electric charge

14. Ohm is the unit of
(a) electric charge
(b) electric resistance
(c) capacitance
(d) none

15. Coulomb is the unit of
(a) electric charge
(b) electric resistance
(c) capacitance
(d) none

16. S.I. unit of area is
(a) m (b) m^2
(c) m^3 (d) cm

17. S.I. unit of volume is
(a) m (b) m^2
(c) m^3 (d) cm

18. S.I. unit of velocity is
(a) m/s (b) cm/s
(c) m/s^2 (d) cm/s^2

19. S.I. unit of acceleration is
(a) m/s (b) cm/s
(c) m/s^2 (d) cm/s^2

20. S.I. unit of density is
(a) g/m^2 (b) kg/m^3
(c) kg/m^2 (d) g/s^2

21. S.I. unit of concentration is
(a) mol/cm^2 (b) mol/m^2
(c) mol/m^3 (d) mol/cm^3

22. S.I. unit of specific volume is
(a) kg/m^3 (b) m^3/kg
(c) kg/m^2 (d) m^2/kg

23. S.I. unit of luminance is
(a) cd/m (b) cd/m^2
(c) cd/m^3 (d) cd/kg

24. S.I. unit of pressure is
(a) pascal (b) newton
(c) hertz (d) all

25. S.I. unit of frequency is
(a) pascal (b) newton
(c) hertz (d) all

26. S.I. unit of capacitance is
(a) ohm (b) coloumb
(c) farad (d) all

27. S.I. unit of conductance is
(a) pascal (b) newton
(c) hertz (d) siemens

28. S.I. unit of dynamic viscosity is
(a) Nm (b) N/m
(c) Pa.s (d) J/kg

29. S.I. unit of surface tension is
(a) Nm (b) N/m
(c) Pa.s (d) J/kg

30. S.I. unit of heat capacity is
(a) J/kg (b) J/K
(c) J/m^3 (d) J/kg

31. S.I. unit of specific heat capacity is
(a) J/kg (b) J/K
(c) J/m^3 (d) J/(kgK)

32. S.I. unit of specific energy is
(a) J/kg (b) J/K
(c) J/m^3 (d) J/(kgK)

33. S.I. unit of thermal conductivity is
(a) J/K (b) W/(mK)
(c) J/m^3 (d) J/(kgK)

34. S.I. unit of energy density is
(a) J/K (b) W/(mK)
(c) J/m^3 (d) J/(kgK)

35. S.I. unit of electric charge density is
(a) C/m^2 (b) C/m^3
(c) $J.m^2$ (d) J/m^3

36. S.I. unit of electric flux density is
(a) C/m^2 (b) C/m^3
(c) $J.m^2$ (d) J/m^3

37. S.I. unit of plane angle is
(a) radian (b) steradian
(c) both (d) none

38. S.I. unit of solid angle is
(a) radian (b) steradian
(c) both (d) none

39. Which of the following is unit of viscosity
(a) poise (b) Pa.s
(c) both (d) none

40. Adiabatic system involves no exchange of
(a) heat (b) mass
(c) temperature (d) all

41. Which of the following is extensive property
(a) mass (b) length
(c) volume (d) all

42. Which of the following is intensive property
(a) temperature (b) pressure
(c) density (d) all

43. Density of the substance is mass per unit____
(a) length (b) area
(c) volume (d) all

44. Which of the following is not a type of density
(a) solid density (b) particle density
(c) bulk density (d) none

45. Solid density of glucose is about ______ kg/m^3
(a) 1500 (b) 1560
(c) 1590 (d) 1400

46. Solid density of glucose is about ______ kg/m^3
(a) 1500 (b) 1560
(c) 1590 (d) 1400

47. Solid density of starch is about ______ kg/m^3
(a) 1500 (b) 1560
(c) 1590 (d) 1400

48. Solid density of globular protein is about ______ kg/m^3
(a) 1500 (b) 1560
(c) 1590 (d) 1400

49. Solid density of fat is about______ kg/m^3
(a) 900-950 (b) 1000
(c) 1540 (d) 2160

50. Solid density of salt is about ______ kg/m^3
(a) 900-950 (b) 1000
(c) 1540 (d) 2160

51. Solid density of citric acid is about ______ kg/m^3
(a) 900-950 (b) 1000
(c) 1540 (d) 2160

52. Solid density of water is about______ kg/m^3
(a) 900-950 (b) 1000
(c) 1540 (d) 2160

53. Bulk density of wheat is about ______ kg/m^3
(a) 770 (b) 800
(c) 320 (d) 673

54. Bulk density of hulled rice is about ______ kg/m^3
(a) 770 (b) 800
(c) 320 (d) 673

55. Bulk density of granulated sugar is about______ kg/m^3
(a) 770 (b) 800
(c) 320 (d) 673

56. Porosity = 1- bulk density/?
(a) particle density (b) solid density
(c) both (d) none

57. Interparticle porosity = 1- bulk density/?

(a) particle density (b) solid density
(c) both (d) none

58. Which of the following involves volume of solution
(a) molarity (b) molality
(c) both (d) none

59. Which of the following is independent of temperature
(a) molality (b) mole fraction
(c) both (d) none

60. S.I. unit of temperature is
(a) °C (b) °F
(c) °K (d) all

61. T (K) = T(°C) +?
(a) 272.73 (b) 273.15
(c) 273.79 (d) 273.95

62. Ice point for Celsius scale is
(a) 0 (b) 4
(c) -4 (d) -5.5

63. Ice point for Fahrenheit scale is
(a) 0 (b) 4
(c) 12 (d) 32

64. Boiling point of water is ____ °F
(a) 100 (b) 272
(c) 212 (d) 373

65. Boiling point of water is ____ °F
(a) 100°C (b) 100.9°C
(c) 102.5 °C (d) 105.9°C

66. Pressure may be expressed as ________ per unit area
(a) work (b) energy
(c) force (d) mass

67. 1 bar = ________ kPa
(a) 0.1 (b) 100
(c) 1000 (d) 10000

68. 1 atmosphere = ________ bar
(a) 1.01325 (b) 101.325
(c) 10.1325 (d) 1013.25

69. 1 atmosphere = ________ kPa
(a) 1.01325 (b) 101.325
(c) 10.1325 (d) 1013.25

70. Enthalpy = Internal Energy + ______
(a) P (b) T
(c) V (d) PV

71. Triple point of water is at
(a) 0.1°C (b) 0.01°C
(c) 0 °C (d) none

72. Phase diagrams are important in studying
(a) crystallization (b) distillation
(c) precipitation (d) all

73. According to first law of thermodynamics, energy of an isolated system remains
(a) variable (b) constant
(c) zero (d) none

74. Potential energy = mg?
(a) v^2 (b) c
(c) h (d) a

75. Kinetic energy = ½ m?
(a) v^2 (b) c
(c) h (d) a

76. When work is done by the system, then w is
(a) –ve (b) +ve
(c) zero (d) none

77. When work is done on- the system, then w is
(a) –ve (b) +ve
(c) zero (d) none

78. Latent heat of fusion for water at 0°C is________ kJ/kg
(a) 323.2 (b) 223.2
(c) 333.2 (d) 523.2

79. Latent heat of vaporization for water at 100°C is________ kJ/kg
(a) 225.706 (b) 2257.06
(c) 325.706 (d) 3257.06

80. 1 horse power = ________ kW
(a) 745.7 (b) 0.7457
(c) 0.07457 (d) 0.547

81. Which of the following is kinetic type pump
(a) centrifugal (b) rotary
(c) reciprocating (d) all

82. Which of the following is a positive displacement type pump
(a) Rotary (b) Reciprocating
(c) both (d) none

83. Stress is force per unit______
(a) length (b) area
(c) volume (d) none

84. When force acting on a surface is perpendicular to it, then it is called
(a) shear stress (b) shear strain
(c) normal stress (d) normal strain

85. When force act parallel to surface, then it is called
(a) shear stress (b) shear strain
(c) normal stress (d) normal strain

86. Gases are ________ fluids
(a) compressible
(b) incompressible
(c) both (d) none

87. Liguids are ________ fluids
(a) compressible
(b) incompressible
(c) both (d) none

88. Approximate viscosity (Pa.s) of air at room temperature is about
(a) 10^{-1} (b) 10
(c) 10^{-3} (d) 10^{-5}

89. Approximate viscosity (Pa.s) of water at room temperature is about
(a) 10^{-1} (b) 10
(c) 10^{-3} (d) 10^{-5}

90. Approximate viscosity (Pa.s) of olive at room temperature is about
(a) 10^{-1} (b) 10
(c) 10^{-3} (d) 10^{-5}

91. Approximate viscosity (Pa.s) of liquid honey at room temperature is about
(a) 10^{-1} (b) 10
(c) 10^{-3} (d) 10^{-5}

92. 1 Pa = ________ dyne/cm²
(a) 10 (b) 100
(c) 1000 (d) 500

93. Which of the following exhibit Newtonian flow
(a) water (b) milk
(c) fruit juice (d) all

94. 1 Pa.s = ________ mPa.s
(a) 10 (b) 100
(c) 1000 (d) 1.0

95. 1 cP = ________ mPa.s
(a) 10 (b) 100
(c) 1000 (d) 1.0

96. 1 Poise = ________ Pa.s
(a) 0.1 (b) 10
(c) 100 (d) 1000

97. Kinematic viscosity = dyanamic viscosity/ ?
(a) mass (b) volume
(c) density (d) none

98. The staright line flow at ________ flow rates is called laminar flow
(a) low (b) intermediate
(c) high (d) none

99. The staright line flow at ________ flow rates is called transistional flow
(a) low (b) intermediate
(c) high (d) none

100. The staright line flow at ________ flow rates is called turbulent flow
(a) low (b) intermediate
(c) high (d) none

101. Ratio of inertial forces to viscous forces is known as
(a) viscosity
(b) Reynolds number
(c) shear stress
(d) none

102. Reynolds number 2100 or less, signifies__________ flow
(a) laminar (b) turbulent
(c) transistional (d) all

103. Reynolds number between 2100-4000, signifies __________ flow
(a) laminar (b) turbulent
(c) transistional (d) all

104. Reynolds number greater than 4000 indicates__________ flow
(a) laminar (b) turbulent
(c) transistional (d) all

105. Energy required to overcome change in elevation during liquid transport is ______energy
(a) pressure (b) kinetic
(c) potential (d) none

106. Which of the following is part of centrifugal pump
(a) impellar (b) eye
(c) vanes (d) all

107. Affinity law is associated with
(a) fluid flow
(b) energy balance
(c) mass balance
(d) none

108. Which of the following is a pressure measuring device
(a) manometer (b) piezometer
(c) both (d) none

109. Pitot tube is used to measure __________ of fluid
(a) pressure (b) velocity
(c) temperature (d) none

110. Canon-Fenske viscometer is __________ type viscsometer
(a) rotational (b) capillary
(c) falling ball (d) none

111. Ostwald viscometer is __________ type viscosmeter
(a) rotational (b) capillary
(c) falling ball (d) none

112. Brookfield viscometer is __________ type viscosmeter
(a) rotational (b) capillary
(c) falling ball (d) none

113. Rotational viscometer measures
(a) force (b) torque
(c) pressure (d) none

114. Which of the following is a non Newtonian fluid
(a) tomato puree (b) soup
(c) salad dressing (d) all

115. Which of the following is a non Newtonian fluid
(a) shear-thinning
(b) shear thickening
(c) both (d) none

116. Apparent viscosity should always be expressed with
(a) shear stress (b) shear rate
(c) both (d) none

117. For shear thinning fliud, with increase in shear rate, apparent viscosity __________
(a) decreases (b) increases
(c) do not change (d) none

118. Shear thinning liquids are also called
(a) elastic (b) plastic
(c) pseudoplastic (d) all

119. Which of the following is a shear thinning liquid
(a) condensed milk (b) fruit puree
(c) soup (d) all

120. For shear thickening fluids, with increase in shear rate, apparent viscosity __________
(a) decreases (b) increases
(c) do not change (d) none

121. Which of the following is a shear thickening fluid
(a) 60 % cornstarch suspension
(b) 60 % wheatstarch suspension
(c) both (d) none

122. Shear thickening fluids are also known as
(a) elastic (b) plastic
(c) dilatent (d) none

123. Tomato ketchup is a type of ______ fluid
(a) dilatent (b) plastic
(c) Bingham plastic (d) none

124. Starch paste is a type of ______ material
(a) Bingham plastic (b) thixotropic
(c) both (d) none

125. Property of granular food directly related to flow is
(a) surface area
(b) angle of repose
(c) surface tension
(d) none

126. Flow of electron is
(a) resistance (b) current
(c) watt (d) none

127. Force causing current flow in an electrical circuit is
(a) current (b) voltage
(c) resistance (d) none

128. Which of the following is a mode of heat transfer
(a) conduction (b) convection
(c) radiation (d) all

129. Mode of heat transfer are of ______ type
(a) two (b) three
(c) four (d) five

130. Which of the following unit operation involves transfer of heat
(a) freezing (b) evaporation
(c) drying (d) all

131. Generally, heat exchangers are classified into ______ group
(a) two (b) three
(c) four (d) single

132. Which of the following is a contact type heat exchanger
(a) steam infusion
(b) steam injection
(c) both (d) none

133. Which of the following is a non-contact type heat exchanger
(a) scraped surface (b) shell & tube
(c) tubular (d) all

134. Plate type heat exchanger comes under ______ type
(a) contact type
(b) non-contact type
(c) both (d) none

135. Plate heat exchangers are suitable for ______ viscosity liquid food
(a) low (b) high
(c) both (d) none

136. Thermal conductivity (W/m°C) of water at 20°C is about
(a) 0.0251 (b) 0.597
(c) 0.173 (d) none

137. Thermal conductivity (W/m°C) of air at 20°C is about
(a) 0.0251 (b) 0.597
(c) 0.173 (d) none

138. Thermal conductivity (W/m°C) of metals at 20°C is about
(a) 0.958 (b) 9.587
(c) 10-50 (d) 50-400

139. Thermal conductivity (W/m°C) of insulating material at 20°C is about
(a) 0.035-0.173 (b) 0.298-0.313
(c) 0.429-0.483 (d) none

140. Approximate convective heat transfer coefficient (W/m^{2}K) of air in free convection is
(a) 5-25 (b) 10-200
(c) 20-100 (d) 50-10000

141. Approximate convective heat transfer coefficient (W/m^{2}K) of air in forced convection is

(a) 5-25 (b) 10-200
(c) 20-100 (d) 50-10000

142. Approximate convective heat transfer coefficient (W/m^{2}K) of water in free convection is
(a) 5-25 (b) 10-200
(c) 20-100 (d) 50-10000

143. Approximate convective heat transfer coefficient (W/m^{2}K) of water in forced convection
(a) 5-25 (b) 10-200
(c) 20-100 (d) 50-10000

144. Approximate convective heat transfer coefficient (W/m^{2}K) of boiling water is
(a) 3000-100000 (b) 10-200
(c) 20-100 (d) 5-25

145. Approximate convective heat transfer coefficient (W/m^{2}K) of condensing water is
(a) 10-200 (b) 20-100
(c) 5-10 (d) 5000-100000

146. Heat transfer coefficient depends on ______ of fluids
(a) density (b) specific heat
(c) viscosity (d) all

147. Heat transfer coefficient depends on
(a) specific heat of fluids
(b) velocity of fluids
(c) geometry of system
(d) all

148. Free convection occurs due to ______ difference in fliud
(a) water activity (b) density
(c) viscosity (d) all

149. Which of the following is a type of fouling
(a) precipitation (b) particulate
(c) corrosion (d) all

150. Microwave frequency generally used in food industry
(a) 915 (b) 2450
(c) both (d) none

151. Mechanism of microwave heating is based on
(a) ionic polarization
(b) dipole rotation
(c) both
(d) none

152. Relative dielectric constant of ice at 2450 (MHz)
(a) 3.2 (b) 0.0029
(c) 12.48 (d) 78

153. Relative dielectric constant of water at 2450 (MHz) at 25°C
(a) 3.2 (b) 0.0029
(c) 12.48 (d) 78

154. Relative dielectric loss constant of ice is
(a) 3.2 (b) 0.0029
(c) 12.48 (d) 78

155. Relative dielectric constant of water at 2450 (MHz) at 25°C
(a) 3.2 (b) 0.0029
(c) 12.48 (d) 78

156. Loss tangent of ice at 2450 MHz is
(a) 3.2 (b) 12.48
(c) 0.16 (d) 0.0009

157. Loss tangent of water at 2450 MHz is at 25°C
(a) 3.2 (b) 12.48
(c) 0.16 (d) 0.0009

158. Which of the following is dimensionless
(a) clean factorc
(b) heat capacity rate ratio
(c) emissivity
(d) all

159. Which of the following is dimensionless
(a) shape factor
(b) friction factor
(c) relative dielectric constant
(d) all

160. Which of the following is dimensionless
(a) Biot number
(b) Grashoff number
(c) Nusselt number
(d) all

161. Which of the following is dimensionless
(a) Prandtl number
(b) Reynolds number
(c) Raleigh number
(d) all

162. Which of the following is dimensionless
(a) absorptivity
(b) transmissivity
(c) mass fraction
(d) all

163. Which of the following is not dimensionless
(a) volume fraction
(b) reflectivity
(c) attenuation factor
(d) none

164. Boiling point of ammonia is
(a) 33.3°C (b) -33.3°C
(c) -43.3°C (d) 43.3°C

165. Boiling point of ammonia at 430.43 kPa pressure
(a) 33.3°C (b) -4.3°C
(c) 0°C (d) 100°C

166. Latent heat of vaporization of a refrigerant should be
(a) high (b) low
(c) both (d) none

167. Freezing temperature of refrigerant should be ________ than evaporator temperature
(a) more (b) less
(c) equal (d) none

168. Critical temperature of refrigerant should be
(a) high (b) low
(c) both (d) none

169. A good refrigerant should be
(a) nontoxic
(b) nonflammable
(c) both
(d) none

170. Molecular weight of dichlorodifluoromethane (refrigerant)
(a) 120.9 (b) 86.5
(c) 102.3 (d) 17.0

171. Boiling point (°C) of dichlorodifluoromethane at 101.3 kPa
(a) -33.3 (b) -26.16
(c) -40.8 (d) -29.8

172. Evaporator pressure (kPa) of dichlorodifluoromethane at -15°C is about
(a) 236.5 (b) 164.0
(c) 296.4 (d) 182.7

173. Condensing pressure (kPa) of dichlorodifluoromethane at 30°C is about
(a) 1203.0 (b) 744.6
(c) 110.1 (d) 1166.5

174. Freezing point (°C) of dichlorodifluoromethane at 101.3 kPa
(a) -77.8 (b) -96.6
(c) -157.8 (d) -160.0

175. Critical temperature (°C) of dichlorodifluoromethane is about
(a) 112.2 (b) 96.1
(c) 101.1 (d) 132.8

176. Critical pressure (kPa) of dichlorodifluoromethane is about
(a) 4060.4 (b) 4115.7
(c) 4936.1 (d) 11423.4

177. Compressure discharge temperature (°C) of dichlorodifluoromethane is about
(a) 43 (b) 55
(c) 98.9 (d) 37.8

178. Compression ratio (30°C/-15°C) of dichlorodifluoromethane is about

(a) 4.07 (b) 4.81
(c) 4.94 (d) 5.06

179. Latent heat of vaporization (kJ/kg) of dichlorodifluoromethane at -15°C

(a) 161.7 (b) 217.7
(c) 209.5 (d) 1314.2

180. Molecular weight of monochlorodifluoromethane (refrigerant) is

(a) 120.9 (b) 86.5
(c) 102.3 (d) 17.0

181. Boiling point (°C) of monochlorodifluoromethane at 101.3 kPa is

(a) -29.8 (b) -26.16
(c) -33.3 (d) -40.8

182. Evaporator pressure (kPa) of monochlorodifluoromethane at-15°C

(a) 296.4 (b) 164.0
(c) 182.7 (d) 236.5

183. Condensing pressure (kPa) of monochlorodifluoromethane at 30°C is about

(a) 744.6 (b) 770.1
(c) 1203.0 (d) 1166.5

184. Freezing point (°C) of monochlorodifluoromethane at 101.3 kPa

(a) -77.8 (b) -96.6
(c) -157.8 (d) -160.0

185. Which of the following is a refrigerant

(a) ammonia
(b) dichlorodifluoromethane
(c) both (d) none

186. Compressure discharge temperature (°C) of monochlorodifluoromethane is about

(a) 43 (b) 55
(c) 98.9 (d) 37.8

187. Compression ratio (30°C/-15°C) of monochlorodifluoromethane is about

(a) 4.07 (b) 4.81
(c) 4.94 (d) 5.06

188. Latent heat of vaporization (kJ/kg) of monochlorodifluoromethane at -15°C

(a) 161.7 (b) 217.7
(c) 209.5 (d) 1314.2

189. Molecular weight of CH_2FCF_3 is

(a) 102.9 (b) 86.5
(c) 102.3 (d) 17.0

190. Boiling point (°C) of CH_2FCF_3 at 101.3 kPa is

(a) -29.8 (b) -26.16
(c) -33.3 (d) -40.8

191. Condensing pressure (kPa) of CH_2FCF_3 at 30°C is about

(a) 744.6 (b) 770.1
(c) 1203.0 (d) 1166.5

192. Evaporator pressure (kPa) of CH_2FCF_3 at-15°C

(a) 296.4 (b) 164.0
(c) 182.7 (d) 236.5

193. Freezing point (°C) of CH_2FCF_3 at 101.3 kPa

(a) -77.8 (b) -96.6
(c) -157.8 (d) -160.0

194. Critical temperature (°C) of CH_2FCF_3 is

(a) 96.1 (b) 101.1
(c) 112.2 (d) 132.8

195. Critical pressure (kPa) of CH_2FCF_3 is

(a) 4115.7 (b) 4936.1
(c) 4060 (d) 11423.4

196. Compressure discharge temperature (°C) of CH_2FCF_3 is about

(a) 43 (b) 55
(c) 98.9 (d) 37.8

197. Compression ratio (30°C/-15°C) of CH_2FCF_3 is about

(a) 4.07 (b) 4.81
(c) 4.94 (d) 5.06

198. Latent heat of vaporization (kJ/kg) of CH_2FCF_3 at -15°C

(a) 161.7 (b) 217.7
(c) 209.5 (d) 1314.2

199. Molecular weight of ammonia is
(a) 120.9 (b) 86.5
(c) 102.3 (d) 17.0

200. Boiling point (°C) of ammonia at 101.3 kPa is
(a) -29.8 (b) -26.16
(c) -33.3 (d) -40.8

201. Evaporator pressure (kPa) of ammonia at-15°C
(a) 296.4 (b) 164.0
(c) 182.7 (d) 236.5

202. Condensing pressure (kPa) of ammonia at 30°C is about
(a) 744.6 (b) 770.1
(c) 1203.0 (d) 1166.5

203. Freezing point (°C) of ammonia at 101.3 kPa
(a) -77.8 (b) -96.6
(c) -157.8 (d) -160.0

204. Critical temperature (°C) of ammonia is
(a) 96.1 (b) 101.1
(c) 112.2 (d) 132.8

205. Critical pressure (kPa) of ammonia is
(a) 4115.7 (b) 4936.1
(c) 4060 (d) 11423.4

206. Compressure discharge temperature (°C) of ammonia is about
(a) 43 (b) 55
(c) 98.9 (d) 37.8

207. Compression ratio (30°C/-15°C) of ammonia is about
(a) 4.07 (b) 4.81
(c) 4.94 (d) 5.06

208. Latent heat of vaporization (kJ/kg) of ammonia at -15°C
(a) 161.7 (b) 217.7
(c) 209.5 (d) 1314.2

209. Which of the following is tetrafluoroethane
(a) $CHClF_2$ (b) CCl_2F_2
(c) CH_2FCF_3 (d) none

210. Chemically ammonia is
(a) halocarbon (b) azeotrope
(c) inorganic compound
(d) none

211. Dichlorodifluoromethane is also known as
(a) frecon (b) freon
(c) both (d) none

212. Which of the following is a component of refrigeration system
(a) condenser (b) compressor
(c) evaporator (d) all

213. Which of the following is a type of condenser
(a) water-cooled (b) air-cooled
(c) evaporative (d) all

214. Unit of cooling load generally used is
(a) ice unit (b) cooling unit
(c) ton of refrigeration (d) none

215. Density of product generally ________ on freezing
(a) increases (b) decreases
(c) remain same (d) none

216. Thermal conductivity of product generally ________ on freezing
(a) increases (b) decreases
(c) remain same (d) none

217. Enthalpy of product generally ________ on freezing
(a) increases (b) decreases
(c) remain same (d) none

218. Plank's equation is used for predicting ________ time
(a) sterilization (b) heating
(c) freezing (d) reaction

219. Which of the following affects freezing time
(a) medium temperature
(b) product size
(c) convective heat transfer coefficient
(d) all

220. Apparent molecular weight of standard dry air is about
(a) 8.314 (b) 8314.41
(c) 28.9645 (d) none

221. Nitrogen (% by volume) is present in standard air
(a) 78.0840 (b) 20.9476
(c) 0.9340 (d) 0.0314

222. Oxygen (% by volume) is present in standard air
(a) 78.0840 (b) 20.9476
(c) 0.9340 (d) 0.0314

223. Argon (% by volume) is present in standard air
(a) 78.0840 (b) 20.9476
(c) 0.9340 (d) 0.0314

224. Carbondioxide (% by volume) is present in standard air
(a) 78.0840 (b) 20.9476
(c) 0.9340 (d) 0.0314

225. Neon (% by volume) is present in standard air
(a) 0.01314 (b) 0.001818
(c) 0.000524 (d) 0.0314

226. Helium (% by volume) is present in standard air
(a) 0.01314 (b) 0.001818
(c) 0.000524 (d) 0.0314

227. Average value of specific heat (kJ/[kg/k]) of dry air is about
(a) 1.005 (b) 2.005
(c) 1.505 (d) 1.795

228. Gas constant ($m^3Pa/kg.mol.K$) of water vapor is about
(a) 261.52 (b) 361.52
(c) 461.52 (d) 561.52

229. Specific heat (kJ/kg.K) of water vapor is about
(a) 0.88 (b) 1.88
(c) 2.88 (d) 2.58

230. Which of the following is associated with psychrometrics
(a) dry bulb temperature
(b) wet bulb temperature
(c) both (d) none

231. Diffusion coefficient ($\times 10^{-9}$ m/s^2) of ammonia gas in water at 20°C
(a) 1.8 (b) 1.6
(c) 1.9 (d) 5.3

232. Diffusion coefficient ($\times 10^{-9}$ m/s^2) of carbondioxide gas in water at 20°C
(a) 1.8 (b) 1.6
(c) 1.9 (d) 5.3

233. Diffusion coefficient ($\times 10^{-9}$ m/s^2) of chlorine gas in water at 20°C
(a) 1.8 (b) 1.6
(c) 1.9 (d) 5.3

234. Diffusion coefficient ($\times 10^{-9}$ m/s^2) of hydrogen gas in water at 20°C
(a) 1.8 (b) 1.6
(c) 1.9 (d) 5.3

235. Diffusion coefficient ($\times 10^{-9}$ m/s^2) of nitrogen gas in water at 20°C
(a) 1.8 (b) 1.6
(c) 1.9 (d) 5.3

236. Diffusion coefficient ($\times 10^{-9}$ m/s^2) of oxygen gas in water at 20°C
(a) 2.1 (b) 1.6
(c) 1.9 (d) 5.3

237. Diffusion coefficient ($\times 10^{-6}$ m/s^2) of ammonia gas in air under standard conditions
(a) 7.7 (b) 10.2
(c) 13.8 (d) 17.0

238. Diffusion coefficient ($\times 10^{-6}$ m/s^2) of benzene in air under standard conditions
(a) 7.7 (b) 10.2
(c) 13.8 (d) 17.0

239. Diffusion coefficient ($\times 10^{-6}$ m/s^2) of CO_2 gas in air under standard conditions
(a) 7.7 (b) 10.2
(c) 13.8 (d) 17.0

240. Diffusion coefficient ($\times 10^{-6}$ m/s^2) of ethyl alcohol in air under standard conditions

(a) 7.7 (b) 10.2
(c) 13.8 (d) 17.0

241. Diffusion coefficient ($\times 10^{-6}$ m/s^2) of hydrogen in air under standard conditions

(a) 61.1 (b) 13.3
(c) 17.8 (d) 10.3

242. Diffusion coefficient ($\times 10^{-6}$ m/s^2) of methyl alcohol in air under standard conditions

(a) 61.1 (b) 13.3
(c) 17.8 (d) 10.3

243. Diffusion coefficient ($\times 10^{-6}$ m/s^2) of nitrogen in air under standard conditions

(a) 61.1 (b) 13.2
(c) 17.8 (d) 10.3

244. Diffusion coefficient ($\times 10^{-6}$ m/s^2) of oxygen in air under standard conditions

(a) 61.1 (b) 13.3
(c) 17.8 (d) 10.3

245. Diffusion coefficient ($\times 10^{-6}$ m/s^2) of sulfurdioxide in air under standard conditions

(a) 61.1 (b) 13.3
(c) 17.8 (d) 10.3

246. Diffusion coefficient ($\times 10^{-6}$ m/s^2) of sulfurtrioxide in air under standard conditions

(a) 9.4 (b) 21.9
(c) 17.8 (d) 10.3

247. Diffusion coefficient ($\times 10^{-6}$ m/s^2) of water vapor in air under standard conditions

(a) 9.4 (b) 21.9
(c) 17.8 (d) 10.3

248. Which of the following is dimensionless

(a) Lewis number
(b) Schmidt number
(c) Sherwood number
(d) all

249. Which of the following has m^2/s unit

(a) mass diffusivity
(b) kinematic viscosity
(c) thermal diffusivity
(d) all

250. Osmotic pressure (kPa) of milk (9% SNF) is

(a) 690 (b) 1587
(c) 2070 (d) 3450

251. Osmotic pressure (kPa) of whey (6% TS) is

(a) 690 (b) 1587
(c) 2070 (d) 3450

252. Osmotic pressure (kPa) of orange juice (11% TS) is

(a) 690 (b) 1587
(c) 2070 (d) 3450

253. Osmotic pressure (kPa) of apple juice (15% TS) is

(a) 690 (b) 1587
(c) 2070 (d) 3450

254. Osmotic pressure (kPa) of grape juice (16% TS) is

(a) 690 (b) 1587
(c) 2070 (d) 3450

255. Osmotic pressure (kPa) of coffee extract (28% TS) is

(a) 690 (b) 1587
(c) 2070 (d) 3450

256. Osmotic pressure (kPa) of lactose (5% w/v) is

(a) 380 (b) 552
(c) 862 (d) 970

257. Osmotic pressure (kPa) of sodium chloride (1% w/v) is

(a) 380 (b) 552
(c) 862 (d) 970

258. Osmotic pressure (kPa) of lactic acid (1% w/v) is
(a) 380 (b) 552
(c) 862 (d) 970

259. Packing density (m^2/m^3) of plate and frame membrane module
(a) 200-400 (b) 300-900
(c) 150-300 (d) 9000-30000

260. Packing density (m^2/m^3) of spiral-wound membrane module
(a) 200-400 (b) 300-900
(c) 150-300 (d) 9000-30000

261. Packing density (m^2/m^3) of tube-in-shell membrane module
(a) 200-400 (b) 300-900
(c) 150-300 (d) 9000-30000

262. Packing density (m^2/m^3) of hollow-fiber membrane module
(a) 200-400 (b) 300-900
(c) 150-300 (d) 9000-30000

263. Permeate flux ($m^3/[m^2day]$) of plate and frame membrane module is
(a) 0.3-1.0 (b) 10.3-11.3
(c) 0.03-0.01 (d) 0.004-0.08

264. Permeate flux ($m^3/[m^2day]$) of spiral-wound membrane module is
(a) 0.3-1.0 (b) 10.3-11.3
(c) 0.03-0.01 (d) 0.004-0.08

265. Permeate flux ($m^3/[m^2day]$) of tube-in-shell membrane module is
(a) 0.3-1.0 (b) 10.3-11.3
(c) 0.03-0.01 (d) 0.004-0.08

266. Permeate flux ($m^3/[m^2day]$) of hollow-fiber membrane module is
(a) 0.3-1.0 (b) 10.3-11.3
(c) 0.03-0.01 (d) 0.004-0.08

267. Flux density ($m^3/[m^3day]$) of plate and frame membrane module is
(a) 60-400 (b) 45-300
(c) 90-900 (d) 36-2400

268. Flux density ($m^3/[m^3day]$) of spiral-wound membrane module is
(a) 60-400 (b) 45-300
(c) 90-900 (d) 36-2400

269. Flux density ($m^3/[m^3day]$) of tube-in-shell membrane module is
(a) 60-400 (b) 45-300
(c) 90-900 (d) 36-2400

270. Flux density ($m^3/[m^3day]$) of hollow-fiber membrane module is
(a) 60-400 (b) 45-300
(c) 90-900 (d) 36-2400

271. Feed channel diameter (mm) of plate and frame membrane module is
(a) 0.1 (b) 1.3
(c) 5 (d) 13

272. Feed channel diameter (mm) of spiral-wound membrane module is
(a) 0.1 (b) 1.3
(c) 5 (d) 13

273. Feed channel diameter (mm) of tube-in-shell membrane module is
(a) 0.1 (b) 1.3
(c) 5 (d) 13

274. Feed channel diameter (mm) of hollow-fiber membrane module is
(a) 0.1 (b) 1.3
(c) 5 (d) 13

275. Which of the following membrane module have low concentration polarization
(a) plate & frame (b) spiral-wound
(c) tube-in-shell (d) hollow-fiber

276. Which of the following is a type of dryer
(a) cabinet (b) tunnel
(c) spray (d) all

277. Which of the following is a type of dryer
(a) puff (b) fluidized-bed
(c) freeze-dryer (d) all

278. Propeller type agitators are suitable for products with viscosity (Pa.s) range of
(a) 500-1000 (b) <3
(c) <100 (d) 50-500

279. Turbine type agitators are suitable for products with viscosity (Pa.s) range of
(a) 500-1000 (b) <3
(c) <100 (d) 50-500

280. Anchor type agitators are suitable for products with viscosity (Pa.s) range of
(a) 500-1000 (b) <3
(c) <100 (d) 50-500

281. Helical and ribbon type agitators are suitable for products with viscosity (Pa.s) range of
(a) 500-1000 (b) <3
(c) <100 (d) 50-500

282. Which of the following is a type of extruder
(a) single-screw (b) twin-screw
(c) both (d) none

283. Which of the following is a function of food package
(a) protection
(b) communication
(c) convenience
(d) all

284. Deci (d) is equal to
(a) 1/10 (b) 1/100
(c) 1/1000 (d) 1/1000000

285. Centi (c) is equal to
(a) 1/10 (b) 1/100
(c) 1/1000 (d) 1/1000000

286. Mili (d) is equal to
(a) 1/10 (b) 1/100
(c) 1/1000 (d) 1/1000000

287. Micro (μ) is equal to
(a) 1/10 (b) 1/100
(c) 1/1000 (d) 1/1000000

288. Nano (n) is equal to
(a) 10^{-3} (b) 10^{-6}
(c) 10^{-9} (d) 10^{-12}

289. Tera (T) is equal to
(a) 10^{12} (b) 10^{9}
(c) 10^{6} (d) 10^{2}

290. Hecto (h) is equal to
(a) 10^{12} (b) 10^{9}
(c) 10^{6} (d) 10^{2}

291. Mega (M) is equal to
(a) 10^{12} (b) 10^{9}
(c) 10^{6} (d) 10^{2}

292. Giga (G) is equal to
(a) 10^{12} (b) 10^{9}
(c) 10^{6} (d) 10^{2}

293. 1 BTU = ________ Joule
(a) 55 (b) 155
(c) 1055 (d) 1155

294. 1 BTU = ________ kJ
(a) 5.5 (b) 1.55
(c) 1.055 (d) 1.155

295. 1 BTU = ________ cal
(a) 4.184 (b) 252.16
(c) 152.16 (d) 15.184

296. 1 kcal = ________ kJ
(a) 4.184 (b) 252.16
(c) 152.16 (d) 15.184

297. 1 Newton = ________ kg m/s²
(a) 1.0 (b) 10
(c) 5 (d) 100

298. 1 Dyne = ____________ kg m/s²
(a) 10^{-2} (b) 10^{-3}
(c) 10^{-5} (d) 10^{-4}

299. 1 Pascal = __________ N/m²
(a) 1.0 (b) 10
(c) 100 (d) 1000

300. 1 Watt = __________ cal/min
(a) 4.34 (b) 14.34
(c) 6.34 (d) 16.34

301. $T_{°C} = (T_{°F} - 32)/?$
(a) 1.8 (b) 5
(c) 18 (d) 8.5

302. Specific heat (kJ/kg K) of canned fish is about
(a) 3.882 (b) 3.35
(c) 1.754 (d) 1.520

303. Specific heat (kJ/kg K) of starch is about
(a) 3.882 (b) 3.35
(c) 1.754 (d) 1.520

304. Specific heat (kJ/kg K) of orange juice is about
(a) 3.882 (b) 3.35
(c) 1.754 (d) 1.520

305. Specific heat (kJ/kg K) of dry milk (non fat) is about
(a) 3.882 (b) 3.35
(c) 1.754 (d) 1.520

306. Specific heat (kJ/kg K) of butter is about
(a) 2.043 (b) 3.831
(c) 3.935 (d) 3.676

307. Specific heat (kJ/kg K) of pasteurized whole milk is about
(a) 2.043 (b) 3.831
(c) 3.935 (d) 3.676

308. Specific heat (kJ/kg K) of skim milk is about
(a) 2.043 (b) 3.831
(c) 3.935 (d) 3.676

309. Specific heat (kJ/kg K) of tomato soup concentrate is about
(a) 2.043 (b) 3.831
(c) 3.935 (d) 3.676

310. Specific heat (kJ/kg K) of egg yolk is about
(a) 2.449 (b) 3.579
(c) 3.634 (d) 3.759

311. Specific heat (kJ/kg K) of apple is about
(a) 2.449 (b) 3.579
(c) 3.634 (d) 3.759

312. Thermal conductivity (W/[m°C]) of apple is about
(a) 0.197 (b) 0.338
(c) 0.577 (d) 0.393

313. Thermal conductivity (W/[m°C]) of butter is about
(a) 0.197 (b) 0.338
(c) 0.577 (d) 0.393

314. Thermal conductivity (W/[m°C]) of egg white is about
(a) 0.197 (b) 0.338
(c) 0.577 (d) 0.393

315. Thermal conductivity (W/[m°C]) of egg yolk is about
(a) 0.197 (b) 0.338
(c) 0.577 (d) 0.393

316. Thermal conductivity (W/[m°C]) of milk is about
(a) 0.220 (b) 0.330
(c) 0.430 (d) 0.530

317. Thermal conductivity (W/[m°C]) of salt is about
(a) 0.047 (b) 0.147
(c) 0.247 (d) 0.347

318. Thermal conductivity (W/[m°C]) of whey is about
(a) 0.241 (b) 0.441
(c) 0.641 (d) 0.841

319. Visosity (Pa.s) of cream (20% fat) at 60°C is
(a) 0.00510 (b) 0.00289
(c) 0.00171 (d) none

320. Visosity (Pa.s) of cream (30% fat) at 60°C is
(a) 0.00510 (b) 0.00289
(c) 0.00171 (d) none

321. Visosity (Pa.s) of cream (40% fat) at 60°C is
(a) 0.00510 (b) 0.00289
(c) 0.00171 (d) none

322. Viscosity (Pa.s) of cornoil at 25°C is
(a) 0.0565 (b) 0.0317
(c) 0.0165 (d) none

323. Viscosity (Pa.s) of cornoil at 38°C is
(a) 0.0565 (b) 0.0317
(c) 0.0165 (d) none

324. Viscosity (Pa.s) of cottonseed oil at 20°C is
(a) 0.0306 (b) 0.0102
(c) 0.0704 (d) none

325. Viscosity (Pa.s) of cottonseed oil at 38°C is
(a) 0.0306 (b) 0.0102
(c) 0.0704 (d) none

326. Viscosity (Pa.s) of peanut oil at 25°C is
(a) 0.0151 (b) 0.0251
(c) 0.0656 (d) none

327. Viscosity (Pa.s) of soybean oil at 30°C is
(a) 0.01 (b) 0.02
(c) 0.03 (d) 0.04

328. Thermal conductivity (W/m°C) of ice at 0°C is about
(a) 1.22 (b) 2.22
(c) 3.22 (d) 4.22

329. Specific heat (kJ/kg°C) of ice at 0°C is about
(a) 1.050 (b) 2.050
(c) 3.050 (d) 4.050

330. Density (kg/m^3) of ice at 0°C is about
(a) 716.2 (b) 816.2
(c) 916.2 (d) 1016.2

331. Highest moisture content is present in
(a) milk (b) apple
(c) egg (d) lettuce

332. Lowest moisture content is present in
(a) wheat (b) rice
(c) peanuts (d) butter

333. Highest protein is present in
(a) egg (b) peanut
(c) lima beans (d) garlic

334. Lowest protein is present in
(a) apple (b) turnip
(c) spinach (d) butter

335. Highest fat content is present in
(a) egg (b) peanut
(c) beef (d) milk

336. Lowest fat content is present in
(a) potatoes (b) peaches
(c) both (d) none

337. Prandtl number of water at saturation pressure at 0°C is
(a) 6.1 (b) 7.0
(c) 5.4 (d) 13.7

338. Prandtl number of water at saturation pressure at 20°C is
(a) 6.1 (b) 7.0
(c) 5.4 (d) 13.7

339. Prandtl number of water at saturation pressure at 25°C is
(a) 6.1 (b) 7.0
(c) 5.4 (d) 13.7

340. Prandtl number of water at saturation pressure at 30°C is
(a) 6.1 (b) 7.0
(c) 5.4 (d) 13.7

341. Prandtl number of water at saturation pressure at 100°C is
(a) 3.55 (b) 2.55
(c) 2.04 (d) 1.75

342. Density (kg/m^3) of water at 0°C is
(a) 997.1 (b) 999.9
(c) 1000 (d) 995.7

343. Density (kg/m^3) of water at 5°C is
(a) 997.1 (b) 999.9
(c) 1000 (d) 995.7

344. Density (kg/m^3) of water at 25°C is
(a) 997.1 (b) 999.9
(c) 1000 (d) 995.7

345. Density (kg/m^3) of water at 30°C is
(a) 997.1 (b) 999.9
(c) 1000 (d) 995.7

346. Density (kg/m^3) of water at 50°C is
(a) 998.1 (b) 999.5
(c) 958.4 (d) 995.7

347. Density (kg/m^3) of water at 100°C is
(a) 998.1 (b) 999.5
(c) 958.4 (d) 995.7

348. Specific heat (kJ/kg°C) of water at 0°C
(a) 4.211 (b) 4.178
(c) 4.226 (d) 4.986

349. Specific heat (kJ/kg°C) of water at 25°C
(a) 4.211 (b) 4.178
(c) 4.226 (d) 4.986

350. Specific heat (kJ/kg°C) of water at 100°C
(a) 4.211 (b) 4.178
(c) 4.226 (d) 4.986

351. Prandtl number of dry air
(a) 1.71 (b) 2.71
(c) 3.71 (d) 0.71

352. Density (kg/m^3) of dry air at 0°C is
(a) 1.252 (b) 1.127
(c) 1.057 (d) 0.916

353. Density (kg/m^3) of dry air at 30°C is
(a) 1.252 (b) 1.127
(c) 1.057 (d) 0.916

354. Density (kg/m^3) of dry air at 100°C is
(a) 1.252 (b) 1.127
(c) 1.057 (d) 0.916

355. Which of the following method is used for measurement of porosity
(a) direct method
(b) optical microscopic method
(c) density method
(d) all

356. Which of the following is a Newtonian fluid
(a) milk (b) honey
(c) corn syrup (d) all

357. Which of the following is non-Newtonian fliud
(a) melted chocolate
(b) dairy cream
(c) fruit puree
(d) all

358. Rheopectic behavior is also reffered to as ________ behavior
(a) plastic
(b) antithixotropic
(c) thixotropic
(d) none

359. Rheology is study of ________ and deformation
(a) texture (b) consistency
(c) flow (d) none

360. Which of the following is associated with rheology
(a) stress (b) strain
(c) shear rate (d) all

361. Which of the following is associated with viscometer
(a) Brookfield
(b) Haake
(c) Epprecht Rheomat
(d) all

362. Which of the following is particularly associated with starch analysis
(a) viscometer (b) rheometer
(c) RVA (d) none

363. G' is also known as
(a) bulk modulus
(b) elastic modulus
(c) both
(d) none

364. G" is also known as
(a) bulk modulus
(b) elastic modulus
(c) both
(d) none

365. Elastic modulus is also known as
(a) storage modulus
(b) loss modulus
(c) both (d) none

366. Bulk modulus is also known as
(a) storage modulus
(b) loss modulus
(c) both
(d) none

367. Storage modulus in rheogram represents _________ behavior
(a) elastic (b) viscous
(c) plastic (d) none

368. Loss modulus in rheogram represents _________ behavior
(a) elastic (b) viscous
(c) plastic (d) none

369. Wheat dough is __________ in nature
(a) elastic (b) viscous
(c) plastic (d) viscoelastic

370. SI unit of magnetic induction is
(a) tesla (b) siemen
(c) farad (d) coulomb

371. SI unit of magnetic potential is
(a) hertz (b) ohm
(c) coulomb (d) volt

372. SI unit of angular velocity is
(a) m/s (b) rad/s
(c) m/s^2 (d) rad/s^2

373. SI unit of energy density
(a) J/m (b) J/m^2
(c) J/m^3 (d) none

374. SI unit of heat flux density
(a) W/m (b) W/m^2
(c) W/m^3 (d) none

375. SI unit of molar energy
(a) J/mol (b) W/mol
(c) m/mol (d) none

376. SI unit of molar heat capacity
(a) J/mol (b) J/K
(c) J/(mol.K) (d) none

377. SI unit of angular acceleration
(a) m/s^2 (b) m/s
(c) rad/s (d) rad/s^2

378. SI unit of current density
(a) A/m (b) A/m^2
(c) A/m^3 (d) all

379. SI unit of electric charge density
(a) C/m (b) C/m^2
(c) C/m^3 (d) all

380. SI unit of entropy is
(a) J/K (b) J/m
(c) J/s (d) none

381. SI unit of heat capacity
(a) J/K (b) J/m
(c) J/s (d) none

382. SI unit of molar entropy
(a) J/mol (b) J/mol.K
(c) both (d) none

383. SI unit of power density
(a) W/m (b) W/m^2
(c) W/m^3 (d) none

384. SI unit of specific energy
(a) J/m (b) J/kg
(c) J/s (d) none

385. SI unit of specific volume
(a) kg/m^3 (b) m^2/kg
(c) m^3/kg (d) kg/m^2

386. SI unit of thermal conductivity
(a) W/m (b) W/m^2
(c) W/m^3 (d) none

387. SI unit of surface tension
(a) N/m (b) N/m^2
(c) N/m^3 (d) none

388. SI unit of specific entropy
(a) J (b) K/J
(c) kg/J (d) none

389. R-744 (refrigerant) is
(a) CO_2 (b) Freon-12
(c) dichloromethane (d) Freon-21

390. R-12 (refrigerant) is
(a) CO_2
(b) dichlorodifluoromethane
(c) dichloromethane
(d) Freon-21

391. R-30 (refrigerant) is
(a) CO_2
(b) dichlorodifluoromethane
(c) dichloromethane
(d) Freon-21

392. R-21 (refrigerant) is
(a) CO_2 (b) Freon-21
(c) Freon-22 (d) Freon-22

393. R-114 (refrigerant) is
(a) dichlorotetrafluoroethane
(b) monochlorodifluoromethane
(c) trichloromonofluoromethane
(d) trichlorotrifluoroethane

394. R-22 (refrigerant) is
(a) dichlorotetrafluoroethane
(b) monochlorodifluoromethane
(c) trichloromonofluoromethane
(d) trichlorotrifluoroethane

395. R-11 (refrigerant) is
(a) dichlorotetrafluoroethane
(b) monochlorodifluoromethane
(c) trichloromonofluoromethane
(d) trichlorotrifluoroethane

396. R-113 (refrigerant) is
(a) dichlorotetrafluoroethane
(b) monochlorodifluoromethane
(c) trichloromonofluoromethane
(d) trichlorotrifluoroethane

397. R-600 (refrigerant) is
(a) ethylene (b) ethane
(c) butane (d) ethylchloride

398. R-170 (refrigerant) is
(a) ethylene (b) ethane
(c) butane (d) ethylchloride

399. R-160 (refrigerant) is
(a) ethylene (b) ethane
(c) butane (d) ethylchloride

400. R-1150 (refrigerant) is
(a) ethylene (b) ethane
(c) butane (d) ethylchloride

401. R-601 (refrigerant) is
(a) propane
(b) methyl formate
(c) isobutane
(d) methyl chloride

402. R-40 (refrigerant) is
(a) propane
(b) methyl formate
(c) isobutane
(d) methyl chloride

403. R-611 (refrigerant) is
(a) propane
(b) methyl formate
(c) isobutane
(d) methyl chloride

404. R-290 (refrigerant) is
(a) propane
(b) methyl formate
(c) isobutane
(d) methyl chloride

405. R-717 (refrigerant) is
(a) ammonia (b) propane
(c) butane (d) ethane

406. Compressor in refrigerator does not start if
(a) blown fuses of voltage supplier
(b) broken electric connection
(c) starting relay not functioning properly
(d) all

407. Compressor starts with jerk after a feqw trips if
(a) starting relay defective
(b) starting winding too weak
(c) both
(d) none

408. Ice melting in the freezer in pending case if
(a) compressor starts after long time
(b) refrigerant leaked out
(c) both
(d) none

409. No proper cooling in refrigerator if
(a) thermostat set on warmer side
(b) too much ice on evaporator coil
(c) too frequent opening of freezer door
(d) all

410. Absolute pressure is normal force per unit ________
(a) length (b) volume
(c) area (d) none

411. International standard value of acceleration due to gravity is
(a) 9.8 ft/s^2 (b) 980.665 ft/s^2
(c) 32.174 ft/s^2 (d) 174.32 ft/s^2

412. Acoustic deals with the study of
(a) hydraulic pressure
(b) temperature
(c) sound
(d) polymer

413. Activated alumina is
(a) aluminium hydroxide
(b) aluminium oxide
(c) aluminium chloride
(d) none

414. Aerology deals with the study of
(a) telescope (b) space
(c) air (d) none

415. Device which increases the electron flow in circuit
(a) voltmeter (b) amplimeter
(c) amplifier (d) all

416. Anemometer measures
(a) air humidity (b) air velocity
(c) air composition (d) none

417. Which of the following is an antifreeze
(a) ethylene glycol (b) glycol
(c) ethyl ether (d) all

418. Azeotrops are mixture of fixed composion of atleast ________ compounds
(a) two (b) three
(c) four (d) none

419. Bar is the unit of
(a) standard (b) pressure
(c) purity (d) none

420. Barometer is used to measure
(a) temperature (b) pressure
(c) humidity (d) all

421. Bernoulli's theorm is about
(a) energy (b) temperature
(c) both (d) none

422. Breeching is________ in steam boiler
(a) metal (b) steam
(c) space (d) none

423. Broiler is atype of
(a) evaporator (b) boiler
(c) chicken (d) none

424. Calorimeter is used to measure
(a) specific heat
(b) heat of combustion
(c) heat of fusion
(d) all

425. Carcass is
(a) refrigerant
(b) slaughter animal body
(c) coolent
(d) none

426. Carrene is composed of
(a) carbon (b) fluorine
(c) chlorine (d) all

427. Cahrle's law shows volume of gas varies linearly with ________
(a) temperature (b) pressure
(c) density (d) all

428. Boyle's law shows relation between volume of gas and ________
(a) temperature (b) pressure
(c) density (d) all

429. Amount of refrigerant in a system is also referred as
(a) density (b) charge
(c) mass (d) none

430. Coefficient of performance = refrigeration effect/?
(a) energy input
(b) energy output
(c) both
(d) none

431. In counter flow, fluid flow in ________ direction
(a) same (b) opposite
(c) both (d) none

432. In co-current flow, fluid flow in ________ direction
(a) same (b) opposite
(c) both (d) none

433. Critical point is the highest ______ at which distinct liquid and gas phases can co-exist
(a) temperature (b) pressure
(c) both (d) none

434. In cross flow, fluid flow in ______ direction
(a) same (b) opposite
(c) perpendicular (d) none

435. Decible is the unit of ______ of sound
(a) pitch (b) load
(c) intensity (d) none

436. Diaphragm is a flexible membrane made of
(a) thin metal (b) rubber
(c) plastic (d) all

437. Dielectric fliud has very high electrical______
(a) resistance (b) conductance
(c) both (d) none

438. Entropy is measure of
(a) randomness (b) disorderness
(c) both (d) none

439. Which of the following is not an enzyme
(a) pepsin (b) amylase
(c) trypsin (d) none

440. Steam point of water in Fahrenheit scale is
(a) 32°F (b) 112°F
(c) 212°F (d) 272°F

441. Ice point of water in Fahrenheit scale is
(a) 32°F (b) 112°F
(c) 212°F (d) 272°F

442. Grain is a unit of
(a) mass (b) volume
(c) density (d) none

443. lb = ________ grains
(a) 3000 (b) 4000
(c) 7000 (d) 11000

444. Halogenated refrigerants contain
(a) fluorine (b) chlorine
(c) bromine (d) all

445. Hydrometer is used to measure
(a) water vapor
(b) water density
(c) vapor pressure
(d) all

446. Lactometer is used to measure ______ ______ of milk
(a) protein (b) moisture
(c) density (d) all

447. Hygrometer is used to measure ______ in air
(a) moisture (b) nitrogen
(c) hydrogen (d) oxygen

448. Glass transition temperature of bread is about
(a) -12°C (b) -42°C
(c) -51°C (d) -34.5°C

449. Glass transition temperature of ice-cream is about
(a) -12°C (b) -42°C
(c) -51°C (d) -34.5°C

450. Texture profile analysis give
(a) cohesiveness (b) springiness
(c) chewiness (d) all

451. Which of the following is measured parameter in texture profile analysis
(a) hardness (b) gumminess
(c) chewiness (d) none

452. Which of the following is calculated parameter in texture profile analysis
(a) cohesiveness (b) adhesiveness
(c) gumminess (d) none

453. Thermal conductivity (W/m.K) of fat and oils at 0°C is about
(a) 0.023 (b) 0.2
(c) 0.45 (d) 2.24

454. Thermal conductivity (W/m.K) of ice at 0°C is about
(a) 0.023 (b) 0.2
(c) 0.45 (d) 2.24

455. Thermal conductivity (W/m.K) of air at 0°C is about
(a) 0.023 (b) 0.2
(c) 0.45 (d) 2.24

456. Thermal conductivity (W/m.K) of carragennan gel (2% aqueous) at 25°C is about
(a) 0.609 (b) 0.164
(c) 0.133 (d) 0.967

457. Thermal conductivity (W/m.K) of olive oil at 25°C is about
(a) 0.609 (b) 0.164
(c) 0.033 (d) 0.967

458. Thermal conductivity (W/m.K) of silicon oil at 25°C is about
(a) 0.609 (b) 0.164
(c) 0.133 (d) 0.967

459. Unit of thermal diffusivity is
(a) m/s (b) m^2/s
(c) m^3/s (d) m/s^2

460. Initial freezing temperature (°C) of apple juice is about
(a) -1.44 (b) -11.33
(c) -1.11 (d) -0.79

461. Initial freezing temperature (°C) of apple juice concentrate is about
(a) -1.44 (b) -11.33
(c) -1.11 (d) -0.79

462. Initial freezing temperature (°C) of carrots is about
(a) -1.44 (b) -11.33
(c) -1.11 (d) -0.79

463. Initial freezing temperature (°C) of chicken is about
(a) -1.44 (b) -11.33
(c) -1.11 (d) -0.79

464. Initial freezing temperature (°C) of grape juice is about
(a) -1.78 (b) -1.44
(c) -1.17 (d) -2.28

465. Initial freezing temperature (°C) of onion is about
(a) -1.78 (b) -1.44
(c) -1.17 (d) -2.28

466. Initial freezing temperature (°C) of orange juice is about
(a) -1.78 (b) -1.44
(c) -1.17 (d) -2.28

467. Initial freezing temperature (°C) of plums is about
(a) -1.78 (b) -1.44
(c) -1.17 (d) -2.28

468. Initial freezing temperature (°C) of tomato pulp is about
(a) -1.22 (b) -1.75
(c) -0.72 (d) -0.56

469. Initial freezing temperature (°C) of spinach is about
(a) -1.22 (b) -1.75
(c) -0.72 (d) -0.56

470. Which of the following is dimensionless
(a) mass fraction
(b) volume fraction
(c) mole fraction
(d) all

471. Which of the following is associated with radiative heat transfer

(a) emissivity (b) reflectivity
(c) absorptivity (d) all

472. Adhesiveness is related to ________ of food

(a) texture (b) color
(c) flavor (d) none

473. Units of Planks constant is

(a) J/s (b) J/m
(c) J.m (d) J.s

474. Units of Boltzmann constant is

(a) J/K (b) J/s
(c) J/m (d) J.m

475. Water activity of baking soda should not exceed

(a) 0.55 (b) 0.45
(c) 0.75 (d) 0.65

476. Water activity of dried egg should not exceed

(a) 0.55 (b) 0.65
(c) 0.75 (d) 0.30

477. Water activity of gelatin should not exceed

(a) 0.55 (b) 0.45
(c) 0.75 (d) 0.65

478. Water activity of hard candies should not exceed

(a) 0.39 (b) 0.30
(c) 0.45 (d) 0.60

479. Water activity of potato flakes should not exceed

(a) 0.11 (b) 0.22
(c) 0.33 (d) 0.44

480. Water activity of flour should not exceed

(a) 0.45 (b) 0.55
(c) 0.65 (d) 0.75

481. Water activity of dried skim milk should not exceed

(a) 0.30 (b) 0.40
(c) 0.50 (d) 0.60

482. Water activity of soluble coffee should not exceed

(a) 0.45 (b) 0.55
(c) 0.65 (d) 0.75

483. Water activity of starch should not exceed

(a) 0.40 (b) 0.50
(c) 0.60 (d) 0.70

484. Water activity of pure fructose should not exceed

(a) 0.63 (b) 0.89
(c) 0.85 (d) 0.92

485. Water activity of pure dextrose should not exceed

(a) 0.63 (b) 0.89
(c) 0.85 (d) 0.92

486. Water activity of pure sucrose should not exceed

(a) 0.63 (b) 0.89
(c) 0.85 (d) 0.92

487. Water activity of pure maltose should not exceed

(a) 0.63 (b) 0.89
(c) 0.85 (d) 0.92

488. Water activity of dehydrated meat should not exceed

(a) 0.52 (b) 0.62
(c) 0.72 (d) 0.82

489. Water activity of dried apples should not exceed

(a) 0.50 (b) 0.60
(c) 0.65 (d) 0.70

490. Water activity of dried apricots should not exceed

(a) 0.50 (b) 0.60
(c) 0.65 (d) 0.70

491. Water activity of dried dates should not exceed

(a) 0.50 (b) 0.60
(c) 0.65 (d) 0.70

492. Water activity of dried peaches should not exceed

(a) 0.50 (b) 0.60
(c) 0.65 (d) 0.70

493. Water activity of dried pears should not exceed

(a) 0.50 (b) 0.60
(c) 0.65 (d) 0.70

494. Water activity of dried plums should not exceed

(a) 0.50 (b) 0.60
(c) 0.65 (d) 0.70

495. Water activity of dried orange powder should not exceed

(a) 0.10 (b) 0.20
(c) 0.30 (d) 0.50

496. Overall heat transfer coefficient for [W/(m^2.K)] for vaccum shelf dryer is about

(a) 5-6 (b) 11-57
(c) 11-85 (d) 28-284

497. Overall heat transfer coefficient for [W/(m^2.K)] for indirect rotary dryer is about

(a) 5-6 (b) 11-57
(c) 11-85 (d) 28-284

498. Overall heat transfer coefficient for [W/(m^2.K)] for jacketed trough dryer is about

(a) 5-6 (b) 11-57
(c) 11-85 (d) 28-284

499. Overall heat transfer coefficient for [W/(m^2.K)] for rotary vaccum dryer is about

(a) 5-6 (b) 11-57
(c) 11-85 (d) 28-284

500. Overall heat transfer coefficient for [W/(m^2.K)] for agitated tray dryer is about

(a) 5-6 (b) 11-57
(c) 28-340 (d) 1135-1700

501. Overall heat transfer coefficient for [W/(m^2.K)] for vaccum dryer is about

(a) 5-6 (b) 11-57
(c) 28-340 (d) 1135-1700

502. Diffusivity ($10^5 \times m^2/s$) of hydrogen gas in air at 0°C at atmospheric pressure is about

(a) 2.60 (b) 1.38
(c) 1.78 (d) 6-11

503. Diffusivity ($10^5 \times m^2/s$) of oxygen gas in air at 0°C at atmospheric pressure is about

(a) 2.60 (b) 1.38
(c) 1.78 (d) 6-11

504. Diffusivity ($10^5 \times m^2/s$) of CO_2 gas in air at 0°C at atmospheric pressure is about

(a) 2.60 (b) 1.38
(c) 1.78 (d) 6-11

505. Diffusivity ($10^5 \times m^2/s$) of water vapors in air at 0°C at atmospheric pressure is about

(a) 2.60 (b) 1.38
(c) 1.78 (d) 6-11

506. Diffusivity ($10^5 \times m^2/s$) of ethanol vapors in air at 0°C at atmospheric pressure is about

(a) 0.71 (b) 0.80
(c) 1.06 (d) 1.33

507. Diffusivity ($10^5 \times m^2/s$) of aceticacid vapors in air at 0°C at atmospheric pressure is

(a) 0.71 (b) 0.80
(c) 1.06 (d) 1.33

508. Diffusivity ($10^5 \times m^2/s$) of ethyl acetate in air at 0°C at atmospheric pressure is about

(a) 0.71 (b) 0.80
(c) 1.06 (d) 1.33

509. Diffusivity ($10^5 \times m^2/s$) of n-Butyl alcohol in air at 0°C at atmospheric pressure is about

(a) 0.70 (b) 0.80
(c) 1.06 (d) 1.33

510. Which of the following has highest diffusivity in dilute aqueous solution at 25°C
(a) CO_2 (b) O_2
(c) acetic acid (d) glucose

511. Which of the following has lowest diffusivity in dilute aqueous solution at 25°C
(a) lactose (b) ethanol
(c) urea (d) catalase

512. Osmotic pressure (kPa) of 0.1 molal NaCl aqueous solution
(a) 248 (b) 253
(c) 259 (d) 462

513. Osmotic pressure (kPa) of 0.1 molal glucose aqueous solution
(a) 248 (b) 253
(c) 259 (d) 462

514. Osmotic pressure (kPa) of 0.1 molal fructose aqueous solution
(a) 248 (b) 253
(c) 259 (d) 462

515. Osmotic pressure (kPa) of 0.1 molal sucrose aqueous solution
(a) 248 (b) 253
(c) 259 (d) 462

516. Osmotic pressure (kPa) of 0.1 molal maltose aqueous solution
(a) 248 (b) 253
(c) 259 (d) 214

517. Osmotic pressure (kPa) of 0.1 molal lactose aqueous solution
(a) 248 (b) 253
(c) 259 (d) 214

518. Osmotic pressure (kPa) of 0.2 molal NaCl aqueous solution
(a) 517 (b) 917
(c) 496 (d) 503

519. Osmotic pressure (kPa) of 0.2 molal glucose aqueous solution
(a) 517 (b) 917
(c) 496 (d) 503

520. Osmotic pressure (kPa) of 0.2 molal fructose aqueous solution
(a) 517 (b) 917
(c) 496 (d) 503

521. Osmotic pressure (kPa) of 0.2 molal sucrose aqueous solution
(a) 517 (b) 917
(c) 496 (d) 503

522. Which of the following is a unit of osmotic pressure
(a) Pa (b) K
(c) T (d) none

523. Osmotic pressure (kPa) of 0.2 molal maltose aqueous solution
(a) 255 (b) 355
(c) 455 (d) 555

524. Osmotic pressure (kPa) of 0.2 molal lactose aqueous solution
(a) 255 (b) 355
(c) 455 (d) 555

525. Osmotic pressure (kPa) of 0.3 molal NaCl aqueous solution
(a) 758 (b) 776
(c) 790 (d) 1372

526. Osmotic pressure (kPa) of 0.3 molal glucose aqueous solution
(a) 758 (b) 776
(c) 790 (d) 1372

527. Osmotic pressure (kPa) of 0.3 molal fructose aqueous solution
(a) 758 (b) 776
(c) 790 (d) 1372

528. Osmotic pressure (kPa) of 0.3 molal sucrose aqueous solution
(a) 758 (b) 776
(c) 790 (d) 1372

529. Osmotic pressure (kPa) of 0.3 molal maltose aqueous solution
(a) 224 (b) 324
(c) 524 (d) 724

530. Osmotic pressure (kPa) of 0.3 molal lactose aqueous solution
(a) 224 (b) 324
(c) 524 (d) 724

531. Osmotic pressure (kPa) of 0.4 molal NaCl aqueous solution
(a) 1013 (b) 1020
(c) 1034 (d) 1820

532. Osmotic pressure (kPa) of 0.4 molal glucose aqueous solution
(a) 1013 (b) 1020
(c) 1034 (d) 1820

533. Osmotic pressure (kPa) of 0.4 molal fructose aqueous solution
(a) 1013 (b) 1020
(c) 1034 (d) 1820

534. Osmotic pressure (kPa) of 0.4 molal sucrose aqueous solution
(a) 1013 (b) 1020
(c) 1034 (d) 1820

535. Osmotic pressure (kPa) of 0.4 molal maltose aqueous solution
(a) 733 (b) 633
(c) 933 (d) 1033

536. Osmotic pressure (kPa) of 0.5 molal NaCl aqueous solution
(a) 1282 (b) 1293
(c) 1307 (d) 2282

537. Osmotic pressure (kPa) of 0.5 molal glucose aqueous solution
(a) 1282 (b) 1293
(c) 1307 (d) 2282

538. Osmotic pressure (kPa) of 0.5 molal fructose aqueous solution
(a) 1282 (b) 1293
(c) 1307 (d) 2282

539. Osmotic pressure (kPa) of 0.5 molal sucrose aqueous solution
(a) 1282 (b) 1293
(c) 1307 (d) 2282

540. Osmotic pressure (kPa) of 1.0 molal NaCl aqueous solution
(a) 2640 (b) 3640
(c) 4640 (d) 5640

541. Osmotic pressure (kPa) of 1.0 molal fructose aqueous solution
(a) 2564 (b) 1564
(c) 3564 (d) 4564

542. Osmotic pressure (kPa) of 1.0 molal sucrose aqueous solution
(a) 2564 (b) 2668
(c) 2764 (d) 2868

543. Electrical conductivity (S/m) of beer at 30°C is about
(a) 0.06 (b) 0.16
(c) 0.26 (d) 0.36

544. Electrical conductivity (S/m) of black coffee at 30°C is about
(a) 0.007 (b) 0.107
(c) 0.207 (d) 0.307

545. Electrical conductivity (S/m) of apple juice at 30°C is about
(a) 0.079 (b) 0.179
(c) 0.214 (d) 0.279

546. Electrical conductivity (S/m) of grape juice at 30°C is about
(a) 0.092 (b) 0.192
(c) 0.292 (d) 0.392

547. Electrical conductivity (S/m) of orange juice at 30°C is about
(a) 0.129 (b) 0.229
(c) 0.329 (d) 0.429

548. Electrical conductivity (S/m) of carrot juice at 30°C is about
(a) 0.282 (b) 0.482
(c) 0.782 (d) 1.282

549. Electrical conductivity (S/m) of tomato juice at 30°C is about
(a) 0.974 (b) 1.974
(c) 2.974 (d) 3.974

550. Electrical conductivity (S/m) of skim milk at 30°C is about
(a) 0.299 (b) 0.399
(c) 0.599 (d) 0.799

551. Electrical conductivity (S/m) of whole milk at 30°C is about
(a) 0.617 (b) 0.717
(c) 0.817 (d) 0.917

552. Electrical conductivity (S/m) of cranberry juice at 30°C is about
(a) 0.052 (b) 0.105
(c) 0.152 (d) 0.205

553. With increase in temperature, electrical conductivity of liquid foods ________
(a) increases (b) decreases
(c) do not change (d) none

554. With increase in salt concentration, electrical conductivity of hydrocolloid solution____
(a) increases (b) decreases
(c) do not change (d) none

555. Radio frequency generally used for food processing
(a) 13.56 MHz (b) 27.12 MHz
(c) 40.68 MHz (d) all

556. Microwave frequency (MHz) generally used in food processing
(a) 915 (b) 5800
(c) 2600 (d) 7412

557. Relaxation frequency (GHz) of water at 0°C is about
(a) 9.007 (b) 12.552
(c) 17.004 (d) 21.862

558. Relaxation frequency (GHz) of water at 10°C is about
(a) 9.007 (b) 12.552
(c) 17.004 (d) 21.862

559. Relaxation frequency (GHz) of water at 20°C is about
(a) 9.007 (b) 12.552
(c) 17.004 (d) 21.862

560. Relaxation frequency (GHz) of water at 30°C is about
(a) 9.007 (b) 12.552
(c) 17.004 (d) 21.862

561. Relaxation frequency (GHz) of water at 40°C is about
(a) 39.690 (b) 27.346
(c) 33.506 (d) 21.862

562. Relaxation frequency (GHz) of water at 50°C is about
(a) 39.690 (b) 27.346
(c) 33.506 (d) 21.862

563. Relaxation frequency (GHz) of water at 60°C is about
(a) 39.690 (b) 27.346
(c) 33.506 (d) 21.862

564. Which food component show low dielectric loss at microwave frequency
(a) bound water (b) protein
(c) starch (d) all

565. Which food component show high dielectric loss at microwave frequency
(a) free water
(b) monosaccharides
(c) ions
(d) all

566. Dielectric constant of PET at 1 kHz is about
(a) 1.5 (b) 2.5
(c) 3.5 (d) 4.5

567. Dielectric loss factor of PET at 1 kHz is about
(a) 0.01 (b) 0.05
(c) 0.10 (d) 0.15

568. Dielectric constant of CPET at 1 kHz is about
(a) 1.3 (b) 3.3
(c) 5.3 (d) 7.3

569. Dielectric loss factor of CPET at 1 kHz is about
(a) 0.02 (b) 0.002
(c) 0.01 (d) 0.032

570. Dielectric constant of polyethylene at 3 GHz is about
(a) 1.25 (b) 2.25
(c) 3.25 (d) 4.25

571. Dielectric constant of plexiglass at 3 GHz is about
(a) 0.6 (b) 1.6
(c) 2.6 (d) 3.6

572. Dielectric constant of teflon at 3 GHz is about
(a) 1.5 (b) 1.7
(c) 2.1 (d) 3.1

573. Dielectric constant of glass (96% SiO_2) at 3 GHz is about
(a) 2.97 (b) 3.84
(c) 4.12 (d) 4.47

574. Dielectric loss factor of polyethylene at 3 GHz is about
(a) 0.26 (b) 0.026
(c) 0.0026 (d) 0.00026

575. Dielectric loss factor of plexiglass at 3 GHz is about
(a) 0.15 (b) 0.015
(c) 0.0015 (d) 0.045

576. Dielectric loss factor of teflon at 3 GHz is about
(a) 0.3 (b) 0.03
(c) 0.003 (d) 0.0003

577. Dielectric loss factor of glass (96% SiO_2) at 3 GHz is about
(a) 0.26 (b) 0.026
(c) 0.0026 (d) 0.00026

578. Dielectric constant of royal grey paper at 1 GHz is
(a) 2.94 (b) 1.94
(c) 3.0 (d) 3.94

579. Dielectric constant of royal grey paper at 3 GHz is
(a) 2.94 (b) 1.94
(c) 3.0 (d) 3.94

580. Dielectric loss factor of royal grey paper at 1 GHz is
(a) 0.216 (b) 0.235
(c) 0.00216 (d) 0.0143

581. Dielectric loss factor of royal grey paper at 3 GHz is
(a) 0.216 (b) 0.235
(c) 0.00216 (d) 0.0143

582. Speed (m/s) of sound in air is
(a) 1482 (b) 5100
(c) 753 (d) 343

583. Speed (m/s) of sound in water is
(a) 1482 (b) 5100
(c) 753 (d) 343

584. Speed (m/s) of sound in aluminium is
(a) 1482 (b) 5100
(c) 753 (d) 343

585. Maximum speed of sound (m/s) in pure water is
(a) 1456 (b) 1556
(c) 1656 (d) 1756

586. Maximum speed of sound (m/s) in 10% sucrose solution is
(a) 1375 (b) 1475
(c) 1575 (d) 1600

587. Maximum speed of sound (m/s) in 20% sucrose solution is
(a) 1375 (b) 1475
(c) 1575 (d) 1600

588. Maximum speed of sound (m/s) in 30% sucrose solution is
(a) 1434 (b) 1678
(c) 1634 (d) 1534

589. Maximum speed of sound (m/s) in 40% sucrose solution is
(a) 1434 (b) 1678
(c) 1634 (d) 1534

590. In sugar solution, speed of sound ______ with increase in sugar concentration
(a) increases (b) decreases
(c) do not change (d) none

591. Attenuation coefficient (Neper/m) of distilled water at 2.1 MHz and 20°C is
(a) 0.9 (b) 5.4
(c) 29.1 (d) 30.1

592. Attenuation coefficient (Neper/m) of castor oil at 2.1 MHz and 20°C is
(a) 0.9 (b) 5.4
(c) 29.1 (d) 30.1

593. Attenuation coefficient (Neper/m) of olive oil at 2.1 MHz and 20°C is
(a) 0.9 (b) 5.4
(c) 29.1 (d) 30.1

594. Attenuation coefficient (Neper/m) of n-hexadecane at 2.1 MHz and 20°C is
(a) 0.9 (b) 5.4
(c) 29.1 (d) 30.1

595. Impedance (10^6kg/m^{2}sec) of distilled water at 2.1 MHz and 20°C is
(a) 1.04 (b) 1.35
(c) 1.45 (d) 1.48

596. Impedance (10^6kg/m^{2}sec) of castor oil at 2.1 MHz and 20°C is
(a) 1.04 (b) 1.35
(c) 1.45 (d) 1.48

597. Impedance (10^6kg/m^{2}sec) of olive oil at 2.1 MHz and 20°C is
(a) 1.04 (b) 1.35
(c) 1.45 (d) 1.48

598. Impedance (10^6kg/m^{2}sec) of n-hexadecane at 2.1 MHz and 20°C is
(a) 1.04 (b) 1.35
(c) 1.45 (d) 1.48

599. Average velocity (m/s) of sound in apple is about
(a) 184.55 (b) 273.55
(c) 383.23 (d) 341.13

600. Average velocity (m/s) of sound in avocado (hard) is about
(a) 184.55 (b) 273.55
(c) 383.23 (d) 341.13

601. Average velocity (m/s) of sound in avocado (soft) is about
(a) 184.55 (b) 273.55
(c) 383.23 (d) 341.13

602. Average velocity (m/s) of sound in carrot (longitudinal) is about
(a) 184.55 (b) 273.55
(c) 383.23 (d) 341.13

603. Average velocity (m/s) of sound in cucumber (longitudinal) is about
(a) 242.81 (b) 267.94
(c) 371.82 (d) 380.04

604. Average velocity (m/s) of sound in melon is about
(a) 242.81 (b) 267.94
(c) 371.82 (d) 380.04

605. Average velocity (m/s) of sound in potato is about
(a) 242.81 (b) 267.94
(c) 371.82 (d) 380.04

606. Average velocity (m/s) of sound in pumpkin is about
(a) 242.81 (b) 267.94
(c) 371.82 (d) 380.04

607. Units of surface tension is dimensionally equal to
(a) J/s (b) J/m
(c) J/m^2 (d) J/m.s

608. Which of the following test is used in rheological analysis
(a) stress relaxation (b) creep
(c) oscillatory test (d) all

609. Which of the following is generally used for determining elastic modulus of material
(a) viscometer
(b) rheometer
(c) texture analyzer
(d) all

610. Rheofermentometer is used for the analysis of
(a) cookies (b) bread
(c) yoghurt (d) all

611. Chewing and swallowing shows a shear rate (sec^{-1}) of about
(a) 10^{-6}-10^{-3} (b) 10^{-1}-10^{1}
(c) 10^{1}-10^{2} (d) 10^{1}-10^{3}

612. Particle sedimentation shows a shear rate (sec^{-1}) of about
(a) 10^{-6}-10^{-3} (b) 10^{-1}-10^{1}
(c) 10^{1}-10^{2} (d) 10^{1}-10^{3}

613. Mixing shows a shear rate (sec^{-1}) of about
(a) 10^{-6}-10^{-3} (b) 10^{-1}-10^{1}
(c) 10^{1}-10^{2} (d) 10^{1}-10^{3}

614. Pipe flow operations shows a shear rate (sec^{-1}) of about
(a) 10^{0}-10^{-3} (b) 10^{2}-10^{3}
(c) 10^{1}-5 x 10^{2} (d) 10^{2}-5 x 10^{4}

615. Plate heat exchanger shows a shear rate (sec^{-1}) of about
(a) 10^{0}-10^{-3} (b) 10^{2}-10^{3}
(c) 10^{1}-5 x 10^{2} (d) 10^{2}-5 x 10^{4}

616. Scrap-surface heat exchanger shows a shear rate (sec^{-1}) of about
(a) 10^{0}-10^{-3} (b) 10^{2}-10^{3}
(c) 10^{1}-5 x 10^{2} (d) 10^{2}-5 x 10^{4}

617. Extruder shows a shear rate (sec^{-1}) of about
(a) 10^{0}-10^{-3} (b) 10^{2}-10^{3}
(c) 10^{1}-5 x 10^{2} (d) 10^{2}-5 x 10^{4}

618. Unit of modulus is
(a) Pa.s (b) Pa/s
(c) Pa/s^2 (d) Pa

619. Rheological properties can be used in
(a) sensory evaluation
(b) molecular conformation
(c) product characterization
(d) all

620. Activation energy (Kcal/mole) for enzymatic reactions is generally about
(a) 5-27 (b) 0-8
(c) 9-40 (d) 10-15

621. Activation energy (Kcal/mole) for chlorophyll degradation is generally about
(a) 5-27 (b) 0-8
(c) 9-40 (d) 10-15

622. Activation energy (Kcal/mole) for non-enzymatic browning is generally about
(a) 5-27 (b) 0-8
(c) 9-40 (d) 10-15

623. Activation energy (Kcal/mole) for hydrolysis of disaccharides is generally about
(a) 5-27 (b) 0-8
(c) 9-40 (d) 10-15

624. Activation energy (Kcal/mole) for carotenoids oxidation is generally about
(a) 5-10 (b) 0-8
(c) 10-22 (d) 60-80

625. Activation energy (Kcal/mole) for lipid oxidation is generally about
(a) 5-10 (b) 0-8
(c) 10-25 (d) 60-80

626. Activation energy (Kcal/mole) for spores destruction is generally about
(a) 5-10 (b) 0-8
(c) 10-25 (d) 60-80

627. Activation energy (Kcal/mole) for vegetables cell destruction is generally about
(a) 10-30 (b) 30-44
(c) 50-150 (d) 180-200

628. Activation energy (Kcal/mole) for protein destruction is generally about
(a) 10-30 (b) 30-50
(c) 50-70 (d) 80-120

629. Denaturation temperature (°C) of actin is about
(a) 63.5 (b) 73.5
(c) 83.5 (d) 68.5

630. Denaturation temperature (°C) of avidin is about
(a) 65 (b) 70
(c) 75 (d) 85

631. Denaturation temperature (°C) of avidin-biotin complex is about
(a) 75 (b) 85
(c) 150 (d) 131

632. Denaturation temperature (°C) of bovine serum albumin is about
(a) 64 (b) 74
(c) 84 (d) 94

633. Denaturation temperature (°C) of conalbumin is about
(a) 76 (b) 71
(c) 81 (d) 61

634. Denaturation temperature (°C) of egg albumin is about
(a) 62 (b) 72
(c) 82 (d) 92

635. Denaturation temperature (°C) of immunoglobulin is about
(a) 62 (b) 72
(c) 82 (d) 92

636. Denaturation temperature (°C) of α-lactalbumin is about
(a) 62 (b) 72
(c) 82 (d) 92

637. Denaturation temperature (°C) of β-lactoglobulin is about
(a) 62 (b) 78
(c) 82 (d) 92

638. Denaturation temperature (°C) of myosin is about
(a) 55 (b) 65
(c) 75 (d) 85

639. Denaturation temperature (°C) of ovalbumin is about
(a) 54 (b) 64
(c) 74 (d) 84

640. Denaturation temperature (°C) of soya protein is about
(a) 63 (b) 93
(c) 83 (d) 73

641. Unfolding of peptide chains in proteins due to heat is
(a) Denaturation (b) coagulation
(c) precipitation (d) none

642. Heat of Denaturation (J/g) of actin is about
(a) 4.5 (b) 10.5
(c) 14.5 (d) 18.5

643. Heat of Denaturation (J/g) of bovine serum albumin is about
(a) 8.2 (b) 12.2
(c) 4.2 (d) 18.2

644. Heat of Denaturation (J/g) of conalbumin is about
(a) 8.2 (b) 12.2
(c) 15.2 (d) 19.2

645. Heat of Denaturation (J/g) of egg globulins is about
(a) 4.8 (b) 8.8
(c) 11.8 (d) 14.8

646. Heat of Denaturation (J/g) of immunoglobulins is about
(a) 4.9 (b) 7.9
(c) 10.9 (d) 13.9

647. Heat of Denaturation (J/g) of α-lactalbumin is about
(a) 9.8 (b) 12.8
(c) 17.8 (d) 22.8

648. Heat of Denaturation (J/g) of β-lactoglobulin is about
(a) 11.9 (b) 13.9
(c) 21.9 (d) 16.9

649. Heat of Denaturation (J/g) of lysozyme is about
(a) 13.2 (b) 21.2
(c) 28.2 (d) 17.2

650. Heat of Denaturation (J/g) of myosin is about
(a) 13.9 (b) 17.9
(c) 20.9 (d) 23.9

651. Heat of Denaturation (J/g) of ovalbumin is about
(a) 11.2 (b) 15.2
(c) 19.2 (d) 24.2

652. Melting temperature (from DSC) of L-arabinose is about
(a) 115 (b) 130
(c) 155 (d) 165

653. Melting temperature (from DSC) of L-fucose is about
(a) 115 (b) 130
(c) 155 (d) 165

654. Melting temperature (from DSC) of L-fructose is about
(a) 115 (b) 130
(c) 155 (d) 165

655. Melting temperature (from DSC) of L-galactose is about
(a) 115 (b) 130
(c) 155 (d) 165

656. Melting temperature (from DSC) of D-galacturonic acid is about
(a) 105 (b) 150
(c) 140 (d) 90

657. Melting temperature (from DSC) of α-D-glucose is about
(a) 105 (b) 150
(c) 140 (d) 90

658. Melting temperature (from DSC) of D-glucuronic acid is about
(a) 105 (b) 150
(c) 140 (d) 90

659. Melting temperature (from DSC) of D-ribose is about
(a) 105 (b) 150
(c) 140 (d) 90

660. Melting temperature (from DSC) of D-xylose is about
(a) 105 (b) 150
(c) 140 (d) 90

661. Melting temperature (from DSC) of cellobiose is about
(a) 155 (b) 185
(c) 195 (d) 220

662. Melting temperature (from DSC) of α-lactose is about
(a) 155 (b) 185
(c) 195 (d) 220

663. Melting temperature (from DSC) of β-lactose is about
(a) 155 (b) 185
(c) 195 (d) 220

664. Melting temperature (from DSC) of lactulose is about
(a) 155 (b) 185
(c) 195 (d) 220

665. Melting temperature (from DSC) of sucrose is about
(a) 155 (b) 185
(c) 195 (d) 220

666. Melting temperature (from DSC) of galactitol is about
(a) 190 (b) 150
(c) 170 (d) 95

667. Melting temperature (from DSC) of D-mannitol is about
(a) 190 (b) 150
(c) 170 (d) 95

668. Melting temperature (from DSC) of maltitol is about
(a) 190 (b) 150
(c) 170 (d) 95

669. Melting temperature (from DSC) of sorbitol is about
(a) 190 (b) 150
(c) 170 (d) 95

670. Melting temperature (from DSC) of xylitol is about
(a) 80 (b) 100
(c) 140 (d) 170

671. Latent heat of melting (J/g) of D-fructose is about
(a) 70 (b) 150
(c) 180 (d) 280

672. Latent heat of melting (J/g) of D-galactose is about

(a) 70 (b) 150
(c) 180 (d) 280

673. Latent heat of melting (J/g) of D-galacturonic acid is about

(a) 70 (b) 150
(c) 180 (d) 280

674. Latent heat of melting (J/g) of α-D-glucose is about

(a) 70 (b) 150
(c) 180 (d) 280

675. Latent heat of melting (J/g) of β-D-glucose is about

(a) 70 (b) 150
(c) 180 (d) 280

676. Latent heat of melting (J/g) of D-xylose is about

(a) 70 (b) 150
(c) 180 (d) 280

677. Latent heat of melting (J/g) of cellobiose is about

(a) 110 (b) 120
(c) 160 (d) 250

678. Latent heat of melting (J/g) of α-lactose is about

(a) 110 (b) 120
(c) 160 (d) 250

679. Latent heat of melting (J/g) of β-lactose is about

(a) 110 (b) 120
(c) 160 (d) 250

680. Latent heat of melting (J/g) of lactulose is about

(a) 110 (b) 120
(c) 160 (d) 250

681. Latent heat of melting (J/g) of sucrose is about

(a) 110 (b) 120
(c) 160 (d) 250

682. Latent heat of melting (J/g) of galactitol is about

(a) 150 (b) 250
(c) 290 (d) 330

683. Latent heat of melting (J/g) of D-mannitol is about

(a) 150 (b) 250
(c) 290 (d) 330

684. Latent heat of melting (J/g) of maltitol is about

(a) 150 (b) 250
(c) 290 (d) 330

685. Latent heat of melting (J/g) of sorbitol is about

(a) 150 (b) 250
(c) 290 (d) 330

686. Latent heat of melting (J/g) of xylitol is about

(a) 150 (b) 250
(c) 290 (d) 330

687. Melting temperature (°C) of inulin is

(a) 100 (b) 150
(c) 200 (d) 250

688. Glass transition temperature (°C) of anhydrous α-casein is about

(a) 148 (b) 152
(c) 156 (d) 197

689. Glass transition temperature (°C) of anhydrous ß-casein is about

(a) 148 (b) 152
(c) 156 (d) 197

690. Glass transition temperature (°C) of collagen is about

(a) 148 (b) 152
(c) 156 (d) 197

691. Glass transition temperature (°C) of gelatin is about

(a) 159 (b) 179
(c) 189 (d) 207

692. Glass transition temperature (°C) of gliadin is about

(a) 159 (b) 179
(c) 189 (d) 207

693. Glass transition temperature (°C) of glutenin is about
(a) 159 (b) 179
(c) 189 (d) 207

694. Glass transition temperature (°C) of elastin is about
(a) 159 (b) 179
(c) 189 (d) 208

695. Glass transition temperature (°C) of α-lactalbumin is about
(a) 149 (b) 159
(c) 169 (d) 179

696. Glass transition temperature (°C) of β-lactalbumin is about
(a) 146 (b) 156
(c) 166 (d) 176

697. Glass transition temperature (°C) of myoglobinis about
(a) 139 (b) 149
(c) 157 (d) 168

698. Glass transition temperature (°C) of ovalbumin is about
(a) 139 (b) 149
(c) 157 (d) 168

699. Glass transition temperature (°C) of zein is about
(a) 135 (b) 145
(c) 155 (d) 165

700. Glass transition temperature (°C) of anhydrous amylose is about
(a) 272 (b) 282
(c) 302 (d) 322

701. Glass transition temperature (°C) of anhydrous amylopectin is about
(a) 294 (b) 314
(c) 304 (d) 284

702. Glass transition temperature (°C) of potato starch is about
(a) 296 (b) 316
(c) 306 (d) 286

703. Glass transition temperature (°C) of waxy maize starch is about
(a) 295 (b) 315
(c) 305 (d) 285

704. Energy flow rate (kJ/sec) in condenser preheater is about
(a) 1451 (b) 1680
(c) 1814 (d) 3500

705. Energy flow rate (kJ/sec) in chest preheater is about
(a) 1451 (b) 1680
(c) 1814 (d) 3500

3.3 ANSWER KEY

Q. A.	Q. A.	Q. A.	Q. A.	Q. A.	Q. A.
1. (b)	2. (d)	3. (a)	4. (c)	5. (d)	6. (c)
7. (d)	8. (d)	9. (d)	10. (d)	11. (c)	12. (c)
13. (c)	14. (b)	15. (a)	16. (b)	17. (c)	18. (a)
19. (c)	20. (b)	21. (c)	22. (b)	23. (b)	24. (a)
25. (c)	26. (c)	27. (d)	28. (c)	29. (b)	30. (b)
31. (d)	32. (a)	33. (b)	34. (c)	35. (b)	36. (a)
37. (a)	38. (b)	39. (c)	40. (a)	41. (d)	42. (d)
43. (c)	44. (d)	45. (b)	46. (c)	47. (a)	48. (d)

Q. A.	Q. A.	Q. A.	Q. A.	Q. A.	Q. A.
49. (a)	50. (d)	51. (c)	52. (b)	53. (a)	54. (c)
55. (b)	56. (b)	57. (a)	58. (a)	59. (c)	60. (c)
61. (b)	62. (a)	63. (d)	64. (c)	65. (a)	66. (c)
67. (b)	68. (a)	69. (b)	70. (d)	71. (b)	72. (d)
73. (b)	74. (c)	75. (a)	76. (b)	77. (a)	78. (c)
79. (b)	80. (b)	81. (a)	82. (c)	83. (b)	84. (c)
85. (a)	86. (a)	87. (b)	88. (d)	89. (c)	90. (a)
91. (b)	92. (a)	93. (d)	94. (c)	95. (d)	96. (a)
97. (c)	98. (a)	99. (b)	100. (c)	101. (b)	102. (a)
103. (c)	104. (b)	105. (c)	106. (d)	107. (a)	108. (c)
109. (b)	110. (b)	111. (b)	112. (a)	113. (b)	114. (d)
115. (c)	116. (b)	117. (a)	118. (c)	119. (d)	120. (b)
121. (c)	122. (c)	123. (c)	124. (b)	125. (b)	126. (b)
127. (b)	128. (d)	129. (b)	130. (d)	131. (a)	132. (c)
133. (d)	134. (b)	135. (a)	136. (b)	137. (a)	138. (d)
139. (a)	140. (a)	141. (b)	142. (c)	143. (d)	144. (a)
145. (d)	146. (d)	147. (d)	148. (b)	149. (d)	150. (c)
151. (c)	152. (a)	153. (d)	154. (b)	155. (c)	156. (d)
157. (c)	158. (d)	159. (d)	160. (d)	161. (d)	162. (d)
163. (c)	164. (b)	165. (c)	166. (a)	167. (b)	168. (a)
169. (c)	170. (a)	171. (d)	172. (d)	173. (b)	174. (c)
175. (a)	176. (b)	177. (d)	178. (a)	179. (a)	180. (b)
181. (d)	182. (a)	183. (c)	184. (c)	185. (c)	186. (b)
187. (d)	188. (b)	189. (c)	190. (b)	191. (b)	192. (b)
193. (b)	194. (b)	195. (c)	196. (a)	197. (b)	198. (c)
199. (d)	200. (c)	201. (d)	202. (d)	203. (a)	204. (d)
205. (d)	206. (c)	207. (c)	208. (d)	209. (c)	210. (c)
211. (b)	212. (d)	213. (d)	214. (c)	215. (b)	216. (a)
217. (b)	218. (c)	219. (d)	220. (c)	221. (a)	222. (b)
223. (c)	224. (d)	225. (b)	226. (c)	227. (a)	228. (c)
229. (b)	230. (c)	231. (a)	232. (a)	233. (b)	234. (d)
235. (c)	236. (a)	237. (d)	238. (a)	239. (c)	240. (b)
241. (a)	242. (b)	243. (b)	244. (c)	245. (d)	246. (a)
247. (b)	248. (d)	249. (d)	250. (a)	251. (a)	252. (b)

Q. A.	Q. A.	Q. A.	Q. A.	Q. A.	Q. A.
253. (c)	254. (c)	255. (d)	256. (a)	257. (c)	258. (b)
259. (a)	260. (b)	261. (c)	262. (d)	263. (a)	264. (a)
265. (a)	266. (d)	267. (a)	268. (c)	269. (b)	270. (d)
271. (c)	272. (b)	273. (d)	274. (a)	275. (d)	276. (d)
277. (d)	278. (b)	279. (c)	280. (d)	281. (a)	282. (c)
283. (d)	284. (a)	285. (b)	286. (c)	287. (d)	288. (c)
289. (a)	290. (d)	291. (c)	292. (b)	293. (c)	294. (c)
295. (b)	296. (a)	297. (a)	298. (c)	299. (a)	300. (b)
301. (a)	302. (b)	303. (c)	304. (a)	305. (d)	306. (a)
307. (b)	308. (c)	309. (d)	310. (a)	311. (d)	312. (d)
313. (a)	314. (c)	315. (b)	316. (d)	317. (c)	318. (c)
319. (c)	320. (b)	321. (a)	322. (a)	323. (b)	324. (c)
325. (a)	326. (c)	327. (d)	328. (b)	329. (b)	330. (c)
331. (d)	332. (c)	333. (b)	334. (a)	335. (b)	336. (c)
337. (d)	338. (b)	339. (a)	340. (c)	341. (d)	342. (b)
343. (c)	344. (a)	345. (d)	346. (a)	347. (c)	348. (c)
349. (b)	350. (a)	351. (d)	352. (a)	353. (b)	354. (d)
355. (d)	356. (d)	357. (d)	358. (b)	359. (c)	360. (d)
361. (d)	362. (c)	363. (b)	364. (a)	365. (a)	366. (b)
367. (a)	368. (b)	369. (d)	370. (a)	371. (d)	372. (b)
373. (c)	374. (b)	375. (a)	376. (c)	377. (d)	378. (b)
379. (c)	380. (a)	381. (a)	382. (b)	383. (b)	384. (b)
385. (c)	386. (a)	387. (a)	388. (d)	389. (a)	390. (b)
391. (c)	392. (b)	393. (a)	394. (b)	395. (c)	396. (d)
397. (c)	398. (b)	399. (d)	400. (a)	401. (c)	402. (d)
403. (b)	404. (a)	405. (a)	406. (d)	407. (c)	408. (c)
409. (d)	410. (c)	411. (c)	412. (c)	413. (b)	414. (c)
415. (c)	416. (b)	417. (d)	418. (a)	419. (b)	420. (b)
421. (a)	422. (c)	423. (c)	424. (d)	425. (b)	426. (d)
427. (a)	428. (b)	429. (b)	430. (a)	431. (b)	432. (a)
433. (c)	434. (c)	435. (c)	436. (d)	437. (a)	438. (c)
439. (d)	440. (c)	441. (a)	442. (a)	443. (c)	444. (d)
445. (b)	446. (c)	447. (a)	448. (a)	449. (d)	450. (d)
451. (a)	452. (c)	453. (b)	454. (d)	455. (a)	456. (a)

Q. A.	Q. A.	Q. A.	Q. A.	Q. A.	Q. A.
457. (b)	458. (c)	459. (b)	460. (a)	461. (b)	462. (c)
463. (d)	464. (a)	465. (b)	466. (c)	467. (d)	468. (c)
469. (d)	470. (d)	471. (d)	472. (a)	473. (d)	474. (a)
475. (b)	476. (d)	477. (b)	478. (b)	479. (a)	480. (c)
481. (a)	482. (a)	483. (c)	484. (a)	485. (b)	486. (c)
487. (d)	488. (c)	489. (d)	490. (c)	491. (b)	492. (d)
493. (c)	494. (b)	495. (a)	496. (a)	497. (b)	498. (c)
499. (d)	500. (c)	501. (d)	502. (d)	503. (c)	504. (b)
505. (a)	506. (c)	507. (d)	508. (a)	509. (a)	510. (b)
511. (d)	512. (d)	513. (c)	514. (b)	515. (a)	516. (d)
517. (d)	518. (b)	519. (a)	520. (c)	521. (d)	522. (a)
523. (c)	524. (c)	525. (d)	526. (b)	527. (c)	528. (a)
529. (d)	530. (d)	531. (d)	532. (c)	533. (a)	534. (b)
535. (c)	536. (d)	537. (b)	538. (c)	539. (a)	540. (c)
541. (a)	542. (b)	543. (b)	544. (c)	545. (d)	546. (a)
547. (d)	548. (d)	549. (b)	550. (c)	551. (a)	552. (b)
553. (a)	554. (a)	555. (b)	556. (a)	557. (a)	558. (b)
559. (c)	560. (d)	561. (b)	562. (c)	563. (a)	564. (d)
565. (d)	566. (c)	567. (a)	568. (b)	569. (b)	570. (b)
571. (c)	572. (c)	573. (b)	574. (c)	575. (b)	576. (d)
577. (c)	578. (c)	579. (a)	580. (a)	581. (b)	582. (d)
583. (a)	584. (b)	585. (b)	586. (c)	587. (d)	588. (c)
589. (b)	590. (a)	591. (a)	592. (d)	593. (b)	594. (a)
595. (d)	596. (c)	597. (b)	598. (a)	599. (a)	600. (b)
601. (c)	602. (d)	603. (c)	604. (a)	605. (d)	606. (b)
607. (c)	608. (d)	609. (b)	610. (b)	611. (c)	612. (a)
613. (d)	614. (a)	615. (b)	616. (c)	617. (d)	618. (d)
619. (d)	620. (b)	621. (a)	622. (c)	623. (d)	624. (c)
625. (c)	626. (d)	627. (c)	628. (d)	629. (c)	630. (d)
631. (d)	632. (a)	633. (d)	634. (d)	635. (b)	636. (a)
637. (b)	638. (a)	639. (d)	640. (b)	641. (a)	642. (c)
643. (b)	644. (c)	645. (c)	646. (d)	647. (c)	648. (d)
649. (c)	650. (a)	651. (b)	652. (c)	653. (b)	654. (a)
655. (d)	656. (a)	657. (b)	658. (c)	659. (d)	660. (b)

Q.	A.	Q.	A.	Q.	A.	Q.	A.	Q.	A.	Q.	A.
661.	(d)	662.	(c)	663.	(d)	664.	(a)	665.	(b)	666.	(a)
667.	(c)	668.	(b)	669.	(d)	670.	(b)	671.	(c)	672.	(d)
673.	(a)	674.	(c)	675.	(b)	676.	(d)	677.	(c)	678.	(d)
679.	(d)	680.	(a)	681.	(b)	682.	(d)	683.	(c)	684.	(a)
685.	(a)	686.	(b)	687.	(b)	688.	(b)	689.	(a)	690.	(d)
691.	(d)	692.	(b)	693.	(c)	694.	(d)	695.	(b)	696.	(a)
697.	(b)	698.	(c)	699.	(d)	700.	(c)	701.	(a)	702.	(b)
703.	(d)	704.	(a)	705.	(c)						

CHAPTER 4

FRUITS & VEGETABLES TECHNOLOGY

4.1 SUBJECTIVE KEYNOTES

1. Fruits and vegetables are among the perishable commodities, important ingredients in the human dietaries. The fruits constitute 80-90% water.
2. Fruits are the mature ovaries of the plants with their seeds. The fleshy portion of pericarp and portion surrounding the seeds is the edible part of most of the fruits.
3. The fruits containing single pits are known as drupes e.g. apricots, cherries, peaches and plum.
4. The fruits containing many pits are known as pomes e.g. apples, quince, pears etc.
5. Fruits are good source of carbohydrates, sugars, vitamins and water, but deficient in proteins. Fruits contain many pigments such as anthocyanin, lycopene, carotene etc.
6. Fresh cut fruits and vegetables undergo enzymatic browning involving enzyme poly phenol oxidase.
7. Tropical and subtropical fruits banana, pineapple, dates, mango, papaya etc.
8. Pectic substances hold the cells of fruits together. In immature fruits, pectin is present while ripened fruits contain pectic acid.
9. The principal organic acids present in fruits are malic, citric, oxalic, benzoic and tartaric acid.
10. The white portion of the skin of citrus fruits is called albedo.
11. The acid content of fruits decreases with ripening whereas sugar content generally increases with ripening of fruits.
12. Avocados are particularly rich in proteins and fat.
13. Blanching is the process of heat treatment to inactivate the natural enzymes of fruits & vegetables. It also aids in further processing of fruits and vegetables
14. The most heat resistant enzymes and indicators of adequacy of blanching are catalase and peoroxidase.
15. Asepsis means preventing the entry of microorganisms.
16. Coagulation of proteins and inactivation of their metabolic enzymes by the application of heat leads to the destruction of microorganisms present in foods.
17. Fruit and tomato products should be heated at 100°C for 30 minutes so that the spore forming bacteria which are sensitive to high acidity may be completely killed.

18. Aseptic canning is a technique in which food is sterilized outside the can and then aseptically placed in previously sterilized cans which are subsequently sealed in an aseptic environment.
19. The best way of preserving pure fruit juice is by freezing.
20. Sulphur dioxide is widely used in the preservation of juice, pulp, nectar, squash, crush, cordial and other products.
21. Potassium metabisulphite is commonly used as a stable source of sulphur dioxide.
22. The maximum amount of sulphur dioxide allowed in fruit juice is 700 ppm, in squash, crush and cordial 350 ppm and in RTS and nectar 100 ppm.
23. Carbonation is the process of dissolving sufficient carbon dioxide in water or beverage so that the product when served gives off the gas as fine bubbles and has a characteristic taste.
24. Syrups containing 66 per cent or more of sugar do not ferment, thus sugar acts as a preservative by osmosis and not as a true poison for microorganisms.
25. Fruit syrup, jam, jelly, marmalade, preserve, candy, crystallized fruit and glazed fruit are preserved by sugar.
26. Decomposition of carbohydrates by microorganisms or enzymes is called 'fermentation'.
27. Salt at a concentration of 15 to 25 per cent is sufficient to preserve most products.
28. Maturity at the time of harvest is the most important factor that determines post harvest life and final quality of the produce.
29. Commercial maturity refers to the stage of a plant organ required by a market. The terms immaturity, optimum maturity and over maturity relates to these requirements, and it has little relation with physiological maturity.
30. Commonly selected factors reflecting horticultural maturity includes firmness, skin color, flesh color, sugar content, soluble solids, pigments, etc.
31. Aerobic respiration is preferred energy producing pathway in fruits and vegetables. Anaerobic respiration produces much less energy than aerobic pathway and elevated CO_2 concentrations have deteriorative effect on the product quality.
32. Dehydro-freezing is a process where freezing is preceded by partial dehydration.
33. Sauerkraut is fermented product of cabbage.
34. Table vinegar which is consumed directly without much modification contain 4% acetic acid whereas vinegar used for pickling can have acetic acid as high as 18%.
35. Aerobic respiration is the major biochemical process supplying energy. It involves the oxidation of certain organic compounds i.e. glucose, maleic acid, stored in the tissues. Aerobic respiration occurs in three phases: glycolysis or Embden-Meyerhoff-Parnas (EMP) pathway, the Krebs or Tri-carboxylic acid cycle and the electron transport chain.
36. Climacteric fruits (apple, melon) have a tendency of sudden increase in respiration rate during their storage from the increased rate of ethylene gas production.
37. To allow even distribution of fruit peel and slices in marmalade and not float to the surface, they are soaked in sugar solution.

38. Marmalade should contain NLT 45% fruit from which it is made.
39. Fruit chutney should contain minimum 40% of fruit in the final product.
40. The tomato paste should have NLT 25% total soluble solids.
41. The tomato puree should have NLT 9% total soluble solids.
42. The brown pigment formed in the maillard reaction is melanoidin.
43. For jelly, fruits with high pectin and high acid are preferred.
44. Papain is the enzyme extracted from papaya, ficin from figs.
45. The process of bringing the fruit and vegetable to room temperature after cold storage is called thawing.
46. Minimally processed foods involve processing operations such as peeling, shredding, cutting etc.
47. Active packaging of fruits and vegetables involve use of scavengers, emitters and moisture barriers.
48. MAP and CAS are used preservation of fruits and vegetables and prolonging their shelf life.
49. Vegetable juices are mostly spoiled by bacteria.
50. Goitrogens are present in cabbage which can cause goitre.
51. Respiration plays a major role in the postharvest life of fresh fruits and vegetables. An enhanced respiration rate is associated with perishability of food.
52. The ratio of moles of CO_2 produced per mole of O_2 consumed is called the respiratory quotient. It is 1 for glucose catabolism.
53. Fruit ripening is a highly co-ordinated, genetically programmed irreversible phenomenon which lead to the development of a soft and edible ripe fruit with desirable quality attributes via series of physiological, biochemical, and organoleptic changes.
54. Climacteric fruits are harvested at full maturity, can be ripened off the parent plant. These produce much larger quantities of ethylene in association with their ripening, and exposure to ethylene treatment will result in faster and more uniform ripening.
55. Non-climacteric fruits are not capable of continuing their ripening process, once they are detached from the parent plant. Also, these fruits produce a very small quantity of endogenous ethylene, and do not respond to external ethylene treatment.
56. Ethylene is a fruit ripening hormone which in minute amounts can trigger many events of cell metabolism including initiation of ripening and senescence. It is produced by unripe fruit in minute quantities (0.01-0.05 ppm)
57. Chemicals that inhibit the synthesis of ethylene (e.g. aminoethoxyvinyl glycine, methoxyvinyl glycine, aminoacetic acid) or its action (e.g. silver thiosulphate, carbon dioxide, nickel, cobalt, methylcyclopropane, norbornadiene) are known as ethylene inhibitors.
58. Fruit ripening involves many complex biochemical changes, including seed maturation, change in colour, abscission from the parent plant, texture softening, production of flavour volatiles, wax development on skin, tissue permeability and change in carbohydrate composition, organic acids and proteins.

59. Q10 value at T + 10°C is defined as the ratio of rate of reaction at T + 10°C to the rate of reaction at T°C.
60. Various fruits and vegetables develop physiological disorders when exposed to temperatures below their optimum storage temperatures, but above freezing points. This type of disorder is known as chilling injury or cold injury.
61. Zero energy cold chambers have capacity to lower 15-30°C temperature than that of atmosphere and can be efficiently used for short-term storage of fruits and vegetables at field.
62. Pre-cooling could be defined as process of storage produces which yields their maximum life. This is often done at temperatures below which physiological disorders (chilling, freezing injuries) are exhibited by the produce or usually just above the temperature that which will cause chilling injury or freezing injury.
63. Equipment used for dry cleaning of fruits and vegetables include aspirators, magnetic separators, screening separators etc.
64. Wet cleaning is more effective than dry cleaning for removing soil from root crops or dust and pesticide residues from soft fruits or vegetables.
65. Chlorination is used to decrease bacterial load of water in the soak tank. High levels of chlorine affect the food, e.g. potatoes develop blackened flesh.
66. Sound waves of frequencies above 16 kHz are used for ultrasonic cleaning.
67. Flotation washing method is based on the differences in buoyancy of the desired and undesired parts of the raw material to be cleaned.
68. Sorting may be regarded as a separation operation based in the differences in physical properties of the raw materials or products.
69. Grading is the quality separation on the basis of an overall assessment of those properties, which affect the acceptance of the raw material for processing, and finished food product for consumer acceptance and safety.
70. Peeling is used in the processing of many fruits and vegetables to remove unwanted or inedible material, and to improve the appearance of the final product. Different types of peeling include knife peeling, abrasion peeling, caustic peeling, flame peeling, flash steam peeling.
71. Size reduction or comminution is the unit operation in which the average size of solid pieces of food is reduced by the application of grinding, compression or impact forces. Different methods of size reduction include- chopping, cutting, slicing and dicing.
72. The cutting operation in minimally processed fruits and vegetables include various operations such as chopping, slicing, shredding, peeling, dicing or sectioning.
73. The shelf life of minimally processed fruits and vegetables is inversely proportional to the increase in respiration rate and ethylene production.
74. Carotenoids, a yellow pigment in fruit and vegetable tissues, are highly susceptible to oxidative breakdown that is catalysed by lipoxygenase enzymes.
75. Senescence is the main reason for yellowing of the tissues in green vegetables and fruits due to the degradation of chlorophyll and is not liked by consumers.
76. The firmness of the flesh of fresh-cut fruit products can be maintained through the application or treatment with calcium compounds.

77. Production of hydrogen (hydrogen swell), CO_2, browning, corrosion of cans due to chemical reactions and liquification, gelation, discoloration of products due to enzymatic reactions are some examples of non-microbial spoilage in canned fruits and vegetables.
78. Sulphite spoilage in canned fruits and vegetables is caused by *Clostridium nigrificans* and *C. bifermentans* uncommonly in low acid food like peas and corn producing H2S.
79. Flat sour spoilage of acid food is caused by *Bacillus coagulans*.
80. Commercially pectin is manufactured from apples or citrus fruit and is available in both powdered and liquid forms.
81. Methods used for judging of end point in jam making are refractometer, boiling point, cold plate test, sheet test.
82. Defects in jam include crystallization, sticky or gummy jam, premature setting, surface graining, shrinkage etc.
83. Defects in jelly include failure to set, cloudiness, crystals, synesresis.
84. The phenomenon of spontaneous release of fluid from a gel is called syneresis.
85. Marmalades are the products made from citrus fruits like oranges and lemons in which shredded peel is used as the suspended material. These are of two types jam marmalades & jelly marmalades.
86. Browning is one problem with marmalades during storage. It can be prevented by addition of small quantity of potassium metabisulphite (KMS). KMS also prevents spoilage due to molds.
87. Theories of gel formation are Fibril theory, Spencer's theory, Hinton's theory and Olsen's theory.
88. TSS in fruit and vegetable chutney should not be less than 50% and 25%, respectively.
89. Drained weight of pickles in brine, oil or vinegar should not be less than 60%.
90. Pickles can be of three types- pickles in vinegar, pickles in citrus juice or brine and pickles in oil.
91. In sauerkraut manufacture, 2.25-2.5% salt is added to shredded fresh cabbage.
92. Different kinds of spoilages occur in pickles are shriveling, bitter taste, blackening, dull or faded appearance, softness or slipperiness, scum formation and cloudiness.
93. Fortified wines are those wines in which distilled spirits such as grape brandy is added to increase the alcohol content in addition to those derived from fermentation. They usually contain 18-22% alcohol. Examples are Sherry, Port, and Champagnes etc.
94. Red wines are those wines in which the skin, stems etc. of the grapes are present during the fermentation to extract the pigment and thereby colouring the fermented juice.
95. **White wines** are fermented grape juice only i.e. without skin, stems etc. The colour of the white wines varies from a pale straw colour to deep gold colour.
96. Winemaking or vinification is the process of wine production starting from the selection of grapes to the bottling of the finished wine.

97. CFTRI developed a fast method for the production of coconut vinegar using 'vinegar generator'. Vinegar generator is a simple device, gives good yield, require no power supply and easy in construction and assemble.

98. In fruits juice industry, enzymes aids in juice extraction, clarification, debittering, cloud stability etc.

99. Thermally processed fruits juices are called sweetened juice when nutritive sweeteners are higher than or equal to 15g/kg.

100. Thermally processed mango pulp/puree TSS NLT 15.0% (sweetened) and TSS NLT 12.0% (unsweetened).

4.2 MULTIPLE CHOICE QUESTIONS

1. The per centage of water in fruits is
(a) 2-5% (b) 10-15%
(c) 40-50% (d) 80-90%

2. Fruits are generally deficient in
(a) proteins
(b) carbohydrates
(c) vitamins
(d) water

3. Avocado contains approximately _____ _____ % fat
(a) 2 (b) 10
(c) 40 (d) 25

4. The prominent organic acid present in grapes is
(a) malic acid (b) citric acid
(c) tartaric acid (d) none

5. Which of the following acid combine with calcium to give calcium oxalate
(a) malic acid b) citric acid
(c) tartaric acid (d) oxalic acid

6. Amla is very good source of
(a) Vitamin A (b) Vitamin B
(c) Vitamin C (d) Vitamin D

7. Yellow fruits contain
(a) chlorophyll (b) carotenoids
(c) anthocyanins (d) none

8. Anthocyanins are found in
(a) capsicum (b) carrots
(c) blueberries (d) banana

9. Carotenoids are _______ soluble
(a) water (b) fat
(c) both (d) none

10. Anthocyanins are _______ soluble
(a) water (b) fat
(c) both (d) none

11. The per centage of fat in amla is
(a) 0.1 (b) 1
(c) 10 (d) 20

12. Among the following, highest amount of calcium is found in
(a) amla (b) carambola
(c) dates (d) apricot

13. Among the following, highest amount of fat is found in
(a) apple (b) avocado
(c) dates (d) cherries

14. Among the following, highest amount of minerals is found in
(a) apple (b) apricot
(c) dates (d) figs

15. Richest source of phosphorus among the following fruits is
(a) avocado (b) banana
(c) dates (d) grapes

16. Richest source of carotene among the following fruits is
(a) amla (b) apricot
(c) banana (d) figs

17. The carotene content in apricot is ______ μg
(a) 9 (b) 0
(c) 2160 (d) 90

18. The vitamin C content in amla is ______ mg
(a) 1 (b) 6
(c) 600 (d) 60

19. Melons contain ______ % water
(a) 95.8 (b) 70
(c) 62.5 (d) 55.6

20. Mango is a rich source of
(a) Vitamin C (b) carotene
(c) calcium (d) phosphorus

21. Mosambi contains ______ μg carotene
(a) 0 (b) 1.5
(c) 2.5 (d) 4.5

22. Wood apple is rich source of
(a) calcium (b) Vitamin C
(c) water (d) none

23. The highest protein content is present in
(a) wood apple (b) pear
(c) prunes (d) sapota

24. The protein content of wood apple is ______ g/100g
(a) 7.1 (b) 0.71
(c) 1.7 (d) 17

25. Strawberries are ______ in color
(a) yellow (b) red
(c) orange (d) purple

26. When a cut apple is exposed to air browning occurs due to presence of ______ substances in fruit
(a) phenolic (b) protein
(c) fat (d) watery

27. The enzyme responsible for enzymatic browning in cut fruits is
(a) PPO (b) catalase
(c) peroxidase (d) lipase

28. Enzymatic browning can be prevented by
(a) sugar (b) lemon juice
(c) both (d) none

29. ______ substances hold the cells of fruits together
(a) protein (b) cellulose
(c) pectic (d) fat

30. In immature fruits, pectin is present as
(a) myopectin (b) protopectin
(c) propectin (d) pectin

31. On ripening of the fruit, protopectin changes to
(a) pectin (b) pectic acid
(c) pectinic acid (d) both (a) & (b)

32. Upon ripening the acid content of fruit
(a) increases
(b) decreases
(c) remains constant
(d) none

33. On ripening the development of ______ occurs
(a) acidity (b) sweetness
(c) sourness (d) none

34. Upon ripening the protein content of fruit
(a) increases
(b) decreases
(c) remains constant
(d) none

35. Upon ripening the nineral content of fruit
(a) increases
(b) decreases
(c) remains constant
(d) none

36. Upon ripening the fat content of fruit
(a) increases (b) decreases
(c) remains constant
(d) none

37. On maturity, vitamin C content of fruit
(a) increases
(b) decreases
(c) remains constant
(d) none

38. A relative humidity of ____% promotes the retention of water in the fruit cells
(a) 10-15 (b) 15-30
(c) 55-60 (d) 85-90

39. The respiration of fruits ______ after harvesting
(a) continues
(b) ceases
(c) increase manifolds
(d) none

40. The metabolic activities of fruit ______ at low temperature
(a) increases (b) decreases
(c) remains constant
(d) none

41. ______ oxygen levels decrease spoilage by retarding ripening process
(a) low (b) high
(c) 0% (d) none

42. ______ CO_2 levels decrease spoilage by retarding ripening process
(a) low (b) high
(c) 0% (d) none

43. The ripening of fruits is hastened by ______ gas
(a) methylene (b) ethylene
(c) CO_2 (d) N_2

44. The ambient storage temperature for bananas is_____ °C
(a) 2 (b) 4
(c) 10 (d) 16

45. Fruits are mature ______ of plants with heir seeds
(a) stigma (b) ovary
(c) receptacle (d) both (b) & (c)

46. Outermost layer of fruit is called
(a) epicarp (b) pericarp
(c) endocarp (d) none

47. Pectins are polymer of
(a) sugar & protein
(b) protein & fat
(c) sugar & acid derivatives
(d) fat & acid derivatives

48. The natural enzyme present in fruit that hydrolyze pectin is
(a) pectin methyl esterase
(b) pectin ethyl esterase
(c) pectin methyl peroxidase
(d) pectin ethyl peroxidase

49. Pectin methyl esterase is deactivated at _____ °C
(a) 5 (b) 20
(c) 40 (d) 80

50. The decrease in viscosity of fresh tomato paste with time is due to
(a) pectin methyl esterase
(b) pectin ethyl esterase
(c) pectin methyl peroxidase
(d) pectin ethyl peroxidase

51. Calcium pectates are
(a) water soluble
(b) water insoluble
(c) fat insoluble
(d) none

52. The pigments and their precursors found in fruits and vegetables occurs in
(a) chloroplast (b) chromoplast
(c) both (d) none

53. Which of the following is a group of pigments in fruits and vegetables
(a) chlorophyll (b) carotenoids
(c) anthocyanins (d) all

54. Chlorophyll is ______ soluble
(a) water (b) oil
(c) both (d) none

55. The metal bound to chlorophyll is
(a) calcium (b) magnesium
(c) phosphorus (d) tin

56. The release of bound Mg in the chlorophyll leads to change of chlorophyll to
(a) pheophytin (b) neophytin
(c) epiphytin (d) eophytin

57. Pheophytin is ________ in color
(a) red (b) brown
(c) purple (d) yellow

58. The addition of ______ salts can prevent the conversion of chlorophyll to pheophytin
(a) calcium (b) magnesium
(c) manganese (d) zinc

59. One molecule of β-carotene is converted to ______ molecules of vitamin A
(a) 1 (b) 2
(c) 3 (d) 4

60. Which of the following is not a precursor of vitamin A
(a) β-carotene (b) α-carotene
(c) chlorophyll (d) all

61. ________ pigments belong to group of plant pigments known as flavonoids
(a) chlorophyll (b) carotenoids
(c) anthocyanin (d) cryptoxanthin

62. Anthocyanins include ____________ pigments
(a) red & blue (b) blue & purple
(c) both (d) none

63. The anthocyanins are most sensitive to
(a) pH (b) water
(c) heat (d) none

64. ________ are responsible for the dark color of the bark of oat trees
(a) anthocyanins (b) chlorophyll
(c) tannins (d) carotenoids

65. Tannins are ________ soluble
(a) water (b) oil
(c) both (d) none

66. The color of tannin complexes is
(a) red (b) purple
(c) black (d) green

67. Important characteristic of tannins is
(a) astringency (b) pungency
(c) sweetness (d) sourness

68. Betalains is present in
(a) apple (b) red beet
(c) papaya (d) grapes

69. Betalains are _______ soluble
(a) water (b) oil
(c) both (d) none

70. During plant photosynthesis, _______ is given out
(a) O_2 (b) CO_2
(c) N_2 (d) all

71. During plant respiration, _______ is given out
(a) O_2 (b) CO_2
(c) N_2 (d) all

72. Upon harvesting, the amount of ________ increase in vegetables
(a) amino acids (b) fats
(c) starch (d) sugar

73. Unripe fruit has
(a) high starch, low sugar
(b) low starch, high sugar
(c) equal levels of starch & sugar
(d) low starch & low sugar

74. Upon harvesting, the water insoluble pectic substance in fruits _____
(a) increase (b) decrease
(c) no change (d) none

75. Upon harvesting, the water soluble pectin substance in fruits _____
(a) increase (b) decrease
(c) no change (d) none

76. The softening of fruits during ripening and storage is due to
(a) pectic substances
(b) pectin
(c) protopectin
(d) none

77. The organic acid contents ______ upon ripening of fruits
(a) increase (b) decrease
(c) no change (d) none

78. The quality decline in stored respiring fruits is termed as
(a) deterioration (b) ripening
(c) senescence (d) none

79. ______ % of oxygen decrease the spoilage by retarding the ripening process of fruits
(a) 0.1-0.2 (b) 1-2
(c) 2-3 (d) 10-20

80. ______ % of CO_2 decrease the spoilage by retarding the ripening process of fruits
(a) 0.03-0.3 (b) 0.3-3
(c) 1-3 (d) 4-5

81. In berry, the edible part is
(a) epicarp (b) pericarp
(c) endocarp (d) thalamus

82. The acid content (malic, citric & benzoic) is ______ ib berries
(a) high (b) low
(c) no acid present (d) none

83. The scientific name of strawberry is
(a) Fragaria virginiana
(b) Embilica officinalisa
(c) *Malus sylvestris*
(d) *Musa paradisiacal*

84. The edible part of strawberry is
(a) receptacle (b) juicy pericarp
(c) epicarp (d) endocarp

85. Cranberry is ______ in color
(a) red (b) blue
(c) purple (d) yellow

86. Logan berries are hybrid of
(a) blue & cran berries
(b) black & blue berries
(c) black & raspberries
(d) none

87. The white portion of citrus fruits is called
(a) albedo (b) malbedo
(c) palbedo (d) nalbedo

88. The scientific name of musambi is
(a) Citrus sinensis
(b) Citrus reticulate
(c) Citrus limon
(d) Citrus parodist

89. Sweet orange or Musambi originated in
(a) India (b) USA
(c) Sri Lanka (d) China

90. Jaffa, Valencias and Washington navel are varieties of
(a) apple (b) sweet orange
(c) lemon (d) guavas

91. Tangerine is also known as
(a) lemon (b) mandarin
(c) sweet orange (d) lime

92. Santra, Kamala and Kinnow are varieties of
(a) mandarin (b) sweet orange
(c) lemon (d) lime

93. A grape fruit is a hybrid between
(a) pummelo & sweet orange
(b) lime & lemon
(c) sour & sweet orange
(d) tangerine & lime

94. Apricots belong to
(a) drupes (b) berries
(c) citrus fruits (d) pomes

95. Apricots, cherries, peaches and plums belong to genus
(a) prunus (b) citrus
(c) rubus (d) none

96. The high alcohol content liquor obtained from sweet cherry is known as
(a) kirschwasser (b) lemonade
(c) sake (d) none

97. Free stones and cling stones are examples of
(a) peach (b) cherry
(c) plums (d) apricots

98. The oldest cultivated plant among the following is
(a) cherry (b) peach
(c) grape (d) plum

99. The cultivation of grapes is called
(a) pisciculture (b) miscipculure
(c) viticulture (d) none

100. Bangalore blue, Bhokri, Gulabi and Thompson seedless are examples of
(a) apricots (b) plum
(c) mangoes (d) grapes

101. The ripe grapes contain
(a) sucrose
(b) glucose & fructose
(c) both
(d) none

102. The principal acid of grapes is
(a) citric & malic
(b) tartaric & citric
(c) tartaric & malic
(d) none

103. The green color of grapes is due to
(a) carotenoids (b) chlorophyll
(c) anthocyanin (d) all

104. The red, blue and black color of grapes is due to
(a) carotenoids (b) chlorophyll
(c) anthocyanin (d) all

105. Melons are commonly eaten in ________ form
(a) processes
(b) heat processed
(c) raw (d) canned

106. Musk, Cantaloupe and Honey dew are varieties of
(a) grapes (b) apple
(c) melon (d) pears

107. The edible portion of pear is
(a) receptacle (b) seed
(c) pericarp (d) thalamus

108. Ambri Kashmiri, Golden Delicious and Baldwin are varieties of
(a) pear (b) apple
(c) plum (d) grapes

109. Principal sugar of apple is
(a) sucrose (b) fructose
(c) glucose (d) all

110. The principal acid of apple is
(a) citric (b) tartaric
(c) malic (d) none

111. Cider, a wine is prepared from
(a) grapes (b) apple
(c) pears (d) plums

112. Indian Gooseberry is also known as
(a) grape (b) amla
(c) peach (d) lime

113. Avocado is also known as
(a) butterfruit (b) acidfruit
(c) malayafruit (d) none

114. Avocados are particularly rich in
(a) fat (b) protein
(c) both (d) none

115. The cholesterol levels of avocados are
(a) 0 (b) high
(c) low (d) none

116. Nira, a sweetend drink is obtained from
(a) grapes (b) dates
(c) pears (d) mango

117. Kapa and Barka are varieties of
(a) mango (b) papaya
(c) jamuns (d) jackfruit

118. The extracts of _____ obtained from barks and seeds are used in treatment of diabetes
(a) jack fruit (b) guava
(c) jambu fruit (d) papaya

119. Mango is a
(a) drupe (b) berry
(c) citrus fruit (d) pome

120. Bombay Green, Chowsa, Pairi, safeda are varieties of
(a) mango (b) grapes
(c) plums (d) papaya

121. Papapya is principal source of enzyme
(a) papain (b) esterase
(c) cathepsins (d) all

122. Papain is commercially used for
(a) tenderization of meat
(b) fixation of red color of meat
(c) removal of fat from meat
(d) all

123. Wood apple is also known as
(a) elephant apple (b) monkey fruit
(c) curd fruit (d) all

124. Kiwi is a rich source of
(a) vitamin A (b) vitamin B
(c) vitamin C (d) vitamin d

125. Paw-paw is another name for
(a) wood apple (b) jack fruit
(c) custard apple (d) none

126. _______ fruit grows on a cactus vine
(a) nectarines (b) Paw-paw
(c) Ximenia (d) Pitahaya

127. Simple fruits are formed of _________
(a) single ovary
(b) multiple ovary
(c) multiple pistil
(d) accessory part of fruit

128. Clusters of flowers which form multiple fruit is called
(a) inflorescence (b) drupelets
(c) pistils (d) none

129. Raspberry is an example of _______ fruit
(a) simple (b) multiple
(c) aggregate (d) none

130. Pineapple is an example of
(a) simple (b) multiple
(c) aggregate (d) none

131. Cucumber is an example of
(a) simple (b) multiple
(c) aggregate (d) none

132. Scientific name of apple is
(a) *Malus sylvestris*
(b) *Musa paradisiacal*
(c) *Magnifera indica*
(d) none

133. Scientific name of banana is
(a) *Malus sylvestris*
(b) *Musa paradisiacal*
(c) *Magnifera indica*
(d) none

134. Scientific name of mango is
(a) *Malus sylvestris*
(b) *Musa paradisiacal*
(c) *Magnifera indica*
(d) none

135. Scientific name of grapes is
(a) *Vitia vinifera*
(b) *Musa paradisiacal*
(c) *Magnifera indica*
(d) none

136. Scientific name of wood apple is
(a) *Malus sylvestris*
(b) Limonia acidissima
(c) *Magnifera indica*
(d) none

137. A fruit formed without fertilization of ovary is called_____ fruit
(a) bisexual
(b) unisexual
(c) parthenocarpic
(d) none

138. In coconut the fruit is called
(a) drupe (b) pome
(c) berry (d) none

139. The seeds of the fruit are actually matured
(a) ovary (b) ovules
(c) stigma (d) calyx

140. The potato belongs to genus
(a) Solanaceae (b) Lilaceae
(c) Asteraceae (d) none

141. The food fro growing plant is stored in
(a) cotlydons (b) radical
(c) plumule (d) hilum

142. Tomatoes are stored and shipped at temperatures between °F
(a) 0-90 (b) 15-20
(c) 50-65 (d) 0-25

143. Only LAB can ferment sugars and nutrients in pickles because they
(a) use a naturally occurring enzyme
(b) are tolerant of salt levels
(c) produce lactic acid
(d) use acetic acid

144. Flavor is sensed by taste buds which are sensory organs located on tongue. The taste buds on the sides of the tongue respond to
(a) sweet (b) bitter
(c) salty (d) sour

145. Lemon juice has a pH of
(a) 7.2 (b) 8.1
(c) 2.3 (d) 6.8

146. Processing loss of following can be as high as 100% in case of
(a) vitamin K (b) vitamin D
(c) vitamin C (d) none

147. Mangoes are stored at temperature ______ °C
(a) 5.5 14 (b) 14-17
(c) 20-25 (d) 25-30

148. Internal browning of apples occur at temperature
(a) below 3°C (b) above 3°C
(c) above 5°C (d) above 10°C

149. ERH stands for
(a) Equillibrium Relative Humidity
(b) Equillibrium Room Humidity
(c) Equal Relative Humidity
(d) Equal Room Humidity

150. The a_w of most of the fresh fruits and vegetables is in the range of
(a) 0.90-0.92 (b) 0.88-0.90
(c) 0.95-0.98 (d) 0.98-1.0

151. Vegetables sown in winter season are known as
(a) rabi (b) kharif
(c) both (d) none

152. Vegetables sown in summer season are known as
(a) rabi (b) kharif
(c) both (d) none

153. The strong flavor of ginger and onion is due to ________ containing volatile compounds
(a) phosphorus (b) sulphur
(c) iron (d) zinc

154. Except for legumes, vegetables are low in
(a) carbohydrates (b) sugar
(c) protein (d) vitamin

155. The fat content of vegetables is
(a) high (b) low
(c) very high (d) none

156. Agathi is the richest source of
(a) calcium (b) iron
(c) phosphorus (d) riboflavin

157. Agathi, curry leaves and drum stick are particularly rich in
(a) water (b) protein
(c) phosphorus (d) iron

158. Among thefollowing, the richest source of vitamin C is
(a) beetroot (b) drumstick
(c) chillies (d) bittergourd

159. Carrots are extremely rich in
(a) calcium (b) iron
(c) phosphorus (d) riboflavin

160. Leaves of colocasia are rich in
(a) calcium (b) iron
(c) carotene (d) thiamin

161. Fenugreek and colocasia leaves are rich in
(a) protein (b) iron
(c) calcium (d) niacin

162. Cabbage belongs to genus
(a) Brassica (b) Daucus
(c) Solanum (d) none

163. Sauerkraut is prepared from
(a) cauliflower (b) cabbage
(c) spinach (d) onion

164. The principal pigment present in red cabbage is
(a) carotenoids
(b) chlorophyll
(c) anthocyanin
(d) cryptoxanthin

165. The Chinese cabbage is known as
(a) Pak-choi (b) Tak-thoi
(c) Cak-phoi (d) none

166. Turnip rooted cabbage is better known as
(a) Knol-khol (b) Khol-rabi
(c) both (d) none

167. Khol-rabi is a german word which means
(a) cabbage-turnip
(b) turnip-cabbage
(c) radish-turnip
(d) none

168. The anticarcinogenic properties of cabbage, cauliflower, broccoli is due to
(a) glucosinolates
(b) isothiocyanates
(c) both
(d) none

169. Cabbage, kale, cauliflower belong to ________ group of crops
(a) root (b) cole
(c) fruit (d) curcubits

170. In radish, the edible portion is
(a) flower (b) bud
(c) leaves (d) root

171. Carrots are rich source of
(a) vitamin A (b) carotene
(c) vitamin D (d) vitamin K

172. In parsnip and turnip, the edible portion is
(a) flower (b) bud
(c) leaves (d) root

173. The scientific name of tomato is
(a) Lycopersicon esculentum
(b) Hibiscus esculentus
(c) Manihot esculenta
(d) none

174. Brinjal is also known as ________ plant
(a) hen (b) egg
(c) cock (d) duck

175. The scientific name of okra is
(a) Hibiscus esculentus
(b) Lycopersicon esculentum
(c) Dioscorea
(d) none

176. Cucumbers, pumpkins & gourds belong to family
(a) Solanum
(b) Brassica
(c) Daucus
(d) Cucurbitaceae

177. Ridged, bottle, bitter, snake and sponge are varieties of
(a) potato (b) radish
(c) gourd (d) carrot

178. The edible portion of agathi is
(a) leaves (b) bud
(c) flower (d) carrot

179. Parsley is a
(a) aromatic herb (b) fruit
(c) tuber (d) cole crop

180. "Eyes" are found in
(a) parsley (b) carrot
(c) potato (d) agathi

181. Potato and tapioca are particularly rich in
(a) protein (b) starch
(c) minerals (d) fat

182. Yams contain _____% starch
(a) 2 (b) 10
(c) 21 (d) 98

183. Onion, garlic and leek belong to ________ group
(a) tuber (b) cole
(c) bulb (d) bud

184. Rhubarb is
(a) vegetable (b) fruit
(c) herb (d) none

185. Rhubarb contain toxic amounts of __________ acid
(a) citric (b) oxalic
(c) malic (d) none

186. Edible mushrooms are
(a) poisonous (b) non-poisonous
(c) both (d) none

187. The skin of vegetables are removed by
(a) hot dipping in alkali solution
(b) lye solution
(c) sun drying
(d) both (a) & (b)

188. The process of heat treatment to inactivate the natural enzyme of vegetables or fruits is called
(a) blanching (b) scalding
(c) pasteurization (d) none

189. Blanching is a _____ treatment
(a) heat
(b) cold
(c) freezing
(d) irradiation

190. The most heat resistance enzymes are
(a) catalase & lipase
(b) lipase & peroxidase
(c) catalase & peroxidase
(d) none

191. Smaller vegetables require ______ blanching times than larger ones
(a) longer (b) shorter
(c) equal (d) none

192. Vegetables are canned in
(a) sugar syrup (b) salt brine
(c) water (d) none

193. Fruits are canned in
(a) sugar syrup (b) salt brine
(c) water (d) none

194. Maturity and ripening are similar
(a) true (b) false
(c) cannot be determined
(d) none

195. Maturity comes _____ than ripeness
(a) sooner (b) later
(c) equal (d) none

196. As an orange ripens on tree
(a) sugar increase, acid decrease
(b) sugar decrease, acid increase
(c) sugar & acid levels are equal
(d) none

197. The TSS of fruit juice is measured by
(a) refractometer
(b) hydrometer
(c) thermometer
(d) both (a) & (b)

198. The per centage of sugars in juice is expressed in
(a) degree (b) °C
(c) brix (d) bricks

199. Higher the brix, _____ is the sugar concentration in juice
(a) higher
(b) lower
(c) no effect on sugar concentration
(d) none

200. _________ is used to stabilize the color of the fresh and processed fruits and vegetables
(a) CO_2 (b) CO
(c) SO_2 (d) N_2

201. Maillard browning in fruits is a ______ reaction
(a) enzymatic
(b) non-enzymatic
(c) temperature induced
(d) none

202. Ascorbic acid in fruits act as a powerful
(a) antidepressant (b) antidiabetic
(c) antioxidant (d) antibacterial

203. Use of SO_2 _________ non-enzymatic maillard browning by reacting with aldehyde groups of sugar so that they are no longer free to combine with amino acids
(a) lower
(b) increase
(c) no effect on browning
(d) none

204. Peels of citrud fruits is particularly rich in
(a) bitter oils (b) citric acid
(c) iron (d) calcium

205. Fruit juices are rich in
(a) vitamin C (b) vitamin E
(c) vitamin B (d) all

206. The firmness, softness, juiciness, chewiness of fruits and vegetables are _______ factors
(a) appearance (b) flavor
(c) textural (d) color

207. Wholeness, damage, gloass, color, transparency of fruit and vegetables are ______ factors
(a) appearance (b) flavor
(c) textural (d) color

208. Sweet, sour, salty, bitter and aromas of fruits and vegetables are ___ factors
(a) appearance (b) flavor
(c) textural (d) color

209. Flavor of a food product is a combination of
(a) taste & smell
(b) taste & appearance
(c) smell & appearance
(d) none

210. The color of fruit extract is measured by
(a) colorimeter
(b) spectrophotometer
(c) tintometer (d) all

211. Consistency is related to _______ of fruit
(a) appearance (b) flavor
(c) texture (d) color

212. The shelf life of a food is the time for which the food is in __________ form
(a) fresh (b) mature
(c) acceptable (d) ripe

213. The major causes of deterioration of fruits and vegetables is
(a) bacteria (b) yeast
(c) molds (d) all

214. Canning of fruits ans vegetables is a ______ process
(a) heat (b) cold
(c) irradiation (d) microwave

215. Blanching of fruits and vegetable is used to
(a) inactivate enzymes
(b) increase sweetness
(c) improve the vitmanin content
(d) none

216. The target organism for canning is
(a) Bacillus cereus (b) *E. coli*
(c) Clostridium botulinum
(d) all

217. The storage life of fruits and vegetables ________ on cold storage
(a) increase (b) decrease
(c) no effect (d) none

218. CAS used for storage of apples and other fruits is called
(a) Controlled Atmosphere Storage
(b) Continued Atmospheric Storage
(c) Cold Atmosphere Storage
(d) none

219. Stored fruits and vegetables ______ oxygen during storage
(a) consume (b) give away
(c) release (d) none

220. CAS ________ the over ripening of fruits and vegetables
(a) prevents (b) allows
(c) has no effect (d) increase

221. In CAS the concentration of oxygen in the package is ______
(a) reduced (b) increased
(c) no effect (d) none

222. In CAS the concentration of oxygen in the package is ______
(a) reduced (b) increased
(c) no effect (d) none

223. In CAS the temperature of the package is ________
(a) reduced (b) increased
(c) no effect (d) none

224. The per centage of O_2 is reduced from 21% to _______%
(a) 0.03 (b) 0.3
(c) 3 (d) 20

225. The per centage of CO_2 is increaed from 0.03% to ______%
(a) 0.03 (b) 0.3
(c) 3 (d) 20

226. The purge gas used in CAS is
(a) CO_2 (b) O_2
(c) N_2 (d) none

227. The CA storage at reduced pressure is known as ________ storage
(a) hyperbaric (b) hypobaric
(c) isobaric (d) none

228. McIntosh apples store well at ____°C
(a) 0.1-1 (b) 1-3
(c) 2-5 (d) 10-20

229. Delicious apples store well at ____°C
(a) 0 (b) 10
(c) 20 (d) 30

230. The injury caused to the fruits and vegetables when they are stored at lower temperature is called ________ injury
(a) chilling (b) freezing
(c) low temperature (d) antifreezing

231. The sweetness of corn ________ on storage
(a) increase (b) decrease
(c) remains same (d) none

232. The sweetness of the corn is reduced cue to conversion of
(a) starch to sugar
(b) sugar to starch
(c) sugars to organic acid
(d) none

233. Upon refrigeration, the vitamin content of fruts and vegetables
(a) increase (b) decrease
(c) remains same (d) none

234. BTU stands for
(a) British Thermal Units
(b) Bureau of Thermal Units
(c) British Technical Units
(d) none

235. 1BTU is equivalent to
(a) 152 cal (b) 100 cal
(c) 252 cal (d) 300 cal

236. The peas are freezed by ____ freezer
(a) still air sharp (b) blast
(c) fluidized bed (d) none

237. The motion of peas during fluidized bed freezing is called
(a) jumping (b) cooling
(c) boiling (d) vibrating

238. The term IQF stands for
(a) Individually Quick Frozen
(b) Ideal Quick Frozen
(c) Individually Quick Fried
(d) none

239. Which of the following is preserved by IQF
(a) cauliflower (b) potatoes
(c) strawberry (d) sapota

240. Food dehydration differs from sun drying in that the former is ______ drying
(a) natural (b) artificial
(c) man-made (d) both (b) & (c)

241. Food dehydration refers to ______ moisture removal from fruits and vegetables
(a) almost complete (b) complete
(c) little (d) no

242. The moisture content in the commodity undergoing food dehydration is ______%
(a) 0.1-0.2 (b) 0.1-0.5
(c) 1-5% (d) 10-15

243. Which of the following is not a function of dehydration
(a) decrease the weight and bulk
(b) improve nutrient availability
(c) preservation
(d) all

244. __________ is known as ripening hormone
(a) methylene (b) ethylene
(c) propene (d) gibberllin

245. ISO stands for
(a) International Organization for Standardization
(b) Indian Standards Organization
(c) Indian Standardization Organization
(d) International Standardization Organization

246. According to ISO, which of the following is not a property of spice
(a) it is dried seed, root or bark
(b) used in bulk quantity
(c) used as an additive
(d) all

247. Which of the following spice is used for pungency
(a) cinnamon (b) clove
(c) pepper (d) saffron

248. Which of the following spice is not used for pungency
(a) ginger (b) chillies
(c) mustard (d) cinnamon

249. Which of the following spice is an aromatic spice
(a) clove (b) turmeric
(c) mustard (d) fenugreek

250. Which of the following spice is not an aromatic spice
(a) cardamom (b) nutmeg
(c) mace (d) paprika

251. Clove consist of
(a) eugenol (b) allicin
(c) menthol (d) none

252. Which of the following spice is used for color
(a) paprika (b) pepper
(c) cassia (d) anise

253. Which of the following spice is not used for color
(a) paprika (b) saffron
(c) turmeric (d) celery

254. Phytotherapy involves the use of
(a) plants (b) animals
(c) chemicals (d) radiation

255. Oleoresins means
(a) oils of spices
(b) carbohydrates of spices
(c) extracted nutrients
(d) none

256. King of spices is
(a) black pepper (b) clove
(c) cinnamon (d) cardamom

257. ________ is the major producer of black pepper in India
(a) Karnataka
(b) Kerala
(c) Tamil Nadu
(d) Andhra Pradesh

258. Black pepper belongs to
(a) drupe (b) pome
(c) berry (d) none

259. The spice value of black pepper is due to
(a) phenols (b) esters
(c) alkaloids (d) none

260. The per centage of alkaloids in black pepper is
(a) 1-2% (b) 2-3%
(c) 4-5% (d) 10-11%

261. White pepper has _____ flavor than black pepper
(a) more (b) less
(c) equal (d) none

262. White pepper has _______ pungency than black pepper
(a) more (b) less
(c) equal (d) none

263. The characteristic odor of pepper is due to _________ present in pericarp of pepper
(a) volatile oil (b) terpenes
(c) alkaloids (d) all

264. Which of the following in not a variety of pepper cultivated in India
(a) white (b) black
(c) green (d) brown

265. Major and Minor are two distinct varieties of
(a) pepper (b) cardamom
(c) cinnamon (d) clove

266. The queen of spices is________
(a) pepper (b) cardamom
(c) cinnamon (d) clove

267. Which of the following countries produces a major fraction of cardamom
(a) India (b) Sri Lanka
(c) Guatemala (d) Tanzania

268. Moisture content of cardamom is _____ %
(a) 7-10 (b) 2-4
(c) 10-15 (d) 15-20

269. Ether extracts of cardamom is _________ %
(a) 2-3.5 (b) 7-15
(c) 25-35 (d) 45-60

270. Carbohydrate content of cardamom is _____ %
(a) 10-19 (b) 29-39
(c) 39-49 (d) 59-69

271. Niacin content of cardamom is _____ mg/100g
(a) 2.3 (b) 20.3
(c) 0.3 (d) 30.3

272. The spice value of cardamom depe-nds on volatile oils present in _______
(a) seeds (b) seed cover
(c) stem (d) leaf

273. The principal constituent of volatile oil in cardamom is
(a) terpenes (b) alkaloids
(c) esters (d) alcohols

274. The quality of cardamom oil is related to ________ content
(a) terpinyl acetate (b) ethyl alcohol
(c) oil (d) starch

275. Ginger is modified
(a) root (b) underground stem
(c) fruit (d) bud

276. The important constituent of ginger oil is
(a) terpenes (b) esters
(c) alkaloids (d) ethers

277. The essential pungent oil of ginger is
(a) volatile (b) non volatile
(c) both (d) none

278. The trade name of oleoresin from ginger is
(a) gingerin (b) gingerlin
(c) gingjeng (d) none

279. The pungent constituents of ginger are
(a) oxymethyl phenols
(b) terpinyl acetate
(c) ethyl alcohol
(d) none

280. Of the two main species Capsicum annum and Capsicum frutescens, the species commonly grown in India is
(a) Capsicum annum
(b) Capsicum frutescens
(c) both
(d) none

281. The pungent principle in chillies is
(a) capsaicin (b) eugenol
(c) both (d) none

282. Chillies are ________ in vitamin C
(a) high (b) low
(c) deficient (d) none

283. Tabasco sauce is prepared from
(a) bird chilli (b) hen chilli
(c) both (d) none

284. Tumeric is a
(a) tuber (b) rhizome
(c) modified root (d) none

285. Tûrmeric belongs to genus
(a) Curcuma (b) Allium
(c) Pimpinella (d) Carum

286. The principal oil present in turmeric is
(a) sesquiterpene ketones
(b) tertiary alcohol
(c) esters
(d) both (a) & (b)

287. Turmeric is an
(a) antidepressant (b) antipyretic
(c) antioxidant (d) all

288. Cinnamon is obtained from
(a) bark (b) fruit
(c) leaves (d) bud

289. The seeds of _____ have hypoglycemic effect
(a) fenugreek (b) ajowan
(c) cinnamon (d) aniseed

290. Garlic is
(a) tuber (b) bud
(c) bulb (d) stem

291. The flavoring comound of garlic is
(a) capsaicin (b) allicin
(c) eugenol (d) none

292. Allicin is a ________ containing compound from garlic
(a) sulfur (b) iron
(c) magnesium (d) calcium

293. White and black are principal types of
(a) mustard (b) garlic
(c) fennel (d) fenugreek

294. The nutmeg oil consists of
(a) terpenes (b) esters
(c) alcohols (d) both (a) & (b)

295. Nutmeg butter is a type of
(a) oil (b) butter
(c) buttermilk (d) all

296. White, yellow and red are varieties of
(a) nutmeg (b) mace
(c) onion (d) asafetida

297. The spice value of onion is due to the presence of ________ compounds
(a) sulfur (b) zinc
(c) iron (d) phosphorus

298. Saffron is obtained from
(a) dried stigma (b) dried ovary
(c) dried pistil (d) dried carpel

299. The coloring principle of saffron is
(a) crocin (b) betalains
(c) anthocyanin (d) all

300. Chemically crocin is
(a) glycoside (b) ester
(c) ether (d) protein

301. Cloves are
(a) tuber (b) bud
(c) bulb (d) stem

302. Principal agent in clove is
(a) eugenol (b) crocin
(c) allicin (d) none

303. The oil obtained from clove is
(a) volatile (b) non volatile
(c) both (d) none

304. Menthol is obtained from
(a) mint (b) nutmeg
(c) mace (d) none

305. The flavoring compound obtained from vanilla is
(a) vanilinin (b) vanin
(c) vanillin (d) valin

306. The flavor of asafoetida is due to ________ compounds
(a) sulfur (b) iron
(c) magnesium (d) calcium

307. The moisture content of coconut is ______ %
(a) 2-3 (b) 30-40
(c) 50-70 (d) >90

308. Good source of protein is
(a) almond
(b) cashewnut
(c) groundnut
(d) all

309. 100 g of groundnut contains ______ g of protein
(a) 2.5 (b) 25
(c) 45 (d) 65

310. Among he following, highest fat content is present in
(a) almond (b) groundnut
(c) walnut (d) cashewnut

311. The fat % in walnut is
(a) 6.4 (b) 16.4
(c) 36.4 (d) 64.4

312. Among the following, the major source of carbohydrates is
(a) almond (b) chestnut
(c) coconut (d) walnut

313. ________ is a rich source of iron and carotene
(a) almond (b) cashewnut
(c) ground nut (d) pistachio nut

314. Nuts are poor source of
(a) fats (b) proteins
(c) vitamin C (d) minerals

315. Acrid juice is obtained from
(a) almond (b) coconut
(c) cashewnut (d) walnut

316. The largest producer of cashewnut in the world is
(a) India (b) Srilanka
(c) Iran (d) Israel

317. The oil of cashewnut is rich in ______ acid
(a) oleic (b) linoleic
(c) palmitic (d) both (a) & (b)

318. The cashew shell oil consists of
(a) anacardic (b) cardol acid
(c) both (d) none

319. The fermented liquor 'kaju' is made from) appl
(a) cashewnut
(b) cashew leaves
(c) cashew apple
(d) none

320. Distilled 'kaju' is known as
(a) feni (b) wine
(c) brandy (d) whisky

321. Which of the following is called 'Kalpa Vriksha'
(a) cashewenut (b) walnut
(c) coconut (d) none

322. As coconut ripens, the thickness of the kernel
(a) increase (b) decrease
(c) remains same (d) none

323. The coconut water is present in
(a) exocarp (b) endosperm
(c) endocarp (d) none

324. Copra is obtained from
(a) cashewnut (b) walnut
(c) coconut (d) peanut

325. Copra is obtained from
(a) ripe coconut
(b) unripe coconut
(c) coconut leaves
(d) none

326. White part of the edible coconut is called
(a) ham (b) meat
(c) sausage (d) hull

327. The coconut water is rich in
(a) potassium (b) calcium
(c) iron (d) phosphorus

328. The coconut water is rich in
(a) ascorbic acid (b) oxalic acid
(c) malic acid (d) folic acid

329. The coconut sap mainly consists of
(a) fructose (b) sucrose
(c) maltose (d) lactose

330. The coconut sap is also known as
(a) peera (b) jeera
(c) neera (d) beera

331. Toddy, a beverage with 4% alcohol is obtained from
(a) cashewnut (b) walnut
(c) coconut (d) peanut

332. A distilled fermented toddy is known as
(a) barrack (b) arrack
(c) gin (d) wine

333. Arrack contains about ____ % alcohol
(a) 10 (b) 25
(c) 35 (d) 45

334. Groundnut is similar to
(a) cashewnut (b) walnut
(c) coconut (d) peanut

335. Peanut originated from
(a) Taiwan (b) India
(c) Srilanka (d) Brazil

336. India ranks ____ in production of groundnut
(a) 1 (b) 2
(c) 3 (d) 4

337. Groundnut contains ____ % protein
(a) 4 (b) 9.7
(c) 20.6 (d) 59

338. Peanut butter has ____ % fat
(a) 18 (b) 28
(c) 38 (d) 48

339. Peanut is not available in ____ form
(a) butter (b) oil
(c) flour (d) distilled liquor

340. Which of the following is bitter almond
(a) *P. amygdalus var amara*
(b) *P. amygdalus var dulcis*
(c) both (a) & (b)
(d) none

341. Which of the following is sweet almond
(a) *P. amygdalus var amara*
(b) *P. amygdalus var dulcis*
(c) both
(d) none

342. __________ is used for extraction of almond oil
(a) amara (b) dulcis
(c) both (d) none

343. In India, most of the almond is cultivated in
(a) Kashmir (b) Kerala
(c) Gujarat (d) Assam

344. The bright golden color of the shell of the almond is due to bleaching by
(a) SO_2 (b) NO_2
(c) CO_2 (d) O_2

345. Almond is found in which of the following forms
(a) oil (b) milk
(c) paste (d) all

346. Almond is not found in which of the following forms
(a) milk (b) oil
(c) flour (d) all

347. The main almond protein is
(a) albumin (b) globulin
(c) zein (d) gluten

348. The bitter almond has ______ protein than sweet almond
(a) higher (b) lower
(c) similar (d) none

349. The bitter almonds are non-edible due to the presence of
(a) amygdalin (b) albumin
(c) globulin (d) none

350. The bitter almond differes from sweet almond in having
(a) amygdalin (b) zein
(c) globulin (d) both (a) & (c)

351. Unlike most fruit nuts, chestnuts are rich in
(a) starch & protein
(b) starch & sugar
(c) protein & fat
(d) starch & fat

352. The fat content of chestnut is
(a) very high (b) high
(c) low (d) absent

353. The chief walnut growing state in India is
(a) Kashmir (b) Bihar
(c) Andhra Pradesh (d) Kerala

354. The young green walnuts are richest sources of
(a) ascorbic acid (b) malic acid
(c) succinic acid (d) none

355. Which of the following oil is used in preparation of 'artist paint'
(a) chestnut (b) groundnut
(c) almond (d) walnut

356. Brazil nut is similar to
(a) walnut (b) hazelnut
(c) almond (d) none

357. Which among the following is best source of selenium
(a) Brazil nut (b) hazelnut
(c) walnut (d) chestnut

358. Brazil nut is extremely rich in
(a) protein
(b) carbohydrates
(c) minerals (d) fats

359. The fat content of Brazil nut is ____ %
(a) 37 (b) 47
(c) 57 (d) 67

360. Cobnut is another name for
(a) Brazil nut (b) hazelnut
(c) walnut (d) chestnut

361. Pecan nuts are
(a) edible (b) non-edible
(c) toxic (d) none

362. Which of the following is the richest source of protein
(a) pine nut (b) hazelnut
(c) Brazil nut (d) none

363. Pine nut has ______ % protein
(a) 31 (b) 21
(c) 41 (d) 51

364. Chestnuts are rich in
(a) protein (b) starch
(c) minerals (d) fat

365. Chestnut have high level of ______ ______ because of which they cannot be eaten raw
(a) tannic acid
(b) ascorbic acid
(c) malic acid
(d) ozalic acid

366. Coffee consists of high amount of
(a) catechins (b) caffeine
(c) proteins (d) fat

367. The largest producer of coffee in the world is
(a) Brazil (b) Srilanka
(c) Paraguay (d) China

368. Coffe occurs in ______ form
(a) berry (b) pome
(c) drupe (d) none

369. Raw or green coffee has ______ flavor or aroma
(a) most (b) low
(c) no (d) none

370. ______ gives aroma to coffee
(a) boiling (b) roasting
(c) frying (d) stewing

371. The caffeine content of coffee is ______ %
(a) 1-2.5 (b) 2-5.5
(c) 5-7.5 (d) 7-9.5

372. Caffeine adds ______ to coffee
(a) sweetness (b) stimulation
(c) bitterness (d) both (b) & (c)

373. One cup of coffee (150 ml) has ______ mg caffeine
(a) 1 (b) 10
(c) 100 (d) 1000

374. The aroma of coffee is due to group of substances present in coffee known as ______
(a) coffeol (b) catechol
(c) caffiol (d) none

375. The astringency of coffee is due to
(a) malic acid
(b) chlorogenic acid
(c) citric acid
(d) oxalic acid

376. The caffeine content in green coffee is ______ than roasted coffee
(a) higher (b) lower
(c) similar (d) none

377. The content chlorogenic acid content in green coffee is ______ than roasted coffee
(a) higher (b) lower
(c) similar (d) none

378. The bitterness of coffee is due to
(a) tannins (b) catechol
(c) caffeine (d) none

379. The best temperature for preparation of coffee is ______ °C
(a) 65-75 (b) 75-85
(c) 85-95 (d) 95-105

380. Which of the following is a type of coffee
(a) drip (b) steeped
(c) French drip (d) all

381. ______ is the substitute for coffee
(a) chicory (b) chicken
(c) chalk (d) none

382. The scientific name of tea is
(a) Camellia sinensis
(b) Coffea liberica
(c) Humulus lupulus
(d) none

383. Tea is obtained from
(a) root (b) bud
(c) leaves (d) stem

384. Which of the following is a type of tea
(a) black (b) green
(c) oolong (d) all

385. Which of the following is a fermented tea
(a) black (b) green
(c) oolong (d) all

386. Which of the following is a semi-fermented tea
(a) black (b) green
(c) oolong (d) all

387. Withering means ______ of tea leaves
(a) drying (b) frying
(c) roasting (d) brewing

388. Which of the following express best quality of tea leaves
(a) orange pekoe (b) pekoe
(c) souchang (d) none

389. Which of the following express poorest quality of tea leaves
(a) orange pekoe (b) pekoe
(c) souchang (d) none

390. Green tea is ______ tea
(a) fermented (b) unfermented
(c) semi-fermented (d) none

391. Which is a type of tea
(a) green (b) white
(c) yellow (d) all

392. Kukicha is a type of
(a) tea (b) coffee
(c) liquor (d) none

393. The flavor of the tea is due to
(a) caffeine (b) polyphenols
(c) essential oils (d) all

394. The caffeine content of tea is ____ %
(a) 1.3 (b) 2.3
(c) 3.3 (d) 4.3

395. The maximum amount of caffeine is present in
(a) bud (b) small leaf
(c) large leaf (d) none

396. Important polyphenols present in tea leaves
(a) catechins
(b) gallocatechins
(c) both
(d) none

397. Enzyme involved in fermentation of tea is
(a) PPO (b) lipase
(c) protease (d) none

398. Coffee and tea have ______ nutririve value
(a) no (b) high
(c) low (d) none

399. The polyphenol compounds of tea contribute to
(a) bitterness (b) aroma
(c) astringency (d) all

400. Which of the following is not a form of tea
(a) hot (b) iced
(c) instant (d) extract

401. The problem faced in ice tea is
(a) flavor (b) cloudiness
(c) astringency (d) all

402. The problem of cloudiness in ice tea is due to
(a) tannins (b) caffeine
(c) starch (d) none

403. Which of the following is an alternative to tea bag
(a) tea sticks (b) tea lollies
(c) tea cups (d) none

404. The chief cocoa producing country is
(a) Ghana (b) India
(c) Brazil (d) Sri Lanka

405. Most of the cocoa in ndia is grown in ________ states
(a) South Indian (b) North Indian
(c) North Eastern (d) N. Western

406. Which of the following is not fermented
(a) coffee (b) tea
(c) cocoa (d) none

407. Cotlydons of ________ are called nibs
(a) coffee (b) tea
(c) cocoa (d) none

408. According to ISI, the fat content of cocoa used for beverage has been fixed to ___ %
(a) 20 (b) 30
(c) 40 (d) 50

409. 'DutchProcess' in processing of cocoa originated in
(a) Holland (b) Netherland
(c) both (d) none

410. Plain chocolate contains ________ % sugar
(a) 20-25 (b) 30-35
(c) 35-40 (d) 40-55

411. Plain chocolate contains ______ % fat
(a) 20-22 (b) 30-32
(c) 32-42 (d) 42-52

412. Milk chocolate contains ______ % sugar
(a) 10-20 (b) 20-30
(c) 35-55 (d) 55-75

413. Milk chocolate contains ______ % fat
(a) 10-20 (b) 28-39
(c) 40-55 (d) 55-75

414. Milk chocolate contains ______ % milk solids
(a) 12 (b) 22
(c) 32 (d) 52

415. Cocoa beans contains ________ % moisture
(a) 2.13 (b) 12.3
(c) 22.3 (d) 33.3

416. Cocoa beans contains ______ % fat
(a) 34.68 (b) 44.68
(c) 54.68 (d) 64.68

417. Cocoa beans contains ________ % pentosans
(a) 11 (b) 21
(c) 31 (d) 41

418. Cocoa beans contains ________ % tannins
(a) 1 (b) 6
(c) 12 (d) 22

419. Cocoa beans contains ________ % theobromine
(a) 2.8 (b) 12.3
(c) 22.3 (d) 33.3

420. Cocoa beans contains ________ % caffeine
(a) 0.6 (b) 6.0
(c) 60.0 (d) 16

421. Which of the following is not a strained form of beverage
(a) coffee (b) tea
(c) cocoa (d) none

422. Cocoa is a rich source of
(a) fat
(b) protein
(c) minerals
(d) carbohydrates

423. Which of the following also yields a butter
(a) coffee (b) tea
(c) cocoa (d) none

424. Cocoa butter is rich in ________ fatty acids
(a) saturated (b) unsaturated
(c) PUFA (d) all

425. Which of the following is an artificial sweetener
(a) aspartame
(b) saccharin
(c) acesulfame potassium
(d) all

426. 'ppb' stands for
(a) parts per billion
(b) parts per barrel
(c) pieces per billion
(d) none

427. Which of the following is a class of soft drink
(a) gin (b) wine
(c) beer (d) none

428. HFCS stands for
(a) high fructose corn syrup
(b) high fat corn syrup
(c) high fructose corn sugar
(d) none

429. The acid mostlyused in soft drinks is
(a) malic (b) citric
(c) tartaric (d) succinic

430. Which of the following is an important constituent of soft drinks
(a) CO (b) O_2
(c) N_2 (d) CO_2

431. Soft drinks are packaged in PET bottles. PET stands for
(a) polyethylene terephthalate
(b) polyester terephthalate
(c) polyether terphthalate
(d) none

432. Strained juice containing moderate quantity of fruit pulp to which sugar is added is called
(a) fruit juice (b) fruit drink
(c) squash (d) cordial

433. _______ is prepared by liquefying the whole fruit
(a) fruit juice (b) fruit drink
(c) squash (d) cordial

434. Fruit squash from which all suspended material os removed and is clear is called
(a) fruit juice (b) fruit drink
(c) squash (d) cordial

435. _______ is prepared by mixing the desired fruit juices at time of serving
(a) fruit juice (b) fruit punch
(c) squash (d) cordial

436. Fruit juice which has been concentrated by removal of water is called
(a) fruit juice concentrate
(b) fruit punch
(c) squash
(d) cordial

437. Cooling drink of sweetened and diluted fruit juice is called
(a) fruit punch (b) sherbet
(c) cordial (d) fruit juice

438. Natural juice pressed out of fruit is known as
(a) fruit juice (b) fruit drink
(c) sherbet (d) fruit punch

439. Finning agents are used in juices to _______________
(a) produce clear juice
(b) add flavor
(c) remove bitterness
(d) all

440. The chemical preservatives added to fruit juices are
(a) sodium benzoate
(b) sodium bisulfate
(c) both (a) & (b)
(d) citric acid

441. Which of the following is distilled liquor
(a) beer (b) whisky
(c) tea (d) wine

442. Which of the following is not a distilled liquor
(a) whisky (b) gin
(c) vodka (d) none

443. Wine is chiefly produced from
(a) grapes (b) apple
(c) pineapple (d) sapota

444. Champagne got its name from distric of Chamagne in
(a) Brazil (b) France
(c) Guatemala (d) Holland

445. Champagne is also called
(a) tingling wine
(b) sparkling wine
(c) sprinkling wine (d) none

446. Port and sherry are types of
(a) brandy (b) beer
(c) whisky (c) fortified wine

447. Which of the following is fortified wine
(a) port (b) sherry
(c) vermouth (d) all

448. In preparation of red wine
(a) both juice & skin are used
(b) only juice
(c) only skin (d) none

449. Dry wine is
(a) sweet (b) not sweet
(c) dehydrated wine (d) none

450. Yeast can tolerate alcohol conten of ____%
(a) 4 (b) 8
(c) 12 (d) 16

451. Most wines have alcohol content __%
(a) 2-4 (b) 4-8
(c) 8-14 (d) none

452. Fortified wines contain ____% alcohol
(a) 2 (b) 5
(c) 10 (d) 20

453. Wines containing less than 14% alcohol are known as
(a) table wine (b) dessert wine
(c) fortified wine (d) none

454. Wines containing more than 14% alcohol are known as
(a) table wine (b) dessert wine
(c) fortified wine (d) both (b) & (c)

455. The juicy pulp of grapes for preparation of wines is called ______
(a) must (b) malt
(c) both (d) none

456. Fermentation of grapes for wine is done with
(a) bacteria (b) yeast
(c) virus (d) none

457. Starter culture for fermentation of grapes is
(a) Saccharomyces ellipsoideus
(b) Candida
(c) Torula
(d) none

458. Cider is fermented wine prepared from
(a) apple (b) sapota
(c) orange (d) pineapple

459. Beer is chiefly prepared from
(a) sorghum (b) barley
(c) wheat (d) maize

460. Ale, stout, lager and porter are types of
(a) wine (b) fortified wine
(c) beer (d) distilled liquor

461. ________ are responsible for bitterness in beer
(a) broken rice
(b) tapioca
(c) hops
(d) cereal adjunct

462. Scientific name of hop is
(a) Humulus lupulus
(b) Persea Americana
(c) Coffea robusta
(d) none

463. Hops are added during beer manufacture to
(a) add bitterness
(b) improve aroma
(c) sterlization of brew
(d) all

464. Ale is prepared by _____ fermentation
(a) top (b) bottom
(c) middle (d) none

465. Lager is prepared by __________ fermentation
(a) top (b) bottom
(c) middle (d) none

466. Yeast used for ale production is
(a) Saccharomyces cerevisiae
(b) Saccharomyces carlsbergensis
(c) both
(d) none

467. Yeast used for lager production is
(a) Saccharomyces carlsbergensis
(b) Saccharomyces cerevisiae
(c) both
(d) none

468. During fermentation of beer
(a) starch-sugar-alcohol
(b) sugar-starch-alcohol
(c) starch-alcohol-sugar
(d) none

469. Which of the following is top fermented beer
(a) ale (b) porter
(c) stout (d) all

470. Ale contains ___% alcohol than lager
(a) higher (b) lower
(c) similar (d) none

471. Beer has_____% alcohol content
(a) 4-5 (b) 1-2
(c) 10-20 (d) >20

472. Saki is _________
(a) beer (b) wine
(c) fruit juice (d) fermented milk

473. Saki is_____ beer
(a) Indian (b) Japanese
(c) African (d) Chinese

474. Saki is prepared from
(a) barley (b) wheat
(c) rice (d) maize

475. Alcohol content of sake is ______%
(a) 15-17 (b) 7
(c) 1-2 (d) 4-5

476. Toddy is prepared from
(a) barley (b) rice
(c) coconut (d) maize

477. Brandy is prepared from distillation of ________
(a) beer (b) wine
(c) toddy (d) saki

478. Rum is prepared from
(a) barley (b) sorghum
(c) molasses (d) rice

479. The alcohol content of distilled liquor is _____%
(a) 10 (b) 20
(c) 30 (d) 40

480. Spirits/ liquors are prepared by
(a) concentration (b) distillation
(c) dehydration (d) filtration

481. Cognac is a
(a) distilled liquor (b) beer
(c) wine (d) toddy

482. Vodka is a
(a) distilled liquor (b) beer
(c) wine (d) toddy

483. Scotch whisky is prepared from
(a) barley malt (b) sorghum
(c) rice (d) wheat

484. Brandy is distilled liquor from
(a) grape (b) apple
(c) sapota (d) coconut

485. Rum has alcohol content of _____%
(a) 10 (b) 20
(c) 30 (d) 40

186. Alcohol content of gin is ____%
(a) 20 (b) 40
(c) 60 (d) 80

487. Popular coffee palnt in India is
(a) Coffea robusta
(b) Coffea arabica
(c) Coffea liberica (d) none

488. Caffeine is not present in
(a) coffee (b) tea
(c) chocolates (d) none

489. Sparkling wine got its name due to the presence of excess of
(a) flavor
(b) carbondioxide
(c) alcohol
(d) none

490. Cereal adjuncts in form of maize, broken rice and raw barley are source of _____ in beer
(a) starch (b) proteins
(c) fat (d) none

491. _____ water is used for beer making
(a) hard (b) soft
(c) both (d) none

492. The extract obtained after steping in beer manufacture is called
(a) port (b) wort
(c) cort (d) sort

493. Beer contains
(a) CO_2 (b) no CO_2
(c) CO (d) none

494. Ale is _______ bitter than lager
(a) more (b) less
(c) equal (d) none

495. Stout has _____ hop taste than porter
(a) stronger (b) weaker
(c) similar (d) none

496. Beer has a calorific value of ______ Kcal/ 100ml
(a) 10 (b) 200
(c) 30 (d) 300

497. Gin is a distilled liquor flavored with
(a) hop (b) juniper
(c) both (d) none

498. CO_2 in solution contributes to
(a) acidity (b) basicity
(c) neutrality (d) none

499. In beverages CO_2 is found in _____ state
(a) solid (b) dissolved
(c) gas (d) none

500. The solubility of CO_2 is more at _____ temperature
(a) high (b) low
(c) not soluble (d) none

501. Chill haze is a defect caused in
(a) beer (b) wine
(c) tea (d) coffee

502. Decaffeinated coffee contains about _______ mg caffeine/ cup
(a) 3 (b) 30
(c) 300 (d) 100

503. India black tea has______ % caffeine
(a) <1 (b) 2-3
(c) 5-10 (d) 10-20

504. India black tea has ______ % tannin
(a) <1 (b) 2-3
(c) 6-10 (d) 10-20

505. India black tea has ______ % soluble solid
(a) <1 (b) 2-3
(c) 15-17 (d) 22-25

506. NHB stands for
(a) National Horticulture Board
(b) Non-Horticulture Board
(c) Nutritional Health Board
(d) none

507. NHB is situated at
(a) Gurgaon (b) Patna
(c) New Delhi (d) Hyderabad

508. IGPB stands for
(a) Indian Grape Producing Board
(b) Indian Grape Processing Board
(c) Indian Grape Production Bureau
(d) none

509. IICPT stands for
(a) Indian Institute of Crop Processing Technology
(b) Indian Institution of Crop Processing Technology
(c) Indian Institute of Crop Production Technology
(d) none

510. The brewer's yeast is
(a) Saccharomyces cerevisiae
(b) Candida
(c) Torulopsis
(d) all

511. Sauerruben is a fermented product of
(a) cucumber (b) olives
(c) cabbage (d) turnip

512. Kimchi is a fermented product of
(a) mixed vegetable in Chinese cabbage
(b) cucumber
(c) vegetables & milk
(d) vegetable & rice

513. Tarhana is a fermented product of
(a) mixed vegetable in Chinese cabbage
(b) cucumber
(c) vegetables & milk
(d) vegetable & rice

514. Kish K is a fermented product of
(a) rice (b) vanilla beans
(c) dough & milk (d) lettuce

515. Sajur asin is a fermented product of
(a) mixed vegetable in Chinese cabbage
(b) cucumber
(c) vegetables & milk
(d) vegetable & rice

516. Poi is a fermented product of
(a) taro (b) milk
(c) cabbage (d) rice

517. Which of the following is yeast fermented product
(a) pickle (b) sauerkraut
(c) beer (d) lettuce kraut

518. Which of the following is bacteria fermented product
(a) dill pickle (b) sauerkraut
(c) tarhana (d) all

519. Sonti is a fermented product of
(a) malt (b) fruit
(c) molasses (d) rice

520. Pulque is a fermented product of
(a) guava (b) agave
(c) molasses (d) malt

521. Pulque is a ______ fermented product
(a) LAB (b) yeast
(c) both a & b
(d) mold

522. Succulometer is used to determine ________ of fruit
(a) texture (b) color
(c) flavor (d) aroma

523. The tenderness of peas is measured by
(a) tenderometer
(b) succulometer
(c) penetrometer
(d) all

524. Bloom gelometer is a type of
(a) tenderometer (b) succulometer
(c) penetrometer (d) none

525. Apples are best stored at _____°C
(a) 1-2 (b) 2-4
(c) 4-7 (d) 0

526. A storage temperature between 0°C and safe temperature of apple
(a) increase quality
(b) cause internal browning
(c) cause soggy breakdown
(d) all

527. Lemons are best stored at_____°C
(a) 1-3 (b) 3-7
(c) 7-10 (d) 13-14

528. Oranges are best stored at _____°C

(a) 1.5-2.5 (b) 3.5-7.5
(c) 7.5-10.5 (d) 13.5-14.5

529. Potatoes are best stored at_____°C

(a) 3 (b) 7
(c) 10 (d) 14

530. Tomatoes are best stored at_____°C

(a) 1-3 (b) 3-7
(c) 7-10 (d) 10-13

531. Pitting is observed in

(a) apples (b) melons
(c) berries (d) bananas

532. Which of the following fruits have the lowest pH

(a) mango (b) pumpkin
(c) pear (d) prunes

533. Blanching is done in case of

(a) fruits (b) vegetables
(c) meat (d) both (a) & (b)

534. Which of the following is a low acid food

(a) pickle (b) tomatoes
(c) sauerkraut (d) carrots

535. Which of the following is a medium acid food

(a) apricots (b) tomatoes
(c) olives (d) carrots

536. Which of the following is a high acid food

(a) lime juice (b) tomatoe soup
(c) sauerkraut (d) figs

537. The fruit juices are pasteurized at

(a) 63°C/30min
(b) 68.5°C /30min
(c) 71°C /30min
(d) 78.5°C /30min

538. Flash pasteurization for juices is done at

(a) 111°C/2 sec (b) 101.1°C /2sec
(c) 121°C /2sec (d) 131°C /2sec

539. Chocolate undergoes following process as one of the production steps from harvest to a finished candy bar

(a) pasteurization
(b) homogenization
(c) fermentation
(d) lyophilization

540. __________ is sweeter than sucrose

(a) fructose (b) lactose
(c) glucose (d) maltose

541. Mustard seeds and oil contains following toxic factor

(a) elaidic acid (b) erucic acid
(c) DHA & EPA (d) mucic acid

542. Canned foods are liable to spoilage for various reasons. Various indica-tions of spoilage are as follows except

(a) hydrogen swell (b) springer
(c) flipper (d) flatter

543. Vinegar contains at least ______ % acetic acid

(a) 4 (b) 6
(c) 8 (d) 10

544. Minimal processing of fruits and vegetables involve

(a) cleaning (b) cutting
(c) grating (d) all

545. Respiration rate of fruits increases to amaximum prior to full ripening in case of _________ fruits

(a) climacteric
(b) non-climacteric
(c) both (d) none

546. Which of the following does not relate to climacteric fruits

(a) increase in respiration rate
(b) increase in ethylene
(c) no increase in respiration rate
(d) none

547. _______ fruit can ripen after harvest

(a) mango (b) lemon
(c) pineapple (d) grapefruit

548. Which of the following is an example of climacteric fruit
(a) pear (b) kiwifruit
(c) sapota (d) pomegranate

549. Which of the following is an example of non-climacteric fruit
(a) papaya (b) passion fruit
(c) strawberry (d) both (a) & (b)

550. Which of the following is not associated with non climacteric fruit
(a) higher production of ethylene
(b) doesnot respond to ethylene treatment
(c) respiration rate doesnot increase after harvest
(d) ripen before harvest

551. Which of the following produces highest ethylene
(a) citrus fruit (b) pine apple
(c) banana (d) passion fruit

552. Which of the following produces lowest ethylene
(a) apricot (b) fig
(c) raspberry (d) pomegranate

553. Which of the following produces highest respiration rate
(a) strawberry (b) banana
(c) avocado (d) apple

554. Which of the following produces lowest respiration rate
(a) pear (b) dates
(c) cherry (d) raspberry

555. The edible part in amaranth is
(a) leaves (b) roots
(c) bulb (d) tuber

556. The edible part in broccoli is
(a) bulb (b) roots
(c) bulb (d) tuber

557. The edible part in celery is
(a) bulb (b) stem
(c) bulb (d) tuber

558. The stage at which a fruit/ vegetable has reached sufficient stage of development that after harvesting, its quality will be acceptable is
(a) maturity (b) ripened
(c) senescent (d) none

559. Maturity indices include
(a) shape & size
(b) color
(c) firmness
(d) all

560. DFFB is a maturity index where DFFB stands for
(a) days from full bloom
(b) do from full bloom
(c) days from fruit bloom
(d) none

561. As the fruit matures, its specific gravity
(a) increases
(b) decreases
(c) remains same
(d) none

562. When a fruit is placed in a tank of water, immature fruit will
(a) float (b) sink
(c) burst (d) none

563. When a fruit is placed in a tank of water, mature fruit will
(a) float (b) sink
(c) burst (d) none

564. Upon maturation, the acidity of fruit
(a) increases (b) decreases
(c) remains same (d) none

565. The oil content _______ with the onset of maturity
(a) increases (b) decreases
(c) remains same (d) none

566. The juice content _______ with the onset of maturity
(a) increases (b) decreases
(c) remains same (d) none

567. Minimum juice content specified for oranges below which they are not harvested
(a) 25% (b) 33%
(c) 30% (d) 35%

568. Ripening and respiration is controlled in fruits and vegetables by spraying
(a) isopropyl N-phenyl carbonate
(b) 2-4-D
(c) maleic hydrazide (d) all

569. Irradiation with ______ krads is used to delay ripening of banana
(a) 25-35 (b) 10-15
(c) 2-10 (d) 40-50

570. Sprouting in potatoes and onions is inhibited by irradiation at _____krads
(a) 25-35 (b) 10-15
(c) 2-5 (d) 5-10

571. The rate of deterioration of fruits and vegetables double for increase in __°C
(a) 1 (b) 5
(c) 10 (d) 15

572. The water loss from fruits can be reduced by _________ relative humidity
(a) increasing (b) decreasing
(c) no change (d) none

573. The recommended relative humidity for storage of fruits and vegetables is ________%
(a) 80-85 (b) 85-90
(c) 90-95 (d) 95-100

574. The storage life of fruits and vegetables can be increased by _____ CO_2
(a) increasing (b) decreasing
(c) no change (d) none

575. The storage life of fruits and vegetables can be increased by _____ O_2
(a) increasing (b) decreasing
(c) no change (d) none

576. The phenolic content ______ with fruit ripening
(a) increases (b) decreases
(c) remains same (d) none

577. The astringency of fruit _______ with ripening
(a) increases (b) decreases
(c) remains same (d) none

578. In banana, the main acids- malic and oxalic acid ________ during ripening
(a) increases (b) decreases
(c) remains same (d) none

579. Monocarpic senescence means
(a) flower once & die
(b) flower twice & die
(c) flower thrice & die
(d) none

580. The dipping of fruit in solution of $CaCl_2$
(a) reduces browning
(b) decrease flesh softening
(c) increase ripening
(d) both (a) & (b)

581. Minimally processed products should be kept at ____
(a) room temperature
(b) 0-3°C
(c) 5-10°C
(d) 10-15°C

582. Treatment of minimally processed products with super atmospheric O_2
(a) inhibits microbial growth
(b) enzymatic coloration
(c) prevent anaerobic fermentation
(d) all

583. E in EMAP stands for
(a) equal (b) equilibrium
(c) easy (d) none

584. OTR stands for
(a) oxygen transfer rate
(b) oxygen transmission rate
(c) oxygen turning rate
(d) none

585. Soft rot in fruits and vegetables is caused due to
(a) Pseudomonas (b) Erwinia
(c) Fusarium (d) Rhizopus

586. Which of the following is hazardous ripening agent
(a) ethylene (b) ethephon
(c) calcium carbide (d) all

587. The transparency of apple juice is measured in
(a) NTU (b) BTU
(c) CTU (d) GTU

588. Blanching is a ______ heat treatment
(a) mild (b) high
(c) not (d) none

589. The adequacy of blanching is indicated by
(a) peroxidase (b) catalase
(c) lipoxygenase (d) both (a) & (b)

590. PPC refers to
(a) para processing contamination
(b) post processing contamination
(c) pre processing contamination
(d) none

591. Blanching involves ______ mode of heat transfer
(a) convection (b) conduction
(c) radiation (d) none

592. The golden bleached appearance of raisins is due to high concentration of ______ content
(a) NO_2 (b) SO_2
(c) $CaCl_2$ (d) none

593. The fruit juice has ______ % solid content
(a) 12 (b) 22
(c) 2 (d) 32

594. ______ enzyme is responsible for cut fruts and vegetables browning
(a) PPO (b) lipase
(c) phosphorylase (d) catalase

595. The oxygen polygalacturonase and pectin methylesterase are involved in
(a) browning
(b) flavor defects
(c) textural changes
(d) taget blanching enzymes

596. Lycopene has __________ provitamin A activity
(a) no (b) very high
(c) similar to beta carotene
(d) low

597. Lycopene is biologically available in ______ form
(a) cis (b) trans
(c) both (d) none

598. IMF has _______ % moisture
(a) 10-20 (b) 20-50
(c) 50-75 (d) 0-20

599. 1 Gray = _______ rads
(a) 1 (b) 10
(c) 100 (d) 1000

600. For prevention sprouts in potatoes and onions _______ kGray radiation dose is used
(a) 0.1-0.5 (b) 0.05-0.15
(c) 2-3 (d) 20-25

601. For mangoes _______ kGray radiation dose is used
(a) 0.075-1.1 (b) 0.05-0.15
(c) 2-3 (d) 20-25

602. For fruits and vegetables _______ kGray radiation dose is used
(a) 0.075-1.0 (b) 0.05-0.15
(c) 1-5 (d) 20-25

603. For fruits and vegetables _______ kGray radiation dose is used
(a) 0.075-1.1 (b) 0.05-0.15
(c) 2-10 (d) 20-25

604. Dm is the radiation dose giving a ___ % reduction inmicrobial population
(a) 50 (b) 70
(c) 80 (d) 90

605. Jam is produced from
(a) fruit juice (b) pulp
(c) clear extracts (d) fruit slices

606. Jelly is produced from
(a) fruit juice (b) pulp
(c) clear extracts (d) fruit slices

607. Marmalade is produced from
(a) fruit slices (b) pulp
(c) clear extracts (d) both (a) & (b)

608. Jams are solid gel which involves
(a) pulp (b) sugar
(c) pectin (d) all

609. The fruit content in jam should be
(a) NLT 40% (b) NMT 40%
(c) NLT 60% (d) NMT 60%

610. In mixed fruit jam the first named fruit should be _______ of the total fruit added
(a) NLT 50% (b) NLT 60%
(c) NLT 40% (d) NLT 70%

611. The total sugar content of jam should not be less than
(a) 40% (b) 58%
(c) 68% (d) 78%

612. The 50°Brix sugar solution can be made by dissolving _____ g sugar in 500ml water
(a) 100 (b) 250
(c) 500 (d) 1000

613. The pectin is present mainly in the
(a) pulp (b) seeds
(c) skin (d) juice

614. Which of the following contains high concentration of pectin
(a) apples (b) strawberries
(c) melon (d) none

615. The pH necessary for making jam is
(a) 5-5.3 (b) 1-1.3
(c) 2-2.3 (d) 3-3.3

616. The acid added to jam to attain desired acidity is
(a) citric (b) malic
(c) tartaric (d) all

617. To allow even distribution of fruit peel or slices in marmalade, they are soaked in
(a) brine solution (b) citric acid
(c) sugar solution (d) malic acid

618. The ideal temperature for filling of jam is ____°C
(a) 62-68 (b) 72-75
(c) 82-85 (d) 98-100

619. For setting of a gel structure, important factor is
(a) sugar (b) acid
(c) pectin (d) all

620. Which of the following foods is nt exempted from food labeling
(a) coffee beans
(b) dehydrated vegetable condiments
(c) instant tea
(d) unpopped popcorn

621. The carbohydrates present in fruits and vegetables supply _____ Kcal/g
(a) 4 (b) 5
(c) 6 (d) 8

622. Solanine is present in
(a) tomato (b) coffee
(c) potato (d) tea

623. Starch is a __________
(a) protein (b) fat
(c) carbohydrate (d) mineral

624. Fruits and vegetables discolor when cut due to
(a) caramelization
(b) sulfating
(c) dehydration
(d) enzymatic browning

625. Fruits and vegetables are primarily composed of
(a) proteins (b) water
(c) carbohydrate (d) fat

626. Fruit juces can be fortified with ________ to increase the nutritional content
(a) antioxidants (b) stabilizers
(c) vitamins (d) all

627. During production of sauerkraut, the cabbage is ________ to contribute to aroma
(a) fermented
(b) homogenization
(c) fried
(d) baked

628. Instant mashed potato flakes are an example of a product that has undergone a ________ process
(a) fermentation
(b) curing
(c) dehydration
(d) caramelization

629. Pressure canners used in canning is known as
(a) steam blanchers (b) retorts
(c) plate exchangers (d) sublimators

630. When peanut is ground to make peanut butter a ________ is added to keep the oil from separating out at the top of the jar during storage
(a) caking agent (b) stabilizer
(c) humectants (d) anti-oxidant

631. Which of the following foods cannot be effectively frozen
(a) broccoli (b) cabbage
(c) carrots (d) lettuce

632. A toxin found in corn and peanuts is
(a) solanine (b) protease
(c) goitrogens (d) aflatoxins

633. Canning is a method for ________ of foods
(a) deterioration (b) preservation
(c) pasteurization (d) none

634. The amount of sulfur dioxide permitted in fruit juice and jam is limited to
(a) 2000ppm (b) 200ppm
(c) 20ppm (d) 2ppm

635. The amount of sulfur dioxide permitted in fruit marmalades is limited to
(a) 4000ppm (b) 400ppm
(c) 40ppm (d) 4 ppm

636. According to PFA, the cardamom seeds obtained from capsule shall contain _____ % volatile oil
(a) NLT 3% (b) NMT 3%
(c) NMT 2% (d) NLT 2%

637. Cinnamon powder is obtained from _______ of *Cinnamomum zeylanicum*
(a) leaves (b) bark
(c) roots (d) seeds

638. According to PFA, the cinnamon powder shall contain _____ % volatile oil
(a) NLT 5% (b) NLT 10%
(c) NLT 15% (d) NLT 20%

639. Coriander powder is obtained by grinding dried ______ of *Coriandrum sativum*
(a) leaves (b) fruits
(c) roots (d) stem

640. Fenugreek is obtained from dried ________ of plant
(a) leaves (b) fruits
(c) roots (d) seeds

641. Kalonji is the common name for
(a) turmeric (b) cumin black
(c) fennel (d) fenugreek

642. Jaepatri is the common name for
(a) fennel (b) fenugreek
(c) nutmeg (d) mace

643. Nutmeg is obtained from dried _____ of Myristica fragrans houtt
(a) seeds (b) stem
(c) bark (d) leaves

644. Poppy is obtained from ________ of *Papaver somniferum* L.
(a) seeds (b) stem
(c) bark (d) leaves

645. Saunf is the common name for
(a) nutmeg (b) fennel
(c) aniseed (d) fenugreek

646. According to PFA, roasted coffee shall have moisture NMT_____%
(a) 1 (b) 2
(c) 5 (d) 10

647. According to PFA, roasted coffee shall have caffeine NLT_____%
(a) 1 (b) 2
(c) 5 (d) 10

648. According to PFA, the acidity of fruit juice shall be NLT_____%
(a) 1 (b) 2
(c) 4 (d) 10

649. According to PFA, cocoa butter shall have saponification value
(a) 10-58 (b) 188- 200
(c) 225-350 (d) none

650. According to PFA, the TSS of sweetened fruit juice shall be NLT_____%
(a) 1 (b) 2
(c) 4 (d) 10

651. According to PFA, the TSS of tomato juice shall be NLT_____%
(a) 1 (b) 2
(c) 5 (d) 10

652. According to PFA, the TSS of fruit syrup shall be NLT_____%
(a) 50 (b) 25
(c) 65 (d) 85

653. According to PFA, minimum % of fruit juice in fruit syrup shall be NLT _____%
(a) 50 (b) 25
(c) 65 (d) 85

654. According to PFA, the TSS of fruit squash shall be NLT_____%
(a) 40 (b) 25
(c) 65 (d) 85

655. According to PFA, minimum % of fruit juice in fruit squash shall be NLT_____%
(a) 50 (b) 25
(c) 65 (d) 85

656. Which of the following is fermented
(a) tomato sauce (b) tomato puree
(c) tomato ketchup (d) none

657. The total acidity in tomato sauce and ketchup is
(a) NLT 1% (b) NLT 20%
(c) NLT 5% (d) NLT 10%

658. According to PFA, the TSS of tomato ketchup & sauce shall be NLT ____%
(a) 50 (b) 25
(c) 65 (d) 85

659. Fumaric acid permissible by PFA in tomato ketchup and sauce is NMT ____%
(a) 30 (b) 3
(c) 0.3 (d) 0.03

660. Jam should have minimum ______ % soluble solids
(a) 28 (b) 48
(c) 68 (d) 88

661. Marmalade is obtained by _______ of fruit juice
(a) boiling (b) fermentation
(c) homogenization (d) irradiation

662. According to PFA, marmalade shall contain NLT _______ % of fruit from which it is made
(a) 25 (b) 45
(c) 65 (d) 85

663. The TSS of marmalade should be
(a) NLT 65% (b) NMT 65%
(c) NLT 45% (d) NMT 45%

664. The TSS of fruit chutney should be
(a) NLT 50% (b) NLT 65%
(c) NLT 35% (d) NLT 75%

665. Minimum per centage of fruit in final product of fruit chutney should be
(a) NLT 40% (b) NLT 65%
(c) NLT 35% (d) NLT 75%

666. Sauce from fruit or vegetable should have TSS
(a) NLT 15% (b) NLT 25%
(c) NLT 35% (d) NLT 55%

667. 667. Minimum per centage of acidity in final product of fruit chutney should be
(a) NLT 5% (b) NLT 1%
(c) NLT 15% (d) NLT 25%

668. The TSS of tomato puree should be
(a) NMT 9% (b) NLT 9%
(c) NMT 25% (d) NLT 25%

669. The TSS of tomato paste should be
(a) NMT 9% (b) NLT 9%
(c) NMT 25% (d) NLT 25%

670. Fruit jelly should have __________% fruit extract
(a) NMT 9% (b) NLT 9%
(c) NMT 45% (d) NLT 45%

671. Fruit jelly should have ____________ % TSS
(a) NLT 40% (b) NLT 65%
(c) NLT 35% (d) NLT 75%

672. The pickle in citrus juice or brine should have __________% salt
(a) NMT 10% (b) NLT 10%
(c) NMT 20% (d) NLT 20%

673. The per centage of fruits and vegetable in the final pickle should be
(a) NLT 40% (b) NLT 60%
(c) NLT 30% (d) NLT 70%

674. Per centage of oil in pickles in oil should be
(a) NMT 10% (b) NLT 10%
(c) NMT 20% (d) NLT 20%

675. For pickles in vinegar, the per centage of vinegar should be _____
(a) NMT 2% (b) NLT 2%
(c) NMT 5% (d) NLT 5%

676. Watermelon seed oil should have _________ moisture
(a) NMT 0.25% (b) NMT 2.5%
(c) NMT 5.5% (d) NMT 7.5%

677. __________ has been the highy cultivated fruit in India in 2010-11
(a) mango (b) banana
(c) papaya (d) guava

678. __________ has been the highy cultivated vegetable in India in 2010-11
(a) potato (b) tomato
(c) onion (d) brinjal

679. The production share of the major fruit banana in 2010-11 is ____%
(a) 20.3 (b) 39.8
(c) 10.0 (d) 25.3

680. The production share of the fruit mango in 2010-11 is______%
(a) 20.3 (b) 3.3
(c) 10.0 (d) 20.3

681. The production share of the potato in 2010-11 is ______%
(a) 20.3 (b) 3.3
(c) 10.0 (d) 28.9

682. The production share of the tomato in 2010-11 is ______%
(a) 20.3 (b) 11.3
(c) 10.0 (d) 20.3

683. The production share of the onion in 2010-11 is ______%
(a) 11.3 (b) 3.3
(c) 30.0 (d) 20.3

684. Major plantation crop in India in 2010-11
(a) coconut (b) cashewnut
(c) cocoa (d) coffee

685. Major production of spice crop in India in 2010-11
(a) garlic (b) chillies
(c) turmeric (d) ginger

686. The production share of the coconut in 2010-11 is ______%

(a) 20.0 (b) 40.0
(c) 10.0 (d) 90.0

687. The production share of cashewnut in 2010-11 is ______%

(a) 5.6 (b) 15.6
(c) 4.6 (d) 14.6

688. The production share of chillies in 2010-11 is ______%

(a) 32.3 (b) 22.87
(c) 15.0 (d) 48.3

689. The production share of turmeric in 2010-11 is ______%

(a) 2.3 (b) 13.12
(c) 18.56 (d) 32.3

690. The production share of ginger in 2010-11 is ______%

(a) 5.3 (b) 13.12
(c) 18.56 (d) 20.3

691. The production share of garlic in 2010-11 is ______%

(a) 19.77 (b) 33.12
(c) 28.56 (d) 10.3

692. The state which reported the highest production of fruits in India in 2010-11 is

(a) Gujarat (b) Rajasthan
(c) Tamil Nadu (d) Punjab

693. The state which reported the highest production of vegetables in India in 2010-11 is

(a) Gujarat (b) West Bengal
(c) Tamil Nadu (d) Punjab

694. The production share of West Bengal in vegetable in 2010-11 is ______%

(a) 2.3 (b) 18.2
(c) 28.56 (d) 32.3

695. The production share of Tamil Nadu in vegetable in 2010-11 is ______%

(a) 13.3 (b) 18.2
(c) 28.56 (d) 32.3

696. The leading cashewnut producing state in India in 2010-11 has been

(a) Maharashtra
(b) Andhra Pradesh
(c) West Bengal
(d) Haryana

697. The leading cocoa producing state in India in 2010-11 has been

(a) Kerala
(b) Andhra Pradesh
(c) West Bengal
(d) Haryana

698. The leading coconut producing state in India in 2010-11 has been

(a) Maharashtra
(b) Andhra Pradesh
(c) Kerala
(d) Haryana

699. The leading spice producing state in India in 2010-11 has been

(a) Maharashtra
(b) Andhra Pradesh
(c) Kerala
(d) Haryana

700. Total Indian horticultural production in 2010-11 in million metric tonnes is

(a) 96.6 (b) 170.8
(c) 211.2 (d) 240.5

701. Total Indian fruits production in 2010-11 in 000, metric tonnes is

(a) 28632 (b) 50867
(c) 68466 (d) 74878

702. Total Indian vegetables production in 2010-11 in 000, metric tonnes is

(a) 58532 (b) 101246
(c) 128449 (d) 146555

703. Total Indian production of plantation crops in 2010-11 in 000, metric tonnes is

(a) 7498 (b) 8051
(c) 11263 (d) 12007

704. Total Indian spices production in 2010-11 in 000, metric tonnes is
(a) 1900 (b) 8051
(c) 5351 (d) 4016

705. According to NHB major apple producing state in India in 2010-11 is
(a) Jammu & Kashmir
(b) Bihar
(c) Tamil Nadu
(d) Punjab

706. According to NHB major banana producing state in India in 2010-11 is
(a) Jammu & Kashmir
(b) Bihar
(c) Tamil Nadu
(d) Punjab

707. According to NHB major orange producing state in India in 2010-11 is
(a) Maharashtra (b) Bihar
(c) Tamil Nadu (d) Punjab

708. According to NHB major citrus fruits producing state in India in 2010-11 is
(a) Andhra Pradesh (b) Bihar
(c) Tamil Nadu (d) Punjab

709. The production share of Tamil Nadu in banana in 2010-11 is ______%
(a) 12.4 (b) 27.7
(c) 38.5 (d) 47.3

710. The production share of Andhra Pradesh in citrus fruits in 2010-11 is ______%
(a) 12.4 (b) 24.2
(c) 38.5 (d) 47.3

711. The production share of Punjab in oranges in 2010-11 is ______%
(a) 12.4 (b) 17.2
(c) 28.5 (d) 47.3

712. According to NHB major grape producing state in India in 2010-11 is
(a) Maharashtra (b) Bihar
(c) Tamil Nadu (d) Punjab

713. According to NHB major mango producing state in India in 2010-11 is
(a) Maharashtra
(b) Uttar Pradesh
(c) Tamil Nadu
(d) Punjab

714. According to NHB major potato producing state in India in 2010-11 is
(a) Maharashtra (b) West Bengal
(c) Tamil Nadu (d) Punjab

715. The production share of Maharashra in grapes in 2010-11 is ______%
(a) 12.5 (b) 62.7
(c) 33.4 (d) 82.3

716. The production share of Uttar Pradesh in mango in 2010-11 is ______%
(a) 23.9 (b) 62.7
(c) 33.4 (d) 82.3

717. The production share of West Bengal in potato in 2010-11 is ______%
(a) 23.9 (b) 62.7
(c) 32.4 (d) 82.3

718. According to NHB major fruit producing country in world in 2010-11 is
(a) USA (b) China
(c) Australia (d) Greece

719. According to NHB major apple producing country in world in 2010-11 is
(a) USA (b) China
(c) Australia (d) Greece

720. According to NHB major banana producing country in world in 2010-11 is
(a) USA (b) China
(c) India (d) SriLanka

721. According to NHB major grapes producing country in world in 2010-11 is
(a) USA (b) China
(c) Australia (d) Greece

722. The production share of China in apple in 2010-11 is ______%
(a) 23.9 (b) 62.7
(c) 47.3 (d) 82.3

723. The production share of India in banana in 2010-11 is ______%
(a) 29.0 (b) 62.7
(c) 33.4 (d) 82.3

724. The production share of China in grapes in 2010-11 is ______%
(a) 23.9 (b) 62.7
(c) 13.4 (d) 82.3

725. According to NHB major mangoes producing country in world in 2010-11 is
(a) USA (b) China
(c) India (d) Greece

726. The production share of India in mangoes in 2010-11 is ______%
(a) 23.9 (b) 62.7
(c) 45.3 (d) 82.3

727. According to NHB major orange producing country in world in 2010-11 is
(a) USA (b) China
(c) Brazil (d) Greece

728. The production share of Brazil in orange in 2010-11 is ______%
(a) 23.9 (b) 62.7
(c) 47.3 (d) 28.3

729. According to NHB major vegetable producing country in world in 2010-11 is
(a) USA (b) China
(c) Australia (d) Greece

730. The production share of China in vegetable in 2010-11 is ______%
(a) 23.9 (b) 62.7
(c) 48.3 (d) 82.3

731. According to NHB major onion producing country in world in 2010-11 is
(a) USA (b) China
(c) Australia (d) Greece

732. The production share of China in apple in 2010-11 is ______%
(a) 27.0 (b) 62.7
(c) 47.3 (d) 82.3

733. According to NHB major potato producing country in world in 2010-11 is
(a) USA (b) China
(c) Australia (d) Greece

734. The production share of China in apple in 2010-11 is ______%
(a) 23.0 (b) 62.7
(c) 47.3 (d) 82.3

735. The production share of India in vegetable in 2010-11 is ______%
(a) 23.9 (b) 62.7
(c) 14.0 (d) 82.3

736. According to NHB major okra producing country in world in 2010-11 is
(a) Iran (b) Brazil
(c) India (d) Sudan

737. The production share of India in okra in 2010-11 is ______%
(a) 23.0 (b) 62.7
(c) 47.3 (d) 73.0

738. The production share of India in potato in 2010-11 is ______%
(a) 13.0 (b) 62.7
(c) 47.3 (d) 73.0

739. Leading wine manufacturing state in India is
(a) Maharashtra (b) Bihar
(c) Haryana (d) Kerala

740. Bangalore blue, Gulabi, Perlette, Sonaka are varieties of
(a) melon (b) banana
(c) grapes (d) lemon

741. Kesar, Rajapuri, Pairi, Malda are varieties of
(a) melon (b) banana
(c) mango (d) lemon

742. Which of the following is a high pectin fruit

(a) gooseberries (b) peaches
(c) pears (d) figs

743. When harvested, banana has ______
(a) high % of starch
(b) high % of sugar
(c) low % of starch (d) none

744. Edible skins of fruits and vegetables has
(a) insoluble fiber (b) soluble fiber
(c) no fiber (d) none

745. Plucking, withering, rolling and fermentation are steps tea ______ manufacturing
(a) tea (b) cocoa
(c) coffee (d) none

746. Stevia is a sweet herb and is ________ times sweeter than sugar
(a) 10 (b) 20
(c) 30 (d) 50

747. Browning occurs due to reaction of PPO with__________
(a) phenols (b) oxygen
(c) protein (d) both (a) & (b)

748. Which of the following does not involves microbes in its processing
(a) coffee (b) pickles
(c) potato chips (d) root bear

749. Maillard reaction is between reducing sugar and __________ group
(a) aldehyde (b) amino
(c) carboxylic (d) ketone

750. Pectin is a polymer of
(a) polygalacturonic acid
(b) maltose
(c) glucose
(d) none

751. Postharvest losses can be minimized by
(a) increasing ethylene
(b) decreasing temperature
(c) increasing temperature
(d) increasing oxygen

752. Fruit Product Order came in
(a) 1954 (b) 1955
(c) 1956 (d) 1957

753. Proanthocyanidins belong to
(a) tea (b) coffee
(c) lemon (d) pineapple

754. Flavoglycosides are present in
(a) tea (b) ginkgo
(c) grape (d) green tea

755. ________ is responsible for antioxidant activity of garlic
(a) allicin (b) caffeine
(c) menthol (d) none

756. NIN stands for
(a) National Institute of Nutrition
(b) National Institutional for Nutrition
(c) National Institute for Neurology
(d) none

757. NIN has revealed that lower intake of fruits & vegetables have been responsible for malnutrition and undernutrition. NIN is situated at
(a) Hyderabad (b) Bangalore .
(c) New Delhi (d) Bhopal

758. Spoilage of foods include
(a) physical damage
(b) action of enzymes present in food
(c) microbial changes
(d) all

759. Asepsis means
(a) removal of mo's
(b) killing of mo's
(c) keepout of mo's
(d) none

760. Self decomposition of fruits and vegetable products can be prevented by
(a) inactivating food enzymes
(b) by keeping them at room temperature
(c) by ripening
(d) none

761. Which of the following is not a method of preservation
(a) picke making
(b) dehydration
(c) modified atmospheric packaging
(d) none

762. Canning involves
(a) low heating
(b) low temperature processing
(c) sterilization
(d) pasteurization

763. Curing, pickling and chutney preparation are examples of
(a) preservation (b) deterioration
(c) fermentation (d) none

764. Tomato crush involves
(a) seeds and pulp (b) pulp
(c) juice (d) skin

765. Sweetened juice contains ________ % TSS
(a) 10 (b) 85
(c) 60 (d) 40

766. Fruit juice consists of
(a) juice and seeds (b) juice and skin
(c) juice and peel (d) juice

767. The fruit juice is sterilized at _____°C
(a) 60-65 (b) 50-55
(c) 70-75 (d) 80-85

768. Commonly added preservative in fruit juice is
(a) potassium metasulfite
(b) sodium benzoate
(c) both
(d) none

769. The bitterness of citrus fruit juice can be reduced by
(a) keeping in open air
(b) by dipping in 2% boiling NaOH
(c) by exposure to high temperature
(d) by keeping at low temperature

770. Fruit syrup consists of ______% TSS
(a) 5 (b) 10
(c) 15 (d) 20

771. Fruits and vegetables fall into the category of ____________ foods
(a) non-perishable
(b) semi-perishable
(c) perishable
(d) dry

772. Apples and potatoes are
(a) non-perishable
(b) semi-perishable
(c) perishable
(d) dry

773. The decomposition of food can be delayed very easily at __________ phase
(a) lag
(b) log
(c) positive acceleration
(d) none

774. In order to preserve food, the log phase needs to be __________
(a) shortened (b) lengthened
(c) destroyed (d) none

775. Which of the following is a method of preservation of fruits and vegetables
(a) HPP (b) irradiation
(c) electric field (d) all

776. The removal of mo's is carried out by
(a) washing (b) filtration
(c) centrifugation (d) all

777. _______ peeling is preferred in fruits and vegetables to remove mo's
(a) cold water (b) hot water
(c) dry (d) lye

778. Centrifugation leads to removal of ____ % micro organisms in fruits and vegetables
(a) 60 (b) 70
(c) 90 (d) 99

779. Low temperature preservation of fruits and vegetables
(a) kills the mo's
(b) slow down mo's activity
(c) no effect on mo's
(d) none

780. Controlled removal of water from fruits and vegetables is called
(a) dehydration (b) sundrying
(c) canning (d) none

781. Facultative bacteria can grow in presence of
(a) O_2 (b) CO_2
(c) N_2 (d) all

782. Fermentation involves the growth of ______ mo's in food product
(a) all (b) selective
(c) thermophiles (d) LAB

783. The addition of chemical preservatives _________ the growth of mo's in food
(a) hinders (b) allows
(c) increase (d) none

784. Preservatives nhibit mo's by
(a) interfering with their cell membrane
(b) interfering with their enzyme activity
(c) interfering with their genetic mechanism
(d) all

785. The pH in pickled and fermented foods is
(a) neutral (b) high
(c) low (d) none

786. Which of the following s not permitted in fruits and vegetables
(a) SO_2
(b) sorbic acid
(c) para hydroxy benzoic acid
(d) nisin

787. The presence of salt for preservation causes
(a) precipitation of enzyme protein
(b) reduction in sulfur linkage in enzyme protein
(c) reaction with aldehyde
(d) none

788. Sodium benzoate is most effective at ____pH
(a) acidic (b) basic
(c) neutral (d) none

789. The SO_2 allowed inhsarbats is ___ ppp
(a) 150 (b) 250
(c) 70 (d) 350

790. _______ causes killing of mo's
(a) chilling (b) irradiation
(c) ohmic heating (d) both (b) & (c)

791. Pasteurization is a heat treatment that kills ______micro organisms
(a) selective (b) all
(c) mo's and toxins (d) none

792. Blanching is done at
(a) temperature>100°C
(b) temperature<100°C
(c) temperature>200°C
(d) none

793. Sterlization is done at
(a) 100°C (b) 121°C
(c) 90°C (d) 150°C

794. The application of high pressure processing in fruits and vegetables processing use pressure________MPa
(a) 100-200 (b) 200-300
(c) 400-600 (d) none

795. When moisture level is ___, there is no enzyme activity
(a) below 1% (b) below 3%
(c) below 5% (d) below 8%

796. Ascorbic acid is a ____ inhibitor of PPO
(a) poor (b) highly effectivo
(c) not (d) none

797. Which of the following are non enzymatic reactions
(a) maillard
(b) caramellization
(c) ascorbic acid oxidation
(d) all

798. The brown pigment formed in maillard reaction is called
(a) melanoidins
(b) metmyoglobin
(c) both
(d) none

799. Optimum moisture for maillard reaction is
(a) 3% (b) 13%
(c) 23% (d) 33%

800. Papain is extracted from
(a) mango (b) papaya
(c) guava (d) tomato

801. Fruit with brix to acid ratio _____ is preferred for juice
(a) 3-9 (b) 13-19
(c) 23-29 (d) 33-39

802. For jelly, fruits with ____ pectin are preferred
(a) high (b) low
(c) no (d) none

803. For jelly, fruits with ____ acid are preferred
(a) high (b) low
(c) no (d) none

804. An antioxidant
(a) prevents oxidation
(b) allows oxidation
(c) does not get consumed
(d) all

805. _____ are involved in chill proofing of beer
(a) carbohydrases (b) proteases
(c) glucosidases (d) pectinases

806. _________ are involved in clarification of fruit juices
(a) carbohydrases (b) proteases
(c) glucosidases (d) pectinases

807. The tenderization of vegetables is done by
(a) cellulase (b) proteases
(c) glucosidases (d) pectinases

808. Darkening of maple sugar upon heating is an example of
(a) maillard reaction
(b) caramellization
(c) enzymatic browning
(d) none

809. Safrole is a toxicant found in
(a) beans (b) spices
(c) almonds (d) spinach

810. Prussic acid is a toxicant found in
(a) beans (b) spices
(c) almonds (d) spinach

811. Oxalic acid is found in
(a) beans (b) spices
(c) almonds (d) spinach

812. Upon digestion, fats yield ____ kcal/g
(a) 1 (b) 4
(c) 9 (d) 12

813. Upon digestion, proteins yield _____ kcal/g
(a) 1 (b) 4
(c) 9 (d) 12

814. The highest cooking losses are reported for
(a) vitamin C (b) vitamin A
(c) vitamin K (d) vitamin B

815. The vitamin good for eyes is
(a) vitamin C (b) vitamin A
(c) vitamin K (d) vitamin B

816. Which of the following has lonest storage life
(a) fruits
(b) leafy vegetables
(c) root crops
(d) dried fruits

817. All molds require ___ for their growth
(a) O_2 (b) CO_2
(c) N_2 (d) NO_2

818. Chill injury occurs at ___ °C
(a) <4 (b) <10
(c) <0 (d) <15

819. CAS stands for
(a) controlled atmospheric storage
(b) canned atmospheric storage
(c) cold atmospheric storage
(d) none

820. There is no microbial activity in food product at a_w
(a) 0.99 (b) 0.77
(c) 0.60 (d) none

821. The most heat resisitant toxin found in canned foods is that of
(a) Clostridium botulinum
(b) Clostridium saccharobutylicum
(c) *E. coli*
(d) none

822. Before freezing, ______ is performed to enhance the storage life of fruits and vegetables
(a) pickling (b) fermentation
(c) blanching (d) all

823. Canned tomato juice is heated by
(a) convection (b) conduction
(c) radiation (d) both (a) & (b)

824. Convection heating is __________ than conduction heating
(a) rapid (b) slower
(c) equal to (d) none

825. For canning, ________ fruits are used
(a) whole (b) cut
(c) peeled and cut (d) all

826. Refrigerated storage of fruits causes _______ of vitamin
(a) no loss (b) full loss
(c) partial loss (d) none

827. The process of bringing the fruit and vegetable product to room temperature after cold storage is called
(a) thawing
(b) fermentation
(c) evaporation
(d) dehydration

828. When the water in the intercellular space of fruits and vegetables freezes rapidly form
(a) no ice crystal
(b) small ice crystal
(c) large ice crystal
(d) none

829. If the moisture is removed more from the outside/ surface than the interior of the food, it leads to
(a) case softening
(b) case hardening
(c) freeze-drying
(d) none

830. Continued dehydration of vegetable dices causes
(a) shrinkage of surface
(b) bulkiness
(c) no effect
(d) none

831. For drying of purees and liquids ________ is used
(a) cabinet driers (b) drum driers
(c) vacuum driers (d) all

832. The blanching time is influenced by
(a) type of vegetable
(b) size of vegetable
(c) temperature
(d) all

833. Enzyme which causes the loss of eating and nutritional quality in fruits and vegetables
(a) lipoxygenase
(b) PPO
(c) polygalacturonase
(d) all

834. The marker enzyme to determine the success of blanching are peroxidase and catalase. The most heat resistant among the two is
(a) peroxidase (b) catalase
(c) both are equal (d) none

835. Function of blanching is
(a) reduce no of contaminating mo's
(b) inactivate enzymes
(c) softens vegetable tissue
(d) all

836. During blanching
(a) no nutrient loss
(b) water soluble components are lost
(c) minerals lost
(d) both (b) & (c)

837. _________ is added to blanching water to maintain firmness in tissues
(a) calcium chloride
(b) magnesium chloride
(c) phosphoric acid
(d) none

838. Pasteurization is a mild heat treatment, thus it is accompanied by
(a) irradiation
(b) low temperature storage
(c) sterilization
(d) both (a) & (b)

839. The cause of color deterioration in fruit juice is
(a) peroxidase (b) PPO
(c) lipoxygenase (d) none

840. UHT processing is used for
(a) fruit juice (b) wine
(c) vegetables (d) all

841. The nutrient losses in UHT processing is _______ compared to retorting
(a) higher (b) lower
(c) equal (d) none

842. UHT stands for
(a) ultra heat treatment
(b) under heat treatment
(c) ultra heating temperature
(d) none

843. In CAS, the composition of the gas arund the respiring food is
(a) controlled (b) uncontrolled
(c) no gas (d) none

844. In MAP, the composition of the gas arund the respiring food is
(a) controlled
(b) uncontrolled
(c) no gas (d) none

845. EMA stands for
(a) equilibrium modified atmosphere
(b) equal modified atmosphere
(c) equilibrium manmade atmosphere
(d) none

846. The normal composition of oxygen in air is
(a) 2% (b) 21%
(c) 41% (d) 71%

847. The normal composition of nitrogen in air is
(a) 2% (b) 21%
(c) 41% (d) 78%

848. Presence of carbon-dioxide in MAP _________ microbial activity
(a) increase (b) decrease
(c) no effect (d) none

849. In MAP, oxygen is ______ by the commodity
(a) consumed (b) produced
(c) neither consumed nor produced
(d) none

850. In MAP, carbon-dioxide is ______ by the commodity
(a) consumed (b) produced
(c) neither consumed nor produced
(d) none

851. The per centage of oxygen for MAP of potatoes
(a) 10 (b) 20
(c) 30 (d) 40

852. The per centage of carbon-dioxide for MAP of potatoes

(a) 10 (b) 20
(c) 30 (d) 40

853. The atmosphere in MAP package is influenced by

(a) permeability of packing material
(b) micro-biological activity
(c) respiration of food
(d) all

854. Active packaging of fruits and vegetable involve

(a) edible moisture barriers
(b) ethylene scavengers
(c) oxygen scavengers
(d) all

855. ___________ is an oxygen absorber in active packaging

(a) sorbate
(b) ferrous carbonate
(c) toluene
(d) none

856. Ethanol _______ in active packaging

(a) decrease microbial growth
(b) increase microbial growth
(c) increase ripening
(d) none

857. The freezing point of most vegetables is _______°C

(a) 1-2 (b) -0.8 to -2.8
(c) -5.8to -0.8.8 (d) 0

858. The freezing point of most fruits is _______°C

(a) 1-2 (b) -0.9 to -2.7
(c) -5.8to -0.8.8 (d) 0

859. Fluidized bed freezers used for freezing of peas have air temperature of _______°C

(a) 0-10 (b) -10 to -20
(c) -25 to -35 (d) -55 to -65

860. The coolant in immersion freezers include

(a) propylene glycol (b) glycerol
(c) calcium chloride (d) all

861. The cellular material of plant cells that leaks out from ruptured cell upon thawing of frozen food is called

(a) drip (b) turgor
(c) both (d) none

862. During freezing of food which of the following takes place

(a) breakdown of chlorophyll
(b) change in pH of fruits
(c) loss of vitamin C
(d) all

863. As the moisture leaves the surface of foods due to cold storage, it produces areas of visible damage known as

(a) chill injury (b) freezer injury
(c) freezer burn (d) heat burn

864. Lyophilization is also called

(a) sublimation (b) freeze drying
(c) cold storage (d) all

865. Lyophilization takes place

(a) under high pressure
(b) under low pressure
(c) under normal pressure
(d) none

866. The texture of the food product which are lyophilized are

(a) poor (b) well
(c) damaged (d) none

867. Fre drying results in _______ of vitamins

(a) no loss (b) high loss
(c) medium loss (d) none

868. Coating of fruits and vegetables

(a) inprove appearance
(b) enhance flavors
(c) modify texture
(d) all

869. Major acid in apple is

(a) malic acid (b) tartaric acid
(c) quinic acid (d) none

870. Which is the major acid in grapes
(a) malic acid (b) tartaric acid
(c) quinic acid (d) none

871. Which is the major acid in lemon
(a) malic acid (b) tartaric acid
(c) quinic acid (d) citric acid

872. Most of the fruits have pH
(a) < 4.5 (b) > 4.5
(c) 7 (d) none

873. The predominant mo's in fruits is
(a) yeast (b) molds
(c) bacteria (d) both (a) & (b)

874. The yeast found in fresh apples belong to genus
(a) Candida (b) Torulopsis
(c) Rhodotorula (d) all

875. The most frequent contaminant of vegetables is
(a) Pseudomonas (b) Enterobacter
(c) Bacillus (d) all

876. The fungi found in onion is
(a) Alternaria (b) Aspergillus
(c) Fusarium (d) Rhizopus

877. The yeast and molds require pH ____ for growth
(a) 1-2 (b) 1.5-8
(c) >9 (d) none

878. Low acid foods have pH
(a) > 5.3 (b) 4.5-5.3
(c) <3.7 (d) none

879. Medium acid foods have pH
(a) > 5.3 (b) 4.5-5.3
(c) <3.7 (d) none

880. High acid foods have pH
(a) > 5.3 (b) 4.5-5.3
(c) <3.7 (d) none

881. Cold sterilization involves use of
(a) x-rays (b) uv rays
(c) β-rays (d) all

882. The spoilage of canned foods occur due to
(a) underprocessing
(b) inadequate cooling
(c) infection from leakage in cans
(d) all

883. When ends of the canned food remains flat and contents become sour, it is called
(a) flipper (b) flat sour
(c) TA spoilage (d) sulfur stinker

884. Flat sour occurs mostly in ________ foods
(a) acid (b) neutral
(c) non acid (d) none

885. Bacteria responsible for flat sour is
(a) *Bacillus stearothermophilus*
(b) *Bacilus coagulans*
(c) both
(d) none

886. TA spoilage is due to
(a) *Bacillus stearothermophilus*
(b) *Bacilus coagulans*
(c) *Clostridium thermosaccharolyticum*
(d) none

887. The swelling and bursting of cans is called
(a) flipper (b) flat sour
(c) TA spoilage (d) sulfur stinker

888. Gas producd in TA spoilage is
(a) O_2 (b) CO_2
(c) H_2 (d) both (b) & (c)

889. Spoiled food in TA has
(a) sour odor (b) sweet odor
(c) no odor (d) none

890. Sulfur stinker spoilage is caused by
(a) *Bacillus stearothermophilus*
(b) *Bacilus coagulans*
(c) *Clostridium thermosaccharolyticum*
(d) *Clostridium nigrificans*

891. Sulfur stinker spoilage occurs in _______ foods

(a) medium acid (b) high acid
(c) low acid (d) none

892. Jam, jeelies and fruit juices are spoiled by
(a) yeast (b) mold
(c) bacteria (d) all

893. Over-filling of can may lead to
(a) flipper (b) springer
(c) both (d) none

894. Metallic contamination in canned foods include formation of ferric tannate which is _____ in color
(a) yellow (b) green
(c) black (d) brown

895. A can with mild positive pressure is called
(a) flipper (b) springer
(c) flat sour (d) buckling

896. Flipper occurs due to
(a) hydrogen swell
(b) overfilling
(c) under exhausting
(d) all

897. A mild swell at one or both ends of can is called
(a) flipper (b) springer
(c) flat sour (d) buckling

898. When both ends remain bulged it is called
(a) soft swell (b) hard swell
(c) both (d) none

899. Sewls may occur due to production of
(a) hydrogen (b) carbondioxide
(c) both (d) none

900. Which of the following canned food is fit for consumption
(a) a) flipper (b) springer
(c) soft swell (d) hard swell

901. Which of the following is a spoilage defect in canned foods
(a) leaker (b) breather
(c) buckling (d) all

902. Vegetable juices are spoiled mostly by
(a) yeast (b) mold
(c) bacteria (d) all

903. The surface of the fruit juice may be contaminated by
(a) yeast (b) mold
(c) bacteria (d) all

904. Fruits and vegetable concentrates are mostly spoiled by
(a) yeast (b) mold
(c) bacteria (d) all

905. Most of the bacteria have their optimum pH at
(a) 1-4 (b) 4-8
(c) 9-12 (d) none

906. Soft drinks are mostly spoiled by
(a) yeast (b) mold
(c) bacteria (d) all

907. Yeast which commonly spoils the soft drinks is
(a) Torulopsis (b) Candida
(c) both (d) none

908. Main type of spoilage of soft drinks
(a) cloudiness (b) ropiness
(c) both (d) none

909. Ropiness in soft drinks occurs due to
(a) Bacillus
(b) LAB
(c) Leuconostoc
(d) Gluconobacter

910. The a_w of jams and jellies is
(a) <0.60 (b) 0.75to 0.86
(c) > 0.99 (d) none

911. Jams, jellies and preserves are mainly spoiled by
(a) osmophilic yeast (b) mold
(c) bacteria (d) all

912. The aw of sun dried dry fruits is
(a) 0.99 (b) >1
(c) 0.85 (d) 0.65

913. No microbial growth but enzymatic browning occurs at aw
(a) 0.99 (b) 0.2-0.6
(c) 0.85 (d) 0.65

914. Sauerkraut is ______ acid fermentation of cabbage
(a) acetic (b) citric
(c) lactic (d) gluconic

915. Soft kraut and black kraut are
(a) varieties of sauerkraut
(b) spoilage
(c) sauerkraut with different acid fermentation
(d) none

916. ________ is caused by red yeasts
(a) pink kraut (b) black kraut
(c) soft kraut (d) ropy kraut

917. Ropy kraut is caused by
(a) Lactobacillus plantarum
(b) *E. coli*
(c) Clostridium botulinum
(d) none

918. Soft pickles is due to
(a) Penicillium (b) Fusarium
(c) both (d) none

919. Black pickles is due to
(a) yeast
(b) bacteria
(c) hydrogen sulfide produced by bacteria
(d) mold

920. Slippery pickles results by action of
a) yeast (b) mold
(c) bacteria (d) all

921. Wine is spoiled by
(a) metals (b) enzyme
(c) both (d) none

922. Cloudiness in wine is due to
(a) tin & copper (b) iron & zinc
(c) Mg & Na (d) none

923. The pH of salad dressing is
(a) 1-2 (b) 3-4
(c) 5-6 (d) 7-8

924. Hydrolytic rancidity is caused by action of
(a) lipase (b) lipoxygenase
(c) peroxidase (d) catalase

925. Salad dressing is prone to spoilage by
(a) yeast & mold (b) bacteria
(c) both (d) none

926. Ketonic rancidity is due to ______ oxidation of lipids
(a) alpha (b) beta
(c) both (d) none

927. Honey has ________% sugar
(a) 20-30 (b) 40-50
(c) 60-70 (d) 70-80

928. The acidity of honey is
(a) 1.2-2.2 (b) 2.2-3.2
(c) 3.2-4.2 (d) 4.2-5.2

929. Chief spoilage of honey is by
(a) osmophilic yeast (b) mold
(c) bacteria (d) all

930. The highest quercetin content is present in
(a) onion (b) capsicum
(c) cranberry (d) tomato

931. Sucrose is cleaved to glucose and fructose by enzyme
(a) invertase (b) zymase
(c) glucosidase (d) peroxidase

932. Highest oxalic acid content is present in
(a) amaranth leaves
(b) spinach
(c) amla
(d) beet root

933. The toxic substance present in Cucurbitaceae family is
(a) goitrogens (b) oxalic acid
(c) cucurbitacins (d) none

934. Goitrogens is present in
(a) tomato (b) cabbage
(c) onion (d) plums

935. Peas, beans and alfalfa have compound that interferes with utilization of
(a) vitamin K (b) vitamin C
(c) vitamin E (d) vitamin D

936. Vasoactive substances are present in
(a) tomatoes (b) onion
(c) garlic (d) ginger

937. The radiation dose in mushrooms is ____kGray
(a) 0.1-0.5 (b) 1.5-3.0
(c) 3.5-5.0 (d) none

938. The radiation dose in strawberries is ___ kGray
(a) 0.1-0.5 (b) 2-3.0
(c) 3.5-5.0 (d) none

939. The radiation dose in bananas is ___ kGray
(a) 0.25-0.75 (b) 1.5-3.0
(c) 3.5-5.0 (d) none

940. Commercial refrigeration temperature is ________°C
(a) 1-2 (b) 3-5
(c) 5-7 (d) 5-9

941. The optimum storage temperature for lemons is
(a) 1-2 (b) 3-5
(c) 5-7 (d) 12-14

942. The optimum storage temperature for apricots is
(a) -0.5-0 (b) 3-5
(c) 5-7 (d) 12-14

943. The optimum storage temperature for melons is
(a) 1-2 (b) 3-4
(c) 4-10 (d) 12-14

944. The optimum storage temperature for onions is
(a) 0 (b) 3-5
(c) 5-7 (d) 12-14

945. The optimum storage temperature for tomatoes is
(a) 1-2 (b) 3-5
(c) 4-10 (d) 12-14

946. Dehydrofreezing of fruits and vegetables involves
(a) only freezing
(b) dehydration before freezing
(c) dehydration after freezing
(d) only dehydration

947. The unfrozen and concentrated solution oozing from packages of fruits and vegetables is called
(a) metacryotic liquid
(b) drip
(c) super critical fluid
(d) none

948. The type of storage for equalization of moisture to a desired level in dried fruits is
(a) sweating (b) chilling
(c) freezing (d) none

949. Jams jellies and honey fall into _________ product category
(a) high moisture
(b) low moisture
(c) intermediate moisture
(d) none

950. The water activity of intermediate moisture foods is
(a) 0.60-0.65 (b) 0.95-0.99
(c) 0.75-0.85 (d) none

951. ________ is used as a preservative in jam. Jellies and pickles
(a) benzoates (b) sorbates
(c) acetate (d) both (a) & (b)

952. Ethanol is germicidal at concentration _____%
(a) 25-45 (b) 45-65
(c) 70-95 (d) none

953. Which of the following rays are not used for irradiation
(a) alpha (b) beta
(c) gamma (d) omega

954. The source of gamma rays is
(a) cobalt 60
(b) phosphorus 32
(c) cesium 137
(d) both a & c

955. The unit of radiation dose is
(a) rad (b) roentgen
(c) rep (d) megarep

956. 1 gray is equivalent to ______ rads
(a) 1 (b) 10
(c) 100 (d) 1000

957. Rad appertization means
(a) radiation sterilization
(b) radiation pasteurization
(c) radiation ionization
(d) none

958. Radurization means
(a) radiation sterilization
(b) radiation pasteurization
(c) radiation ionization
(d) none

959. Radicidation is
(a) radiation sterilization
(b) radiation pasteurization
(c) radiation ionization
(d) none

960. Radiation dose for spices is ____ kgray
(a) 1 (b) 10-30
(c) 50-70 (d) >90

961. Microwaves used in food industry is ______ megacycles
(a) 915 (b) 2450
(c) both (d) none

962. ________% of salt is added to prepare sauerkraut
(a) 1.25-2.25 (b) 2.25-3.25
(c) 2.25-2.50 (d) none

963. 100° salometer reading corresponds to ______% salt
(a) 10.5 (b) 26.5
(c) 36.5 (d) 95.5

964. A breather in the can
(a) allows the air to pass through
(b) allows the mo's to pass through
(c) both (d) none

965. Perry is wine from
(a) peaches (b) plums
(c) pears (d) cherries

966. Mead is wine from
(a) peaches (b) plums
(c) honey (d) cherries

967. Good sauerkraut has acidity _____
(a) 0.7% (b) 1.7%
(c) 2.7% (d) 3.7%

968. UV radiation dose is higher for
(a) gram –ve bacteria
(b) gram +ve bacteria
(c) molds (d) yeasts

969. β-rays are
(a) electromagnetic waves
(b) produced from cobalt and cesium
(c) stream of electrons
(d) all

970. Cathode rays are similar to
(a) X-rays (b) gamma rays
(c) beta rays (d) none

971. _______ is used to label foods with low level ionization radiation
(a) radappertization (b) radurization
(c) radicidation (d) picowaved

972. x-, β- and γ-rays have ________ penetration
(a) poor (b) good
(c) no (d) none

973. Among x-, β-, γ- and cathode rays, rays having less penetration is
(a) x rays (b) beta rays
(c) gamma rays (d) cathode rays

974. The rays used for commercial sterilization are
(a) x-rays (b) gamma rays
(c) both (d) none

975. Gamma rays are effective upto ___ cm in foods
(a) 2 (b) 10
(c) 20 (d) 15

976. Cathode rays are effective upto ___ cm in foods
(a) 0.5 (b) 0.05
(c) 5.0 (d) 50.0

977. which of the following is true about cathode rays
(a) good penetration
(b) directional
(c) penetrate upto 5 cm
(d) none

978. The bacterial efficacy of dose of radiation depends on
(a) kind of mo's
(b) number of mo's
(c) composition of food
(d) all

979. The optimum dose for killing virus is _______ kgray
(a) 1-10 (b) 10-40
(c) 0.1-0.4 (d) >40

980. The optimum dose for killing yeast is ______ kgray
(a) 4-9 (b) 10-40
(c) 0.1-0.4 (d) >40

981. The optimum dose for killing mold is _______ kgray
(a) 1.3-11 (b) 12-42
(c) 0.1-0.4 (d) >40

982. The optimum dose for killing bacteria is ______ kgray
(a) 1-4.2 (b) 10-40
(c) 0.1-0.4 (d) >40

983. The optimum dose for killing bacterial spores is _______ kgray
(a) 1-3 (b) 3.1-37
(c) 0.1-0.4 (d) >40

984. The optimum dose for killing spores of *Clostridium botulinum* A is ______ k gray
(a) 1-10 (b) 19-37
(c) 0.1-0.4 (d) >40

985. The optimum dose for killing spores of *Clostridium botulinum* E is ______ kgray
(a) 1-10 (b) 15-18
(c) 0.1-0.4 (d) >40

986. High radiation dose cause
(a) non adverse effect
(b) side reaction
(c) secondary changes
(d) both (b) & (c)

987. The irradiation dose for fresh fruits is _____ k gray
(a) 1 (b) 3.1-37
(c) 0.1-0.4 (d) >40

988. The irradiation dose for dry or dehydrated fruits is _____ k gray
(a) 1 (b) 30
(c) 0.1-0.4 (d) >40

989. Microwaves are electromagnetic waves between
(a) visible and infrared
(b) gamma & x rays
(c) infrared & radio
(d) none

990. Microwaves cause _________ of food
(a) cold sterilization
(b) heating
(c) both
(d) none

991. Most of the micro-organisms are present on _________ of food
(a) surface (b) interior
(c) seeds (d) flesh

992. At storage temperature below 2.2-4.4°C, potatoes
(a) turn sweet
(b) have more sugar
(c) have more starch
(d) both (a) & (b)

993. Scalding of vegetables is a
(a) freezing treatment
(b) heat treatment
(c) irradiation
(d) none

994. *E. coli* in vegetabe is indication of ______ sanitary conditions
(a) good (b) poor
(c) not related to (d) none

995. Most of the vegetables require ______ heat treatment than fruits
(a) more (b) less
(c) equal (d) none

996. The common class I preservative is
(a) sugar (b) salt
(c) both (d) none

997. Brine is
(a) salt solution
(b) sugar solution
(c) solution of chemical preservatives
(d) all

998. Irradiation has been used to
(a) delay sprouting
(b) stop insect reproduction
(c) complete destruction of mo's
(d) all

999. Asepsis means
(a) keeping out mo's
(b) destruction of mo's
(c) propagation of mo's
(d) pasteurization

1000. Which of the following decrease the microdial load
(a) washing (b) trimming
(c) dipping (d) all

1001. Central Potato Ressearch Institute is situated at
(a) Bhopal (b) Solan
(c) New Delhi (d) Gwalior

4.3 ANSWER KEY

Q.	A.	Q.	A.	Q.	A.	Q.	A.	Q.	A.	Q.	A.
1.	(d)	2.	(a)	3.	(d)	4.	(c)	5.	(d)	6.	(c)
7.	(b)	8.	(c)	9.	(b)	10.	(a)	11.	(a)	12.	(c)
13.	(b)	14.	(c)	15.	(a)	16.	(b)	17.	(c)	18.	(c)
19.	(a)	20.	(b)	21.	(a)	22.	(a)	23.	(a)	24.	(a)
25.	(b)	26.	(b)	27.	(a)	28.	(c)	29.	(c)	30.	(b)
31.	(d)	32.	(b)	33.	(b)	34.	(c)	35.	(c)	36.	(c)
37.	(a)	38.	(d)	39.	(a)	40.	(b)	41.	(a)	42.	(b)
43.	(b)	44.	(d)	45.	(d)	46.	(a)	47.	(c)	48.	(a)
49.	(d)	50.	(a)	51.	(b)	52.	(c)	53.	(d)	54.	(b)
55.	(b)	56.	(a)	57.	(b)	58.	(b)	59.	(b)	60.	(c)
61.	(c)	62.	(c)	63.	(a)	64.	(c)	65.	(a)	66.	(c)
67.	(a)	68.	(b)	69.	(a)	70.	(a)	71.	(b)	72.	(c)

Q. A.	Q. A.	Q. A.	Q. A.	Q. A.	Q. A.
73. (a)	74. (b)	75. (a)	76. (b)	77. (b)	78. (c)
79. (c)	80. (d)	81. (b)	82. (a)	83. (a)	84. (a)
85. (a)	86. (c)	87. (a)	88. (a)	89. (d)	90. (b)
91. (b)	92. (a)	93. (a)	94. (a)	95. (a)	96. (a)
97. (a)	98. (c)	99. (c)	100. (d)	101. (b)	102. (c)
103. (b)	104. (d)	105. (c)	106. (c)	107. (a)	108. (b)
109. (b)	110. (c)	111. (b)	112. (b)	113. (a)	114. (c)
115. (a)	116. (b)	117. (d)	118. (c)	119. (a)	120. (a)
121. (a)	122. (a)	123. (d)	124. (c)	125. (c)	126. (d)
127. (a)	128. (a)	129. (c)	130. (c)	131. (b)	132. (a)
133. (b)	134. (c)	135. (a)	136. (b)	137. (c)	138. (a)
139. (b)	140. (a)	141. (a)	142. (c)	143. (b)	144. (d)
145. (c)	146. (c)	147. (a)	148. (a)	149. (a)	150. (d)
151. (a)	152. (b)	153. (b)	154. (c)	155. (b)	156. (a)
157. (b)	158. (b)	159. (c)	160. (c)	161. (b)	162. (a)
163. (b)	164. (c)	165. (a)	166. (c)	167. (a)	168. (c)
169. (b)	170. (d)	171. (b)	172. (d)	173. (a)	174. (b)
175. (a)	176. (b)	177. (c)	178. (a)	179. (a)	180. (c)
181. (b)	182. (c)	183. (c)	184. (a)	185. (b)	186. (b)
187. (d)	188. (a)	189. (a)	190. (c)	191. (b)	192. (b)
193. (a)	194. (b)	195. (a)	196. (a)	197. (d)	198. (c)
199. (a)	200. (c)	201. (b)	202. (c)	203. (a)	204. (a)
205. (d)	206. (c)	207. (a)	208. (b)	209. (a)	210. (d)
211. (c)	212. (c)	213. (d)	214. (c)	215. (a)	216. (c)
217. (a)	218. (a)	219. (a)	220. (a)	221. (a)	222. (b)
223. (a)	224. (c)	225. (c)	226. (c)	227. (b)	228. (c)
229. (a)	230. (a)	231. (b)	232. (b)	233. (b)	234. (a)
235. (c)	236. (c)	237. (c)	238. (a)	239. (c)	240. (d)
241. (a)	242. (c)	243. (b)	244. (b)	245. (a)	246. (b)
247. (c)	248. (d)	249. (d)	250. (d)	251. (a)	252. (a)
253. (d)	254. (a)	255. (a)	256. (a)	257. (b)	258. (c)
259. (c)	260. (c)	261. (b)	262. (b)	263. (d)	264. (d)
265. (b)	266. (b)	267. (a)	268. (a)	269. (b)	270. (c)
271. (a)	272. (a)	273. (a)	274. (a)	275. (b)	276. (a)

Q. A.	Q. A.	Q. A.	Q. A.	Q. A.	Q. A.
277. (b)	278. (a)	279. (a)	280. (a)	281. (a)	282. (a)
283. (a)	284. (b)	285. (a)	286. (d)	287. (c)	288. (a)
289. (a)	290. (c)	291. (b)	292. (a)	293. (a)	294. (d)
295. (a)	296. (c)	297. (a)	298. (a)	299. (a)	300. (a)
301. (b)	302. (a)	303. (a)	304. (a)	305. (c)	306. (a)
307. (b)	308. (d)	309. (b)	310. (c)	311. (d)	312. (b)
313. (d)	314. (c)	315. (c)	316. (a)	317. (d)	318. (c)
319. (c)	320. (a)	321. (c)	322. (a)	323. (b)	324. (c)
325. (a)	326. (b)	327. (a)	328. (a)	329. (b)	330. (c)
331. (c)	332. (b)	333. (c)	334. (d)	335. (d)	336. (a)
337. (d)	338. (d)	339. (d)	340. (a)	341. (b)	342. (a)
343. (a)	344. (a)	345. (d)	346. (c)	347. (b)	348. (a)
349. (a)	350. (a)	351. (b)	352. (c)	353. (a)	354. (a)
355. (d)	356. (d)	357. (a)	358. (d)	359. (d)	360. (b)
361. (a)	362. (a)	363. (a)	364. (b)	365. (a)	366. (b)
367. (a)	368. (a)	369. (c)	370. (b)	371. (a)	372. (d)
373. (c)	374. (a)	375. (b)	376. (a)	377. (a)	378. (a)
379. (c)	380. (d)	381. (a)	382. (a)	383. (c)	384. (d)
385. (a)	386. (c)	387. (a)	388. (a)	389. (c)	390. (b)
391. (d)	392. (a)	393. (d)	394. (d)	395. (a)	396. (c)
397. (a)	398. (a)	399. (a)	400. (d)	401. (b)	402. (a)
403. (a)	404. (a)	405. (a)	406. (a)	407. (c)	408. (a)
409. (c)	410. (d)	411. (c)	412. (c)	413. (b)	414. (a)
415. (a)	416. (c)	417. (a)	418. (b)	419. (a)	420. (a)
421. (c)	422. (a)	423. (c)	424. (a)	425. (d)	426. (a)
427. (d)	428. (a)	429. (b)	430. (d)	431. (a)	432. (c)
433. (b)	434. (d)	435. (b)	436. (a)	437. (b)	438. (a)
439. (a)	440. (c)	441. (b)	442. (d)	443. (a)	444. (b)
445. (b)	446. (c)	447. (d)	448. (a)	449. (b)	450. (d)
451. (c)	452. (d)	453. (a)	454. (d)	455. (a)	456. (b)
457. (a)	458. (a)	459. (b)	460. (c)	461. (c)	462. (a)
463. (d)	464. (a)	465. (b)	466. (a)	467. (a)	468. (a)
469. (d)	470. (a)	471. (a)	472. (a)	473. (b)	474. (c)
475. (a)	476. (c)	477. (b)	478. (c)	479. (d)	480. (b)

Q.	A.	Q.	A.	Q.	A.	Q.	A.	Q.	A.	Q.	A.
481.	(a)	482.	(a)	483.	(a)	484.	(a)	485.	(d)	486.	(c)
487.	(a)	488.	(d)	489.	(b)	490.	(a)	491.	(a)	492.	(b)
493.	(a)	494.	(a)	495.	(a)	496.	(c)	497.	(b)	498.	(a)
499.	(b)	500.	(b)	501.	(a)	502.	(a)	503.	(b)	504.	(c)
505.	(d)	506.	(a)	507.	(a)	508.	(a)	509.	(a)	510.	(a)
511.	(d)	512.	(a)	513.	(c)	514.	(c)	515.	(d)	516.	(a)
517.	(c)	518.	(d)	519.	(d)	520.	(b)	521.	(b)	522.	(a)
523.	(d)	524.	(c)	525.	(d)	526.	(d)	527.	(d)	528.	(a)
529.	(c)	530.	(d)	531.	(b)	532.	(d)	533.	(d)	534.	(d)
535.	(a)	536.	(a)	537.	(a)	538.	(c)	539.	(c)	540.	(a)
541.	(b)	542.	(d)	543.	(a)	544.	(d)	545.	(a)	546.	(c)
547.	(a)	548.	(d)	549.	(d)	550.	(a)	551.	(d)	552.	(d)
553.	(a)	554.	(b)	555.	(a)	556.	(c)	557.	(b)	558.	(a)
559.	(d)	560.	(a)	561.	(b)	562.	(a)	563.	(b)	564.	(b)
565.	(a)	566.	(a)	567.	(d)	568.	(a)	569.	(a)	570.	(b)
571.	(c)	572.	(a)	573.	(c)	574.	(a)	575.	(b)	576.	(b)
577.	(b)	578.	(a)	579.	(a)	580.	(d)	581.	(c)	582.	(d)
583.	(b)	584.	(b)	585.	(b)	586.	(c)	587.	(a)	588.	(a)
589.	(d)	590.	(b)	591.	(a)	592.	(b)	593.	(a)	594.	(a)
595.	(c)	596.	(a)	597.	(a)	598.	(b)	599.	(c)	600.	(b)
601.	(a)	602.	(c)	603.	(c)	604.	(d)	605.	(b)	606.	(c)
607.	(d)	608.	(d)	609.	(a)	610.	(a)	611.	(c)	612.	(c)
613.	(c)	614.	(a)	615.	(d)	616.	(d)	617.	(c)	618.	(c)
619.	(d)	620.	(d)	621.	(a)	622.	(c)	623.	(c)	624.	(d)
625.	(b)	626.	(c)	627.	(a)	628.	(c)	629.	(b)	630.	(b)
631.	(d)	632.	(d)	633.	(b)	634.	(a)	635.	(c)	636.	(a)
637.	(b)	638.	(c)	639.	(b)	640.	(d)	641.	(b)	642.	(d)
643.	(a)	644.	(a)	645.	(c)	646.	(c)	647.	(a)	648.	(c)
649.	(b)	650.	(d)	651.	(c)	652.	(c)	653.	(b)	654.	(a)
655.	(b)	656.	(d)	657.	(a)	658.	(b)	659.	(c)	660.	(c)
661.	(b)	662.	(b)	663.	(a)	664.	(a)	665.	(a)	666.	(a)
667.	(b)	668.	(b)	669.	(d)	670.	(d)	671.	(b)	672.	(b)
673.	(b)	674.	(b)	675.	(a)	676.	(b)	677.	(a)	678.	(b)
679.	(d)	680.	(d)	681.	(b)	682.	(a)	683.	(a)	684.	(b)

Q. A.	Q. A.	Q. A.	Q. A.	Q. A.	Q. A.
685. (d)	686. (a)	687. (b)	688. (c)	689. (b)	690. (a)
691. (c)	692. (b)	693. (b)	694. (a)	695. (a)	696. (a)
697. (c)	698. (b)	699. (d)	700. (d)	701. (d)	702. (d)
703. (c)	704. (a)	705. (c)	706. (d)	707. (a)	708. (b)
709. (b)	710. (c)	711. (a)	712. (b)	713. (b)	714. (b)
715. (a)	716. (c)	717. (b)	718. (b)	719. (c)	720. (b)
721. (c)	722. (a)	723. (c)	724. (c)	725. (c)	726. (c)
727. (d)	728. (b)	729. (c)	730. (b)	731. (a)	732. (b)
733. (a)	734. (c)	735. (c)	736. (c)	737. (a)	738. (a)
739. (c)	740. (c)	741. (a)	742. (a)	743. (a)	744. (a)
745. (c)	746. (d)	747. (c)	748. (b)	749. (a)	750. (b)
751. (b)	752. (a)	753. (b)	754. (a)	755. (a)	756. (a)
757. (d)	758. (c)	759. (a)	760. (d)	761. (c)	762. (a)
763. (a)	764. (a)	765. (d)	766. (d)	767. (c)	768. (b)
769. (c)	770. (c)	771. (b)	772. (a)	773. (b)	774. (d)
775. (d)	776. (d)	777. (d)	778. (b)	779. (a)	780. (d)
781. (b)	782. (a)	783. (d)	784. (c)	785. (c)	786. (a)
787. (a)	788. (d)	789. (d)	790. (a)	791. (b)	792. (b)
793. (c)	794. (a)	795. (b)	796. (d)	797. (a)	798. (b)
799. (b)	800. (b)	801. (a)	802. (a)	803. (a)	804. (b)
805. (d)	806. (a)	807. (b)	808. (b)	809. (c)	810. (d)
811. (c)	812. (b)	813. (a)	814. (b)	815. (d)	816. (a)
817. (a)	818. (a)	819. (c)	820. (a)	821. (c)	822. (d)
823. (a)	824. (d)	825. (c)	826. (a)	827. (b)	828. (b)
829. (a)	830. (d)	831. (d)	832. (d)	833. (a)	834. (d)
835. (d)	836. (a)	837. (d)	838. (b)	839. (d)	840. (b)
841. (a)	842. (a)	843. (b)	844. (a)	845. (b)	846. (d)
847. (b)	848. (a)	849. (b)	850. (a)	851. (a)	852. (d)
853. (d)	854. (b)	855. (a)	856. (b)	857. (b)	858. (c)
859. (d)	860. (a)	861. (d)	862. (c)	863. (d)	864. (b)
865. (b)	866. (c)	867. (d)	868. (a)	869. (b)	870. (d)
871. (a)	872. (d)	873. (d)	874. (d)	875. (b)	876. (b)
877. (a)	878. (b)	879. (c)	880. (d)	881. (d)	882. (b)
883. (c)	884. (c)	885. (c)	886. (c)	887. (d)	888. (a)

Q.	A.	Q.	A.	Q.	A.	Q.	A.	Q.	A.	Q.	A.
889.	(d)	890.	(c)	891.	(a)	892.	(c)	893.	(c)	894.	(a)
895.	(d)	896.	(b)	897.	(c)	898.	(c)	899.	(b)	900.	(d)
901.	(c)	902.	(b)	903.	(a)	904.	(b)	905.	(a)	906.	(c)
907.	(c)	908.	(a)	909.	(b)	910.	(a)	911.	(d)	912.	(b)
913.	(c)	914.	(b)	915.	(a)	916.	(a)	917.	(c)	918.	(c)
919.	(a)	920.	(c)	921.	(a)	922.	(b)	923.	(a)	924.	(a)
925.	(c)	926.	(d)	927.	(c)	928.	(a)	929.	(a)	930.	(a)
931.	(a)	932.	(c)	933.	(b)	934.	(c)	935.	(a)	936.	(b)
937.	(b)	938.	(a)	939.	(c)	940.	(d)	941.	(a)	942.	(c)
943.	(a)	944.	(c)	945.	(b)	946.	(a)	947.	(a)	948.	(c)
949.	(c)	950.	(d)	951.	(c)	952.	(d)	953.	(d)	954.	(a)
955.	(c)	956.	(a)	957.	(b)	958.	(b)	959.	(b)	960.	(c)
961.	(c)	962.	(b)	963.	(a)	964.	(c)	965.	(c)	966.	(b)
967.	(c)	968.	(d)	969.	(c)	970.	(d)	971.	(b)	972.	(d)
973.	(c)	974.	(c)	975.	(a)	976.	(b)	977.	(d)	978.	(b)
979.	(a)	980.	(a)	981.	(a)	982.	(b)	983.	(b)	984.	(b)
985.	(d)	986.	(a)	987.	(b)	988.	(c)	989.	(b)	990.	(a)
991.	(d)	992.	(b)	993.	(b)	994.	(a)	995.	(c)	996.	(a)
997.	(c)	998.	(a)	999.	(d)	1000.	(b)				

CHAPTER 5

DAIRY TECHNOLOGY

5.1 SUBJECTIVE KEYNOTES

1. Major constituents of milk are water, fat, protein, lactose and minerals.
2. Minor constituents of milk are phospholipids, sterols, vitamins, enzymes, pigments etc.
3. True constituents of milk are milk fat, casein and lactose.
4. Milk proteins are considered complete proteins as they contain all essential amino acids i.e. tryptophan, phenylalanine, lysine, threonine, valine, methionine, leucine, arginine and histidine
5. Milk is an excellent source of calcium and phosphorus.
6. Milk is low in iron, copper and iodine.
7. Various types of milk depending on fat and SNF content are: full cream milk, standardized milk, recombined milk, toned milk, double toned milk and skim milk.
8. Different types of processed milk includes pasteurized milk, flavored milk, sterilized milk and UHT milk.
9. Jersey breed cow milk contains highest fat content among cow breeds.
10. Murrah breed buffalo milk contains highest fat content.
11. Pearson's square method is used for standardization of fat and SNF in processed milk.
12. Phosphatase test is used to check adequacy of pasteurization.
13. Lactose is the sugar found only in milk and is 1/6th as sweet as sucrose. Lactose accounts for approximately 54% of the total solids-not-fat content of milk
14. Carotene (fat soluble) is responsible for yellow color of milk, ghee, cream, butter and other dairy products while riboflavin gives a greenish yellow color to whey.
15. The TA of cow milk varies from 0.13 to 0.14% and 0.14 to 0.15% in buffalo milk.
16. Pasteurization of milk is done at 63°C/30 minutes, 72°C/15sec.
17. Standardised milk should contain a minimum of 4.5% fat and 8.5% SNF.
18. Toned milk should contain a minimum of 3.0% fat and 8.5% SNF.
19. Double toned milk should contain a minimum of 1.5% fat and 9.0% SNF.
20. Cream should contain NLT 25% milk fat.
21. Butter must contain NMT 1.5% by weight curd, NMT 3.0% by weight of common salt and NLT 80% by weight of milk fat.

22. The flavor of butter is chiefly attributed by diacetyl content.
23. The color of butter is due to added coloring agent annatto.
24. Butteroil refers to fat concentrate obtained mainly from butter or cream by removal of practically all the water and SNF.
25. The per centage of butter fat in butteroil is 99.5-99.8%.
26. Ice cream is frozen dairy product and should contain NLT 10% milk fat, 3.5% protein, 36% total solids, NMT 0.5% stabilizers and emulsifiers by weight.
27. Cheese is a product obtained by draining after the coagulation of milk and may contain coagulating agent, sodium chloride and calcium chloride not exceeding 0.02% by weight.
28. Unsweetened condensed milk should contain NLT 8% milk fat and NLT 26% milk solids.
29. Sweetened condensed milk should contain NLT 9% milk fat, NLT 31% total milk solids and NLT40% canesugar.
30. Unsweetened condensed skim milk should contain NLT 20% total milk soilds and NMT 0.5% fat by weight.
31. Sweetened condensed skim milk should contain NLT 26% total milk soilds, NLT 40% cansugar and NMT 0.0% fat content by weight.
32. Whole milk powder should contain NMT 5% moisture, NLT 26% fat and NMT 1.2% total acidity.
33. Skim milk powder should contain NMT 1.5% milk fat, NMT 5% moisture and NMT 1.5% total acidity.
34. Milk powders are drum dried/roller dried and spray dried.
35. The maximum solubility index should be 15 for roller dried and 2 for spray dried powder.
36. Khoa, mawa, rabri and kulfi are concentrated whole milk products.
37. Dahi, shrikhand, paneer and chhana are coagulated milk products.
38. Makkhan, ghee, lassi and ghee residue are products of clarified butter fat industry.
39. Khoa should contain NLT 20% milk fat.
40. Pindi, dhar, danedar are varieties of khoa.
41. Chhana or paneer should contain NMT 70% moisture and NLT 50% milk fat of dry matter.
42. Ghee is the pure clarified fat derived solely from milk or from desi butter and should contain NLT 99-99.5% milk fat and NMT 0.5% moisture.
43. Ghee residue refers to the charred light to dark brown residue which is obtained after the filtration of ghee.
44. Ghee residue is the by-product of ghee industry.
45. Diacetyl content in butter should be NMT 4ppm.
46. India is the leading producer of milk in world with annual milk production of 155.5 million tonnes (2015-16).
47. Per capita availability of milk in India in 2015-16 was 337 g/day.

48. Sanitizers and cleaning agents form an important part of the dairy industry for hygiene maintenance.
49. The common adulterants of milk are urea, starch, sucrose, water, neutralizers and detergents.
50. Dr. V. Kurien is known as "Father of White Revolution".
51. Excellent Indian cow breeds include Sahiwal, Rathi, Gir and Red Sindhi.
52. Murrah, Mehsani and Jaffarbadi are some excellent Indian buffalo breeds.
53. Milk production is growing 7% by volume and about 5% by value.
54. The first large-scale and systematic breakthrough in dairy co-operatives in India was made in 1948 by the Kaira District Co-operatives Milk producers Union Limited at Anand in 1946, referred to as "Anand pattern". The Kaira Cooperative Union established a marketing agency named Gujarat Cooperative Milk Marketing Federation (GCMMF).
55. The National Dairy Development Board (NDDB), set up at Anand in 1965, drew up a program known as "Operation flood" (commenced in 1970) to replicate the Anand pattern in 18 areas of milk production in the milk sheds of Bombay, Delhi, Calcutta and Madras.
56. Operation Flood was one of the world's largest rural development programs which ran for 26 years and eventually helped India to emerge as the world"s largest milk producer.
57. Milk is processed and marketed by 170 milk producer cooperative unions, which federate into 15 State Cooperative Milk Marketing Federations.
58. Over the years, several brands have been created by cooperatives like Amul (Gujarat), Verka (Punjab), Vita (Haryana), Vijaya (Andhra Pradesh), Saras (Rajasthan), Nandini (Karnataka), Milma (Kerala) and Gokul (Kolhapur).
59. In 1992 the MMPO was promulgated under the essential commodities Act 1955 to regulate milk and milk products production in the country.
60. Milk contains proteins in the form of casein and whey proteins. Of the total protein in milk, about 80% is casein and 20% is whey protein.
61. Casein can be fractionated into four major components: alpha-, beta-, gamma-, and kappa-casein.
62. Whey protein consists predominantly of beta-lactoglobulin and alpha-lactalbumin.
63. The first stage at which raw milk is received is known as milk reception dock or platform or raw milk receiving dock (RMRD).
64. Reception of raw milk include operations such as unloading, grading, conveying, sampling, testing, weighing or measuring and recording, dumping and pumping.
65. At RMRD, rigorous examination of raw milk is carried out which include organoleptic evaluation, physical and chemical tests to assess the quality and to decide for acceptance or rejection of raw milk. All these tests are known as "platform tests".
66. Standardization of milk refers to the adjustment, i.e. raising or lowering, of the fat and/or solids-not-fat percentages of milk to a desired value, so as to conform to the legal or other requirements prescribed.

67. The term 'commercial sterilization' refers to the absence of any microorganisms capable of reproducing in foods under non-refrigerated conditions of storage and distribution, instead of the absolute absence of every existing form of microbial life.
68. The first aseptically processed milk in metal cans was developed in Denmark by Jonas Nielsen in 1913.
69. Hydrogen peroxide (H_2O_2) is the most popular sterilization method for aseptic packaging, because it is relatively fast and efficient.
70. Defects in UHT milk include cooked and stale flavor, chalky, astringent, sediment, fat separation, gelation etc.
71. The Lactoperoxidase system (LP system) of raw milk preservation is currently the only approved method of raw milk preservation, apart from refrigeration, in Codex.
72. The LP-system operates by the reactivation of the enzyme lactoperoxidase, which is naturally present in raw milk, by the addition of thiocyanate (10 ppm) and a source of peroxide (8.5 ppm of sodium percarbonate). This results in a blocking of bacterial metabolism thereby preventing the multiplication of bacteria present in the milk.
73. Thermization is a mild pasteurization operation sometimes applied to raw milk that is intended to be held after intake for extended periods of refrigerated storage before manufacture of certain products. Thermization is typically heating milk to 57-68°C for a period of 15 sec, which will inactivate psychrotrophic bacteria and thereby prevent enzyme production.
74. Chhana is a base material for sweetmeats like Rasogolla, Sandesh, etc.
75. Bactofugation is a process in which a specially designed centrifuge called a 'Bactofuge' is used to separate micro-organisms from milk (for cheese making or UHT milk). The optimal bactofugation temperature is 55–62°C.
76. Lactose in milk can be hydrolyzed by adding an enzyme β-D-galactosidase into milk, which breaks down lactose into glucose and galactose.
77. The milk serum has higher specific gravity (1.036) than that of fat (0.9) which results in cream and skim milk separation in cream separator.
78. Cream is defined as the fat rich portion of milk obtained by gravity or mechanical method of separation. The fat content of cream varies widely in the range of 18- 85% depending upon the method of separation.
79. Fat % in low, medium & high fat cream should not be less than 25,40 & 60%, respectively.
80. Minimum 76% fat content should be present in desi butter as per FSSAI standards.
81. The steps involved in preparation of cream for buttermaking are standardization of cream (38-41% fat), pasteurization (80oC/16 sec.), cooling (~ 5oC) and ageing (5-8oC/6-12 h) of cream.
82. The word ghee is evolved from Sanskrit word 'ghrut'. Ghee is a common Indian name for clarified butter fat.
83. Ghee may contain butylated hydroxyl anisole (BHA) at concentration not exceeding 0.02%. Baudouin test should be negative. Positive test indicates presence of dalda (hydrogenated vegetable fat) as an adulterant in ghee.
84. In India, ghee is marketed from organized dairies under Agmark certification, wherein there are three grades based on the Free Fatty Acids (FFA) contents: Special

(Max. 1.4% FFA), General (Max. 2.5% FFA) and Standard (Max. 3.0% FFA).

85. Cow milk fat has a dense yellow colour due to presence of carotenoids (ß-carotene).

86. Ghee has a melting range of 28 to 44°C. Its Butyro-refractometer (BR) reading is 40 to 45 at 40°C. The saponification number is not less than 220. The iodine value varies from 26 to 38. The Polenske value (PV) of cow ghee (2.0-3.0) is higher than for buffalo ghee (1.0-1.5).

87. Anhydrous milk fat (AMF) is a milk fat-based dairy product in its purest form. It is the fatty products derived exclusively from milk and/ or products obtained from milk by means of process which results in almost total removal of water and milk solids not fat.

88. Chhana is heat and acid-coagulated milk product, also known as India's traditional soft, cottage cheese.

89. Chhana making is essentially a process involving destabilization of casein particle by acidification of milk with dilute acid at relatively higher temperature.

90. Rasogolla is chhana based sweet & was made for the first time in 1868 in Kolkata by Mr. Nobin Chandra Das.

91. *Khoa* is an important indigenous milk product. It is conventionally prepared by heating, evaporating and desiccation of milk in an open kettle at atmospheric pressure accompanied by continuous stirring until dough like consistency is achieved. Three varieties of khoa are dhap, pindi and danedar.

92. Shrikhand is a popular fermented, sweetened, indigenous dairy product having semi-solid consistency. It has typical sweetish-sour taste. Shrikhand is fermented milk product which is very close to flavoured quarg of Germany.

93. In fruit Shrikhand milk fat on dry matter basis not less than 7.0% and protein on dry matter not less than 9.0%.

94. Kefir is characterized by foamy, effervescent milk product. It is fermented milk product which involves mixed lactic acid and alcoholic type of fermentation.

95. Koumiss is a mare's milk fermented product similar to kefir and is very popular in Russia and Central Asia.

96. Yoghurt is a coagulated milk product obtained by lactic acid fermentation through the action of *S. thermophilus* and *Lb. bulgaricus*.

97. The cheese may contain maximum of 3000 ppm sorbic acid or its Na, K or Ca salts calculated as sorbic acid, and or 12.5 ppm of nisin, either singly or in combination. Natamycin may be used for surface treatment only.

98. All cheeses, whether rennet or acid set, can be classified as soft, semi-soft (semi-hard), hard, or very hard, depending on their moisture content.

99. Sweetened condensed milks are milk products which can be obtained by the partial removal of water from milk with the addition of sugar, or by any other process which leads to a product of the same composition and characteristics. It may sometimes contain added refined lactose, calcium chloride, citric acid and sodium citrate, sodium salts of orthophosphoric acid and polyphosphoric acid not exceeding 0.3 per cent by weight of the finished product.

100. Evaporated milk is sterilized, concentrated, homogenized milk. The product can be kept without refrigeration and has a long shelf life.

5.2 MULTIPLE CHOICE QUESTIONS

1. The highest producer of milk in world is
(a) China (b) India
(c) Pakistan (d) USA

2. The highest producer of milk in India is
(a) Gujarat (b) Haryana
(c) Rajasthan (d) Uttar Pradesh

3. The minimum per centage of milk fat in standardized milk is
(a) 4.5 (b) 4
(c) 3.5 (d) 3

4. Which among the following has the highest milk protein content
(a) cow (b) buffalo
(c) human (d) goat

5. Which among the following is a minor component of milk
(a) water (b) fat
(c) lactose (d) phospholipids

6. The principal sugar/carbohydrate in milk is
(a) glucose (b) sucrose
(c) lactose (d) maltose

7. Milk is deficient in vitamin
(a) vitamin A (b) vitamin B
(c) vitamin D (d) vitamin C

8. The average size (micron) of fat globule is
(a) 0.1-1.0 (b) 0.01-0.1
(c) 2-5 (d) 20-50

9. The principal milk protein is
(a) Casein (b) lactoglobulin
(c) lactalbumin (d) gliadin

10. Casein is precipitated at __________ pH in milk
(a) 4.0 (b) 4.6
(c) 4.8 (d) 5.0

11. Sandiness in ice cream is due to
(a) Casein (b) lactose
(c) phospholipids (d) calcium

12. Which of the following is exclusively found in milk
(a) lactose (b) phospholipids
(c) vitamin D (d) sucrose

13. The milk pigment riboflavin is
(a) fat soluble (b) water soluble
(c) both (d) none

14. The yellow color of milk products like ghee is due to
(a) cholesterol (b) riboflavin
(c) carotene (d) xanthophyll

15. The greenish yellow color of whey is due to
(a) riboflavin (b) carotene
(c) xanthophylls (d) casein

16. Which of the following enzyme is basis for checking the adequacy of pasteurization
(a) lipase (b) phosphatase
(c) peroxidase (d) catalase

17. Natural milk is ______ in nature
(a) acidic (b) basic
(c) highly acidic (d) neutral

18. The per cent TA of cow milk ranges from
(a) 0.15-0.16 (b) 0.13-0.14
(c) 0.17-0.18 (d) 0.18-0.19

19. The per cent TA of buffalo milk ranges from
(a) 0.12-0.14 (b) 0.18-0.20
(c) 0.14-0.15 (d) 0.10-0.11

20. The developed acidity of milk is due to formation of
(a) citric acid (b) lactic acid
(c) formic acid (d) glutamic acid

21. The titratable acidity of milk is referred as
(a) % of citric acid
(b) % of formic acid
(c) % of lactic acid
(d) % of glutamic acid

22. pH of freshly drawn cow milk is
(a) 6.8-7.0 (b) 6.2-6.4
(c) 6.4-6.6 (d) 6.0-6.2

23. pH of freshly drawn buffalo milk is
(a) 6.7-6.8 (b) 6.5-6.6
(c) 6.9-7.0 (d) 7.1-7.2

24. Higher pH values of milk is due to
(a) casein
(b) mastitis
(c) bacterial action
(d) light exposure

25. Lower pH values of milk is due to
(a) casein
(b) mastitis
(c) bacterial action
(d) light exposure

26. The average specific gravity of cow milk ranges from
(a) 1.000-1.002 (b) 1.028-1.030
(c) 1.050-1.060 (d) 1.042-1.062

27. The average specific gravity of buffalo milk ranges from
(a) 1.035-1.037 (b) 1.040-1.043
(c) 1.028-1.031 (d) 1.047-1.050

28. The specific gravity of milk is expressed at_________°C
(a) 10.6 (b) 12.6
(c) 15.6 (d) 17.6

29. Which among the following has highest specific gravity
(a) whole milk
(b) buffalo milk
(c) skim milk
(d) standardized milk

30. Which among the following has highest specific gravity
(a) fat (b) protein
(c) lactose (d) salt

31. Addition of water to milk _________ the specific gravity
(a) decreases (b) increases
(c) do not affect (d) none

32. The specific gravity of milk is increased upon addition of
(a) fat (b) water
(c) both (a) & (b) (d) skim milk

33. Milk freezes at a temperature
(a) similar to water
(b) less than water
(c) greater than water
(d) milk does not freeze

34. The average freezing point depression of cow milk is
(a) 0.647 (b) 0.347
(c) 0.547 (d) 0.447

35. The average freezing point depression of buffalo milk is
(a) 0.649 (b) 0.349
(c) 0.549 (d) 0.449

36. Mastitis milk shows a decreased _________ level
(a) potassium (b) lactoferrin
(c) both (d) none

37. Full form of NDDB is
(a) National Dairy Development Board
(b) National Dairy Department Bureau
(c) National Draught Area Development
(d) none

38. Head quarter of NDDB is situated at
(a) New Delhi (b) Anand
(c) Ahmedabad (d) Banglore

39. Chairman of NDDB is
(a) Dr. Amrita Patel (b) Dr. P.R. Patel
(c) Dr. A. Shah (d) K.P. Singh

40. GCMMF is associated with
(a) Goa (b) Gujarat
(c) Gangtok (d) None

41. Milk SNF refers to
(a) salts not fat (b) solids not fat
(c) soluble neutral fats
(d) none

42. Pasteurization of milk is done at
(a) 63°C for 30 min
(b) 63°C for 30 sec
(c) 71°C for 30 min
(d) 59°C for 30 sec

43. The heat stability of milk is judged by
(a) MBRT (b) COB
(c) lactometer (d) freezing point

44. The detection of adulteration of milk with water is done by
(a) T.A.
(b) ethanol test
(c) lactometer
(d) Alcohol-Alizarin test

45. Bactofugation is a process to remove bacteria by
(a) sedimentation
(b) centrifugation
(c) precipitation
(d) homogenization

46. COB stands for
(a) clot on boiling
(b) crumble on boiling
(c) clotting of bacteria
(d) crumling of bacteria

47. CIP stands for
(a) clot in pasteurization
(b) cleaning in plant
(c) cells in pasteurizer
(d) cleaning in place

48. COP stands for
(a) Clinical Operating Procedure
(b) Cleaning Out of Place
(c) Cumulative Operating Procedure
(d) none

49. Milk is stored at ________°C upon receiving
(a) 0 (b) 5
(c) 20 (d) room temperature

50. Preheating of milk is done at _____°C before homogenization
(a) 35-40 (b) 40-45
(c) 45-50 (d) 50-55

51. Pearson's square method is used to
(a) detect adulteration in milk
(b) bacteria removal
(c) form a standardized milk
(d) none

52. HTST of milk is done at
(a) 72°C/15sec (b) 82°C/10sec
(c) 79°C/30sec (d) 75°C/15sec

53. Pateurization of milk causes destruction of
(a) pathogenic and non-pathogenic bacteria
(b) all bacteria and toxins
(c) only toxins
(d) both both (b) & (c)

54. Pasteurization ________ the cream volume
(a) decreases (b) increases
(c) no effect (d) none

55. Homogenization of milk is carried out at
(a) 2500 psi (b) 2000 psi
(c) 1500 psi (d) 3000 psi

56. Vacreation refers to pasteurization of cream/milk under
(a) high pressure
(b) reduced pressure
(c) high temperature
(d) low temperature

57. Time/temperature combination of UHT milk
(a) 100°C/10sec (b) 150°C/1sec
(c) 150°C/5sec (d) 100°C/30sec

58. Metal which aid in oxidation of milk
(a) P & Ca (b) Cu & Fe
(c) Co & Mg (d) Calcium

59. The sequence of steps involved in process of milk are
(a) clarification, homogenization, preheating, pasteurization, cooling
(b) preheating, homogenization, clarification, pasteurization, cooling
(c) cooling, homogenization, preheating, clarification, pasteurization
(d) clarification, preheating, homogenization, pasteurization, cooling

60. Acidophillus milk microorganism
(a) Lactobacillus acidophilus
(b) lactococcus acidophilus
(c) Streptococcus acidophilus
(d) streptococcus lactis

61. Kumiss is a type of
(a) Ice cream
(b) cheese
(c) fermented milk
(d) fermented soy product

62. Yoghurt microorganism is
(a) Streptococcus thermophillus
(b) Lactobacillus bulgaricus
(c) both
(d) none

63. According to FSSA, standardized milk should contain
(a) 4.5%fat & 8.5% SNF
(b) 5.0%fat & 8.5% SNF
(c) 4.5%fat & 9.5% SNF
(d) 5.0%fat & 9.5% SNF

64. According to FSSA, recombined milk should contain
(a) 4.0%fat & 8.5% SNF
(b) 3.0%fat & 8.5% SNF
(c) 3.5%fat & 9.0% SNF
(d) none

65. According to FSSA, toned milk should contain
(a) 4.0% fat & 8.5% SNF
(b) 3.0%fat & 8.5% SNF
(c) 3.5%fat & 9.0% SNF
(d) none

66. According to FSSA, double toned milk should contain
(a) 1.0% fat & 9.0% SNF
(b) 2.0% fat & 9.0% SNF
(c) 1.5% fat & 8.5% SNF
(d) 1.5% fat & 9.0% SNF

67. In milk, whey proteins are present in
(a) true solution
(b) suspended particles
(c) colloidal solution
(d) none

68. Which of the following has highest particle size in milk
(a) fat globules (b) casein
(c) whey protein (d) lactose

69. Milk is ________ type emulsion
(a) oil in water (b) water in oil
(c) not emulsion (d) both (a) & (b)

70. The normal pH of milk is
(a) 6.6-6.8 (b) 7.8-8.8
(c) 2.4 (d) 5-5.5

71. Most abundant milk proteins are
(a) whey proteins
(b) albumin
(c) casein
(d) fat globule membrane proteins

72. Which of the following is not a type of milk casein
(a) alpha (b) beta
(c) gama (d) omega

73. Milk serum proteins are also called
(a) whey protein (b) casein
(c) both (d) none

74. α-lactalbumin and β-lactoglobulin comprises the
(a) casein (b) whey proteins
(c) both (d) none

75. Which of the following enzyme is not found in milk system
(a) peroxidase (b) catalase
(c) phosphatase (d) diastage

76. Lactose is split into ________& ________ by lactase
(a) glucose & fructose
(b) galactose & glucose
(c) galactose & fructose
(d) none

77. The sunlight flavor of milk is due to oxidation of
(a) fat (b) carbohydrate
(c) protein (d) vitamin

78. Milk fat mainly consists of
(a) free fatty acids
(b) TGA
(c) mono & diglycerides
(d) phospholipid

79. Fresh milk is
(a) slightly basic
(b) extremely acidic
(c) slighltly acidic
(d) extremely basic

80. Which of the following is highest in milk
(a) phospholipid (b) viatmin
(c) minerals (d) lactose

81. Which of the following casein is highest in milk
(a) α-casein (b) β-casein
(c) k-casein (d) none

82. ________ reach the maximum level when green feed is in plenty
(a) carotene (b) vitamin B_2
(c) both (a) & (b) (d) none

83. Yieth ration in milk is __________ for the presence of lactose: protein: mineral
(a) 9: 13: 2 (b) 13: 9: 2
(c) 2: 13: 9 (d) 13:2:9

84. The SPC for milk has been specified as
(a) NLT 200000 (b) NMT 200000
(c) NMT 20000 (d) NLT 20000

85. Commercially vitamin D in milk can be enhanced through
(a) irradiation (b) fortification
(c) feed (d) all

86. Alcohol test for raw milk is used to find out its stability during
(a) heating of milk
(b) freezing of milk
(c) transit
(d) none of the above

87. Developed acidity in milk is due to the accumulation of
(a) salts (b) lactic acid
(c) calcium ions (d) phosphates

88. Souring and gassiness in milk is due to
(a) LAB (b) Str. Thermophilus
(c) *E. coli* (d) yeast

89. The SPC/ml of pasteurized milk should be
(a) NMT 30000 (b) NLT 30000
(c) NMT 20000 (d) NLT 20000

90. Lactometer is used to check
(a) adulteration of milk with water
(b) for heat stability
(c) for fat estimation
(d) for freezing point determination

91. HTST stands for
(a) heating temperature short time
(b) high temperature short time
(c) high temperature small time
(d) heating temperature small time

92. Standardization refers to _______ of milk
(a) adjusting the fat content
(b) adjusting the acidity
(c) adjusting SNF
(d) both (a) & (c)

93. Pasteurization was invented by
(a) Dr. Soxhlet
(b) Nicolas Appert
(c) Louis Pasteur
(d) Robert Koch

94. FDV stands for ________ in milk processing
(a) flow direct valve
(b) first direct valve
(c) flow divert valve
(d) flow divert vacreation

95. Mother Dairy has its headquarter at
(a) New Delhi (b) Mumbai
(c) Anand (d) Gandhinagar

96. Bitter taste in milk is due to enzymatic breakdown of
(a) fats
(b) carbohydrates
(c) proteins
(d) calcium

97. Cowy flavor of milk is due to
(a) acetonemia (b) mastitis
(c) ketosis (d) both (a) & (b)

98. Rancid flavor in milk is primarily due to
(a) caproic acid (b) butyric acid
(c) palmitic acid (d) stearic acid

99. The president of IDA is
(a) NR Bhasin (b) Amrita Patel
(c) HN Patel (d) SK Shah

100. Head quarter of IDA is situated at
(a) Mumbai (b) New Delhi
(c) Bhopal (d) Banglore

101. According to FSSA/PFA, cream should contain per cent milk fat
(a) NLT 25 b) NMT 25
(c) NMT 20 (d) NLT 20

102. Maximum clumping of fat in cream in cream take place at ________ °C
(a) 5 b) 7
(c) 9 (d) 10

103. As the size of fat globules increases, the rate at which cream rises/ separates
(a) decreases (b) increases
(c) no effect (d) none

104. Lower the temperature of milk ________ will be the fat% of cream
(a) lower (b) equivalent
(c) higher (d) no effect

105. 100 kg milk testing 8% fat, cream produced 15 kg testing 50% fat, skimming efficiency is
(a) 93.75 (b) 93.75 %
(c) 98.7 (d) 98.7%

106. Temperature (°C) of cream during addition of neutralizer should be
(a) 38-42 (b) 29-32
(c) 20-23 (d) 42-45

107. Plastic cream consists of ________ % milk fat
(a) 20-25 (b) 30-40
(c) 65-85 (d) 55

108. High milk fat % is present in ________ cream
(a) sour (b) plastic
(c) butter (d) sweet

109. Higher the temperature ________ will be the viscosity of cream
(a) lower (b) higher
(c) no effect (d) none

110. Higher the temperature of separation ________ the viscosity of cream
(a) lower (b) higher
(c) no effect (d) none

111. Single stage homogenization ______ viscosity
(a) decrease (b) increase
(c) do not effect (d) none

112. ________ % fat is ideal for whipping of cream
(a) 25-30 (b) 30-35
(c) 35-40 (d) 40-45

113. The specific gravity of cream ________ with increase in fat %
(a) decreases (b) increases
(c) do not effect (d) none

114. The density of milk fat at 16°C is
(a) 1.013 (b) 0.53
(c) 0.93 (d) 0.83

115. The velocity of rise of fat globules is ________ related to size of fat globule
(a) directly
(b) inversely
(c) no relation
(d) cannot be determined

116. Creaming occurs faster due to ________ size of fat globule
(a) smaller (b) larger
(c) average (d) no effect

117. Lower temperature of milk during separation, ________ the fat % of cream
(a) lower (b) higher
(c) no effect (d) none

118. Skimming efficiency of cream refers to
(a) % total fat from milk recovered in cream
(b) % total fat recovered in butter
(c) both (a) & (b)
(d) none

119. Higher the acidity of milk ________ the efficiency of separation
(a) lower (b) higher
(c) no effect (d) none

120. Pasteurization of cream is done at
(a) 71°C/40 min (b) 71°C/3 min
(c) 71°C/20 min (d) 71°C/15 sec

121. According to PFA, table butter should contain ______ % milk fat
(a) NLT 80% (b) NMT 80%
(c) NMT 70% (d) NLT 70 %

122. The per centage of common salt in table butter is
(a) NMT 1% (b) NMT 2%
(c) NMT 3% (d) NMT 4%

123. The diacetyl content in butter should be
(a) NMT 4 ppm (b) NLT 4 ppm
(c) NLT 10 ppm (d) NMT 10 ppm

124. The flavor of butter is due to
(a) ethanol (b) esters
(c) diacetyl (d) propionates

125. Moisture content (%) in butter is generally
(a) 10 (b) 6
(c) 11 (d) 16

126. The yellow color of butter is due to
(a) alizarin (b) sunset yellow
(c) annatto (d) none

127. Overrun of butter is expressed in
(a) ratio (b) per centage
(c) kg (d) none

128. The ageing of cream for butter preparation is done at
(a) 10-15°C (b) 5-10°C
(c) 0°C (d) none

129. The starter culture (%) cream fro ripening
(a) 0.05-1.0 (b) 0.1-0.5
(c) 0.5-2.0 (d) none

130. Milk fat is a ________ fat
(a) true (b) mixed
(c) both (d) none

131. Milk fats with low melting point usually have ________ fatty acids
(a) stearin (b) palmitin
(c) olein (d) both (a) & (b)

132. Milk fats with high melting point usually have __________ fatty acids
(a) stearin (b) palmitin
(c) tributyrin (d) both (a) & (b)

133. Smaller size fat globules ________ the churning period
(a) smaller (b) longer
(c) do not affect (d) none

134. Greater the viscosity of cream ______ the churning period
(a) longer (b) smaller
(c) do not affect (d) none

135. The optimum temperature for churning is
(a) 20-30°C (b) 9-11°C
(c) 18-20°C (d) none

136. Higher churning temperature results in ________ body of butter
(a) strong (b) weak
(c) both (d) none

137. Higher the fat % of cream________ the churning period
(a) higher (b) lower
(c) no effect (d) none

138. Greater sized fat globules results in _______ fat loss
(a) higher (b) lower
(c) no effect (d) none

139. Which is the method of salting in butter
(a) dry salting (b) wet salting
(c) brine salting (d) all

140. The salt used for salting of butter is
(a) NaCl (b) $CaCl_2$
(c) $MgCl_2$ (d) KCl

141. The commercial cold storage of butter is done at______°C
(a) -5 to -10 (b) -10 to -20
(c) -23 to -29 (d) -30 to -40

142. The overrun in butter can be
(a) NMT 10 % (b) NMT 25%
(c) NMT 35 % (d) NMT 80 %

143. Butter is _______ emulsion
(a) oil in water (b) water in oil
(c) not an emulsion (d) none

144. Tempering of butter is done at °C
(a) 2 (b) 10
(c) 16 (d) 20

145. Lipase activity in butter preparation causes_______flavor in butter
(a) sour (b) sweet
(c) bitter (d) none

146. Proteolytic bacterial activity in butter preparation causes ______ flavor in butter
(a) cheesy (b) cooked
(c) rancid (d) yeasty

147. Crumbly texture in butter arises due to
(a) over working
(b) under working
(c) high melting point fats
(d) both (a) & (b)

148. Gummy texture in butter arises due to
(a) over working
(b) under working
(c) high melting point fats
(d) both (a) & (b)

149. Overworking of butter causes
(a) rancidity
(b) yeasty flavor
(c) greasy texture
(d) gummy texture

150. Mottled is a kind of __________ defect in butter
(a) texture (b) flavor
(c) color (d) none

151. Mealy is a kind of __________ defect in butter
(a) texture (b) flavor
(c) color (d) none

152. The spreading range of butter is
(a) 2-8°C (b) 16-32°C
(c) 16-24°C (d) 10-14°C

153. Higher the Cu & Fe in butter ________ will be the keeping quality
(a) higher (b) lower
(c) no effect (d) none

154. Higher the acidity of butter_______ will be the keeping quality
(a) lower (b) higher
(c) no effect (d) none

155. Higher the salt content of butter ________ will be the keeping quality
(a) higher (b) lower
(c) no effect (d) none

156. Higher the curd content of butter ________ will be the keeping quality
(a) lower (b) higher
(c) no effect (d) none

157. Exposure to light ________
(a) increases the vitamin D in butter
(b) improves keeping quality
(c) decreases keeping quality
(d) none

158. Unsalted butter has ________ keeping quality
(a) higher (b) lower
(c) no effect (d) none

159. Normally worked butter has air content of __________ per 100g
(a) 5-10 ml (b) 0.5-10 ml
(c) 10-20 ml (d) 0.1-20 ml

160. Normal butter has a curd content of
(a) 0.6-1.0 (b) 1-6%
(c) 0.1-0.6% (d) none

161. Greater the size of fat globules ________ the fat loss in buttermilk
(a) higher (b) lower
(c) no effect (d) none

162. Increased proportion of soft fats ________ churning period
(a) longer (b) shorter
(c) no effect (d) none

163. Greater the viscosity of cream ________ the churning period
(a) longer (b) shorter
(c) no effect (d) none

164. The fat globules carry a net ______ charge
(a) +ve (b) –ve
(c) zero (d) none

165. The ripening of cream is done at ________°C
(a) 31 (b) 11
(c) 21 (d) 15

166. The standardization of cream for butter making is done to
(a) 20-25% fat (b) 25-35% fat
(c) 35-40% fat (d) 10-15% fat

167. Pasteurization of cream is done at
(a) 63°C/30 min
(b) 82-88°C/no hold
(c) 70-72°C/15 sec
(d) none

168. Butter is a rich source of ______ soluble vitamin
(a) water (b) fat
(c) both (d) none

169. The standard color amount is ________ /100kg of butter
(a) 0-250ml (b) 250-500ml
(c) 500-750ml (d) 750-1000ml

170. Which of the following is not a property of annatto
(a) non toxic (b) have flavor
(c) water soluble (d) oil soluble

171. The emulsifier and stabilizer content in ice cream should be
(a) NMT 0.5% (b) NLT 0.5%
(c) NMT 1% (d) NLT 1%

172. The milk fat content in ice cream should be
(a) 15% (b) 8%
(c) 10% (d) 16%

173. The milk protein content (%) in ice cream should be
(a) 2.5 (b) 3.5
(c) 4.5 (d) 5.5

174. Total solids content (%) in ice cream should be
(a) 26 (b) 36
(c) 20 (d) 30

175. Starch content in ice cream should be
(a) NMT 5% (b) NLT 5%
(c) NMT 10% (d) NLT 10%

176. Softy ice cream is
(a) hardened (b) not hardened
(c) rich in fat (d) none

177. Sugar content (%) in icecream is
(a) 10 (b) 15
(c) 20 (d) 30

178. Minimum weight (g/litre) of ice cream should be
(a) 585 (b) 525
(c) 550 (d) 545

179. Acidity of ice cream should be
(a) 0.5% (b) 0.25%
(c) 1.0% (d) 1.5%

180. SPC in ice cream should be
(a) NMT 250000 (b) NLT 250000
(c) NMT 15000 (d) NLT 150000

181. Coliform/g in ice cream should be
(a) NMT 90 (b) NLT 90
(c) NMT 45 (d) NLT 45

182. Phosphatase test for good quality milk should be
(a) –ve
(b) +ve
(c) not done for milk
(d) none

183. Ice cream is a ____ source of nutrients
(a) rich (b) poor
(c) no nutrient (d) none

184. Riboflavin is also called
(a) vitamin C (b) vitamin K
(c) vitamin G (d) vitamin D

185. Milk SNF does not include
(a) protein (b) sugar
(c) mineral (d) fat

186. Sandiness is caused due to high % of
(a) proteins (b) fat
(c) SNF (d) none

187. Basic and apparent are types of ________ of ice cream
(a) viscosity
(b) pasteurization technique
(c) homogenization technique
(d) none

188. Higher stabilizer content ________ the viscosity in ice cream
(a) increases (b) decreases
(c) no effect (d) none

189. Higher temperature ______ the viscosity of ice cream
(a) increases (b) decreases
(c) no effect (d) none

190. Higher fat content ______ the viscosity of ice cream
(a) increases (b) decreases
(c) no effect (d) none

191. Higher sugar content ______ the viscosity of ice cream
(a) increases (b) decreases
(c) no effect (d) none

192. Presence of salts ______ the viscosity of ice cream
(a) increases (b) decreases
(c) no effect (d) none

193. Pasteurization (batch/holding) ______ the viscosity of ice cream
(a) increases (b) decreases
(c) no effect (d) none

194. Ageing ________ the viscosity of ice cream
(a) increases (b) decreases
(c) no effect (d) none

195. As TSS increases, viscosity of ice cream
(a) increases (b) decreases
(c) no effect (d) none

196. Higher the acidity ________ the viscosity of ice cream
(a) lower (b) higher
(c) no effect (d) none

197. Specific gravity of ice cream is
(a) 1.05-1.12 (b) 0.5-0.75
(c) 0.25-0.50 (d) none

198. Specific gravity of ice cream is measured at ________°C
(a) 25 (b) 16
(c) 24 (d) 14

199. Freezing point (°C) of ice cream mix is about
(a) -2.5 (b) -3.5
(c) -1.5 (d) -4.5

200. The sequence of processing steps in manufacture of ice cream
(a) pasteurization, homogenization, freezing, hardening
(b) homogenization, pasteurization, freezing, hardening
(c) pasteurization, homogenization, hardening, freezing
(d) homogenization, pasteurization, hardening, freezing

201. Corn syrup is mixture of
(a) dextrose+maltose
(b) dextrose+lactose
(c) dextrose+galactose
(d) maltose+maltose

202. Gelatin is obtained from
(a) collagen (b) myosin
(c) egg (d) none

203. Addition of sugar ________ freezing point of ice cream
(a) increases (b) decreases
(c) no effect (d) none

204. Saccharin is a ________ sweetener
(a) artificial (b) natural
(c) both (d) none

205. Which of the following is not a stabilizer in ice cream
(a) guar gum (b) gelatin
(c) sodium alginate (d) polysorbate

206. CMC satnds for
(a) carbon methyl cellulose
(b) carboxy methyl cellulose
(c) cellulose of methyl carbon
(d) none

207. The principal flavoring material in vanilla is
(a) villin (b) vanillin
(c) vaniella (d) vanole

208. The nuts are generally added to ice cream at concentration
(a) 3-5% (b) 1-3%
(c) 0.3-0.5% (d) 0.1-0.3%

209. Pasteurization of ice cream mix is done at
(a) 68.5°C/30 min (b) 80°C/25 sec
(c) both (d) none

210. Pasteurization temperature of ice cream is ______ than milk
(a) lower (b) higher
(c) similar (d) none

211. Homogenization temperature (°C) of ice cream is
(a) 63-77 (b) 77-88
(c) 88-96 (d) 105-110

212. ________ stage homogenization is done for ice cream
(a) one (b) two
(c) three (d) four

213. The ageing of ice cream is done at______°C
(a) 5 (b) 10
(c) 20 (d) 15

214. Rapid freezing is preferred over slow freezing in ice cream
(a) true
(b) false
(c) cannot be determined
(d) none

215. Continuous freesing takes ________ time than batch freezing
(a) longer (b) shorter
(c) equal (d) none

216. Overrun in ice cream is due to
(a) addition of air
(b) addition of stabilizer
(c) both
(d) none

217. Over run (%) in ice cream is gener-ally
(a) 70-80 (b) 80-90
(c) 50-60 (d) none

218. Which of the following is not a package of ice cream
(a) multiservice (b) single service
(c) dual service (d) none

219. The over run (%) in softy is
(a) 50-70 (b) 30-50
(c) 70-90 (d) 150-200

220. Hardening of ice cream is done at________°C
(a) -28 (b) -18
(c) -8 (d) -38

221. Slow hardening is preferred over fast hardening
(a) true (b) false
(c) no effect (d) none

222. The freezing point (°C) of solid CO_2 is
(a) -78 (b) -88
(c) -98 (d) -108

223. Dry ice is also known as
(a) solid O_2 (b) solid N_2
(c) solid CO_2 (d) Freon

224. Softy consists of ________ fat
(a) 1-3% (b) 3-6%
(c) 6-9% (d) 9-12%

225. Which of the following is not a common contaminant of ice cream
(a) bacteria
(b) yeast & mold
(c) osmophillic yeast
(d) viruses

226. Microbial contaminant of dry fruit ice cream include
(a) thermophillic bacteria
(b) yeasts & mold
(c) coliforms
(d) osmophilic yeast

227. Excessive stabilizer can lead to
(a) fast melting (b) slow melting
(c) no melt (d) none

228. Curdy meltdown in ice cream is due to
(a) addition of curd
(b) high acidity of mix
(c) low acidity of mix
(d) none

229. Sogginess in ice cream may be due to
(a) high overrun
(b) low sugar content
(c) excessive stabilizer
(d) none

230. Fluffy texture in ice cream is due to
(a) low overrun
(b) high solid content
(c) low emulsifier
(d) low milk solids

231. Icy texture of ice cream is due to
(a) fast hardening
(b) fast freezing
(c) high stabilizer
(d) low solid content

232. Packaging of ice cream is done prior to freezing
(a) true
(b) false
(c) cannot be determined
(d) none

233. Hardening of ice cream is done after packaging
(a) true
(b) false
(c) cannot be determined
(d) none

234. Overrun is achieved during
(a) homogenization (b) hardening
(c) freezing (d) packaging

235. Stabilizer gelatin is of________ origin
(a) plant (b) animal
(c) man made (d) none

236. Mono and diglycerides are examples of
(a) emulsifier (b) stabilizer
(c) sweetener (d) none

237. Fresh fruits can be added to ice cream at ________%
(a) 10-12 (b) 7-8
(c) 2-3 (d) 4-5

238. Flavors are added after _______ in ice cream manufacture
(a) mixing
(b) homogenization
(c) pasteurization
(d) none

239. Fruits are added in ice cream at ________ stage
(a) mixing
(b) homogenization
(c) pasteurization
(d) freezing

240. Which of the following is not an example of emulsifier
(a) glycerides
(b) lecithin
(c) egg white
(d) sodium alginate

241. Sandiness defect is mainly due to
(a) sucrose (b) sucralose
(c) galactose (d) lactose

242. Ageing of ice cream mix is done after freezing
(a) true
(b) false
(c) cannot be determined
(d) none

243. Whipping ___________the overrun of ice cream
(a) increases (b) decreases
(c) do not affect (d) none

244. Average ice cream mix has water content (%)
(a) 61 (b) 71
(c) 81 (d) 91

245. Normal surface tension (dyne/cm^2) value of ice cream mix
(a) 38-43 (b) 48-53
(c) 58-63 (d) 68-73

246. Glucose ________ the freezing point of ice cream
(a) increases (b) decreases
(c) do not affect (d) none

247. At 20°C, the surface tension of milk is
(a) 20-40 (b) 40-60
(c) 60-80 (d) 80-100

248. Fresh dairy product has ________ acidity than stale products
(a) higher (b) lower
(c) similar (d) none

249. The neutralizer of ice cream is
(a) SBC (b) NaOH
(c) KOH (d) CaOH

250. Ice cream is not a good source of
(a) vitaminA (b) vitamin C
(c) vitamin D (d) calcium

251. Ices have _______% overrun
(a) 10-15 (b) 15-20
(c) 20-25 (d) none

252. Lactose is ___________ sweeter than sucrose
(a) more (b) less
(c) equal (d) none

253. Egg yolk solids added to ice cream mix are
(a) NMT 0.5% (b) NLT 0.5%
(c) NMT 5% (d) NLT 5%

254. In ice cream manufacturing ageing is done
(a) post freezing of mix
(b) at the time of freezing
(c) before freezing
(d) after hardening

255. As TSS in water increases, freezing point
(a) increases (b) decreases
(c) do not affect (d) none

256. The quantity of emulsifier added to ice cream is
(a) 2-4% (b) 6-8%
(c) 4-6% (d) none

257. In icre cream manufacture, fast freezing is required to get
(a) more number of ice crystals
(b) homogeneous nucleation
(c) decrease ice crystal
(d) all

258. IICMA stands for
(a) Indian Ice Cream Manufacturer's Association
(b) Indian Ice Cream Maker's Association
(c) Indian Ice Cream Manufacturer's Agency
(d) none

259. AMF stands for
(a) any milk fat
(b) anhydrous milk fat
(c) animal milk fat
(d) none

260. Milk fat is delivered in the form of
(a) liquid fat (b) AMF
(c) milk (d) none

261. Hardening of ice cream is done at ______ °C
(a) -2 to -4 (b) -8 to -10
(c) -18 to -20 (d) -29 to -31

262. Average moisture content in very hard cheese is
(a) NLT 25% (b) NMT 25%
(c) NLT 35% (d) NMT 35%

263. Average moisture content in hard cheese is
(a) 15-25 (b) 25-36%
(c) 36-45% (d) 2-5%

264. Average moisture content in soft cheese is
(a) 10% (b) 20%
(c) 30% (d) 40%

265. Which of the following is example of hard cheese
(a) camembert (b) blue
(c) roquefort (d) swiss

266. For cheese making, standardization of casein to fat ratio in cheese milk should be
(a) 0.50-0.52 (b) 0.68-0.70
(c) 0.58-0.60 (d) 0.48-0.50

267. Pasteurization temperature for cheese milk is
(a) 63°C/30min (b) 71°C/15sec
(c) 78°C/15min (d) both (a) & (b)

268. The amount of starter added to milk for cheese is
(a) 0.5-1% (b) 0.25-0.30%
(c) 1-1.5% (d) 1.5-2%

269. The temperature at the time of starter addition to the cheese milk is
(a) 20-21°C (b) 30-31°C
(c) 40-41°C (d) 50-51°C

270. The amount of color added to cheese is
(a) 0.3-30ml (b) 30-200ml
(c) 200-333ml (d) none

271. Rennet enzyme is composed of
(a) rennin and trypsin
(b) rennin and pepsin
(c) trypsin and pepsin
(d) none

272. The bitterness in cheese is due to
(a) pepsin (b) trypsin
(c) rennin (d) none

273. The optimum pH for rennin is
(a) 21°C (b) 31°C
(c) 41°C (d) 51°C

274. The amount of liquid rennet added to milk is
(a) 5-10ml/100 lit of milk
(b) 10-15 ml/100 lit of milk
(c) 15-25 ml/100 lit of milk
(d) none

275. The amount of salt added to cheese curd is
(a) 1-2% (b) 0.1-0.2%
(c) 3-5% (d) 0.3-0.5%

276. The curing of cheese is also known as
(a) ripening (b) souring
(c) maturing (d) all

277. Green cheese refers to
(a) green color cheese
(b) unripened cheese
(c) not fully ripened
(d) both both (b) & (c)

278. Example of soft cheese is
(a) cheddar (b) gouda
(c) emmental (d) camembert

279. Eye cheese is
(a) cheddar (b) gouda
(c) swiss (d) camembert

280. Cheddaring is done to
(a) improve texture
(b) improve taste
(c) to improve shelf life
(d) none

281. Milling of cheese is done for
(a) size reduction
(b) texture development
(c) improve shelf life
(d) none

282. The eyes results from ________ acid fermentation
(a) lactic acid (b) acetic acid
(c) propionic acid (d) malic acid

283. The average composition of cholesterol in cheese is
(a) traces (b) absent
(c) 100mg (d) 1000mg

284. Cheese is ________ in protein
(a) deficient (b) rich source
(c) traces (d) none

285. The lactose content of cheese is
(a) deficient (b) rich source
(c) traces (d) none

286. Rouquefort cheese originated in
(a) Italy (b) Austria
(c) France (d) Norway

287. Emmental cheese originated in
(a) Italy (b) Austria
(c) Switzerland (d) Norway

288. IDF stands for
(a) International Dairy Federation
(b) Indian Dairy Federation
(c) International Dietary Foundation
(d) none

289. Fat in extra hard cheese is
(a) <60% (b) 65%
(c) 70% (d) 80%

290. Fat in hard cheese is
(a) <60% (b) 65%
(c) 70% (d) 80%

291. Fat in soft cheese is
(a) 10-50% (b) 65%
(c) 70% (d) 80%

292. Cottage cheese is
(a) semi hard (b) semi soft
(c) soft (d) hard

293. Cottage cheese is
(a) unripened (b) mold ripened
(c) surface ripened (d) none

294. Mozarella cheese originated in
(a) France (b) UK
(c) USA (d) Italy

295. Pizza cheese is
(a) Limburger (b) Mozarella
(c) Ricotta (d) Cottage

296. Schardinger's enzyme is also known as which catalyses the oxidation of aldehyde
(a) phosphatase
(b) xanthine oxidase
(c) alkaline phosphatase
(d) esterase

297. For satisfactory coagulation ________ % of $CaCl_2$ is added
(a) 0.2% (b) 0.02%
(c) 0.4% (d) 0.04%

298. FMFS stands for
(a) fat in moisture free substance
(b) fatty acid in moisture free substance
(c) fat in modified form in substance
(d) none

299. WFFS stands for
(a) water in fat free substance
(b) water in fully fat substance
(c) water in fatty acid free source
(d) none

300. WFFS value for cheddar cheese is
(a) 34% (b) 44%
(c) 54% (d) 64%

301. WFFS value for cottage cheese is
(a) 55% (b) 65%
(c) 75% (d) 85%

302. Rind rot is due to
(a) low acidity (b) less moisture
(c) high acidity (d) none

303. The emulsifie & stabilizersr in processed cheese is
(a) NMT 3% (b) NMT 4%
(c) NMT 5% (d) NMT 6%

304. The moisture in processed cheese should be
(a) 37% (b) 47%
(c) 57% (d) 67%

305. Nisin level in processed cheese is
(a) 1% (b) 0.01%
(c) 0.1% (d) 2%

306. Fat content in processed cheese is
(a) 40% (b) 50%
(c) 60% (d) 70%

307. Rennin is a ________ containing enzyme
(a) sulfur (b) iron
(c) manganese (d) cobalt

308. High acidity in milk ________ the clotting rate of rennin
(a) increase (b) decrease
(c) no effect (d) none

309. Mastitis milk coagulates ______ with rennin
(a) rapidly (b) slowly
(c) no effect (d) none

310. Heating of milk ________ the coagulation by rennin
(a) improves (b) favors
(c) lowers (d) both (a) & (b)

311. Syneresis of cheese curd is called
(a) expulsion of whey
(b) contraction of curd
(c) weeping of curd
(d) all

312. Function of salting in cheese is
(a) removal of whey
(b) hardening of curd
(c) retards formation of lactic acid
(d) all

313. Parafinning is done to
(a) improve the flavor of cheese
(b) improve the nutrient quality
(c) reduce loss of moisture
(d) none

314. An increase in coagulation period ______ the moisture content of cheese
(a) increase (b) decrease
(c) no effect (d) none

315. An increase in curd particle size for cooking ______ the moisture content of cheese
(a) increase (b) decrease
(c) no effect (d) none

316. Curing of cheese is done at ______°C
(a) 0-16 (b) 20-30
(c) 30-40 (d) 40-50

317. Cottage cheese is prepared from
(a) whole milk (b) skim milk
(c) toned milk
(d) double toned milk

318. Excessive gassy fermentation in cheese leads to
(a) seamy defect
(b) rind rot
(c) huffed
(d) crumbled body

319. ____ acid development in cheese curd leads to rubbery cheese
(a) lower (b) higher
(c) no effect (d) none

320. Pin hole defect in cheese occurs due to
(a) addition of pins
(b) excess fat
(c) contamination by mo's
(d) high acid

321. Swiss holes is due to
(a) propionic bacteria
(b) lactic acid bacteria
(c) yeast
(d) mold

322. Cracked paraffin leads to
(a) appearance defect
(b) increase in moisture content of cheese
(c) increase contamination by mo's
(d) all

323. The successful manufacture of condensed milk was started by
(a) Appert (b) Pasteur
(c) Borden (d) Koch

324. In condensed milk ________ of water has been removed
(a) no (b) part
(c) complete (d) none

325. The ratio of concentration of milk solids is about
(a) 1:3.5 (b) 1:2.5
(c) 1:4.5 (d) 1:0.5

326. Milk solids in condensed milk
(a) NLT 31% (b) NMT 31%
(c) NLT 41% (d) NMT 41%

327. Milk solids in skim sweetened condensed milk
(a) NLT 36% (b) NMT 36%
(c) NLT 26% (d) NMT 26%

328. Fat per centage in condensed milk is
(a) NLT 9% (b) NMT 9%
(c) NLT 19% (d) NMT 19%

329. Fat per centage in skim sweetened condensed milk is
(a) NLT 0.5% (b) NMT 0.5%
(c) NLT 5% (d) NMT 5%

330. Sucrose per centage in condensed milk is
(a) 20% (b) 30%
(c) 40% (d) 50%

331. Acidity % as lactic acid in condensed milk is
(a) 0.25 (b) 0.35
(c) 0.45 (d) 0.55

332. Bacterial count/ g of condensed milk is
(a) 50 (b) 500
(c) 5000 (d) 50000

333. Coliform count/g of condensed milk is
(a) 50 (b) 90
(c) negative (d) 10

334. Yeast and mold count/g of condensed milk is
(a) 1 (b) 10
(c) 100 (d) 1000

335. Unsweetened condensed skim milk should contain ______ milk solids
(a) NLT 20% (b) NLT 30%
(c) NLT 40% (d) NLT 50%

336. The specific gravity of condensed milk at 16°C is
(a) 1.082 (b) 1.355
(c) 1.303 (d) 1.506

337. Highest specific gravity at 16°C is for ______ milk
(a) condensed
(b) evaporated
(c) skim & sweetened
(d) none

338. The brown pigment causing discoloration of condensed milk is
(a) alizarin (b) melanoidin
(c) melano (d) riboflavin

339. The cooked flavor is due to ______ groups
(a) SH (b) C=O
(c) CHO (d) none

340. The amount of browning is increased with ______ in total solids
(a) increase (b) decrease
(c) no effect (d) none

341. The amount of browning is increased with ______ in pH
(a) increase (b) decrease
(c) no effect (d) none

342. The amount of browning is increased with ______ in storage temperature
(a) increase (b) decrease
(c) no effect (d) none

343. The amount of browning is increased with ______ in oxygen
(a) increase (b) decrease
(c) no effect (d) none

344. The amount of browning is increased with ______ in reducing sugar
(a) increase (b) decrease
(c) no effect (d) none

345. Maillard browning in condensed milk is due to ______ reaction
(a) fat-fat (b) amino-sugar
(c) amino-fat (d) fat-sugar

346. The viscosity of condensed milk ______ with lowering of the pH
(a) increase (b) decrease
(c) no effect (d) none

347. The flavor of the condensed milk is chiefly due to
(a) milk proteins (b) mineral
(c) milk fat (d) salts

348. A stable milk has ______ alcohol index
(a) 1 (b) 3
(c) 7 (d) 10

349. Alcohol-alizarin test is used for
(a) heat stability of milk
(b) fat determination
(c) amino acid determination
(d) mineral determination

350. The ratio of fat: snf for milk for condensed milk has been specified to be
(a) 1:2.44 (b) 2.44:1
(c) 2:2.44 (d) 2.44:2

351. The baume hydrometer is used for
(a) fat test (b) density test
(c) protein test (d) none

352. Condensed milk undergoes a ______ step homogenization
(a) 1 (b) 2
(c) 3 (d) 4

353. Crystallization of lactose of condensed milk is most rapid at
(a) 10°C (b) 20°C
(c) 30°C (d) 40°C

354. The crystal defect in condensed milk is chiefly dur to
(a) sucrose (b) lactose
(c) maltose (d) glucose

355. The introduction of lactose in very fine powder form during cooling process to provide nuclei for crystallization is known as

(a) feeding (b) seeding
(c) heeding (d) beading

356. Seeding results in the formation of ________ sized lactose crystal

(a) large (b) small
(c) no effect (d) none

357. The seed lactose usage dose is ______ g/1000kg of fluid milk

(a) 375-500 (b) 475-600
(c) 575-700 (d) 675-800

358. Evaporated milk differs from condensed milk in

(a) having more sugar
(b) no added sugar
(c) no water (d) none

359. Additive step in manufacture of evaporated milk is

(a) homogenization
(b) pasteurization
(c) sterilization
(d) cooling

360. The sterilization test for evaporated milk is known as

(a) air sterilization test
(b) pilot sterilization test
(c) lactose sterilization test
(d) none

361. The apparent viscosity for evapor-ated milk determined by pilot sterilization test is

(a) 50°R (b) 100°R
(c) 150°R (d) 200°R

362. The sterilization temperature for evaporated milk is

(a) 63°C (b) 71°C
(c) 121°C (d) 118°C

363. The storage temperature of evaporated milk is

(a) 5-16°C (b) 16-32°C
(c) 32-40°C (d) 40-50°C

364. The heat coagulation of milk is due to

(a) proteins (b) salt balance
(c) fat (d) both (a) & (b)

365. For optimum heat stability of milk, the per centage of calcium and magnesium ions/ phosphate and citrate ions in cow milk is

(a) 0.20 (b) 0.27
(c) 0.30 (d) 0.37

366. For optimum heat stability of milk, the per centage of calcium and magnesium ions/phosphate and citrate ions in buffalo milk is

(a) 0.20 (b) 0.27
(c) 0.30 (d) 0.39

367. Milk with high acidity is ____ stable

(a) more (b) less
(c) no effect (d) none

368. Milk with high total solids is ____ stable

(a) more (b) less
(c) no effect (d) none

369. Homogenization of milk ________ heat stability of milk

(a) increase (b) decrease
(c) no effect (d) none

370. For heat stability, balance of minerals required is

(a) Ca+Mg/P+Mn
(b) Mg+Mn/P+Ca
(c) Ca+Mg/P+citrate
(d) none

371. Bloat defect in condensed milk is due to

(a) bacteria (b) yeast
(c) virus (d) mold

372. The addition of finely powdered lactose causes _______ crystallization

(a) mass (b) low
(c) fat (d) none

373. WMP stands for
(a) whole milk powder
(b) white milk powder
(c) white milk paste
(d) whole milk paste

374. SMP stands for
(a) skim milk powder
(b) small milk powder
(c) skim milk paste
(d) small milk paste

375. NFDM stands for
(a) non fat dry milk
(b) new fat dry milk
(c) non fat dairy milk
(d) non fat dried milk

376. Milk powder should contain ______% moisture
(a) NMT 5% (b) NLT 5%
(c) NMT 2% (d) NLT 2%

377. Whole milk powder should contain maximum ________% fat
(a) 16% (b) 26%
(c) 2% (d) 36%

378. Whole milk powder should contain ________% acidity
(a) NMT 0.2% (b) NMT 1.2%
(c) NMT 3.2% (d) NMT 4.2%

379. Maximum solubility index for roller dried milk powder is
(a) 5 (b) 10
(c) 15 (d) 20

380. Maximum solubility index for spray dried milk powder is
(a) 1 (b) 2
(c) 3 (d) 4

381. Skim milk powder should contain ________% fat
(a) NMT 0.5% (b) NMT 1.5%
(c) NMT 3.5% (d) NMT 4.5%

382. .The milk is condensed to _________% total solids for milk powder
(a) 5-10% (b) 10-15%
(c) 16-18% (d) 24-28%

383. 1μ is equivalent to how many millimeters
(a) 0.001mm (b) 0.01mm
(c) 0.1mm (d) 1mm

384. WPN stands for
(a) whey protein nitrogen
(b) whey protein nitrate
(c) white powder nitrogen
(d) none

385. Instantization refers to
(a) making powder soluble
(b) making powder insoluble
(c) restricting powder solubility
(d) none

386. For instantization, the desired particle size is
(a) 10 μM (b) 20 μM
(c) 30 μM (d) 40 μM

387. _______ dried milk powder is desirable
(a) drum (b) spray
(c) sun (d) none

388. The density of SMP is
(a) 1-1.2g/ml
(b) 1.44-1.48 g/ml
(c) 2.44-2.48 g/ml
(d) 2-2.2 g/ml

389. The density of WMP is
(a) 1.26-1.32g/ml
(b) 1.44-1.48 g/ml
(c) 2.44-2.48 g/ml
(d) 2-2.2 g/ml

390. ____________ dried powder has more flowability
(a) drum (b) spray
(c) sun (d) none

391. Example of free flowing agents in milk powder is
(a) NaOH (b) NaCl
(c) calcium silicate (d) none

392. Amount of calcium silicate added to milk powder as free flowing agent is ______ %
(a) 0.5 (b) 1.5
(c) 2.5 (d) 3.5

393. ______ the acidity, higher will be the oxidation of powder
(a) higher (b) lower
(c) no effect (d) none

394. Permitted antioxidant added to milk powder is
(a) phospholipids
(b) BHA
(c) magnesium chloride
(d) none

395. The permitted level of antioxidant in milk powder is______%/g of product
(a) 0.1 (b) 0.01
(c) 1 (d) 0.001

396. Tallowy is a kind of ______ defect
(a) flavor (b) color
(c) body (d) none

397. Ancidity is caused by ______ enzyme
(a) phosphatase (b) lipase
(c) lactase (d) peroxidase

398. Burnt flavor is due to______groups
(a) SH (b) NH_2
(c) CHO (d) C=O

399. The buttermilk powder contains ______% fat
(a) 5-6 (b) 6-10
(c) 10-15 (d) 15-20

400. Cream powder consists of ______ % fat
(a) 10 (b) 30
(c) 60 (d) 90

401. Butter powder contains ________ % fat
(a) 81.98 (b) 51.6
(c) 61.6 (d) 71.6

402. Which of the following is not an Indian dairy product
(a) kulfi (b) dahi
(c) paneer (d) none

403. Which of the following is a traditional Indian dairy product
(a) buttermilk (b) butter
(c) ghee (d) ice cream

404. Khoa is similar to
(a) condensed milk
(b) evaporated milk
(c) clotted cream
(d) none

405. Kheer or basundi is similar to
(a) condensed milk
(b) evaporated milk
(c) clotted cream
(d) none

406. Rabri is similar to
(a) condensed milk
(b) evaporated milk
(c) clotted cream
(d) none

407. Kulfi is similar to
(a) condensed milk
(b) evaporated milk
(c) clotted cream
(d) ice cream

408. Dahi is similar to
(a) yoghurt (b) cheese
(c) butter (d) buttermilk

409. Paneer is similar to
(a) yoghurt (b) cheese
(c) butter (d) buttermilk

410. Chhana is similar to
(a) yoghurt (b) green cheese
(c) butter (d) buttermilk

411. Makkhan is similar to
(a) yoghurt (b) cheese
(c) butter (d) buttermilk

412. Ghee is similar to
(a) yoghurt (b) cheesc
(c) butteroil (d) buttermilk

413. Lassi is similar to
(a) yoghurt (b) cheese
(c) butter (d) buttermilk

414. Which of the following is a fermented milk product
(a) makkhan (b) kheer
(c) khoa (d) chhana

415. Which of the following is a fermented milk product
(a) kheer (b) khoa
(c) chhana (d) chakka

416. Which of the following is a fermented milk product
(a) kheer (b) khoa
(c) chhana (d) lassi

417. Which of the following is a fermented milk product
(a) shrikhand (b) khoa
(c) chhana (d) kheer

418. Which of the following is a milk concentrate
(a) kheer (b) khoa
(c) mawa (d) all

419. Which of the following is a milk concentrate
(a) butter (b) kulfi
(c) chhana (d) paneer

420. Kheer is also known as
(a) shrikhand (b) rabri
(c) chhana (d) basundi

421. Khoa is also known as
(a) shrikhand (b) mawa
(c) chhana (d) rabri

422. Moisture content of kheer is about ______%
(a) 47 (b) 57
(c) 67 (d) 77

423. Fat content of kheer is about ______%
(a) 8 (b) 18
(c) 28 (d) 30

424. Protein content of kheer is about ______%
(a) 6 (b) 16
(c) 22 (d) 24

425. Lactose content of kheer is about ______%
(a) 2.5 (b) 8.5
(c) 16.5 (d) 20

426. Average shelf life of kheer at room temperature is about ______days
(a) 2 (b) 5
(c) 10 (d) 15

427. Average shelf life of kheer at refrigerated temperature is about ______days
(a) 2 (b) 5
(c) 10 (d) 15

428. Khoa is a ______ whole milk product
(a) dehydrated (b) partially dehydrated
(c) both (d) none

429. Milk fat content of khoa should not be less than ____%
(a) 5 (b) 10
(c) 15 (d) 30

430. Which of the following is type of khoa
(a) pindi (b) dhap
(c) danedaar (d) all

431. Fat content in pindi khoa is ______%
(a) 2-5 (b) 5-12
(c) 10-16 (d) 21-26

432. Fat content in dhap khoa is ______%
(a) 2-5 (b) 5-12
(c) 10-16 (d) 20-23

433. Fat content in danedaar khoa is ______%
(a) 2-5 (b) 5-12
(c) 10-16 (d) 20-25

434. Pindi type khoa is used for making
(a) plain burfi (b) pista burfi
(c) peda (d) all

435. Dhap type khoa is used for making
(a) burfi (b) gulabjamun
(c) peda (d) all

436. Danedaar type khoa is used for making
(a) gulabjamun (b) kalakand
(c) peda (d) all

437. Moisture content of khoa prepared from cow milk is ________%
(a) 10 (b) 20
(c) 25 (d) 30

438. Fat content of khoa prepared from cow milk is ______%
(a) 10 (b) 20
(c) 25 (d) 30

439. Protein content of khoa prepared from cow milk is ______%
(a) 17.8 (b) 19.2
(c) 25.3 (d) 28.7

440. Moisture content of khoa prepared from buffalo milk is ______%
(a) 10 (b) 20
(c) 25.5 (d) 30

441. Lactose content of khoa prepared from cow milk is ______%
(a) 10 (b) 20
(c) 25.5 (d) 30

442. Fat content of khoa prepared from buffalo milk is ______%
(a) 37.1 (b) 20
(c) 25.5 (d) 30

443. Protein content of khoa prepared from buffalo milk is ______%
(a) 17.8 (b) 19.2
(c) 25.3 (d) 28.7

444. Lactose content of khoa prepared from buffalo milk is ______%
(a) 10 (b) 18.2
(c) 22.1 (d) 30

445. Iron content of khoa prepared from cow milk is _______ppm
(a) 53 (b) 103
(c) 253 (d) 153

446. Which of the following occurs during khoa making from milk
(a) change in state
(b) change in color intensity
(c) homogenization of milk fat
(d) all

447. Which of the following occurs during khoa making from milk
(a) heat coagulation of milk proteins
(b) free fat formation
(c) partial precipitation of milk salts
(d) all

448. Color of cow milk khoa is
(a) pale yellow (b) whitish
(c) both (d) none

449. Color of buffalo milk khoa is
(a) pale yellow (b) whitish
(c) both (d) none

450. Body of cow milk khoa is
(a) slightly hard (b) soft
(c) both (d) none

451. Body of buffalo milk khoa is
(a) slightly hard (b) soft
(c) both (d) none

452. Texture of buffalo milk khoa is
(a) smooth
(b) granular
(c) both (a) & (b)
(d) slightly sandy

453. Texture of cow milk khoa is
(a) smooth
(b) granular
(c) both (a) & (b)
(d) slightly sandy

454. Which of the following is a characteristic of buffalo milk khoa
(a) slightly salty (b) slightly sweet
(c) both (d) none

455. Which of the following is a characteristic of cow milk khoa
(a) slightly salty (b) slightly sweet
(c) both (d) none

456. Which of the following khoa is highly suitable for sweets
(a) cow milk khoa
(b) buffalo milk khoa
(c) both (d) none

457. Which of the following spoilage is associated with low temperature storage of khoa
(a) rancid flavor (b) sour flavor
(c) stale (d) none

458. Which of the following spoilage is associated with room temperature storage of khoa
(a) rancid flavor (b) sour flavor
(c) stale (d) none

459. Use of high acidity milk for khoa making results in
(a) rancid flavor (b) sour flavor
(c) stale (d) none

460. Rancid flavor of khoa during storage is due to action of
(a) amylase (b) protease
(c) lipase (d) all

461. Which of the following is responsible for hard body of khoa
(a) excessive low fat content of milk
(b) excessive low moisture content
(c) milk adulteration with starch
(d) all

462. Gritty texture of khoa is due to
(a) sand like particles
(b) lactose crystals
(c) both (d) none

463. Moldy surface of khoa may be due to
(a) excessive high moisture content of khoa
(b) excessive high humidity of storage room
(c) air leaky packaging of khoa
(d) all

464. Coarse texture of khoa is due to
(a) high acidity of milk
(b) low fat content of milk
(c) high temperature of dehydration
(d) all

465. Khurchan is a ______ whole milk product
(a) concentrated (b) sweetened
(c) both (d) none

466. Moisture content of khurchan is about ______ %
(a) 17.9 (b) 27.9
(c) 37.9 (d) 47.9

467. Fat content of khurchan is about ______ %
(a) 3.6 (b) 13.6
(c) 23.6 (d) 33.6

468. Protein content of khurchan is about ______ %
(a) 5 (b) 10
(c) 15 (d) 20

469. Lactose content of khurchan is about ______ %
(a) 5 (b) 10
(c) 15 (d) 20

470. Sugar content of khurchan is about ______ %
(a) 5 (b) 10
(c) 15 (d) 20

471. Rabri is ______ whole milk product
(a) concentrated (b) sweetened
(c) both (d) none

472. Moisture content of khurchan is about ______ %
(a) 10 (b) 20
(c) 30 (d) 40

473. Protein content of khurchan is about ______ %
(a) 10 (b) 20
(c) 30 (d) 40

474. Fat content of khurchan is about _____%
(a) 10 (b) 20
(c) 30 (d) 40

475. Lactose content of khurchan is about _____%
(a) 7 (b) 17
(c) 37 (d) 47

476. Indian curd is also known as
(a) dahi (b) rabri
(c) malai (d) none

477. Which of the following culture is used in sweet dahi preparation
(a) Streptococcus lactis
(b) S. diacetilactic
(c) S. cremoris (d) all

478. Which of the following culture is used in sour dahi preparation
(a) Lactobacillus bulgaricus
(b) Streptococcus thermophilus
(c) both
(d) none

479. The acidity of sweet dahi as lactic acid is________%
(a) 0.7 (b) 1.0
(c) 1.2 (d) 1.5

480. The acidity of sour dahi as lactic acid is ________%
(a) 0.7 (b) 1.0
(c) 1.2 (d) 1.5

481. Phosphatase test for sweet dahi should be
(a) –ve (b) +ve
(c) both (d) none

482. Water content of whole milk dahi is ________%
(a) 55 (b) 65
(c) 75 (d) 85

483. Fat content of whole milk dahi is ________%
(a) 5-8 (b) 10-12
(c) 13.5 (d) 14-18

484. Protein content of whole milk dahi is ________%
(a) 1.2 (b) 2.2
(c) 3.2 (d) 4.2

485. Lactose content of whole milk dahi is ________%
(a) 5 (b) 10
(c) 15 (d) 20

486. Ash content of whole milk dahi is ________%
(a) 0.5 (b) 0.7
(c) 1.0 (d) 1.2

487. Which of the following is a type of sweetened dahi
(a) misti dahi (b) lal dahi
(c) payodhi dahi (d) all

488. Which of the following is a characteristic of misti dahi
(a) cooked flavor
(b) caramelized flavor
(c) firm body
(d) all

489. Which of the following is used as aroma producer in dahi
(a) Str. Lactis (b) Str. Cremoris
(c) Str. Diacetilactis (d) all

490. Dahi can be used for the preparation of
(a) chakka (b) shrikhand
(c) makkhan (d) all

491. Skim milk dahi is beneficial for
(a) heart patient (b) liver patient
(c) brain patient (d) all

492. Shrikhand is generally prepared from
(a) skim milk (b) whole milk
(c) both (d) none

493. Basic ingredient for shrikhand is
(a) chakka (b) makkhan
(c) ghee (d) all

494. Moisture content of chakka is ______%
(a) 32 (b) 42
(c) 52 (d) 63

495. Panir is generally associated with
(a) rennet coagulation
(b) soft cheese
(c) both
(d) none

496. Moisture content of surati panir is about
(a) 51 (b) 61
(c) 71 (d) 81

497. Total soilds content of surati panir is about
(a) 18.8 (b) 28.8
(c) 38.8 (d) 42.2

498. Fat content of surati panir is about
(a) 7.5 (b) 13.5
(c) 21.5 (d) 25.5

499. Moisture content of *chhena* from cow milk is ______%
(a) 33.4 (b) 53.4
(c) 18.6 (d) 28.4

500. Fat content of *chhena* from cow milk is ______%
(a) 14.8 (b) 24.8
(c) 10.4 (d) 32.8

501. Protein content of *chhena* from cow milk is _______%
(a) 7.4 (b) 11.4
(c) 17.4 (d) 14.4

502. Lactose content of *chhena* from cow milk is _______%
(a) 1.1 (b) 2.1
(c) 3.1 (d) 4.1

503. Ash content of *chhena* from cow milk is _______%
(a) 1.1 (b) 2.1
(c) 3.1 (d) 4.1

504. Moisture content of *chhena* from buffalo milk is ______%
(a) 41.6 (b) 51.6
(c) 61.6 (d) 71.6

505. Protein content of *chhena* from buffalo milk is ______%
(a) 7.4 (b) 11.4
(c) 14.4 (d) 18.4

506. Fat content of chhana from buffalo milk is ______%
(a) 20.8 (b) 24.8
(c) 29.4 (d) 11.5

507. Chhana is a good source of
(a) fat (b) protein
(c) vitamin A & D (d) all

508. Amount of citric acid required to coagulate 1 kg of fresh milk
(a) 1-1.5g (b) 2-2.5g
(c) 3.5-4g (d) 5g

509. Amount of lactic acid required to coagulate 1 kg of fresh milk
(a) 1-1.5g (b) 2-2.5g
(c) 3.5-4g (d) 5g

510. For desirable body and texture in chhana, pH of coagulation should be around
(a) 3.4 (b) 4.4
(c) 5.4 (d) 4.9

511. For desirable body and texture in chhana, temperature of coagulation should be around
(a) 62°C (b) 72°C
(c) 82°C (d) 92°C

512. Which of the following affect the body and texture of chhana
(a) fat content of milk
(b) acidity of milk
(c) adulteration of milk
(d) all

513. Which of the following is flavor defect in chhana
(a) smoky (b) sour
(c) rancid (d) all

514. Which of the following is body and texture defect in chhana
(a) hard body (b) coarse texture
(c) both (d) none

515. Which of the following is color & appearance defect in chhana
(a) dry surface (b) moldy surface
(c) foreign matter (d) all

516. Chhana is used in the preparation of
(a) rossogulla (b) sandesh
(c) pantooa (d) all

517. Which of the following is a traditional Indian dairy product
(a) ghee (b) makkhan
(c) both (d) none

518. Makkhan is also known as
(a) ghee (b) desi ghee
(c) desi butter (d) all

519. Moisture content (%) of makkhan is about
(a) 10 (b) 20
(c) 30 (d) 40

520. Fat content (%) of makkhan is about
(a) 50 (b) 60
(c) 70 (d) 80

521. According to PFA, makkhan should contain not less than ______% milk fat by weight
(a) 62 (b) 66
(c) 76 (d) 82

522. For makkhan preparation, amount of starter culture required in winter season
(a) 0.5-1.0% (b) 1.0-1.5%
(c) 2.0-2.5% (d) 3%

523. For makkhan preparation, amount of starter culture required in summer season
(a) 0.5-1.0% (b) 1.0-1.5%
(c) 2.0-2.5% (d) 3%

524. Amount of buffalo milk required to produce 1 kg of makkhan
(a) 7 (b) 14
(c) 22 (d) 19

525. Amount of cow milk required to produce 1 kg of makkhan
(a) 7 (b) 14
(c) 22 (d) 19

526. According to PFA, maximum limit of free fatty acid (as oleic acid) in ghee is
(a) 3 (b) 9
(c) 13 (d) 19

527. According to PFA, maximum moisture content (%) in ghee is
(a) 0.3 (b) 0.9
(c) 1.3 (d) 1.9

528. Butyro-refractometer reading of ghee at 40°C is about
(a) 30-35 (b) 40-45
(c) 50-55 (d) 60-65

529. Reichert value of ghee is generally about
(a) 11-15 (b) 21-28
(c) 31-36 (d) 41-48

530. Baudouin test for ghee should be
(a) positive (b) negative
(c) both (d) none

531. Fat content (%) in ghee is about
(a) 80-85.5 (b) 88.5-89.5
(c) 91.5-95.5 (d) 99-99.5

532. Carotene content (µg/g) of cow ghee is about
(a) 3.2-7.4 (b) 11.2-14.5
(c) 19.8-23.2 (d) 25

533. Vitamin A (IU/g) of cow ghee is about
(a) 10-12 (b) 19-34
(c) 40-43 (d) 49-67

534. Tocopherol content (µg/g) of cow ghee is about
(a) 6-18 (b) 21-23
(c) 26-48 (d) 62-87

535. Vitamin A (IU/g) of buffalo ghee is about
(a) 11-15 (b) 17-38
(c) 42-57 (d) 62-79

536. Tocopherol content (μg/g) of buffalo ghee is about
(a) 2-9 (b) 11-15
(c) 18-37 (d) 46-58

537. Which of the following is present as traces in ghee
(a) charred casein (b) copper salts
(c) iron salts (d) all

538. Polenske value of ghee should normally be not more than
(a) 2 (b) 4
(c) 5 (d) 7

539. Saponfication value of ghee should normally be not less than
(a) 110 (b) 160
(c) 220 (d) 135

540. Iodine value of ghee normally ranges
(a) 10-14 (b) 16-21
(c) 26-38 (d) 45-53

541. Which of the following is the richest source of milk fat
(a) ice cream (b) butter
(c) ghee (d) buttermilk

542. Which of the following is associated with ghee making
(a) milk (b) curd
(c) butter (d) all

543. Which of the following is a method of ghee making
(a) desi method
(b) creamery butter method
(c) direct cream method
(d) all

544. Which of the following is small scale ghee making method
(a) desi method
(b) creamery butter method
(c) direct cream method
(d) all

545. Which of the following is associated with continuous ghee making method
(a) large scale
(b) high fat recovery
(c) in-place cleaning
(d) all

546. Which of the following is a by-prod-uct of desi method of ghee making
(a) butter milk (b) ghee residue
(c) both (d) none

547. Which of the following is a by-product of creamery butter method of ghee making
(a) butter milk (b) ghee residue
(c) skim milk (d) all

548. Which of the following is a by-product of direct cream method of ghee making
(a) whole milk (b) ghee residue
(c) both (d) none

549. Which of the following is a by-product of continuous method of ghee making
(a) butter milk (b) skim milk
(c) both (d) none

550. Which of the following method give strong aroma of ghee
(a) pre-stratification method
(b) direct cream method
(c) continuous method
(d) desi method

551. Which of the following is a by-product of prestratification method of ghee making
(a) butter milk (b) ghee residue
(c) skim milk (d) all

552. Large grains in ghee is produced by
(a) desi method
(b) creamery butter method
(c) direct cream method
(d) all

553. Which of the following method produce ghee with greasy texture

(a) desi method
(b) creamery butter method
(c) continuous method
(d) all

554. Which of the following method produce ghee with greasy texture
(a) desi method
(b) creamery butter method
(c) direct cream method
(d) all

555. Clarification temperature (°C) of ghee is generally about
(a) 80-90 (b) 110-120
(c) 95-100 (d) 135-150

556. Higher clarification temperature (°C) of ghee leads to
(a) reduce vitamin A
(b) dark color
(c) good keeping quality
(d) all

557. Lower clarification temperature (°C) of ghee leads to
(a) improve color
(b) decrease keeping quality
(c) higher residual moisture
(d) all

558. Granular texture of ghee is developed during
(a) clarification (b) cooling
(c) both (d) none

559. Melting point (°C) of cow ghee is about
(a) 25 (b) 29
(c) 31 (d) 37

560. Melting point (°C) of buffalo ghee is about
(a) 25 (b) 29
(c) 31 (d) 37

561. Suitable temperature (°C) of ghee packing is
(a) 25-25 (b) 27-29
(c) 30-32 (d) 35-37

562. Color of cow ghee by desi method is
(a) deep yellow (b) white
(c) light yellow (d) none

563. Color of buffalo ghee by desi method is white with ________tinge
(a) yellowish (b) greenish
(c) both (d) none

564. In desi method of ghee making, item of fat loss is
(a) lassi (b) ghee residue
(c) handling (d) all

565. In creamery butter method of ghee making, item of fat loss is
(a) butter milk (b) ghee residue
(c) skim milk (d) all

566. In direct cream method of ghee making, item of fat loss is
(a) skim milk (b) ghee residue
(c) handling (d) all

567. % fat recovery in desi method of ghee making is about
(a) 70-75 (b) 60-70
(c) 80-85 (d) 90-95

568. % fat recovery in creamery butter method of ghee making is about
(a) 70-75 (b) 88-92
(c) 60-70 (d) 80-85

569. % fat recovery in direct cream (40-50% fat) method of ghee making is about
(a) 70-75 (b) 60-70
(c) 85-88 (d) 90-95

570. % fat recovery in direct cream (65-85% fat) method of ghee making is about
(a) 70-75 (b) 60-70
(c) 85-88 (d) 90-95

571. Higher fat recovery in ghee is due to ________ fat loss in ghee residue
(a) lower (b) higher
(c) both (d) none

572. Lower fat loss in ghee residue is due to its _____ non-fatty solid s content
(a) higher (b) lower
(c) both (d) none

573. Higher the temperature of storage _____ the keeping quality of ghee
(a) lower (b) higher
(c) both (d) none

574. Higher the initial moisture content _____ the keeping quality of ghee
(a) higher (b) lower
(c) both (d) none

575. Higher the initial acidity _____ the keeping quality of ghee
(a) lower (b) higher
(c) both (d) none

576. Higher the initial sediment content _____ the keeping quality of ghee
(a) higher (b) lower
(c) both (d) none

577. Higher the copper content _____ the keeping quality of ghee
(a) higher (b) lower
(c) both (d) none

578. Higher the iron content _____ the keeping quality of ghee
(a) lower (b) higher
(c) both (d) none

579. Copper and iron in ghee causes
(a) protein hydrolysis
(b) fat hydrolysis
(c) fat oxidation (d) all

580. BHA content (%) in ghee should not be more than
(a) 0.02 (b) 0.2
(c) 0.3 (d) 0.5

581. Which of the following is a ghee adulterant
(a) vanaspati
(b) refined vegetables oil
(c) animal body fat
(d) all

582. Presence of vanspati in ghee can be identified by _____ test
(a) starch (b) iodine
(c) Baudouin (d) none

583. AG in AGMARK represents
(a) agronomy (b) acidic ghee
(c) agricultural (d) none

584. AGMARK act was passed in _____ year
(a) 1937 (b) 1954
(c) 1957 (d) 1964

585. The AGMARK ghee grading scheme was initiated in
(a) 1937 (b) 1938
(c) 1954 (d) 1964

586. According to AGMARK, phytosterol acetate test of ghee should be
(a) positive (b) negative
(c) both (d) none

587. Accoring to AGMARK, FFA (% oleic) content of special grade ghee should not exceed
(a) 0.5 (b) 1.4
(c) 2.5 (d) 3.5

588. Accoring to AGMARK, FFA (% oleic) content of general grade ghee should not exceed
(a) 0.5 (b) 1.4
(c) 2.5 (d) 3.5

589. AGMARK red label of ghee is related to
(a) special grade (b) general grade
(c) both (d) none

590. AGMARK green label of ghee is related to
(a) special grade (b) general grade
(c) both (d) none

591. When high grade ghee is melted, it should be
(a) clear (b) transparent
(c) sediment free (d) all

592. Ghee is used in
(a) cooking
(b) confectionery
(c) direct consumption
(d) all

593. Lassi is also known as
(a) chhas (b) matha
(c) both (d) none

594. Lassi is by-product of ____________ making
(a) butter (b) desi-butter
(c) ice cream (d) dahi

595. Water content (%) of lassi is generally about
(a) 76.2 (b) 86.2
(c) 96.2 (d) 82.4

596. Total solids (%) of lassi is generally about
(a) 3.8 (b) 5.8
(c) 7.8 (d) 9.8

597. Fat content (%) of lassi is generally about
(a) 0.8 (b) 2.2
(c) 3.8 (d) 4.2

598. SNF content (%) of lassi is generally about
(a) 1 (b) 3
(c) 5 (d) 7

599. Protein content (%) of lassi is generally about
(a) 0.3 (b) 0.7
(c) 1.3 (d) 2.7

600. Lactose content (%) of lassi is generally about
(a) 02 (b) 1.2
(c) 2.2 (d) 3.2

601. Lactic acid content (%) in lassi is generally about
(a) 0.11 (b) 0.44
(c) 0.7 (d) 0.85

602. Ash content (%) of lassi is generally about
(a) 0.4 (b) 0.9
(c) 1.4 (d) 1.9

603. Lassi contains appreciable amount of
(a) milk proteins (b) phospholipids
(c) both (d) none

604. Greater the dilution of curd ________ will be the yield of lassi
(a) lower (b) higher
(c) same (d) none

605. Method/technique used for recovery of ghee from ghee residues
(a) centrifugal (b) pressure
(c) both (d) none

606. Which of the following is a by-product of dairy industry
(a) skim milk (b) butter milk
(c) whey (d) all

607. Which of the following is a by-product of cream industry
(a) skim milk (b) butter milk
(c) whey (d) all

608. Which of the following is a by-product of butter industry
(a) skim milk (b) butter milk
(c) whey (d) all

609. Which of the following is a by-product of cheese industry
(a) skim milk (b) butter milk
(c) whey (d) all

610. Which of the following is a by-product of casein industry
(a) skim milk (b) butter milk
(c) whey (d) all

611. Which of the following is a by-product of ghee industry
(a) SMP (b) ghee residue
(c) both (d) none

612. Whey is a by-product of ________ industry
(a) cheese (b) casein
(c) paneer (d) all

613. Cow skim milk generally has water content (%) of about
(a) 80.6 (b) 85.6
(c) 90.6 (d) 95.6

614. Cow skim milk generally has fat content (%) of about
(a) 0.1 (b) 1.0
(c) 1.5 (d) 2.0

615. Cow skim milk generally has protein content (%) of about
(a) 1.6 (b) 3.6
(c) 5.6 (d) 7.9

616. Cow skim milk generally has lactose content (%) of about
(a) 2 (b) 5
(c) 9 (d) 11

617. Cow skim milk generally has ash content (%) of about
(a) 0.1 (b) 0.7
(c) 1.2 (d) 1.6

618. Cow butter milk (from sweet cream) generally has water content (%) of about
(a) 81 (b) 87
(c) 91 (d) 96

619. Cow butter milk (from sweet cream) generally has fat content (%) of about
(a) 0.1 (b) 0.4
(c) 0.9 (d) 1.6

620. Cow butter milk (from sweet cream) generally has protein content (%) of about
(a) 2.4 (b) 3.4
(c) 5.4 (d) 7.4

621. Cow butter milk (from sweet cream) generally has lactose content (%) of about
(a) 2.5 (b) 6.5
(c) 8.5 (d) 4.5

622. Cow butter milk (from sweet cream) generally has ash content (%) of about
(a) 0.1 (b) 0.2
(c) 0.4 (d) 0.7

623. Cow milk lassi (from sweet cream) generally has water content (%) of about
(a) 90.6 (b) 91.0
(c) 88.4 (d) 96.2

624. Cow milk lassi (from sweet cream) generally has fat cotntent (%) of about
(a) 0.1 (b) 0.4
(c) 0.8 (d) 1.5

625. Cow milk lassi (from sweet cream) generally has protein content (%) of about
(a) 1.4 (b) 2.4
(c) 3.4 (d) 4.2

626. Cow milk lassi (from sweet cream) generally has lactose content (%) of about
(a) 0.5 (b) 1.2
(c) 4.5 (d) 5.0

627. Cow milk lassi (from sweet cream) generally has ash content (%) of about
(a) 0.1 (b) 0.2
(c) 0.4 (d) 0.7

628. Cow milk ghee residue (from sweet cream butter) generally has water content (%)
(a) 9.7 (b) 19.7
(c) 49.7 (d) 99.7

629. Cow milk ghee residue (from sweet cream butter) generally has fat content (%)
(a) 21.4 (b) 41.4
(c) 61.4 (d) 81.4

630. Cow milk ghee residue (from sweet cream butter) generally has protein content (%)
(a) 0.4 (b) 12.6
(c) 18.2 (d) 24.8

631. Cow milk ghee residue (from sweet cream butter) generally has ash content (%)
(a) 0.4 (b) 0.7
(c) 1.5 (d) 4.1

632. Cow milk chhana whey generally has water content (%)
(a) 9.7 (b) 71.0
(c) 93.6 (d) 80.6

633. Cow milk chhana whey generally has fat content (%)
(a) 0.5 (b) 1.5
(c) 2.5 (d) 4.5

634. Cow milk chhana whey generally has protein content (%)
(a) 0.1 (b) 0.4
(c) 2.4 (d) 4.1

635. Cow milk chhana whey generally has lactose content (%)
(a) 1.1 (b) 5.1
(c) 7.1 (d) 11.2

636. Cow milk chhana whey generally has ash content (%)
(a) 0.1 (b) 0.4
(c) 0.9 (d) 1.5

637. Cow milk cheese whey generally has water content (%)
(a) 76.1 (b) 88.1
(c) 83.1 (d) 93.1

638. Cow milk cheese whey generally has fat content (%)
(a) 0.02 (b) 0.3
(c) 0.8 (d) 1.2

639. Cow milk cheese whey generally has protein content (%)
(a) 0.2 (b) 0.4
(c) 0.9 (d) 1.5

640. Cow milk cheese whey generally has lactose content (%)
(a) 1.2 (b) 2.9
(c) 3.2 (d) 4.9

641. Cow milk cheese whey generally has ash content (%)
(a) 0.1 (b) 0.4
(c) 0.6 (d) 1.3

642. Cow milk acid casein whey generally has water content (%)
(a) 98.1 (b) 93.1
(c) 99.5 (d) 85.1

643. Cow milk acid casein whey generally has fat content (%)
(a) 0.1 (b) 0.5
(c) 1.0 (d) 1.5

644. Cow milk acid casein whey generally has protein content (%)
(a) 1.0 (b) 2.0
(c) 3.0 (d) 4.0

645. Cow milk acid casein whey generally has lactose content (%)
(a) 2.1 (b) 3.1
(c) 5.1 (d) 3.8

646. Cow milk acid casein whey generally has ash content (%)
(a) 0.1 (b) 0.2
(c) 0.4 (d) 0.7

647. Skim milk can be utilized fro the preparation of
(a) flavored milk
(b) acidophilus milk
(c) Bulgarian buttermilk
(d) all

648. Skim milk can be utilized fro the preparation of
(a) dried skim milk
(b) cottage cheese
(c) casein
(d) all

649. Buttermilk can be utilized fro the preparation of
(a) condensed buttermilk
(b) dried buttermilk
(c) soft cheese
(d) all

650. Whey can be utilized fro the preparation of
(a) whey beverage
(b) Ricotta cheese
(c) Primost cheese
(d) all

651. Ghee residue can be utilized fro the preparation of
(a) ghee toffee (b) sweetmeat
(c) both (d) none

652. Which of the following is a type of cheese
(a) Quarg (b) Gammelost
(c) Sapsago (d) all

653. Which of the following is a type of casein
(a) acid casein (b) rennet casein
(c) both (d) none

654. Which of the following is a continuous process for acid casein
(a) Sheffield process
(b) Universal process
(c) both
(d) none

655. Browning defect in casein may be due to
(a) overheating of casein during drying
(b) presence of extra fat in casein during drying
(c) presence of sulphate during drying
(d) all

656. Poor solubility of casein may be due to
(a) excess high temperature of cooking curd
(b) overheating of casein during drying
(c) molding of curd during drying
(d) all

657. Low adhesive strength of casein is due
(a) high ash content in casein
(b) molding of curd during drying
(c) both (d) none

658. Rennet casein is used in
(a) plastic buttons
(b) plastic jewellery
(c) both (d) none

659. Acid casein is used in
(a) adhesive (b) paint
(c) bristles (d) all

660. Acid casein is used in
(a) paper coating (b) fiber
(c) film (d) all

661. According to ISI, moisture content of edible casein is NMT _____%
(a) 5 (b) 10
(c) 15 (d) 20

662. According to ISI, ash content of edible casein is NMT _____%(dry basis)
(a) 2.5 (b) 1.5
(c) 3.5 (d) 5.5

663. According to ISI, acid insoluble ash content of edible casein is NMT ___%
(a) 0.1 (b) 0.3
(c) 0.5 (d) 1.0

664. According to ISI, fat content of edible casein is NMT _____%
(a) 2.5 (b) 1.5
(c) 3.5 (d) 5.5

665. According to ISI, nitrogen content of edible casein is NMT _____%
(a) 4.5 (b) 8.5
(c) 14.5 (d) 18.5

666. According to ISI, bacterial count (per gram)
(a) NMT 50000 (b) NMT 100000
(c) NMT 500000 (d) none

667. According to ISI, coliform count (per gram)
(a) NMT 10 (b) NMT 50
(c) NMT 150 (d) NMT 200

668. According to ISI, mold count (per gram)
(a) NMT 10 (b) NMT 50
(c) NMT 150 (d) NMT 200

669. Which of the following is a common grade of lactose
(a) crude (b) edible
(c) USP (d) all

670. Lactose content in crude grade of lactose is _____%

(a) 95.4 (b) 98.4
(c) 99.0 (d) 98.85

671. Moisture content in crude grade of lactose is _____%

(a) 0.1 (b) 0.3
(c) 0.5 (d) 1.5

672. Protein content in crude grade of lactose is _____%

(a) 0.01 (b) 0.2
(c) 0.4 (d) 0.8

673. Ash content in crude grade of lactose is _____%

(a) 0.03 (b) 0.2
(c) 0.4 (d) 0.8

674. Acidity (as lactic acid) in crude grade of lactose is _____%

(a) 0.04 (b) 0.06
(c) 0.4 (d) 0.9

675. Lipid content in crude grade of lactose is _____%

(a) 0.1 (b) 0.3
(c) 0.5 (d) 1.5

676. Lactose content in edible grade of lactose is _____%

(a) 95.4 (b) 98.4
(c) 99.0 (d) 98.85

677. Moisture content in edible grade of lactose is _____%

(a) 0.1 (b) 0.3
(c) 0.5 (d) 1.5

678. Protein content in edible grade of lactose is _____%

(a) 0.01 (b) 0.1
(c) 0.4 (d) 0.8

679. Ash content in edible grade of lactose is _____%

(a) 0.03 (b) 0.2
(c) 0.4 (d) 0.8

680. Acidity (as lactic acid) in edible grade of lactose is _____%

(a) 0.04 (b) 0.06
(c) 0.4 (d) 0.9

681. Lipid content in edible grade of lactose is _____ %

(a) 0.1 (b) 0.3
(c) 0.5 (d) 1.5

682. Lactose content in USP grade of lactose is _____ %

(a) 95.4 (b) 98.4
(c) 99.0 (d) 98.85

683. Moisture content in USP grade of lactose is _____ %

(a) 0.1 (b) 0.3
(c) 0.5 (d) 1.5

684. Protein content in USP grade of lactose is _____ %

(a) 0.01 (b) 0.1
(c) 0.4 (d) 0.8

685. Ash content in USP grade of lactose is _____ %

(a) 0.03 (b) 0.2
(c) 0.4 (d) 0.8

686. Acidity (as lactic acid) in USP grade of lactose is _____ %

(a) 0.04 (b) 0.06
(c) 0.4 (d) 0.9

687. Lipid content in USP grade of lactose is _____ %

(a) 0.01 (b) 0.3
(c) 0.5 (d) 1.5

688. According to ISI, maximum limit for total ash in commercial lactose is __ %

(a) 0.5 (b) 1.0
(c) 1.5 (d) 2.0

689. According to ISI, maximum limit for fat in commercial lactose is _____%

(a) 0.5 (b) 1.0
(c) 1.5 (d) 2.5

690. According to ISI, maximum limit for nitrogen in commercial lactose is ____%

(a) 0.05 (b) 1.0
(c) 1.5 (d) 2.5

691. According to ISI, minimum limit for lactose in commercial lactose is ________ %

(a) 70 (b) 80
(c) 90 (d) 99

692. According to ISI, maximum limit for specific rotation in commercial lactose is _____%

(a) 26 (b) 36
(c) 42 (d) 52

693. According to ISI, maximum limit for lead in commercial lactose is __________ ppm

(a) 0.5 (b) 1.0
(c) 1.5 (d) 2.5

694. According to ISI, maximum limit for arsenic in commercial lactose is __________ ppm

(a) 0.5 (b) 1.0
(c) 1.5 (d) 2.5

695. Which of the following is a cultured milk product

(a) yoghurt (b) kefir
(c) koumiss (d) all

696. Which of the following is a cultured milk product

(a) kishk
(b) cultured buttermilk
(c) both
(d) none

697. Which of the following is formed during fermentation of milk

(a) CO_2 (b) acetic acid
(c) diacetyl (d) all

698. Yoghurt is a ________ type of milk product

(a) acidified (b) coagulated
(c) both (d) none

699. Which of the following is similar to yoghurt

(a) jugurt (b) leban
(c) both (d) none

700. Which of the following is atype of yoghurt

(a) set type (b) stirred type
(c) drink type (d) all

701. Energy obtained from plain low fat yoghurt is about ______Kcal

(a) 40 (b) 61
(c) 91 (d) 119

702. Energy obtained from fortified yoghurt is about ______Kcal

(a) 40 (b) 61
(c) 91 (d) 119

703. Protein content of plain low fat yoghurt is about ______%

(a) 4.5 (b) 5
(c) 5.5 (d) 6

704. Protein content of fortified yoghurt is about ______%

(a) 4.5 (b) 5
(c) 5.5 (d) 6

705. Lactose content of yoghurt is about ______%

(a) 1-2 (b) 4-5
(c) 8-10 (d) 0.5-1

706. Fat content of low fat yoghurt is about ______%

(a) 6-8 (b) 4-5
(c) 8-10 (d) 0.5-1.2

707. Fat content of yoghurt is about ______%

(a) 6-8 (b) 4
(c) 8-10 (d) 0.5-1.2

708. Yoghurt should have minimum SNF content of ______%

(a) 5.5 (b) 6.5
(c) 8.5 (d) 2.5

709. Which of the following factor affect the quality of yoghurt

(a) ingredients
(b) heat treatment
(c) emulsification
(d) all

710. Which of the following factor affect the quality of yoghurt
(a) choice of milk base
(b) culture
(c) both (d) none

711. Which of the following milk based ingredient is used in yoghurt making
(a) Lactobacillus bulgaricus
(b) Streptococcus thermophilus
(c) both (d) none

712. Yoghurt is similar to
(a) dahi (b) paneer
(c) condensed milk (d) lassi

713. Which of the following is used as a sweetener in yoghurt
(a) sucrose (b) glucose
(c) maltose (d) all

714. When sucrose is added before fermentation in yoghurt, it inhibits the growth of
(a) osmophilic yeast
(b) osmophilic molds
(c) both
(d) none

715. Which of the following is not a non-calorific sweetener used in yoghurt
(a) saccharin (b) maltose
(c) cyclamate (d) aspartame

716. Cyclamate is ______times sweeter than sucrose
(a) 15-20 (b) 30-80
(c) 110-220 (d) 240-350

717. Saccharin is _____ times sweeter than sucrose
(a) 15-20 (b) 30-80
(c) 110-220 (d) 240-350

718. Which of the following is used as stabilizer in yoghurt
(a) starch (b) alginate
(c) carrageenan (d) all

719. Main function of stabilizer in yoghurt is
(a) maintain viscosity
(b) maintain texture
(c) prevent serum separation
(d) all

720. Stabilizers (gums) are generally added maximum ______ % in yoghurt
(a) 0.5 (b) 1.0
(c) 2.0 (d) 5

721. Starch and pectin may be added max to ___________% in yoghurt
(a) 0.5 (b) 1.0
(c) 2.0 (d) 5.0

722. Maximum SO_2 (mg/kg) that can be added in fruit yoghurt
(a) 60 (b) 120
(c) 150 (d) 300

723. Maximum benzoic acid (mg/kg) that can be added in fruit yoghurt
(a) 60 (b) 120
(c) 150 (d) 300

724. Maximum sorbic (mg/kg) that can be added in fruit yoghurt
(a) 60 (b) 120
(c) 150 (d) 300

725. Maximum methylhydroxy benzoate (mg/kg) that can be added in fruit yoghurt
(a) 60 (b) 120
(c) 150 (d) 300

726. Homogenization of milk reduces fat globule size to less than
(a) 2 micron (b) 5 micron
(c) 20 micron (d) 25 micron

727. Diameter (micron) of fat globules in milk ranges
(a) 0.25-0.5 (b) 1.0-20
(c) 21-25 (d) 25-35

728. Homogenization used for yoghurt manufacture is
(a) single stage (b) double stage
(c) both (d) none

729. In yoghurt manufacture, first stage homogenization is done at ______ MPa

(a) 4 (b) 10
(c) 15 (d) 20

730. In yoghurt manufacture, second stage homogenization is done at ______ MPa

(a) 4 (b) 10
(c) 15 (d) 20

731. In yoghurt manufacture, milk is heated to 60-80°C prior to fermentation which helps in

(a) pasteurization
(b) hydration of proteins
(c) maximum coagulation of protein
(d) all

732. Inoculation temperature (°C) of set type yoghurt for short duration is generally

(a) 32 (b) 42
(c) 25 (d) 48

733. Inoculation temperature (°C) of set type yoghurt for longer incubation is generally

(a) 32 (b) 42
(c) 25 (d) 48

734. In yoghurt, starter culture is generally added at a level of

(a) 2% (b) 4%
(c) 4.5% (d) 6%

735. Isoelectric point of casein is about

(a) 3.6 (b) 4.6
(c) 5.6 (d) 6.5

736. Fermntation in yoghurt is generally arrested at approximate pH of

(a) 2.5 (b) 3.0
(c) 4.0 (d) 5.5

737. Cooling after fermentation in yoghurt manufacture is done to develop

(a) color (b) texture
(c) flavor (d) both (b) & (c)

738. Kefir can be made from ______ milk

(a) goat (b) sheep
(c) cow (d) all

739. Which of the following is present in kefir

(a) Lactobacilli (b) Lactococci
(c) Streptococci (d) all

740. Which of the following is present in larger amount in kefir

(a) Lactobacilli (b) Lactococci
(c) Streptococci (d) all

741. Koumiss was traditionally made from______ milk

(a) goat (b) sheep
(c) mare (d) none

742. Generally, Koumiss has ______ flavor

(a) acidic (b) yeasty
(c) both (d) none

743. Kishk is made from

(a) low fat yoghurt (b) salt
(c) burghol (d) all

744. Burghol is

(a) parboiled wheat
(b) parboiled rice
(c) parboiled pearl millet
(d) none

745. Which of the following is a culture dairy product

(a) Quarg (b) fromagefrais
(c) both (d) none

746. Which of the following starter is used in quarg making

(a) Lactococcus lactis
(b) Lactococcus cremoris
(c) Lactococcus diacetylactis
(d) all

747. Quarg is generally made from

(a) skim milk (b) whole milk
(c) both (d) none

748. Fromage frais is also known as ______ _____ cheese

(a) fresh (b) stale
(c) white (d) none

749. Total solids (%) in quarg is generally about

(a) 7.5 (b) 17.5
(c) 27.5 (d) 37.5

750. Moisture (%) in quarg is generally about

(a) 45.5 (b) 50.5
(c) 62.5 (d) 82.5

751. Protein (%) in quarg is generally about

(a) 2.5 (b) 12.5
(c) 18.5 (d) 22.5

752. Lactose (%) in quarg is generally about

(a) 1.5 (b) 2.5
(c) 3.5 (d) 4.5

753. Fat (%) in quarg is generally about

(a) 0.2 (b) 1.2
(c) 2.0 (d) 2.8

754. Ash (%) in quarg is generally about

(a) 0.2 (b) 0.7
(c) 1.2 (d) 1.5

755. pH of quarg is generally about

(a) 3.5 (b) 2.8
(c) 4.5 (d) 5.8

756. Total solids (%) of creamy fromage frias is about

(a) 13 (b) 15
(c) 19 (d) 25

757. Moisture conent (%) of creamy fromage frias is about

(a) 71 (b) 77
(c) 81 (d) 87

758. Fat content (%) of creamy fromage frias is about

(a) 0.5 (b) 2.5
(c) 5.0 (d) 8.0

759. pH of creamy fromage frias is about

(a) 3.55 (b) 3.65
(c) 4.55 (d) 4.65

760. Total solids (%) of very low fat fromage frias is about

(a) 13 (b) 15
(c) 19 (d) 25

761. Moisture (%) of very low fat fromage frias is about

(a) 71 (b) 77
(c) 81 (d) 87

762. Fat (%) of very low fat fromage frias is about

(a) 0.5 (b) 1.5
(c) 2.5 (d) 5.0

763. pH of very low fat fromage frias is about

(a) 3.55 (b) 3.65
(c) 4.55 (d) 4.65

764. Rennet contain which of the following enzyme

(a) Chymosin (b) pepsin
(c) both (d) none

765. Minimum fat content in cream shoul be

(a) 20 (b) 30
(c) 40 (d) 50

766. Which of the following is a property of cultured cream

(a) slight acidic
(b) mild cheest flavor
(c) viscosity
(d) all

767. Which of the following is atype of bacteriocin

(a) nisin (b) cytolysin
(c) pediocin (d) all

768. Which of the following is atype of bacteriocin

(a) lacticin (b) enterocin
(c) both (d) none

769. Which of the following has antimicrobial property
(a) actadardine (b) cinamycin
(c) lactococcin (d) all

770. Which of the following probiotic can be added to yoghurt
(a) insulin
(b) oligosaccharides
(c) lactulose
(d) all

771. Yoghurt is a type of weak ______ gel
(a) elastic (b) viscous
(c) viscoelastic (d) none

772. Mouthfeel of yoghurt is
(a) smooth (b) viscous
(c) slimy (d) all

773. Which of the following affect the quality of dairy drinking products
(a) additives
(b) fat content of milk
(c) homogenization
(d) all

774. Which of the following stabilizer is used in dairy drinking products
(a) guar (b) pectin
(c) CMC (d) all

775. Which of the following is a concentrated yoghurt
(a) syuzma (b) chakka
(c) skyr (d) all

776. Which of the following is a concentrated yoghurt
(a) torba (b) tan
(c) mastou (d) all

777. Talmjolk is a traditional fermented milk product from
(a) Finland (b) Sweden
(c) Norway (d) all

778. Which of the following is a traditional fermented milk product
(a) Tafmjolk (b) surmjolk
(c) filbunke (d) all

779. Dielectric contant of milk fat is about
(a) 3.1 (b) 5.1
(c) 13.1 (d) 7.9

780. Plasmin is a ________ enzyme
(a) lipolytic (b) proteolytic
(c) carbohydrate hydrolyzing
(d) all

781. Plasmin attacks
(a) α-casein (b) β-casein
(c) both (d) none

782. Which of the following vitamin is generally present in highest amount in milk
(a) pantothenic acid (b) biotin
(c) folic acid (d) vitamin K

783. Freezing point of milk is about ___°C
(a) -0.532 (b) -0.222
(c) -0.952 (d) 0

784. Milk is a good source of ______
(a) water (b) protein
(c) fat (d) all

785. Cowmilk provide ________ cal/100g of milk
(a) 75 (b) 100
(c) 150 (d) 200

786. Buffalo milk provide ______ cal/100g of milk
(a) 75 (b) 100
(c) 150 (d) 200

787. Which of the following is highest in amount in milk
(a) protein (b) fat
(c) minerals (d) water

788. Which of the following is used as a milk adulterant in India
(a) neutralizer (b) starch
(c) urea (d) all

789. Which of the following increase nitrogen content in milk
(a) neutralizer (b) starch
(c) urea (d) all

790. pH of natural milk is around
(a) 5.8 (b) 6.8
(c) 8.8 (d) 10.2

791. pH of synthetic milk is around
(a) 5.8 (b) 6.8
(c) 8.8 (d) 10.2

792. Taste of synthetic milk is
(a) palatable (b) bitter
(c) sour (d) sweet

793. Synthetic milk become ______ on heating
(a) white (b) light green
(c) yellowish (d) purple

794. Synthetic milk become ______ on storage
(a) white (b) light green
(c) yellowish (d) purple

795. Which of the following gives positive test of urea
(a) natural milk
(b) synthetic milk
(c) both (a) & (b)
(d) none

796. Synthetic milk gives _____ test of urea
(a) strongly positive (b) negative
(c) both (d) none

797. Natural milk gives ________ test of urea
(a) weakly positive (b) negative
(c) both (d) none

798. Fat globules in milk are visible with
(a) microscope (b) naked eye
(c) ultramicroscope (d) none

799. Casein in milk is visible with
(a) microscope (b) naked eye
(c) ultramicroscope (d) none

800. Lipoprotein particles in milk are visible with
(a) microscope (b) naked eye
(c) ultramicroscope
(d) electron microscope

801. Density of fat globules in milk at 20°C is ________ kg/m^3
(a) 920 (b) 1100
(c) 1300 (d) 1700

802. Density of casein micelles in milk at 20°C is ________ kg/m^3
(a) 920 (b) 1100
(c) 1300 (d) 1700

803. Density of globular protein in milk at 20°C is ________ kg/m^3
(a) 920 (b) 1100
(c) 1300 (d) 1700

804. Density of lipoprotein in milk at 20°C is ________ kg/m^3
(a) 920 (b) 1100
(c) 1300 (d) 1700

805. Which of the following is not a part of milk plasma
(a) fat globules
(b) casein micelles
(c) globular protein (d) lipoprotein

806. Which of the following is not a part of milk serum phase
(a) fat globules
(b) casein micelles
(c) globular protein
(d) lipoprotein

807. Casein micelles consists of
(a) water (b) protein
(c) salts (d) all

808. Particle diameter of fat globule in milk is ______
(a) 0.1-10 μm (b) 20-400 nm
(c) 3-6 nm (d) 10 nm

809. Particle diameter of casein micelles in milk is ______
(a) 0.1-10 μm (b) 20-400 nm
(c) 3-6 nm (d) 10 nm

810. Particle diameter of globular protein in milk is______
(a) 0.1-10 μm (b) 20-400 nm
(c) 3-6 nm (d) 10 nm

811. Particle diameter of lipoprotein in milk is _______
(a) 0.1-10 μm (b) 20-400 nm
(c) 3-6 nm (d) 10 nm

812. Which of the following is a milk protein
(a) casein (b) lactoferrin
(c) serum albumin (d) all

813. Water activity of milk is about
(a) 0.785 (b) 0.825
(c) 0.895 (d) 0.995

814. Which of the following affects the crystallization of milk
(a) crystallization temperature
(b) faster cooling
(c) globule size (d) all

815. Which of the following is not a soluble at pH 4.6
(a) casein (b) globular protein
(c) lipoprotein (d) all

816. Which of the following leads to clotting by rennet
(a) casein
(b) globular protein
(c) lipoprotein (d) all

817. Which of the following can be denatured by heating
(a) casein
(b) globular protein
(c) lipoprotein (d) all

818. Hydrophobicity of α-lactalbumin is ____ %
(a) 22 (b) 49
(c) 39 (d) 28

819. Hydrophobicity of β-lactglobulin is ____%
(a) 22 (b) 49
(c) 39 (d) 29

820. Hydrophobicity of serum albumin is ____%
(a) 24 (b) 49
(c) 39 (d) 28

821. Lactoferrin inhibits which of the following bacteria
(a) Bacillus stearothermoplilus
(b) Bacillus subtilis
(c) both
(d) none

822. Lactoferrin inhibits bacteria by removal of
(a) calcium (b) copper
(c) iron (d) none

823. Protein efficiency ratio of milk casein is
(a) 2.5 (b) 3.5
(c) 1 (d) 10

824. Protein efficiency ratio of milk serum protein is
(a) 2.5 (b) 3.0
(c) 1 (d) 10

825. Protein efficiency ratio of milk protein is
(a) 2.5 (b) 3.3
(c) 1 (d) 10

826. Which of the following enzyme is present in milk
(a) xanthine oxidase
(b) sulfhydryl oxidase
(c) catalase
(d) all

827. Which of the following enzyme is present in milk
(a) lactoperoxidase
(b) superoxidismutase
(c) lipoprotein lipase
(d) all

828. Which of the following enzyme is present in fat globule membrane of milk
(a) xanthine oxidase
(b) alkaline phosphates
(c) both
(d) none

829. Which of the following enzyme is present in plasma portion of milk
(a) sulfhydryl oxidase
(b) superoxidismutase
(c) both
(d) none

830. Which of the following enzyme is present in casein micelles of milk
(a) lipoprotein lipase (b) plasmin
(c) both (d) none

831. Which of the following enzyme is present in serum portion of milk
(a) lactopoeroxidase (b) ribonuclease
(c) both (d) none

832. Immunoglobulin content in first colostrums is ________%
(a) 2 (b) 7
(c) 15 (d) 25

833. Colostrum has high amount of
(a) somatic cells (b) Cu
(c) Fe (d) all

834. Which of the following is not a lyophobic colloid in milk
(a) fat globules (b) gas bubbles
(c) fat crystals
(d) casein micelles

835. Which of the following is not a type of fat globule aggregate
(a) floccules (b) clusters
(c) granules (d) none

836. Which of the following is associated with milk casein micelle
(a) submicelle (b) peptide chain
(c) nanocluster (d) all

837. Refractive index of milk is generally determined at wavelength of
(a) 523nm (b) 553nm
(c) 573nm (d) 589nm

838. Which of the following donot contribute to the refractive index of milk
(a) fat globule (b) air bubbles
(c) lactose crystals (d) all

839. Which of the following affect the skimming efficiency
(a) temperature (b) viscosity
(c) fat globule size (d) all

840. Fat globule size depends on
(a) homogenization pressure
(b) temperature
(c) type of surfactant
(d) all

841. Aggregation of homogenized globules may be caused by
(a) renneting (b) souring
(c) heating (d) all

842. Water activity of evaporated milk is
(a) 0.83 (b) 0.97
(c) 0.986 (d) 0.995

843. Water activity of icecream mix is
(a) 0.83 (b) 0.97
(c) 0.986 (d) 0.995

844. Water activity of sweetened condensed milk is
(a) 0.83 (b) 0.97
(c) 0.986 (d) 0.995

845. Water activity of skim milk powder (4.5% water) is
(a) 0.02 (b) 0.1
(c) 0.2 (d) 0.5

846. Water activity of skim milk powder (3% water) is
(a) 0.02 (b) 0.1
(c) 0.2 (d) 0.5

847. Water activity of skim milk powder (1.5 % water) is
(a) 0.02 (b) 0.1
(c) 0.2 (d) 0.5

848. Water activity of cheese is
(a) 0.76-0.80 (b) 0.94-0.98
(c) 0.88-0.92 (d) 0.82-0.85

849. Which of the following decrease the water activity of milk
(a) concentration (b) evaporation
(c) both (d) none

850. Which of the following is not a type of membrane processing
(a) ultrafiltration
(b) ultracentrifugation
(c) reverse osmosis
(d) electrodialysis

851. Which of the following is associated with milk processing
(a) thermalization (b) separation
(c) homogenization (d) all

852. Which of the following process involves agitation or beating
(a) working
(b) standardization
(c) churning
(d) separation

853. Which of the following is most important in describing rheology of butter
(a) elastic modulus
(b) yield stress
(c) intrinsic viscosity
(d) relative viscosity

854. Butter is a________ material
(a) elastic (b) plastic
(c) viscoelastic (d) none

855. Consistency of butter depends on
(a) temperature
(b) storage condition
(c) working
(d) all

856. Water content in AMF should not exceed
(a) 0.1% (b) 0.4%
(c) 0.8% (d) 1.5%

857. Water content in buttermilk should not exceed
(a) 0.1% (b) 0.4%
(c) 0.8% (d) 1.5%

858. Hydrogenation of fat _____ the high melting proportion of fat
(a) decrease (b) increase
(c) no effect (d) none

859. According to American standards, fat content of evaporated milk is ____%
(a) 0.1 (b) 4
(c) 9 (d) 7.8

860. According to American standards, SNF content of evaporated milk is _____%
(a) 0.1 (b) 4
(c) 18.1 (d) 7.8

861. According to British standards, fat content of evaporated milk is _____%
(a) 0.1 (b) 4
(c) 9 (d) 7.8

862. According to British standards, SNF content of evaporated milk is _____%
(a) 20 (b) 22
(c) 18.1 (d) 7.8

863. Lactose content of whole milk powder is _____%
(a) 38 (b) 46
(c) 51 (d) 72

864. Casein content of whole milk powder is _____%
(a) 38 (b) 19.5
(c) 26 (d) 72

865. Lactose content of skim milk powder is _____ %
(a) 38 (b) 46
(c) 51 (d) 72

866. Casein content of whole milk powder is _____ %
(a) 38 (b) 46
(c) 51 (d) 27

867. Which of the following additive is used to improve the flowing property of milk powder
(a) sodium silicate
(b) calcium phosphate
(c) both (d) none

868. Porosity of whole milk powder ranges from
(a) 0.40-0.55 (b) 0.45-0.70
(c) 0.75-0.95 (d) 0.85-0.98

869. Porosity of skim milk powder ranges from
(a) 0.40-0.55 (b) 0.45-0.70
(c) 0.75-0.95 (d) 0.85-0.98

870. Which of the following is soluble in water
(a) rennet casein (b) lactalbumen
(c) both (d) none

871. Starter culture used for Bulgarian buttermilk is
(a) 1% (b) 2%
(c) 3% (d) 4%

872. Starter culture used for Acidophilus milk is
(a) 1% (b) 0.5%
(c) 2.5% (d) 3%

873. Roba is fermented milk product from
(a) Egypt (b) Hungary
(c) India (d) Russia

874. Tarho is fermented milk product from
(a) Egypt (b) Hungary
(c) India (d) Russia

875. Yaourt is fermented milk product from
(a) Egypt (b) Hungary
(c) India (d) Russia

876. Zabady is fermented milk product from
(a) Egypt (b) Hungary
(c) India (d) Russia

877. Alcohol content in kefir is about
(a) 0.05-1% (b) 1-1.5%
(c) 2% (d) 3.5%

878. Lactic acid content in kefir is about
(a) 0.25-0.5% (b) 0.7-1.0
(c) 2% (d) 3.5%

879. Which of the following is not a sparkling drink
(a) kefir (b) kumiss
(c) both (d) none

880. Which of the following is associated with cheese making
(a) salting (b) curing
(c) ripening (d) all

881. Which of the following milk has highest fat content
(a) buffalo milk (b) cow milk
(c) human (d) none

882. ________ % of milk lipase is left after pasteurization
(a) 1-2 (b) 2-5
(c) 5-10 (d) 10-15

883. Which of the following is degradation product of amino acid
(a) ammonia (b) amine
(c) H_2S (d) all

884. Which of the following is an off flavor in cheese
(a) yeasty (b) burnt
(c) bitter (d) all

885. Which of the following factors affect the consistency of cheese
(a) temperature (b) fat content
(c) water content (d) all

886. Butyric acid fermentation in cheese depends on
(a) lactic acid content (b) NaCl
(c) nitrite addition (d) all

887. White mild is associated with ______ cheese
(a) Roquefort (b) Gorgonzola
(c) Camembert (d) Swiss

888. Blue mold is associated with ________ cheese
(a) Roquefort (b) Gorgonzola
(c) both (d) none

889. Protein content of Emmentater cheese is ______%
(a) 9 (b) 19
(c) 29 (d) 39

890. Rennet is used in processing of
(a) ice cream (b) butter
(c) cheese (d) all

891. Water content of Emmentater cheese is ____%

(a) 10 (b) 15
(c) 25 (d) 35

5.3 ANSWER KEY

Q.	A.	Q.	A.	Q.	A.	Q.	A.	Q.	A.	Q.	A.
1.	(b)	2.	(d)	3.	(a)	4.	(b)	5.	(d)	6.	(c)
7.	(d)	8.	(c)	9.	(a)	10.	(b)	11.	(b)	12.	(a)
13.	(b)	14.	(c)	15.	(a)	16.	(b)	17.	(a)	18.	(b)
19.	(c)	20.	(b)	21.	(c)	22.	(c)	23.	(a)	24.	(b)
25.	(c)	26.	(b)	27.	(a)	28.	(c)	29.	(c)	30.	(d)
31.	(a)	32.	(d)	33.	(b)	34.	(c)	35.	(c)	36.	(c)
37.	(a)	38.	(b)	39.	(a)	40.	(b)	41.	(b)	42.	(a)
43.	(b)	44.	(c)	45.	(b)	46.	(a)	47.	(d)	48.	(b)
49.	(b)	50.	(a)	51.	(c)	52.	(a)	53.	(a)	54.	(a)
55.	(a)	56.	(b)	57.	(b)	58.	(b)	59.	(d)	60.	(a)
61.	(c)	62.	(c)	63.	(a)	64.	(b)	65.	(b)	66.	(d)
67.	(c)	68.	(a)	69.	(a)	70.	(a)	71.	(c)	72.	(d)
73.	(a)	74.	(b)	75.	(d)	76.	(b)	77.	(a)	78.	(b)
79.	(c)	80.	(d)	81.	(a)	82.	(a)	83.	(b)	84.	(b)
85.	(d)	86.	(a)	87.	(b)	88.	(c)	89.	(a)	90.	(a)
91.	(b)	92.	(d)	93.	(c)	94.	(c)	95.	(a)	96.	(c)
97.	(d)	98.	(b)	99.	(a)	100.	(b)	101.	(a)	102.	(b)
103.	(b)	104.	(c)	105.	(b)	106.	(b)	107.	(c)	108.	(b)
109.	(a)	110.	(a)	111.	(b)	112.	(b)	113.	(a)	114.	(c)
115.	(a)	116.	(b)	117.	(b)	118.	(a)	119.	(a)	120.	(c)
121.	(a)	122.	(c)	123.	(a)	124.	(c)	125.	(d)	126.	(c)
127.	(b)	128.	(b)	129.	(c)	130.	(b)	131.	(c)	132.	(d)
133.	(b)	134.	(a)	135.	(b)	136.	(b)	137.	(b)	138.	(a)
139.	(d)	140.	(a)	141.	(c)	142.	(b)	143.	(b)	144.	(c)
145.	(c)	146.	(a)	147.	(b)	148.	(c)	149.	(c)	150.	(c)
151.	(a)	152.	(c)	153.	(b)	154.	(a)	155.	(b)	156.	(a)
157.	(c)	158.	(b)	159.	(b)	160.	(a)	161.	(b)	162.	(b)
163.	(a)	164.	(b)	165.	(c)	166.	(c)	167.	(b)	168.	(b)

Q. A.	Q. A.	Q. A.	Q. A.	Q. A.	Q. A.
169. (a)	170. (c)	171. (a)	172. (c)	173. (b)	174. (b)
175. (a)	176. (b)	177. (b)	178. (b)	179. (b)	180. (a)
181. (a)	182. (a)	183. (a)	184. (c)	185. (d)	186. (c)
187. (a)	188. (a)	189. (b)	190. (a)	191. (b)	192. (a)
193. (b)	194. (a)	195. (a)	196. (b)	197. (a)	198. (b)
199. (a)	200. (a)	201. (a)	202. (a)	203. (b)	204. (a)
205. (d)	206. (b)	207. (b)	208. (b)	209. (c)	210. (b)
211. (a)	212. (b)	213. (a)	214. (a)	215. (b)	216. (c)
217. (a)	218. (c)	219. (b)	220. (b)	221. (b)	222. (a)
223. (c)	224. (b)	225. (d)	226. (b)	227. (b)	228. (b)
229. (c)	230. (d)	231. (d)	232. (b)	233. (a)	234. (c)
235. (b)	236. (a)	237. (b)	238. (c)	239. (d)	240. (d)
241. (d)	242. (b)	243. (a)	244. (a)	245. (b)	246. (b)
247. (b)	248. (b)	249. (a)	250. (b)	251. (c)	252. (b)
253. (a)	254. (c)	255. (b)	256. (d)	257. (d)	258. (a)
259. (b)	260. (b)	261. (a)	262. (a)	263. (a)	264. (d)
265. (d)	266. (b)	267. (d)	268. (a)	269. (b)	270. (b)
271. (b)	272. (a)	273. (c)	274. (c)	275. (a)	276. (d)
277. (d)	278. (d)	279. (c)	280. (a)	281. (a)	282. (c)
283. (c)	284. (b)	285. (c)	286. (c)	287. (c)	288. (a)
289. (a)	290. (a)	291. (a)	292. (c)	293. (a)	294. (d)
295. (b)	296. (b)	297. (b)	298. (a)	299. (a)	300. (c)
301. (b)	302. (c)	303. (b)	304. (b)	305. (c)	306. (a)
307. (a)	308. (a)	309. (b)	310. (c)	311. (d)	312. (d)
313. (c)	314. (a)	315. (a)	316. (a)	317. (b)	318. (c)
319. (a)	320. (c)	321. (a)	322. (d)	323. (c)	324. (b)
325. (b)	326. (a)	327. (c)	328. (a)	329. (b)	330. (c)
331. (b)	332. (b)	333. (c)	334. (b)	335. (a)	336. (c)
337. (c)	338. (b)	339. (a)	340. (a)	341. (a)	342. (a)
343. (a)	344. (a)	345. (b)	346. (a)	347. (c)	348. (c)
349. (a)	350. (a)	351. (b)	352. (b)	353. (c)	354. (b)
355. (b)	356. (b)	357. (a)	358. (b)	359. (c)	360. (b)
361. (c)	362. (d)	363. (a)	364. (d)	365. (d)	366. (d)
367. (b)	368. (b)	369. (b)	370. (c)	371. (b)	372. (a)

Q. A.	Q. A.	Q. A.	Q. A.	Q. A.	Q. A.
373. (a)	374. (a)	375. (a)	376. (a)	377. (b)	378. (b)
379. (c)	380. (b)	381. (b)	382. (c)	383. (a)	384. (a)
385. (a)	386. (b)	387. (b)	388. (b)	389. (a)	390. (b)
391. (c)	392. (b)	393. (a)	394. (b)	395. (b)	396. (a)
397. (b)	398. (a)	399. (a)	400. (c)	401. (a)	402. (d)
403. (c)	404. (b)	405. (a)	406. (c)	407. (d)	408. (a)
409. (b)	410. (b)	411. (c)	412. (c)	413. (d)	414. (a)
415. (d)	416. (d)	417. (a)	418. (d)	419. (b)	420. (d)
421. (b)	422. (c)	423. (a)	424. (a)	425. (b)	426. (a)
427. (c)	428. (b)	429. (d)	430. (d)	431. (d)	432. (d)
433. (d)	434. (d)	435. (b)	436. (b)	437. (c)	438. (c)
439. (b)	440. (b)	441. (c)	442. (a)	443. (a)	444. (c)
445. (b)	446. (d)	447. (d)	448. (a)	449. (b)	450. (a)
451. (b)	452. (c)	453. (d)	454. (b)	455. (a)	456. (b)
457. (b)	458. (a)	459. (b)	460. (c)	461. (d)	462. (c)
463. (d)	464. (d)	465. (c)	466. (b)	467. (c)	468. (c)
469. (c)	470. (c)	471. (c)	472. (c)	473. (a)	474. (b)
475. (b)	476. (a)	477. (d)	478. (c)	479. (a)	480. (b)
481. (a)	482. (d)	483. (a)	484. (c)	485. (a)	486. (b)
487. (d)	488. (d)	489. (c)	490. (d)	491. (a)	492. (b)
493. (a)	494. (d)	495. (c)	496. (c)	497. (b)	498. (b)
499. (b)	500. (b)	501. (c)	502. (b)	503. (b)	504. (b)
505. (c)	506. (d)	507. (b)	508. (b)	509. (c)	510. (c)
511. (d)	512. (d)	513. (c)	514. (d)	515. (d)	516. (c)
517. (c)	518. (b)	519. (d)	520. (c)	521. (c)	522. (a)
523. (b)	524. (b)	525. (a)	526. (a)	527. (b)	528. (b)
529. (b)	530. (d)	531. (a)	532. (b)	533. (c)	534. (b)
535. (c)	536. (d)	537. (a)	538. (c)	539. (c)	540. (c)
541. (d)	542. (d)	543. (a)	544. (d)	545. (c)	546. (d)
547. (b)	548. (c)	549. (d)	550. (d)	551. (a)	552. (a)
553. (c)	554. (b)	555. (d)	556. (d)	557. (b)	558. (b)
559. (c)	560. (c)	561. (a)	562. (c)	563. (d)	564. (d)
565. (d)	566. (c)	567. (b)	568. (c)	569. (d)	570. (a)
571. (b)	572. (a)	573. (b)	574. (a)	575. (b)	576. (b)

Q. A.	Q. A.	Q. A.	Q. A.	Q. A.	Q. A.
577. (a)	578. (c)	579. (a)	580. (d)	581. (c)	582. (c)
583. (a)	584. (b)	585. (b)	586. (b)	587. (c)	588. (a)
589. (b)	590. (d)	591. (d)	592. (c)	593. (b)	594. (c)
595. (a)	596. (a)	597. (b)	598. (c)	599. (b)	600. (b)
601. (a)	602. (c)	603. (b)	604. (c)	605. (d)	606. (a)
607. (b)	608. (c)	609. (c)	610. (b)	611. (d)	612. (c)
613. (a)	614. (b)	615. (b)	616. (b)	617. (c)	618. (b)
619. (b)	620. (d)	621. (d)	622. (d)	623. (c)	624. (a)
625. (b)	626. (c)	627. (a)	628. (c)	629. (d)	630. (d)
631. (c)	632. (a)	633. (b)	634. (b)	635. (b)	636. (d)
637. (b)	638. (c)	639. (d)	640. (c)	641. (b)	642. (a)
643. (a)	644. (c)	645. (d)	646. (d)	647. (d)	648. (d)
649. (d)	650. (c)	651. (d)	652. (c)	653. (c)	654. (d)
655. (d)	656. (c)	657. (c)	658. (d)	659. (d)	660. (b)
661. (a)	662. (a)	663. (b)	664. (c)	665. (a)	666. (a)
667. (b)	668. (d)	669. (b)	670. (b)	671. (d)	672. (c)
673. (c)	674. (a)	675. (c)	676. (c)	677. (b)	678. (b)
679. (b)	680. (a)	681. (d)	682. (a)	683. (a)	684. (a)
685. (a)	686. (a)	687. (c)	688. (d)	689. (a)	690. (c)
691. (d)	692. (d)	693. (b)	694. (d)	695. (c)	696. (d)
697. (c)	698. (c)	699. (d)	700. (b)	701. (d)	702. (d)
703. (c)	704. (b)	705. (d)	706. (b)	707. (c)	708. (d)
709. (c)	710. (c)	711. (a)	712. (d)	713. (c)	714. (b)
715. (b)	716. (d)	717. (d)	718. (d)	719. (a)	720. (b)
721. (a)	722. (b)	723. (d)	724. (b)	725. (a)	726. (b)
727. (b)	728. (c)	729. (a)	730. (d)	731. (b)	732. (a)
733. (a)	734. (b)	735. (c)	736. (d)	737. (d)	738. (d)
739. (a)	740. (c)	741. (c)	742. (d)	743. (a)	744. (c)
745. (d)	746. (a)	747. (a)	748. (b)	749. (d)	750. (b)
751. (c)	752. (a)	753. (b)	754. (c)	755. (c)	756. (c)
757. (d)	758. (d)	759. (a)	760. (d)	761. (a)	762. (c)
763. (c)	764. (a)	765. (d)	766. (d)	767. (c)	768. (d)
769. (d)	770. (c)	771. (d)	772. (d)	773. (d)	774. (d)
775. (d)	776. (d)	777. (d)	778. (a)	779. (b)	780. (c)

Q. A.	Q. A.	Q. A.	Q. A.	Q. A.	Q. A.
781. (a)	782. (a)	783. (d)	784. (a)	785. (b)	786. (d)
787. (d)	788. (c)	789. (b)	790. (d)	791. (b)	792. (c)
793. (c)	794. (c)	795. (a)	796. (a)	797. (a)	798. (c)
799. (d)	800. (a)	801. (b)	802. (c)	803. (b)	804. (a)
805. (b)	806. (d)	807. (a)	808. (b)	809. (c)	810. (d)
811. (d)	812. (d)	813. (d)	814. (a)	815. (a)	816. (b)
817. (d)	818. (d)	819. (a)	820. (c)	821. (c)	822. (a)
823. (b)	824. (b)	825. (d)	826. (d)	827. (c)	828. (c)
829. (c)	830. (c)	831. (b)	832. (d)	833. (d)	834. (d)
835. (d)	836. (d)	837. (d)	838. (d)	839. (d)	840. (d)
841. (c)	842. (b)	843. (a)	844. (c)	845. (b)	846. (a)
847. (b)	848. (c)	849. (b)	850. (d)	851. (c)	852. (b)
853. (b)	854. (d)	855. (a)	856. (b)	857. (b)	858. (d)
859. (c)	860. (c)	861. (b)	862. (a)	863. (b)	864. (c)
865. (a)	866. (c)	867. (b)	868. (a)	869. (d)	870. (b)
871. (c)	872. (a)	873. (b)	874. (d)	875. (a)	876. (a)
877. (b)	878. (d)	879. (d)	880. (a)	881. (d)	882. (d)
883. (d)	884. (d)	885. (d)	886. (c)	887. (c)	888. (c)
889. (c)	890. (d)				

CHAPTER 6

CEREALS TECHNOLOGY

6.1 SUBJECTIVE KEYNOTES

1. The cereal crops that are grown for their edible fruit are generally called grain, but botanically referred to as caryopsis.
2. The cereal seed consists of two major components, the endosperm and embryo or germ.
3. The endosperm encompass the bulk of the seed and is the energy source of stored food.
4. A semi permeable layer under the pericarp, which is called testa, surrounds the embryo and is derived from the inner ovary wall.
5. Wheat is a single-seeded fruit, 4- to 10-mm long, consisting of a germ and endosperm enclosed by an epidermis and a seed coat
6. The endosperm cells are packed with starch granules, which lie embedded in a matrix that is largely protein.
7. The corn seed is a single fruit called the kernel. It includes an embryo, endosperm, aleurone, and pericarp.
8. 80% of dry matter of cereals is carbohydrate out of which starch is the most important carbohydrate in all cereals.
9. The fibre constituents are cellulose, hemicellulose and pentosans,
10. Among cereals, whole wheat, ragi and bajra contain high amount of fibre.
11. Rice contains less amount of protein compared to other cereals.
12. Within the endosperm the concentration of protein increases from the center to the periphery.
13. The types of protein present in cereals are albumins, globulins, prolamines (gliadins) and glutelins.
14. The gliadins and glutelins are known as gluten forming proteins.
15. The gluten has unique elasticity and flow properties which are used for baking bread and other products.
16. Cereals contain 6-12% protein, which is generally deficient in lysine.
17. Milling operation in cereals generally removes much of the bran (fibrous portion) and fat from grains.
18. Lipids are present to the extent of 1-2% in wheat and rice, and 3% in maize.

19. Wheat germ contains lipids 6-11% and bran 3-5% and endosperm 0.8-1.5%.
20. About 95% of minerals in cereals are the phosphates and sulphates of potassium, magnesium and calcium.
21. Phytates present in cereals decrease the absorption of iron.
22. Cereals are poor sources of calcium and iron particularly rice is a very poor source of these two elements.
23. Ragi is a rich source of calcium and iron. Millets (ragi, bajra, jowar) are rich in minerals and fibre.
24. Parboiling which includes soaking in water and steaming of paddy results in seeping of vitamins present in outer layer into the grain.
25. Cereals do not contain either vitamin A or C except maize which contains small amount of carotenes.
26. Certain grains contain many enzymes and of these the amylases, proteases, lipases and oxido-reductases are of importance.
27. The 1000 grain weight is a good indicator of the grain size, which can vary relative to growing conditions and maturity.
28. The bulk density of cereal grains is determined by measuring the weight of a grain sample of known volume.
29. Freshly harvested paddy normally has a moisture content of 18-25%.
30. For long storage, the grain should be dried to 13% moisture or less.
31. One-fourth the weight of paddy consists of husk, which makes up the largest milling by-product of rice.
32. Rice husk has a high ash content, about 20%, the major portion of which (over 90%) consists of silica.
33. Fully burnt white ash of husk contains 95% silica and can be used for manufacturing sodium silicate, silica gel, insulating bricks, etc.
34. Rice bran is the most valuable by-product of rice milling industry.
35. Rice bran contains 12-15% protein, 14-20% oil if the paddy is raw and 18-25% if parboiled, and is a rich source of B-group vitamins.
36. Parboiling is a hydrothermal treatment followed by drying before milling for the production of milled parboiled grain.
37. Water soluble B-vitamins and other water soluble nutrients diffuse into the endosperm during parboiling and hence the loss of nutrients is less in parboiled rice even after polishing the presence of vitamin E is particularly noted in parboiled rice.
38. Wheat types may be classified as hard or soft wheat and strong or weak.
39. Hardness is related to the degree of adhesion between the starch granules and surrounding protein.
40. The botanical name of corn is Zea mays and used as food and feed.
41. Corn-germ contains about 21.0 per cent proteins and 32 per cent fat and since the germ constitutes about 13 per cent of the grain, it is an important source of protein and fat.

42. Barley is used to a small extent as a food in the form of pot barley or pearl barley or barley flour and mostly used for preparation of malt.
43. The oat, grain consists of husk, aleurone layer, germ and endosperm.
44. The important oat products of commerce are rolled oats and oat flakes.
45. Oats free from the husk are called groats.
46. Apart from maize and sorghum, the major millet crops of India are Pearl millet called bajra and finger millet known as ragi.
47. Minor millets are grown and they are: the common millet or proso millet, foxtail millet and kodo millet.
48. Barley malt is used in brewing and in the preparation of malt extract for pharmaceutical purposes and in the preparation of, malted milk powder.
49. Ready-to-serve breakfast cereal foods include flaked breakfast cereals, puffed breakfast cereals, shredded and granular breakfast cereals and puffing by extrusion.
50. The total dietary fiber content of oat is considerable with the hemicellulose content of the oat hull is between 30 and 50%.
51. Soft wheat flour is undesirable for bread but for cakes, pastries and cookies soft wheat flour is required.
52. For proper gluten development glutenin and gliadin should be in a ratio of 1:1 of in the flour. This is the reason why only wheat flour is suitable for almost all bakery products.
53. Patent flour is the highest quality of all commercial grades of white flour as it consists of the innermost part of the endosperm and is essentially free of bran and germ. This makes patent flour lowest in ash and whitest in colour.
54. Clear flour is the lowest quality of all commercial grades of flour as it consists of the outer part of the endosperm and contains germ and bran. The flour streams that remain after patent flour are the clear flour.
55. Straight flour is milled from the entire endosperm. It is made by combining all usable streams of flour from the milling process and contains bran and germ particles that are not easily separated from the endosperm.
56. All-purpose flour is often made from a blend of hard and soft wheat and contains 9.5–11.5 percent protein. This flour comes bleached or unbleached, and may contain added malted barley flour.
57. When the flour contains all the three components of wheat kernel- bran, germ and endosperm then it is called as whole meal flour.
58. Composite flour is prepared by blending wheat flour with other flours (cereal, legumes etc.) to lower the use of wheat by partial substitution for economic reasons.
59. The germ enriched flour should contain minimum 10% (calculated on dry basis) cooked germ.
60. There are some functional ingredients termed as 'flour improver' which are combined with flour to improve processing and baking quality. These are also called as dough conditioners, dough improver and bread improver. Examples include bleaching agents, maturing agents, oxidizing agents, reducing agents, enzymes and emulsifiers.

61. Bakery shortening is fat portion of bakery products used to weaken or shorten the dough by weakening its gluten network resulting in soft texture of bakery products.
62. Sugar undergoes a series of complex browning reactions above 160°C imparting the brown crust colour to many baked goods.
63. Leavening is of four types i.e. chemical, biological, mechanical & steam leavening.
64. Mixing is a term used to describe combined process of stirring, beating and blending. Proper mixing of ingredients is a prerequisite for good quality bakery product.
65. Transfer of energy to the dough during mixing helps to increase gas retention and increase bread volume.
66. Bread dough mixing can be carried out using horizontal, vertical, reciprocating or continuous mixers.
67. In over-mixing of bread dough, the developed gluten structure is broken down resulting into wet, sticky dough known as 'broken down dough'.
68. In under-mixing, the bread dough is underdeveloped and hence is less elastic. The volume of bread will be low and at times the bread will collapse in proofing or in the oven as gluten will not have the proper extensibility to hold the gas in the dough.
69. The final proofing is done at temperature of 30-35°C and RH of 85 % for 55-65 minutes.
70. During the fermentation of bread dough the starch is converted into sugar which is used by yeast to produce CO_2 gas and ethanol. The carbon dioxide gas formed is entrapped by the gluten film resulting into expansion of bread loaf volume.
71. After the final proofing, the leavened bread dough is baked at temperature of 220°C for 20-30 minutes in baking oven.
72. During baking as the temperature of the bread dough increased, it passes three phases-oven spring, starch gelatinization and browning and aroma formation.
73. The Chorleywood bread process (CBP) method, a mechanical dough development was developed in 1958 by British Baking Industries Research Association at Chorleywood, UK. In 1950s the first mechanical dough development process 'Do-maker' was successfully developed.
74. In the activated dough development method, the commonly used active ingredients are reducing agents like L-cysteine hydrochloride (35 ppm), proteolytic anzymes and ascorbic acid (more than 50 ppm) that reduce mixing time of flour, oxidizing agents, fat or an emulsifier, yeast and salt.
75. Celiac disease is a severe form of a continuum of disorders called gluten sensitivity and is due to a permanent intolerance to gluten in genetically susceptible individuals.
76. Sourdough is a mixture of flour and water fermented with LAB and yeasts. Traditionally sourdough is added as an ingredient to unmodified flour of wheat for bread making. Generally, rye bread is prepared by sourdough process.
77. Transglutaminase catalyses either protein cross-linking, binding of free amines or the deamidation of glutamine residues to form glutamate, depending on the available substrates and conditions in the system.
78. According to UK Classification, biscuits are classified based on the formulation as hard dough and short dough biscuits.

79. According to USA Classification, biscuits are classified based on the method of processing and the shape of biscuits such as cutting type, rotary molding type, wire cutting type and depositing type cookies.
80. The process of manufacturing of biscuits mainly involves the following three steps- mixing, shaping or forming and baking.
81. Wafers represent a specialized type of low-moisture, light-textured biscuit, generally made from cereals. The thin, crisp, precisely shaped wafers are available in variety of shapes including flat, hollow, molded cones, rolled cones and sticks.
82. The product density of wafer ranges from 0.10 to 0.25 g/cm3.
83. Cakes are sweet baked products characterized by high level of sugar in the formulation and hence are foods of high calorific value. Ingredient used in cakes include flour, sugar, shortening and egg are the essential ingredients for cake manufacture.
84. Corn syrup, commercial name of glucose syrup, has largely replaced invert sugar as a confectionery ingredient.
85. The most common type of glucose syrup in sugar confectionery is 42 DE and is also referred to as confectioner's glucose.
86. As the DE of corn syrup increases, the properties like browning, flavour, freezing point depression, humectancy, hygroscopy, osmotic pressure, sugar crystallization, sweetness, etc increases while some properties like cohesiveness, foam stability, viscosity decreases.
87. If the substrate fat is butterfat then at low degree of hydrolysis, it gives a 'buttery' or 'creamy' flavour. As the FFA content is increased, the 'cheesy' flavour becomes pronounced.
88. Bulking sweeteners includes polyols like sorbitol, lactitol, polydextrose, maltitol, isomalt, erythritol, etc.
89. Deck ovens are used on small scale. Tunnel ovens are used for continuous baking operation.
90. Vital wheat gluten is the insoluble protein portion of wheat flour obtained by separation, washing and drying so that it contains about 75 to 80 percent protein.
91. Oxidising agents act on the gluten in flour to increase its strength and produce larger, more-uniform finished products. Natural ascorbic acid and chemicals like potassium bromate are used as oxidising agents.
92. Reducing agents act on the gluten in flour to increase extensibility, reduce elasticity, and shorten mix time. These include L-cysteine and non-leavening yeast, which is a natural source of glutathione.
93. Fillers are used to standardize the strength of dough conditioners and make concentrated ingredients easier to scale. Flour, starch, and calcium sulphate are most frequently used as fillers.
94. Baking soda is also known as sodium bicarbonate (NaHCO3). It produces carbon dioxide on heating.
95. Residual sodium carbonate (washing soda) can cause a bitter, "soapy" taste and a yellow colour in the baked product.

96. The efficiency of the acids used in baking powder can be compared by their neutralising value which is the number of parts of the acidic material required to neutralise 100 parts of baking soda.

97. Ammonium carbonate or bicarbonate is used as leavening agent in biscuits and crackers etc. that are to be baked at low moisture.

98. Yeast is an oval shaped, single celled and colourless fungus that produces carbon dioxide gas through fermentation.

99. The shortening to be used in bakery should have bland flavor, white appearance, good plasticity and flavor & oxidative stability.

100. The types of sugar used in bakery products includes sanding sugar, granulated sugar, icing sugar & brown sugar.

6.2 MULTIPLE CHOICE QUESTIONS

1. Cereals are regarded as ___________ source of energy
   (a) richest (b) cheapest
   (c) both (d) none
2. Cereals are used for
   (a) source of energy
   (b) animal feed
   (c) industrial purpose (d) all
3. Cereal grains belong to
   (a) monocots family (b) gramineae
   (c) both (d) none
4. Jowar is common name for
   (a) oats (b) maize
   (c) sorghum (d) millet
5. The chief cereal of human diet is
   (a) rice (b) wheat
   (c) maize (d) all
6. The cereal grains are obtained from
   (a) stem (b) flowers
   (c) leaves (d) shoot
7. Which of the following is called naked grain
   (a) wheat (b) oats
   (c) maize (d) all
8. Which of the following is called covered caryopsis
   (a) rice (b) oats
   (c) barley (d) all
9. Paddy is the name given to
   (a) polished rice
   (b) rice with husk
   (c) millet
   (d) sorghum
10. The grain generally comprises of
    (a) seed coat (b) germ
    (c) endosperm (d) all
11. The outer layer of cereals is called
    (a) pericarp (b) seed coat
    (c) aleurone (d) endosperm
12. The endosperm is surrounded by one or more layers of cells called
    (a) aleurone (b) seed coat
    (c) hypoderm (d) epidermis
13. The aleurone layer & all the covering external to it comprise the
    (a) bran (b) seed coat
    (c) epidermis (d) hypodermis
14. The endosperm mainly consists of
    (a) starch (b) protein
    (c) minerals (d) both (a) & (b)
15. Germ is separated from embryo by
    (a) scutellum (b) endosperm
    (c) aleurone layer (d) none
16. Most of the B vitamins are present in
    (a) endosperm (b) scutellum
    (c) aleurone layer (d) germ

17. Germ is rich in
(a) fat (b) protein
(c) carbohydrates (d) both (a) & (b)

18. Which of the following is absent in cereal grain
(a) carbohydrates (b) protein
(c) vitamins (d) none

19. Crude, soluble and insoluble are types of
(a) vitamins (b) fiber
(c) protein (d) germ fat

20. The fiber consists of
(a) cellulose
(b) hemicelluloses
(c) pentsans (d) all

21. Which of the following is absent in cereals
(a) sucrose (b) maltose
(c) raffinose (d) none

22. Removal of the husk of rice increases its
(a) protein content (b) soluble fiber
(c) insoluble fiber (d) none

23. The protein content in barley is ______ g/100g
(a) 11.5 (b) 10.6
(c) 9.6 (d) 8.6

24. The fat content in barley is ______ g/100g
(a) 1.3 (b) 2.3
(c) 3.3 (d) 4.3

25. The fiber content in barley is ______ g/100g
(a) 2.8 (b) 3.9
(c) 4.2 (d) 5.5

26. The mineral content in barley is ______ g/100g
(a) 2.2 (b) 0.2
(c) 1.2 (d) 0.8

27. The protein content in maize is ______ g/100g
(a) 11.1 (b) 10.6
(c) 9.6 (d) 8.6

28. The fat content in maize is ___ g/100g
(a) 1.3 (b) 2.3
(c) 3.6 (d) 4.3

29. The fiber content in maize is ______ g/100g
(a) 2.7 (b) 1.7
(c) 4.2 (d) 5.5

30. The protein content in oats is ______ g/100g
(a) 13.6 (b) 7.3
(c) 7.5 (d) 12.1

31. The protein content in ragi is ______ g/100g
(a) 13.6 (b) 7.3
(c) 7.5 (d) 12.1

32. The protein content in rice is ______ g/100g
(a) 13.6 (b) 7.3
(c) 7.5 (d) 12.1

33. The protein content in rye is ____ g/100g
(a) 13.6 (b) 7.3
(c) 7.5 (d) 12.1

34. The fat content of oat is ______ g/100g
(a) 7.6 (b) 1.3
(c) 0.5-1 (d) 1.7

35. The fat content of ragi is ______ g/ 100g
(a) 7.6 (b) 1.3
(c) 0.5-1 (d) 1.7

36. The fat content of rice is ______ g/100g
(a) 7.6 (b) 1.3
(c) 0.5-1 (d) 1.7

37. The fat content of oat is ______ g/100g
(a) 7.6 (b) 1.3
(c) 0.5-1 (d) 1.7

38. The mineral content of oats is ______ g/100g
(a) 1.8 (b) 2.7
(c) 5.2 (d) 7.2

39. The mineral content of ragi is ______ g/100g
(a) 1.8 (b) 2.7
(c) 5.2 (d) 7.2

40. The fiber content of oats is ______ g/100g
(a) 3.5 (b) 6.3
(c) 0.2-0.6 (d) 2.0

41. The fiber content of ragi is ______ g/100g
(a) 3.6 (b) 6.3
(c) 0.2-0.6 (d) 2.0

42. The fiber content of rice is ______ g/100g
(a) 3.5 (b) 6.3
(c) 0.2-0.6 (d) 2.0

43. The fiber content of rye is ______ g/100g
(a) 3.5 (b) 6.3
(c) 0.2-0.6 (d) 2.0

44. The protein content of sorghum is ______g/100g
(a) 10.4 (b) 10.8-14
(c) 7.2 (d) 4.6

45. The protein content of wheat is ______g/100g
(a) 10.4 (b) 10.8-14
(c) 7.2 (d) 4.6

46. The fat content of sorghum is ______ g/100g
(a) 1.9 (b) 1.50
(c) 2.3 (d) 3.5

47. The fat content of wheat is ______ g/100g
(a) 1.9 (b) 1.50
(c) 2.3 (d) 3.5

48. The fiber content of sorghum is ______g/100g
(a) 1.6 (b) 1.2-1.9
(c) 2.3 (d) 3.5

49. The fiber content of wheat is ______g/100g
(a) 1.6 (b) 1.2-1.9
(c) 2.3 (d) 3.5

50. The mineral content of wheat is ______ g/100g
(a) 1.5 (b) 1.2-1.9
(c) 2.3 (d) 3.5

51. Which of the following has highest protein
(a) oat (b) rye
(c) wheat (d) rice

52. Which of the following has highest fat content
(a) oat (b) rye
(c) wheat (d) rice

53. Which of the following has highest fiber content
(a) oat (b) rye
(c) barley (d) bajra

54. ICAR stands for
(a) Indian College of Agricultural Research
(b) International Council of Agricultural Research
(c) Indian Council of Agricultural Research
(d) none

55. ICAR is situated at
(a) New Delhi (b) Bangalore
(c) Chandigarh (d) Karnal

56. DWR stands for
(a) Director of Wheat Research
(b) Directorate of Wheat Research
(c) Distance Institute of Wheat Research
(d) none

57. DWR is situated at
(a) New Delhi (b) Bangalore
(c) Chandigarh (d) Karnal

58. Which of the following has lowest protein
(a) oat (b) rye
(c) wheat (d) rice

59. In the endosperm, the concentration of protein ________ from center to the periphery
(a) increases (b) decreases
(c) no change (d) none

60. Which of the following is not a type of protein in cereals
(a) albumen (b) globulin
(c) gliadin (d) zein

61. The gluten protein is present in
(a) maize (b) wheat
(c) rice (d) barley

62. Cereals are generally ________ in lysine
(a) rich (b) poor
(c) deficient (d) none

63. Highest protein content is present in
(a) germ (b) aleurone
(c) endosperm (d) both (a) & (b)

64. Mostly lipids are present in
(a) germ (b) bran
(c) endosperm (d) both (a) & (b)

65. ________ is particularly rich in lipids
(a) germ (b) aleurone
(c) scutellum (d) bran

66. Major constituent of ash in rice husk is
(a) silica (b) alumina
(c) magnesium (d) phosphorus

67. ______% silica is present in rice husk ash
(a) 46 (b) 66
(c) 96 (d) 86

68. Phosphorus is present in cereals in ________ form
(a) nicotinic acid (b) phosphates
(c) phytin (d) phytic acid

69. Which of the following is rich in iron
(a) oats (b) rye
(c) maize (d) bajra

70. Alpha amylase is present in
(a) barley (b) wheat
(c) both (d) none

71. Beta amylase is present in
(a) barley (b) wheat
(c) rye (d) all

72. The cereal proteases are present in
(a) aleurone layer (b) germ
(c) endosperm (d) bran

73. The spoilage of cereals is mainly by
(a) bacteria (b) yeast
(c) molds (d) virus

74. A moisture of less than_____% is considered safe for storage
(a) <14 (b) <15
(c) <18 (d) <20

75. The commonly grown variety of wheat is
(a) Triticum aestivum
(b) Triticum durum
(c) Triticum dicoccum (d) none

76. The bread wheat is
(a) Triticum aestivum
(b) Triticum durum
(c) Triticum dicoccum (d) none

77. The macaroni wheat is
(a) Triticum aestivum
(b) Triticum durum
(c) Triticum dicoccum (d) none

78. The club wheat is
(a) Triticum aestivum
(b) Triticum durum
(c) Triticum compactum
(d) none

79. The increase in the wheat production in India is due to
(a) White revolution
(b) Blue revolution
(c) Green revolution
(d) Yellow revolution

80. The father of Green revolution in India is
(a) Norman Borlaug
(b) Kurian
(c) both
(d) none

81. Most of wheat grown in India is consumed as
(a) chapati (b) bread
(c) biscuits (d) pasta

82. The wheat grain is composed of
(a) bran (b) endosperm
(c) germ (d) all

83. The endosperm portion of wheat is rich in
(a) starch (b) soluble sugar
(c) lipid (d) mineral

84. The bran constitutes of
(a) cellulose
(b) hemicelluloses
(c) both
(d) none

85. Most of the niacin is present in
(a) aleurone (b) endosperm
(c) scutellum (d) embryo

86. Most of the pantothenic acid is present in
(a) aleurone (b) endosperm
(c) scutellum (d) embryo

87. Thiamin is present in
(a) aleurone (b) endosperm
(c) scutellum (d) embryo

88. Riboflavin is mainly present in
(a) aleurone (b) endosperm
(c) scutellum (d) embryo

89. Starch is mostly present in
(a) aleurone (b) endosperm
(c) scutellum (d) germ

90. Sugar is mostly present in
(a) aleurone (b) endosperm
(c) scutellum (d) germ

91. Cellulose is mostly present in
(a) aleurone (b) endosperm
(c) bran (d) embryo

92. Hemicellulose is mostly present in
(a) aleurone (b) endosperm
(c) bran (d) embryo

93. Albumins and globulins are generally regarded as ________ proteins
(a) non gluten (b) gluten
(c) non wheat (d) none

94. Gliadins and glutenins are generally regarded as ________ proteins
(a) non gluten (b) gluten
(c) non wheat (d) none

95. Gluten proteins are rich in ______ amino acid
(a) glutamine (b) proline
(c) tryptophan (d) both (a) & (b)

96. Non gluten proteins are rich in
(a) glutamine (b) proline
(c) tryptophan (d) both (a) & (b)

97. Albumin proteins are _____ soluble
(a) water (b) salt
(c) alcohol (d) dilute acid

98. Globulin proteins are _____ soluble
(a) water (b) salt
(c) alcohol (d) dilute acid

99. Gliadin proteins are _____ soluble
(a) water (b) salt
(c) alcohol (d) dilute acid

100. Glutenin proteins are _____ soluble
(a) water (b) salt
(c) alcohol
(d) dilute acid & alkalis

101. Gluten is ________ in nature
(a) viscous (b) elastic
(c) both (d) none

102. Viscous nature of wheat dough is due to ________ protein
(a) gliadin (b) glutenin
(c) zein (d) hordein

103. Elasticity of wheat dough is due to ________ protein

(a) gliadin (b) glutenin
(c) zein (d) hordein

104. Bran constitute ________ % of wheat grain

(a) 10-15 (b) 2-3
(c) 80-85 (d) 40

105. Germ constitute ______ % of wheat grain

(a) 10-15 (b) 2-3
(c) 80-85 (d) 40

106. Endosperm constitute ______ % of wheat grain

(a) 10-15 (b) 2-3
(c) 80-85 (d) 40

107. Steely, glassy, mealy and horny are ________ of wheat endosperm

(a) flavor (b) texture
(c) appearance (d) none

108. Which of the following is with vitreous kernel

(a) macaroni wheat
(b) emmer wheat
(c) einkorn wheat
(d) all

109. Which of the following has mealy kernel

(a) bread wheat (b) club wheat
(c) both (d) none

110. Vitreous kernels are

(a) translucent (b) bright
(c) dull (d) both (a) & (b)

111. Mealy kernels are

(a) opaque (b) dark
(c) translucent (d) both (a) & (b)

112. Terms 'hard' and 'soft' are associated with _______ of wheat

(a) varieties (b) flour
(c) germ (d) bran

113. Terms 'strong' and 'weak' are associated with _______ of wheat

(a) varieties (b) flour
(c) germ (d) bran

114. Vitreous grains are

(a) hard (b) weak
(c) soft (d) none

115. Mealy grains are

(a) hard (b) weak
(c) soft (d) both b & c

116. Hardness of grain is related to degree of adhesion between ________ & ________

(a) bran & endosperm
(b) starch & protein
(c) starch & germ
(d) protein & germ

117. Hard wheats yield ______ flour

(a) coarse (b) gritty
(c) soft (d) both (a) & (b)

118. Soft wheats yield _________ flour

(a) coarse (b) gritty
(c) fine (d) both (a) & (b)

119. The process of obtaining flour from wheat is called

(a) tempering (b) milling
(c) conditioning (d) all

120. The moisture content of wheat at time of harvest is ____%

(a) 8-10 (b) 10-12
(c) 12-14 (d) 14-16

121. Which of the following operation leads to increase in moisture content of wheat grains

(a) tempering (b) conditioning
(c) milling (d) both (a) & (b)

122. Conditioning and tempering operation ______ the yield of flour from wheat grains

(a) increase (b) decrease
(c) no effect (d) none

123. The common name for semolina is
(a) rava (b) suji
(c) dalia (d) both (a) & (b)

124. Which pigment is present in wheat flour
(a) anthocyanin (b) lycopene
(c) xanthophyll (d) chlorophyll

125. The chemical used for bleaching of wheat flour is
(a) chlorine
(b) benzoyl peroxide
(c) acetone peroxide
(d) all

126. Flour high in gluten and protein content is called ______ flour
(a) strong (b) weak
(c) soft (d) hard

127. Flour low in gluten & protein content is called ______ flour
(a) strong (b) weak
(c) soft (d) hard

128. Strong flour is recommended for
(a) bread (b) cookies
(c) cakes (d) biscuits

129. Weak flour is recommended for
(a) biscuits (b) cookies
(c) cakes (d) all

130. Pasta/macaroni is prepared from
(a) whole wheat flour
(b) refined flour
(c) semolina (d) none

131. Wheat is used for
(a) pasta (b) bread
(c) beer (d) all

132. The wheat belong to genus
(a) Triticum (b) Oryza
(c) Secale (d) Pennisetum

133. The husk in rice form ______ % of seed grain
(a) 21 (b) 2.25
(c) 90.75 (d) 60

134. The germ in rice form ______ % of seed grain
(a) 21 (b) 2.25
(c) 90.75 (d) 60

135. The starchy endosperm in rice form ______ % of seed grain
(a) 21 (b) 2.25
(c) 90.75 (d) 60

136. The pericarp in rice form ______ % of seed grain
(a) 21 (b) 2.25
(c) 90.75 (d) 6.0

137. Japonica and indica are varieties of
(a) rice (b) wheat
(c) rye (d) millet

138. The main carbohydrate of rice is
(a) soluble sugar
(b) insoluble sugar
(c) starch
(d) galactomannan

139. Parboiled rice mean
(a) partially boiled rice
(b) complete boiled rice
(c) milled rice
(d) home pounded

140. Which sugar is present in rice
(a) glucose (b) fructose
(c) sucrose (d) all

141. The principal protein of rice is
(a) oryzenin (b) zein
(c) glutenin (d) gliadin

142. The rice proteins are deficient in
(a) arginine (b) lysine
(c) threonine (d) both (a) & (b)

143. Polished rice is poor in
(a) calcium (b) iron
(c) both (d) none

144. The sticky consistency of rice after cooking is due to
(a) alpha amylase (b) lipase
(c) oxidase (d) protease

145. The parboiling of rice includes
(a) steeping (b) heating
(c) drying (d) all

146. Which of the following is a property of parboiled rice
(a) color remains same
(b) nutritive value of rice decreases
(c) rice does not form glutinous mass upon cooking
(d) all

147. Parboiling of rice ________ nutrients
(a) increase (b) decrease
(c) no effect (d) none

148. Milling yield of rice is _____%
(a) 60 (b) 55
(c) 70 (d) 85

149. Brown rice is
(a) rice with husk
(b) kernel without hull
(c) kernel without bran
(d) kernel without germ

150. Which of the following has higher protein content
(a) cereals (b) legumes
(c) oilseeds (d) none

151. Polished rice is _______ colored
(a) white (b) brown
(c) yellowish (d) none

152. Milled parboiled rice has ________ resistance to insects
(a) increased (b) decreased
(c) no effect (d) none

153. The nutritive value of parboiled rice is
(a) high (b) low
(c) similar to normal rice
(d) none

154. During parboiling of rice
(a) nutrients from endosperm travel to bran & hull
(b) nutrients from bran & hull travel to endosperm
(c) no movement (d) none

155. Polishing of rice results in _____of vitamin B
(a) increase (b) decrease
(c) no effect (d) none

156. Parboiling of rice leads to
(a) change of grain color
(b) development of unpleasant smell
(c) fermentative off flavor
(d) all

157. Polishing of rice is carried out by
(a) HR conversion process
(b) Malek process
(c) Solvent extraction process
(d) all

158. The well stored rice grains swell due to cooking white fresh grains donot. This is due to
(a) alpha amylase in active form
(b) alpha amylase in inactive form
(c) both (d) none

159. Excessive use of polished rice leads o
(a) night blindness (b) beri beri
(c) anaemia (d) scurvy

160. Protein content of rice flour is _____%
(a) 1-2 (b) 30.5
(c) 5-9 (d) 9-12

161. Which of the following is prepared from parboiled rice
(a) rice flour (b) parched rice
(c) puffed rice (d) flaked rice

162. Saki, an alcoholic beverage is prepared from
(a) wheat (b) barley
(c) rice (d) maize

163. The fermenting organism for Saki is
(a) Saccharomyces
(b) Aspergillus
(c) Lactobacillus
(d) Candida

164. Rice bran contains ______% protein
(a) 2 (b) 5
(c) 8 (d) 12

165. Rice bran contains ______% fat
(a) 6 (b) 9
(c) 12 (d) 15

166. Rice bran contains ______% ash
(a) 22 (b) 33
(c) 11 (d) 44

167. Husk of rice has high __________ content
(a) magnesium (b) silica
(c) phosphorus (d) calcium

168. Rice husk contains 12 % ______, thus used for production of furfural
(a) carbohydrates (b) pentosans
(c) phytic acid (d) proteins

169. Which of the rice donot occur
(a) red (b) brown
(c) black (d) green

170. Parimal rice in India is grown in
(a) Bihar (b) Punjab
(c) Uttar Pradesh (d) Haryana

171. Sona Masuri is rice type grown in
(a) Andhra Pradesh
(b) Karnataka
(c) Tamil Nadu
(d) both (a) & (b)

172. The scientific name of corn is
(a) Zea mays
(b) Oryza sativa
(c) Hordeum vulgare
(d) none

173. The flint, dent, sweet, waxy are varieties of
(a) corn (b) wheat
(c) rice (d) barley

174. Waxy corn has higher proportion of __________
(a) amylose
(b) amylopectin
(c) similar proportion of amylose and amylopectin
(d) none

175. Which of the following has highest grain size
(a) rice (b) corn
(c) wheat (d) rye

176. The predominant portion of maize is
(a) protein (b) starch
(c) fat (d) minerals

177. Which of the following sugar is present in corn
(a) fructose (b) sucralose
(c) galactose (d) lactose

178. The principal protein of maize is
(a) zein (b) glutenins
(c) gliadins (d) hordein

179. The major protein of maize germ is
(a) zein (b) glutelin
(c) albumin (d) globulin

180. Of the total fat, ______% is found in endosperm of corn
(a) 1 (b) 15
(c) 2 (d) 5

181. Maize is particularly deficient in
(a) thiamin (b) niacin
(c) riboflavin (d) carotene

182. The biologically unavailable form of niacin in corn is
(a) niacytin (b) niacytonin
(c) niatin (d) niacinatylin

183. The people solely feeding on maize develop
(a) beri beri (b) scurvy
(c) night blindness (d) pellagra

184. Wet and dry milling are used to process
(a) maize (b) rice
(c) oats (d) wheat

185. The major product of dry maize milling is
(a) grits (b) coarse meal
(c) fine meal (d) germ

186. The major carbohydrates of sorghum is
(a) starch (b) sugars
(c) cellulose (d) none

187. The cyanogenic glycoside formed in sorghum upon germination is
(a) dhurrin (b) niacytin
(c) phytic acid (d) nicotinic acid

188. Dhurrin is _______ substance
(a) toxic (b) nutritive
(c) flavor enhancer (d) none

189. The principal protein of sorghum is
(a) albumin (b) globulin
(c) prolamine (d) glutelin

190. The low palatability and digestibility of sorghum based products is
(a) TGA (b) tannins
(c) prolamine (d) anthocyanins

191. The dry milling of sorghum has highest proportion of
(a) grits (b) germ
(c) fiber (d) bran

192. Pearling is a milling process for
(a) barley (b) corn
(c) sorghum (d) oats

193. Kafir beer, containing 3% alcohol is prepared from malt of
(a) barley (b) sorghum
(c) corn (d) millet

194. Bajra is the common name for
(a) pearl millet (b) sorghum
(c) proso millet (d) kodo millet

195. Ragi is the common name for
(a) pearl millet (b) finger millet
(c) proso millet (d) kodo millet

196. Pearl and ragi are varieties of
(a) sorghum (b) millet
(c) oats (d) rye

197. The coarse grains include
(a) millets (b) maize
(c) sorghum (d) all

198. About 50% of phosphorus of bajra is present in ______ form
(a) phosphorus (b) phosphate
(c) phytin (d) phytic acid

199. The common variety of millet grown in India is
(a) pearl (b) kodo
(c) finger (d) proso

200. The major protein of pearl millet is
(a) prolamine (b) globulin
(c) albumin (d) glutelin

201. Pearl millet has ______ proportion of lysine
(a) high (b) low
(c) absent (d) none

202. The major carbohydrates of peral millet is
(a) starch (b) sugar
(c) pentosans (d) hemicellulose

203. The starch of pearl millet has _______ proportion of amylose than amylopectin
(a) higher (b) lower
(c) similar (d) none

204. The proportion of amylopectin in pearl millet is
(a) 68% (b) 50%
(c) 37% (d) 45%

205. Pearling is done for
(a) sorghum (b) bajra
(c) rice (d) both (a) & (b)

206. The major proteins of ragi are
(a) prolamine (b) glutelin
(c) albumin (d) both (a) & (b)

207. Broom, hog and white millet are common names of
(a) pearl millet (b) finger millet
(c) proso millet (d) kodo millet

208. Barley belongs to genus
(a) Panicum (b) Elcusine
(c) Triticum (d) Hordeum

209. The six rowed barley is
(a) H. vulgare (b) H. distichon
(c) both (d) none

210. The two rowed barley is
(a) H. vulgare (b) H. distichon
(c) both (d) none

211. The hull content of hulled barley is ______%
(a) 3 (b) 6
(c) 10 (d) 13

212. The mineral found in high proportion in barley is
(a) magnesium (b) calcium
(c) phosphorus (d) iron

213. The barley starch has ______ % amylose
(a) 10 (b) 20
(c) 40 (d) 60

214. The barley starch has ______ % amylopectin
(a) 20 (b) 40
(c) 60 (d) 80

215. The major proportion of protein in barley is
(a) albumin (b) globulin
(c) prolamin (d) glutelin

216. The lysine content of barley is
(a) very high (b) very low
(c) absent (d) none

217. Barley consumption ________ blood glucose level in body
(a) reduce (b) increase
(c) no effect (d) none

218. Barley is used for
(a) beer (b) whisky
(c) malt (d) all

219. Malting of barley involves
(a) soaking (b) germination
(c) drying (d) all

220. Malts used in brewing must have ________ nitrogen content
(a) high (b) low
(c) no (d) none

221. Oats belong to genus
(a) Avena (b) Oryza
(c) Hordeum (d) none

222. The hull proportion of oats is _____% of grain
(a) 2-3 (b) 10-12
(c) 20-30 (d) 50-60

223. Oatmeal flour is ________ for bread making
(a) highly recommended
(b) not suitable
(c) not determined
(d) none

224. In India, the commonly grown oat variety is
(a) A. sativa
(b) A. byzantina
(c) not grown in India
(d) none

225. The kernel of oat is called
(a) caryopsis (b) groat
(c) grit (d) none

226. The average kernel content of oat is
(a) 60% (b) 65%
(c) 70% (d) 75%

227. The husk of oat is ________
(a) tough & fibrous
(b) not easily removed
(c) inedible (d) both (a) & (b)

228. The milling yield of oatmeal is
(a) 10% (b) 15-20%
(c) 35-40% (d) 55-60%

229. Rye belong to genus
(a) Secale (b) Avena
(c) Hordeum (d) none

230. _______is used for cultivating ergot
(a) rye (b) oats
(c) sorghum (d) barley

231. Rye contains ______vitamin B than wheat
(a) more (b) less
(c) equal (d) none

232. Oralmat is an extract of
(a) oats (b) rye
(c) barley (d) sorghum

233. Triticale is a hybrid between
(a) wheat & oat
(b) wheat & sorghum
(c) wheat & rye
(d) wheat & millet

234. The grains of triticale have protein content of _____%
(a) 4-8 (b) 14-18
(c) 24-28 (d) 34-38

235. Farina is another name for
(a) suji (b) semolina
(c) grits (d) both (a) & (b)

236. Bulger is obtained from
(a) wheat (b) rye
(c) oats (d) sorghum

237. Which of the following forms gluten
(a) rice (b) millet
(c) wheat (d) all

238. Hominy grits is obtained from
(a) barley (b) maize
(c) millets (d) rye

239. Another cereal which form gluten to a lesser extent is
(a) oats (b) sorghum
(c) maize (d) rye

240. Gluten forming property is due to the presence of
(a) albumin & globulin
(b) gliadin & glutenin
(c) albumin & glutenin
(d) globulin & gliadin

241. During formation of wheat dough ______ bonds are cleaved to _____ bonds
(a) SS to SH (b) SH to SS
(c) SS to NH_2 (d) all

242. Which of the following require strong gluten characteristics
(a) bread (b) cookies
(c) biscuits (d) cakes

243. Which of the following require weak gluten characteristics
(a) bread (b) cookies
(c) biscuits (d) both b & c

244. Bread flours should have _______ protein content
(a) no (b) high
(c) low (d) none

245. Cookies flour should have _______ protein content
(a) no (b) high
(c) low (d) none

246. Glutenin gives ________ to gluten
(a) viscous nature (b) elasticity
(c) extensibility (d) all

247. Glutenin gives ________ to gluten
(a) viscous nature (b) elasticity
(c) extensibility (d) both a & c

248. Cookies require ____% protein content
(a) 2-3 (b) 7-9
(c) 14-15 (d) >16

249. Bread require _____% protein content
(a) 2-3 (b) 7-9
(c) 12-14 (d) >16

250. The moisture content of wheat flour should be
(a) NMT 13% (b) NMT 9%
(c) NLT 20% (d) NLT 17%

251. Leavening agents cause ________in volume
(a) increase (b) decrease
(c) no change (d) none

252. Common leavening agent is
(a) SBC (b) ABC
(c) yeast (d) all

253. A slight excess of sodium bicarbonate give rise to ________smell
(a) no (b) unpleasant
(c) pleasant (d) none

254. The biscuit flour should have more
(a) extensibility (b) elasticity
(c) both (d) none

255. Higher the extensibility of dough ____ will be the spread ratio of cookies
(a) higher (b) lower
(c) no effect (d) none

256. Higher the elasticity of dough ______ will be the spread ratio of cookies
(a) higher (b) lower
(c) no effect (d) none

257. The bread flour should have more ________
(a) elasticity (b) extensibility
(c) both (d) none

258. Presence of higher glutenin result in ______ bread quallity
(a) poor (b) good
(c) no effect (d) none

259. Presence of higher gluten result in ______ bread quallity
(a) poor (b) good
(c) no effect (d) none

260. HMWGS and LMWGS are subunits of
(a) albumin (b) globulin
(c) glutenin (d) gliadin

261. α-, β-, γ- and ω- are subclasses of
(a) albumin (b) globulin
(c) glutenin (d) gliadin

262. Glutenin have __________ number of disulfide bonds
(a) even (b) odd
(c) no (d) none

263. Glutenin forms _____ disulfide bonds
(a) intramolecular
(b) intermolecular
(c) both (d) none

264. Mostly gliadin have ______ number of disulfide bonds
(a) even (b) odd
(c) no (d) none

265. Gliadin forms ______ disulfide bonds
(a) intramolecular
(b) intermolecular
(c) both
(d) none

266. The viscous nature of gliadins is due to ________bonding
(a) intramolecular
(b) intermolecular
(c) both
(d) none

267. The function of tempering of wheat grain is to
(a) mellow the endosperm
(b) toughen the bran
(c) both
(d) none

268. Which of the following has higher molecular weight
(a) albumin (b) globulin
(c) gliadin (d) glutenin

269. The average molecular weight of glutenin is _____ kDa
(a) 120-200 (b) 30-70
(c) 20-40 (d) 80-90

270. The average molecular weight of gliadin is ________ kDa
(a) 120-200 (b) 30-70
(c) 20-40 (d) 80-90

271. Which of the following has highest molecular weight
(a) HMWGS (b) LMWGS
(c) alpha gliadin (d) beta gliadin

272. Cake flour is ______ strength flour
(a) high (b) low
(c) medium (d) none

273. Chapati flour is _____ strength flour
(a) high (b) low
(c) medium (d) none

274. The extensibility of biscuit flour can be increased by
(a) SBC
(b) ABC
(c) reducing agents
(d) oxidizing agents

275. Hard wheats are tempered to _____ _____ % moisture
(a) 16.5 (b) 15.5
(c) 14.5 (d) 13.5

276. Soft wheats are tempered to _______ % moisture
(a) 16.5 (b) 15.5
(c) 14.5 (d) 13.5

277. Tempering is _____ to conditioning of wheat grains
(a) similar (b) different
(c) done prior to (d) none

278. Temperature is a criterion in
(a) tempering
(b) conditioning
(c) milling
(d) none

279. The correct order is
(a) harvesting-cleaning-tempering-milling-flour
(b) cleaning-harvesting-tempering-milling-flour
(c) tempering-harvesting-cleaning-milling-flour
(d) none

280. Pastry flour is _____ in protein
(a) low (b) high
(c) very high (d) none

281. Which of the following is a baked product
(a) biscuit (b) cookies
(c) cake (d) all

282. Gluten is formed of
(a) gliadin & glutenin
(b) gliadin & glutenin & starch
(c) glutenin & germ
(d) none

283. According to FDA, enriched flour has added
(a) thiamin (b) iron
(c) folic acid (d) all

284. Leavening is a result of
(a) air (b) water vapor
(c) CO_2 (d) all

285. The flour mixes with egg white has _____ air than those with egg yolk
(a) more (b) less
(c) equal (d) none

286. The chemical leavening agent perform by producing
(a) CO_2 (b) O_2
(c) NO_2 (d) all

287. Baking soda is
(a) SBC
(b) ABC
(c) sodium carbonate
(d) all

288. Baking powder is
(a) SBC + acid component
(b) ABC + acid component
(c) sodium carbonate+acid component
(d) none

289. The yeast used for wheat based baked products is
(a) Saccharomyces (b) Candida
(c) both (d) none

290. The optimum temperature for yeast activity in dough is _____°C
(a) 30 (b) 20
(c) 10 (d) 5

291. The optimum pH for yeast activity in dough is _____
(a) 1-2 (b) 3
(c) 4-6 (d) 6-8

292. Yeast forms ________ & ________ from sugar
(a) alcohol & CO_2
(b) ketone & CO_2
(c) ketone & CO_2
(d) all

293. One molecule of glucose gives ______ mole of alcohol
(a) 2 (b) 3
(c) 4 (d) 5

294. One molecule of glucose gives ______ mole of carbon dioxide
(a) 2 (b) 3
(c) 4 (d) 5

295. Malt is added to wheat flour to increase ________ activity
(a) invertase (b) amylase
(c) zymase (d) all

296. Compressed and active are forms of
(a) yeast (b) SBC
(c) baking powder (d) all

297. Compressed yeast has ______ moisture content than active dry yeast
(a) higher (b) lower
(c) similar (d) none

298. Compressed yeast has ________ % moisture content
(a) 50 (b) 60
(c) 70 (d) 80

299. Active dry yeast has ____ %moisture content
(a) 8 (b) 2
(c) 20 (d) 30

300. Which of the following is made from fermented flour
(a) idli (b) dosa
(c) dhokla (d) all

301. ______ is prepared from rice flour and Bengal gram flour
(a) idli (b) dosa
(c) dhokla (d) all

302. Which of the following has a pancake appearance
(a) idli (b) dosa
(c) dhokla (d) all

303. Which of the following is made from rice flour
(a) dosa (b) appam
(c) pizza (d) both (a) & (b)

304. Drop, bar, rolled are types of
(a) cakes (b) cookies
(c) bread (d) pasta

305. The composition of flour, fat and water in pastry is
(a) 5:1:0.5 (b) 6:2:1
(c) 4:2:1 (d) 4:8:4

306. Macaroni, spaghetti, vermicelli and noodles are examples of
(a) pasta products
(b) fermented products
(c) both (d) none

307. Pasta products are made from ______ of durum wheat
(a) fine flour
(b) straight grade flour
(c) semolina (d) grits

308. Commonly chapati is prepared from
(a) whole wheat flour
(b) refined flour
(c) semolina (d) none

309. The wheat variety commonly used for chapati making is
(a) T. aestivum (b) T.durum
(c) T. dicoccum (d) none

310. The ideal protein content of wheat flour for chapati is ____%
(a) 5-8 (b) 8-11
(c) 10-13 (d) 14-17

311. The per centage of total sugars in wheat flour for chapatis is
(a) 0.5% (b) 1%
(c) 2.5% (d) 5%

312. Flours with ______ diastatic activities are desirable for chapati
(a) high (b) low
(c) no (d) none

313. Pulses are _____ of pod bearing plants
(a) fruits (b) seeds
(c) stem (d) both (a) & (b)

314. The pulses belong to family ______
(a) Triticum (b) Leguminosae
(c) Juglandaceae (d) none

315. Pulses have ______ protein content
(a) low (b) high
(c) no (d) none

316. The average protein content in pulses ranges from _____%
(a) 2-5 (b) 5-10
(c) 10-20 (d) 20-40

317. Botanical name of chick pea
(a) Cajanus cajan
(b) Lathyrus sativus
(c) Pisum sativum (d) none

318. The protein content of bengal gram is _____g/100g
(a) 17.1 (b) 24
(c) 28.2 (d) 25.1

319. The protein content of black gram is _____g/100g
(a) 17.1 (b) 24
(c) 28.2 (d) 25.1

320. The protein content of cow pea is _____g/100g
(a) 17.1 (b) 24
(c) 28.2 (d) 25.1

321. The protein content of green gram is _____g/100g
(a) 17.1 (b) 24
(c) 28.2 (d) 25.1

322. The protein content of kesari dhal is _____g/100g
(a) 17.1 (b) 24
(c) 28.2 (d) 25.1

323. The protein content of lentil is _____ g/100g
(a) 17.1 (b) 24
(c) 28.2 (d) 25.1

324. The fat content of bengal gram is _____ g/100g
(a) 5.3 (b) 1.4
(c) 1.0 (d) 2.5

325. The fat content of black gram is _____ g/100g
(a) 5.3 (b) 1.4
(c) 1.0 (d) 2.5

326. The fat content of green gram is _____ g/100g
(a) 5.3 (b) 1.4
(c) 1.0 (d) 2.5

327. The mineral content of black gram is _____ g/100g
(a) 3.5 (b) 2.3
(c) 3.2 (d) 4.6

328. The mineral content of green gram is _____ g/100g
(a) 3.5 (b) 2.3
(c) 3.2 (d) 4.6

329. The mineral content of kesari dhal is _____ g/100g
(a) 3.5 (b) 2.3
(c) 3.2 (d) 4.6

330. The mineral content of soybean is _____ g/100g
(a) 3.5 (b) 2.3
(c) 3.2 (d) 4.6

331. The fiber content of bengal gram is _____ g/100g
(a) 3.9 (b) 0.9
(c) 4.1 (d) 2.3

332. The mineral content of cow pea is _____ g/100g
(a) 3.5 (b) 2.3
(c) 3.2 (d) 4.6

333. The fiber content of black gram is _____ g/100g

(a) 3.9 (b) 0.9
(c) 4.1 (d) 2.3

334. The fiber content of green gram is _____ g/100g

(a) 3.9 (b) 0.9
(c) 4.1 (d) 2.3

335. The fiber content of lentil is _____ g/100g

(a) 3.9 (b) 0.7
(c) 4.1 (d) 2.3

336. The fiber content of pea is _____g/100g

(a) 3.9 (b) 0.9
(c) 4.5 (d) 2.3

337. The fiber content of soybean is _____ g/100g

(a) 2.9 (b) 0.9
(c) 4.1 (d) 3.7

338. The fiber content of kesari dhal is _____ g/100g

(a) 3.9 (b) 0.9
(c) 4.1 (d) 2.3

339. The protein content of peas is _____ g/100g

(a) 19.7 (b) 24
(c) 28.2 (d) 25.1

340. The protein content of red gram is _____ g/100g

(a) 17.1 (b) 22.3
(c) 28.2 (d) 25.1

341. The protein content of soybean is _____ g/100g

(a) 17.1 (b) 24
(c) 42.2 (d) 25.1

342. The fat content of red gram is _____ g/100g

(a) 1.7 (b) 2.4
(c) 0.7 (d) 5.1

343. The fat content of soybean is _____ g/100g

(a) 19.5 (b) 25.4
(c) 30.7 (d) 5.1

344. The mineral content of soybean is ___ _____ g/100g

(a) 3.5 (b) 2.3
(c) 4.6 (d) 4.6

345. Which of the following has highest protein content

(a) red gram
(b) peas
(c) soybean
(d) Bengal gram

346. Which of the following has highest fat content

(a) red gram
(b) peas
(c) soybean
(d) Bengal gram

347. Which of the following has highest mineral content

(a) red gram (b) peas
(c) soybean (d) Bengal gram

348. The chief pulse proteins are

(a) albumins (b) globulins
(c) gluten (d) all

349. Pulse protein is ___________ in sulfur containing amino acids

(a) deficient (b) rich
(c) none (d) low

350. Pulse protein s deficient in

(a) methionine (b) tryptophan
(c) both (d) none

351. The flatulence causing sugar are

(a) raffinose (b) stachyose
(c) verbascose (d) all

352. The flatulence causing sugars escape digestion due to

(a) absence of α-galactosidase
(b) presence of α- galactosidase
(c) presence of β- galactosidase
(d) none

353. The phosphorus of pulses is present in ______ form
(a) phytate phosphorus
(b) phytic acid
(c) native form
(d) none

354. The availability of proteins and minerals of pulses is increased by
(a) cooking (b) soaking
(c) germination (d) all

355. Dry legumes are ________ source of ascorbic acid
(a) rich
(b) poor
(c) completely absent
(d) none

356. The process of removal of skin of pulse sedds is
(a) soaking (b) germination
(c) decortication (d) none

357. Soaking ______ the amount of phytic acid in pulses
(a) reduce (b) increase
(c) no effect (d) none

358. The vitamin content of pulses ______ upon germination
(a) increase (b) decrease
(c) no change (d) none

359. Folic acid content of pulses ________ upon germination
(a) increase (b) decrease
(c) no change (d) none

360. Pantothenic acid content of pulses ________ upon germination
(a) increase (b) decrease
(c) remain same (d) none

361. During germination of pulses
(a) conversion of starch to sugar
(b) sugar to starch
(c) vitamin decrease
(d) all

362. The germination process ________ the anti nutritional factor in pulses
(a) increase (b) decrease
(c) remain same (d) none

363. ANF stands for
(a) anti nutritive factors
(b) anti nutritional factors
(c) anti nutritional functions
(d) none

364. Cooking of pulses
(a) destroys enzyme inhibitors
(b) improves nutritional quality
(c) improves palatability
(d) all

365. Longer cooking time of pulses ____________ the lysine content
(a) increase (b) decrease
(c) remain same (d) none

366. The fermentation of pulses cause
(a) increased digestibility
(b) increased palatability
(c) high nutritive value
(d) all

367. Which of the following pulse has protein value similar to animal protein
(a) red gram (b) lentil
(c) soybean (d) black gram

368. ________ form lather or foam when shaken with water
(a) haemagglutinins
(b) saponin
(c) cyanogenic glycosides
(d) all

369. Goitre can be caused by
(a) saponin (b) goitrogen
(c) alkaloid (d) all

370. Lathyrism is caused by
(a) kesari dal (b) black gram
(c) chick pea (d) all

371. Favism is caused by
(a) Vicia faba (b) chick pea
(c) lentil (d) kesari dal

372. Lathyrism causes
(a) nervous disorder (b) paralysis
(c) blood disease (d) none

373. In lathyrism
(a) normal formation of collagen fiber is interfered
(b) normal RBC is not formed
(c) water begins to form in lungs
(d) none

374. Favism is also known as
(a) hemolytic anaemia
(b) paralysis
(c) nervopus disorder
(d) all

375. Favism occurs due to the decreased activity of enzyme
(a) glucose 6 phosphate dehydrogenase
(b) alkaline phosphatase
(c) peroxidase
(d) none

376. The problem of favism is severed by intake of
(a) ground nut (b) faba beans
(c) rice (d) none

377. The popular, Mysore pak is prepared from
(a) red gram (b) bengal gram
(c) green gram (d) none

378. The Gujurati dish, Khaman prepared from
(a) red gram (b) bengal gram
(c) green gram (d) none

379. Because of its high net protein utilization value ________ is the best pulse protein
(a) Bengal gram (b) black gram
(c) red gram (d) green gram

380. Which of the following is also known as pigeon pea
(a) Bengal gram (b) black gram
(c) red gram (d) green gram

381. The south Indian dish Sambhar is prepared from
(a) Bengal gram (b) black gram
(c) red gram (d) green gram

382. The chief proteins of black gram are
(a) albumins (b) globulins
(c) prolamine (d) both (a) & (b)

383. The albumins and globulins constitute ________ % of protein of black gram
(a) 15-20 (d) 25-30
(c) 45-55 (d) 55-56

384. Papad is mainly prepared from
(a) Bengal gram (b) black gram
(c) red gram (d) green gram

385. Mung is the common name for
(a) Bengal gram (b) black gram
(c) red gram (d) green gram

386. Masur dal is the common name for
(a) Bengal gram (b) black gram
(c) lentil (d) green gram

387. Kulthi is the common name for
(a) Bengal gram (b) black gram
(c) horse gram (d) green gram

388. Cluster bean is the another name for
(a) guar gum
(b) locust bean gum
(c) gum arabic (d) none

389. Cluster bean is used as
(a) as dal (b) thickener
(c) leavener (d) all

390. Haba, Burma, Hibbert and Rangoon are names of
(a) lima bean (b) cow pea
(c) kidney bean (d) velvet

391. Goa bean is common name for
(a) winged bean (b) velvet
(c) cow pea (d) kidney bean

392. Soybean is a
(a) pulse (b) oilseed
(c) both (d) none

393. TVP stands for
(a) textured vegetable protein
(b) tasty vegetable protein
(c) textured vegetable paste
(d) none

394. Nutrinuggetes are prepared from
(a) soybean (b) kidney bean
(c) velvet bean (d) winged bean

395. Soybean has ______ effect, so used by diabetic patients
(a) hypoglycaemic
(b) hyperglycaemic
(c) no effect (d) none

396. Soybean is used in preparation of
(a) flour (b) oil
(c) meal (d) all

397. Tofu is prepared from
(a) soybean (b) wheat
(c) curd (d) millet

398. Tofu is a fermented
(a) pancake (b) beverage
(c) curd (d) pickle

399. The soya sauce is prepared from fermentation by
(a) Saccharomyces (b) Aspergillus
(c) Candida (d) Lactobacillus

400. Soya sauce has moisture content of _____ %
(a) 20-30 (b) 40-45
(c) 50-55 (d) 60-67

401. The paste of soybean is called
(a) miso (b) tofu
(c) tempeh (d) none

402. The fermenting agent in soya paste is
(a) Aspergillus oryzae
(b) Saccharomyces cerevisiae
(c) both
(d) none

403. The miso has ________ % protein
(a) 10 (b) 20
(c) 30 (d) 40

404. Tempeh is prepared from
(a) cluster bean (b) soybean
(c) gren gram (d) milk

405. The fermenting agent in tempeh is
(a) Aspergillus oryzae
(b) Saccharomyces cerevisiae
(c) Rhizopus oryzae
(d) none

406. Tempeh is an ______ product
(a) Japanese (b) Indonesian
(c) Chinese (d) Korean

407. Natto is a __________ product
(a) Japanese (b) Indonesian
(c) Chinese (d) Korean

408. Fermenting agent in natto is
(a) Aspergillus oryzae
(b) Saccharomyces cerevisiae
(c) Rhizopus oryzae
(d) Bacillus subtilis

409. Natto is chiefly eaten with
(a) chapati (b) rice
(c) bread (d) none

410. Hamanatto is a ______ product
(a) Japanese (b) Indonesian
(c) Chinese (d) Korean

411. Hamanatto is a product made from
(a) soybean (b) lentil
(c) red gram (d) green gram

412. The bran is much easily separated from endosperm for
(a) hard wheat (b) soft wheat
(c) both (d) none

413. Which wheat has higher extraction rate
(a) hard (b) soft
(c) equal (d) none

414. Triticum aestivum is ______ wheat
(a) diploid (b) haploid
(c) hexaploid (d) tetraploid

415. Triticum durum is ______ wheat
(a) diploid (b) haploid
(c) hexaploid (d) tetraploid

416. The highest content of protein is present in ______
(a) bran
(b) aleurone layer
(c) subaleurone layer
(d) endosperm

417. Towards the center of the grain, protein content ______
(a) increases (b) decreases
(c) remains same (d) none

418. Stains used for starch is
(a) periodic acid-schiff reaction
(b) iodine or potassium iodide
(c) both (d) none

419. Stains used for protein is
(a) fast green (b) ponceauze
(c) lectin stains (d) all

420. Stains used for lipid is
(a) fast green (b) ponceauze
(c) lectin stains (d) sudan IV

421. SEM stands for
(a) standard electron microscopy
(b) scanning electron microbiology
(c) scanning electron microscopy
(d) none

422. TEM stands for
(a) time dependent electron microscopy
(b) tandem electron microscopy
(c) transmission electron microscopy
(d) none

423. Gluten protein, also called storage protein account for ______ % of whole quantity of proteins in grain
(a) 80 (b) 75
(c) 60 (d) 65

424. Sprout-damaged wheat has high ______ activity
(a) amylase (b) lipase
(c) peroxidase (d) none

425. Too much of alpha amylase activity cause ______ dough
(a) non sticky (b) sticky
(c) crumbled (d) none

426. Alpha amylase activity is measured in unit
(a) falling number
(b) amylase number
(c) both
(d) none

427. Higher falling number indicates ______ amylase activity
(a) low (b) high
(c) no (d) none

428. Lower falling number indicates ______ amylase activity
(a) low (b) high
(c) no (d) none

429. Very damaged wheat sample will have falling number
(a) 60 (b) 250
(c) 400 (d) none

430. Very sound wheat sample will have falling number
(a) 60 (b) 250
(c) 400 (d) none

431. Minimum falling number generally accepted for bread is
(a) 60 (b) 250
(c) 400 (d) none

432. Hard wheat when milled, yields ______ level of damaged starch
(a) low (b) high
(c) no (d) none

433. Soft wheat when milled, yields ______ level of damaged starch
(a) low (b) high
(c) no (d) none

434. High levels of damaged starch is suitable for
(a) biscuits (b) bread
(c) cookies (d) none

435. Low levels of damaged starch is suitable for
(a) biscuits (b) bread
(c) cookies (d) both (a) & (c)

436. Flour obtained from hard wheat is
(a) high flowability
(b) easier to handle
(c) both
(d) none

437. Hardness of grain is controlled by gene located on short arm of chromosome
(a) 5A (b) 5B
(c) 5C (d) 5D

438. T. aestivum has ______% protein
(a) 7-12 (b) 12-14
(c) 14-16 (d) none

439. T. durum has ______% protein
(a) 7-12 (b) 12-14
(c) 14-16 (d) none

440. T. compactum has ______% protein
(a) 7-12 (b) 12-14
(c) 14-16 (d) none

441. T. aestivum is used for preparation of ______
(a) bread (b) cookies
(c) pasta (d) all

442. T.durum is used for preparation of ______
(a) bread (b) cookies
(c) pasta (d) all

443. T. compactum is used for preparation of ______
(a) bread (b) cookies
(c) pasta (d) all

444. The protein present on the surface of starch granule in wheat & responsible for hardness of grain is
(a) globulin (b) glutenin
(c) gliadin (d) friabilin

445. Friabilin is __________ soluble
(a) water (b) fat
(c) dilute acid (d) alcohol

446. Hard kernels has _______ density than soft kernels
(a) higher (b) lower
(c) similar (d) none

447. The rate of movement of moisture into grain is rapid in ______ grains
(a) soft
(b) hard
(c) same for both types of grain
(d) none

448. Moisture level tolerance with respect to flour yield is _______ in hard grain
(a) more (b) less
(c) same (d) none

449. The separation of endosperm from bran is better in ______ grains
(a) hard (b) soft
(c) same (d) none

450. The reduction of endosperm to flour is better in _______ grains
(a) hard (b) soft
(c) same (d) none

451. Sieving behavior of flour is _______ in hard variety
(a) good
(b) poor
(c) same as soft wheat
(d) none

452. The aleurone granules are rich in ____________ amino acids
(a) acidic (b) basic
(c) neutral (d) none

453. The wheat germ is composed of
(a) embryo (b) scutellum
(c) endosperm (d) both (a) & (b)

454. A and B granules are types of _____ _____ granule in wheat
(a) starch (b) protein
(c) lipid (d) all

455. The larger sized starch granules are _____
(a) elliptical shaped
(b) A granule
(c) B granule
(d) both (a) & (b)

456. Smaller sized type B starch granules are _____ shaped
(a) elliptical (b) ovoid
(c) round (d) none

457. The content of pentosans in wheat is _____%
(a) 3-5 (b) 2-3
(c) 5-7 (d) 7-9

458. Pentosans have _____ water absorption capacity
(a) poor (b) good
(c) very good (d) none

459. The water absorption capacity of pentosans is _____ times their mass
(a) 2 (b) 4
(c) 6 (d) 10

460. The protein content in grain coat and endosperm in wheat is _____%
(a) 23-33 (b) 36-42
(c) 9-14 (d) 60-70

461. The protein content in embryo in wheat is ____%
(a) 23-33 (b) 36-42
(c) 9-14 (d) 60-70

462. The protein content in endosperm in wheat is ____%
(a) 23-33 (b) 36-42
(c) 9-14 (d) 60-70

463. Starch is absent in
(a) grain coat (b) embryo
(c) endosperm (d) both (a) & (b)

464. The starch content in whole wheat grain is _______%
(a) 23-33 (b) 36-42
(c) 9-14 (d) 60-70

465. The starch content in endosperm of wheat grain is _______%
(a) 23-33 (b) 36-42
(c) 78-84 (d) 60-70

466. The sugar content in whole wheat grain is ________%
(a) 23-33 (b) 3-6
(c) 9-14 (d) 60-70

467. The sugar content in grain coat and endosperm of wheat is _______%
(a) 23-33 (b) 3-5
(c) 9-14 (d) 60-70

468. The sugar content in germ of wheat is _____%
(a) 22-28 (b) 3-6
(c) 9-14 (d) 60-70

469. The sugar content in endosperm of wheat is _______%
(a) 23-33 (b) 3-4
(c) 9-14 (d) 60-70

470. The pentosan content in grain coat and endosperm of wheat is ____%
(a) 30-40 (b) 7-8.5
(c) 12-16 (d) 0.5-0.7

471. The fat content in grain coat and endosperm of wheat is ____%
(a) 30-40 (b) 7-8.5
(c) 12-16 (d) 0.5-0.7

472. The fat content in germ of wheat is ____%
(a) 30-40 (b) 7-8.5
(c) 12-16 (d) 0.5-0.7

473. The fat content in endosperm of wheat is ____%
(a) 30-40 (b) 7-8.5
(c) 12-16 (d) 0.5-0.7

474. Highest minerals are present in
(a) grain coat and endosperm
(b) germ
(c) endosperm
(d) none

475. Whole wheat grain has the largest component
(a) starch (b) protein
(c) fiber (d) lipids

476. The largest component of endosperm of wheat grain is
(a) starch (b) protein
(c) fiber (d) lipids

477. The largest component of germ of wheat grain is
(a) starch (b) protein
(c) fiber (d) lipids

478. The largest component of aleurone layer of wheat grain is
(a) starch (b) protein
(c) fiber (d) lipids

479. The largest component of bran of wheat grain is
(a) starch (b) protein
(c) fiber (d) lipids

480. 95% of world wide production of wheat is
(a) hexaploid (b) tetraploid
(c) diploid (d) none

481. Amylase is _______ digesting enzyme
(a) fat (b) starch
(c) protein (d) all

482. Denaturation is related to__________
(a) fat (b) starch
(c) protein (d) all

483. Gelatinization is related to_________.
(a) fat (b) starch
(c) protein (d) all

484. Retrogradation is related to__________
(a) fat (b) starch
(c) protein (d) all

485. Durum wheat is used for pasta product because
(a) Double concentration of xanthophyll
(b) semolina gives good texture
(c) high mineral content
(d) all

486. Durum wheat has _________ sub-unit, hence good for pasta products
(a) gliadin 45 (b) gliadin 42
(c) gliadin 35 (d) gliadin 32

487. Mineral content of durum wheat is _____ times greater than bread wheat
(a) 0.5 (b) 1.0
(c) 1.5 (d) 2.5

488. Which of the following test is not for grain analysis
(a) test weight (b) kernel width
(c) AWRC
(d) hectoliter weight

489. Test weight indicates the ___ of grain
(a) density (b) plumpness
(c) both (d) none

490. The unit of hectoliter weight is
(a) kg/hL (b) kg/100L
(c) g/100L (d) both (a) & (b)

491. Immature and shriveled grains have _______test weight
(a) higher (b) lower
(c) equal (d) none

492. Grains with lower test weight gives _________ yield of flour
(a) high (b) poor
(c) good (d) none

493. Weight in gram of 1000 kernel is called
(a) test weight (b) TKW
(c) HW (d) all

494. TKW is an important parameter for
(a) milling yield (b) protein
(c) endosperm content
(d) all

495. Wheat varieties with higher TKW gives ________milling yield
(a) higher (b) lower
(c) similar (d) none

496. General range of thousand kernel weight is ________ for bread wheat
(a) 30-45g (b) 50-65g
(c) 70-95g (d) 110-125g

497. General range of thousand kernel weight is ______ for durum wheat
(a) 35-55g (b) 50-65g
(c) 70-95g (d) 110-125g

498. Small, medium and bold are categories of
(a) wheat kernels
(b) starch granules
(c) protein
(d) penosans

499. Small wheat grain generally show TKW
(a) <12.0g (b) <22.0g
(c) <32.0g (d) none

500. Medium wheat grain generally show TKW
(a) 12-18 (b) 22-28
(c) 32-38g (d) 42-48g

501. Bold wheat grain generally show TKW
(a) >28g (b) >38g
(c) >48g (d) >58g

502. The aleurone layer remains attached with endosperm in ________ grains
(a) hard (b) soft
(c) both (d) none

503. A larger grinding force is required in case of ________ grains
(a) hard (b) soft
(c) both (d) none

504. The molecular weight of friabilin is________kDa
(a) 15 (b) 30
(c) 45 (d) 60

505. The presence of friabilin makes the grain ______
(a) hard (b) soft
(c) both (d) none

506. Which of the following measures hardness of grain
(a) pearling resistance
(b) grinding resistance
(c) particle size index
(d) all

507. NIR stands for
(a) Near Infrared Refraction
(b) Near Infraction Reflection
(c) Near Induction Refraction
(d) none

508. SKCS stands for
(a) Single Kernel Characterization System
(b) Single Kernel Characteristics System
(c) Single Kernel Size Characterization System
(d) none

509. Air oven method and Karl Fischer method is used to
(a) moisture (b) ash
(c) fat (d) protein

510. Stronger flour leads to cookies with ________ quality
(a) superior (b) inferior
(c) poor (d) both b & c

511. Standard cookies flour is generally soft flour having protein content
(a) 1-2 (b) 7-8
(c) 9-12 (d) 12-15

512. Proteins are measured by
(a) Kjeldahl (b) NIR
(c) Lowry (d) all

513. Baking test is conducted on ________ gluten
(a) dry (b) wet
(c) powdered (d) all

514. Baking test is done to assess ______

(a) gluten quality
(b) gluten quantity
(c) protein quantity
(d) all

515. In baking test, gluten balls are baked at ______°C

(a) 180 (b) 200
(c) 220 (d) 240

516. Rapeseed displacement method is used to measure

(a) volume of gluten
(b) quantity of gluten
(c) test weight
(d) hectoliter weight

517. Good quality of gluten is shown by

(a) high volume (b) low volume
(c) no volume (d) none

518. Gluten performing the best in baking test will be good for

(a) bread (b) cookies
(c) cakes (d) pastries

519. The gluten performing the best in baking test will have volume ______cm³

(a) 90 (b) 100
(c) 125 (d) 250

520. Indian wheat flours contain sufficient amount of ______amylase

(a) α- (b) β-
(c) γ- (d) ω-

521. Ash content reflects the quantity of ______ present in flour

(a) mineral
(b) protein & mineral
(c) lipid & protein
(d) lipid & mineral

522. High ash content of flour indicates ______ bran contamination

(a) lower (b) higher
(c) no (d) none

523. Milled products containing higher levels of ash content are ____ in color

(a) lighter (b) darker
(c) no effect (d) none

524. The flour obtained after milling comprises of

(a) endosperm
(b) endosperm & germ
(c) endosperm, germ & bran
(d) none

525. The extraction rate of commercial grade flour is ____%

(a) 50 (b) 60
(c) 70 (d) 80

526. ______ rolls open the wheat grains from its crease

(a) break (b) reduction
(c) both (d) none

527. ______ rolls grind the large endosperm particles into flour

(a) break (b) reduction
(c) both (d) none

528. ______ rolls are fluted

(a) break (b) reduction
(c) both (d) none

529. ______ rolls are smooth

(a) break (b) reduction
(c) both (d) none

530. The ratio of speeds of two rolls- break and reduction is called

(a) differential (b) average
(c) compression (d) all

531. Shrunken kernels have ______ ratio of bran to endosperm than sound kernels

(a) more (b) less
(c) similar (d) none

532. Tempering is done to adjust ______ content of wheat grains

(a) protein (b) moisture
(c) ash (d) lipid

533. Tempering of wheat grains is done at _____ temperature
(a) high (b) low
(c) room temperature (d) none

534. Tempering is done for _____ duration
(a) 1 h (b) 4 h
(c) 18-72 h (d) 10 h

535. Conditioning is done to adjust _____ content of wheat grains
(a) moisture (b) protein
(c) lipid (d) ash

536. Conditioning is done at _____ temperature
(a) elevated (b) low
(c) room (d) none

537. Conditioning is done for _____ duration
(a) 1 (b) 8-18 h
(c) 18-72 h (d) 5-8 h

538. The efficiency of tempering depends upon
(a) initial moisture content of grains
(b) temperature of process
(c) time of process
(d) all

539. The ideal temperature for tempering is _____°C
(a) 5 (b) 10
(c) 25 (d) 40

540. The average tempering time for hard wheat is _____ h
(a) 10 (b) 16
(c) 30 (d) 36

541. The average tempering time for soft wheat is _____ h
(a) 10 (b) 16
(c) 30 (d) 36

542. The average tempering time for durum wheat is _____ h
(a) 10 (b) 6
(c) 30 (d) 36

543. Combination of all streams from the break and reduction rolls gives
(a) straight grade flour
(b) baker's patent flour
(c) patent flour (d) none

544. _____ flour is the whitest and contains lowest ash content
(a) straight grade flour
(b) baker's patent flour
(c) patent flour
(d) none

545. Patent flour is about _____ % of total flour
(a) 50 (b) 60
(c) 75 (d) 85

546. The first clear flour is _____ % of total flour
(a) 20 (b) 30
(c) 50 (d) 80

547. The second clear flour is _____ % of total flour
(a) 2 (b) 3
(c) 5 (d) 8

548. Flour extraction rate is expressed in
(a) dimensionless (b) kg/h
(c) % (d) kg

549. As the extraction rate of flour increases, the ash content in it _____
(a) increase (b) decrease
(c) remains same (d) none

550. As the extraction rate of flour increases, the protein content in it _____
(a) increase (b) decrease
(c) remains same (d) none

551. As the extraction rate of flour increases, the starch content in it _____
(a) increase (b) decrease
(c) remains same (d) none

552. As the extraction rate of flour increases, the lipid content in it _____
(a) increase (b) decrease
(c) remains same (d) none

553. Flour with 70% extraction rate has _____% ash
(a) 0.2-0.4 (b) 0.4-0.45
(c) 0.6-0.65 (d) 2-2.5

554. Flour with 70% extraction rate has _____% starch
(a) 25 (b) 55
(c) 65 (d) 75

555. Flour with 70% extraction rate has _____% lipids
(a) 1.0-1.5 (b) 2.0-2.5
(c) 3.0-3.5 (d) 4.0-4.5

556. A plansifter consists of
(a) breaking rolls
(b) reduction rolls
(c) sieves (d) all

557. Soft wheat endosperm is
(a) amorphous (b) crystalline
(c) both (d) none

558. Hard wheat endosperm is
(a) amorphous (b) crystalline
(c) both (d) none

559. _______ wheat endosperm breaks into smaller particles
(a) hard (b) soft
(c) both (d) none

560. _______ wheat endosperm breaks into larger particles
(a) hard (b) soft
(c) both (d) none

561. Soft wheat produces flour with _______ damgesd starch
(a) high (b) low
(c) no (d) none

562. Hard wheat produces flour with _______ damgesd starch
(a) high (b) low
(c) no (d) none

563. Commercial flour particle granulati on is _________ μ
(a) 0-150 (b) 150-200
(c) 200-250 (d) 250-300

564. The tempering time for durum wheat is _______ than hard and soft wheat
(a) longer (b) shorter
(c) equal (d) none

565. Flours with smaller size 1-17μ contain _____% of protein
(a) high (b) low
(c) no (d) none

566. Flours with smaller size 17-40μ contain ______% of starch
(a) high (b) low
(c) no (d) none

567. Flours with smaller size 17-40μ contain ______% of protein
(a) high (b) low
(c) no (d) none

568. As the tempering moisture _______, flour yield decreases
(a) increase (b) decrease
(c) no change (d) none

569. As the tempering moisture increases, ash content _______
(a) increase (b) decrease
(c) no change (d) none

570. As the tempering moisture _______, flour color
(a) deteriorates (b) improve
(c) remains same (d) none

571. Hard wheat offers ______ tolerance to added water than soft wheat
(a) more (b) less
(c) similar (d) none

572. The tempering time is influenced by
(a) kernel hardness
(b) protein per centage
(c) initial moisture content
(d) all

573. As the temperature increase, the rate of moisture penetration ______
(a) increase (b) decrease
(c) remains same (d) none

574. As the temperature increase, the tempering time
(a) increase (b) decrease
(c) remains same (d) none

575. Soft wheat should be fed into mill _____ than hard wheat as they sift less readily than hard wheat
(a) rapidly (b) slowly
(c) vigorously (d) none

576. The ease with which endosperm can be separated from bran is called
(a) extraction rate
(b) milling efficiency
(c) milling separation index
(d) none

577. The physical separation of flour into fractions of higher protein and starch content is
(a) air classification
(b) extraction
(c) tempering
(d) milling

578. Osborne classified the wheat proteins into four fractions –albumins, globulins, gliadins and glutenins in year
(a) 1907 (b) 1807
(c) 1917 (d) 1817

579. _______ can be extracted with 0.5M NaCl
(a) albumins (b) globulins
(c) gliadins (d) glutenins

580. Gluten proteins account for _______% of total flour proteins
(a) 20-30 (b) 30-40
(c) 60-70 (d) 80-90

581. Non gluten proteins account for _______% of total flour proteins
(a) 1-2 (b) 10-20
(c) 30-40 (d) 40-50

582. Gliadin was coined by
(a) Osborne (b) Shewry
(c) Taddei (d) none

583. Glutenin was coined by
(a) Osborne (b) Shewry
(c) Taddei (d) none

584. _________ proteins have property of foaming
(a) non gluten (b) gluten
(c) gliadin (d) none

585. The classification of gluten proteins into sulfur poor, rich and HMW prolamins was given by
(a) Osborne (b) Shewry
(c) Taddei (d) none

586. The term monomeric is generally associated with
(a) albumin (b) globulin
(c) gliadin (d) glutenin

587. The term polymeric is generally associated with
(a) albumin (b) globulin
(c) gliadin (d) glutenin

588. α-, β-, γ- and ω- are subclassifications of
(a) albumin (b) globulin
(c) gliadin (d) glutenin

589. Low molecular weight and high molecular weight subunits are subclassification of
(a) albumin (b) globulin
(c) gliadin (d) glutenin

590. Which of the following falls into the category of S-poor prolamins
(a) ω-gliadin
(b) low molecular weight subunits
(c) α-gliadins
(d) all

591. Which of the following falls into the category of S-rich prolamins
(a) γ-gliadin
(b) low molecular weight subunits
(c) α-gliadins
(d) all

592. The gliadins are classified into α-, β-, γ- and ω-gliadins in ______ order of electrophoretic mobility under acidic conditions
(a) increasing (b) decreasing
(c) similar (d) none

593. The relative molecular mass of gliadin subfractions is
(a) α-< β-< γ-< ω- (b) α-> β-> γ-> ω-
(c) α-< β-> γ- <ω- (d) none

594. HPLC stands for
(a) High Performance Liquid Chromatography
(b) High Pressure Liquid Chromato graphy
(c) Hot pressure Liquid Chromatography
(d) none

595. RP-HPLC stands for
(a) Reverse Pressure HPLC
(b) Red Phase HPLC
(c) Reverse Protein HPLC
(d) none

596. Which of the following are structurally similar, so kept under I group
(a) α-, β- (b) α-, γ-
(c) γ-, β- (d) α-, ω-

597. α- and γ-gliadins are ______ in sulfur containing amino acids
(a) rich (b) poor
(c) no (d) none

598. ω-gliadins is ____________ in sulfur containing amino acids
(a) rich (b) poor
(c) no (d) none

599. The molecular weight of α- type gliadins is
(a) 10000-20000 (b) 20000-30000
(c) 30000-45000 (d) 50000-65000

600. The molecular weight of ω- type gliadins is
(a) 10000-20000 (b) 20000-30000
(c) 30000-45000 (d) 44000-80000

601. Which of the following has highest molecular weight
(a) α- (b) β-
(c) γ- (d) ω-

602. PAGE stands for
(a) Poly Acrylamide Gel Electrophoresis
(b) Protein Acrylamide Gel Electrphoresis
(c) Poly Acrylamide Gel Electroplating
(d) none

603. SDS PAGE is
(a) sodium dodecyl sulfate PAGE
(b) sodium decyl sulfate PAGE
(c) sodium dodecyl sodium PAGE
(d) none

604. PAGE used for separating polypeptides components of glutenin is
(a) SDS-PAGE (b) Acid-PAGE
(c) PAGE (d) none

605. PAGE used for separating polypeptides components of gliadin is
(a) SDS-PAGE (b) Acid-PAGE
(c) PAGE (d) none

606. The subunits of glutenin are subdivided into
(a) A, B, C (b) C, D, E
(c) C (d) all

607. The largest polypeptide of glutenin are
(a) A (b) B
(c) C (d) all

608. The molecular weight of A subunit of glutenin is
(a) 10000-20000 (b) 20000-45000
(c) 80000-160000 (d) 5000-9000

609. Which of the following is a larger polypeptide
(a) X (b) Y
(c) both similar (d) none

610. Molecular weight of X-type polypeptides is
(a) 83-88000 (b) 5000-500000
(c) 10-110 (d) none

611. Molecular weight of Y-type polypeptides is
(a) 67-74000 (b) 5000-500000
(c) 10-110 (d) none

612. The low molecular weight subunits of glutenin is divided into
(a) B & C (b) A & C
(c) C & D (d) Y & Z

613. Which of the following is a longer polypeptide
(a) B (b) C
(c) both similar (d) none

614. Molecular weight of B-subunits
(a) 42-51000 (b) 30-40000
(c) 83-88000 (d) none

615. Molecular weight of B-subunits
(a) 42-51000 (b) 30-40000
(c) 83-88000 (d) none

616. Principal bonds of gluten polypeptides is
(a) hydrogen (b) disulfide
(c) C-C (d) none

617. Disulfide bond is ________ type bond
(a) covalent (b) ionic
(c) vander wall's (d) none

618. Ionic bond is ____ than covalent bond
(a) stronger (b) weaker
(c) similar (d) none

619. α type and γ-gliadins have ________ disulfide linkages
(a) inter molecular
(b) intramolecular
(c) no
(d) all

620. ω-type giladins have ______ disulfide linkages
(a) inter molecular
(b) intramolecular
(c) no
(d) all

621. Gluten proteins are ________ in polar amino acids
(a) rich (b) poor
(c) no (d) all

622. Gluten proteins have ________ charge density
(a) high (b) low
(c) no (d) none

623. Gluten has low charge density due to
(a) low level of basic amino acid
(b) amino acid present in amide form
(c) both
(d) none

624. Gliadins have ________ number of cysteine residues
(a) odd (b) even
(c) both (d) none

625. Glutenins have ________ number of cysteine residues
(a) odd (b) even
(c) both (d) none

626. When total protein of grain increases, then amount gluten proteins ______
(a) increases (b) decreases
(c) remain same (d) none

627. When total protein of grain increases, then amount albumins & globulins ________
(a) increases (b) decreases
(c) remain same (d) none

628. Mostly gliadins are extracted in ________% ethanol
(a) 80 (b) 95
(c) 70 (d) 60

629. Too elastic dough has higher
(a) glutenins (b) Gliadins
(c) albumins (d) globulins

630. Too elastic dough give ______ bread making performance
(a) poor (b) good
(c) both (d) none

631. Extensible doughs have ______ mixing time
(a) longer (b) shorter
(c) excessively long (d) none

632. Extensible doughs with shorter mixing time are recommended for
(a) bread (b) cookies
(c) cakes (d) both (b) & (c)

633. Extensible doughs with longer mixing time are recommended for
(a) bread (b) cookies
(c) cakes (d) both b & c

634. Each bread wheat cultivar contains ______ high molecular weight subunits
(a) 3-5 (b) 5-8
(c) 8-11 (d) 11-13

635. HMWGS are encoded at Glu1 loci on the ______arms of the chromosomes 1A, 1B & 1D
(a) long (b) short
(c) both (d) none

636. SDS in SDS sedimentation
(a) nullifies the charge on proteins
(b) provides acidic medium
(c) swell the glutenin
(d) none

637. Lactic acid in SDS sedimentation test
(a) nullifies the charge on proteins
(b) provides acidic medium
(c) swell the glutenin
(d) none

638. High sedimentation volume is recommended for
(a) bread (b) biscuits
(c) both (d) none

639. High sedimentation volume is recommended for
(a) cookies (b) biscuits
(c) cakes (d) all

640. The Glu-1 score of bread wheat range from 3 to ______
(a) 4 (b) 10
(c) 100 (d) 50

641. The Glu-1 score for 5+10 subunit is
(a) 4 (b) 3
(c) 2 (d) 1

642. The Glu-1 score for 1 subunit is
(a) 4 (b) 3
(c) 2 (d) 1

643. The Glu-1 score for 17+18 subunit is
(a) 4 (b) 3
(c) 2 (d) 1

644. The Glu-1 score for 2* subunit is
(a) 4 (b) 3
(c) 2 (d) 1

645. The Glu-1 score for 7+8 subunit is
(a) 4 (b) 3
(c) 2 (d) 1

646. The Glu-1 score for 13+16 subunit is
(a) 4 (b) 3
(c) 2 (d) 1

647. The Glu-1 score for 7+9 subunit is
(a) 4 (b) 3
(c) 2 (d) 1

648. The Glu-1 score for 2+12 subunit is
(a) 4 (b) 3
(c) 2 (d) 1

649. The Glu-1 score for 3+12 subunit is
(a) 4 (b) 3
(c) 2 (d) 1

650. The Glu-1 score for Null subunit is
(a) 4 (b) 3
(c) 2 (d) 1

651. The Glu-1 score for 7 subunit is
(a) 4 (b) 3
(c) 2 (d) 1

652. The Glu-1 score for 4+12 subunit is
(a) 4 (b) 3
(c) 2 (d) 1

653. The Glu-1 score for 6+8 subunit is
(a) 4 (b) 3
(c) 2 (d) 1

654. The Glu-1 score for 20 subunit is
(a) 4 (b) 3
(c) 2 (d) 1

655. A score of 10 indicates
(a) good bread making quality
(b) good cookie quality
(c) good biscuits
(d) all

656. Glucose is ___________ related with bread making quality
(a) positively (b) negatively
(c) no relation (d) none

657. Which of the following is present in good bread wheat
(a) 5+10 (b) 2+12
(c) 6+8 (d) 20

658. Which of the following is present in poor bread wheat
(a) 5+10 (b) 1
(c) 2* (d) 2+12

659. The superiority of 17+18 over 20 is
(a) number of cysteine residues
(b) position of cysteine residues
(c) both (a) & (b)
(d) none

660. The strength of gluten is attributed to
(a) gliadin (b) glutenin
(c) both (d) globulin

661. Gliadins are ________ in nature
(a) viscous (b) globular
(c) smaller (d) all

662. Extensibility of dough is desirable for
(a) cookies (b) bread
(c) buns (d) all

663. Wheat gluten is ______ protein
(a) water soluble
(b) water insoluble
(c) salt soluble
(d) none

664. The protein content of gluten is generally
(a) 20 (b) 40
(c) 50 (d) 80

665. Commercial gluten powder generally has a particle size (μm)
(a) <50 (b) <150
(c) < 200 (d) >250

666. Commercial starch powder generally has a particle size (μm)
(a) <50 (b) <150
(c) < 200 (d) <250

667. Maximum ash content (%) of gluten powder is generally
(a) 2 % (b) 4%
(c) 6% (d) 10%

668. Maximum fat % of gluten powder is generally
(a) 2 % (b) 4%
(c) 6% (d) 10%

669. Maximum fiber content (%) of gluten powder is generally
(a) 1.5 % (b) 2.5%
(c) 3.5% (d) 4%

670. Wheat is unique due to presence of
(a) gluten
(b) fat
(c) protein content
(d) albumin & globulin

671. Which of the following is a property of gluten
(a) viscoelasticity (b) film forming
(c) thermosetting (d) all

672. Most of the enzymes of wheat are present in
(a) outer aleurone (b) bran
(c) germ (d) all

673. The chief wheat enzyme important in baking industry is
(a) alpha amylase (b) beta amylase
(c) hemicellulase (d) glucoamylase

674. The arabinoxylan content in wheat flour is
(a) 2-3% (b) 0.2-0.3%
(c) 1-2% (d) 3-5%

675. The non-polar lipids of wheat flour ________the mixing time of flour
(a) increases (b) decreases
(c) no effect (d) none

676. Defatted wheat flour is ________ in appearance
(a) darker (b) yellowish
(c) whiter (d) none

677. ______ lipids improve baking quality of wheat flour
(a) polar (b) non-polar
(c) both (d) none

678. ______lipids have detrimental effect on baking quality
(a) polar (b) non-polar
(c) both (d) none

679. Polar lipids contribute
(a) high loaf volume
(b) low loaf volume
(c) higher bread softness
(d) both (a) & (c)

680. Which of the following unsaturated fatty acid is most detrimental for bread quality
(a) oleic (b) linoleic
(c) linoleinic (d) all

681. Among polar lipids _______ are most effective in improving bread quality
(a) glycolipids (b) phospholipids
(c) both (d) none

682. Which of the following act as gas cell stabilizer
(a) glycolipids (b) non-polar lipids
(c) phospholipids (d) both (a) & (c)

683. Gas cell breakers are
(a) glycolipids
(b) non-polar lipids
(c) phospholipids
(d) both (a) & (c)

684. Which of the following cookie parameter is desirable
(a) more spread
(b) cracked surface
(c) both
(d) none

685. Lipids have _________effect on cookie quality
(a) positive (b) negative
(c) no effect (d) none

686. The size (μm) of starch granules ranges from
(a) 1-30 (b) 15-45
(c) 55-65 (d) none

687. Average diameter (μm) of A-type granule of starch
(a) 4 (b) 8
(c) 14 (d) none

688. Average diameter (μm) of B-type granule of starch
(a) 4 (b) 8
(c) 14 (d) none

689. Per centage of starch in A-type granule
(a) 25 (b) 50
(c) 75 (d) 90

690. Per centage of starch in B-type granule
(a) 25 (b) 50
(c) 75 (d) 90

691. % total of number of granule A (less)
(a) 10% (b) 20%
(c) 30% (d) 90%

692. % total of number of granule B (more)
(a) 10% (b) 20%
(c) 30% (d) 90%

693. Which starch granule produces higher loaf volume in bread
(a) A (b) B
(c) both (d) none

694. Which starch granule produces lower loaf volume in bread
(a) A (b) B
(c) both (d) none

695. The proportion of amylase and amylopectin in wheat starch is
(a) 30:70 (b) 50:50
(c) 40:60 (d) 60:40

696. The formation of gel or paste of starch at elevated temperature is called
(a) gel formation (b) gelatinization
(c) jelly (d) all

697. Wheat flour with higher proportion of damaged starch has ______ activity
(a) higher (b) lower
(c) no effect (d) none

698. The desired level (mg/10g flour) of maltose value is
(a) 2.5-3.5 (b) 5-10
(c) 12-17 (d) 20-25

699. The water holding capacity of damaged starch is __________ than normal starch
(a) higher (b) lower
(c) similar (d) none

700. The water holding capacity (g water/ g of dry starch) of normal starch
(a) 0.44 (b) 0.98
(c) 1.12 (d) 1.35

701. The water holding capacity (g water/ g of dry starch) of damaged starch
(a) 0.44 (b) 0.98
(c) 2.0 (d) 4.35

702. Desired level of damaged for bread
(a) 7-9% (b) 2-37%
(c) 15-16% (d) 1-2%

703. Cookie level of damaged starch
(a) <1% (b) <7%
(c) <10% (d) <12%

704. Bread staling/retrogradation is due to
(a) starch (b) protein
(c) lipids (d) fiber

705. Staling of bread is maximum at ________°C
(a) 1.0 (b) 4.0
(c) 7.0 (d) 10

706. The water level of ________ % is required for retrogradation to occur
(a) 10-20 (b) 20-30
(c) 30-50 (d) 50-70

707. The moisture content of bread is
(a) 10% (b) 2%
(c) 40% (d) 70%

708. Starch retrogradation changes ____ to _____ pattern of starch
(a) A to B (b) B to A
(c) no change (d) none

709. The water holding capacity of B pattern of starch is
(a) high (b) poor
(c) similar to A (d) none

710. Retrogradation is reversible at ___________ temperature °C
(a) >45 (b) >10
(c) >2 (d) <10

711. Diastatic activity is mg of _______ produced per 10g of flour per hour at 30°C
(a) glucose (b) maltose
(c) sucrose (d) fructose

712. The optimum water content % of bread dough
(a) 65% (b) 45%
(c) 50% (d) 35%

713. The loaf volume is _________ related with water content in dough
(a) inversely (b) directly
(c) no relation (d) none

714. Which of the following increases the water absorption capacity of flour
(a) damaged starch (b) protein
(c) pentosans (d) all

715. The gelation of starch begins at ___°C
(a) 80 (b) 50
(c) 30 (d) 90

716. Moisture content (%) of freshly baked cookies and biscuits is
(a) upto 5% (b) upto 10%
(c) upto 15% (d) upto 25%

717. _________ is known as baker's yeast
(a) S. cerevisiae (b) A. niger
(c) S. uvarum (d) all

718. The spongy structure of bread is due to
(a) yeast activity (b) CO_2
(c) gluten (d) all

719. The optimum pH for CO_2 production by yeast in bread is due to
(a) 3.5 (b) 4.5
(c) 5.5 (d) 7.5

720. Salt is added to dough to
(a) enhance gluten development
(b) toughen gluten
(c) increase mixing time
(d) all

721. The optimum size (μ) of sugar for good cookie spread is
(a) 50-75 (b) 60-120
(c) 150-180 (d) 200-250

722. The function of shortening in dough is
(a) reduce toughness of dough
(b) high machinability
(c) enhance flavor
(d) all

723. Which of the following is an oxidizing agent used in baking industry
(a) azodicarbonamide
(b) potassium iodate
(c) ascorbic acid
(d) all

724. The emulsifier used in bakery is
(a) calcium steroyl-2 lactylate
(b) sodium steroyl-2 lactylate
(c) diacetyl tartaric esters of monoglycerides
(d) all

725. Cookie dough with emulsifiers result in __________ spread
(a) higher (b) lower
(c) no effect (d) none

726. Ropiness in bakery products is prevented by
(a) calcium propionate
(b) potassium sorbate
(c) both (d) none

727. Fisher units used to measure
(a) flour particle size
(b) damaged starch
(c) flour color
(d) diastase activity

728. Flour coarser than ________FU is considered to be granular
(a) 22 (b) 32
(c) 12 (d) 2

729. Falling number below 150 indicates
(a) normal wheat
(b) sprouted wheat
(c) both
(d) none

730. AWRC stands for
(a) alkaline water retention capacity
(b) acid water retention capacity
(c) alkaline water reserve capacity
(d) acid water reserve capacity

731. Good bread flour has wet gluten in range
(a) 5-10% (b) 12-25%
(c) 30-36% (d) 45-49%

732. Good biscuits flour has wet gluten in range
(a) 5-10% (b) 12-25%
(c) 30-36% (d) 45-49%

733. Good bread flour has dry gluten in range

(a) 10-12% (b) 15-25%
(c) 30-36% (d) 45-49%

734. Goodcookie flour has dry gluten in range

(a) 8-9% (b) 12-25%
(c) 30-36% (d) 45-49%

735. Higher sedimentation indicates

(a) higher HMW glutenin
(b) higher gliadin
(c) higher albumins
(d) higher globulins

736. Flour with sedimentation volume ______ ml is suitable for cookies

(a) less than 20 (b) 20-40
(c) more than 40 (d) none

737. Flour with sedimentation volume ___ ml is suitable for fermented products

(a) less than 20 (b) 20-40
(c) more than 40 (d) none

738. Flour with sedimentation volume _______ml is suitable for bread

(a) less than 20 (b) 20-40
(c) more than 40 (d) none

739. AWRC is test used for judging quality of flour for making of

(a) bread (b) cookies
(c) pasta (d) cakes

740. The higher value of AWRC indicate ______ cookie diameter

(a) higher (b) lower
(c) no effect (d) none

741. The study of deformation and flow of matter is called

(a) rheology (b) physics
(c) chemistry (d) none

742. The farinograph measures the resistance of dough to mixing blades in _____ units

(a) fischer (b) brabender
(c) farinograph (d) all

743. The basis for gluten viscoelasticity is

(a) gliadin (b) glutenin
(c) both (a) & (b) (d) globulins

744. The elastic modulus is indicated by

(a) G' (b) G"
(c) Tan θ (d) none

745. The viscous modulus is indicated by

(a) G' (b) G"
(c) Tan θ (d) none

746. The presence of β-spiral is responsible for ______ of dough

(a) viscosity (b) elasticity
(c) extensibility (d) all

747. Which of the following domain of glutenins is rich in β-turns

(a) C
(b) N
(c) central repetitive domain
(d) all

748. Carbondioxide is released by yeast during

(a) mixing (b) fermentation
(c) panning (d) all

749. _______ protein allow the gas cells to expand during bread making

(a) gliadin (b) glutenin
(c) albumin (d) globulin

750. ______ proteins provide strength to prevent gas cells from over expanding

(a) gliadin (b) glutenin
(c) albumin (d) globulin

751. There is a rapid loss of carbondioxide as a result of high ______ of dough

(a) elasticity (b) viscous
(c) both (d) none

752. PDR stands for

(a) peak dough result
(b) peak dough resistance
(c) protein dough resistance
(d) none

753. The isoelectric point of gliadin proteins is
(a) >6.5 (b) <6.5
(c) both (d) none

754. The isoelectric point of gliadin proteins is
(a) >6.5 (b) <6.5
(c) both (d) none

755. Yeast produce ____ g of carbondioxide from 1 g of glucose
(a) 0.5 (b) 1
(c) 2.5 (d) 5

756. Yeast produce ______ ml volume of carbondioxide from 1 g of glucose
(a) 100 (b) 175
(c) 200 (d) 275

757. What is the per centage of yeast required to produce acceptable bread
(a) 1-2 (b) 2-3
(c) 3-5 (d) 5-7

758. The sticky nature of dough is due to high number of ____ groups
(a) SH (b) SS
(c) HH (d) OH

759. Extensibility in the dough is increased by _______ groups
(a) SH (b) SS
(c) HH (d) OH

760. The elasticity in the dough is increased by ______groups
(a) SH (b) SS
(c) HH (d) OH

761. The addition of sugar _______ the gelatinization temperature of starch
(a) increase (b) decrease
(c) no effect (d) none

762. Emulsifiers in bread industry belong to
(a) polar lipids
(b) non polar lipids
(c) both
(d) none

763. GMP stands for
(a) glucose mono phosphate
(b) glutenin macro polymer
(c) gliadin macro polymer
(d) none

764. Alpha amylase acts on the ______ bonds
(a) α(1-4) (b) α(1-6)
(c) β(1-4) (d) β(1-6)

765. Which of the following is not a type of bread
(a) pan (b) sweet
(c) sour (d) flat

766. The optimum stage of bread dough is when it is
(a) elastic (b) smooth
(c) silky (d) all

767. The yeast fermentation rate is doubled for each degree rise in temperature range between ______°C
(a) 20-40 (b) 40-60
(c) 10-20 (d) none

768. The temperature iat which the yeast cells get killed is ____°C
(a) >40 (b) >60
(c) >80 (d) >50

769. Yeast activity is inactivated at pH
(a) <3 (b) 5
(c) 6 (d) 7

770. Puching of the dough in between the fermentation period is called
(a) dough make up (b) proving
(c) knock back (d) panning

771. The process of dividing the dough into individual pieces for bread making is called
(a) dough making (b) scaling
(c) knock back (d) panning

772. The proving of bread is done at _____°C
(a) 20-25 (b) 30-35
(c) 40-45 (d) 50-55

773. The proving of bread is done at ____% relative humidity
(a) 65 (b) 75
(c) 85 (d) 95

774. The baking of bread is done at ____ °C for 20 minutes
(a) 180 (b) 200
(c) 220 (d) 240

775. CBP stands for
(a) Central Bread Process
(b) Central Bread Procedure
(c) Chorleywood Bread Process
(d) none

776. During fermentation, pH of bread-dough is reduced from
(a) 4.7 to 3.3 (b) 5.5 to 2
(c) 5.5 to 4.7 (d) 6.5 to 5.5

777. During fermentation, the pH of bread dough
(a) increase (b) decrease
(c) remains same (d) none

778. The diastatic enzymes convert starch into
(a) maltose (b) glucose
(c) sucrose (d) fructose

779. The rapid rise of dough during first few minutes in oven is called
(a) oven spring
(b) crust formation
(c) scaling
(d) proving

780. Yeast is inactivated at temperature ____°C
(a) 45 (b) 55
(c) 65 (d) 75

781. β-amylase is denatured at____°C
(a) 47-51 (b) 57-71
(c) 67-81 (d) 71-87

782. α-amylase is denatured at temperature ____°C
(a) 45-55 (b) 55-65
(c) 65-95 (d) 75-105

783. The browning reaction involves
(a) reducing sugars (b) proteins
(c) both (d) none

784. The colored compounds formd as a result of browning is
(a) melanoidins (b) melanin
(c) both (d) none

785. The browning reaction starts at ___°C
(a) 100 (b) 120
(c) 140 (d) 160

786. Hearth bread involves __________ fermentation
(a) lactic acid (b) acetic acid
(c) oxalic acid (d) all

787. Flat bread is most popular in
(a) Asian countries (b) Europe
(c) USA (d) none

788. Rolls have _______ level sugar than other bread
(a) higher (b) lower
(c) no sugar (d) none

789. WVTR stands for
(a) water vapour transparency rate
(b) water vapour transmission rate
(c) water vapour travelling rate
(d) none

790. The WVTR of packaging material of bread should be
(a) high (b) low
(c) optimum (d) none

791. Milk bread should contain atleast ____% milk solids
(a) 2 (b) 6
(c) 10 (d) 20

792. Wheat germ bread must contain atleast ______% germ
(a) 10 (b) 20
(c) 30 (d) 40

793. Sugar free bread have sugar ____%
(a) 0.2 (b) 1
(c) 2 (d) none

794. Coeliac disease is caused due to
(a) zein (b) oryzenin
(c) gluten (d) hordein

795. The presence of rope in bread is due to breakdown of ______ and protein
(a) starch (b) lipids
(c) sugars (d) dextrins

796. Ropiness occurs at temperature ______ °C
(a) <35 (b) >35
(c) <10 (d) <4

797. Causative organism for rope in bread is
(a) Bacillus mesentericus
(b) Saccharomyces cereviseae
(c) *Aspergillus niger*
(d) none

798. The ropiness in bread can be prevented by
(a) sodium hydroxide
(b) acetic acid
(c) vinegar
(d) both (b) & (c)

799. The mold inhibitors are
(a) acetic acid (b) propionates
(c) both (d) none

800. Which of the following is not a bread defect
(a) ropiness (b) staling
(c) oven spring (d) mold growth

801. POEMS stands for
(a) Polyoxyethylene monostearate
(b) Polyoxymethylene monostearate
(c) Paraoxyethylene monostearate
(d) none

802. POEMS is a
(a) surfactant
(b) mold inhibitor
(c) ropiness inhibitor
(d) all

803. Polysorbate 60 is a
(a) antistaling agent
(b) dough conditioner
(c) both
(d) none

804. The recrystallization of starch fraction after baking is called
(a) retrogradation
(b) crust staling
(c) ropiness
(d) all

805. The word biscuit has been derived from
(a) danis biscytis
(b) danis biscoctus
(c) danis discoctus
(d) none

806. The term cookie has originated from
(a) cookie (b) koekje
(c) mokeje (d) none

807. Cookies and biscuits are
(a) similar (b) different
(c) bread type (d) none

808. The level of sugar in cookies is ________ than biscuits
(a) higher (b) lower
(c) similar (d) none

809. The level of shortening in cookies is ________than biscuits
(a) higher (b) lower
(c) similar (d) none

810. The level of salt in cookies is ________ than biscuits
(a) higher (b) lower
(c) similar (d) none

811. Cookies are made from
(a) weaker dough
(b) developed dough
(c) fermented dough
(d) all

812. Crackers are made from
(a) weaker dough
(b) developed dough
(c) fermented dough
(d) all

813. Higher spread in cookies is achieved with________sugar
(a) fine (b) coarse
(c) dextrose (d) none

814. Fast and slow acting are types of
(a) baking soda (b) emulsifiers
(c) baking powder (d) sweeteners

815. Fast acting release carbondioxide at
(a) room temperature
(b) low temperature
(c) elevated temperature
(d) none

816. Slow acting release carbondioxide at
(a) room temperature
(b) low temperature
(c) elevated temperature
(d) none

817. Cookies are ________ products
(a) fermented
(b) chemically leavened
(c) non-leavened
(d) none

818. Which of the following is atype of cookie
(a) deposit (b) rotary molded
(c) wirecut (d) all

819. Which of the following has lowest sugar levels
(a) deposit (b) rotary molded
(c) wirecut (d) cutting

820. Which of the following has highest shortening
(a) deposit (b) rotary molded
(c) wirecut (d) cutting

821. Soft wheat flour with particles less than ________ μ is most desireable for cookies
(a) 38 (b) 10
(c) 20 (d) 28

822. Creaming and all-in-one are methods of ________ preparation
(a) cookies (b) bread
(c) pasta (d) flat bread

823. In creaming, first ________ and ________ are added
(a) sugar and fat
(b) sugar and flour
(c) fat and flour
(d) milk and sugar

824. ________ spread ratio is desired for cookies
(a) low (b) high
(c) no (d) none

825. ________ prevent rancidity in cookies
(a) BHA (b) BHT
(c) TBHQ (d) all

826. Which of the following takes least time to cook
(a) brown rice (b) parboiled rice
(c) white rice (d) none

827. Which of the following takes most time to cook
(a) brown rice (b) parboiled rice
(c) white rice (d) none

828. The milling yield of white rice is
(a) 66-70% (b) 70-80%
(c) 80-90% (d) 100%

829. Starch of rice comprises of________% of rice solid
(a) 70 (b) 80
(c) 90 (d) 100

830. Which of the following has highest lysine content in rice
(a) albumin (b) glutelin
(c) globulin (d) prolamins

831. Major proportion of lipid in rice is obtained from
(a) bran (b) brown rice
(c) polished rice (d) milled rice

832. Lowest proportion of lipid in rice is obtained from
(a) bran (b) brown rice
(c) polished rice (d) milled rice

833. Major rice lipid constitute
(a) TGA (b) phospholipids
(c) glycolipids (d) waxes

834. Bran of rice is rich in ________lipids
(a) polar (b) non –polar
(c) neutral (d) none

835. Endosperm of rice is rich in ________ lipids
(a) polar (b) non –polar
(c) neutral (d) none

836. Waxy rice has _____% of amylose
(a) high
(b) low
(c) similar to amylopectin
(d) none

837. Milling yield of rice________after parboiling
(a) deteriorate (b) improves
(c) no change (d) none

838. Rice i.e. not dehusked is called
(a) paddy (b) cargo rice
(c) white rice (d) parboiled rice

839. Rice i.e. only dehusked is called
(a) paddy (b) cargo rice
(c) white rice (d) parboiled rice

840. Which of the following is longest
(a) head rice (b) brokens
(c) small brokens (d) none

841. The germ of corn is_______% of the total kernel
(a) 11.5 (b) 82.3
(c) 18.5 (d) none

842. The endosperm of corn is_______% of the total kernel
(a) 11.5 (b) 82.3
(c) 18.5 (d) none

843. Most of the corn protein is present in
(a) endosperm (b) germ
(c) both (d) none

844. The per centage of protein in germ of corn is
(a) 11.5 (b) 82.3
(c) 18.5 (d) none

845. Most of the sugar of corn is present in
(a) endosperm (b) germ
(c) both (d) none

846. The per centage of sugar in corn germ is
(a) 11.0 (b) 82.3
(c) 18.5 (d) none

847. The per centage of sugar in whole corn kernel is
(a) 11.5 (b) 1.94
(c) 18.5 (d) none

848. Upon corn milling, the highest proportion is that of
(a) grits (b) coarse meal
(c) fine meal (d) flour

849. The per centage yield of grits in corn milling is
(a) 40 (b) 5
(c) 14 (d) 11

850. The per centage yield of flour in corn milling is
(a) 40 (b) 5
(c) 14 (d) 11

851. The per centage yield of germ in corn milling is
(a) 40 (b) 5
(c) 14 (d) 11

852. The per centage yield of hominy feed in corn milling is
(a) 40 (b) 5
(c) 14 (d) 11

853. The major protein of barley is
(a) glutelins (b) albumin
(c) globulin (d) hordein

854. Hordeins form ________ % of total protein in barley
(a) 20 (b) 30
(c) 40 (d) 50

855. The hordein proteins are ____ soluble
(a) fat (b) alcohol
(c) salt (d) water

856. Majority of the lipids of barley are present in
(a) embryo
(b) aleurone layer
(c) both (d) none

857. Which of the following is a milled product of barley
(a) pearls (b) pearling dust
(c) feed meal (d) all

858. The moisture % of malt is
(a) 1-2 (b) 2-4
(c) 4-6 (d) 8-10

859. The per centage of sugar is higher in
(a) barley (b) malt
(c) both (d) none

860. The per centage of sugar in malt is
(a) 8-10 (b) 12-18
(c) 18-24 (d) 24-30

861. Diastatic power is hgher of
(a) barley (b) malt
(c) equal (d) none

862. Diastatic power of malt is
(a) 100-250°L (b) 50-100°L
(c) 300-450°L (d) none

863. The pericarp forms ____% of sorghum
(a) 2 (b) 6
(c) 84 (d) 10

864. The endosperm forms __________% of sorghum
(a) 2 (b) 6
(c) 84 (d) 10

865. The germ forms _______% of sorghum
(a) 2 (b) 6
(c) 84 (d) 10

866. The largest part of sorghum is
(a) pericarp (b) endosperm
(c) germ (d) none

867. The pericarp forms ________ % of pearl millet
(a) 8.4 (b) 25
(c) 75 (d) 16.5

868. The endosperm forms ________ % of pearl millet
(a) 8.4 (b) 25
(c) 75 (d) 16.5

869. The germ forms ___________% of pearl millet
(a) 8.4 (b) 25
(c) 75 (d) 16.5

870. The average starch content of sorghum is
(a) 8.4 (b) 25
(c) 75 (d) 69.5

871. Major proportion of starch in sorghum is that of
(a) amylose (b) amylopectin
(c) both equal (d) none

872. The per centage of amylopectin in sorghum is
(a) 20-40 (b) 40-50
(c) 60-70 (d) 70-80

873. The per centage of amylose in sorghum is
(a) 20-30 (b) 40-50
(c) 60-70 (d) 70-80

874. Waxy sorghum has
(a) 100% amylopectin
(b) 100% amylose
(c) both
(d) none

875. The fat content is highest in __________ millet
(a) pearl (b) finger
(c) foxtail (d) common

876. Decortication of grain is a method to remove________
(a) lipids
(b) proteins
(c) fiber
(d) carbohydrates

877. Total dietary fiber in pearl millet is _______%
(a) 20.4 (b) 18.6
(c) 14.2 (d) 17.2

878. Total dietary fiber in finger millet is _______%
(a) 20.4 (b) 18.6
(c) 14.2 (d) 17.2

879. Total dietary fiber in sorghum is _______ %
(a) 20.4 (b) 18.6
(c) 14.2 (d) 17.2

880. Total dietary fiber in wheat is _______%
(a) 20.4 (b) 18.6
(c) 14.2 (d) 17.2

881. Total dietary fiber in rice is _______%
(a) 20.4 (b) 18.6
(c) 14.2 (d) 8.3

882. Which of the following has highest dietary fiber
(a) pearl millet (b) sorghum
(c) wheat (d) rice

883. Goitrogens are present in
(a) pearl millet (b) sorghum
(c) wheat (d) rice

884. Enrichment and fortification are _________ terms
(a) similar (b) different
(c) not determined (d) none

885. A process in which the lost nutrient are added back to the product is called
(a) enrichment (b) fortification
(c) both (d) none

886. A process of adding 10% or more of a daily value for a specific nutrient to a product is
(a) enrichment (b) fortification
(c) both (d) none

887. The major source of dietary proteins in India is
(a) cereals (b) pulses
(c) milk (d) meat

888. The seed coat of legumes ranges from _________% of total seed
(a) 15-30 (b) 20-45
(c) 30-45 (d) none

889. Most of the pulses have _______ lysine content compared to that present in egg protein
(a) more (b) less
(c) equal (d) none

890. Major proteins of legume is
(a) globulin (b) albumin
(c) glutelin (d) gluten

891. Digestibility of legume protein is
(a) good (b) high
(c) poor (d) none

892. The digestibility of legume proteins can be improved by
(a) cooking (b) soaking
(c) roasting (d) all

893. The poor digestibility of legume proteins is due to the presence of
(a) protease inhibitor
(b) polyphenols
(c) both
(d) none

894. Major constituent of pulses is
(a) protein (b) carbohydrate
(c) lipid (d) minerals

895. Principal carbohydrate of pulses is
(a) starch
(b) sucrose
(c) sucrosyoligosaccharide
(d) all

896. Which of the following is associated with flatulence
(a) raffinose (b) stachyose
(c) verbascose (d) all

897. Which of the following has highest protein content
(a) soybean (b) French bean
(c) faba bean (d) winged bean

898. Which of the following has highest lipid content
(a) soybean (b) French bean
(c) faba bean (d) winged bean

899. Which of the following is not a legume
(a) horse gram (b) lentil
(c) rice bean (d) maize bean

900. The RDI level for fiber is ______ grams
(a) 1-25 (b) 25-50
(c) 50-75 (d) 75-100

6.3 ANSWER KEY

Q.	A.	Q.	A.	Q.	A.	Q.	A.	Q.	A.	Q.	A.
1.	(b)	2.	(d)	3.	(c)	4.	(c)	5.	(d)	6.	(b)
7.	(d)	8.	(d)	9.	(b)	10.	(d)	11.	(a)	12.	(a)
13.	(a)	14.	(d)	15.	(a)	16.	(b)	17.	(d)	18.	(d)
19.	(b)	20.	(d)	21.	(d)	22.	(a)	23.	(a)	24.	(a)
25.	(b)	26.	(c)	27.	(a)	28.	(c)	29.	(a)	30.	(a)
31.	(b)	32.	(c)	33.	(d)	34.	(a)	35.	(b)	36.	(c)
37.	(d)	38.	(a)	39.	(b)	40.	(a)	41.	(a)	42.	(c)
43.	(d)	44.	(a)	45.	(b)	46.	(a)	47.	(b)	48.	(a)
49.	(b)	50.	(a)	51.	(a)	52.	(a)	53.	(c)	54.	(c)
55.	(a)	56.	(b)	57.	(d)	58.	(d)	59.	(a)	60.	(a)
61.	(b)	62.	(b)	63.	(d)	64.	(d)	65.	(a)	66.	(a)
67.	(c)	68.	(c)	69.	(a)	70.	(c)	71.	(d)	72.	(b)
73.	(c)	74.	(a)	75.	(a)	76.	(a)	77.	(b)	78.	(c)
79.	(c)	80.	(a)	81.	(a)	82.	(d)	83.	(a)	84.	(c)
85.	(a)	86.	(b)	87.	(c)	88.	(a)	89.	(d)	90.	(d)
91.	(c)	92.	(c)	93.	(a)	94.	(b)	95.	(d)	96.	(c)
97.	(a)	98.	(b)	99.	(c)	100.	(d)	101.	(c)	102.	(a)
103.	(b)	104.	(a)	105.	(b)	106.	(c)	107.	(b)	108.	(d)
109.	(c)	110.	(d)	111.	(d)	112.	(a)	113.	(b)	114.	(a)
115.	(d)	116.	(b)	117.	(d)	118.	(c)	119.	(b)	120.	(a)
121.	(d)	122.	(a)	123.	(d)	124.	(c)	125.	(d)	126.	(a)
127.	(b)	128.	(a)	129.	(d)	130.	(c)	131.	(d)	132.	(a)

Q. A.	Q. A.	Q. A.	Q. A.	Q. A.	Q. A.
133. (a)	134. (b)	135. (c)	136. (d)	137. (a)	138. (c)
139. (a)	140. (d)	141. (a)	142. (d)	143. (c)	144. (a)
145. (d)	146. (c)	147. (a)	148. (c)	149. (b)	150. (c)
151. (a)	152. (a)	153. (a)	154. (b)	155. (b)	156. (d)
157. (c)	158. (b)	159. (b)	160. (c)	161. (d)	162. (c)
163. (b)	164. (d)	165. (d)	166. (a)	167. (b)	168. (b)
169. (d)	170. (a)	171. (d)	172. (a)	173. (a)	174. (b)
175. (b)	176. (b)	177. (a)	178. (a)	179. (b)	180. (b)
181. (b)	182. (a)	183. (d)	184. (a)	185. (a)	186. (a)
187. (a)	188. (a)	189. (c)	190. (b)	191. (a)	192. (c)
193. (b)	194. (a)	195. (b)	196. (b)	197. (d)	198. (c)
199. (a)	200. (a)	201. (b)	202. (a)	203. (b)	204. (a)
205. (d)	206. (d)	207. (c)	208. (d)	209. (a)	210. (b)
211. (d)	212. (d)	213. (b)	214. (d)	215. (d)	216. (b)
217. (a)	218. (d)	219. (d)	220. (b)	221. (a)	222. (c)
223. (b)	224. (b)	225. (b)	226. (d)	227. (d)	228. (d)
229. (a)	230. (a)	231. (a)	232. (b)	233. (c)	234. (b)
235. (d)	236. (a)	237. (c)	238. (b)	239. (d)	240. (b)
241. (a)	242. (a)	243. (d)	244. (b)	245. (c)	246. (b)
247. (d)	248. (b)	249. (c)	250. (a)	251. (a)	252. (d)
253. (b)	254. (a)	255. (a)	256. (b)	257. (a)	258. (b)
259. (b)	260. (c)	261. (d)	262. (b)	263. (b)	264. (a)
265. (a)	266. (a)	267. (c)	268. (d)	269. (a)	270. (b)
271. (a)	272. (c)	273. (c)	274. (c)	275. (a)	276. (b)
277. (b)	278. (b)	279. (a)	280. (a)	281. (d)	282. (a)
283. (d)	284. (d)	285. (a)	286. (a)	287. (a)	288. (a)
289. (a)	290. (a)	291. (c)	292. (a)	293. (a)	294. (a)
295. (b)	296. (a)	297. (a)	298. (c)	299. (a)	300. (d)
301. (c)	302. (d)	303. (d)	304. (b)	305. (b)	306. (a)
307. (c)	308. (a)	309. (a)	310. (c)	311. (c)	312. (a)
313. (d)	314. (b)	315. (b)	316. (d)	317. (c)	318. (a)
319. (b)	320. (b)	321. (b)	322. (c)	323. (d)	324. (a)
325. (b)	326. (b)	327. (c)	328. (a)	329. (b)	330. (d)
331. (a)	332. (c)	333. (b)	334. (a)	335. (b)	336. (c)

Q. A.	Q. A.	Q. A.	Q. A.	Q. A.	Q. A.
337. (d)	338. (d)	339. (a)	340. (b)	341. (c)	342. (a)
343. (a)	344. (c)	345. (c)	346. (c)	347. (c)	348. (b)
349. (a)	350. (c)	351. (d)	352. (a)	353. (a)	354. (d)
355. (b)	356. (c)	357. (a)	358. (a)	359. (b)	360. (c)
361. (a)	362. (b)	363. (b)	364. (d)	365. (b)	366. (d)
367. (c)	368. (b)	369. (b)	370. (a)	371. (a)	372. (b)
373. (a)	374. (a)	375. (a)	376. (b)	377. (b)	378. (b)
379. (a)	380. (a)	381. (c)	382. (d)	383. (d)	384. (b)
385. (d)	386. (c)	387. (c)	388. (a)	389. (b)	390. (a)
391. (a)	392. (c)	393. (a)	394. (a)	395. (a)	396. (d)
397. (a)	398. (c)	399. (b)	400. (d)	401. (a)	402. (a)
403. (a)	404. (b)	405. (c)	406. (b)	407. (a)	408. (d)
409. (b)	410. (d)	411. (a)	412. (a)	413. (a)	414. (c)
415. (d)	416. (c)	417. (b)	418. (c)	419. (d)	420. (d)
421. (c)	422. (c)	423. (a)	424. (a)	425. (b)	426. (a)
427. (a)	428. (b)	429. (a)	430. (c)	431. (b)	432. (b)
433. (a)	434. (b)	435. (d)	436. (c)	437. (d)	438. (c)
439. (b)	440. (a)	441. (a)	442. (c)	443. (b)	444. (d)
445. (a)	446. (a)	447. (a)	448. (a)	449. (a)	450. (a)
451. (a)	452. (b)	453. (d)	454. (a)	455. (d)	456. (c)
457. (b)	458. (c)	459. (d)	460. (a)	461. (b)	462. (c)
463. (d)	464. (d)	465. (c)	466. (b)	467. (b)	468. (a)
469. (b)	470. (a)	471. (b)	472. (c)	473. (d)	474. (a)
475. (a)	476. (a)	477. (b)	478. (c)	479. (c)	480. (a)
481. (b)	482. (c)	483. (b)	484. (b)	485. (d)	486. (a)
487. (c)	488. (c)	489. (c)	490. (d)	491. (b)	492. (b)
493. (b)	494. (d)	495. (a)	496. (a)	497. (a)	498. (a)
499. (c)	500. (c)	501. (b)	502. (b)	503. (a)	504. (a)
505. (b)	506. (d)	507. (a)	508. (a)	509. (a)	510. (d)
511. (b)	512. (d)	513. (b)	514. (a)	515. (c)	516. (a)
517. (a)	518. (a)	519. (c)	520. (b)	521. (a)	522. (b)
523. (b)	524. (a)	525. (c)	526. (a)	527. (b)	528. (a)
529. (b)	530. (a)	531. (a)	532. (b)	533. (c)	534. (c)
535. (a)	536. (a)	537. (b)	538. (d)	539. (c)	540. (d)

Q. A.	Q. A.	Q. A.	Q. A.	Q. A.	Q. A.
541. (a)	542. (b)	543. (a)	544. (c)	545. (c)	546. (a)
547. (b)	548. (c)	549. (a)	550. (a)	551. (b)	552. (a)
553. (b)	554. (d)	555. (a)	556. (c)	557. (a)	558. (b)
559. (b)	560. (a)	561. (b)	562. (a)	563. (a)	564. (b)
565. (a)	566. (a)	567. (b)	568. (b)	569. (b)	570. (b)
571. (a)	572. (d)	573. (a)	574. (b)	575. (b)	576. (c)
577. (a)	578. (a)	579. (b)	580. (d)	581. (b)	582. (c)
583. (a)	584. (a)	585. (b)	586. (c)	587. (d)	588. (c)
589. (d)	590. (a)	591. (d)	592. (b)	593. (a)	594. (a)
595. (a)	596. (a)	597. (a)	598. (b)	599. (c)	600. (d)
601. (d)	602. (a)	603. (a)	604. (a)	605. (b)	606. (a)
607. (a)	608. (c)	609. (a)	610. (a)	611. (a)	612. (a)
613. (a)	614. (a)	615. (b)	616. (b)	617. (a)	618. (b)
619. (b)	620. (c)	621. (a)	622. (b)	623. (c)	624. (b)
625. (a)	626. (a)	627. (c)	628. (c)	629. (a)	630. (a)
631. (b)	632. (d)	633. (a)	634. (a)	635. (a)	636. (a)
637. (c)	638. (a)	639. (d)	640. (b)	641. (a)	642. (b)
643. (b)	644. (b)	645. (b)	646. (b)	647. (c)	648. (c)
649. (c)	650. (d)	651. (d)	652. (d)	653. (d)	654. (d)
655. (a)	656. (a)	657. (a)	658. (d)	659. (c)	660. (b)
661. (d)	662. (a)	663. (b)	664. (d)	665. (d)	666. (d)
667. (a)	668. (a)	669. (a)	670. (a)	671. (d)	672. (d)
673. (a)	674. (a)	675. (a)	676. (c)	677. (a)	678. (b)
679. (d)	680. (b)	681. (a)	682. (d)	683. (b)	684. (c)
685. (b)	686. (a)	687. (c)	688. (a)	689. (c)	690. (a)
691. (a)	692. (d)	693. (a)	694. (b)	695. (a)	696. (b)
697. (a)	698. (a)	699. (a)	700. (a)	701. (c)	702. (a)
703. (b)	704. (a)	705. (b)	706. (b)	707. (c)	708. (a)
709. (b)	710. (a)	711. (b)	712. (a)	713. (b)	714. (d)
715. (b)	716. (a)	717. (a)	718. (d)	719. (c)	720. (d)
721. (d)	722. (d)	723. (d)	724. (d)	725. (a)	726. (c)
727. (a)	728. (a)	729. (b)	730. (a)	731. (c)	732. (b)
733. (a)	734. (a)	735. (a)	736. (a)	737. (b)	738. (c)
739. (b)	740. (b)	741. (a)	742. (b)	743. (c)	744. (a)

Q. A.	Q. A.	Q. A.	Q. A.	Q. A.	Q. A.
745. (b)	746. (b)	747. (c)	748. (b)	749. (a)	750. (b)
751. (a)	752. (b)	753. (a)	754. (b)	755. (a)	756. (d)
757. (c)	758. (a)	759. (a)	760. (b)	761. (b)	762. (a)
763. (b)	764. (a)	765. (b)	766. (d)	767. (a)	768. (a)
769. (a)	770. (c)	771. (b)	772. (b)	773. (c)	774. (c)
775. (c)	776. (c)	777. (b)	778. (a)	779. (a)	780. (b)
781. (b)	782. (c)	783. (c)	784. (a)	785. (d)	786. (a)
787. (a)	788. (a)	789. (b)	790. (c)	791. (b)	792. (a)
793. (d)	794. (c)	795. (a)	796. (b)	797. (a)	798. (d)
799. (c)	800. (c)	801. (a)	802. (a)	803. (c)	804. (a)
805. (b)	806. (b)	807. (b)	808. (a)	809. (a)	810. (c)
811. (a)	812. (b)	813. (b)	814. (c)	815. (a)	816. (c)
817. (b)	818. (d)	819. (d)	820. (b)	821. (a)	822. (a)
823. (a)	824. (b)	825. (d)	826. (c)	827. (a)	828. (a)
829. (c)	830. (a)	831. (a)	832. (d)	833. (a)	834. (c)
835. (a)	836. (b)	837. (b)	838. (a)	839. (b)	840. (a)
841. (a)	842. (b)	843. (b)	844. (c)	845. (b)	846. (a)
847. (b)	848. (a)	849. (a)	850. (b)	851. (c)	852. (d)
853. (a)	854. (b)	855. (b)	856. (c)	857. (d)	858. (c)
859. (b)	860. (a)	861. (b)	862. (a)	863. (b)	864. (c)
865. (d)	866. (b)	867. (a)	868. (c)	869. (d)	870. (d)
871. (b)	872. (d)	873. (a)	874. (a)	875. (a)	876. (c)
877. (a)	878. (b)	879. (c)	880. (d)	881. (d)	882. (a)
883. (a)	884. (b)	885. (a)	886. (b)	887. (a)	888. (a)
889. (c)	890. (a)	891. (c)	892. (d)	893. (c)	894. (b)
895. (a)	896. (d)	897. (a)	898. (a)	899. (d)	900. (b)

CHAPTER 7

MEAT FISH AND POULTRY TECHNOLOGY

7.1 SUBJECTIVE KEYNOTES

1. In India goat, sheep, pig, buffalo and cattle meat constitutes majority of the total meat production.
2. The skeletal muscle is the principal muscle tissue in meat, although very little of smooth tissue is also present.
3. The main connective tissue types are adipose tissue (fat), bone and connective tissue proper.
4. Smooth and cardiac muscles are involuntary in nature.
5. Skeletal and cardiac muscles are sometimes referred as striated muscles due to their specific microscopic appearance.
6. Each muscle is surrounded by a sheath of connective tissue known as epimysium.
7. Skeletal muscle fibres are long, narrow, almost tubular multinucleated cells which may extend from one end to the other.
8. The individual fibre may also be classified as red, intermediate and white.
9. Red muscle fibres have smaller diameter, lower glycolytic metabolism and ATPase activity but higher oxidative metabolism as compared to white muscle fibres.
10. Sarcomere is the functional unit of myofibril.
11. The myofibril itself is seen to be composed of a number of thick (myosin) and thin (actin) filaments.
12. The tendons attach muscle with bone whereas ligaments connect two bones or support organs.
13. Muscle tissue contains approximately 75% water and 25% solids, of which 19% are proteins. Lipids constitute about 2.5 to 5% of muscle.
14. Myofibrillar proteins constitute contractile part of the muscle and make up about 60% of the total protein in the skeletal muscle.
15. Myofibrillar proteins contribute to approximately 95% of the water holding capacity, 75% of the emulsifying capacity and to a large extent the tenderness of meat.
16. Collagen is the main fibrous protein the muscles and significantly influences the meat toughness.

17. Collagen fibres shrink or shorten at a temperature of 600°C but higher temperatures or boiling cause transformation to water soluble gelatin.
18. The smallest unit of collagen molecular structure is tropo collagen.
19. Lipid is a major component of the carcass of a meat animal and is inversely proportional to the moisture content.
20. Immediately after slaughter, muscle normally contains a very small amount (nearly 1%) of glycogen.
21. The rate and amount of glycolysis influence the color, tenderness and water holding capacity of meat.
22. Most of the total body inorganic material is located in skeletal tissue primarily as salts of calcium and phosphorus and some other minerals especially magnesium.
23. The inorganic elements play an important role in the conversion of muscle to meat and cause the development of rigor mortis and alteration of fluid balance which cause a drop in pH and water holding capacity.
24. Water-soluble vitamins are localized in lean tissues whereas fat-soluble vitamins in fatty tissues.
25. Pork contains 5-10 times more thiamine content as compared to mutton.
26. Stress and fatigue lower the quality of meat in several ruminant species due to depletion of glycogen in the muscle
27. Due to low acid production, the ultimate pH of the muscle remains high causing a condition called dark cutting meat or dark, firm and dry (DFD) meat.
28. In pigs, low ultimate pH is achieved within 45 minutes due to rapid glycolysis and results in pale, soft and exudative (FSE) pork that has higher drip and cooking losses.
29. Stunning is a process employed to create s state of immobility or unconsciousness at the time of slaughter to 1 minute.
30. Meat cutting refers to the skill of separation of carcass into wholesale primal cuts in order to facilitate requirements of meat trade, cater to the consumer preference and convenient handling by the butchers.
31. Postmortem inspection involves visual perception and palpation of organs and tissues, incisions where necessary and laboratory tests.
32. Primal and sub-primal cuts of fresh meat are overwrapped with thermoplastic films having excellent optical properties.
33. Post slaughter, the accumulation of lactic acid due to anaerobic glycolysis lowers down the muscle pH which is an important postmortem change during the conversion of muscle to meat.
34. There is decline in postmortem pH of the meat from 7 to ultimately 5.5-5.7.
35. Rigor mortis refers to stiffening of muscles after death and is another important postmortem change in the process of conversion of muscle to meat.
36. The onset of rigor mortis is enhanced at ambient temperature above 200°C.
37. Pre-rigor meat is quite tender but its toughness keeps on increasing until rigor mortis is completed.
38. The onset of rigor mortis is also accompanied by a decrease in water holding capacity.

39. A rapid decline in muscle pH also causes denaturation of collagenous connective tissue.
40. The principal meat pigment is myoglobin.
41. Myoglobin content of more active species and muscles is higher than the passive ones.
42. In intact meat, iron in the heme ring of myoglobin exists in the reduced form.
43. Upon cutting, grinding or exposure to air, myoglobin is oxygenated to form oxymyoglobin which has a bright red color.
44. In conditions of less oxygen, partial vacuum or semi permeable package, myoglobin as well as oxymyoglobin is oxidized to brown colored metmyoglobin.
45. The capacity of meat to retain its water during the application of physical forces in known as water holding capacity (WHC).
46. Marbling refers to the intramuscular fat which can be visibly detected when the muscle surface is cut.
47. Fresh meat with a good water holding capacity is less prone to shrinkage during storage.
48. When prerigor meat is fast chilled below 150°C, there is shortening of muscles. This cold shortening or contraction is more at 0.0°C and still more at -2.0°C.
49. Cold shortening can be avoided by keeping the meat above 140°C for sufficient time to pass the rigor stage.
50. When prerigor meat is frozen, a severe type of rigor mortis ensues during thawing called thaw rigor, where the shortening so produced may be 60 to 80% of the original length of the unrestrained muscle.
51. Meat and poultry production in India has been about 4.6 Mt per year with goats and sheep contributing 54%, buffalo and cattle 26%, poultry 13% and pig 7%..
52. India produces 3.6% of global egg production, i.e., 61 million tonnes. The annual growth rate of egg is 5 to 8%.
53. India has the world's largest cattle population (1/2 of the total buffalo population and 1/6th of the total goat population of the world).
54. India's fish production is only 1/10th that of China (the world leader). India is second largest fish producer.
55. Indian seafood industry is the 3rd largest seafood market in the world.
56. India ranks 2nd in aquaculture production but currently only 6% of the coastline is utilized for this purpose.
57. Shechita is Jewish laws of slaughtering animals.
58. Dhabihah is Islamic law of slaughtering animals.
59. Homeostasis is the process of adaption of body's internal environment with the external environment to cope up the extreme variation conditions like change in temperature, daylight, oxygen deficiency, energy supply and other conditions that stimulates the internal body changes.
60. A sharp decline in postmortem pH even before the dissipation of body heat through carcass chilling may cause denaturation of muscle proteins.

61. High pH meat is darker but less consistently tender than normal pH meat.
62. The pH of living animals is around pH 7, but after death the sugars in the muscles are converted to lactic acid, lowering the pH. A normal, non - stressed animal has muscle pH of around pH 5.5 after death.
63. There are a number of vegetable enzymes such as papain (papaya), bromelin (pineapple), and ficin (fig) used to tenderize meat.
64. Grinding is a very popular means of increasing tenderness of meat, especially beef.
65. Colour of meat is dependent on the oxidation-reduction potential of the environment, maturity, and chemical state of haemoglobin and myoglobin in meat.
66. Cured meats such as ham and bacon should be frozen for a short period of time because the salt in them hastens rancidity.
67. Freshly slaughtered meat carcasses or primal cuts need to be cooled to below 40°F within 24 hours to prevent souring or spoiling.
68. Meat can be cut into roasts, rolled roasts, steaks, chops, stew meat, ground meat, etc.
69. 'Fat fish' includes varieties such as mullet, mackerel, trout, tuna and salmon.
70. 'Lean fish' includes flounder, cod, whiting, redfish, croakers, snapper, grouper, sheep head and most freshwater fish.
71. Fat fish should be dipped in an ascorbic acid solution (20 sec) to control rancidity and flavour change.
72. Lean fish should be dipped in a brine solution (20 sec) to firm the fish and to decrease drip loss from thawing.
73. Red colour of meat is more stable at lower temperatures because the rate of oxidation of the pigment decreases.
74. Sodium nitrite has the ability to react with the red meat pigment to form the heat stable red curing colour.
75. Dry curing of meat involves rubbing of curing mix on meat surface. Curing mix is prepared with nitrite curing salt, together with spices and other additives.
76. Wet curing (pickling) involves the application of curing brine to the meat.
77. An alternative and quick way of wet curing is to accelerate the diffusion of the curing substances by pumping brine into the meat tissue. It is also known as injection curing.
78. The most known undesirable effect of smoking of meat is the risk of residues of benzopyrene in smoked products which can be carcinogenic.
79. The optimal temperature in "cold" smoking of meat is 15 to 18°C.
80. Hot smoking of meat is carried out at temperatures of 60 to 80°C.
81. Cold smoking is used for fermented meat products and precooked sausage. Hot smoking is used for a range of raw-cooked sausages, bacon and cooked ham products.
82. Dry heat cooking of meat involves roasting, broiling or sauteing.
83. Moist heat cooking of meat involves braising, steaming or poaching.
84. Dried meat products are the result of the simple dehydration or drying of lean meat in natural conditions or in an artificially created environment.

85. Dried meats can be classified as dry and semi-dry sausages, dried (i.e., dry cured) whole muscle meats, and whole muscle meat snacks.
86. Shelf-stable semi-dry sausages must have a moisture protein ratio of 3.1:1 or less and a pH of 5.0 or less.
87. Dried (mostly dry cured) whole muscle products include prosciutto, parma and country hams as well as pancetta, cappocola, coppa, bresaola, etc.
88. Electrical stunning is achieved by passing a current through the bird that causes immediate loss of consciousness and sensibility to pain.
89. Concussion stunning may be achieved by a blow to the head using a captive-bolt instrument or a weighted stick.
90. Scalding is the process of submerging the birds/carcass in hot water to loosen the feather so that they can be easily separated further. The birds can either be dipped in hot water or sprayed with steam.
91. Semi-scald or slack scald is the name given to scalding for 30-60 s in 125-130 °F water. Sub-scald is the use of water at 138-140 °F for 30-75 seconds. Hard-scald or full scald requires a water temperature of 140-150 °F.
92. Picking is the process of removal of feathers from the poultry.
93. Evisceration is the process for removal of edible and inedible viscera from the carcass.
94. Aging or maturing of meat is the process of holding meat for a duration of time between death and deboning and is done because the muscle need to develop rigor mortis before deboning in order to prevent toughening.
95. Rigor mortis is largely complete in 4 hours in chickens.
96. The ropy, twisted strands of albumen that anchor the yolk to the centre of the thick egg white are known as chalaza.
97. The membrane surrounding the egg yolk and attached to the chalazae is known as vitelline membrane.
98. Fish muscle is organized as tissue blocks of myomeres attached to myocommata. This structure makes the flesh inherently soft during cooking.
99. The amount of total volatile base (TVB) is measured by distilling a fish extract, and determining the base concentration by titration against acid.
100. Fish protein isolate (FPI) is fish protein which has been purified to a protein content of at least 90% of the dry material.

7.2 MULTIPLE CHOICE QUESTIONS

1. The muscles of warm blooded animals like cattle, sheep & pigs is called
(a) meat (b) beef
(c) ham (d) none

2. Meat is a ________ source of nutrients required by humans
(a) poor (b) rich
(c) no nutrients (d) none

3. Meat is rich in protein having ____%
(a) 1-10 (b) 15-20
(c) 30-40 (d) 40-50

4. Meat has __________ amono acids
(a) limited
(b) all
(c) all essential
(d) none

5. Liver is a rich source of vitamin ______
(a) A (b) C
(c) D (d) K

6. The surface of the muscle is called ______
(a) sarcolemma (b) sarcoplasm
(c) myofibrils (d) none

7. The semifluid material within the muscle cell is called
(a) sarcolemma (b) sarcoplasm
(c) myofibrils (d) none

8. Each muscle fiber is composed of many small
(a) sarcolemma (b) sarcoplasm
(c) myofibrils (d) none

9. The dark bands of muscle is known as ______ bands
(a) isotropic (b) anisotropic
(c) Z line (d) H zone

10. The light bands of muscle is known as ______ bands
(a) isotropic (b) anisotropic
(c) Z line (d) H zone

11. The dark line in the center of each of the I band is called____line
(a) Z (b) Y
(c) H (d) D

12. The centre of A band is called ______ zone
(a) H (b) B
(c) Z (d) D

13. M line occurs in the centre of
(a) Z line (b) H zone
(c) B zone (d) H line

14. The contractile unit of muscle is called
(a) sarcolemma (b) sarcoplasm
(c) myofibrils (d) sarcomere

15. Sarcomere is the material located between two
(a) H zone (b) Z line
(c) M line (d) none

16. A-band consists of ______ band
(a) thick (b) thin
(c) both (d) none

17. I-band consists of ______ band
(a) thick (b) thin
(c) both (d) none

18. During contraction, the length of ______ remains the same
(a) A-band (b) I-band
(c) H zone (d) all

19. During contraction, the length of ______ shorten
(a) A-band (b) I-band
(c) H zone (d) both (b) & (c)

20. Protein of muscle include ______ proteins
(a) contractile (b) soluble
(c) insoluble (d) all

21. The contractile proteins are soluble in
(a) water (b) salt solution
(c) dilute acid (d) none

22. Enzymes are ______ soluble
(a) water (b) salt
(c) insoluble (d) both (a) & (b)

23. ______ forms the major protein of thick filaments
(a) myosin (b) actin
(c) actomyosin (d) tropomyosin

24. 60% of myofibrils are composed of
(a) myosin (b) actin
(c) actomyosin (d) tropomyosin

25. The major protein of thin filaments is
(a) myosin (b) actin
(c) actomyosin (d) tropomyosin

26. 15-30% of myofibrils is composed of
(a) myosin (b) actin
(c) actomyosin (d) tropomyosin

27. ______ consists of two fractions F and G
(a) myosin (b) actin
(c) actomyosin (d) tropomyosin

28. F in F-actin stands for
(a) fibrous (b) fibrillar
(c) first (d) none

29. G in G-actin stands for
(a) green (b) gel
(c) globular (d) none

30. ________ form bind with ATP in presence of magnesium to form F form
(a) G (b) N
(c) R (d) M

31. ATP stands for
(a) Adenosine Tri Phosphate
(b) Adenaline Tri Phosphate
(c) Active Tri Phosphate
(d) none

32. Actomyosin is formed of
(a) myosin (b) actin
(c) both (d) tropomyosin

33. Myosin requires ______ for its activity
(a) Mg (b) Ca
(c) P (d) Na

34. Myosin is inhibited by
(a) Mg (b) Ca
(c) P (d) Na

35. Tropomyosin B is a contractile protein found in ______ filaments
(a) thick (b) thin
(c) both (d) none

36. Which of the following is not a form of tropomyosin
(a) A (b) B
(c) C (d) none

37. ________ inhibits the contraction activity
(a) tropnin C (b) troponin I
(c) both (d) none

38. When calcium ions are removed from the sarcoplsm, ________ occurs
(a) contraction (b) relaxation
(c) muscle function stops
(d) none

39. Which of the following is fibrous protein
(a) collagen (b) elastin
(c) both (a) & (b) (d) myosin

40. Gelatin is obtained from
(a) collagen (b) elastin
(c) actin (d) myosin

41. Which is tougher
(a) collagen (b) elastin
(c) both are equal (d) none

42. The structural unit of collagen is
(a) elastin (b) actin
(c) troponin (d) trocollagen

43. The quantity of collagen ________ upon ageing of animal
(a) increase (b) decrease
(c) no change (d) none

44. There is difficulty in chewing of meat of older animals than that of younger anmals because of
(a) higher collagen
(b) lower collagen
(c) higher elastin
(d) lower elastin

45. ________ is the constituent of tendons
(a) elastin (b) collagen
(c) actin (d) myosin

46. ______ is the constituent of ligaments
(a) elastin (b) collagen
(c) actin (d) myosin

47. Elastin
(a) denature on heating
(b) forms gelatin
(c) is heat resistant
(d) none

48. Lean meat contains ______% proteins
(a) 10 (b) 20
(c) 40 (d) 60

49. The fat content of meat depends on
(a) breed (b) age
(c) type (d) all

50. The fat content of meat varies from _____%
(a) 1-5 (b) 5-40
(c) 40-50 (d) 50-70

51. Meat fats are rich in ____ fatty acids
(a) saturated (b) unsaturated
(c) PUFA (d) none

52. The cholesterol content of meat is _____ mg/100g
(a) 2 (b) 25
(c) 50 (d) 75

53. Which of the following is found in small quantity in meat
(a) carbohydrates (b) protein
(c) fat (d) none

54. Which of the following is found in small quantity in meat
(a) glucose (b) glycogen
(c) fructose (d) both (a) & (b)

55. Meat is a good source of
(a) iron (b) phosphorus
(c) vitamin B (d) all

56. Most of the __________ are present in lean portion of meat
(a) vitamin (b) minerals
(c) both (d) none

57. Color of the meat is due to
(a) haemoglobin
(b) metmyoglobin
(c) myoglobin (d) none

58. The energy required for contraction is provided by
(a) ATP (b) calcium
(c) magnesium (d) none

59. In glycolysis, glycogen is finally converted to
(a) pyruvate (b) citric acid
(c) malic acid (d) glucose

60. Beef is meat of ________
(a) cattle (b) buffalo
(c) pig (d) fish

61. Beef has ______ g/100g moisture
(a) 24.3 (b) 44.3
(c) 74.3 (d) 94.3

62. Beef has ______ g/100g protein
(a) 22.6 (b) 36.6
(c) 46.6 (d) 56.6

63. Beef has ______ g/100g minerals
(a) 0.1 (b) 1
(c) 10 (d) 20

64. Beef has ______ g/100g fat
(a) 1.6 (b) 2.6
(c) 15.6 (d) 22.6

65. Mutton has ______ g/100g moisture
(a) 31.5 (b) 61.5
(c) 71.7 (d) 81.5

66. Mutton has ______ g/100g protein
(a) 18.5 (b) 13.3
(c) 1.3 (d) 77.4

67. Mutton has ______ g/100g fat
(a) 18.5 (b) 13.3
(c) 1.3 (d) 77.4

68. Mutton has ______ g/100g minerals
(a) 18.5 (b) 13.3
(c) 1.3 (d) 77.4

69. Pork has ______ g/100g moisture
(a) 18.7 (b) 4.4
(c) 1.0 (d) 77.4

70. Pork has ______ g/100g protein
(a) 18.7 (b) 4.4
(c) 1.0 (d) 77.4

71. Pork has ______ g/100g fat
(a) 18.7 (b) 4.4
(c) 1.0 (d) 77.4

72. Pork has ______ g/100g minerals
(a) 18.7 (b) 4.4
(c) 1.0 (d) 77.4

73. Which of the following has highest protein
(a) beef (b) mutton
(c) pork (d) none

74. Which of the following has highest fat
(a) beef (b) mutton
(c) pork (d) none

75. Which of the following has highest minerals
(a) beef (b) mutton
(c) pork (d) none

76. Lactic acid is produced as the end product of glycolysis n
(a) presence of O_2 (b) absence of O_2
(c) always (d) none

77. The rigor mortis is also called ________ of muscle
(a) stiffness (b) relaxation
(c) swelling (d) none

78. Rigor mortis occurs
(a) after death
(b) before death
(c) in presence of oxygen
(d) none

79. Rigor mortis occurs due to
(a) accumulation of lactic acid
(b) decrease in pH of muscle
(c) absence of ATP
(d) all

80. The water binding capacity of muscle at rigor mortis
(a) increase (b) decrease
(c) remains same (d) none

81. Tenderization of muscleprotein is due to enzyme
(a) cathepsins (b) peroxidase
(c) lipase (d) amylase

82. Ageing of meat is done at _______°C
(a) 4 (b) 0.5 to -2.0
(c) 1-2 (d) 4-7

83. The best flavor and tenderness is developed in meat aged for
(a) 2-4 weeks (b) 2-4 months
(c) 2-4 years (d) 2-4 days

84. Which of the following is aged commonly
(a) beef (b) mutton
(c) pork (d) all

85. Which of the following is used for tenderizing meat
(a) pounding (b) cutting
(c) pinning (d) all

86. Tenderization can be brought about by placing in ________ leaves
(a) banana (b) papaya
(c) mango (d) coconut

87. Bromelin is obtained from
(a) papaya (b) fig
(c) guava (d) pineapple

88. The enzyme present in papaya responsible for tenderizing of meat is
(a) papain (b) bromelin
(c) cathepsin (d) ficin

89. Ficin is obtained from
(a) papaya (b) fig
(c) guava (d) pineapple

90. Salts __________ the water holding capacity of muscle fibres
(a) increase (b) decrease
(c) no effect (d) none

91. Salts added to muscle fibres ________ tenderness
(a) increase (b) decrease
(c) no effect (d) none

92. Salts added to muscle fibres ________ juiciness
(a) increase (b) decrease
(c) no effect (d) none

93. Which of the following salt is used for tenderization
(a) sodium chloride
(b) sodium bicarbonate
(c) potassium phosphate
(d) all

94. The tenderness of meat in increased upon __________ in pH
(a) increase (b) decrease
(c) no effect (d) none

95. Soaking beef in concentrated vinegar will __________ its tenderness and juiciness
(a) increase (b) decrease
(c) no effect (d) none

96. Curing of meat is done to
(a) increase protein
(b) increase fat
(c) increase tenderness
(d) increase minerals

97. The curing agents include
(a) sodium chloride
(b) sodium nitrite
(c) sugar and spices
(d) all

98. ________ fix the red color of myoglobin
(a) sodium chloride
(b) sodium nitrite
(c) suagr and spices
(d) all

99. The common process of curing is __________ of salts on animal meat
(a) rubbing (b) immersion
(c) injection (d) all

100. Smoking of meat is a method of a
(a) preservation
(b) flavor addition
(c) coagulation of proteins
(d) all

101. Smoking prevents
(a) mo's (b) rancidity
(c) both (d) none

102. Carcass means
(a) whole animal
(b) animal without organs
(c) animal with organs
(d) none

103. The meat of cattle slaughtered 3-4 weeks after birth is called
(a) veal (b) beef
(c) pork (d) mutton

104. The meat of cattle slaughtered over 1 year is called
(a) veal (b) beef
(c) pork (d) mutton

105. A bovine male animal castrated at a very young age is called
(a) veal (b) beef
(c) steer (d) mutton

106. A female bovine animal that has not borne a calf is called
(a) veal (b) heifer
(c) steer (d) mutton

107. A male bovine animal that is castrated after maturing is called
(a) veal (b) heifer
(c) steer (d) stag

108. A male or female bovine animal upto 12 months of age, generally from 3-8 months of age is called
(a) veal (b) heifer
(c) calf (d) stag

109. The flesh of young bovine animal (sheep) of both sexes whose age is 12 months or under is called
(a) lamb (b) heifer
(c) steer (d) stag

110. Carcasses of young sheep usually from 12-20 months is called
(a) yearling mutton (b) heifer
(c) steer (d) stag

111. The flesh of both the males & females of ovine species that are 20 months in age at time of slaughter is called
(a) mature mutton
(b) yearling mutton
(c) steer
(d) stag

112. The meat of swine is called
(a) mutton (b) beef
(c) veal (d) pork

113. Good quality pork is obtained from animal between ________ months
(a) 3-12 (b) 12-24
(c) 24-36 (d) 36-48

114. The cut from the belly portion of hog carcass is called
(a) mutton (b) bacon
(c) veal (d) ham

115. People eating pork are susceptible to
(a) trichionosis (b) pneumonia
(c) AIDS (d) typhoid

116. Trichionosis is caused by
(a) Wuchherria
(b) Anopheles
(c) Trichinella spiralis
(d) none

117. The first and second stomach of the ruminant forms the
(a) sweet bread (b) heart
(c) tripe (d) none

118. ______ are made from ground or minced meat
(a) sausage (b) carcass
(c) stag (d) mutton

119. The amount and distribution of fat in meat is called
(a) demarbling (b) marbling
(c) both (d) none

120. Over cooking of meat can cause
(a) contraction of muscle fibres
(b) shrinkage
(c) toughness
(d) all

121. Braising, stewing and pressure cooking are ______ methods of cooking meat
(a) dry heat (b) moist heat
(c) frying (d) none

122. Roasting, broiling, panboiling and frying are ____________ methods of cooking meat
(a) dry heat (b) moist heat
(c) frying (d) none

123. Which of the following is better for cooking of meat
(a) low temperature/ long time
(b) high temperature/ short time
(c) both (d) none

124. The recommended temperature in the centre of meat cut for welldone stages for lamb is ______°C
(a) 42 (b) 52
(c) 72 (d) 82

125. The recommended temperature in the centre of meat cut for welldone stages for beef is ______°C
(a) 42 (b) 52
(c) 77 (d) 92

126. The recommended temperature in the centre of meat cut for welldone stages for pork is ______°C
(a) 42 (b) 52
(c) 77 (d) 92

127. Which of the following involve direct heating of meat in open fire of gas flame
(a) roasting (b) broiling
(c) frying (d) all

128. Broiling is carried out at atemerature of ______°C
(a) 176 (b) 220
(c) 196 (d) 256

129. Which of the following produces a more juicy and tender meat
(a) roasting (b) broiling
(c) frying (d) all

130. At higher temperature, the meat fat decompose to
(a) fatty acids (b) acrolein
(c) trans fats (d) none

131. Heating _____ the water holding capacity of meat
(a) increase (b) decrease
(c) no effect (d) none

132. Collagen is _______ helix structure
(a) double (b) triple
(c) single (d) none

133. Gelatin is _______ degraded collagen
(a) non (b) completely
(c) partially (d) none

134. Broiling is applied to tender cuts tht are atleast _____ com thick
(a) 1cm (b) 2-5cm
(c) 3.5cm (d) 4.5cm

135. The demineralized bone is called
(a) ossein (b) casein
(c) ligament (d) tendon

136. Gelatin is used in
(a) sponges (b) cream salads
(c) jellies (d) all

137. Gelatin is a protein of _______ biological value
(a) high (b) low
(c) no (d) none

138. Gelatin is _______ in essential amino acids
(a) deficient (b) rich
(c) low (d) none

139. A gel of good texture is obtained with _____ % gelatin
(a) 0.5-1.5 (b) 1.5-2.0
(c) 2.5-3.0 (d) 3.5-4.0

140. High acidity _______ the gel forming ability of gelatin
(a) increase (b) decrease
(c) no effect (d) none

141. _________ sugar increases the setting time of gel
(a) small amount
(b) high amount
(c) no effect of sugar
(d) none

142. _____ should not be used with gelatin
(a) mango (b) banana
(c) pineapple (d) all

143. _________ enzyme present in pineapple hydrolyzes gelatin
(a) bromelin (b) ficin
(c) cathepsin (d) papain

144. Fat marbling makes the muscle
(a) tender (b) tough
(c) non-juicy (d) none

145. Increased toughening of meat refers to
(a) ageing of meat (b) rigor mortis
(c) slaughtering (d) all

146. Veal has _______% protein
(a) 18.8 (b) 14.0
(c) 66.0 (d) 27.7

147. Veal has _______% fat
(a) 18.8 (b) 14.0
(c) 66.0 (d) 27.7

148. Veal has _______% water
(a) 18.8 (b) 14.0
(c) 66.0 (d) 27.7

149. Lamb has _______% protein
(a) 18.8 (b) 15.7
(c) 66.0 (d) 27.7

150. Lamb has _______% fat
(a) 18.8 (b) 14.0
(c) 66.0 (d) 27.7

151. Lamb has _______% moisture
(a) 18.8 (b) 14.0
(c) 56.0 (d) 27.7

152. Horse has _______% protein
(a) 20.0 (b) 14.0
(c) 66.0 (d) 27.7

153. Horse has _______% fat
(a) 18.8 (b) 4.0
(c) 66.0 (d) 27.7

154. Horse has _______% moisture
(a) 18.8 (b) 14.0
(c) 74.0 (d) 27.7

155. Which one has the lowest protein
(a) beef (b) pork
(c) lamb (d) horse

156. Which one has the highest carbohydrate
(a) beef (b) pork
(c) lamb (d) horse

157. Which one has the lowest fat
(a) beef (b) pork
(c) lamb (d) horse

158. Which one has the highest fat
(a) beef (b) pork
(c) lamb (d) horse

159. Which one has the highest protein
(a) beef (b) pork
(c) lamb (d) horse

160. Horse meat has _____% carbohydrate
(a) 1 (b) 10
(c) 50 (d) 80

161. Ageing/ ripening of meat is done at ________°C
(a) 2 (b) 7
(c) 10 (d) 25

162. Bologna and frankfurters are example of
(a) bacon (b) ham
(c) sausage (d) all

163. Oxymyoglobin is responsible for ________ color of meat
(a) red (b) purple
(c) brown (d) green

164. Intact muscle has ________ color
(a) red (b) purple
(c) brown (d) green

165. Upon exposure to air, the muscle color changes to
(a) bright red (b) purple
(c) brown (d) green

166. On prolonged exposure to air, oxymyoglobin changes to
(a) myoglobin
(b) metmyoglobin
(c) hemoglobin
(d) none

167. Metmyoglobin is ________ in color
(a) red (b) purple
(c) brown (d) green

168. Meat cured with nitrites has ______ color
(a) red (b) purple
(c) brown (d) green

169. Nitrite combines with myoglobin to produce nitric oxide myoglobin which is _______ colored
(a) red (b) purple
(c) brown (d) pink

170. On cooking, nitric oxide myoglobin converts to
(a) nitrosohemochrome
(b) metmyoglobin
(c) oxymyoglobin
(d) none

171. During smoking, the internal temperature of meat should reach ______°C
(a) 12 (b) 32
(c) 52 (d) 82

172. Smoking is done for
(a) 18-24 hrs (b) 1-2 hrs
(c) 1-2 days (d) 1-2 months

173. For pork, the internal temperature should reach ________ for destruction of Trichinosis parasite ______°C
(a) 12 (b) 32
(c) 58 (d) 82

174. Metmyoglobin has iron in _____ state
(a) Fe^{2+} (b) Fe^{3+}
(c) Fe^{4+} (d) none

175. Oxymyoglobin has iron in ______ state
(a) Fe^{2+} (b) Fe^{3+}
(c) Fe^{4+} (d) none

176. Irradiation of metmyoglobin in presence of H_2O_2 makes
(a) ferrylmyoglobin
(b) cholemyoglobin
(c) sulfmyoglobin (d) none

177. Ferrylmyoglobin has Fe in ______ state
(a) Fe^{2+} (b) Fe^{3+}
(c) Fe^{4+} (d) none

178. Ferrylmyoglobin is ______ in color
(a) bright red (b) purple
(c) brown (d) green

179. The reduction of myoglobin changes myoglobin to
(a) ferrylmyoglobin
(b) cholemyoglobin
(c) sulfmyoglobin
(d) none

180. Cholemyoglobin is ______ in color
(a) bright red (b) purple
(c) brown (d) green

181. The oxidation of myoglobin in presence of sulfide makes
(a) ferrylmyoglobin
(b) cholemyoglobin
(c) sulfmyoglobin (d) none

182. Sulfmyoglobin is ______ in color
(a) bright red (b) purple
(c) brown (d) green

183. Which of the following are of similar color
(a) ferrylmyoglobin & myoglobin
(b) cholemyoglobin & sulfmyoglobin
(c) sulfmyoglobin & metmyoglobin
(d) none

184. Nitrosomochrome is ______ in color
(a) pink (b) purple
(c) brown (d) green

185. Myoglobin has iron in ______ state
(a) Fe^{2+} (b) Fe^{3+}
(c) Fe^{4+} (d) none

186. Nitric-oxide myoglobin has ______ color
(a) bright pink (b) purple
(c) brown (d) green

187. Oxidation of sulfmyoglobin results in
(a) porphyrins
(b) cholemyoglobin
(c) sulfmyoglobin (d) none

188. Porphyrins are ______ in color
(a) brown (b) yellow
(c) colorless (d) all

189. Metmyoglobin is ______ in color
(a) brown (b) yellow
(c) colorless (d) red

190. Casings are usually associated with
(a) sausage (b) ham
(c) hog (d) bacon

191. Frankfurters are
(a) cooked (b) smoked
(c) both (d) none

192. Salami is ______ type of sausage
(a) fermented (b) smoked
(c) cooked (d) all

193. Natural casings are made from
(a) intestine (b) liver
(c) stomach (d) spleen

194. The casings are used to hold the ______ meat
(a) minced (b) ground
(c) restructured (d) all

195. Frankfurters are made from ______ meat
(a) mutton (b) beef
(c) pork (d) all

196. Trichinosis parasite is present in
(a) mutton (b) beef
(c) pork (d) all

197. ______ % O_2 is recommended for MAP of fresh red meats
(a) 20-50 (b) 2-5
(c) 1-2 (d) 10-15

198. Upon thawing of meat, the liquid that comes out is called
(a) drip (b) blot
(c) drop (d) none

199. The white portions of the muscle refers to
(a) muscle (b) fat
(c) minerals (d) none

200. The dark portions of the muscle refers to
(a) muscle (b) fat
(c) minerals (d) none

201. The meat is most susceptible to microbial spoilage because of
(a) high moisture (b) neutral pH
(c) high nutrients (d) all

202. Keeping away of mo's from meat is called
(a) asepsis (b) sepsis
(c) slaughtering (d) none

203. Removal of organs from the dead animal body is called
(a) asepsis (b) sepsis
(c) evisceration (d) none

204. The commercially sterile meat is known as
(a) shelf stable
(b) non shelf stable
(c) refrigerated
(d) none

205. The meat that are heated enough to kill part of spoilage mo's are called
(a) shelf stable
(b) non shelf stable
(c) refrigerated
(d) both (b) & (c)

206. The processing temperature for shelf stable canned cured meat is ____C
(a) 28 (b) 48
(c) 98 (d) 158

207. The processing temperature for non-shelf stable canned cured meat is ______°C
(a) 28 (b) 65
(c) 98 (d) 158

208. Chilling temperature for preserv-ation of meat is _______°C
(a) 0 (b) -1.4 to 2.2
(c) 3-5 (d) 6-9

209. The preservation of frozen meat is done at temperature __________°C
(a) 0 (b) -1.4 to 2.2
(c) -12.2 to -28.9 (d) 6-9

210. Irradiation dose for meat preserva-tion is _______ kGray
(a) 1-2 (b) 20-70
(c) 0.1-0.5 (d) 70-90

211. The cover pickle used for immersing the meat may contain ____ % of NaCl
(a) 5 (b) 10
(c) 15 (d) 20

212. Pumping pickle used for injecting into the meat may contain _______% NaCl
(a) 12 (b) 24
(c) 36 d 48

213. The real color fixative in meat is
(a) sodium nitrite
(b) sodium nitrate
(c) nitric oxide (d) nitrous oxide

214. Nitrosyl hemochrome is __________ in color
(a) pink (b) red
(c) brown (d) purple

215. Nitrosyl hemochrome has Fe in _______ state
(a) Fe^{2+} (b) Fe^{3+}
(c) Fe^{4+} (d) Fe^{5+}

216. Denatured metmyoglobin has _____ ______ color
(a) pink (b) red
(c) brown (d) purple

217. Denatured metmyoglobin has Fe in ______ state
(a) Fe^{2+} (b) Fe^{3+}
(c) Fe^{4+} (d) Fe^{5+}

218. The curing temperature with a pickling solution for a meat is ______ °C
(a) 2.2-3.3 (b) 10.3-15.3
(c) 15.3-19.3 (d) none

219. Salami and bologna are examples of
(a) sausage (b) ham
(c) hog (d) bacon

220. Which of the following antibiotics is recommended for preservation of meat
(a) chlortetracycline
(b) oxytetracycline
(c) nisin
(d) all

221. The pH of raw meat is
(a) 1.2-3.4 (b) 5.7-7.2
(c) 7.9-10.1 (d) none

222. An increase in pH ________ of meat
(a) favors spoilage
(b) inhibits spoilage
(c) no-effect
(d) none

223. Aerobic bacteria is mostly present on
(a) surface of meat
(b) interior of meat
(c) inintramuscular fat
(d) none

224. Putrefaction occurs
(a) in presence of oxygen
(b) in absence of oxygen
(c) in absence of CO_2
(d) none

225. Surface slime in an example of meat spoilage due to
(a) aerobic condition
(b) anaerobic condition
(c) both
(d) none

226. Bloom is
(a) red color of meat
(b) green color f meat
(c) brown color of meat
(d) none

227. Phosphorescence is a defect caused by
(a) phosphorescent bacteria
(b) luminous bacteria
(c) both
(d) none

228. The musty or earthy flavor of meat is due to
(a) actinomycetes
(b) flavobacterium
(c) chromobacterium
(d) none

229. The white fuzzy growth caused by molds on meat is called
(a) black spot (b) white spot
(c) whiskers (d) green patches

230. Practically all the eggs present in the market are
(a) fertile (b) infertile
(c) boiled (d) poached

231. Eggs have ______ nutritive value
(a) poor (b) no
(c) excellent (d) none

232. Protein content of eggs is ___%
(a) 12-14 (b) 10-12
(c) 1 (d) 90

233. The lipid content of eggs is _______%
(a) 12-14 (b) 10-12
(c) 1 (d) 90

234. Eggs contain high per centage of ________ fatty acids
(a) unsaturated (b) saturated
(c) no fatty acids (d) none

235. The calorie per egg is ______ kcal
(a) 20 (b) 40
(c) 60 (d) 80

236. Most of the calories of egg is due to
(a) albumen (b) yolk
(c) egg white (d) none

237. The minerals of egg are _______ %
(a) 12-14 (b) 10-12
(c) 1 (d) 90

238. The yolk content of iron is _______ mg/egg
(a) 100 (b) 150
(c) 250 (d) 350

239. Which of the following is present in egg
(a) albumen (b) albumin
(c) both (d) none

240. The iron content of yolk is _______ mg/egg
(a) 0.42 (b) 1.42
(c) 14.2 (d) 142.2

241. The shell of the egg is made from
(a) sodium chloride
(b) carcium carbide
(c) calcium carbonate
(d) all

242. Which of the membrane of egg shell is thicker
(a) inner (b) outer
(c) both are same (d) none

243. Which of the membrane of egg shell is thinner
(a) inner (b) outer
(c) both are same (d) none

244. The thickness of outer membrane of shell is ______ μm
(a) 12 (b) 24
(c) 48 (d) 56

245. The thickness of inner membrane of shell is ______ μm
(a) 12 (b) 22
(c) 48 (d) 56

246. A freshly laid egg has ________ air cells
(a) no (b) small number
(c) large number (d) none

247. The egg yolk is enclosed in sac called
(a) vitelline membrane
(b) albuminous sac
(c) cuticle
(d) none

248. The yolk is connected to the albumen with
(a) germ
(b) chalazae
(c) vitelline membrane
(d) cuticle

249. The germinal disc or germ spot is present in
(a) albumen (b) yolk
(c) albumin (d) globulin

250. The ________ develops into a chick
(a) germ spot (b) chalazae
(c) yolk (d) albumen

251. The white column beneath the germ spot is called
(a) vertebra
(b) chalazae
(c) latebra
(d) vitelline membrane

252. The per centage of shell in egg is ______
(a) 8-11 (b) 56-61
(c) 27-32 (d) none

253. The per centage of egg white in egg is ______
(a) 8-11 (b) 56-61
(c) 27-32 (d) none

254. The per centage of egg yolk in egg is ________
(a) 8-11 (b) 56-61
(c) 27-32 (d) none

255. Major portion of egg is that of
(a) albumen (b) yolk
(c) shell (d) all

256. The per centage of water in egg is ______
(a) 8-11 (b) 56-61
(c) 27-32 (d) 65

257. The per centage of ash in egg is ______
(a) 8-11 (b) 56-61
(c) 27-32 (d) 1

258. Which has higher mineral content in egg
(a) egg white (b) egg yolk
(c) both (d) none

259. The per centage of ash in egg yolk is ______
(a) 8-11 (b) 56-61
(c) 27-32 (d) 1.5

260. The per centage of ash in egg white is ______
(a) 0.82 (b) 56-61
(c) 27-32 (d) 1.5

261. The per centage of fat in egg yolk is ______
(a) 8-11 (b) 56-61
(c) 31-32 (d) 1.5

262. The per centage of fat in egg white is ______
(a) 0.05 (b) 56-61
(c) 27-32 (d) 1.5

263. The per centage of protein in egg white is ______
(a) 10 (b) 56-61
(c) 27-32 (d) 1.5

264. The per centage of protein in egg white is ______
(a) 16.7 (b) 56-61
(c) 27-32 (d) 1.5

265. The per centage of water in egg yolk is ______
(a) 49 (b) 56-61
(c) 27-32 (d) 1.5

266. The per centage of water in egg white is ______
(a) 8-11 (b) 56-61
(c) 27-32 (d) 87.77

267. Major per centage of water is present in
(a) yolk (b) albumen
(c) both (d) none

268. Major per centage of lipid is present in
(a) yolk (b) albumen
(c) both (d) none

269. The major lipid of egg yolk is
(a) triglycerides
(b) monoglycerides
(c) diglycerides
(d) none

270. Egg yolk has ______ % triglycerides
(a) 28 (b) 38
(c) 58 (d) 68

271. Egg yolk has ______ % phospholipids
(a) 28 (b) 38
(c) 58 (d) 68

272. Egg yolk has ______ % cholesterol
(a) 5 (b) 38
(c) 58 (d) 68

273. Yolk is rich in ______ vitamins
(a) B (b) C
(c) D (d) both (a) & (b)

274. Air cells are mostly present on
(a) narrower end (b) broader end
(c) middle (d) none

275. Total fat in quail egg is ______%
(a) 11.1 (b) 13.8
(c) 28.3 (d) 21.8

276. Total fat in duck egg is ______%
(a) 11.1 (b) 13.8
(c) 28.3 (d) 21.8

277. Total fat in goose egg is ______%
(a) 11.1 (b) 13.3
(c) 28.3 (d) 21.8

278. Highest cholesterol is present in ____ ________ egg
(a) turkey (b) goose
(c) quail (d) hen

279. Cholesterol in chicken is ___ mg/100g
(a) 844 (b) 423
(c) 933 (d) none

280. Cholesterol in quail is _____ mg/100g
(a) 844 (b) 423
(c) 933 (d) none

281. Cholesterol in duck is _____mg/100g
(a) 884 (b) 423
(c) 933 (d) none

282. Cholesterol in turkey is ____mg/100g
(a) 844 (b) 423
(c) 933 (d) none

283. Cholesterol in goose is _____mg/100g
(a) 852 (b) 423
(c) 933 (d) none

284. All egg white proteins are
(a) globular (b) fibrous
(c) both (d) none

285. The foaming properties of egg is due to
(a) egg yolk (b) egg white
(c) both (d) none

286. Which of the following is egg white protein
(a) ovomucoid (b) phosvitin
(c) livetin (d) lipovitellins

287. The main egg white protein is
(a) ovalbumin (b) conalbumin
(c) ovomucoid (d) ovomucin

288. The per centage of ovalbumin of total albumins is
(a) 25 (b) 35
(c) 45 (d) 55

289. Ovalbumin is further composed of A1, A2 and A3. Which among the following is in highest proportion?
(a) A1 (b) A2
(c) A3 (d) none

290. The per centage of conalbumin of total albumins is
(a) 13 (b) 33
(c) 43 (d) 53

291. The per centage of ovomucoid of total albumins is
(a) 10 (b) 20
(c) 30 (d) 40

292. Which of the following is trypsin inhibitor
(a) ovalbumin (b) conalbumin
(c) ovomucoid (d) ovomucin

293. The thickness of albumen is due to
(a) ovalbumin (b) conalbumin
(c) ovomucoid (d) ovomucin

294. The enzyme capable of dissolving the cell wall of bacteria is
(a) peroxidase (b) amylase
(c) lysozyme (d) lipase

295. Lysozyme has ____________ number of disulfide linkage
(a) 1 (b) 2
(c) 3 (d) 4

296. Avidin is ________ protein
(a) egg yolk (b) egg white
(c) non (d) none

297. _______ binds biotin and makes it unavailable
(a) ovomucin (b) avidin
(c) ovomucoid (d) none

298. Ovoglobulin, ovoinhibitor, flavo-protein are part of
(a) white (b) yolk
(c) not egg proteins (d) none

299. Which of the following has excellent foaming properties
(a) ovoglobulin (b) ovoinhibitor
(c) lipoprotein (d) flavoprotein

300. The egg yolk solids consist of
(a) lipovitellins (b) phosvitin
(c) lipoprotein (d) all

301. The major portion of egg yolk solids is
(a) lipovitellins (b) phosvitin
(c) lipoprotein (d) all

302. The per centage of lipovitellins is ________ % of egg yolk solids
(a) 50 (b) 60
(c) 70 (d) 80

303. The iron carrier of egg yolk is
(a) lipovitellins (b) phosvitin
(c) lipoprotein (d) all

304. Lipovittelins are ____________ density lipoproteins
(a) high (b) low
(c) no (d) none

305. The ratio of neutral lipid and phospholipid in lipovitellin
(a) 20:80 (b) 30:70
(c) 40:60 (d) 50:50

306. Under anerobic conditions, glycogen forms
(a) glucose (b) lactic acid
(c) pyruvate (d) all

307. The ratio of neutral lipid and phospholipid in low density lipoprotein
(a) 20:80 (b) 26:74
(c) 40:60 (d) 50:50

308. The color of egg yolk is due to
(a) carotenoids
(b) xanthophylls
(c) both (a) & (b)
(d) anthcyanin

309. Pigments are present in
(a) egg white (b) egg yolk
(c) egg shell (d) none

310. Most abundant mineral present in egg is
(a) phosphorus (b) calcium
(c) magnesium (d) iron

311. Most of the calcium inegg is present in
(a) shell (b) yolk
(c) egg white (d) none

312. Most abundant mineral present in yolk is
(a) phosphorus (b) calcium
(c) magnesium (d) iron

313. The iron present in egg is
(a) unavailable
(b) freely assimiable
(c) not present
(d) none

314. The size of hen egg depends in
(a) body size of hen
(b) stage of laying
(c) season of laying
(d) all

315. Mostly the average weight of hen egg is
(a) 45g (b) 75g
(c) 65g (d) 85g

316. Smaller sized eggs contains ________ proportion of egg yolk
(a) higher (b) lower
(c) similar (d) none

317. A high porosity of egg shell is
(a) desirable (b) undesirable
(c) no effect (d) none

318. As the quality of egg deteriorates, the size of air cell
(a) increase (b) decrease
(c) no change (d) none

319. As the quality of egg deteriorates, the egg white beomes
(a) thinner (b) thicker
(c) no change (d) none

320. A healthy egg has yolk at ________ position
(a) extreme left (b) extreme right
(c) centre (d) none

321. The process for evaluating the quality of egg is
(a) candling (b) handling
(c) scalding (d) stewing

322. Candling reveals ______ in egg
(a) cracks
(b) size of air cell
(c) firmness of albumen
(d) all

323. When dropped in water, a health egg
(a) floats (b) sinks
(c) breaks (d) none

324. When dropped in water, a poor quality egg
(a) floats (b) sinks
(c) breaks (d) none

325. Fresh eggs have ______ yolk rather than flat yolk
(a) high (b) low
(c) liquid (d) none

326. Yolk index is
(a) height/ diameter of yolk
(b) diameter/ height of yolk
(c) height of yolk
(d) diameter of yolk

327. Albumen index is
(a) height of thick albumen
(b) diameter of thick albumen
(c) height/ diameter of albumen
(d) none

328. Fresh eggs have ______ amount of thick white albumen
(a) larger (b) smaller
(c) no amount (d) none

329. A stale egg would spread ______ over amount area than fresh egg
(a) more (b) less
(c) no spreading (d) none

330. In India, the eggs are graded according to
(a) weight (b) shell color
(c) shape (d) yolk color

331. An egg is graded as extra large if its weight is ______ g
(a) > 60 (b) > 50
(c) > 40 (d) 55

332. An egg is graded as large if its weight is ______ g
(a) 53-59 (b) 38-44
(c) 45-52 (d) 60-65

333. An egg is graded as medium if its weight is ______ g
(a) 53-59 (b) 38-44
(c) 45-52 (d) 60-65

334. An egg is graded as small if its weight is ______ g
(a) 53-59 (b) 38-44
(c) 45-52 (d) 60-65

335. In India eggs are graded according to weight into ______ classes
(a) 4 (b) 5
(c) 6 (d) 7

336. Fresh layed egg has pH of albumen
(a) 1.2 (b) 6
(c) 7.9 (d) 9

337. The pH of the egg ______ as the CO_2 of albumen escapes through the pores of the shell as it is laid
(a) increase (b) decrease
(c) remains same (d) none

338. As the alkalinity increases the egg white becomes
(a) thicker (b) thinner
(c) no effect (d) none

339. The pH of thr yolk as it is laid
(a) 7.9 (b) 9.3
(c) 8.8 (d) 6.3

340. The pH of egg increases from 7.9 to ______ as it is laid due to escape of CO_2 from the shell pores
(a) 7.9 (b) 9.3
(c) 8.8 (d) 6.3

341. The most common contaminant of hen egg is
(a) Salmonella (b) Clostridium
(c) Bacillus (d) Candida

342. Most of the egg production takes place in
(a) winter (b) summer
(c) rainy (d) spring

343. Eggs can be stored for long periods without any reduction in its quality at
(a) room temperature
(b) 10°C
(c) -1°C
(d) 20°C

344. Mineral oil is sprayed over egg which
(a) prevents loss of moisture
(b) prevents loss of moisture
(c) closes pores of shell
(d) all

345. Thermostabilization is a method for prolonging storage life of egg. It involves heating egg at
(a) 54°C/15 min
(b) boiling water/5 sec
(c) 68°C/30 min
(d) both (a) & (b)

346. When the egg is frozen, its shell
(a) cracks (b) contracts
(c) no effect (d) none

347. Pasteurization of egg can be carried out at
(a) 61.5°C/ 3.5-4 min
(b) 54°C/15 min
(c) boiling water/5 sec
(d) 68°C/30 min

348. Gelation is the condition of
(a) yolk (b) albumen
(c) both (d) none

349. Gelation occurs upon
(a) heat treatment (b) freezing
(c) storage (d) none

350. Egg white contain traces of ______ which combines with egg proteins to cause browning
(a) glucose (b) lactose
(c) galactose (d) fructose

351. Egg white proteins begin to coagulate at
(a) 52°C (b) 32°C
(c) 22°C (d) 12°C

352. Egg yolk proteins begin to coagulate at
(a) 65°C (b) 32°C
(c) 22°C (d) 12°C

353. The yolk coagulates at a temperature ______ than egg white
(a) lower (b) higher
(c) similar (d) none

354. The presence of salt ______ the coagulation temperature
(a) increase (b) decrease
(c) no effect (d) none

355. The thin egg white produces ______ volume than thick white
(a) more (b) less
(c) similar (d) none

356. A greenish grey layer at the interface of yolk and egg white in hand cooked eggs is due to
(a) ferrous sulfide (b) sulfite
(c) hydrogen sulfide (d) all

357. Soft and hard cooked eggs are those prepared
(a) in shell (b) outside shell
(c) frying egg (d) none

358. The eggs prepared by sliding the content of a good quality egg into a pan of water heated to simmering temperature is called
(a) poached
(b) fried
(c) scrambled
(d) French omelette

359. Meringues are foams of
(a) egg white
(b) egg yolk
(c) chemical foaming agents
(d) none

360. _______ meringue has more sugar
(a) soft
(b) hard
(c) both have similar sugar
(d) none

361. _______ is a good emulsifying agent
(a) egg yolk (b) egg white
(c) albumen (d) none

362. Designer eggs are those which have
(a) unique shapes
(b) different vitamin content
(c) lower cholesterol
(d) both (a) & (b)

363. An average egg shell has _______ mg calcium
(a) 100 (b) 200
(c) 2200 (d) 5800

364. An average egg shell has _______ mg phosphorus
(a) 100 (b) 20
(c) 2200 (d) 5800

365. An average egg shell has _________ calcium than phosphorus
(a) more (b) less
(c) similar (d) none

366. The yolk is formed in _____ of hen
(a) ovaries (b) oviduct
(c) lower oviduct (d) none

367. Albumen is formed in _____ of hen
(a) ovaries (b) oviduct
(c) lower oviduct (d) none

368. Calcium or shell is formed in _______ of hen
(a) ovaries (b) oviduct
(c) lower oviduct (d) none

369. The fertile eggs are formed when sperm travels the oviduct and fertile the
(a) yolk
(b) yolk & albumen
(c) yolk, albumen & shell
(d) none

370. The outermost covering of the egg is
(a) cuticle
(b) chalaziferous layer
(c) vitelline membrane
(d) shell

371. A large air cell is an indication of _______egg
(a) fresh (b) stale
(c) boiled (d) none

372. The shell has _______ % calcium carbonate
(a) 94 (b) 1
(c) 10 (d) 45

373. The shell has _______ % magnesium carbonate
(a) 94 (b) 1
(c) 10 (d) 45

374. The shell has _______ % calcium phosphate
(a) 94 (b) 1
(c) 10 (d) 45

375. The loss of carbon dioxide from the pores of egg makes the egg
(a) acidic (b) alkaline
(c) neutral (d) none

376. _______ is removed prior to drying of egg white
(a) glucose (b) sucrose
(c) cholesterol (d) none

377. Poultry refers to fowl reared for their
(a) flesh (b) egg
(c) feathers (d) all

378. Chicken of 8-10 weeks with tender meat is called

(a) broiler (b) rooster
(c) stag (d) none

379. Chicken of 3-5 months of age is called

(a) broiler (b) rooster
(c) stag (d) none

380. A mature chicken cock under 10 months of age is called

(a) broiler (b) rooster
(c) stag (d) none

381. A mature chicken more than 10 months of age is called

(a) stewing chicken
(b) fowl
(c) stag
(d) both (a) & (b)

382. A mature male chicken more than 10 months of age is called

(a) cock (b) fowl
(c) stag (d) both (a) & (b)

383. The birds are generally not fed for ________ before slaughter

(a) 12hrs (b) 12 days
(c) 12 months (d) none

384. Which of the following is a hot water treatment

(a) roasting (b) scalding
(c) frying (d) none

385. The birds are scalded after bleeding at

(a) 60°C/45 sec (b) 60°C/45 min
(c) 90°C/45 sec (d) 90°C/45 min

386. Scalding is done for

(a) cooking (b) defeathering
(c) bleeding (d) all

387. The best temperature for prolonging the shelf life of poultry is

(a) room temperature
(b) refrigeration
(c) freezing
(d) none

388. The protein content of poultry is

(a) 1-2% (b) 18-25%
(c) 26-36% (d) > 40%

389. Poultry meat has ________ protein to fat ratio

(a) high (b) low
(c) equal (d) none

390. Raw chicken has ________ flavor

(a) no (b) little
(c) strong (d) both (a) & (b)

391. Older and tougher birds are cooked by

(a) moist heat (b) dry heat
(c) frying (d) none

392. Young and tender birds are cooked by

(a) moist heat (b) dry heat
(c) frying (d) none

393. The turkey meat has ______ % protein

(a) 20.5 (b) 8.3
(c) 10.1 (d) 18.1

394. The turkey meat has ________ % fat

(a) 20.5 (b) 8.3
(c) 10.1 (d) 18.1

395. The fowl meat has ________ % protein

(a) 20.5 (b) 8.3
(c) 10.1 (d) 18.1

396. The fowl meat has ________ % fat

(a) 20.5 (b) 8.3
(c) 10.1 (d) 18.1

397. The broiler meat has ______ % protein

(a) 20.5 (b) 8.3
(c) 10.1 (d) 18.1

398. The broiler meat has ________ % fat

(a) 14.3 (b) 8.3
(c) 10.1 (d) 18.1

399. The duck meat has ________ % protein

(a) 20.5 (b) 8.3
(c) 10.1 (d) 15.1

400. The duck meat has ________ % fat

(a) 20.5 (b) 26.5
(c) 10.1 (d) 18.1

401. The goose meat has _______ % protein
(a) 20.5 (b) 8.3
(c) 10.1 (d) 15.7

402. The goose meat has ________ % fat
(a) 31.0 (b) 8.3
(c) 10.1 (d) 18.1

403. Lowest moisture is present in _______ meat
(a) goose (b) duck
(c) turkey (d) fowl

404. Highest protein is present in ________ meat
(a) goose (b) duck
(c) turkey (d) fowl

405. Lowest protein is present in ________ meat
(a) goose (b) duck
(c) turkey (d) fowl

406. Highest fat is present in ________ meat
(a) goose (b) duck
(c) turkey (d) fowl

407. Lowest fat is present in ________ meat
(a) goose (b) duck
(c) turkey (d) fowl

408. During broiling, the internal temperature of the meat should reach
(a) 24°C (b) 34°C
(c) 54°C (d) 74°C

409. Roasting of the bird should continue till the internal temperature reach
(a) 24°C (b) 34°C
(c) 85°C (d) 74°C

410. Tandoor chicken is
(a) barbecued (b) boiled
(c) fried (d) poached

411. Which of the following involves the use of more water
(a) braising (b) stewing
(c) both (d) none

412. Finfish and shell fish are
(a) inedible (b) edible
(c) not eatable (d) none

413. Molluse and crustaceans are examples of
(a) finfish (b) shell fish
(c) both (d) non

414. The poultry should be frozen at temperature ________
(a) 0°C (b) -2°C
(c) -17°C (d) -70°C

415. The radiation odse for chicken carcasses to destroy salmonella is _____ kgray
(a) 2.5 (b) 5
(c) 7.5 (d) 10

416. Oysters, clams, scallops are examples of
(a) finfish (b) mollusks
(c) crustaceans (d) none

417. Lobsters, crabs, shrimp are examples of
(a) finfish (b) mollusks
(c) crustaceans (d) none

418. Example of fresh water fishes are
(a) sardine (b) mullet
(c) catfish (d) all

419. Lean fishes have fat
(a) < 2% (b) 2-5%
(c) > 5% (d) > 10%

420. Medium fishes have fat
(a) < 2% (b) 2-5%
(c) > 5% (d) > 10%

421. Fat fishes have fat
(a) < 2% (b) 2-5%
(c) > 5% (d) > 10%

422. Example of lean fish are
(a) Bombay duck (b) salmon
(c) sardine (d) tuna

423. Which of the following is not a lean fish
(a) halibut (b) cod
(c) perch (d) salmon

424. Example of fat fish is
(a) sardine (b) salmon
(c) tuna (d) all

425. The fishes are mainly categorized based on their
(a) carbohydrates (b) fat
(c) protein (d) minerals

426. Majority of the fish fat is located in
(a) liver
(b) pyrolic caecum
(c) mesenteria (d) all

427. Mass of the eggs of fish is called
(a) mutt (b) roe
(c) caecum (d) none

428. The principal component of fish oils is
(a) triglycerides
(b) monoglycerides
(c) diglycerides (d) none

429. Majority of the fatty acids in the fish are
(a) saturated (b) unsaturated
(c) hydrogenated (d) none

430. Fish fat from fish eggs is high in
(a) cholesterol (b) lecithin
(c) waxes (d) none

431. Fish is a _______ source of protein
(a) poor (b) rich
(c) deficient (d) none

432. The protein content of fin fish is
(a) 10% (b) 20%
(c) 30% (d) 40%

433. The protein content of shel fish is ________ than fin fish
(a) higher (b) lower
(c) similar (d) none

434. Saltwater fishes have ________ iron than freshwater fishes
(a) more (b) less
(c) similar (d) none

435. The iodine content of ______ fish is exceptionally high
(a) salt water (b) fresh water
(c) both (d) none

436. The iodine content of salt water fish is exceptionally ____________ than fresh water fish
(a) higher (b) lower
(c) similar (d) none

437. The iodine content of marine fish per 100g of fish meat is
(a) 0.01-0.5mg (b) 0.5-1mg
(c) 1-2 (d) 2.3

438. Iodine is necessary for prevention of
(a) scurvy (b) goitre
(c) rickets (d) pellagra

439. Fish oils are richest source of vitamin
(a) A (b) D
(c) K (d) both (a) & (b)

440. FPC stands for
(a) fat protein concentrate
(b) fish protein concentrate
(c) fish putrid concentrate
(d) none

441. AMD stands for
(a) age mediated disease
(b) age related macular degeneration
(c) age mediated deficiency
(d) none

442. Omega 3 oils found in fish helps in prevention of
(a) AMD (b) scurvy
(c) goiter (d) pellagra

443. PCB stands for
(a) polychlorinated biphenyls
(b) poly chlorine biphenyls
(c) poultry chicken and birds
(d) none

444. 4. Shrimp, clams and oysters contain ________ level of mercury
(a) high (b) low
(c) extremely high (d) none

445. Higher level of mercury is found in
(a) tuna (b) grouper
(c) lobster (d) cod

446. Fishes are ________ perishable
(a) highly (b) semi
(c) non (d) none

447. Rigor mortis in fish is relatively ________ as compared to other mammals
(a) longer (b) shorter
(c) similar (d) none

448. The fishy odor is due to
(a) trimethyl amine
(b) oxidation of fat
(c) both (d) none

449. The removal of internal organs of fish is known as
(a) gutting (b) putting
(c) mutting (d) cutting

450. The length wise cuts of fin fish is called
(a) fillets (b) steaks
(c) sticks (d) none

451. Cross cut section of fin fish is called
(a) fillets (b) steaks
(c) sticks (d) none

452. Foshes are preserved by
(a) drying (b) salting
(c) canning (d) all

453. Shell fish are ________ perishable than finfish
(a) more (b) less
(c) equally (d) none

454. Broiling, baking and frying are examples of ________ heat treatment
(a) moist (b) dry
(c) smoking (d) none

455. Fish meal has ______ % proteins
(a) 10-20 (b) 30-50
(c) 50-70 (d) 70-90

456. Fish meal has ______ % fat
(a) 2-5 (b) 5-10
(c) 10-20 (d) 20-30

457. Fish meal has ______ % minerals
(a) 10-12 (b) 30-50
(c) 50-70 (d) 70-90

458. Fish meal has ______ % moisture
(a) 6-12 (b) 30-50
(c) 50-70 (d) 70-90

459. Fish flour contains ________ % protein
(a) 6-12 (b) 30-50
(c) 50-70 (d) 85-90

460. Rearing of fish is called
(a) pisciculture (b) sericulture
(c) agriculture (d) none

461. Fish found in the surface and middle water layer of sea is called
(a) pelagic (b) demersal
(c) both (d) none

462. Fish found in the bottom water layer of sea is called
(a) pelagic (b) demersal
(c) both (d) none

463. Pelagic fishes are _____ in fat
(a) rich (b) poor
(c) no (d) none

464. Demersal fish are ______ in fat
(a) rich (b) poor
(c) no (d) none

465. Fish proteins are ________ digestible
(a) highly (b) poorly
(c) non (d) none

466. Shell fish are ______ in vitaminB than fin fish
(a) rich (b) poor
(c) no (d) none

467. Which of the following meat is referred as white meat
(a) goat (b) buffalo
(c) sheep (d) all

468. Mehsan, surti, gaddi and Marwari are names of
(a) cow (b) buffalo
(c) goat (d) pig

469. Meat of goat is called
(a) chevon (b) veal
(c) mutton (d) beef

470. Nellore, madras red, TrichyBlack are examples of
(a) goat (b) buffalo
(c) sheep (d) pig

471. The meat of sheep is called
(a) chevon (b) veal
(c) mutton (d) beef

472. Meat from sheep of less than 12 months of age is called
(a) lamb (b) chevon
(c) mutton (d) veal

473. Meat from sheep of more than 12 months of age is called
(a) lamb (b) chevon
(c) mutton (d) veal

474. Male sheeps castrated early in life are called
(a) ether (b) gimmer
(c) both (d) none

475. Pigs are slaughtered at ________ of age
(a) 6-7 months (b) 6-7 yrs
(c) 12-24 months (d) 3-4 yrs

476. Ideal age for slaughter of buffalo calves for mat is
(a) 6-7 months (b) 6-7 yrs
(c) 18-24 months (d) 3-4 yrs

477. Female sheep which has bot yet borne a lamb is called
(a) ether (b) gimmer
(c) both (d) none

478. Bulk of the meat is formed of
(a) skeletal muscles
(b) cardiac muscles
(c) unstriated muscles
(d) none

479. Skeletal muscles are ______ muscle
(a) striated (b) unstriated
(c) both (d) none

480. Each muscle is surrounded by a sheath called
(a) epimysium (b) perimysium
(c) endomysium (d) none

481. The functional unit of myofibril is
(a) sarcoplasm (b) sarcomere
(c) muscle fibre (d) A-band

482. Only ______ filaments are present in H zone
(a) thick (b) thin
(c) neither (d) none

483. Tendons attach
(a) muscle with bone
(b) bone to muscle
(c) muscle to heart
(d) none

484. Ligament attach
(a) muscle with bone
(b) bone to muscle
(c) muscle to heart
(d) none

485. The smallest unit of collagen molecular structure is
(a) reticulin (b) tropocollagen
(c) elastin (d) actin

486. Insulin deficiency results in ______ tissue glycogen
(a) higher (b) lower
(c) no effect (d) none

487. DFD stands for
(a) dark firm and dry
(b) dry firm and dark
(c) dry fat and dark
(d) dark fat and dry

488. DFD occurs at ________ pH
(a) higher (b) lower
(c) neutral (d) none

489. PSE stands for
(a) pale soft and exudate
(b) poly soft and exudate
(c) protein soft and exudate
(d) none

490. PSE occurs in __________pH
(a) higher (b) lower
(c) neutral (d) none

491. Antemortem examination is done ________ to slaughter
(a) prior (b) after
(c) at time of (d) none

492. All animals to be slaughtered should be rested for atleast _____ hrs
(a) 1 (b) 6
(c) 12 (d) 24

493. The concentration of carbondioxide usd for chemicalstunning of pigs is
(a) 25% (b) 50%
(c) 65% (d) 95%

494. Jewish slaughter is called
(a) schechita (b) kosher
(c) schochet (d) terepha

495. Carcass fit for Jewish consumption is
(a) schechita (b) kosher
(c) schochet (d) terepha

496. Carcass unfit for Jewish consumption is
(a) schechita (b) kosher
(c) schochet (d) terepha

497. Muslim method of slaughter is
(a) schechita (b) kosher
(c) halal (d) terepha

498. In Jhatka method of slaughter _____ _______ is damaged
(a) medulla oblongata
(b) back bone
(c) stomach
(d) leg

499. Rigor mortis occurs due to
(a) depletion of ATP
(b) permanent actomyosin cross bridges
(c) both
(d) none

500. At rigor mortis
(a) meat is tender
(b) meat is tough
(c) meat water absorption capacity is low
(d) both (b) & (c)

7.3 ANSWER KEY

Q. A.	Q. A.	Q. A.	Q. A.	Q. A.	Q. A.
1. (a)	2. (b)	3. (b)	4. (c)	5. (a)	6. (a)
7. (b)	8. (c)	9. (b)	10. (a)	11. (a)	12. (a)
13. (b)	14. (d)	15. (b)	16. (c)	17. (b)	18. (a)
19. (d)	20. (d)	21. (b)	22. (d)	23. (a)	24. (a)
25. (b)	26. (b)	27. (a)	28. (b)	29. (c)	30. (a)
31. (a)	32. (c)	33. (a)	34. (b)	35. (b)	36. (c)
37. (b)	38. (b)	39. (c)	40. (a)	41. (b)	42. (d)

Q. A.	Q. A.	Q. A.	Q. A.	Q. A.	Q. A.
43. (a)	44. (a)	45. (b)	46. (a)	47. (c)	48. (b)
49. (d)	50. (b)	51. (a)	52. (d)	53. (a)	54. (d)
55. (d)	56. (c)	57. (c)	58. (a)	59. (b)	60. (a)
61. (c)	62. (a)	63. (b)	64. (b)	65. (c)	66. (a)
67. (b)	68. (c)	69. (d)	70. (a)	71. (b)	72. (c)
73. (a)	74. (b)	75. (b)	76. (b)	77. (a)	78. (a)
79. (d)	80. (b)	81. (a)	82. (b)	83. (a)	84. (a)
85. (d)	86. (b)	87. (d)	88. (a)	89. (b)	90. (a)
91. (a)	92. (a)	93. (d)	94. (a)	95. (a)	96. (c)
97. (d)	98. (b)	99. (d)	100. (d)	101. (c)	102. (b)
103. (a)	104. (b)	105. (c)	106. (b)	107. (d)	108. (c)
109. (a)	110. (a)	111. (a)	112. (d)	113. (a)	114. (b)
115. (a)	116. (c)	117. (c)	118. (a)	119. (b)	120. (d)
121. (b)	122. (a)	123. (a)	124. (d)	125. (c)	126. (c)
127. (b)	128. (a)	129. (a)	130. (b)	131. (b)	132. (b)
133. (c)	134. (b)	135. (a)	136. (d)	137. (b)	138. (a)
139. (b)	140. (b)	141. (a)	142. (c)	143. (a)	144. (a)
145. (b)	146. (a)	147. (b)	148. (c)	149. (b)	150. (a)
151. (c)	152. (a)	153. (b)	154. (c)	155. (b)	156. (d)
157. (d)	158. (b)	159. (d)	160. (a)	161. (a)	162. (c)
163. (a)	164. (b)	165. (a)	166. (b)	167. (c)	168. (a)
169. (d)	170. (a)	171. (c)	172. (a)	173. (c)	174. (b)
175. (a)	176. (a)	177. (c)	178. (a)	179. (b)	180. (d)
181. (c)	182. (d)	183. (b)	184. (a)	185. (a)	186. (a)
187. (a)	188. (d)	189. (a)	190. (a)	191. (c)	192. (a)
193. (a)	194. (d)	195. (a)	196. (c)	197. (a)	198. (a)
199. (b)	200. (a)	201. (d)	202. (a)	203. (c)	204. (a)
205. (d)	206. (c)	207. (b)	208. (b)	209. (c)	210. (b)
211. (c)	212. (b)	213. (c)	214. (a)	215. (d)	216. (c)
217. (b)	218. (a)	219. (a)	220. (d)	221. (b)	222. (a)
223. (a)	224. (b)	225. (a)	226. (a)	227. (c)	228. (a)
229. (c)	230. (b)	231. (c)	232. (a)	233. (b)	234. (a)
235. (d)	236. (b)	237. (c)	238. (c)	239. (a)	240. (b)
241. (c)	242. (b)	243. (a)	244. (c)	245. (b)	246. (a)

Q. A.	Q. A.	Q. A.	Q. A.	Q. A.	Q. A.
247. (a)	248. (b)	249. (b)	250. (a)	251. (c)	252. (a)
253. (b)	254. (c)	255. (a)	256. (d)	257. (d)	258. (b)
259. (d)	260. (a)	261. (c)	262. (a)	263. (a)	264. (a)
265. (a)	266. (d)	267. (b)	268. (a)	269. (a)	270. (d)
271. (a)	272. (a)	273. (c)	274. (b)	275. (a)	276. (b)
277. (b)	278. (a)	279. (b)	280. (a)	281. (a)	282. (c)
283. (a)	284. (a)	285. (b)	286. (a)	287. (a)	288. (d)
289. (a)	290. (a)	291. (a)	292. (d)	293. (d)	294. (c)
295. (d)	296. (b)	297. (b)	298. (a)	299. (a)	300. (d)
301. (d)	302. (c)	303. (b)	304. (a)	305. (c)	306. (b)
307. (b)	308. (c)	309. (b)	310. (b)	311. (a)	312. (a)
313. (b)	314. (d)	315. (a)	316. (a)	317. (b)	318. (a)
319. (a)	320. (c)	321. (a)	322. (d)	323. (b)	324. (a)
325. (a)	326. (a)	327. (a)	328. (a)	329. (a)	330. (a)
331. (a)	332. (a)	333. (c)	334. (b)	335. (a)	336. (c)
337. (a)	338. (b)	339. (d)	340. (b)	341. (a)	342. (d)
343. (c)	344. (d)	345. (d)	346. (a)	347. (a)	348. (c)
349. (b)	350. (a)	351. (a)	352. (a)	353. (b)	354. (b)
355. (a)	356. (a)	357. (a)	358. (a)	359. (a)	360. (b)
361. (a)	362. (d)	363. (c)	364. (b)	365. (a)	366. (a)
367. (b)	368. (c)	369. (a)	370. (a)	371. (b)	372. (a)
373. (b)	374. (b)	375. (b)	376. (a)	377. (d)	378. (a)
379. (b)	380. (c)	381. (d)	382. (a)	383. (a)	384. (b)
385. (a)	386. (b)	387. (c)	388. (b)	389. (a)	390. (d)
391. (a)	392. (b)	393. (a)	394. (b)	395. (a)	396. (c)
397. (d)	398. (a)	399. (d)	400. (b)	401. (d)	402. (a)
403. (a)	404. (d)	405. (b)	406. (b)	407. (c)	408. (d)
409. (c)	410. (a)	411. (b)	412. (b)	413. (b)	414. (c)
415. (a)	416. (b)	417. (c)	418. (d)	419. (a)	420. (b)
421. (c)	422. (a)	423. (d)	424. (d)	425. (b)	426. (d)
427. (b)	428. (a)	429. (b)	430. (b)	431. (b)	432. (b)
433. (b)	434. (a)	435. (a)	436. (a)	437. (a)	438. (b)
439. (a)	440. (b)	441. (b)	442. (a)	443. (a)	444. (b)
445. (b)	446. (a)	447. (b)	448. (c)	449. (a)	450. (a)

Q. A.	Q. A.	Q. A.	Q. A.	Q. A.	Q. A.
451. (b)	452. (d)	453. (a)	454. (b)	455. (c)	456. (a)
457. (a)	458. (a)	459. (d)	460. (a)	461. (a)	462. (b)
463. (a)	464. (b)	465. (a)	466. (a)	467. (d)	468. (c)
469. (a)	470. (c)	471. (c)	472. (a)	473. (c)	474. (a)
475. (a)	476. (c)	477. (b)	478. (a)	479. (a)	480. (a)
481. (b)	482. (a)	483. (a)	484. (b)	485. (b)	486. (b)
487. (a)	488. (a)	489. (a)	490. (b)	491. (a)	492. (d)
493. (c)	494. (a)	495. (b)	496. (d)	497. (c)	498. (a)
499. (c)	500. (d)				

CHAPTER 8

BIO PROCESS TECHNOLOGY

8.1 SUBJECTIVE KEYNOTES

1. Bioprocess technology is the study of use of biological organisms and biological materials such as enzymes for food processing and preservation.
2. Fermentation is the process of conversion of simpler molecules into complex one using microorganisms under specific condition.
3. Bioprocesses helps in industrial production of enzymes, amino acids, antibiotics, yeast, SCP, organic acids etc.
4. Beer, wine, ethanol, yoghurt, cheese are examples of fermented products made using microorganisms.
5. Lactic acid bacteria is widely used in preparation of fermented dairy products.
6. Single cell proteins (SCP) term was coined by C.L. Wilson.
7. Single cell proteins can be prepared using bacteria, actinomycetes, molds & higher fungi and yeasts.
8. Yeasts used for SCP production are *Candida* sp., *Debaryomyces* sp., *Hansenula* sp., *Kluveryomyces* sp. etc.
9. Algae used for the SCP production are Chlorella & Spirulina.
10. Bacteria used for the SCP production are Acinetobacter, Cellulomonas, Methylococcus, Methylomonas, Pseudomonas etc.
11. Molds used for the SCP production are Agaricus, Aspergillus, Morchella, Trichoderma etc.
12. Production of bakers yeast is multistage, aerobic and fedbatch process.
13. S. cerevisiae & S. carlsbergensis are generally used as brewing yeasts.
14. Dried yeast can be grown on molasses, wood sugar, whey, ethanol & hydrocarbon.
15. Invertase and Lactase enzyme can be extracted from yeast commercially.
16. Corynebacterium glutamicum is generally used for commercial production of glutamic acid and lysine.
17. Examples of antibiotics include penicillin, streptomycin, neomycin, cephalosporin, erythromycin etc.
18. Penicillin can be produced by Penicillium chrysogenum and Penicillium notatum.
19. Alexander Fleming discovered penicillin antibiotic.

20. Fleming, Florey and Chain received noble prize for discovery of penicillin.
21. Downstream processing involves separation, concentration, filtration, purification and drying of final products of fermentation.
22. Nucleosides and nucleotides can also be produced microbially.
23. Organic acid produced by microbial fermentation includes citric acid, lactic acid, gluconic acid etc.
24. Citric acid can be produced using surface and submerged fermentation processes.
25. *Aspergillus niger* is generally used for microbial production of citric acid.
26. Commercial polysaccharide produced from microbial processes includes xanthan gum, dextrans, pullulans etc.
27. Pullulan is a fungal polysaccharide produced by Aureobasidium pullulans.
28. Xanthan gum is a bacterial polysaccharide produced by Xanthomonas campestris.
29. Dextrans are bacterial polysaccharide produced by Leuconostoc mesenteroides, L. dextranicum and L. citrovorum.
30. Pigments like carotenes and lycopenes can also be produced by microbial processes.
31. Hops contribute characteristic aroma and bitter flavor to beer.
32. Vodka, whiskies, rum and tequila are distilled alcoholic beverages.
33. Mold modified foods include soy sauce, Miso, Sufu, Tempeh, Ang-Kak etc.
34. Industrial alcohol can be produced from molasses as well as wheat starch.
35. Castor oil is used as antifoaming agent in industrial production of alcohol.
36. Alcohol is purified in industrial process by distillation.
37. Downstream processing is the most expensive unit operation in bioprocesses.
38. Microencapsulation of enzyme can be used to extend the shelf life of enzymes.
39. Microorganisms can also be used for waste treatment.
40. Fermenter design plays a crucial role in the yield of final product.
41. Recently bioprocesses are used to produce bioactive compounds that can be utilized as nutraceuticals.
42. Conversion of laboratory level process to plant level process is known as scale up.
43. Computer applications can be used in fermentation industries for data logging, data analysis, process modeling and process control.
44. Isolation of solid fermentation products can be done using filtration, centrifugation, coagulation and flocculation.
45. Primary separation of fermentation products involves liquid extraction, dissociation extraction, sorption processing and precipitation.
46. Purification of fermented products involves fractional liquid-liquid extraction, chromatography, crystallization and drying.
47. Immobilization of microbial cells can be done by carrier binding, cross-linking and entrapping methods.
48. Fermentation plant design includes pumps and piping, heat exchange, instrumentation & control and foam control.

49. Microbial bioprocesses can also be used for bio-fuel production.
50. Biotechnological techniques are used for strain improvement of culture microorganisms.
51. Activated Sludge System is a biochemical process for treating industrial wastewater that uses oxygen and microorganisms to biologically oxidize remaining organic pollutants.
52. Biomass is organic material derived from recently living organisms including plants, animals and their byproducts.
53. Biogas is the gas produced by the breakdown of organic matter during the anaerobic digestion process.
54. Carbon sequestration is the process of capture and long-term storage of carbon dioxide or other forms of carbon through biological (i.e. plants) chemical or physical processes.
55. Eutrophication is a process where water bodies receive excess nutrients that stimulate excessive plant growth.
56. Flocculent is a material that is added to a liquid to encourage the clumping together of fine particulates into a floc. The floc may then float to the top of the liquid surface, settle to the bottom of the liquid (sedimentation), or be readily filtered from the liquid.
57. Renewable energy is the energy that comes from resources which are continually replenished such as sunlight, wind, and biomass.
58. Sugarcane shoots that grow from the stubble left behind from harvesting are known as ratoons.
59. Residue left after starch fermentation to ethanol is known as stillage.
60. Tail Gas is the residual biogas from the process of converting biogas to bio-methane process.
61. Vapour Recovery Units are used to remove and recover vapours from storage tanks.
62. The liquid residue left in the distillation of ethanol from sugarcane is known as vinnase.
63. Bio-insecticide is any material used in insect control that is derived from living organisms, such as bacteria, plant cells, or animal cells.
64. Adjuvant is Insoluble material that increases the formation and persistence of antibodies when injected with an immunogen.
65. Affinity chromatography is a technique used in bioprocess engineering and analytical biochemistry for separation and purification of almost any biomolecule, but typically a protein, on the basis of its biological function or chemical structure. The molecule to be purified is specifically and reversibly adsorbed by a complementary binding substance (ligand) that is immobilized on a matrix, the matrix usually being in the form of beads. The matrix then is washed to remove contaminants, and the molecule of interest is dissociated from the ligand and is recovered from the matrix in purified form by changing the experimental conditions.
66. Anabolism is related to synthesis reactions whereas catabolism is related to breakdown reactions.

67. Antibody is a protein produced by the immune system in response to an antigen (a molecule that is perceived to be foreign). Antibodies bind specifically to their target antigen to help the immune system destroy the foreign entity.
68. Anti-nutrients are the substances that act in direct competition with or otherwise inhibit or interfere with the use or absorption of a nutrient.
69. Callus is a cluster of undifferentiated plant cells that can, for some species, be induced to form the whole plant.
70. Cell line refers to cells which grow and replicate continuously in cell culture outside the living organism.
71. Liquid chromatography is an analytical technique in which substances are separated based on their differential movement within a liquid stream. A common form of liquid chromatography is column chromatography in which the dissolved substances may bind from the liquid differentially to a column of solid material with different affinities and subsequently be released thus be carried from the column into the at different speeds by the liquid through the column liquid, thus creating a basis for separation.
72. Mass spectrometry is an analytical technique by which compounds in a vacuum compartment are ionized, eventually fragmented, accelerated, and detected based upon the mass-dependent behavior of the ionized compounds or their fragments in response to the application of a magnetic or electric field in a vacuum.
73. Micronutrients are the substances such as a vitamin or trace element, essential for healthy growth and development but required only in minute amounts.
74. Protoplast is the cellular material that remains after the cell wall has been removed.
75. In vitro growth of only one type of microorganism is known as pure culture.
76. Secondary metabolites are the chemical substances within a biological organism that are not necessary for primary cellular functions.
77. The N-terminal sequence of a secreted protein, which is required for transport through the cell membrane is known as signal sequence.
78. Trypsin inhibitors are the anti-nutrient proteins present in plants such as soybeans that inhibit the digestive enzyme, trypsin if not inactivated by heating or other processing methods.
79. In 1928, penicillin was discovered as an antibiotic by Alexander Fleming.
80. In 1949, Pauling shows that sickle cell anaemia is a molecular disease resulting from a mutation in the protein molecule haemoglobin.
81. In 1963, new wheat varieties developed by Norman Borlaug increase yields by 70 percent.
82. In 1969, an enzyme was synthesized in vitro for the first time.
83. Three basic types of biological reactor systems used to treat waste gases are bioscrubbers, biotrickling filters and biofilters.
84. Water phase in bioscrubbers and biotrickling filters is continuously moving, whereas it is stationary in biofilters.
85. Bioscrubber consists of a srubber unit and a regeneration unit.

86. Biofiltration is biological removal of air contaminants from waste gases by means of aerobic microorganisms that are immobilized on a porous solid support.

87. Humulin is synthesized in a special non-disease producing laboratory strain of Escherichia coli bacteria that has been genetically altered to produce human insulin.

88. In 1979, Cooney calculated theoretical yields for penicillin G biosynthesis on the basis of material and energy balances using a biosynthetic pathway based on reaction stoichiometry.

89. Continuous culture systems can be operated as Turbidostats in which cell density is controlled by increasing or decreasing the flow of culture medium into the growth vessel.

90. Chemostat consists of a vessel in which the culture is grown.

91. Helmstetter Cumming Technique (HCT) is an excellent technique to obtain synchrony and it is based on selective separation methodology.

92. A bioreactor can be sterilized by destroying the organisms with some lethal agent such as heat, radiations or chemicals.

93. Microorganisms can be removed from fluids by mechanical methods such as filtration, centrifugation, flotation, or electrostatically.

94. During batch fermentations the population of microorganisms goes through several distinct growth phases: lag, acceleration, exponential growth, deceleration, stationary and death.

95. In the lag phase of microbial growth virtually no growth occurs and the microbial population remains relatively constant.

96. During exponential growth, when all nutrients are supplied in excess and are therefore non-limiting, there is a direct relationship between cell numbers and biomass concentration, assuming that mean cell size is constant.

97. The specific growth rate of the microorganism continues decelerating until all of the available limiting substrate is metabolized.

98. Cell concentration in a single chemostat can be increased by recycling biomass from the product stream back to the reactor.

99. Bioreactors are operated continuously in a few bioprocess industries such as brewing, production of bakers yeast and waste treatment; enzyme conversions can also be carried out using continuous systems.

100. The classic mixed reactor is the stirred tank; however mixed reactors can also be of bubble column, airlift or other configuration as long as concentrations of substrate, product and catalyst inside the vessel are uniform.

8.2 MULTIPLE CHOICE QUESTIONS

1. SCP stands for
   (a) single cell protein
   (b) single chain protein
   (c) both
   (d) none

2. Which of the following is used as protein source in industrial fermentation
   (a) soybean meal
   (b) corn steep liquor
   (c) both (d) none

3. In microbial growth curve, increase in rate of growth is checked in______ phase
(a) acceleration (b) deceleration
(c) death phase (d) none

4. Fermenter is mainly used for the mass culture of
(a) microorganisms
(b) plant & animal cells
(c) both (d) none

5. Bioreactor is mainly used for the mass culture of
(a) microorganisms
(b) plant & animal cells
(c) both (d) none

6. Secondary metabolites are produced in
(a) anaphase (b) idiophase
(c) both (d) none

7. Which of the following is secondary metabolite
(a) penicillin (b) erythromycin
(c) both (d) none

8. Citric acid is commercially produced from
(a) *Aspergillus niger*
(b) *Lactobacillus* sp.
(c) *Streptococcus* sp.
(d) none

9. Production of structurally similar compound from a particular one during fermentation is
(a) bioconversion
(b) biotransformation
(c) bioreaction
(d) all

10. Introduction of genes from any organism into microorganism is
(a) genetics
(b) recombinant DNA technology
(c) electrophoresis
(d) none

11. Alexander Fleming discovered
(a) penicillin (b) streptomycin
(c) neomycin (d) all

12. Alexander Fleming was working on ______ when he discovered penicillin
(a) Lactobacillus (b) Aspergillus
(c) Staphylococcus (d) none

13. Who won noble prize for discovery of penicillin
(a) Fleming (b) Florey
(c) Chain (d) all

14. Which of the following is a continuous type reactor
(a) Homogenously mixed reactor
(b) plug flow reactor
(c) both
(d) none

15. Which of the following is a lactam antibiotic
(a) penicillin (b) cephalosporin
(c) both (d) none

16. Acidic proteases are found in
(a) animal cells (b) moulds
(c) yeasts (d) all

17. Which of the following produces neutral proteases
(a) *Clostridium* sp.
(b) *Streptococcus* sp.
(c) *Bacillus* sp.
(d) all

18. Which of the following is a type of protease
(a) serine protease
(b) cysteine proteases
(c) aspartic protease
(d) all

19. Microbial proteases can be produced by
(a) solid state fermentation
(b) submerged fermentation
(c) both
(d) none

20. Directed evolution of enzymes approach uses
(a) DNA shuffling
(b) random priming recombination
(c) staggered extension process
(d) all

21. Metagenomics is the culture-independent analysis of a mixture of microbial genomes using
(a) expression (b) sequencing
(c) both (d) none

22. Which of the following is an example of bioprocess
(a) cheese making
(b) soy sauce making
(c) brewing (d) all

23. Which of the following is a type of biosensor
(a) electrochemical biosensor
(b) ion channel switch biosensor
(c) piezoelectric biosensor
(d) all

24. Which of the following is a type of biosensor
(a) potentiometer (b) amperometer
(c) calorimeter (d) all

25. Which of the following is a type of biosensor
(a) optical (b) acoustic
(c) immunosensor (d) all

26. Which of the following is most expensive in SCP production
(a) substrate (b) fermentation
(c) recovery (d) none

27. SCP may contain protein content (%) upto
(a) 10 (b) 80
(c) 30 (d) 20

28. Which of the following molasses is used in SCP production
(a) cane (b) beet
(c) corn (d) all

29. Which of the following is used for SCP production from cane molasses
(a) xxx (b) xxx
(c) xxx (d) xxx

30. Saccharomyces cerevisiae
(a) C. utilis (b) xxx
(c) both (d) none

31. Molasses require pretreatment for SCP production to reduce
(a) mineral content
(b) suspended solids
(c) both
(d) none

32. Which of the following impurty in methane causes problem in SCP ferementation process
(a) ethane (b) propane
(c) H_2S (d) all

33. Which of the following substrate is used in SCP production
(a) methane (b) methanol
(c) ethanol (d) all

34. Maximum specific growth rate of Methylophilus methylotrophus
(a) $0.52h^{-1}$ (b) $1.85h^{-1}$
(c) $5.59h^{-1}$ (d) $20.2h^{-1}$

35. Saturation constant of Methylophilus methylotrophus
(a) 0.1 mg/L (b) 1.0 mg/L
(c) 10mg/L (d) 100mg/L

36. Respiratory quotient of Methylophilus methylotrophus
(a) 0.43 (b) 0.13
(c) 2.73 (d) 5.78

37. Oxidation quotient of Methylophilus methylotrophus
(a) 0.11 (b) 0.90
(c) 3.92 (d) 5.17

38. Acinetobacter gives highest yield of SCP with
(a) gas oil (b) n-hcxanc
(c) both (d) none

39. Which of the following can be used in SCP production
(a) Achromobacter
(b) Acinetobacter
(c) Bacillus
(d) all

40. Which of the following can be used in SCP production
(a) Brevibacterium
(b) *Cellulomonas* sp.
(c) *Methylococcus* sp.
(d) all

41. Which of the following can be used in SCP production
(a) Methylococcus capsulatus
(b) Methylomonas clara
(c) Methylomonas methanolica
(d) all

42. Which of the following can be used in SCP production
(a) Methylophilus (b) Nocardia
(c) Pseudomonas (d) all

43. Which of the following give highest SCP yield (g/g substrate)
(a) Methylomonas
(b) Methylophilus
(c) Nocardia
(d) Brevibacterium

44. Which of the following give lowest SCP yield (g/g substrate)
(a) Thermomonospora
(b) Nocardia
(c) Pseudomonas
(d) Methylococcus

45. Which of the following is used in SCP production
(a) Candida
(b) Debaryomyces
(c) Kluveromyces
(d) all

46. Which of the following is used in SCP production
(a) Hansenula (b) Rhodotorula
(c) Torulopsis (d) all

47. Which of the following is used in SCP production
(a) Candida utilis
(b) Candida tropicalis
(c) Candida rigida
(d) all

48. Which of the following is used in SCP production
(a) Candida lypolytica
(b) Candida pseudotropicalis
(c) both
(d) none

49. Which of the following is used in SCP production
(a) Rhodotorula glutinis
(b) Rhodotorula rubra
(c) both
(d) none

50. Which of the following is used in SCP production
(a) Torulopsis methanosorba
(b) Torulopsis methanolovescens
(c) both
(d) none

51. On which of the following substarte Candida lypolytica give lowest SCP yield
(a) n-alkanes (b) gas oil
(c) ethanol (d) all

52. Candida ethanothermophilum require ______°C temperature for SCP production
(a) 30 (b) 32
(c) 40 (d) 35

53. Candida tropicalis require ______°C temperature for SCP production
(a) 30 (b) 32
(c) 40 (d) 35

54. Which of the following is used in SCP production

(a) Agaricus blazei
(b) Agaricus campestris
(c) both
(d) none

55. Which of the following is used in SCP production
(a) Aspergillus (b) Boletus
(c) Calvatia (d) all

56. Which of the following is used in SCP production
(a) Lentinus edodes (b) Morchella
(c) Aspergillus (d) all

57. Which of the following is used in SCP production
(a) Paecilomyces (b) Trichoderma
(c) Tricholoma (d) all

58. Which of the following is used in SCP production
(a) Trichoderma viride
(b) Trichoderma harzianum
(c) both
(d) none

59. Which of the following component is highest in Chlorella sorokiniana
(a) protein (b) fat
(c) carbohydrate (d) ash

60. Which of the following component is highest in Spirulina maxima
(a) protein (b) fat
(c) carbohydrate (d) ash

61. Which of the following yeast is used in brewing industry
(a) Saccharomyces cerevisiae
(b) S. uvarum
(c) S. sake
(d) all

62. Which of the following is used as food yeast
(a) Candida utilis
(b) Candida tropicalis
(c) Candida lypolytica
(d) all

63. Zymomonas mobilis has ethanol tolerance of_________ g/L
(a) 130 (b) 290
(c) 376 (d) 679

64. One gram of glucose generally yields _________gram of ethanol
(a) 0.968 (b) 0.768
(c) 0.511 (d) 0.128

65. In citric acid production, optimum iron concentration for pure sucrose is
(a) 2ppm (b) 20ppm
(c) 40ppm (d) 120ppm

66. Submerged fermented of citric acid is done in
(a) stirred reactor (b) airlift reactor
(c) both (d) none

67. Which of the following is generally used as strain in citric acid submerged fermentation
(a) *Aspergillus niger*
(b) Saccharomyces
(c) Kluyveromyces
(d) none

68. Citric acid crystallizes as anhydrous acid at ________°C
(a) 5 (b) 40
(c) 150 (d) 240

69. Isolation of Corynebacterium glutamicum was done in
(a) 1957 (b) 1984
(c) 1864 (d) 1823

70. Sugar cane molasses has a biotin content of _________mg/kg
(a) 0.2-0.8 (b) 0.02-0.08
(c) 1.4-2.8 (d) 3.5

71. Epsom salt is
(a) $MgSO_4.7H_2O$ (b) $CaSO_4$
(c) Na_2SO_4 (d) none

72. 6-APA consists of
(a) thiazolidine ring
(b) beta lactam ring
(c) both (d) none

73. Aeration rate for tetracycline production is
(a) 1.0 vvm (b) 10 vvm
(c) 100 vvm (d) 150 vvm

74. Streptomycin is produced by using strain of
(a) Pencillium (b) Streptomyces
(c) Candida (d) none

75. Who discovered penicillin antibiotic
(a) Rudolph Virchow
(b) Robert Hooke
(c) Alexander Fleming
(d) none

76. Lag phase in growth curve is also known as
(a) death phase
(b) period of adaptation
(c) telophase
(d) none

77. Mass culture of microorganisms is carried out in
(a) fermentor (b) reactor
(c) petriplate (d) none

78. Dextran is commercially produced by
(a) Lactobacillus bulgaricus
(b) Leuconostoc mesenteroides
(c) Bacillus cereus
(d) none

79. Commercially ethanol is produced by
(a) Saccharomyces cerevisiae
(b) Kluyveromyces fragilis
(c) both
(d) none

80. Secondary metabolites are generally produced in
(a) idiophase (b) tropophase
(c) both (d) none

81. Primary metabolites are generally produced in
(a) idiophase (b) tropophase
(c) both (d) none

82. Cephalosporium acrimonium produces
(a) penicillin (b) streptomycin
(c) cephalosporin (d) none

83. __________ increases production of compound in cultures
(a) elicitors (b) pH
(c) antibiotics (d) none

84. Process that brings together genes from different sources
(a) PCR (b) recombination
(c) both (d) none

85. Fed batch system is generally used for production
(a) primary metabolites
(b) secondary metabolites
(c) both
(d) none

86. HMR in bioprocesses stands for
(a) high metabolic rate
(b) homogenously mixed reactor
(c) high mass ratio
(d) none

87. Which of the following is an advantage of batch system
(a) short time
(b) optimum product recovery
(c) both
(d) none

88. Down stream processing in bio-processes is related to
(a) fermentation
(b) culture
(c) substrate
(d) product recovery

89. Organisms used in solid substrate fermentation can tolerate
(a) low water activity
(b) high water activity
(c) both
(d) none

90. Penicillin was discosvered in
(a) 1901 (b) 1928
(c) 1935 (d) 1941

91. Citric acid is a primary metabolic product of
(a) TCA cycle
(b) glycolysis
(c) fatty acid metabolism
(d) amino acid metabolism

92. The science of fermentation is known as
(a) fermentology
(b) zymology
(c) biotechnology
(d) phycology

93. The term fermentation is derived from the Latin word
(a) fervere (b) frevene
(c) fermeni (d) none

94. Beta lactam antibiotics combine specifically with
(a) PBP (b) PCP
(c) PDP (d) PCB

95. Which of the following is not an application of antibiotics
(a) control plant diseases
(b) treatment of tumers
(c) food preservative
(d) none

96. Which of the following is an application of antibiotics
(a) biochemistry tool
(b) molecular biology
(c) both
(d) none

97. Specialized enrichment technique used for isolation of microorganisms
(a) soil treatment
(b) selective inhibitors
(c) nutritional variation
(d) all

98. Strain improvement can be done by
(a) mutant selection
(b) recombination
(c) recombinant DNA technology
(d) all

99. Which of the following is a beta-lactam antibiotics
(a) penicillin (b) cephalosporin
(c) both (d) none

100. Beta lactam antibiotics are specific inhibitors of bacterial ________ synthesis
(a) cell wall
(b) cell membrane
(c) ribosome
(d) none

101. Earlier penicillin was produced from
(a) Penicillium chrysogenum
(b) Penicillium notatum
(c) both
(d) none

102. 6-aminopenicillanic acid (APA) consists of
(a) thiazolidine ring
(b) beta-lactum ring
(c) both
(d) none

103. Now-a days, penicillin is produced from
(a) Penicillium chrysogenum
(b) Penicillium notatum
(c) both
(d) none

104. Molecular weight of sodium penicillin G is ________ kDa
(a) 372.4 (b) 473.8
(c) 356.4 (d) 783.5

105. Molecular weight of penicillin V is ________ kDa
(a) 372.4 (b) 473.8
(c) 456.4 (d) 783.5

106. Which of the following is used as nitrogen source in penicillin production
(a) corn steep liquor
(b) cottonseed flour
(c) soybean meal
(d) all

107. Penicillium chrysogenum strain is used by most of penicillin manufacturers is
(a) NRRL 1951 (b) WIS Q 176
(c) both (d) none

108. Penicillin is a ________ metabolite
(a) primary (b) secondary
(c) both (d) none

109. The required aeration rate during penicillin production is
(a) 0.5-1.0 vvm (b) 1.0-2.0 vvm
(c) 5.0 vvm (d) 10 vvm

110. Fermentation temperature (°C) during penicillin production is about
(a) 20-22 (b) 25-28
(c) 32-35 (d) 40-43

111. Recovery process of penicillin include
(a) filteration (b) extraction
(c) adsorption (d) all

112. Recovery process of penicillin include
(a) crystallization (b) drying
(c) both (d) none

113. Solvent extraction of penicillin is done at
(a) high pH (b) low pH
(c) neutral pH (d) none

114. Solvent extraction of penicillin is done using
(a) amyl acetate (b) butyl acetate
(c) both (d) none

115. Chlorotetracycline was first isolated in
(a) 1945 (b) 1956
(c) 1962 (d) 1979

116. Chlorotetracycline was first isolated from culture of
(a) Streptomyces (b) Penicillin
(c) Candida (d) none

117. Which of the following strain is used for producing tetracycline
(a) Streptomyces aureus
(b) S. antibioticus
(c) S. feofaciens (d) all

118. Which of the following strain is used for producing tetracycline
(a) Streptomyces aureofaciens
(b) S. flavus
(c) both (d) none

119. Tetracycline consists of fused ________ ________ ring system
(a) three (b) four
(c) five (d) two

120. Tetracyclines are effective against
(a) garm +ve (b) gram –ve
(c) both (d) none

121. Which of the following is most widely used antibiotic
(a) Penicillin
(b) Cephalosporin
(c) both
(d) none

122. Highest yield for tetracycline in microbial processes is about
(a) 25g/litre (b) 150g/litre
(c) 165g/litre (d) 185g/litre

123. Streptomycin is effective against
(a) garm +ve (b) gram –ve
(c) both (d) none

124. Streptomycin was discovered by
(a) Alexander Fleming
(b) Selman Waksman
(c) Ernst Chain
(d) none

125. Streptomycin was discovered in
(a) 1944 (b) 1934
(c) 1924 (d) 1974

126. Selman Waksman received nobel prize for discovery of streptomycin in
(a) 1944 (b) 1952
(c) 1964 (d) 1974

127. Maximum production of streptomycin is achieved at pH range
(a) 7.6-8.0 (b) 4.0-5.5
(c) 5.0-6.2 (d) none

128. Which of the following is a fermenter component
(a) air inlet (b) steam inlet
(c) motor (d) all

129. Which of the following is a fermenter component
(a) pH control reservoir
(b) steam outlet
(c) nutrient outlet (d) all

130. Which of the following is a fermenter component
(a) cold water jacket
(b) temperature recorder
(c) temperature controller
(d) all

131. Which of the following is a fermenter component
(a) harvest line (b) culture inlet
(c) paddle wheel (d) all

132. Construction material of fermenter should be
(a) non toxic
(b) corrosion proof
(c) smooth surface (d) all

133. Which of the following is a type of agitator used in fermenter
(a) disc turbine (b) vaned disc
(c) both (d) none

134. Which of the following is a type of agitator used in fermenter
(a) open turbine
(b) variable pitch
(c) marine propeller
(d) all

135. Scaba 6 SRGT is a type of
(a) agitator (b) fermenter
(c) rheometer (d) all

136. Which of the following is a type of fermenter
(a) continuous stirred tank
(b) air-lift
(c) tower fermenter
(d) all

137. Down stream processing involves
(a) separation
(b) cell disruption
(c) extraction
(d) all

138. Down stream processing involves
(a) concentration (b) purification
(c) drying (d) all

139. Separation of particle cells in down stream processing involves
(a) filtration
(b) centrifugation
(c) flocculation
(d) all

140. Cell disruption in down stream processing is done by
(a) agitation
(b) homogenization
(c) high pressure pumping
(d) all

141. Non mechanical method for cell disruption in down stream processing is done by
(a) osmotic shock
(b) freezing
(c) enzymatic digestion
(d) all

142. Concentration of product in DSP is done by
(a) evaporation
(b) ultrafiltration
(c) adsorption
(d) all

143. Product purification in DSP is done by
(a) crystallization
(b) chromatography
(c) both (d) none

144. Which of the following product is obtained using fermentation
(a) antibiotics (b) SCP
(c) ethanol (d) all

145. TCA cycle is also known as
(a) citric acid cycle (b) Krebs cycle
(c) both (d) none

146. Airlift fermenters do not have
(a) impellar blades (b) shaft
(c) both (d) none

147. Which of the following is a microbial source of α-amylase
(a) Aspergillus oryzae
(b) Bacillus licheniformis
(c) both (d) none

148. Which of the following is a microbial source of β-amylase
(a) Bacillus cereus
(b) Bacillus megaterium
(c) Bacillus polymyxa
(d) all

149. Which of the following is a microbial source of cellulase
(a) *Aspergillus niger*
(b) Trichoderma reesei
(c) both (d) none

150. Which of the following is a microbial source of glucoamylase
(a) *Aspergillus niger*
(b) *Rhizopus* sp.
(c) both (d) none

151. Which of the following is a microbial source of lactase
(a) *Aspergillus niger*
(b) Kluyveromyces fragilis
(c) Kluyveromyces lactis
(d) all

152. Which of the following is a microbial source of lipase
(a) *Aspergillus niger*
(b) Candida lipolytica
(c) Geotrichum candidum
(d) all

153. Which of the following is a microbial source of rennet
(a) Endothia parasitica
(b) Mucor pusillus
(c) both
(d) none

154. Which of the following can be produced using bioprocesses
(a) amino acid (b) nucleotide
(c) ethanol (d) all

155. Which of the following can be produced using bioprocesses
(a) citric acid (b) acetic acid
(c) glutamic acid (d) all

156. Which of the following is a microbial polysaccharide
(a) pullulans (b) xanthan
(c) dextran (d) all

157. Generally modes of fermentation are of ________ types
(a) three (b) four
(c) five (d) six

158. Which of the following is a mode of fermentation
(a) batch (b) fed batch
(c) continuous (d) all

159. Which of the following is not a disadvantage of batch fermentation
(a) high cost
(b) optimum product recovery
(c) time loss
(d) wastage of unused nutrients

160. Which of the following is a type of homogenously mixed reactor
(a) chemostat (b) turbidostat
(c) both (d) none

161. Which of the following is a natural penicillin
(a) benzyl penicillin
(b) propicillin
(c) methicillin (d) all

162. Which of the following is a natural penicillin
(a) propicillin
(b) amylopenicillin
(c) methicillin (d) all

163. Which of the following is a semisynthetic penicillin
(a) propicillin
(b) amylopenicillin
(c) benzylpenicillin
(d) all

164. Which of the following is a semisynthetic penicillin
(a) amylpenicillin (b) methicillin
(c) benzylpenicillin (d) all

165. Penicillin may cure
(a) pneumonia (b) scarlet fever
(c) tetanus (d) all

166. Which of the following can produce cephalosporin
(a) Cephalosporium chrysogenum
(b) Cephalosporium acremonium
(c) both (d) none

167. Which of the following can produce cephalosporin
(a) Emericellopsis (b) Paecilomyces
(c) both (d) none

168. Which of the following antibiotic is used as food preservative
(a) nisin (b) pimarisin
(c) both (d) none

169. Which of the following antibiotic is used as antitumor antibiotics
(a) actinomysin (b) mitomysin C
(c) both (d) none

170. Which of the following is associated with enzyme immobilization
(a) cross-linked enzyme
(b) carrier bound enzyme
(c) gel enclosures
(d) all

171. Which of the following is a method of enzyme immobilization
(a) adsorption
(b) covalent binding
(c) both
(d) none

172. Which of the following is a method of enzyme immobilization
(a) entrapment
(b) membrane confinement
(c) both
(d) none

173. Active form of bakers yeast is
(a) compressed yeast
(b) active yeast
(c) cream yeast
(d) all

174. Cream yeast contain water content around
(a) 85% (b) 50%
(c) 35% (d) 15%

175. Compressed yeast contain water content around
(a) 27% (b) 40%
(c) 70% (d) 90%

176. Cream yeast contain dry matter content around
(a) 27% (b) 40%
(c) 70% (d) 90%

177. Production of ethanol from glucose occurs by
(a) EM pathway (b) TCA cycle
(c) glycolysis (d) none

178. Yeast use din manufacture of alcoholic beverage is
(a) S. cerevisiae
(b) S. carlbergensis
(c) both (d) none

179. Dry adjunct used in beer making is
(a) corn (b) rice
(c) sorghum (d) all

180. Which of the following is related with beer manufacturing
(a) mash (b) wort
(c) both (d) none

181. Which of the following has highest per capita consumption of beer
(a) Germany (b) Italy
(c) France (d) Sweden

182. Which of the following has lowest per capita consumption of beer
(a) Canada (b) Italy
(c) Austria (d) Denmark

183. Hops provide ________ flavor in beer
(a) bitter (b) salty
(c) sour (d) none

184. Hops are obtained from
(a) Cyamopsis tetragonolobus
(b) Humulus lupulus
(c) Cicer aretinum
(d) none

185. Which of the following is a bottom fermenting yeast
(a) S. uvarum
(b) S. carlsbergensis
(c) both
(d) none

186. Which of the following is a top fermenting yeast
(a) S. uvarum
(b) S. carlsbergensis
(c) S. cerevisiae (d) none

187. EM pathway ferementsugar to
(a) ethanol (b) CO_2
(c) both (d) none

188. The wine containers can be made of
(a) wood
(b) metal
(c) glass-rein forced polyester
(d) all

189. For red wines, fermentation takes 3-5 days at ________ °C
(a) 20 (b) 25-30
(c) 35-40 (d) none

190. Which of the following is a distilled beverage
(a) brandy (b) rum
(c) whisky (d) all

191. Industrial alcohol can be prepared from
(a) molasses (b) wheat starch
(c) cassava starch (d) all

192. Which of the following is used for industrial alcohol production
(a) Saccharomyces cerevisiae
(b) Zymomonas mobilis
(c) both
(d) none

193. Which of the following enzyme is used in industrial alcohol production
(a) amylase (b) glucoamylase
(c) both (d) none

194. Which of the following homoferm-entative LAB occurs in sourdough
(a) *L. acidophilus* (b) *L. casei*
(c) L. delbrueckii (d) all

195. Which of the following homoferm-entative LAB occurs in sourdough
(a) L.farciminis (b) L. plantarum
(c) both (d) none

196. Which of the following heteroferme-ntative LAB occurs in sourdough
(a) *L. brevis* (b) *L. buchneri*
(c) *L. fermentum* (d) all

197. Which of the following heteroferm-entative LAB occurs in sourdough
(a) L. fructovorans
(b) L. sanfrancisco
(c) both
(d) none

198. Which of the following yeast occurs in sourdough

(a) Saccharomyces cerevisiae
(b) Saccharomyces exiguus
(c) Pichia saitoi
(d) all

199. Which of the following yeast occurs in sourdough
(a) Candida crusei
(b) Torulopsis holmii
(c) both (d) none

200. Trickling generator is used in production of
(a) ethanol (b) acetic acid
(c) yoghurt (d) all

201. Acetic acid is industrially produced from
(a) Acetobacter (b) Gluconobacter
(c) both (d) none

202. Acetic acid is industrially produced from
(a) Acetobacter aceti
(b) Acetobacter pasteurianus
(c) Acetobacter peroxidans
(d) all

203. Acetic acid is industrially produced from
(a) Acetobacter peroxidans
(b) Gluconobacter oxydans
(c) both (d) none

204. Citric acid is industrially produced from
(a) *Aspergillus niger*
(b) Candida lypolytica
(c) both (d) none

205. Aspergillus grow on ________ for citric acid production
(a) sucrose (b) molasses
(c) both (d) none

206. Candida lypolytica grow on ________ for citric acid production
(a) paraffin substrate
(b) sucrose
(c) molasses (d) none

207. Which of the following process is used for citric acid production
(a) surface (b) submerged
(c) both (d) none

208. Submerged process for citric acid production can be done by
(a) batch
(b) fed batch
(c) continuous process
(d) all

209. Which of the following is used for lactic acid production
(a) Lactobacillus delbrueckii
(b) *L. bulgaricus*
(c) L. pentosus
(d) all

210. Which of the following is used for malic acid production
(a) *Aspergillus niger*
(b) A. flavus
(c) A. oryzae
(d) all

211. Which of the following polysaccharide can be obtained from microorganisms
(a) alginate (b) cellulose
(c) chitosan (d) all

212. Which of the following polysaccharide can be obtained from microorganisms
(a) curdlan (b) dextran
(c) gellan gum (d) all

213. Which of the following polysaccharide can be obtained from microorganisms
(a) pullulan (b) xanthan
(c) both (d) none

214. Xanthan gum is composed of
(a) D-glucose (b) D-mannose
(c) D-glucuronic acid (d) all

215. The ratio of D-glucose, D-mannose and D-glucuronic acid in xanthan is respectively
(a) 1:1:1 (b) 2:2:1
(c) 2:1:2 (d) 1:1:2

216. Xanthan gum is commercially produced from
(a) Xanthomonas (b) Pseudomonas
(c) Lactobacillus (d) none

217. Microbial alginate can be produced from
(a) Azotobacter (b) Pseudomonas
(c) both (d) none

218. Dextran can be produced by
(a) Leuconostoc (b) Streptococcus
(c) Aerobacter (d) all

219. Gellan gum is composed of
(a) glucose (b) rhamnose
(c) glucuronic acid (d) all

220. Gellan gum can be obtained from
(a) Pseudomonas elodea
(b) Alcaligenes faecalis
(c) both (d) none

221. Curdlan can be produced from
(a) Pseudomonas elodea
(b) Alcaligenes faecalis
(c) both
(d) none

222. Pullulan can be obtained from
(a) Aureobasidum (b) Azotobacter
(c) Arthrobacter (d) all

223. Which of the following can be obtained from *Aspergillus* sp.
(a) pectinase (b) cellulose
(c) lactase (d) all

224. Which of the following can be obtained from *Aspergillus* sp.
(a) α-amylase (b) glucoamylase
(c) both (d) none

225. Which of the following can be obtained from *Bacillus* sp.
(a) α-amylase
(b) glucose isomerase
(c) both
(d) none

226. Whch of the following is associated with immobilized enzyme
(a) bound
(b) microencapsulation
(c) entrapped
(d) all

8.3 ANSWER KEY

Q.	A.	Q.	A.	Q.	A.	Q.	A.	Q.	A.	Q.	A.
1.	(a)	2.	(c)	3.	(b)	4.	(a)	5.	(b)	6.	(b)
7.	(c)	8.	(a)	9.	(b)	10.	(b)	11.	(a)	12.	(c)
13.	(d)	14.	(c)	15.	(c)	16.	(d)	17.	(d)	18.	(d)
19.	(c)	20.	(d)	21.	(c)	22.	(d)	23.	(d)	24.	(d)
25.	(d)	26.	(c)	27.	(b)	28.	(d)	29.	(c)	30.	(c)
31.	(d)	32.	(d)	33.	(b)	34.	(b)	35.	(a)	36.	(b)
37.	(b)	38.	(d)	39.	(d)	40.	(d)	41.	(d)	42.	(c)
43.	(a)	44.	(d)	45.	(d)	46.	(d)	47.	(c)	48.	(c)
49.	(c)	50.	(b)	51.	(c)	52.	(a)	53.	(c)	54.	(d)
55.	(d)	56.	(d)	57.	(c)	58.	(a)	59.	(a)	60.	(d)
61.	(d)	62.	(a)	63.	(c)	64.	(a)	65.	(c)	66.	(a)

Q.	A.	Q.	A.	Q.	A.	Q.	A.	Q.	A.	Q.	A.
67.	(b)	68.	(a)	69.	(b)	70.	(a)	71.	(c)	72.	(a)
73.	(b)	74.	(c)	75.	(b)	76.	(a)	77.	(b)	78.	(c)
79.	(a)	80.	(b)	81.	(c)	82.	(a)	83.	(b)	84.	(b)
85.	(b)	86.	(b)	87.	(d)	88.	(a)	89.	(b)	90.	(a)
91.	(b)	92.	(a)	93.	(a)	94.	(d)	95.	(c)	96.	(d)
97.	(d)	98.	(c)	99.	(a)	100.	(b)	101.	(c)	102.	(a)
103.	(c)	104.	(a)	105.	(d)	106.	(b)	107.	(b)	108.	(a)
109.	(b)	110.	(d)	111.	(c)	112.	(b)	113.	(c)	114.	(a)
115.	(a)	116.	(d)	117.	(c)	118.	(b)	119.	(c)	120.	(c)
121.	(a)	122.	(c)	123.	(b)	124.	(a)	125.	(b)	126.	(a)
127.	(d)	128.	(d)	129.	(d)	130.	(d)	131.	(d)	132.	(c)
133.	(d)	134.	(a)	135.	(d)	136.	(d)	137.	(d)	138.	(d)
139.	(d)	140.	(d)	141.	(d)	142.	(c)	143.	(d)	144.	(c)
145.	(c)	146.	(c)	147.	(d)	148.	(c)	149.	(c)	150.	(d)
151.	xxx	152.	(d)	153.	(c)	154.	(d)	155.	(d)	156.	(d)
157.	(a)	158.	(d)	159.	(b)	160.	(c)	161.	(a)	162.	(b)
163.	(a)	164.	(b)	165.	(d)	166.	(c)	167.	(c)	168.	(c)
169.	(c)	170.	(d)	171.	(c)	172.	(c)	173.	(d)	174.	(a)
175.	(c)	176.	(a)	177.	(a)	178.	(c)	179.	(d)	180.	(c)
181.	(a)	182.	(b)	183.	(a)	184.	(b)	185.	(c)	186.	(c)
187.	(c)	188.	(d)	189.	(b)	190.	(d)	191.	(d)	192.	(c)
193.	(c)	194.	(d)	195.	(c)	196.	(d)	197.	(c)	198.	(d)
199.	(c)	200.	(b)	201.	(c)	202.	(d)	203.	(c)	204.	(c)
205.	(c)	206.	(a)	207.	(c)	208.	(d)	209.	(d)	210.	(d)
211.	(d)	212.	(d)	213.	(c)	214.	(d)	215.	(b)	216.	(a)
217.	(c)	218.	(d)	219.	(d)	220.	(a)	221.	(b)	222.	(d)
223.	(d)	224.	(c)	225.	(c)	226.	(d)				

CHAPTER 9

FOODS AND NUTRITION

9.1 SUBJECTIVE KEYNOTES

1. Anything that we eat and is beneficial for our body is food.
2. Nutrition is a science which tells us about what happens to the food when it enters our body and thereafter.
3. Nutrients are invisible chemicals in the foods which are necessary for keeping the body healthy.
4. The protein from animal sources is known as animal protein and the protein from plant sources is known as plant protein.
5. One gram of protein gives 4 kcal. An adult needs 1 gm of protein for every kilogram of body weight.
6. One gram of carbohydrate gives 4 kcal. An adult person needs 400-420 gm of carbohydrates every day.
7. One gram of fat gives you 9 kcal.
8. Animal fats increase the chances of heart related diseases.
9. An adult needs only 50-60 gm of fat every day.
10. The minerals calcium and phosphorus are found in bones and teeth.
11. Minerals do not provide energy & are not destroyed during food preparation.
12. Calcium and phosphorus are known as macro or major minerals, as they are present in the body in large amounts especially in bones and teeth.
13. Milk and its products like butter and ghee are the best sources of calcium.
14. These are the substances which occur in very small amounts in foods, and are essentials for proper functioning of the body.
15. Vitamins cannot be manufactured by the body, therefore must be supplied by the diet.
16. Those vitamins which are soluble in fats are known as fat soluble vitamins. These are Vitamins A, D, E, and K.
17. Those Vitamins which are soluble in water are known as water soluble vitamins. These are Vitamins of B Group and C.
18. Dietary fibre, also known as roughage, is the undigestible carbohydrate present in food.
19. Fibre is present in foods of plant origin only.

20. Deficiency disease is a condition which occurs in the body when a particular nutrient in daily meals has been missing for a prolonged period.
21. When there is lack of proteins and carbohydrates in the diet for a long time the deficiency disease known as Protein Energy Malnutrition (PEM) occurs.
22. PEM deficiency occurs mainly in children below 5 years of age.
23. When the protein is deficient but calories or energy from carbohydrates is sufficient, the condition known as Kwashiorkor occurs.
24. When protein and energy both are insufficient for a prolonged period then a disease condition known as marasmus occurs.
25. Deficiency of Vitamin A leads to dryness of conjunctiva (the white portion of the eye).
26. Beri Beri is quite prevalent in areas where polished rice was the staple food.
27. Vitamin C deficiency disorder known as scurvy is endemic to people who do not eat fresh fruits and vegetables.
28. Deficiency of iodine leads to a number of disorders like goitre, cretinism, mental retardation and deaf-mutism.
29. The only solution to goitre is to eat iodised salt.
30. Cereals form the bulk of our daily meals & are the main and also the cheapest source of energy.
31. Sprouting cereals helps in improving their vitamin C content.
32. Flesh foods including meat, poultry and fish are rich in proteins, vitamin A and Vitamin B.
33. Fruits especially citrus fruits, amla and guava, are good sources of vitamin C.
34. The 5-carbon monosaccharide ribose is an important component of coenzymes (e.g., ATP, FAD, and NAD).
35. Catabolism is the metabolic reaction cells undergo to extract energy.
36. Glycolysis is the name given to the series of biochemical reactions in which glucose is converted into pyruvate.
37. Trigylcerides have a glycerol backbone bonded to three fatty acids.
38. In most naturally occurring unsaturated fatty acids, all are cis bonds.
39. Two essential fatty acids are linoleic acid (LA) and alpha-linolenic acid (ALA).
40. Amino acids are molecules containing an amine group, a carboxylic acid group and a side chain that varies between different amino acids.
41. In enzymatic reactions, the molecules at the beginning of the process are called substrates, and they are converted into different molecules, called the products.
42. There are actually eight separate vitamins in the B family: thiamin (vitamin B1), riboflavin (vitamin B2), niacin, vitamin B6, folate, vitamin B12, biotin, and pantothenic.
43. Meat fats are rich in saturated fatty acids. The cholesterol content of meat is 75 mg / 100 gm.
44. Nuts are a rich source of protein and fat and a good source of B – Vitamin and antioxidant vitamin E.

45. The unsaturated glycerides in oil can be converted to more saturated glycerides by the addition of hydrogen through the process known as hydrogenation.
46. Smoking temperature is defined as the lowest temperature at which visible fumes consisting of volatile gaseous products of decomposition are evolved.
47. Mannitol is used in sugar free chewing gum and xylitol is used in pastries, jam, ice cream.
48. Alitame is a dipeptide based amide. It is 2000 times sweeter than sucrose.
49. Aspartame is made by combining two amino acids – aspartic acid and phenyl alanine.
50. Saccharin is sodium orthobenzene sulphonamide or its calcium salt & is 300 times sweeter than sucrose.
51. Allergy or allergic reactions are the immunologically induced tissue response to a foreign substance.
52. Antioxidants are the group of substances that prevent the damage caused by the oxidation of fatty acids and proteins by oxygen free radicals.
53. Atherosclerosis is the gradual deposition of fatty materials and fibrous tissues in the inner lining of the arteries with eventual obstruction of blood flow.
54. Body Mass Index is the body weight in relation to height. Body weight in kilograms divided by height in metres^2.
55. Weaning foods are the foods which are used during gradual transition of the infant from breastfeeding to a normal diet.
56. The diet containing all essential macro and micro nutrients in optimum quantities and in appropriate proportions that meet the requirements.
57. Anaboilsm is the process by which complex materials in tissues and organs are built up from simple substances.
58. A group of substances that prevent the damage caused by the oxidation of fatty acids and proteins by oxygen free radicals.
59. Beta-Carotene is a yellow - orange plant pigment which yields vitamin A by oxidation in the body.
60. A substance in human milk which stimulates the growth of a *Lactobacillus bifidus* in the infants' intestine is known as Bifidus factor.
61. Unit used to indicate the energy value of foods is known as calorie.
62. Process of breakdown of complex organic constituents in the body is known as catabolism.
63. Cataract is the opacity of the lens of the eye resulting in impaired vision.
64. Cholesterol is a lipid constituent of blood and tissues derived from diet as well as from synthesis within the body.
65. Inflammation and scarring of liver tissues is known as cirrhosis. It results in impaired liver function.
66. The milk produced by mammals during the first few days after delivery is known as colostrums. It is high in immunoglobulins.

67. Disease of the heart due to inadequate blood supply as a result of narrowing/obstruction of coronary arteries which nourish heart muscle is known as coronary heart disease.
68. Diabetes mellitus is a disease in which the blood glucose is increased and the body tissues cannot use glucose properly.
69. Fatty acids such as linoleic acid and alpha linolenic acid which are not made in the human body and must be supplied through the diet are known as essential fatty acids.
70. Structural parts of plant tissues which are resistant to the human digestive enzymes are known as dietary fibres.
71. Foods are classified into different groups for exchange. Each "exchange list" includes a number of measured foods of similar nutritive value that can be substituted interchangeably in meal plans.
72. Free radicals are the highly reactive oxygen-derived species formed in the body during normal metabolic processes. They have the capacity to damage cellular components by oxidation.
73. High-density lipoproteins transport cholesterol from the extra-hepatic tissues to the liver. They are anti-atherogenic.
74. Substances produced by a gland (endocrine) which are secreted directly into the blood stream to produce a specific effect on another organ are known as hormones.
75. An increase in the concentration of blood lipids is known as hyperlipidemia.
76. Ischaemia is lack of blood supply to an organ or tissue resulting in reduced oxygen supply, caused either by constriction or obstruction of the blood vessel.
77. Lactose intolerance is a disorder resulting from improper digestion of milk sugar called lactose, due to lack of an enzyme, lactase, in the intestinal mucosa.
78. Linoleic acid is a fatty acid containing 18 carbon atoms and two double bonds. The first double bond is on the sixth carbon atom from the methyl end. Therefore it is called n-6 fatty acid and is abbreviated as 18:2 n-6.
79. Technical term used for fats is lipids . They are important dietary constituents. The group includes triglycerides, steroids, cholesterol and other complex lipids.
80. Low-density lipoproteins transport cholesterol from the liver to tissues. High blood levels indicate that more cholesterol is being transported to tissues.
81. Macrocytic anaemia is a type of anaemia characterized by red blood cells which are larger than normal.
82. Microcytic anaemia is a type of anaemia characterized by red blood cells which are smaller than normal.
83. Micronutrients are the nutrients which are required in small quantities, such as vitamins and trace elements.
84. Abnormal porousness or thinning of bones is known as osteoporosis.
85. Recommended Dietary Allowances (RDA) are the amounts of dietary energy and nutrients considered sufficient for maintaining good health by the people of a country.
86. The foods which have been processed to improve their appearance, colour, taste, odour or keeping quality are known as refined foods.

87. Stroke is a term used for cerebro-vascular disease, a sudden condition that arises from blocking or bleeding of blood vessels in the brain, resulting in paralysis.
88. The condition in which the blood changes from a liquid to a semi-solid state and produces a blood clot (thrombus) which blocks blood flow is known as thrombosis.
89. Feeling of satisfaction after food intake is termed as satiety.
90. Trans-fatty acids are mainly produced during hydrogenation of oils; a few also occur naturally in very small quantities.
91. Triglycerides is a complex of fatty acids and glycerol. It is major type of dietary fat and the principal form in which energy is stored in the body.
92. Bitot's spots is the clinical sign of vitamin A deficiency, characterized by dryness of the eyes accompanied by foamy accumulations on the conjunctiva that often appear near the outer edge of the iris.
93. Theobromine is an alkaloid compound with a molecular structure similar to caffeine. Theobromine has a mild stimulating effect on humans, and is found in certain foods, such as cocoa and chocolate.
94. Albumen index is a measure of egg quality which compares the ratio of the height of a mass of egg white to its diameter as it sits on a flat surface, with a higher index indicating higher quality.
95. Amyloid is a non-nitrogenous food consisting mostly of starch.
96. Barlow's disease is a disease in children caused by a vitamin C deficiency, analogous to scurvy in adults.
97. Bomb calorimeter is a device for measuring the amount of heat released during chemical combustion.
98. Cholestasis is the failure of bile to reach the stomach, owing to an obstruction.
99. Chymorrhoea is the flow of chyme from the stomach to the duodenum.
100. Uric acid is a chemical compound that is formed from nitrogen in waste products from the body and that also forms crystals in the joints of people who have gout.

9.2 MULTIPLE CHOICE QUESTIONS

1. The largest single source of energy in diet is

(a) carbohydrates (b) protein
(c) fat (d) minerals

2. Carbohydrates have empirical formula

(a) $Cx(H_2O)y$ (b) Cx(CHO)y
(c) Cx(CO)y (d) Cx(OH)y

3. The building bloacks of carbohydrates are

(a) monosaccharides
(b) disaccharides
(c) polysaccharides
(d) oligosaccharides

4. Disaccharides have ___ units of sugar

(a) 1 (b) 2
(c) 3 (d) 4

5. The formula for sucrose is

(a) $C_{12}H_{22}O_{11}$ (b) $C_{11}H_{24}O_{12}$
(c) $C_{12}H_{28}O_{16}$ (d) none

6. Oligosaccharides have ________ units of sugar

(a) 2 (b) 3-11
(c) 15 (d) 1

7. Carbohydrates that can be digested and absorbed in human small intestine is referred to as
(a) noncarbohydrates
(b) glycaemic carbohydrates
(c) nonglycaemic carbohydrates
(d) all

8. Plant polysaccharides in foods resistant to hydrolysis by human diges-tive system are called
(a) dietary fiber
(b) soluble fiber
(c) insoluble fiber
(d) all

9. Which are glycaemic carbohydrates
(a) glucose (b) raffinose
(c) stachyose (d) all

10. Which are non glycaemic carbohydrates
(a) fructose (b) glucose
(c) raffinose (d) none

11. Stachyose and verbascose are
(a) glycaemic sugar
(b) non-glycaemic sugar
(c) polysaccharides
(d) none

12. Glycaemic refers to ability to rise blood _________ levels after digestion and absorption
(a) glucose (b) cholesterol
(c) protein (d) all

13. RDS stands for
(a) Rapidly Digestible Starch
(b) Role of Dietary Starch
(c) Rapidly Digestible Sugars
(d) none

14. SDS stands for
(a) Slowly Digestible Starch
(b) Sugar of Dietary Starch
(c) Slowly Digestible Sugars
(d) none

15. RS stands for
(a) Resistant Sugar
(b) Resistant Starch
(c) Role of Starch
(d) none

16. When all glucose is released in under 20 minutes, starch is referred as
(a) RDS (b) SDS
(c) RS (d) none

17. When all glucose is released in 20-120 minutes, starch is referred as
(a) RDS (b) SDS
(c) RS (d) none

18. When starch is not hydrolysed after 120 minutes, starch is referred as
(a) RDS (b) SDS
(c) RS (d) none

19. DP stands for
(a) Degree of Polymer
(b) Degree of Polymerization
(c) Degree of Protein
(d) none

20. Which of the following are synthetic carbohydrates
(a) polydextrose (b) neosugar
(c) maltodextrins (d) all

21. Polydextrose has a texture similar to
(a) fat (b) sucrose
(c) lactose (d) none

22. Total dietary fiber intake per day for adult males is
(a) 10g (b) 25g
(c) 50g (d) 100g

23. Total dietary fiber intake per day for adult females is
(a) 19g (b) 29g
(c) 39g (d) 49g

24. Pectins and gums are
(a) soluble fiber
(b) insoluble fiber
(c) resistant starch (d) none

25. Which of the following is largely fermented in intestine
(a) soluble fiber
(b) insoluble fiber
(c) resistant starch
(d) none

26. Lactulose is a disaccharide with
(a) galactose & fructose
(b) lactose & fructose
(c) galactose & glucose
(d) lactose & glucose

27. The glycaemic index of glucose is
(a) 100 (b) 50
(c) 25 (d) 0

28. White bread glycaemic index is
(a) 100 (b) 50
(c) 25 (d) 75

29. Glycaemic index is expressed as
(a) g/min (b) kg/day
(c) mg/day (d) none

30. Rice glycaemic index is
(a) 100 (b) 50
(c) 25 (d) 75

31. Legume glycaemic index is
(a) 100 (b) < 48
(c) < 25 (d) 75

32. Maltose glycaemic index is
(a) 105 (b) 50
(c) 25 (d) 75

33. Fructose glycaemic index is
(a) 100 (b) 50
(c) 20 (d) 75

34. Sucrose glycaemic index is
(a) 109 (b) 59
(c) 29 (d) 79

35. In the presence of oxygen glucose is oxidized to
(a) CO_2 (b) lactate
(c) CHO (d) alcohol

36. The excess of carbohydrates in the diet are converted into storage form
(a) glycogen (b) galactogen
(c) starch (d) none

37. The glycogen reserves are stored in
(a) pancreas (b) liver
(c) spleen (d) stomach

38. The glycogen stores in an adult human is _____g
(a) 10-100 (b) 150-200
(c) 250-500 (d) 500-750

39. An adult brain requires _______ g glucose/day
(a) 40 (b) 140
(c) 20 (d) 200

40. Fructosuria is due to _____ of enzyme fructokinase
(a) absence (b) hyperactivity
(c) slow activity (d) none

41. Fructokinase is involved in
(a) synthesis of fructose
(b) breakdown of fructose
(c) excretion of fructose
(d) none

42. Increased amount of dietary fiber in diet, ________ the chances of obesity
(a) increase (b) decrease
(c) no effect (d) none

43. The major portion of dietary lipids is
(a) phospholipids (b) TGA
(c) lipoproteins (d) none

44. Saturated fatty acid contains
(a) C-C (b) C=C
(c) C≡C (d) none

45. The principal fatty acid of butter is
(a) palmitic acid (b) butyric acid
(c) acetic acid (d) stearic acid

46. The principal fatty acid of peanut is
(a) arachdic acid (b) butyric acid
(c) acetic acid (d) stearic acid

47. The principal fatty acid of coconut oil is
(a) palmitic acid (b) butyric acid
(c) lauric acid (d) stearic acid

48. Cholesterol is present in
(a) plant food (b) animal food
(c) both (d) none

49. Commonly occurring fatty acids are in ________ form
(a) trans (b) cis
(c) both (d) none

50. Bile acids are
(a) polar (b) non polar
(c) amphipathic (d) none

51. Which of the following lipoproteins has lowest density
(a) chylomicrons
(b) very low density lipoprotein
(c) low density lipoprotein
(d) high density lipoprotein

52. Most of the cholesterol (70%) is carried by ________ in the plasma
(a) chylomicrons
(b) very low density lipoprotein
(c) low density lipoprotein
(d) none

53. 1 kcal = ________ kJ
(a) 4.184 (b) 4.814
(c) 4.418 (d) 4.148

54. Respiratory quotient is the ratio of
(a) CO_2 produced/ O_2 consumed
(b) O_2 consumed/CO_2 produced
(c) CO_2 consumed/O_2 produced
(d) O_2 produced/ CO_2 consumed

55. Fat metabolism require ________ oxygen than carbohydrates
(a) higher (b) lower
(c) equal (d) none

56. Respiratory quotient for carbohydrates is
(a) 1 (b) < 1
(c) > 1 (d) 0

57. Respiratory quotient for protein is
(a) 1 (b) < 1
(c) > 1 (d) 0

58. Respiratory quotient for fats is
(a) 1 (b) < 1
(c) > 1 (d) 0

59. Respiratory quotient for organic acids is
(a) 1 (b) < 1
(c) > 1 (d) 0

60. BMR stands for
(a) Basal Metabolic Reaction
(b) Body Metabolic Reaction
(c) Basal Metabolic Rate
(d) Body Metabolic Rate

61. BMR is affected by
(a) age (b) sex
(c) physiological state (d) all

62. BMR is measured in
(a) kJ/min (b) mJ/day
(c) both (d) none

63. TEM stands for
(a) Thermal Effect of Meal
(b) Total Effect of Meal
(c) Thermal Effect of Meat
(d) Total Effect of Meat

64. Salivary amylase is also known as
(a) α-amylase (b) glucoamylase
(c) β-amylase (d) none

65. α-amylase of saliva acts on
(a) protein (b) fat
(c) carbohydrates (d) none

66. The gastric juices are
(a) highly basic (b) highly acidic
(c) neutral (d) none

67. Pepsinogen is an ________ form of enzyme
(a) active
(b) inactive
(c) not a form of enzyme
(d) none

68. The pH in the stomach is
(a) 1-2.5 (b) 3.5-5
(c) 5.5-7 (d) 7.5-9

69. Trypsinogen is converted to trypsin by
(a) pepsinogen
(b) rennin
(c) enterokinase
(d) chymotrypsinogen

70. PDCAAS stands for
(a) Protein Digestibility Corrected Amino Score
(b) Protein Digestibility Common Amino Score
(c) Protein Dietary Common Amino score
(d) all

71. PDCAAS is used for analyzing
(a) protein quantity
(b) protein quality
(c) lipoprotein quality
(d) none

72. DIT stands for
(a) Diet Induced Thermogenesis
(b) Diet Induced Therpy
(c) Diet Therpy
(d) none

73. SEF stands for
(a) specific effect of food
(b) specific enthalpy of food
(c) standard efficiency
(d) none

74. Sattvic foods are also known as
(a) fresh (b) processed
(c) IMF (d) meat foods

75. Rajsic foods are also known as
(a) fresh (b) processed
(c) IMF (d) meat foods

76. Tamasic foods are also known as
(a) fresh (b) processed
(c) decayed (d) meat foods

77. Cholesterol free means________of cholesterol
(a) < 2mg (b) <5mg
(c) <10mg (d) <20mg

78. Low cholesterol means ________ of cholesterol
(a) <20mg (b) <40mg
(c) <60mg (d) <80mg

79. Reduced/less cholesterol refers to atleast
(a) 25% less (b) 50% less
(c) 75% less (d) 80% less

80. Sugar free refers to
(a) <0.5g/serving (b) <5g/serving
(c) <10g/serving (d) none

81. Reduced sugar refers to atleast ________% sugar/serving
(a) <2.5% (b) <5%
(c) <10% (d) none

82. High fiber refers to
(a) >0.5g/serving (b) >1g/serving
(c) >2g/serving (d) >5g/serving

83. Added fiber refers to atleast ________ g more fiber/serving
(a) 2.5 (b) 0.5
(c) 1.0 (d) 1.5

84. Calorie free refers to ________ cal/serving
(a) <1 (b) <2
(c) <4 (d) <5

85. Low calorie refers to ________ cal/serving
(a) <10 (b) <20
(c) <30 (d) <40

86. Fat free refers to ________ g/serving
(a) <2.5 (b) <0.5
(c) <1.0 (d) <1.5

87. Low fat refers to ________ g/serving
(a) <1 (b) <2
(c) <3 (d) <5

88. Sodium free refers to ___ mg/ serving
(a) < 2.5 (b) < 0.5
(c) < 1.0 (d) < 5.0

89. Low sodium refers to ___ mg/ serving
(a) < 30 (b) < 50
(c) < 80 (d) < 140

90. PEM refers to
(a) protein energy malnutrition
(b) protein efficiency metabolism
(c) protein efficiency malnutrition
(d) protein energy malnutrition

91. PEM leads to
(a) kwashiorkar (b) marasmus
(c) both (d) none

92. NIN stands for
(a) National Indian Nutrition
(b) National Institute of Nutrition
(c) Non Ionic Nutrition
(d) none

93. NIN is situated at
(a) New Delhi (b) Chennai
(c) Hyderabad (d) Patna

94. Cirhossis is caused due to
(a) alcohol consumption
(b) fatty foods
(c) high fiber
(d) high protein

95. The organ related to cirhossis
(a) small intestine (b) liver
(c) pancreas (d) spleen

96. The peak concentration of alcohol in blood is ________ g/dl
(a) 0.015 (b) 0.005
(c) 0.025 (d) 0.05

97. The permitted limit of blood alcohol for driving is
(a) 0.015 (b) 0.005
(c) 0.025 (d) 0.05

98. Metabolism of alcohol occurs in
(a) small intestine (b) liver
(c) pancreas (d) spleen

99. Chief alcohol metabolizing enzyme
(a) ADH (b) CDH
(c) AAD (d) CDE

100. ADH stands for
(a) alcohol dehydrogenase
(b) aldehyde
(c) adenine hydrolysate
(d) none

101. Death can occurs at blood alcohol concentration of__________g/dl
(a) 0.45-0.6 (b) 0.08-0.15
(c) 0.2-0.4 (d) none

102. Alcohol _________ the basal metabolic rate (BMR)
(a) increases (b) decreases
(c) no effect (d) none

103. Alcohol hepatitis is caused by
(a) virus
(b) over consumption of alcohol
(c) bacteria
(d) all

104. Women drinking alcohol during pregnancy give birth to a baby with
(a) unusual facial appearance
(b) prenatal growth impairment
(c) postnatal growth impairment
(d) all

105. Ethanol used _____________ for its metabolism in body
(a) niacin
(b) ascorbic acid
(c) thiamin (d) tocopherol

106. Wernicke's encephalopathy occurs due to acute________ deficiency in alcohol drinkers
(a) thiamin (b) vitamin C
(c) lysozyme (d) none

107. CHD stands for
(a) Coronary Heart Disease
(b) Chronic Health Disease
(c) Chronic Heart Disorder
(d) none

108. Moderate alcohol consumption ______ ______ the chances of CHD
(a) enhances (b) reduces
(c) no effect (d) none

109. Recommended levels of alcohol/day for men is
(a) 0-5g (b) 5-10g
(c) 10-12g (d) 20-30g

110. Recommended levels of alcohol/day for women is
(a) 0-5g (b) 5-10g
(c) 10-12g (d) 20-30g

111. 1 g of NaCl contains ________ mmol of sodium which is the principal cation in most extracellular fliud
(a) 17.1 (b) 11.1
(c) 1.1 (d) 5.1

112. Vitamin D deficiency in children causes
(a) rickets (b) marasmus
(c) both (d) none

113. Majority of the iron in the body is in
(a) haemoglobin of RBC
(b) liver
(c) spleen
(d) pancreas

114. Recommended iron intake by FAO/ WHO for men is ________ mg/day
(a) 1.0 (b) 6.0
(c) 11 (d) 35

115. RDA stands for
(a) Recommended Dietary allowances
(b) Role of Dieatry analysis
(c) both
(d) none

116. RDA depends on
(a) age (b) sex
(c) physical work (d) all

117. BMR of males is ________ than females
(a) higher (b) lower
(c) similar (d) none

118. Sedentory workers require ________ calories than hardworking people
(a) higher (b) lower
(c) similar (d) none

119. RDA of protein for infant 0-6 months is ________ g/kg body weight
(a) 2 (b) 4
(c) 6 (d) 8

120. RDA of calcium for infant 0-6 months is _____ mg/kg body weight
(a) 200 (b) 400
(c) 500 (d) 800

121. RDA of folic acid for infant 0-6 months is _____ mcg/kg body weight
(a) 25 (b) 45
(c) 65 (d) 85

122. Human milk has ________ essential amino acids
(a) no (b) all
(c) few (d) none

123. RDA of iron for infant 0-6 months is _____ mg/kg body weight
(a) 1 (b) 5
(c) 10 (d) 15

124. Nutrition word is derived from
(a) nutricus (b) nutritious
(c) nutra (d) nutrias

125. Nutricus means
(a) essential
(b) suckle at the breast
(c) eat healthy
(d) none

126. The yellowish fluid that comes from the mammary gland is called
(a) closstrict (b) proteinaceous
(c) colostrum (d) none

127. The % daily value is based on a ______ ______ calorie diet
(a) 2000 (b) 2500
(c) 3000 (d) 3500

128. Which of the following maynot be used a claim on a food label
(a) calorie free (b) low calorie
(c) sugar free (d) low sugar

129. Which of the following foods is not exempted from food labeling
(a) whole coffee beans
(b) dehydrated vegetables
(c) plain instant tea
(d) unpopped popcorn

130. The basal metabolic rate of a human being is not affected by
(a) diet (b) size
(c) age (d) sex

131. A calorie is the amount of energy required to raise _______ of water by one degree celcius
(a) 1 ounce (b) 1g
(c) 1kg (d) none

132. In food, carbohydrates supply _______ kcal/g
(a) 1 (b) 2
(c) 3 (d) 4

133. In food, proteins supply ______ kcal/g
(a) 1 (b) 2
(c) 3 (d) 4

134. In food, fat supply _______ kcal/g
(a) 1 (b) 4
(c) 3 (d) 9

135. Protein is required for
(a) bacterial inhibition
(b) proper bowel function
(c) production of antibodies
(d) absortion of water

136. Which of the following work together to maintain chemical, fluid and electrical balance between tissue cells and blood
(a) Ca & P
(b) Na & K
(c) Fe & Vitamin C
(d) Ca & Vitamin D

137. Which of the following cannot be digested, absorbed but looks, feels and behaves like fat
(a) olestra (b) simplesse
(c) saccharin (d) none

138. Which of the following is a macronutrient needed by our bodies to maintain health
(a) Cu (b) Zn
(c) Mg (d) Fe

139. A fatty acid does not contain which of the following elements
(a) C (b) N
(c) O (d) H

140. Starch is a
(a) protein (b) carbohydrates
(c) fat (d) mineral

141. Which of the food components is primarily derived from fruits, vegetables and grains
(a) protein (b) carbohydrates
(c) fat (d) mineral

142. Fats and oils are part of a family of compounds called
(a) protein (b) carbohydrates
(c) lipids (d) mineral

143. Glucose melts at ______°C
(a) 50 (b) 100
(c) 150 (d) 200

144. ______ reacts with amino acid when milk is heated to contribute to the tan color and caremellized flavor of cooked milk products
(a) casein (b) lactose
(c) whey (d) milk fat

145. A process that changes the shape of a protein molecule without breaking in covalent bonds is
(a) denaturation
(b) coagulation
(c) agglutination
(d) saturation

146. The use of food additives in US is regulated by
(a) FDA (b) USDA
(c) both (d) none

147. A food containing 8g fat, 4g carbohydrates and 5g protein, gives total calorie
(a) 88 (b) 108
(c) 93 (d) 113

148. Vitamin D is added to milk to prevent a condition called
(a) scurvy (b) pellagra
(c) rickets (d) beriberi

149. Energy lost when water molecules from ice crystals is
(a) specific heat
(b) latent heat
(c) heat of fusion
(d) heat of vaporization

150. Fruits and vegetables are primarily composed of
(a) carbohydrates (b) proteins
(c) water (d) mineral

151. Water activity of pure water is _____
(a) 1 (b) 2
(c) 3 (d) 4

152. Inorganic elements essential for human growth and health are
(a) vitamins (b) mineral
(c) protein (d) fiber

153. _____ is the time a food product can be stored before deteriorating
(a) retail life
(b) refrigeration life
(c) quality life
(d) shelf life

154. Food heats up in microwave oven primarily due to vibration of _______ molecules
(a) water (b) fat
(c) protein (d) carbohydrates

155. GRAS is
(a) Generally Regarded As Safe
(b) Generally Recognized As Safe
(c) Green Reagent And Safe
(d) none

156. Building block of protein is
(a) amino acid (b) TGA
(c) peptides (d) none

157. An addition of the nutrient to food such as vitmin D to milk is called
(a) irradiation (b) fermentation
(c) nutrification (d) fortification

158. Processing losses of following can be as high as 100% in case of
(a) Vitamin K (b) Vitamin D
(c) Vitamin C (d) none

159. Catabolism is
(a) synthesis of new compounds needed for used in body
(b) breakdown of compounds in body
(c) interconversion of compounds
(d) none

160. RDA of thiamin in adult is _____ mg/day
(a) 1 (b) 2
(c) 3 (d) 4

161. Bitot's spots is caused due to deficiency of
(a) Vitamin K (b) Vitamin D
(c) Vitamin C (d) Vitamin A

162. 6μg of β-carotene gives rise to ______ of Vitamin A
(a) 1 μg (b) 5 μg
(c) 10 μg (d) 100 μg

163. Water soluble vitamin is
(a) B (b) C
(c) D (d) both (a) & (b)

164. Pellagra is caused due to deficiency of
(a) Vitamin B_3 (b) Vitamin B_1
(c) Vitamin B_2 (d) Vitamin B_6

165. Pellagra means
(a) sour mouth (b) sour skin
(c) reduced vision (d) none

166. ________ is the disease of 3Ds
(a) pellagra (b) anaemia
(c) nightblindness (d) none

167. The Ds in pellagra are
(a) dermatitis (b) diarhhoea
(c) dermentia (d) all

168. Vitamin B_6 occurs in nature in _______ form
(a) pyridoxine (b) pyridoxal
(c) pyridoxamine (d) all

169. The human requirement of biotin is______ µg/day
(a) 10 (b) 20
(c) 30 (d) 40

170. RDA for folate in adults is ______ µg/day
(a) 100 (b) 200
(c) 300 (d) 400

171. Megaloblastic anaemia is caused due to deficiency of
(a) folic acid (b) ascorbic acid
(c) fatty acid (d) amino acid

172. Dehydroascorbic acid is ______ form of ascorbic acid
(a) oxidized (b) reduced
(c) native form (d) none

173. Antisterlity vitamin is ______
(a) vitamin C (b) vitamin E
(c) vitamin K (d) vitamin D

174. The chemical name of Vitamin E is
(a) tocopherol (b) glutathione
(c) pyridoxine (d) none

175. BMI stands for
(a) Body Measurement Index
(b) Body Mass Index
(c) Bodily Measurement Index
(d) Body Measurement Indication

176. Unit of BMI is
(a) kg/m^2 (b) $kg.m^3$
(c) kg/m (d) none

177. BMI is an index of
(a) adiposity
(b) metabolic rate
(c) energy requirement
(d) none

178. Underweights have BMI _____ kg/m^2
(a) 1 (b) <18.5
(c) 18.5-24.9 (d) >30

179. Normal adults have BMI _____ kg/m^2
(a) 1 (b) <18.5
(c) 18.5-24.9 (d) >30

180. Overweight have BMI _____ kg/m^2
(a) 1 (b) <18.5
(c) 18.5-24.9 (d) >30

181. Obese adult have BMI _____ kg/m^2
(a) 1 (b) <18.5
(c) 18.5-24.9 (d) >30

182. Marasmus means
(a) wasting (b) utilizing
(c) weak (d) none

183. Oedema is present in
(a) marasmus (b) kwashiorkor
(c) both (d) none

184. Diabetes mellitus is characterized by
(a) hyperglycemia
(b) frequent urination
(c) hypoglycemia
(d) all

185. IDDM
(a) Insulin Dependent Diabetes Mellitus
(b) Insulin Deferring Diabetes Mellitus
(c) Insulin Diet Diabetes Mellitus
(d) Insulin Dependent Disease Mellitus

186. Insulin is produced by
(a) α-cells of pancreas
(b) β-cells of pancreas
(c) gall bladder
(d) liver

187. Which of the following constitute the macronutrients
(a) carbohydrates (b) protein
(c) fat (d) all

188. Which of the following constitute the micronutrients
(a) carbohydrates (b) protein
(c) fat (d) vitamins

189. UNICEF means
(a) United Nations International Children Educational Fund
(b) United National Intelligent Children Educational Fund
(c) United Nations International Children Educational Function
(d) none

190. Which of the following element forms the major proportion of body weight
(a) oxygen (b) carbon
(c) hydrogen (d) nitrogen

191. Our body proteins varies from _______% body weight
(a) 1-5 (b) 5-10
(c) 10-15 (d) 15-20

192. Our body minerals varies from _______% body weight
(a) 3-5 (b) 5-10
(c) 10-15 (d) 15-20

193. Our body water varies from_______% body weight
(a) 10-42 (b) 42-70
(c) 70-80 (d) 80-90

194. Which of the following are non proteinogenic amino acid which have nutritional value
(a) taurine (b) ornithine
(c) citrulline (d) all

195. AIDS is
(a) Acquired Immuno Deficiency Virus
(b) Agent of Immuno Deficiency Virus
(c) Acquired Immuno Defective Virus
(d) none

196. Formation of proteins from amino acid is
(a) catabolism (b) anabolism
(c) metabolism (d) none

197. Formation of urea from proteins is
(a) catabolism (b) anabolism
(c) metabolism (d) none

198. Which of the following is essential amino acid in humans
(a) lysine (b) methionine
(c) valine (d) all

199. PDCAAS is expressed as the _____ requirement on relation to the total protein requirement
(a) amino acid (b) protein
(c) calorie (d) none

200. Which has the highest lysine content among the following
(a) legumes (b) cereals
(c) nuts (d) fruits

201. Which has the highest threonine content among the following
(a) legumes (b) animal food
(c) nuts (d) fruits

202. PDCAAS for wheat is
(a) 40 (b) 60
(c) 80 (d) 100

203. PDCAAS for rice is
(a) 46 (b) 56
(c) 76 (d) 96

204. PDCAAS for beef is
(a) 40 (b) 60
(c) 80 (d) >100

205. The limiting amino acid in rice, wheat, maize and sorghum is
(a) methionine (b) lysine
(c) cysteine (d) all

206. The limiting amino acid in beef is
(a) methionine
(b) lysine
(c) cysteine
(d) both (b) & (c)

207. DP means
(a) degree of polymerization
(b) data of polymerization
(c) degree of polymer
(d) data of polymer

208. DP of oligosaccharides is
(a) 1-2 (b) 3-9
(c) >9 (d) none

209. PDCAAS for maize is
(a) 43 (b) 53
(c) 73 (d) 93

210. DP of polysaccharides is
(a) 1-2 (b) 3-9
(c) >9 (d) none

211. DP of monosaccharides is
(a) 1 (b) 3-9
(c) >9 (d) none

212. 95% of carbohydrates are digested in
(a) mouth
(b) stomach
(c) small intestine
(d) large intestine

213. The ductless glands are
(a) endocrine (b) exocrine
(c) both (d) none

214. Hormones are ________ chemicals
(a) nutritive (b) non nutritive
(c) fatty acids (d) amino acids

215. Pituitary hormones are released from
(a) hypothalamus (b) thymus
(c) pancreas (d) all

216. Oxytocin and vasopressin are released from
(a) anterior pituitary
(b) posterior pituitary
(c) both (d) none

217. Melanocyte stimulating hormone is released from
(a) anterior pituitary
(b) posterior pituitary
(c) pars intermedia (d) all

218. Prolactin, thyroid stimulating hormone, lutenising hormone is released from
(a) anterior pituitary
(b) posterior pituitary
(c) pars intermedia (d) all

219. __________ is known as anti-diuretic hormone
(a) vasopressin (b) prolactin
(c) oxytocin (d) thyroid

220. ______ acts on the smooth musclesof our body and stimulate their contraction
(a) vasopressin (b) prolactin
(c) oxytocin (d) thyroid

221. _______ induces ovulation in females
(a) vasopressin
(b) prolactin
(c) oxytocin
(d) lutenising hormone

222. ________ stimulates the growth and development of ovarian follicles in females
(a) follicle stimulating hormone
(b) prolactin
(c) oxytocin
(d) thyroid

223. Melatonin which influences metabolism and pigmentation & menstrual cycle is released from
(a) pituitary (b) pineal
(c) thymus (d) pancreas

224. Deficiency of iodine in our diet results in
(a) hypothyroidism
(b) hyperthyroidism
(c) both
(d) none

225. During goitre, there is ________ of thyroid gland
(a) enlargement (b) shrinkage
(c) no change (d) none

226. The emergency hormones are
(a) adrenaline
(b) nor adrenaline
(c) both
(d) none

227. Adrenaline and nor adrenaline are released from
(a) adrenal medulla
(b) adrenal cortex
(c) posterior pituitary
(d) thymus

228. Glucocorticoids which stimulate gluconeogenesis, lipolysis and proteolysis are relased by
(a) adrenal medulla
(b) adrenal cortex
(c) posterior pituitary
(d) thymus

229. Glucagon is released by
(a) α-cells of pancreas
(b) β-cells of pancreas
(c) both
(d) none

230. Glucagon is ________ hormone
(a) peptide
(b) carbohydrates
(c) fatty acid
(d) none

231. The blood glucose levels are maintained by
(a) glucagon (b) thyroid
(c) adrenaline (d) melatonin

232. Glucagon is ________ hormone
(a) hyperglycemic (b) hypoglycemic
(c) both (d) none

233. Insulin is released from
(a) α-cells of pancreas
(b) β-cells of pancreas
(c) both
(d) none

234. Insulin is a ____________ hormone
(a) peptide
(b) carbohydrates
(c) fatty acid
(d) none

235. Insulin stimulates conversion of ________
(a) amino acid to protein
(b) glucose to glycogen
(c) glycogen to glucose
(d) none

236. Insulin is a cause of
(a) hyperglycemia (b) hypoglycemia
(c) both (d) none

237. The glucose homeostasis in blood is maintained by
(a) insulin & glucagon
(b) FSH & LH
(c) vasopressin
(d) prolactin

238. Prolonged hyperglycemia leads to diabetes
(a) incidipus (b) mellitus
(c) both (d) none

239. ANF
(a) Atrial Natriuretic Factor
(b) Anti Natriuretic Factor
(c) Atrial Nutrition Factor
(d) none

240. ANF ________ the blood pressure in human body
(a) decrease (b) increase
(c) no effect (d) none

241. The formation of RBC is called
(a) erythropoietin (b) erythropoiesis
(c) both (d) none

242. Which of the following are protein hormones
(a) insulin (b) hypothalamic
(c) pituitary (d) all

243. Which of the following are steroid hormones
(a) cortisol (b) estradiol
(c) testosterone (d) all

244. Iodo thyronines are
(a) thyroid hormone
(b) pituitary hormone
(c) pancreas hormones
(d) none

245. Amino acid derivatives are
(a) epinephrine (b) oxytocin
(c) cortisol (d) testosterone

246. Largest gland of the body
(a) heart (b) liver
(c) brain (d) stomach

247. In stomach, the food mixes with acidic gastric juice and is called
(a) chyme (b) chyle
(c) bile (d) none

248. Bile has
(a) enzymes (b) no enzymes
(c) amylase (d) protease

249. A blood group has _____ antigen on RBC's
(a) A (b) B
(c) AB (d) nil

250. B blood group has _____ antigen on RBC's
(a) A (b) B
(c) AB (d) nil

251. AB blood group has _____ antigen on RBC's
(a) A (b) B
(c) both (d) nil

252. O blood group has _____ antigen on RBC's
(a) A (b) B
(c) AB (d) nil

253. A blood group has _____ antibodies in plasma
(a) anti A (b) anti B
(c) AB (d) nil

254. B blood group has _____ antibodies in plasma
(a) anti A (b) anti B
(c) AB (d) nil

255. AB blood group has _____ antibodies in plasma
(a) anti A (b) anti B
(c) AB (d) nil

256. O blood group has _____ antibodies in plasma
(a) anti A (b) anti B
(c) both (d) nil

257. A blood group individual can donate blood to
(a) A,O (b) B,O
(c) AB (d) O

258. B blood group individual can donate blood to
(a) A,O (b) B,O
(c) AB (d) O

259. AB blood group individual can receive blood from
(a) A,O (b) B,O
(c) AB (d) all

260. O blood group individual can donate blood to
(a) A,O (b) B,O
(c) both (a) & (b) (d) none

261. AB are
(a) universal recipients
(b) universal donors
(c) both
(d) none

262. O are
(a) universal recipients
(b) universal donors
(c) both (d) none

263. A normal heart beats _____ times a minute
(a) 62 (b) 72
(c) 82 (d) 92

264. The duration of cardiac cycle is_____ seconds
(a) 1 (b) 0.5
(c) 0.8 (d) 1.5

265. Hypertension is a term used for blood pressure higher than
(a) 120/80 (b) 110/80
(c) 100/80 (d) 90/80

266. The term lipid was introduced by
(a) Chevreul (b) Bloor
(c) Vorel (d) none

267. Phospholipids were discovered by
(a) Chevreul (b) Bloor
(c) Thudicum (d) none

268. Myristic is a ________ fatty acid
(a) saturated
(b) unsaturated
(c) poly unsaturated
(d) none

269. Elaidic is a ________ fatty acid
(a) saturated
(b) monounsaturated
(c) poly unsaturated
(d) none

270. Arachidonic a ________ fatty acid
(a) saturated
(b) unsaturated
(c) poly unsaturated
(d) none

271. Short chain fatty acids are
(a) water soluble
(b) water insoluble
(c) both (d) none

272. PUFA stands for
(a) Poly Unsaturated Fatty Acid
(b) Poly Unused Fatty Acid
(c) Poly Unsaturated Fat & Acids
(d) none

273. Linoleate and α-linolenate are in _________ form
(a) cis (b) trans
(c) saturated (d) none

274. Partial dehydrogenation of fatty acids during processing of oils leads to formation of
(a) trans fats (b) cis fatty acid
(c) PUFA (d) none

275. Triglycerides are composed of 1 molecule of glycerol and _____ fatty acids
(a) 1 (b) 2
(c) 3 (d) 4

276. Bile acids are stored in
(a) liver (b) gall bladder
(c) stomach (d) speen

277. Hydrogenation bring ________ in fatty acid
(a) unsaturation (b) saturation
(c) polyunsaturation (d) none

278. Trans fatty acid ________ LDL blood cholesterol
(a) increase (b) decrease
(c) no effect (d) none

279. Trans fatty acid ________ HDL blood cholesterol
(a) increase (b) decrease
(c) no effect (d) none

280. Trans fatty acid ________ the risk of cardiovascular disease
(a) increase (b) decrease
(c) no effect (d) none

281. Brain lipids contain ______ proportion of cholesterol
(a) low (b) high
(c) no (d) none

282. The naturally occurring fatty acids have ________ bond
(a) cis (b) trans
(c) both (d) none

283. DRI stands for
(a) Dietary Reference Intake
(b) Diet Reference Intake
(c) Daily Reference Intake
(d) none

284. An important blood clotting vitamin is
(a) K (b) B
(c) C (d) D

285. Chemical name of Vitamin A is
(a) Retinol (b) Calciferol
(c) Tocopherol (d) Phylloquinone

286. Chemical name of Vitamin D is
(a) Retinol (b) Calciferol
(c) Tocopherol (d) Phylloquinone

287. Chemical name of Vitamin E is
(a) Retinol (b) Calciferol
(c) Tocopherol (d) Phylloquinone

288. Chemical name of Vitamin K is
(a) Retinol (b) Calciferol
(c) Tocopherol (d) Phylloquinone

289. Chemical name of Vitamin B_2 is
(a) Riboflavin (b) Calciferol
(c) Tocopherol (d) Phylloquinone

290. Chemical name of Vitamin B_3 is
(a) Retinol (b) Niacin
(c) Tocopherol (d) Phylloquinone

291. Chemical name of Vitamin B_{12} is
(a) Retinol (b) Calciferol
(c) Cobalamin (d) Phylloquinone

292. Chemical name of Vitamin H is
(a) Retinol (b) Calciferol
(c) Tocopherol (d) Biotin

293. Pernicious anaemia is due to deficiency of
(a) Vit B_2 (b) Vit B_{12}
(c) Vit B_3 (d) Vit B_6

294. Impaired fat and carbohydrates metabolism is due to deficiency of
(a) Retinol (b) Calciferol
(c) Tocopherol (d) Biotin

295. Which vitamin is formed in the skin by UV radiation
(a) K (b) B
(c) C (d) D

296. Due to deficiency of vitamin D, ____________ can occur in older people
(a) rickets (b) osteomalacia
(c) PEM (d) none

297. Due to deficiency of vitamin D, _______ can occur in younger children
(a) rickets (b) osteomalacia
(c) PEM (d) none

298. The precursor of vitamin D in the skin is
(a) 7-dehydrocholesterol
(b) 6-dehydrocholesterol
(c) 5-dehydrocholesterol
(d) 4-dehydrocholesterol

299. Hemorrhagic disease can arise due to deficiency of vitamin
(a) K (b) B
(c) C (d) D

300. Vitamin C deficiency cause
(a) diabetes (b) scurvy
(c) rickets (d) pellagra

301. Niacin forms an important part of
(a) NAD (b) NADP
(c) FAD (d) both (a) & (b)

302. Riboflavin forms an important part of
(a) NAD (b) NADP
(c) FAD (d) both (a) & (b)

303. Chemical name of Vitamin B_5 is
(a) Retinol
(b) Niacin
(c) Tocopherol
(d) Pantothenic acid

304. ________ increases the absorption of iron in the body

(a) Vitamin C (b) Vitamin B
(c) Vitamin K (d) Vitamin D

305. ________ is referred to as vitamin Q

(a) ubiquinone (b) quinine
(c) inositol (d) none

306. Which of the following is a rich source of calcium

(a) cheese (b) eggs
(c) beet (d) bread

307. NFSM stands for

(a) National Food Security Mission
(b) Natural Foods Security Mission
(c) Nation Food Security Mission
(d) none

308. The BMR ________ with age

(a) increase (b) decrease
(c) no effect (d) none

309. BMR is the number of calories a person's body uses while at

(a) moderate activity
(b) at rest
(c) hypoactivity
(d) standing

310. The RDI daily intake of calcium is ______mg

(a) 600 (b) 1200
(c) 1800 (d) 2400

311. Body mass index is figured by dividing

(a) weight (kg)/height (m^2) × 100
(b) height (m^2)/ weight (kg) × 100
(c) both
(d) none

312. Higher value of BMI indicate

(a) no fat (b) low fat
(c) more fat (d) none

313. Lower values of BMI are related to

(a) shorter life (b) longer life
(c) no correlation (d) none

314. BMI for normal men & women shoud be in range of ______ kg/m^2

(a) 19-27 (b) 9-20
(c) 1-9 (d) 30-45

315. A BMI of < 17 indicates

(a) PEM (b) obesity
(c) both (d) none

316. A BMI of < 19 indicates

(a) PEM (b) obesity
(c) underweight (d) none

317. A BMI of 19-27 indicates

(a) PEM (b) obesity
(c) acceptable weight (d) none

318. A BMI of >27 indicates

(a) PEM (b) obesity
(c) both (d) none

319. Normal BMI for infants is ________ kg/m^2

(a) 13 (b) 26
(c) 6 (d) 20

320. Omega 6 fatty acid are considered as

(a) essential fatty acid
(b) semi-dispensable fatty acid
(c) non-essential fatty acid
(d) none

321. Omega 6 fatty acids are unsaturated fatty acid having carbon carbon double bond in the ______ position

(a) n-6 (b) n-2
(c) n-3 (d) n-9

322. ________ promotes the formulation of RBC and serves a component of vitamin B_{12}

(a) chromium (b) cobalt
(c) zinc (d) iron

323. Too much cholesterol in blood causes

(a) coronary heart disease
(b) goiter
(c) scurvy
(d) none

324. Which of the following is referred to as bad cholesterol
(a) LDL (b) HDL
(c) both (d) none

325. Which of the following is referred to as good cholesterol
(a) LDL (b) HDL
(c) both (d) none

326. TFA stands for
(a) total fatty acid
(b) trans fatty acid
(c) transitional fatty acid
(d) none

327. Cis form of fatty acid is
(a) bent form
(b) straight form
(c) mixture of bent and straight forms
(d) none

328. Trans form of fatty acid is
(a) bent form
(b) straight form
(c) mixture of bent and straight forms
(d) none

329. High intake of trans fatty acid causes ____ in cholesterol
(a) increase (b) decrease
(c) no effect (d) none

330. Which of the following is a powerful anti oxidant
(a) Vitamin E (b) Vitamin D
(c) Vitamin B (d) Vitamin K

331. Consumption of saturated fat in diet ____the bad cholesterol
(a) increase (b) decrease
(c) no effect (d) none

332. Diets withhighr mount of mono unsaturated fatty acid and PUFA _______the bad cholesterol levels
(a) increase (b) decrease
(c) no effect (d) none

333. Omega 3 and 6 fatty acids have _____ effect on cardiovascular health
(a) positive (b) negative
(c) no effect (d) none

334. GL stands for
(a) glycemic load (b) glucose load
(c) glucose ladder (d) none

335. Cholesterol levels acceptable are ______ mg/dL
(a) < 200 (b) > 200
(c) < 400 (d) > 400

336. High cholesterol levels are ______ mg/dL
(a) < 200 (b) > 240
(c) < 180 (d) > 600

337. LD in LD50 stands for
(a) large dose (b) lethal dose
(c) low dose (d) none

338. LC in LC 50 stands for
(a) large concentration
(b) lethal concentration
(c) low concentration
(d) none

339. LD 50 is the amount of the material given all at once, which causes the death of _______ % of group of test animals
(a) 25 (b) 50
(c) 75 (d) 100

340. LC value is concentration of chemical in air that kills ______ % of test animals in agiven time
(a) 25 (b) 50
(c) 75 (d) 100

341. LD 50 was invented by
(a) JW Trevan (b) Appert
(c) Robert Koch (d) none

342. LDLO stands for
(a) Low dose causing lethality
(b) lethal dose causing lethality
(c) lowest dose causing lethality
(d) none

343. TDLO stands for
(a) Low dose causing toxicity
(b) lethal dose causing total loss
(c) lowest dose causing toxic effect
(d) none

344. Smaller the LD 50 value ______ the toxic chemical is
(a) more (b) less
(c) no relation (d) none

345. Larger the LD 50 value ______ the toxic chemical is
(a) more (b) less
(c) no relation (d) none

346. Are irradiated foods nutritious
(a) yes (b) no
(c) cannot say (d) none

347. Neutraceutical word originate from
(a) nutrition (b) pharmaceutical
(c) both (d) none

348. Neitraceutical are available as
(a) part of food (b) capsule
(c) supplement (d) both (b) & (c)

349. WHO stands for
(a) World Health Organization
(b) Whole Health Organization
(c) World Heart Organization
(d) none

350. The nutritional status can be assessed by
(a) nutrional anthropometry
(b) clinical examination
(c) biochemical tests
(d) all

351. Anthropometry involves the measurement of
(a) weight (b) height
(c) head and chest (d) all

352. The height of an individual is measured by
(a) stadiometer (b) barometer
(c) inch tape (d) all

353. Cheilosis occurs due to deficiency of
(a) Vitamin A (b) riboflavin
(c) thiamine (d) niacin

354. Bleeding gum occurs due to deficiency of
(a) Vitamin C (b) riboflavin
(c) thiamine (d) niacin

355. Enlargement of thyroid gland is due to deficiency of
(a) Vitamin A (b) riboflavin
(c) iodine (d) niacin

356. Estimation of hemoglobin in blood detects ______ deficiency in body
(a) iron (b) potassium
(c) calcium (d) iodine

357. Which of the following is a 5 carbon carbohydrates
(a) erythrose (b) ribose
(c) glucose (d) all

358. Which of the following is a 4 carbon carbohydrates
(a) erythrose (b) ribose
(c) glucose (d) all

359. Fructose, a carbohydrates contain ___ group
(a) ketone (b) aldehyde
(c) phenol (d) both (b) & (c)

360. Which of the following is a heteropolysaccharide
(a) heparin (b) starch
(c) cellulose (d) all

361. Ptyalin is another name for
(a) salivary amylase
(b) pancreatic amylase
(c) trypsin
(d) enterokinase

362. Lactase enzyme converts lactose into
(a) glucose + glucose
(b) glucose + fructose
(c) glucose + galactose
(d) none

363. The conversion of glucose into glycogen in body is called
(a) glycogenesis
(b) glycolysis
(c) glyconeogenesis
(d) lipolysis

364. The conversion of glucose into fat in body is called
(a) glycogenesis
(b) glycolysis
(c) glyconeogenesis
(d) lipogenesis

365. Breakdown of glucose releasing energy is called
(a) glycogenesis
(b) glycolysis
(c) glyconeogenesis
(d) lipolysis

366. Conversion of glycogen to glucose is
(a) glycogenolysis
(b) glycolysis
(c) glyconeogenesis
(d) lipolysis

367. Glycogen is usually stored in the
(a) liver (b) spleen
(c) thyroid (d) pancreas

368. In India. ______% of day energy needs are met from carbohydrates
(a) 20-40 (b) 30-50
(c) 60-80 (d) 100

369. Dietary fibre is ______ digested by gastrointestinal secretion
(a) not (b) readily
(c) resist to being (d) none

370. Which of the following are soluble fibers
(a) gums (b) pectins
(c) mucilages (d) all

371. Which of the following are insoluble fibers
(a) cellulose (b) hemicellulose
(c) lignins (d) all

372. Extra calories from glucose are deposited in body as
(a) TGA (b) proteins
(c) minerals (d) resistant fibers

373. The formation of glycogen from non carbohydrates sources is called
(a) glucogenolysis
(b) glycolysis
(c) gluconeogenesis
(d) lipolysis

374. The starch splitting enzyme is
(a) amylase (b) protease
(c) lipase (d) all

375. The hormone ____________ secreted by cells of islets of langerhans helps in utilization of sugar by tissues
(a) insulin (b) thyroxin
(c) LH (d) melatonin

376. Proteins are composed of
(a) carbon (b) hydrogen
(c) nitrogen (d) all

377. Essential amino acids are also known as
(a) indispensable
(b) semi-dispensable
(c) both (d) none

378. None essential amino acids are also known as
(a) indispensable
(b) semi-dispensable
(c) dispensable
(d) none

379. Threonine, leucine and isoleucine are
(a) essential amino acid
(b) non essential amino acid
(c) semi dispensable amino acids
(d) none

380. Cysteine & glutamic acid are
(a) essential amino acid
(b) non essential amino acid
(c) semi dispensable amino acids
(d) none

381. A complete protein is one which has all ________ in relatively the same amount as human beings required to promote & maintain normal growth
(a) essential amino acid
(b) non essential amino acid
(c) semi dispensable amino acids
(d) none

382. Egg protein is a ______ protein and has highest bio value
(a) complete (b) incomplete
(c) semi complete (d) none

383. Proteins of animal foods are categorized as _____ protein
(a) good quality (b) poor quality
(c) no (d) none

384. Proteins of plant foods are categorized as _____ protein
(a) good quality (b) poor quality
(c) no (d) none

385. The limiting amino acid in wheat germ is
(a) lysine (b) tryptophan
(c) methionine (d) all

386. The biological value of egg proteins is
(a) 96 (b) 90
(c) 74 (d) 80

387. The biological value of milk proteins is
(a) 96 (b) 90
(c) 74 (d) 80

388. The biological value of meat proteins is
(a) 96 (b) 90
(c) 74 (d) 80

389. The biological value of rice proteins is
(a) 96 (b) 90
(c) 74 (d) 80

390. The biological value of wheat proteins is
(a) 66 (b) 90
(c) 74 (d) 80

391. The biological value of maize proteins is
(a) 96 (b) 50
(c) 74 (d) 80

392. The biological value of bengal gram proteins is
(a) 96 (b) 90
(c) 74 (d) 80

393. The biological value of ground nut proteins is
(a) 96 (b) 55
(c) 74 (d) 80

394. The biological value of gingelly proteins is
(a) 96 (b) 90
(c) 62 (d) 80

395. The biological value of fish proteins is
(a) 96 (b) 90
(c) 74 (d) 80

396. Animal foods are _______ source of protein
(a) moderate (b) poor
(c) rich (d) none

397. Cereals, millets nd legumes are ______ source of protein
(a) moderate (b) poor
(c) rich (d) none

398. Protein digesting enzyme are present in
(a) mouh
(b) stomach
(c) small intestine
(d) both (b) & (c)

399. ICMR stands for
(a) International Council of Medical Research
(b) Indian Council of Medical Research
(c) International Corporation of Medical Research
(d) none

400. ICMR recommends _____ g/kg/dayas safe level of intake in terms of dietary protein for Indians
(a) 1 (b) 2
(c) 5 (d) 10

401. Proteins are distinguished from carbohydrates and fat by presence of
(a) nitrogen (b) carbon
(c) oxygen (d) hydrogen

402. Amino acids are linked together by _____ bond
(a) hydrogen (b) peptide
(c) carbon-carbon (d) none

403. _____ protein is reference protein
(a) cereal (b) egg
(c) hydrogen (d) meat

404. Deficiency of protein is manifested as
(a) PEM (b) obesity
(c) goiter (d) cirrhosis

405. Waxes and oils are examples of _____ lipids
(a) simple (b) compound
(c) derived (d) none

406. Phospholipids, glycolipids and lipolipids are examples of _____ lipids
(a) simple (b) compound
(c) derived (d) none

407. Sterols and fatty acids are examples of _____ lipids
(a) simple (b) compound
(c) derived (d) none

408. Cholesterol is of _______ origin
(a) animal (b) plant
(c) both (d) none

409. Linolenic, linoleic and arachidonic acid are _____ fatty acids
(a) essential (b) non essential
(c) semi essential (d) none

410. Palmitic, oleic and butyric acid are example of
(a) essential (b) non essential
(c) semi essential (d) none

411. Vitamin A, K, E and D are _____ soluble vitamins
(a) fat (b) water
(c) both (d) none

412. The breakdown of fat to yield energy is called
(a) lipolysis (b) glycolysis
(c) proteolysis (d) none

413. The synthesis of cholesterol and phospholipids in body is called
(a) lipogenesis (b) lipolytic
(c) glycogenesis (d) glycolysis

414. Lipogenesis occurs in _____ in body
(a) liver (b) spleen
(c) pancreas (d) none

415. Hydrogenation causes _______ of fats
(a) solidification
(b) liquidification
(c) no effect
(d) none

416. _____ is produced by hydrogenation of vegetable oils
(a) vanaspati (b) butter
(c) ghee (d) refined oil

417. Toad skin or phrynoderma is due to
(a) excess of fat
(b) deficiency of fat
(c) excess of carbohydrates
(d) deficiency of proteins

418. _______ energy is used to maintain transport of nutrients
(a) osmotic (b) electrical
(c) chemical (d) thermal

419. ____________ energy is used for transmission of nerve impulse
(a) osmotic (b) electrical
(c) chemical (d) thermal

420. _______ energy is used for synthesis of new compounds
(a) osmotic (b) electrical
(c) chemical (d) thermal

421. ________ energy is used for heat regulation

(a) osmotic (b) electrical
(c) chemical (d) thermal

422. 1KJ is equal to ______ kcal

(a) 0.239 (b) 0.439
(c) 0.889 (d) 0.139

423. The instrument for measuring heat output of the body or energy value of food is

(a) calorimeter
(b) colorimeter
(c) thermometer
(d) none

424. The energy measured using a bomb calorimeter for 1 g of carbohydrates is ________ Kcal

(a) 4.1 (b) 5.1
(c) 6.1 (d) 7.1

425. The energy measured using a bomb calorimeter for 1 g of fats is ______Kcal

(a) 9.45 (b) 10.45
(c) 11.45 (d) 8.45

426. The energy measured using a bomb calorimeter for 1 g of protein is ______ Kcal

(a) 4.1 (b) 5.65
(c) 6.1 (d) 7.1

427. The coefficient of digestibility is the proportion of an ingested nutrient that ultimately becomes available to body cells. The coefficient of digestibility for carbohydrates is

(a) 1.98 (b) 0.98
(c) 2.98 (d) 3.98

428. The coefficient of digestibility for fat is

(a) 1.95 (b) 0.95
(c) 2.95 (d) 3.95

429. The coefficient of digestibility for protein is

(a) 1.92 (b) 0.92
(c) 2.92 (d) 3.92

430. The more lean bod mass ______ the BMR

(a) higher (b) lower
(c) non effect (d) none

431. Fever ________ the BMR by 7% for each degree rise in temperature in Fahrenheit

(a) increases (b) decreases
(c) no effect (d) none

432. Stree ________ the BMR

(a) increases (b) decreases
(c) no effect (d) none

433. Smoking and high intake of caffeine ______ BMR

(a) increases (b) decreases
(c) no effect (d) none

434. Children have ___________ BMR than adult

(a) higher (b) lower
(c) non effect (d) none

435. Starvation or fasting ________ BMR

(a) increases (b) decreases
(c) no effect (d) none

436. According to ICMR, sedentary workers require ________ kcal/hr

(a) 80-180 (b) 170-240
(c) 250-350 (d) none

437. According to ICMR, moderate workers require ________ kcal/hr

(a) 80-180 (b) 170-240
(c) 250-350 (d) none

438. According to ICMR, heavy workers require ________ kcal/hr

(a) 80-180 (b) 170-240
(c) 250-350 (d) none

439. Protein is not digested completely due to ________

(a) absence of nitrogen
(b) presence of nitrogen
(c) presence of carbon
(d) none

440. Macrominerals are those which are required in diet by more than ______ mg per day
(a) 100 (b) 50
(c) 20 (d) 10

441. Microminerals are those which are required in diet by more than______ mg per day
(a) 100 (b) 50
(c) 20 (d) 10

442. High per centage of calcium and phosphate is present in
(a) body fluid (b) bones
(c) nerves (d) glands

443. The ratio of calcium and phosphate in our body is
(a) 1:2 (b) 2:1
(c) 3:1 (d) 4:1

444. In skeletal systm calcium and phosphate are present in form of
(a) calcium
(b) hydroxyapatite crystals
(c) both
(d) none

445. RDA for calcium is ________ mg/day for an adult
(a) 100 (b) 200
(c) 300 (d) 400

446. RDA for calcium is ________ mg/day for pregnant and lactating women
(a) 1000 (b) 2000
(c) 3000 (d) 4000

447. ______ occurs when calcium in blood drops below critical level
(a) osteomalacia (b) tetany
(c) both (d) none

448. The total bod iron in adults is _____g
(a) 4 (b) 8
(c) 12 (d) 16

449. Iron forms a part of
(a) hemoglobin (b) myoglobin
(c) enzyme (d) all

450. RDA for iron in normal adult male is ________ mg/day
(a) 20 (b) 30
(c) 40 (d) 50

451. RDA for iron in normal adult female is _____mg/day
(a) 20 (b) 30
(c) 40 (d) 50

452. RDA for iron in lactating women is _____ mg/day
(a) 20 (b) 30
(c) 40 (d) 50

453. RDA for iron in pregnant women is _____ mg/day
(a) 20 (b) 30
(c) 40 (d) 50

454. Anemia occurs when hemoglobin falls below _________ g/dl in adult man and woman
(a) 10 (b) 8
(c) 12 (d) 14

455. Spoon shapes nails develop due to deficieny of
(a) calcium (b) iron
(c) zinc (d) potassium

456. ______ mg of iodine is present in an adult human body
(a) 10-15 (b) 15-23
(c) 23-28 (d) 28-33

457. Iodine is essential for synthesis of ______ thyroid hormones
(a) T_3 (b) T_4
(c) both (d) none

458. The RDA for iodine is for an adult is _______ µg/day
(a) 50 (b) 100
(c) 150 (d) 200

459. Cretinism occurs in children due to deficiency of
(a) calcium (b) iron
(c) iodine (d) potassium

460. Goitrogens are the substances present in food which ______ goiter
(a) prevents (b) causes
(c) no effect (d) none

461. The total amount of zinc in body is ______ g
(a) 0.3 (b) 1.3
(c) 2.3 (d) 4.3

462. ______ is the constituent of hormone insulin
(a) zinc (b) iodine
(c) potassium (d) calcium

463. RDA requirement for zinc in adult is ______ mg/day
(a) 10.5 (b) 15.5
(c) 20.5 (d) 25.5

464. ______ vitamins are accumulated in the body
(a) water soluble (b) water insoluble
(c) both (d) none

465. Xerosis, in which the outer lining of eye ball loses its usual moisture and becomes white, dry and wrinkled is caused by inadequate intake of
(a) Vitamin A (b) Vitamin B
(c) Vitamin K (d) Vitamin C

466. ______ facilitates the calcium absorption and deposition in the bone
(a) Vitamin A (b) Vitamin B
(c) Vitamin K (d) Vitamin D

467. RDA requirement for vitamin D in adult is ______ µg/day
(a) 100 (b) 200
(c) 300 (d) 400

468. ______ is essential for the activation of prothrombin
(a) Vitamin A (b) Vitamin B
(c) Vitamin K (d) Vitamin D

469. ______ is an excellent anti-oxidant which combines with pre radicals oxidizing them to harmless substances that can be excreted
(a) Vitamin A (b) Vitamin B
(c) Vitamin K (d) Vitamin C

470. RDA requirement for vitamin C in adult is ______ mg/day
(a) 20 (b) 40
(c) 60 (d) 80

471. RDA requirement for vitamin C in children is ______ mg/day
(a) 20 (b) 40
(c) 60 (d) 80

472. RDA requirement for vitamin C in lactating women is ______ mg/day
(a) 20 (b) 40
(c) 60 (d) 80

473. Scorbutic rosary, a beaded appearance of the ribs results from inadequate intake of vitamin
(a) Vitamin A (b) Vitamin B
(c) Vitamin K (d) Vitamin C

474. Which of the following is not a type of beri beri
(a) dry (b) wet
(c) infantile (d) adult

475. The inflammation of the tongue is called
(a) glossitis (b) tonsils
(c) dermatitis (d) none

476. ____ is converted to niacin in humans
(a) tyrosine (b) tryptophan
(c) methionine (d) serine

477. RDA requirement for vitamin C in adult is ______ mg/day
(a) 20 (b) 40
(c) 60 (d) 100

478. RDA requirement for vitamin C in pregnant women is ______ mg/day
(a) 200 (b) 400
(c) 600 (d) 800

479. RDA requirement for vitamin C in children is ______ mg/day
(a) 20-40 (b) 50-70
(c) 70-100 (d) 110-140

480. Anti pernicious anaemia factor has two component, the vitamin B_{12} and cobalamine is

(a) extrinsic factor (b) intrinsic factor
(c) co factor (d) none

481. RDA requirement for B_{12} in adult is ______ mg/day

(a) 1 (b) 2
(c) 3 (d) 4

482. Nutraceutical is related to

(a) nutrition food (b) tasty food
(c) medicinal food (d) none

483. Water is a ________ compound

(a) colorless (b) calorieless
(c) both (d) none

484. Saliva is about ______% water

(a) 95 (b) 75
(c) 65 (d) 55

485. When pH falls below 7, it is termed

(a) acidity (b) alkalinity
(c) neutrality (d) none

486. When pH is above 7, it is termed

(a) acidity (b) alkalinity
(c) neutrality (d) none

487. The gastric juice of stomach is

(a) acidic (b) basic
(c) neutral (d) none

488. pH of gastric juice is

(a) 2 (b) 4
(c) 6 (d) 8

489. pH of urine is

(a) 2 (b) 4
(c) 6 (d) 8

490. A person should take enough water to excrete ______ ml of urine per day

(a) 600-1000 (b) 1200-1500
(c) 1500-1800 (d) 1800-2000

491. ______ ml of water is used up for respiration per day

(a) 100 (b) 200
(c) 300 (d) 400

492. ECF stands for

(a) Extra Cellular Fluid
(b) Extraneous Cellular Fluid
(c) Energy Circulating Fluid
(d) none

493. ICF stands for

(a) Intra Cellular Fluid
(b) Intrinsic Cellular Fluid
(c) Intake of Circulating Fluid
(d) none

494. Dehydration occurs due to excessive ______ of water

(a) loss (b) consumtion
(c) no effect (d) none

495. UNICEF formula consist of ________ g/lit water of NaCl

(a) 1.5 (b) 2.5
(c) 3.5 (d) 4.5

496. UNICEF formula consist of ________ g/lit water of $NaHCO_3$

(a) 1.5 (b) 2.5
(c) 3.5 (d) 4.5

497. UNICEF formula consist of ________ g/lit water of KCl

(a) 1.5 (b) 2.5
(c) 3.5 (d) 4.5

498. UNICEF formula consist of ________ g/lit water of glucose

(a) 20 (b) 30
(c) 40 (d) 50

499. Administration of water or fluid orally to correct dehydration is called

(a) rehydration (b) hydration
(c) both (d) none

500. Water which is safe and wholesome is called ________ water

(a) potable (b) distilled
(c) ground (d) none

501. The normal intake if water per day in adults is ________ ml

(a) 1400-2000 (b) 2400-3000
(c) 3400-4000 (d) 4400-5000

9.3 ANSWER KEY

Q.	A.	Q.	A.	Q.	A.	Q.	A.	Q.	A.	Q.	A.
1.	(a)	2.	(a)	3.	(a)	4.	(b)	5.	(a)	6.	(b)
7.	(b)	8.	(d)	9.	(a)	10.	(c)	11.	(b)	12.	(a)
13.	(a)	14.	(a)	15.	(b)	16.	(a)	17.	(b)	18.	(c)
19.	(b)	20.	(d)	21.	(a)	22.	(b)	23.	(a)	24.	(a)
25.	(a)	26.	(a)	27.	(a)	28.	(d)	29.	(d)	30.	(b)
31.	(b)	32.	(a)	33.	(c)	34.	(b)	35.	(a)	36.	(a)
37.	(b)	38.	(c)	39.	(b)	40.	(a)	41.	(b)	42.	(b)
43.	(b)	44.	(a)	45.	(b)	46.	(a)	47.	(c)	48.	(b)
49.	(b)	50.	(c)	51.	(a)	52.	(c)	53.	(a)	54.	(a)
55.	(a)	56.	(a)	57.	(b)	58.	(b)	59.	(c)	60.	(c)
61.	(d)	62.	(c)	63.	(a)	64.	(a)	65.	(c)	66.	(b)
67.	(b)	68.	(a)	69.	(c)	70.	(a)	71.	(b)	72.	(a)
73.	(a)	74.	(a)	75.	(b)	76.	(c)	77.	(a)	78.	(a)
79.	(a)	80.	(a)	81.	(a)	82.	(d)	83.	(a)	84.	(d)
85.	(d)	86.	(b)	87.	(c)	88.	(d)	89.	(d)	90.	(a)
91.	(c)	92.	(b)	93.	(c)	94.	(a)	95.	(b)	96.	(c)
97.	(d)	98.	(b)	99.	(a)	100.	(a)	101.	(a)	102.	(a)
103.	(b)	104.	(d)	105.	(c)	106.	(a)	107.	(a)	108.	(b)
109.	(d)	110.	(c)	111.	(a)	112.	(a)	113.	(a)	114.	(c)
115.	(a)	116.	(d)	117.	(a)	118.	(a)	119.	(a)	120.	(c)
121.	(a)	122.	(b)	123.	(a)	124.	(a)	125.	(b)	126.	(c)
127.	(a)	128.	(d)	129.	(d)	130.	(a)	131.	(b)	132.	(d)
133.	(d)	134.	(d)	135.	(c)	136.	(b)	137.	(a)	138.	(c)
139.	(b)	140.	(b)	141.	(b)	142.	(c)	143.	(c)	144.	(b)
145.	(a)	146.	(a)	147.	(b)	148.	(c)	149.	(b)	150.	(c)
151.	(a)	152.	(b)	153.	(d)	154.	(a)	155.	(a)	156.	(a)
157.	(d)	158.	(c)	159.	(b)	160.	(a)	161.	(d)	162.	(a)
163.	(d)	164.	(a)	165.	(b)	166.	(a)	167.	(d)	168.	(d)
169.	(c)	170.	(b)	171.	(a)	172.	(a)	173.	(b)	174.	(a)
175.	(b)	176.	(a)	177.	(a)	178.	(b)	179.	(c)	180.	(c)
181.	(d)	182.	(a)	183.	(b)	184.	(a)	185.	(a)	186.	(b)
187.	(d)	188.	(d)	189.	(a)	190.	(a)	191.	(c)	192.	(a)

Q. A.	Q. A.	Q. A.	Q. A.	Q. A.	Q. A.
193. (b)	194. (d)	195. (a)	196. (b)	197. (a)	198. (d)
199. (a)	200. (a)	201. (b)	202. (a)	203. (b)	204. (d)
205. (b)	206. (d)	207. (a)	208. (b)	209. (a)	210. (c)
211. (a)	212. (c)	213. (a)	214. (b)	215. (a)	216. (b)
217. (c)	218. (a)	219. (a)	220. (c)	221. (d)	222. (a)
223. (b)	224. (a)	225. (a)	226. (c)	227. (a)	228. (b)
229. (a)	230. (a)	231. (a)	232. (a)	233. (b)	234. (a)
235. (b)	236. (b)	237. (a)	238. (b)	239. (a)	240. (a)
241. (b)	242. (d)	243. (d)	244. (a)	245. (a)	246. (b)
247. (a)	248. (b)	249. (a)	250. (b)	251. (c)	252. (d)
253. (b)	254. (a)	255. (d)	256. (c)	257. (a)	258. (b)
259. (d)	260. (c)	261. (a)	262. (b)	263. (b)	264. (c)
265. (a)	266. (b)	267. (c)	268. (a)	269. (b)	270. (c)
271. (a)	272. (a)	273. (a)	274. (a)	275. (c)	276. (b)
277. (b)	278. (a)	279. (b)	280. (a)	281. (b)	282. (a)
283. (a)	284. (a)	285. (a)	286. (b)	287. (c)	288. (d)
289. (a)	290. (b)	291. (c)	292. (d)	293. (b)	294. (d)
295. (d)	296. (b)	297. (a)	298. (a)	299. (a)	300. (b)
301. (d)	302. (c)	303. (d)	304. (a)	305. (a)	306. (a)
307. (a)	308. (b)	309. (b)	310. (b)	311. (a)	312. (c)
313. (b)	314. (a)	315. (a)	316. (c)	317. (c)	318. (b)
319. (a)	320. (a)	321. (c)	322. (b)	323. (a)	324. (a)
325. (b)	326. (b)	327. (a)	328. (b)	329. (a)	330. (a)
331. (b)	332. (a)	333. (a)	334. (a)	335. (b)	336. (b)
337. (b)	338. (b)	339. (b)	340. (a)	341. (c)	342. (c)
343. (a)	344. (b)	345. (a)	346. (c)	347. (d)	348. (a)
349. (d)	350. (d)	351. (a)	352. (b)	353. (a)	354. (c)
355. (a)	356. (b)	357. (a)	358. (a)	359. (a)	360. (a)
361. (c)	362. (a)	363. (d)	364. (b)	365. (a)	366. (a)
367. (c)	368. (c)	369. (d)	370. (d)	371. (a)	372. (c)
373. (a)	374. (a)	375. (d)	376. (a)	377. (c)	378. (a)
379. (b)	380. (a)	381. (a)	382. (a)	383. (b)	384. (b)
385. (a)	386. (b)	387. (c)	388. (d)	389. (a)	390. (b)
391. (c)	392. (b)	393. (c)	394. (d)	395. (c)	396. (a)

Q.	A.	Q.	A.	Q.	A.	Q.	A.	Q.	A.	Q.	A.
397.	(d)	398.	(b)	399.	(a)	400.	(a)	401.	(b)	402.	(b)
403.	(a)	404.	(a)	405.	(b)	406.	(c)	407.	(a)	408.	(a)
409.	(b)	410.	(a)	411.	(a)	412.	(a)	413.	(a)	414.	(a)
415.	(a)	416.	(b)	417.	(a)	418.	(b)	419.	(c)	420.	(d)
421.	(a)	422.	(a)	423.	(a)	424.	(a)	425.	(b)	426.	(b)
427.	(b)	428.	(b)	429.	(a)	430.	(a)	431.	(a)	432.	(a)
433.	(a)	434.	(b)	435.	(a)	436.	(b)	437.	(c)	438.	(b)
439.	(a)	440.	(c)	441.	(b)	442.	(b)	443.	(b)	444.	(d)
445.	(a)	446.	(b)	447.	(a)	448.	(d)	449.	(a)	450.	(b)
451.	(b)	452.	(c)	453.	(c)	454.	(b)	455.	(b)	456.	(c)
457.	(c)	458.	(c)	459.	(b)	460.	(c)	461.	(a)	462.	(b)
463.	(b)	464.	(a)	465.	(d)	466.	(d)	467.	(c)	468.	(d)
469.	(b)	470.	(b)	471.	(d)	472.	(d)	473.	(d)	474.	(a)
475.	(b)	476.	(d)	477.	(b)	478.	(c)	479.	(a)	480.	(a)
481.	(c)	482.	(c)	483.	(a)	484.	(a)	485.	(b)	486.	(a)
487.	(a)	488.	(c)	489.	(b)	490.	(d)	491.	(a)	492.	(a)
493.	(a)	494.	(c)	495.	(b)	496.	(a)	497.	(a)	498.	(a)
499.	(a)	500.	(b)	501.	(§§§)						

CHAPTER 10

FOOD ADDITIVES

10.1 SUBJECTIVE KEYNOTES

1. Food additives are the substances other than basic food ingredient used for processing, preservation and storage of food products.
2. Food additives are generally used to solve product specific problems.
3. Fortification and enrichment are also related to food additives.
4. Antioxidants are the substances added to fats and fat containing foods (such as snacks) to retard oxidation and prolong keeping time.
5. BHA, BHT, PG & TBHQ are used as antioxidants in food products.
6. ADI value for BHA is upto 0.5mg/kg body weight and for BHT is 0.3mg/kg body weight.
7. Ascorbic acid and tocopherols are naturally occurring antioxidants.
8. Antioxidants prevent lipid oxidation by interrupting free radical chain mechanism.
9. Antioxidants prevent browning in cut fruits and vegetables.
10. Chelating agents are also known as sequestrants.
11. Chelating agents are the compounds that form complexes with metal ions and thus prevent oxidative rancidity and discoloration reactions.
12. Examples of chelating agents includes sorbic acid, poly carboxylic acid, polyphosphates and EDTA.
13. EDTA is the most common chelating agent used in foods.
14. Chelating agents are not antioxidants as they only serve as scavengers of metals which catalyze oxidation.
15. Color agents used in foods are of two types: natural and synthetic colors.
16. Terpenoids, flavonoids, sulfur compounds are used as flavoring compounds.
17. Four basic types of tastes are: sour, sweet, salty and bitter.
18. Umami is considered as the fifth taste.
19. Nitrite and nitrates are used as curing agent.
20. Curing agent contribute flavor and color stability to meat.
21. Nitrites inhibit the growth of Clostridium and Streptococcus.
22. Polyphosphates such as sodium tripolyphosphate is also used in meat curing.

23. Emilsifiers are used to stabilize the emulsion.
24. Lecithin, monoglycerides, polysorbate are used as emulsifier.
25. Mono-sodium glutamate and nucleotides can be used as flavor enhancers.
26. In yeast extract, flavor enhancing substances in ribonucleotides.
27. Benzoyl peroxide, chlorine gas and nitrosyl chloride are used as flour improvers.
28. Calcium silicate, tri calcium phosphate is generally used as anti-caking agent.
29. Potassium bromated, potassium iodate and calcium peroxide are used as dough improver.
30. Humectants are moisture retention agents.
31. Functions of humectants include control of viscosity, texture, moisture retention, reduction of water activity and control of crystallization.
32. Anti-caking agents are used to prevent lump formation.
33. Magnesium carbonate is generally used as anticaking agent in baking powder and table salt.
34. Thiamine, nicotinic acid, iron and calcium are added to the flour as nutrient supplements.
35. Cyclamate, saccharin, aspartame, acesulfame K, sucralose and neotame are used as non-nutritive sweeteners.
36. Citric acid is used as acidulant in various foods.
37. Sodium chloride, sugar, sulphur dioxide, nitrate and nitrite, sorbic acid, acetic acid, propionic acid, benzoic acid, parabens are used as preservative in various food products.
38. Acetic acid in the form of vinegar (4% acetic acid) is used as preservative in various foods like pickles.
39. Benzoic acid is widely used as anti-microbial agent.
40. Epoxides are cyclic ethers that destroy micro-organisms.
41. Nisin, pimaracin and chlorotetracycline are used as antimicrobiotics.
42. Stabilizers and thickeners are used to stabilize emulsion and foams, improve texture, inhibit crystallization and encapsulate flavors.
43. Examples of stabilizers and thickeners include gum Arabica, guar gum, carrageenan, alginic acid, starch, CMC and pectin.
44. Stabilizers and thickeners are hydrophilic in nature.
45. Stabilizer and thickeners are termed as hydrocolloids.
46. Enzymes are also used as food additives such as rennin pectinase.
47. Clarifying agents such as bentonite, gelatin and synthetic resins are used to remove haziness or sediment in fruit juices, beers and wines.
48. Aluminium sulphates and calcium salts are used as firming agent.
49. Red pigment bixin is derived from the seed coat of Bixa orellana.
50. Citric and phosphoric acids are used as acidulants in soft drink and beverages.

51. Sensitisation is a Part of the allergic process in which an individual becomes hypersensitive to a substance and suffers an allergic reaction when next exposed to it.
52. The “E” number system for food additives intended to assist as a short code for some of the lengthier chemical names and to indicate common European safety approval.
53. Antioxidants reduce the oxidative deterioration that leads to rancidity, loss of flavour, colour and nutritive value of foodstuffs.
54. Some antioxidants actually remove oxygen by self-oxidation, e.g. ascorbic acid, whilst others interfere in the mechanism of oxidation, e.g. tocopherols, gallic acid esters, BHA and BHT.
55. Monosodium glutamate is widely blamed for an intolerance reaction that became known as ‘Chinese Restaurant Syndrome’.
56. Acceptable Daily Intake (ADI)is the amount of the substance or additive that is considered safe for consumption on daily basis.
57. Foods that contain aspartame must be labelled with the words ‘contains a source of phenylalanine’.
58. Foods that contain more than 10% added polyols must carry the indication ‘excessive consumption may produce laxative effects’.
59. A processing aid is defined as a substance that is not consumed as a food ingredient by itself; is intentionally used in the processing of raw materials, foods or their ingredients, to fulfil technological purposes during treatment or processing; and is capable of resulting in the unintended but technically unavoidable presence of its residues or its derivatives in the finished product, and the residues of which do not present any risk to human health and do not have any technological effect on finished products.
60. A flavouring is a material used or intended for use in or on food to impart odour, taste or both.
61. ‘Miscellaneous additive’ refers to any food additive that is used or intended to be used primarily as an acid, acidity regulator, anti-caking agent, antifoaming agent, antioxidant, bulking agent, carrier, carrier solvent, emulsifier, emulsifying salt, firming agent, flavour enhancer, flour treatment agent, foaming agent, gelling agent, glazing agent, humectant, modified starch, packaging gas, preservative, propellant, raising agent, sequestrant, stabiliser or thickener; but does not include use as a processing aid or any enzyme except invertase or lysozyme.
62. An extraction solvent is any solvent used or intended to be used in an extraction procedure, including, in any particular case further to its use in such a procedure, any substance other than such a solvent derived exclusively from such a solvent.
63. A ‘flavour precursor’ is defined as a product, not necessarily having flavouring properties itself, intentionally added to food for the sole purpose of producing flavour by breaking down or reacting with other components during food processing.
64. The Food and Agriculture Organization (FAO) of the United Nations, and the World Health Organization (WHO) established the Codex Alimentarius Commission in 1963 after recognising a need for international industry standards for the food industry worldwide, in order to protect the health of consumers and to make international trade in food easier.

65. The International Numbering System (INS) has been developed for the purpose of identifying food additives in ingredients lists as an alternative to the declaration of the specific name, which can be lengthy and complex.
66. It is the role of the Joint FAO/WHO Expert Committee on Food Additives (JECFA) to evaluate food additives in terms of their toxicology and to prepare specifications for each additive (including ADI).
67. ADI for saccharin is 8mg/kg body weight. ADI for aspartame is 40mg/kg body weight.
68. ADI for acesulfame K is 15mg/kg body weight.
69. A food additive may be direct, secondary direct or indirect, depending on how it is used.
70. Direct food additives are divided into eight categories: food preservatives; coatings, films and related substances; special dietary and nutritional additives; anticaking agents; flavouring agents and related substances; gums, chewing-gum bases and related substances; other specific usage additives; and multipurpose additives.
71. Secondary direct food additives are components used in ingredients of processed foods that may become additives in the final food. These are divided into four categories: polymer substances and polymer adjuvants for food treatment; enzyme preparations and microorganisms; solvents, lubricants, release agents and related substances; and specific usage additives.
72. Indirect food additives are materials that may become part of a food as a component of packaging material, adhesives, food-processing equipment, surfaces and containers used for food handling, and certain production aids and sanitisers.
73. Curcumin is the principal pigment of turmeric, a spice that is obtained from the rhizomes of *Curcuma longa*. It is obtained by solvent extraction from the plant to produce an oleoresin, which is then purified by crystallisation. It provides a bright lemon-yellow colour when used in foods.
74. Tartrazine is a water-soluble synthetic dye commercially available as the sodium salt of the dye. Tartrazine is also available as the aluminium lake, which is waterinsoluble. Tartrazine is classed as a monoazo dye.
75. It has good stability to heat and light and has been allocated an ADI of 7.5 mg/kg body weight by JECFA.
76. Sunset yellow is a water-soluble synthetic dye commercially available as the sodium salt of the dye. Sunset yellow is also available as the aluminium lake, which is water insoluble. Sunset yellow is classed as a monoazo dye.
77. Bixin is the principal carotenoid pigment extracted from the seeds of Bixa Orellana. Alkaline hydrolysis of bixin converts it from an oilsoluble colour to the water-soluble pigment norbixin.
78. Annatto is a naturally derived colour that is used to provide orange shades in both lipid and aqueous food phases. The pure pigment has a high tinctorial strength with characteristic absorption maxima of 470–502 nm for bixin and 452–482 nm in respect of norbixin.
79. Paprika extract contains the oil-soluble carotenoid pigments capsorubin and capsanthin.

80. Lycopene is an oil-soluble pigment, the commercial preparations of which consist of a mixture of carotenoids, with lycopene being the principal constituent. The pure pigment has a high tinctorial strength with an absorption maximum in the region of 472 nm when measured in hexane.

81. Major pigment is beetroot red is betanin which shows absorption maximum is in the region of 535 nm when measured in aqueous solution at pH 5.0.

82. The ADI value of sorbic acid and derivatives is 25 mg/kg body weight. Typical usage levels are 2000 ppm in solid food, 300 ppm in beverages.

83. Benzoic acid is produced by the oxidation of toluene. The ADI is calculated as a total for all benzoates and is 5 mg/kg, expressed as benzoic acid.

84. Nisin is an antimicrobial peptide (small protein) or "bacteriocin" produced by certain strains of the lactic acid bacterium *Lactococcus lactis* subsp. *lactis*.

85. Natamycin is a natural antimicrobial produced by *Streptomycetes*. It shows strong activity against yeasts and moulds, but shows no activity against bacteria or viruses. Its effect is predominantly fungicidal. There are no reported allergic reactions to natamycin in food and the product has GRAS status in the U.S. The ADI of natamycin is 0.25-0.3 mg/kg of body weight/day.

86. Dimethyl dicarbonate is used for the cold sterilisation of beverages.

87. Sodium nitrate is found in nature as Chile saltpetre and is used as preservative in food products. The nitrates are permitted only in cured meats, pickled fish and a number of cheeses.

88. Erythorbic acid and sodium erythorbate are stereoisomers of ascorbic acid and sodium ascorbate, respectively. the erythorbates are limited in their applications to preserved and cured meat up to 500 mg/kg and also in preserved and semi-preserved fish products and frozen and deep frozen fish with red skin, at a maximum use level of 1500 mg/kg

89. Butylated hydroxyanisole is a mixture of the 3-tert-butyl and 2-tert-butyl derivatives of 4-hydroxyanisole (also called 4-methoxyphenol). It is soluble in fats, oils, alcohol and ether, but insoluble in water. BHA is stable to heat and mildly alkaline conditions that makes it suitable in baked and fried foods.

90. Butylated hydroxytoluene is an antioxidant and is added to delay or prevent rancidity in fats and oils in foodstuffs. BHT is less effective than BHA.

91. Lecithin is a mixture or fraction of phospholipids, which are obtained from soya and egg by physical processes. Phospholipids are the active ingredients of lecithin. Lecithin is used on the surface of powders to improve "instant" properties. In bakery applications, lecithin is used to increase the extensibility of the gluten in bread making, and in batters to improve the overall distribution of ingredients in cakes and to assist the release of wafers from hot iron moulds.

92. Citric acid is one of the most widely used of all food additives. Its clean, tart taste has found wide application in soft drinks (carbonates, squashes, nectars and powdered beverages), sugar confectionery, jams, jellies, preserves, soups and sauces.

93. Monocalcium phosphate is added to flour to reduce the risk of growth of the bacteria that lead to the spoilage condition known as "rope". It is used as a source of calcium to improve the structure from low-gluten flours, to increase the rate of gelling of some

milk-based desserts, and to increase the firmness of canned vegetables such as carrots and tomatoes.

94. Alginates are the principal structural polysaccharide component of brown seaweeds. Alginate is present in seaweed as a mixed salt of sodium, potassium, calcium and magnesium. Extraction involves ion exchange in an alkaline medium followed by precipitation, purification and recovery of the alginic acid. Alginic acid is a copolymer of mannuronic acid and guluronic acid.

95. Propylene glycol alginate (PGA) is cold-water-soluble, and functions in food as a thickener, suspending agent and stabiliser.

96. Carrageenans are extracted from certain red seaweeds of the class *Rhodophyceae*. They are used as gelling agents, thickening agents and stabilisers. Carrageenans can interact with casein protein in dairy products.

97. Xanthan gum is used to increase viscosity in sauces and dressings, drinks and cakes.

98. Karaya gum is as a thickener in sauces, in particular brown sauce. It is also used in coatings, fillings, toppings and chewing gum.

99. Gellan gum is an extracellular polysaccharide secreted by the microorganism *Sphingomonas elodea*, previously referred to as *Pseudomonas elodea*.

100. Polysorbate 40 is widely used as a surfactant, often in combination with other emulsifiers for forming oil-in-water emulsions. It is used to stabilise sauces and dressings and in bakery margarine to improve aeration, cake volume and texture. It is also used in whipped cream and coffee whiteners.

10.2 MULTIPLE CHOICE QUESTIONS

1. Which of the following is not a food additive
   (a) sugar (b) flour
   (c) salt (d) none

2. Which of the following is a type of food additive
   (a) preservative (b) acidulants
   (c) antimicrobials (d) all

3. Which of the following is a type of food additive
   (a) antioxidants (b) stabilizer
   (c) thickener (d) all

4. Which of the following is a type of food additive
   (a) sequestrants
   (b) flavor enhancer
   (c) sweetener
   (d) all

5. Which of the following is a type of food additive
   (a) chelating agent
   (b) flour maturing agent
   (c) enzymes
   (d) all

6. Which of the following is a type of food additive
   (a) coloring agent
   (b) flavoring agent
   (c) anti-caking agent
   (d) all

7. Acetic acid can be used as
   (a) Acidulant (b) preservative
   (c) both (d) none

8. Which of the following is an acidulant
   (a) acetic acid (b) adipic acid
   (c) caprylic acid (d) all

9. Which of the following is an acidulant
(a) citric acid
(b) dehydroacetic acid
(c) fumaric acid
(d) all

10. Which of the following is an acidulant
(a) Glucono-delta-lactone
(b) lactic acid
(c) malic acid
(d) all

11. Which of the following is an acidulant
(a) phosphoric acid
(b) propionic acid
(c) sodium diacetate
(d) all

12. Which of the following is an acidulant
(a) tartaric acid (b) succinic acid
(c) both (d) none

13. Which of the following is not used as acidulant in food
(a) benzoic acid (b) tartaric acid
(c) lactic acid (d) all

14. Which of the following is not used as acidulant in food
(a) citric acid (b) ascorbic acid
(c) malic acid (d) all

15. Which of the following is not used as acidulant in food
(a) acetic acid (b) nitric acid
(c) citric acid (d) all

16. Which of the following is not used as acidulant in food
(a) adipic acid (b) fatty acid
(c) fumaric acid (d) all

17. Which of the following is not used as acidulant in food
(a) glycerol (b) benzene
(c) sugar (d) all

18. pKa of an acid is the negative log of its ________ constant
(a) association (b) dissociation
(c) both (d) none

19. Larger the dissociation constant ________ is the pKa of acid
(a) lower (b) higher
(c) do not effect (d) none

20. pKa of acetic acid is about
(a) 2.75 (b) 4.75
(c) 6.75 (d) 8.75

21. pKa_1 of lactic acid is about
(a) 3.08 (b) 6.86
(c) 9.86 (d) 12.86

22. pKa of gluconic acid is about
(a) 1.6 (b) 3.6
(c) 5.6 (d) 9.6

23. Which of the following food acid is generally a hygroscopic solid
(a) tartaric acid
(b) sodium acid sulfate
(c) malic acid
(d) all

24. Which of the following food acid is generally a non-hygroscopic solid
(a) lactic acid (b) fumaric acid
(c) acetic acid (d) all

25. Which of the following food acid is generally used in liquid form
(a) lactic acid (b) fumaric acid
(c) malic acid (d) all

26. Which of the following food acid is generally used in liquid form
(a) lactic acid (b) acetic acid
(c) phosphoric acid (d) all

27. Buffer reduces which of the following function in foods
(a) flavor variation
(b) color variation
(c) gel-strength variation
(d) all

28. Level of sourness in food depends on
(a) type of acid
(b) pH of food
(c) concentration of acid
(d) all

29. If the pH of food is below 4.0 then __________ acid should be used for sourness
(a) phosphoric acid (b) fumaric acid
(c) lactic acid (d) all

30. If the pH of food is above 4.0 then ____ acid should be used for sourness
(a) sodium acid sulfate
(b) tartaric acid
(c) phosphoric acid
(d) all

31. If the pH of food is above 4.0 then__________acid should be used for sourness
(a) fumaric acid (b) lactic acid
(c) tartaric acid (d) all

32. Which of the following acid is more sour (per unit weight basis)
(a) fumaric acid (b) lactic acid
(c) tartaric acid (d) none

33. Which of the following is more hydrophilic in nature
(a) acetic acid (b) citric acid
(c) lactic acid (d) malic acid

34. Acetic acid contribute which of the following flavor effect
(a) volatile (b) pungent
(c) both (d) none

35. Citirc acid contribute which of the following flavor effect
(a) clean sourness
(b) bright sourness
(c) refreshing sourness
(d) all

36. Fumaric acid contribute which of the following flavor effect
(a) clean sourness
(b) persistent sourness
(c) both
(d) none

37. Lactic acid contribute which of the following flavor effect
(a) smooth sourness
(b) persistent sourness
(c) mild dairy aroma
(d) all

38. Malic acid contribute which of the following flavor effect
(a) mellow sourness
(b) fruit flavors
(c) natural flavor profile
(d) all

39. Role of acidulants is
(a) flavoring
(b) lowering pH
(c) buffering agent
(d) all

40. Which of the following is function of citric acid
(a) antioxidant in fats & oils
(b) preserve meats
(c) prevent browning in salads
(d) all

41. Which of the following is a source of citric acid
(a) lemon
(b) *Aspergillus niger*
(c) both
(d) none

42. Acid found in vinegar is
(a) lactic (b) acetic
(c) citric (d) all

43. Which of the following contribute to cola flavor
(a) phosphoric acid (b) malic acid
(c) fumaric acid (d) all

44. Which of the following is widely used in pickle industry
(a) malic (b) acetic
(c) fumaric (d) all

45. Acetic acid has a role in
(a) pH control (b) flavor
(c) shelf life (d) all

46. Main function of lactic acid is
(a) provide flavor
(b) moderate chelator
(c) pH adjustment
(d) all

47. Main function of lactic acid is
(a) antimicrobial activity
(b) coagulating agent
(c) both
(d) none

48. Malic acid is naturally found in
(a) apples (b) pears
(c) tomatoes (d) all

49. Main function of mailc acid is
(a) provide flavor
(b) moderate chelator
(c) pH adjustment
(d) all

50. Which of the following is considered as first food acidulant
(a) tartaric acid (b) malic acid
(c) lactic acid (d) all

51. Buffering capacity of malic acid is about
(a) 1.5-2.5 (b) 2.5-3.0
(c) 3.5-4.5 (d) 5.0-6.0

52. Adipic acid is used in
(a) gelatin powder
(b) baking powder
(c) processed cheese
(d) all

53. Succinic acid in bread dough modifies
(a) plasticity (b) gas formation
(c) both (d) none

54. Which of the following acid is used as acidulant in beverage
(a) citric (b) malic
(c) both (d) none

55. Which of the following is a sweetener
(a) E950 (b) E960
(c) E968 (d) all

56. Which of the following is a flour treatment agent
(a) E1100 (b) E1101
(c) E925 (d) all

57. Which of the following is a glazing agent
(a) E901 (b) E915
(c) E907 (d) all

58. Which of the following is a flavor enhancer
(a) E621 (b) E633
(c) E642 (d) all

59. Which of the following is a thickener
(a) E400 (b) E412
(c) E419 (d) all

60. Tamarind contains good quantity of ________ acid
(a) tartaric (b) malic
(c) phosphoric (d) all

61. Apples contains good quantity of ________ acid
(a) tartaric (b) malic
(c) phosphoric (d) all

62. Cherry contains good quantity of ________ acid
(a) tartaric (b) malic
(c) phosphoric (d) all

63. Cola contains good quantity of ________ acid
(a) tartaric (b) malic
(c) phosphoric (d) all

64. Grapes contains good quantity of ________ acid
(a) tartaric (b) malic
(c) both (d) none

65. Synonym used for acetic acid is
(a) ethanoic acid
(b) ethylic acid
(c) monocarboxylic acid
(d) all

66. Chemical formula of acetic acid is
(a) CH_2COOH (b) CH_3COOH
(c) C_2H_5COOH (d) none

67. Molecular mass of acetic acid is
(a) 50 (b) 60
(c) 70 (d) 40

68. Boiling point (°C) of acetic acid is about
(a) 97.9 (b) 107.0
(c) 117.9 (d) 87.9

69. Melting temperature (°C) of acetic acid is about
(a) 16.6 (b) 26.6
(c) 36.6 (d) 46.6

70. Flash point (°C) of acetic acid is about
(a) 33 (b) 43
(c) 53 (d) 63

71. Acetic acid is generally used in
(a) sauces
(b) salad dressings
(c) mayonnaise (d) all

72. Maximum level (%) of acetic acid can be used in bakery products
(a) 0.25 (b) 0.50
(c) 0.75 (d) 1.0

73. Maximum level (%) of acetic acid can be used in meat products
(a) 0.2 (b) 0.6
(c) 0.9 (d) 1.2

74. Maximum level (%) of acetic acid can be used in condiments
(a) 3 (b) 6
(c) 9 (d) 15

75. Synonym used for adipic acid
(a) acifoloctin (b) adipinic acid
(c) hexanedoic acid (d) all

76. Adipic acid is ______ times more soluble than fumaric acid
(a) two (b) four
(c) seven (d) ten

77. pK_1 value of adipic acid is about
(a) 3.43 (b) 4.43
(c) 5.43 (d) 6.43

78. pK_2 value of adipic acid is about
(a) 3.41 (b) 4.44
(c) 5.42 (d) 6.43

79. Acceptable daily intake of adipic acid for humans (mg/kg body weight)
(a) 5.0 (b) 50
(c) 150 (d) 200

80. pK_1 value of caprylic acid is about
(a) 3.89 (b) 4.89
(c) 5.89 (d) 6.89

81. Synonym used for citric acid
(a) boxylic acid (b) citretten
(c) aciletten (d) all

82. Molecular mass of citric acid is about
(a) 137 (b) 192
(c) 156 (d) 176

83. Flash point (°C) of citric acid is about
(a) 100 (b) 125
(c) 150 (d) 175

84. Citric acid used as acidity regulator in
(a) cultured milk (b) margarine
(c) canned fruits (d) all

85. Citric acid used as acidity regulator in
(a) cheese (b) jam
(c) meat products (d) all

86. Maximum level (%) of citric acid used in milk products by weight
(a) 0.15 (b) 0.25
(c) 0.35 (d) 0.5

87. Maximum level ppm) of citric acid used in pork by weight
(a) 250 (b) 400
(c) 650 (d) 800

88. Melting point (°C) anhydrous form of citric acid is about
(a) 123 (b) 153
(c) 173 (d) 193

89. pK_1 value of citric acid is about
(a) 3.14 (b) 4.77
(c) 6.39 (d) 8.42

90. pK_2 value of citric acid is about
(a) 3.14 (b) 4.77
(c) 6.39 (d) 8.42

91. pK_3 value of citric acid is about
(a) 3.14 (b) 4.77
(c) 6.39 (d) 8.42

92. Synonym used for fumaric acid is
(a) allomaleic acid (b) boletic acid
(c) lichenic acid (d) all

93. Molecular mass of fumaric acid is
(a) 96 (b) 106
(c) 116 (d) 136

94. pK_1 value of fumaric acid is about
(a) 3.03 (b) 3.72
(c) 4.44 (d) 4.72

95. pK_2 value of fumaric acid is about
(a) 3.03 (b) 3.72
(c) 4.44 (d) 4.72

96. ADI (mg/kg body weight) of fumaric acid for human is unconditionally about
(a) 6-10 (b) 10-15
(c) 15-20 (d) 21-27

97. Synonym used for lactic acid is
(a) acetonic acid (b) milk acid
(c) both (d) none

98. Molecular mass of lactic acid is
(a) 70 (b) 80
(c) 90 (d) 110

99. Calcium lactate is used as
(a) firming agent
(b) flavoring agent
(c) gelling agent
(d) all

100. Malic acid is used in_______ products
(a) bakery (b) dairy
(c) fruits & vegetable
(d) all

101. Molecular mass of malic acid is
(a) 124 (b) 134
(c) 174 (d) 184

102. pK_1 value of malic acid is about
(a) 3.4 (b) 4.22
(c) 5.11 (d) 6.26

103. pK_2 value of malic acid is about
(a) 3.4 (b) 4.22
(c) 5.11 (d) 6.26

104. Acceptable daily intake of malic acid for human is ____ mg/kg body weight
(a) 20 (b) 50
(c) 100 (d) 200

105. Phosphoric acid is used as acidulant in ______ products
(a) bakery (b) dairy
(c) meat (d) all

106. Molecular mass of phosphoric acid is
(a) 68 (b) 98
(c) 128 (d) 148

107. pK_1 value of phosphoric acid is about
(a) 2.12 (b) 5.32
(c) 7.21 (d) 12.67

108. pK_2 value of phosphoric acid is about
(a) 2.12 (b) 5.32
(c) 7.21 (d) 12.67

109. pK_3 value of phosphoric acid is about
(a) 2.12 (b) 5.32
(c) 7.21 (d) 12.67

110. ADI value of phosphoric acid for human is ________mg/kg body weight
(a) 20 (b) 40
(c) 70 (d) 110

111. Synonym used for carboxylic acid is
(a) carboxyethane
(b) ethanecarboxylic acid
(c) ethylformic acid
(d) all

112. Molecular mass of propionic acid is
(a) 54 (b) 74
(c) 94 (d) 84

113. Boiling point (°C) of propionic acid at
(a) 4.6 (b) 85.8
(c) 122.0 (d) 141.1

114. Melting point (°C) of propionic acid is
(a) -21.5 (b) -11.5
(c) 21.5 (d) 32.7

115. Flash point (°C) of propionic acid is
(a) 38 (b) 58
(c) 78 (d) 88

116. Propionic acid can be used as ______ agent in cheese
(a) flavoring (b) antifungal
(c) both (d) none

117. Maximum recommended level (%) of propionic acid in cheese products is about
(a) 0.3 (b) 0.9
(c) 1.5 (d) 2.0

118. Maximum recommended level (%) of propionic acid in white bread is about
(a) 0.12 (b) 0.32
(c) 0.60 (d) 1.0

119. Maximum recommended level (%) of propionic acid in whole wheat products is about
(a) 0.18 (b) 0.38
(c) 0.78 (d) 1.8

120. pK_1 value of propionic acid is about
(a) 2.87 (b) 3.87
(c) 4.87 (d) 5.87

121. Sodium diacetate is used as
(a) chelating agent (b) acidulant
(c) preservative (d) all

122. Sodium diacetate is used in
(a) edible oil
(b) sugar preserves
(c) dairy products
(d) all

123. Molecular mass of sodium diacetate is
(a) 142 (b) 152
(c) 162 (d) 172

124. Boiling point (°C) of sodium diacetate is
(a) 135 (b) 175
(c) 235 (d) 275

125. Melting point (°C) of sodium diacetate is
(a) 58 (b) 68
(c) 78 (d) 88

126. Sodium diacetate inhibits growth of ______ in bakery products
(a) bread mould
(b) rope forming bacteria
(c) both
(d) none

127. Sodium diacetate inhibits mold growth in malt syrups at ______ %
(a) 0.5 (b) 1.0
(c) 1.5 (d) 2.0

128. Maximum recommended level (%) of sodium diacetate in snack food is
(a) 0.05 (b) 0.25
(c) 0.5 (d) 1.0

129. Maximum recommended level (%) of sodium diacetate in oil is
(a) 0.01 (b) 0.1
(c) 0.5 (d) 1.0

130. Maximum recommended level (%) of sodium diacetate in gravies and sauces is
(a) 0.05 (b) 0.25
(c) 0.5 (d) 1.0

131. Maximum recommended level (%) of sodium diacetate in baked goods is
(a) 0.1 (b) 0.2
(c) 0.3 (d) 0.4

132. pK_1 value of sodium diacetate is about
(a) 2.75 (b) 4.75
(c) 3.75 (d) 5.75

133. ADI value of sodium diacetate for human is ______mg/kg body weight
(a) 5 (b) 15
(c) 25 (d) 35

134. Succinic acid can be used as
(a) acidulant (b) preservative
(c) flour additive (d) all

135. Synonym used for succinic acid is
(a) amber acid
(b) butanedioic acid
(c) ethanedicaboxylic acid
(d) all

136. Molecular mass of succinic acid is
(a) 98 (b) 118
(c) 128 (d) 138

137. Boiling point (°C) of succinic acid is about
(a) 135 (b) 175
(c) 235 (d) 275

138. Maximum recommended level (%) of succinic acid in meat is
(a) 0.0061 (b) 0.084
(c) 2.6 (d) none

139. Maximum recommended level (%) of succinic acid in condiments and relishes is
(a) 0.0061 (b) 0.084
(c) 2.6 (d) none

140. Succinic acid can be used in
(a) meat (b) condiments
(c) jam (d) all

141. pK_1 value of succinic acid is about
(a) 3.26 (b) 4.16
(c) 5.61 (d) 6.42

142. pK_2 value of succinic acid is about
(a) 3.26 (b) 4.16
(c) 5.61 (d) 6.42

143. Which of the following is used in bakery products
(a) reducing agent (b) acidulant
(c) both (d) none

144. Which of the following is used in bakery products
(a) ABC (b) SBC
(c) both (d) none

145. Which of following product involve enzyme in their manufacturing process
(a) biscuits (b) cheese
(c) fruit jiuce (d) all

146. Tartaric acid can be used as
(a) stabilizer (b) acidulant
(c) chelating agent (d) all

147. Tartaric acid can be used as
(a) acidulant (b) humectant
(c) firming agent (d) all

148. Molecular mass of tartaric acid
(a) 100 (b) 150
(c) 200 (d) 250

149. Melting point (°C) of tartaric acid is about
(a) 120 (b) 140
(c) 170 (d) 210

150. pK_1 value of tartaric acid is about
(a) 2.98 (b) 3.37
(c) 4.34 (d) 5.59

151. pK_2 value of succinic acid is about
(a) 2.98 (b) 3.37
(c) 4.34 (d) 5.59

152. ADI value of tartaric acid for human is _________mg/kg body weight
(a) 10 (b) 20
(c) 30 (d) 40

153. Which of the following is an antioxidant
(a) vitamin E (b) vitamin C
(c) beta carotene (d) all

154. RDA value for vitamin E is about
(a) 15mg (b) 25mg
(c) 35mg (d) 40mg

155. RDA value (mg) for vitamin C for adult women is about
(a) 55 (b) 75
(c) 90 (d) none

156. RDA value (mg) for vitamin C for adult men is about
(a) 55 (b) 75
(c) 90 (d) none

157. Which of the following is a natural antioxidant
(a) vitamin E (b) vitamin C
(c) both (d) none

158. Which of the following is a source of vitamin E
(a) vegetables oil (b) walnuts
(c) almonds (d) all

159. Which of the following is a source of vitamin E
(a) peanuts (b) olives
(c) avocado (d) all

160. Which of the following is a source of vitamin C
(a) citrus fruits (b) tomatoes
(c) peppers (d) all

161. Which of the following is a source of vitamin C
(a) green leafy vegetables
(b) potatoes
(c) strawberries
(d) all

162. Which of the following is a common source of beta carotene
(a) mango (b) papaya
(c) spinach (d) all

163. Which of the following is an anti-oxidant
(a) ascorbyl palmitate
(b) sodium ascorbate
(c) calcium ascorbate
(d) all

164. Which of the following is an antioxidant
(a) BHA (b) BHT
(c) ethoxyquin (d) all

165. Which of the following is an antioxidant
(a) propyl galate (b) TBHQ
(c) tocopherol (d) all

166. Which of the following is an antioxidant
(a) citric acid (b) tartaric acid
(c) EDTA (d) all

167. Molecular mass of ascorbyl palmitate is
(a) 314 (b) 414
(c) 514 (d) 470

168. Ascorbyl palmitate in food may functions as
(a) antioxidant (b) sequestrant
(c) vitamin C source (d) all

169. Sodium ascorbate in food is used as
(a) antioxidant
(b) color preservative
(c) both
(d) none

170. Molecular mass of sodium ascorbate is
(a) 198 (b) 228
(c) 248 (d) 278

171. Melting point (°C) of sodium ascorbate is
(a) 100 (b) 200
(c) 300 (d) 250

172. E number of sodium ascorbate is
(a) E301 (b) E302
(c) E303 (d) E304

173. E number of calcium ascorbate is
(a) E301 (b) E302
(c) E303 (d) E304

174. Molecular mass of BHA is about
(a) 160 (b) 180
(c) 200 (d) 220

175. Melting point (°C) of BHA is
(a) 55 (b) 75
(c) 85 (d) 105

176. Molecular mass of BHT is about
(a) 160 (b) 180
(c) 200 (d) 220

177. Molecular mass of ethoxyquin is about
(a) 167 (b) 197
(c) 217 (d) 237

178. Boilingpoint (°C) of ethoxyquin is about
(a) 100 (b) 125
(c) 150 (d) 175

179. FDA approved ________ppm of ethoxyquin in chilli powder
(a) 50 (b) 100
(c) 200 (d) 400

180. FDA approved ________ ppm of ethoxyquin in paprika
(a) 50 (b) 100
(c) 200 (d) 400

181. FDA approved ________ ppm of ethoxyquin in uncooked meat fat
(a) 5 (b) 15
(c) 25 (d) 45

182. FDA approved ________ ppm of ethoxyquin in uncooked poultry fat
(a) 3 (b) 9
(c) 15 (d) 25

183. FDA approved ________ ppm of ethoxyquin in eggs
(a) 0.1 (b) 0.5
(c) 3.5 (d) 5.0

184. Molecular mass of propyl gallate is about
(a) 112 (b) 152
(c) 212 (d) 252

185. Melting point (°C) of propyl gallate is
(a) 125 (b) 150
(c) 175 (d) 195

186. Propyl gallate is not used in
(a) bakery products
(b) maragarine
(c) dry sausages
(d) all

187. Propyl gallate is generally used at ______% in chewing gum base
(a) 0.1 (b) 0.3
(c) 0.5 (d) 0.75

188. Propyl gallate is generally used at ______% in non-alcoholoic beverages
(a) 0.1 (b) 0.3
(c) 0.5 (d) 0.75

189. Molecular mass of TBHQ is about
(a) 133 (b) 166
(c) 183 (d) 196

190. Boiling point (°C) of TBHQ is
(a) 125 (b) 195
(c) 295 (d) 325

191. TBHQ addition causes some sidcoloration in presence of ________ pH
(a) acidic (b) alkaline
(c) neutral (d) none

192. Molecular mass of alpha-tocopherol is about
(a) 430 (b) 470
(c) 456 (d) 496

193. Tocopherol is generally used as
(a) antioxidant
(b) vitamin supplement
(c) both
(d) none

194. Tocopherol is used as antioxidant in
(a) infant food (b) mayonnaise
(c) salad dressing (d) all

195. Which of the following method is employed for addition of tocopherol
(a) direct addition (b) spraying
(c) dipping (d) all

196. Tocopherol should not exceed ______ ppm in UK & Europe in foods
(a) 200 (b) 300
(c) 500 (d) 700

197. Synonym used for beta-carotene is
(a) provitamin A
(b) food oranges
(c) natural yellow 26
(d) all

198. Molecular mass of beta-carotene is about
(a) 536 (b) 436
(c) 336 (d) 236

199. Melting point (°C) of beta-carotene is
(a) 156 (b) 179
(c) 198 (d) 223

200. Beta-carotene act as ________ in food
(a) antioxidant (b) nutrient
(c) colorant (d) all

201. Which of the following is a source of antioxidant
(a) rosemary extract (b) tea extract
(c) both (d) none

202. Molecular mass of EDTA is about
(a) 262 (b) 292
(c) 312 (d) 342

203. EDTA in food is used as
(a) Chelator (b) antioxidant
(c) both (d) none

204. Which of the following is a form of EDTA
(a) calcium disodium EDTA
(b) sodium dihydrogen EDTA
(c) disodium EDTA
(d) all

205. Boiling point (°C) of EDTA is
(a) 238 (b) 278
(c) 298 (d) 312

206. Which of the following is a food colorant
(a) annatto
(b) anthocyanin
(c) beta-apo-8'-carotenal
(d) all

207. Which of the following is a food colorant
(a) Brilliant Blue FCF
(b) canthaxanthin
(c) caramel
(d) all

208. Which of the following is a food colorant
(a) carmine (b) carminic acid
(c) carmoisine (d) all

209. Which of the following is a food colorant
(a) beta-carotene (b) chlorophyll
(c) chlorophyllin (d) all

210. Which of the following is a food colorant
(a) curcumin (b) indigotine
(c) litholrubine (d) all

211. Which of the following is a food colorant
(a) Brilliant Black
(b) quinoline yellow
(c) sunset yellow
(d) all

212. Which of the following is a food colorant
(a) tartrazine
(b) titanium dioxide
(c) lutein
(d) all

213. Which of the following is a food colorant
(a) lycopene
(b) paprika extract
(c) riboflavin
(d) all

214. Which of the following is a food colorant
(a) alkanet (b) allura red
(c) both (d) none

215. Allura red color is used in ________ products
(a) bakery (b) dairy
(c) beverages (d) all

216. Molecular mass of Allura Red AC is about
(a) 396 (b) 496
(c) 296 (d) 356

217. Which of the following is a property of Allura Red AC
(a) heat stable
(b) light stable
(c) yellowish-red color
(d) all

218. Which of the following is a property of amaranth
(a) water soluble (b) bluish red color
(c) both (d) none

219. Amaranth color fades in presence of
(a) SO_2 (b) heat
(c) light (d) none

220. Wavelength (nm) of visible range of human eye is
(a) 28-70 (b) 250-350
(c) 380-770 (d) none

221. Which of the following is not a function of food colorant
(a) impart color
(b) give good appearance
(c) decoration
(d) preservation

222. FD&C colors are certifiable for use in coloring
(a) food (b) drugs
(c) cosmetics (d) all

223. Which of the following is natural color
(a) carotenoids (b) annatto
(c) anthocyanin (d) all

224. Which of the following is natural color
(a) betalains (b) riboflavin
(c) chlorophyll (d) all

225. Which of the following is insect color
(a) cochineal (b) carmine
(c) both (d) none

226. Which of the following is natural identical color
(a) beta-carotene
(b) beta-apo-8'-carotenal
(c) canthaxanthin
(d) all

227. Which of the following is an advantage of synthetic color
(a) stability (b) high purity
(c) cheapness (d) all

228. Dyes are ______ in nature
(a) soluble (b) insoluble
(c) both (d) none

229. Lakes are ______ in nature
(a) soluble (b) insoluble
(c) both (d) none

230. Which of the following has high pure color content
(a) dyes (b) lakes
(c) both (d) none

231. Dyes contains________ % of pure color content
(a) 10-40 (b) 40-60
(c) 60-76 (d) 90-93

232. Lakes contains________ % of pure color content
(a) 10-40 (b) 40-60
(c) 60-76 (d) 90-93

233. Dyes are used at rate (%) of
(a) 0.01-0.03 (b) 0.1-0.3
(c) 1.0-3.0 (d) none

234. Lakes are used at rate (%) of
(a) 0.01-0.03 (b) 0.1-0.3
(c) 1.0-3.0 (d) none

235. Which of the following is used as food additive
(a) aroma substances
(b) vitamins
(c) flavor enhancer
(d) all

236. Which of the following is a flavor enhancer
(a) MSG (b) 5'-nucleotide
(c) maltol (d) all

237. Maltol has ________ like odor
(a) caramel (b) pungent
(c) both (d) none

238. Which of the following can be used as sugar substitute
(a) manniitol (b) xylitol
(c) sorbitol (d) all

239. Which of the following can be used as sweetener
(a) saccharin (b) cyclamate
(c) monellin (d) all

240. ADI value of saccharin is about ______mg/kg body weight
(a) 0-2.5 (b) 5-10
(c) 15-21 (d) 22-28

241. Cyclamate is marketed as ________ salt of cyclohexane sulfamic acid
(a) Na (b) Ca
(c) both (d) none

242. ADI value of cyclamate is about ______mg/kg body weight
(a) 0-11 (b) 15-21
(c) 24-32 (d) none

243. Synthesis of cyclamate is based on ________ of cyclohexyl amine
(a) chlorination (b) sulfornation
(c) both (d) none

244. Monellin is ________ in nature
(a) carbohydrate (b) protein
(c) lipid (d) none

245. Which of the following is used as sweetener
(a) thaumatin (b) curculin
(c) miraculin (d) all

246. Which of the following has sweetening ability
(a) stevioside (b) osladin
(c) phyllodulcin (d) all

247. Thoumatin is ________ in nature
(a) carbohydrate (b) protein
(c) lipid (d) none

248. Curculin is__________in nature
(a) carbohydrate (b) protein
(c) lipid (d) none

249. Miraculin is a type of ______
(a) glycoprotein (b) lipoprotein
(c) both (d) none

250. ADI value for potassium salt of acesufame is about ________mg/kg body weight
(a) 0.5-2.0 (b) 0-9
(c) 11-15 (d) 20-30

251. ADI value of aspartame is about ______ mg/kg body weight
(a) 0-40 (b) 40-60
(c) 55-79 (d) 80-100

252. Which of the following can be used as food colorant
(a) tartrazine (b) curcumin
(c) zeaxanthin (d) all

253. Which of the following can be used as food colorant
(a) bixin (b) lycopene
(c) canthaxanthin (d) all

254. Which of the following can be used as food colorant
(a) carmoisine (b) amaranth
(c) ponceau 4R (d) all

255. Which of the following can be used as food colorant
(a) carmine (b) anthocyanidin
(c) erythrosine (d) all

256. Which of the following can be used as food colorant
(a) indigocarmine
(b) patent blue V
(c) brillian blue FCF
(d) all

257. Which of the following can be used as food colorant
(a) chlorophyll (b) chlorophyllin
(c) green S (d) all

258. Which of the following give yellow color
(a) tartrazine (b) riboflavin
(c) curcumin (d) all

259. Which of the following give yellow color
(a) curcumin (b) zeaxanthin
(c) both (d) none

260. Sunset yellow FCF gives ________ color in water
(a) yellow (b) orange
(c) red (d) green

261. Which of the following give orange color
(a) sunset yellow FCF
(b) beta-carotene
(c) bixin
(d) all

262. Which of the following give orange color
(a) lycopene
(b) canthaxanthin
(c) astaxanthin
(d) all

263. Which of the following give orange color
(a) sunset yellow F
(b) beta-apo-8'-carotenal
(c) both
(d) none

264. Which of the following give red color with bluish tint
(a) carmoisine (b) amaranth
(c) both (d) none

265. Which of the following give scarlet-red color
(a) ponceau 4R (b) black BN
(c) indigocarmine (d) all

266. Which of the following give bright red color
(a) indigocarmine (b) carmine
(c) anthocyanidin (d) all

267. Which of the following give purple blue color
(a) indigocarmine (b) carmine
(c) anthocyanidin (d) all

268. Which of the following give red violet color
(a) indigocarmine (b) carmine
(c) anthocyanidin (d) all

269. Which of the following give cherry red color
(a) erythrosine
(b) patent blue FCF
(c) lycopene (d) all

270. Which of the following give blue color with greenish tint
(a) patent blue V
(b) brilliant blue FCF
(c) both (d) none

271. Which of the following give green color
(a) astaxanthin (b) chlorophyll
(c) bixin (d) all

272. Which of the following give green color
(a) chlorophyllin (b) erythrosine
(c) riboflavin (d) all

273. Which of the following give green color
(a) curcumin (b) tartrazine
(c) green S (d) all

274. Which of the following give violet color with bluish tint
(a) zeaxanthin (b) black BN
(c) indigocarmine (d) all

275. λ_{max} for tartrazine in water (solvent) is about
(a) 426 nm (b) 445nm
(c) 485nm (d) 516nm

276. λ_{max} (nm) for riboflavin in water (solvent) is about
(a) 426 nm (b) 445nm
(c) 485nm (d) 516nm

277. λ_{max} (nm) for sunset yellow FCF in water (solvent) is about
(a) 426 nm (b) 445nm
(c) 485nm (d) 516nm

278. λ_{max} (nm) for carmoisine in water (solvent) is about
(a) 426 nm (b) 445nm
(c) 485nm (d) 516nm

279. λ_{max} (nm) for amaranth in water (solvent) is about
(a) 520 (b) 505
(c) 527 (d) 532

280. λ_{max} (nm) for ponceau 4R in water (solvent) is about
(a) 520 (b) 505
(c) 527 (d) 532

281. λ_{max} (nm) for erythrosine in water (solvent) is about
(a) 520 (b) 505
(c) 527 (d) 532

282. λ_{max} (nm) for red G in water (solvent) is about
(a) 520 (b) 505
(c) 527 (d) 532

283. λ_{max} (nm) for indigocarmine in water (solvent) is about
(a) 610 (b) 630
(c) 638 (d) 405

284. λ_{max} (nm) for patent blue V in water (solvent) is about
(a) 610 (b) 630
(c) 638 (d) 405

285. λ_{max} (nm) for brilliant blue FCF in water (solvent) is about
(a) 610 (b) 630
(c) 638 (d) 405

286. λ_{max} (nm) for chlorophillin in water (solvent) is about
(a) 610 (b) 630
(c) 638 (d) 405

287. λ_{max} (nm) for green S in water (solvent) is about
(a) 632 (b) 570
(c) 602 (d) 513

288. λ_{max} (nm) for black BN in water (solvent) is about
(a) 632 (b) 570
(c) 602 (d) 513

289. λ_{max} (nm) for bixin in chloroform (solvent) is about
(a) 471 (b) 503
(c) both (d) none

290. λ_{max} (nm) for canthaxanthin in chloroform (solvent) is about
(a) 485 (b) 503
(c) 570 (d) 602

291. λ_{max} (nm) for astaxanthin in chloroform (solvent) is about
(a) 503 (b) 488
(c) 602 (d) 570

292. λ_{max} (nm) for chlorophyll in chloroform (solvent) is about
(a) 412 (b) 512
(c) 485 (d) 575

293. Which of the following is used as alkaline salt in food processing
(a) NaOH (b) $NaHCO_3$
(c) Na_2CO_3 (d) all

294. Which of the following is used as alkaline salt in food processing
(a) $MgCO_3$ (b) MgO
(c) $Ca(OH)_2$ (d) all

295. Which of the following is used as alkaline salt in food processing
(a) Na_2HPO_4 (b) sodium citrate
(c) both (d) none

296. Brilliant blue FCF is FD&C blue number ________
(a) 1 (b) 2
(c) both (d) none

297. Indigocarmine is FD&C blue number ________
(a) 1 (b) 2
(c) both (d) none

298. Fast green FCF is FD&C green number________
(a) 1 (b) 2
(c) 3 (d) none

299. Erythrosine FCF is FD&C red number ________
(a) 1 (b) 2
(c) 3 (d) none

300. Tartrazine FCF is FD&C yellow number ________
(a) 5 (b) 6
(c) 3 (d) none

301. Tartrazine is insoluble in
(a) ethanol (b) glycerol
(c) glycols (d) all

302. Sunset yellow FCF is FD&C yellow number________
(a) 5 (b) 6
(c) 4 (d) none

303. Brilliant blue FCF is generally soluble in
(a) water (b) glycerol
(c) glycols (d) all

304. Fast green FCF is generally soluble in
(a) water (b) glycerol
(c) glycols (d) all

305. Erythrosine is insoluble in
(a) acid media (b) glycol
(c) glycerol (d) all

306. Iron is present in Fe^{2+} state in
(a) myoglobin (b) oxymyoglobin
(c) nitrisomyoglobin (d) all

307. Iron is present in Fe^{3+} state in
(a) metmyoglobin (b) myoglobin
(c) oxymyoglobin (d) all

308. Myoglobin gives ________ color
(a) brown (b) bright red
(c) purplish red (d) all

309. Oxymyoglobin gives ________ color
(a) brown (b) bright red
(c) purplish red (d) all

310. Metmyoglobin gives ________ color
(a) brown (b) bright red
(c) purplish red (d) all

311. Nitrosomyoglobin gives ________ color
(a) pink (b) bright red
(c) both (d) none

312. Carotenoid is present in
(a) paprika (b) tomato
(c) egg yolk (d) all

313. Natural carotenoids have ________ structure
(a) aliphatic (b) alicyclic
(c) both (d) none

314. Chlorophyll ________is generally found in green plants
(a) chl-a (b) chl-b
(c) both a & b (d) chl-c

315. Chlorophyll ________is generally found in marine algae
(a) chl-a (b) chl-b
(c) both a & b (d) chl-c

316. Caramel is produced generally from sugars by heating at________°C
(a) 70 (b) 90
(c) 121 (d) 190

317. Caramel can be produced from
(a) corn syrup (b) glucose syrup
(c) molasses (d) all

318. Caramel is generally used in
(a) bakery (b) confectionery
(c) spirits (d) all

319. Color of paprika is due to
(a) carotenoids (b) chlorophyll
(c) riboflavin (d) all

320. Yellow color of turmeric is due to
(a) carotene (b) curcumin
(c) lycopene (d) all

321. Curcumin is soluble in
(a) ethanol (b) ether
(c) glacial acetic acid (d) all

322. Curcumin gives ________ color in acidic pH
(a) bright yellow (b) reddish-brown
(c) both (d) none

323. Curcumin gives ________ color in alkaline pH
(a) bright yellow
(b) reddish-brown
(c) both
(d) none

324. Which of the following can act as chelating agent
(a) oxalic acid (b) succinic acid
(c) lactic acid (d) all

325. Which of the following can act as chelating agent
(a) citric acid (b) malic acid
(c) polyphosphoric acid
(d) all

326. ________ salt of orthophosphoric acid is used as chelating agent
(a) Na (b) K
(c) Ca (d) all

327. Which of the following can act as surfactant
(a) proteins (b) hydrocolloids
(c) phospholipids (d) all

328. HLB-value of oleic acid is about
(a) 1.0 (b) 2.1
(c) 3.4 (d) 4.7

329. HLB-value of sorbitol tristearate is about
(a) 1.0 (b) 2.1
(c) 3.4 (d) 4.7

330. HLB-value of stearyl monoglyceride is about
(a) 1.0 (b) 2.1
(c) 3.4 (d) 4.7

331. HLB-value of sorbitol monostearte is about
(a) 1.0 (b) 2.1
(c) 3.4 (d) 4.7

332. HLB-value of sorbitol monolaurate is about
(a) 8.6 (b) 9.8
(c) 10.5 (d) 14.9

333. HLB-value of gelatin is about
(a) 8.6 (b) 9.8
(c) 10.5 (d) 14.9

334. HLB-value of polyoxyethylene sorbitol tristearate is about
(a) 8.6 (b) 9.8
(c) 10.5 (d) 14.9

335. HLB-value of methyl cellulose is about
(a) 8.6 (b) 9.8
(c) 10.5 (d) 14.9

336. HLB-value of polyoxyethylene sorbitol monostearate is about
(a) 8.6 (b) 9.8
(c) 10.5 (d) 14.9

337. HLB-value of polyoxyethylene sorbitol monooleateis about
(a) 15 (b) 18
(c) 20 (d) 25

338. HLB-value of sodium oleate is about
(a) 15 (b) 18
(c) 20 (d) 25

339. HLB-value of potassium oleate is about
(a) 15 (b) 18
(c) 20 (d) 25

340. HLB-value of 3-6 is related to
(a) w/o emulsifiers
(b) humectants
(c) o/w emulsifiers
(d) turbidity stabilization

341. HLB-value of 7-9 is related to
(a) w/o emulsifiers
(b) humectants
(c) o/w emulsifiers
(d) turbidity stabilization

342. HLB-value of 8-18 is related to
(a) w/o emulsifiers
(b) humectants
(c) o/w emulsifiers
(d) all

343. HLB-value of 15-18 is related to
(a) w/o emulsifiers
(b) humectants
(c) turbidity stabilization
(d) all

344. Which of the following act as bleaching agent
(a) Cl_2 (b) ClO_2
(c) NOCl (d) all

345. Which of the following act as bleaching agent
(a) NO_2 (b) N_2O_4
(c) both (d) none

346. Synonym used for allura red AC
(a) FD&C red 40 (b) E129
(c) both (d) none

347. Molecular mass of allura red is about
(a) 496 (b) 480
(c) 512 (d) 526

348. Which of the following is a property of allura red AC
(a) heat stable
(b) light stable
(c) yellowish red color
(d) all

349. Aluminium used as food color for surface coating of
(a) sugar (b) confectionery
(c) both (d) none

350. Molecular mass of aluminium is about
(a) 23.98 (b) 26.98
(c) 33.98 (d) 36.98

351. Boiling point of aluminium is about _______°C
(a) 400 (b) 800
(c) 1800 (d) 225

352. Amaranth is used as food color in
(a) confectionery
(b) desserts
(c) alcoholic beverages
(d) all

353. Annatto is used as food color in
(a) butter (b) confectionery
(c) cheese (d) all

354. Molecular mass of bixin is about
(a) 380 (b) 394
(c) 280 (d) 294

355. Molecular mass of nor-bixin is about
(a) 380 (b) 394
(c) 280 (d) 294

356. Alternatives for annatto can be
(a) carminic acid (b) sunset yellow
(c) beta carotene (d) all

357. Which of the following is problem with annatto
(a) oxidation (b) heat stability
(c) light stability (d) all

358. Synonym used for anthocyanin
(a) grapefruit extract
(b) cyanidin
(c) malvidin
(d) all

359. Synonym used for anthocyanin
(a) peonidin (b) delphinidin
(c) petunidin (d) all

360. Molecular formula of cyanidin is
(a) $C_{15}H_{11}O_6Cl$ (b) $C_{16}H_{13}O_6Cl$
(c) both (d) none

361. Molecular formula of peonidin is
(a) $C_{15}H_{11}O_6Cl$ (b) $C_{16}H_{13}O_6Cl$
(c) both (d) none

362. Molecular formula of malvinidin is
(a) $C_{15}H_{11}O_6Cl$ (b) $C_{16}H_{13}O_6Cl$
(c) $C_{17}H_{15}O_5Cl$ (d) none

363. Molecular formula of delphinidin is
(a) $C_{15}H_{11}O_7Cl$ (b) $C_{16}H_{13}O_6Cl$
(c) $C_{17}H_{15}O_5Cl$ (d) none

364. Molecular formula of petunidin is
(a) $C_{16}H_{11}O_6Cl$ (b) $C_{16}H_{13}O_7Cl$
(c) $C_{17}H_{15}O_5Cl$ (d) none

365. Molecular mass of cyanidin is about
(a) 322 (b) 394
(c) 280 (d) 294

366. Molecular mass of peonidin is about
(a) 380 (b) 336
(c) 280 (d) 294

367. Molecular mass of malvidin is about
(a) 380 (b) 394
(c) 366 (d) 294

368. Molecular mass of delphinidin is about
(a) 380 (b) 394
(c) 340 (d) 294

369. Molecular mass of petunidin is about
(a) 380 (b) 352
(c) 280 (d) 294

370. Molecular mass of pelargonidin is about
(a) 306 (b) 394
(c) 280 (d) 294

371. A standard grape skin extract contains ____ % total anthocyanin
(a) 1 (b) 5
(c) 10 (d) 15

372. Alternative for anthocyanin is
(a) carmine (b) erythrosine
(c) allura red (d) all

373. Anthocyanin require _____ environment to exhibit red color
(a) acidic (b) basic
(c) both (d) none

374. Synonym used for beta-apo-8 carotene
(a) food orange 6 (b) E160e
(c) both (d) none

375. Molecular mass of beta-apo-8-carotenal
(a) 316 (b) 416
(c) 322 (d) 387

376. Alternative to beta-apo-8-carotenal is
(a) paprika (b) sunset yellow
(c) annatto (d) all

377. Beta-apo-8-carotenal is stable towards
(a) pH
(b) acidic conditions
(c) both
(d) none

378. Synonyms used for Brilliant Blue FCF is
(a) FD&C Blue No. 1
(b) Food Blue 2
(c) both
(d) none

379. Molecular mass of Brilliant blue FCF is about
(a) 792 (b) 816
(c) 834 (d) 873

380. Alternative for Brilliant blue FCF is
(a) patent blue V (b) green
(c) indigotine (d) all

381. Brilliant blue FCF is stable towars
(a) SO_2 (b) sodium benzoate
(c) heat & light (d) all

382. Alternative to brown FK is
(a) brown HT (b) caramels
(c) both (d) none

383. Brown FK is permitted in
(a) USA (b) UK & Europe
(c) Canada (d) Australia

384. Synonym used for brown HT is
(a) E155 (b) food brown 3
(c) CI 20285 (d) all

385. Brown HT is a type of
(a) flavor (b) color
(c) preservative (d) none

386. Alternative to brown HT is
(a) Brown FK (b) caramels
(c) both (d) none

387. Brown HT is permitted with restrictions in
(a) USA (b) Australia
(c) Canada (d) all

388. Calcium carbonate used as food color in
(a) sugar confectionery
(b) dairy creamer
(c) both
(d) none

389. Synonym used for calcium carbonate is
(a) pigment white 18 (b) chalk
(c) E170 (d) all

390. Molecular mass of calcium carbonate is
(a) 90 (b) 100
(c) 150 (d) 200

391. Alternative used for calcium carbonate is
(a) titanium dioxide (b) CO_2
(c) lycopene (d) all

392. Canthaxanthin is used as food color in
(a) sauces (b) confectionery
(c) salad dressing (d) all

393. Synonym for canthaxanthin is
(a) CI 40850
(b) Food Orange 8
(c) E 161g (d) all

394. Molecular mass of canthaxanthin is
(a) 564 (b) 600
(c) 650 (d) 200

395. Alternative used for canthaxanthin is
(a) carmine (b) ponceau 4R
(c) erythrosine (d) all

396. Canthaxanthin exhibit good stability to
(a) heat (b) light
(c) acid (d) all

397. Canthaxanthin is not permitted in
(a) USA (b) Canada
(c) UK (d) all

398. Caramel is used as food color in
(a) bakery (b) confectionery
(c) desserts (d) all

399. Caramel is not permitted in
(a) USA (b) Canada
(c) UK (d) none

400. Which of the following is a type of food caramel color
(a) plain caramel
(b) ammonia caramel
(c) caustic sulfite caramel
(d) all

401. Alternative for beet root red is
(a) anthocyanin (b) carmine
(c) amaranth (d) all

402. Alternative to carmine is
(a) anthocyanin (b) allura red
(c) erythrosine (d) all

403. Carmine is stable to
(a) heat (b) light
(c) oxidation (d) all

404. Synonym used for carminic acid is
(a) cochineal (b) E 120
(c) beetroot red (d) all

405. Alternative used ofr carminic acid is
(a) sunset yellow (b) carmin
(c) beetroot red (d) all

406. Which of the following belong to azo class
(a) Ponceau 4R
(b) erythrosine
(c) brilliant blue FCF
(d) all

407. Which of the following belong to azo class
(a) Fast green FCF
(b) carmoisine
(c) brilliant blue FCF
(d) all

408. Which of the following belong to azo class
(a) sunset yellow
(b) indigo carmine
(c) both
(d) none

409. Which of the following belong to triarylmethane class
(a) brilliant blue FCF
(b) Fast green FCF
(c) both
(d) none

410. Which of the following belong to indigoid class
(a) indigo carmine
(b) erythrosine
(c) both
(d) none

411. Which of the following belong to xanthene class
(a) indigo carmine (b) erythrosine
(c) both (d) none

412. Which of the following belong to pyrazolone class
(a) tartrazine
(b) indigo carmine
(c) Fast green FCF
(d) none

413. Chemical class of sunset yellow FCF is
(a) azo (b) xanthenes
(c) triarylmethane (d) all

414. Maximum limit (ppm) of saccharin sodium in carbonated water is above
(a) 10 (b) 100
(c) 450 (d) 500

415. Maximum limit (ppm) of saccharin sodium in supari is above
(a) 100 (b) 400
(c) 1200 (d) 4000

416. Maximum limit (ppm) of saccharin sodium in pan masala is above
(a) 100 (b) 400
(c) 1200 (d) 8000

417. Maximum limit (ppm) of saccharin sodium in chocolate (plain) is above
(a) 100 (b) 500
(c) 1200 (d) 8000

418. Maximum limit (ppm) of saccharin sodium in chewing gum is above
(a) 100 (b) 400
(c) 1200 (d) 3000

419. Maximum limit (ppm) of aspartame in carbonated water is
(a) 700 (b) 400
(c) 1200 (d) 8000

420. Maximum limit (ppm) of aspartame in bread and biscuits is
(a) 700 (b) 1400
(c) 2200 (d) 8000

421. Maximum limit (ppm) of aspartame in sweets is
(a) 700 (b) 200
(c) 800 (d) 1000

422. Maximum limit (ppm) of of saccharin sodium in chocolate (plain) is above
(a) 100 (b) 500
(c) 1200 (d) 8000

423. Maximum limit (ppm) of aspartame in jams and jellies is
(a) 700 (b) 200
(c) 800 (d) 1000

424. Maximum limit (ppm) of aspartame in chocolate is
(a) 700 (b) 2000
(c) 800 (d) 1000

425. Maximum limit (ppm) of aspartame in chewing gum is
(a) 700 (b) 200
(c) 800 (d) 10000

426. Maximum limit (ppm) of aspartame in custard powder mix is
(a) 700 (b) 200
(c) 800 (d) 1000

427. Maximum limit (ppm) of aspartame in icecream is
(a) 700 (b) 200
(c) 800 (d) 1000

428. Maximum limit (ppm) of aspartame in fruit nectar is
(a) 600 (b) 200
(c) 800 (d) 1000

429. Maximum limit (ppm) of aspartame in flavored milk is
(a) 600 (b) 200
(c) 800 (d) 1000

430. Maximum limit (ppm) of aspartame in yoghurt is
(a) 600 (b) 200
(c) 800 (d) 1000

431. Maximum limit (ppm) of aspartame in gulab jamun is
(a) 600 (b) 200
(c) 800 (d) 1000

432. Maximum limit (ppm) of aspartame in ready to eat cereals is
(a) 600 (b) 200
(c) 800 (d) 1000

433. Maximum limit (ppm) of acesulfame K in carbonated water is
(a) 600 (b) 200
(c) 300 (d) 400

434. Maximum limit (ppm) of acesulfame K in bread and biscuits is
(a) 1600 (b) 1200
(c) 1300 (d) 1000

435. Maximum limit (ppm) of acesulfame K in cakes and pastries is
(a) 1600 (b) 1200
(c) 1300 (d) 1000

436. Maximum limit (ppm) of acesulfame K in sweets is
(a) 500 (b) 1200
(c) 1300 (d) 1000

437. Maximum limit (ppm) of acesulfame K in chocolate (milk) is
(a) 500 (b) 1200
(c) 1300 (d) 1000

438. Maximum limit (ppm) of acesulfame K in plain chocolate is
(a) 500 (b) 1200
(c) 1300 (d) 1000

439. Maximum limit (ppm) of acesulfame K in sugarfree confectionery is
(a) 7000 (b) 6000
(c) 3500 (d) 4000

440. Maximum limit (ppm) of acesulfame K in ready to serve coffee and beverage is
(a) 600 (b) 1200
(c) 1300 (d) 1000

441. Maximum limit (ppm) of acesulfame K in ice candy is
(a) 1600 (b) 1200
(c) 1300 (d) 800

442. Maximum limit (ppm) of acesulfame K in cereal based beverages is
(a) 500 (b) 1200
(c) 1300 (d) 800

443. Maximum limit (ppm) of acesulfame K in chewing gum is
(a) 500 (b) 1200
(c) 1300 (d) 5000

444. Maximum limit (ppm) of acesulfame K in fruit nectar is
(a) 300 (b) 500
(c) 700 (d) 1000

445. Maximum limit (ppm) of acesulfame K in noncarbonated water based beverages (non-alcoholic) is
(a) 300 (b) 500
(c) 700 (d) 1000

446. Maximum limit (ppm) of sucralose in carbonated water is
(a) 100 (b) 200
(c) 300 (d) 500

447. Maximum limit (ppm) of sucralose in softdrink concentrate is
(a) 100 (b) 200
(c) 300 (d) 500

448. Maximum limit (ppm) of sucralose in biscuits and breads is
(a) 250 (b) 500
(c) 750 (d) 1000

449. Maximum limit (ppm) of sucralose in cakes and pastries is
(a) 250 (b) 500
(c) 750 (d) 1000

450. Maximum limit (ppm) of sucralose in sweets is
(a) 750 (b) 1500
(c) 1000 (d) 2000

451. Maximum limit (ppm) of sucralose in gulabjamun is
(a) 250 (b) 500
(c) 750 (d) 1000

452. Maximum limit (ppm) of sucralose in yoghurt is
(a) 300 (b) 400
(c) 900 (d) 1000

453. Maximum limit (ppm) of sucralose in sweetened buttermilk is
(a) 300 (b) 400
(c) 900 (d) 1000

454. Maximum limit (ppm) of sucralose in ice cream is
(a) 300 (b) 400
(c) 900 (d) 1000

455. Maximum limit (ppm) of sucralose in jam and jellies is
(a) 300 (b) 450
(c) 900 (d) 1000

456. Maximum limit (ppm) of sucralose in chutney is
(a) 300 (b) 400
(c) 800 (d) 1000

457. Maximum limit (ppm) of sucralose in cake mixes is
(a) 300 (b) 400
(c) 700 (d) 1000

458. Maximum limit (ppm) of sucralose in ready to serve coffee beverage is
(a) 300 (b) 400
(c) 700 (d) 1000

459. Maximum limit (ppm) of sucralose in confectionery is
(a) 300 (b) 400
(c) 900 (d) 1500

460. Maximum limit (ppm) of sucralose in ice candy is
(a) 300 (b) 400
(c) 800 (d) 1000

461. Maximum limit (ppm) of sucralose in vegetable nectar is
(a) 250 (b) 400
(c) 900 (d) 1000

462. Maximum limit (ppm) of sucralose in no alcoholic beverages is
(a) 100 (b) 300
(c) 900 (d) 1000

463. Which of the following is a class I preservative
(a) common salt (b) sugar
(c) dextrose (d) all

464. Which of the following is a class I preservative
(a) glucose syrup (b) spices
(c) vinegar (d) all

465. Which of the following is a class I preservative
(a) edible vegetable oil
(b) honey
(c) both
(d) none

466. Which of the following is a class II preservative
(a) benzoic acid
(b) sulphurous acid
(c) nitrate
(d) all

467. Which of the following is a class II preservative
(a) nitrites (b) sorbic acid
(c) propionic acid (d) all

468. Which of the following is a class II preservative
(a) lactic acid (b) nisin
(c) calcium propionate
(d) all

469. Maximum limit of SO_2 in meat sausages is about _______ ppm
(a) 450 (b) 1000
(c) 1500 (d) 2000

470. Maximum limit of SO_2 in fruit pulp is about _______ ppm
(a) 450 (b) 1000
(c) 1500 (d) 2000

471. Maximum limit of SO_2 in fruit juice concentrate is about _______ ppm
(a) 450 (b) 1000
(c) 1500 (d) 2000

472. Maximum limit of SO_2 in non alcoholic wine is about _______ ppm
(a) 350 (b) 1000
(c) 1500 (d) 2000

473. Maximum limit of SO_2 in fruit cordials is about _______ ppm
(a) 350 (b) 1000
(c) 1500 (d) 2000

474. Maximum limit of SO_2 in jam is about _______ ppm
(a) 40 (b) 100
(c) 150 (d) 200

475. Maximum limit of SO_2 in khandsari is about _______ ppm
(a) 70 (b) 140
(c) 210 (d) 280

476. Maximum limit of SO_2 in gur is about _______ ppm
(a) 70 (b) 140
(c) 210 (d) 280

477. Maximum limit of SO_2 in jaggery is about _______ ppm
(a) 70 (b) 140
(c) 210 (d) 280

478. Which of the following is used as preservative
(a) salt (b) sugar
(c) both (d) none

479. Maximum limit of SO_2 in refined sugar is about _______ ppm
(a) 40 (b) 100
(c) 450 (d) 1000

480. Maximum limit of SO_2 in corn flour is about _______ ppm
(a) 40 (b) 100
(c) 450 (d) 1000

481. Maximum limit of SO_2 in corn syrup is about _______ ppm
(a) 40 (b) 100
(c) 450 (d) 1000

482. Maximum limit of SO_2 in canned rasgulla is about _______ ppm
(a) 40 (b) 100
(c) 450 (d) 1000

483. Maximum limit of SO_2 in gelatine is about _______ ppm
(a) 40 (b) 100
(c) 450 (d) 1000

484. Maximum limit of SO_2 in beer is about _______ ppm
(a) 40 (b) 70
(c) 200 (d) 450

485. Maximum limit of SO_2 in cider is about _______ ppm
(a) 40 (b) 70
(c) 200 (d) 450

486. Maximum limit of SO_2 in alcoholic wines is about _______ ppm
(a) 40 (b) 70
(c) 200 (d) 450

487. Maximum limit of SO_2 in ready to serve beverages is about _______ ppm
(a) 40 (b) 70
(c) 200 (d) 450

488. Maximum limit of SO_2 in dehydrated vegetables is about _______ ppm
(a) 350 (b) 500
(c) 1000 (d) 2000

489. Maximum limit of SO_2 in syrup and sharbats is about _______ ppm
(a) 350 (b) 500
(c) 1000 (d) 2000

490. Maximum limit of SO_2 in dried ginger is about _______ ppm
(a) 350 (b) 500
(c) 1000 (d) 2000

491. Maximum limit of benzoic acid in ready to serve beverages is about _______ ppm
(a) 70 (b) 120
(c) 220 (d) 270

492. Maximum limit of benzoic acid in brewed ginger bear is about _______ ppm
(a) 70 (b) 120
(c) 220 (d) 270

493. Maximum limit of benzoic acid in tomato sauce is about _______ ppm
(a) 250 (b) 500
(c) 750 (d) 1000

494. Maximum limit of benzoic acid in tomato puree & paste is about ___ ppm
(a) 350 (b) 650
(c) 550 (d) 750

495. Maximum limit of benzoic acid in syrups & sharbats is about ____ ppm
(a) 300 (b) 600
(c) 900 (d) 1000

496. Maximum limit of benzoic acid in jam is about _______ ppm
(a) 200 (b) 400
(c) 600 (d) 800

497. Maximum limit of benzoic acid in marmalade is about _______ ppm
(a) 200 (b) 400
(c) 600 (d) 800

498. Maximum limit of benzoic acid in fat spread is about _______ ppm
(a) 300 (b) 600
(c) 900 (d) 1000

499. Maximum limit (ppm) of sodium nitrite in pickled meat is about
(a) 200 (b) 100
(c) 300 (d) 400

500. Maximum limit (ppm) of sodium nitrite in bacon is about
(a) 200 (b) 100
(c) 300 (d) 400

501. Maximum limit (ppm) of sodium nitrite in beef is about
(a) 200 (b) 100
(c) 300 (d) 400

502. Maximum limit (ppm) of sorbic acid in cheese is about
(a) 1000 (b) 2000
(c) 3000 (d) 5000

503. Maximum limit (ppm) of sorbic acid in flour confectionery is about
(a) 500 (b) 1500
(c) 3000 (d) 5000

504. Maximum limit (ppm) of sorbic acid in preserved chappatis is about
(a) 500 (b) 1500
(c) 3000 (d) 5000

505. Maximum limit (ppm) of sorbic acid in panir is about
(a) 1500 (b) 2000
(c) 3000 (d) 5000

506. Maximum limit (ppm) of sorbic acid in chhana is about
(a) 1500 (b) 2000
(c) 3000 (d) 5000

507. Maximum limit (ppm) of sorbic acid in fat spread is about
(a) 1000 (b) 2000
(c) 3000 (d) 5000

508. Maximum limit (ppm) of propionates in flour for baked foods is about
(a) 1200 (b) 2200
(c) 3200 (d) 4200

509. Maximum limit (ppm) of propionic acid in paneer is about
(a) 1000 (b) 2000
(c) 3000 (d) 4000

510. Maximum limit (ppm) of sorbic acid in fruit juice is about
(a) 50 (b) 100
(c) 200 (d) 500

511. Maximum limit (ppm) of sorbic acid in fruit juice concentrate with preservative is about
(a) 50 (b) 100
(c) 200 (d) 500

512. Maximum limit (ppm) of sorbic acid in nectars is about
(a) 50 (b) 100
(c) 200 (d) 500

513. Maximum limit (%) of ethyl gallate (as antioxidant) in fats and oil is about
(a) 0.01 (b) 0.02
(c) 0.05 (d) none

514. Maximum limit (%) of propyl gallate (as antioxidant) in fats and oil is about
(a) 0.01 (b) 0.02
(c) 0.05 (d) none

515. Maximum limit (%) of BHA (as antioxidant) in fats and oil is about
(a) 0.01 (b) 0.02
(c) 0.05 (d) none

516. Maximum limit (%) of citric acid (as antioxidant) in fats and oil is about
(a) 0.01 (b) 0.02
(c) 0.05 (d) none

517. Maximum limit (%) of tartaric acid (as antioxidant) in fats and oil is about
(a) 0.01 (b) 0.02
(c) 0.05 (d) none

518. Maximum limit (%) of TBHQ (as antioxidant) in fats and oil is about
(a) 0.01 (b) 0.02
(c) 0.05 (d) none

519. MSG is not allowed in
(a) pasteurized cream
(b) butter
(c) cut fruits
(d) all

520. MSG is not allowed in
(a) frozen vegetables
(b) flaked grains
(c) cereal flours
(d) all

521. MSG is not allowed in
(a) pasta (b) syrups
(c) processed fish (d) all

522. MSG is not allowed in
(a) salt (b) spices
(c) weaning food (d) all

523. MSG is not allowed in
(a) packaged water (b) coffee
(c) wines (d) all

524. MSG is not allowed in
(a) margarine (b) fat spread
(c) carbonated water
(d) all

525. MSG is not allowed in
(a) baking powder (b) arrowroot
(c) sago (d) all

526. MSG is not allowed in
(a) ice candies (b) icecream
(c) cocoa butter (d) all

527. MSG is not allowed in
(a) bread (b) vinegar
(c) toffee (d) all

528. MSG is not allowed in
(a) chocolate
(b) pan masala
(c) alcoholic beverages
(d) all

529. Maximum level (ppm) of calcium gluconate in confections is about
(a) 2500 (b) 600
(c) 10000 (d) 40000

530. Maximum level (ppm) of calcium carbonate as neutralizer in various food is about
(a) 2500 (b) 600
(c) 10000 (d) 40000

531. Maximum level (ppm) of calcium oxide (neutralizer) in specified dairy products is about
(a) 2500 (b) 600
(c) 10000 (d) 40000

532. Maximum level (ppm) of phosphoric acid in beverages is about
(a) 2500 (b) 600
(c) 10000 (d) 40000

533. Maximum level (ppm) of phosphoric acid in beverages is about
(a) 2500 (b) 600
(c) 10000 (d) 40000

534. Maximum level (ppm) of phosphoric acid in bread is about
(a) 2500 (b) 600
(c) 10000 (d) 40000

535. Maximum level (ppm) of phosphoric acid in milk preparations is about
(a) 50 (b) 600
(c) 4000 (d) 10000

536. Maximum level (ppm) of phosphoric acid in cake mixes is about
(a) 50 (b) 600
(c) 4000 (d) 10000

537. Maximum level (ppm) of phosphoric acid in protein foods is about
(a) 50 (b) 600
(c) 4000 (d) 10000

538. Maximum level (ppm) of tartaric acid as acidulant is about
(a) 50 (b) 600
(c) 4000 (d) 10000

539. Maximum level (ppm) of EDTA in emulsion foods is about
(a) 50 (b) 600
(c) 4000 (d) 10000

540. Maximum level (ppm) of EDTA in salad dressing is about
(a) 50 (b) 600
(c) 4000 (d) 10000

541. Maximum level (ppm) of EDTA in fat spread is about
(a) 50 (b) 600
(c) 4000 (d) 10000

542. Maximum level (ppm) of fumaric acid as acidulant is about
(a) 50 (b) 600
(c) 3000 (d) 10000

543. Maximum level (ppm) of ammonium carbonate in baked foods is about
(a) 50 (b) 600
(c) 5000 (d) 10000

544. Maximum level (ppm) of ammonium bicarbonate in baked foods is about
(a) 50 (b) 600
(c) 5000 (d) GMP

545. Maximum level (ppm) of baking powder in baked foods is about
(a) 50 (b) 600
(c) 5000 (d) GMP

546. Maximum level (ppm) of ammonium phosphate monobasic in bread is about
(a) 50 (b) 600
(c) 2500 (d) 10000

547. Maximum level (ppm) of calcium phosphate in bread is about
(a) 50 (b) 600
(c) 2500 (d) 10000

548. Maximum level (ppm) of calcium carbonate in bread is about
(a) 50 (b) 600
(c) 5000 (d) 10000

549. Maximum level (ppm) of potassium bromate in bread is about
(a) 50 (b) 600
(c) 5000 (d) 10000

550. Maximum level (ppm) of potassium iodate in bread is about
(a) 50 (b) 600
(c) 5000 (d) 10000

551. Maximum level (ppm) of ammonium chloride in bread is about
(a) 50 (b) 500
(c) 5000 (d) 10000

552. Maximum level (ppm) of fungal alpha amylase in bread is about
(a) 100 (b) 600
(c) 5000 (d) 10000

553. Maximum level (ppm) of sodium stearoyl-2-lactylate in bread is about
(a) 50 (b) 600
(c) 5000 (d) 10000

554. Maximum level (ppm) of calcium steroyl-2-lactylate in bread is about
(a) 50 (b) 600
(c) 5000 (d) 10000

555. Maximum level (ppm) of L-cysteine monohydrochloride in bread is about
(a) 50 (b) 600
(c) 5000 (d) 90

556. Maximum level (ppm) of benzoyl peroxide in flour for baking is about
(a) 40 (b) 20
(c) 200 (d) 440

557. Maximum level (ppm) of potassium bromate in flour for baking is about
(a) 40 (b) 20
(c) 200 (d) 440

558. Maximum level (ppm) of ascorbic acid in flour for baking is about
(a) 40 (b) 20
(c) 200 (d) 440

559. Maximum level (ppm) of gluconodelta lactone in cured meat is about
(a) 50 (b) 600
(c) 5000 (d) 10000

560. Maximum level (ppm) of chlorine in flour for bakery is about
(a) 500 (b) 2000
(c) 5000 (d) 10000

561. Maximum level (ppm) of ascorbic acid in canned chicken/mutton is about
(a) 500 (b) 2000
(c) 5000 (d) 10000

562. Maximum level (ppm) of phosphates in cooked ham is about
(a) 500 (b) 2000
(c) 5000 (d) 8000

563. Minimum dye content (%) in tartrazine should be
(a) 17 (b) 47
(c) 65 (d) 87

564. Maximum limit (%) of water insoluble matter in tartrazine is
(a) 0.2 (b) 0.5
(c) 1 (d) 10

565. Maximum limit (mg/kg) of lead in tartrazine is
(a) 10 (b) 50
(c) 40 (d) 85

566. Maximum limit (mg/kg) of arsenic in tartrazine is
(a) 10 (b) 3
(c) 40 (d) 85

567. Maximum limit (mg/kg) of heavy metals in tartrazine is
(a) 10 (b) 50
(c) 40 (d) 85

568. Minimum dye content (%) in sunset yellow should be
(a) 17 (b) 47
(c) 65 (d) 87

569. Beta carotene is soluble in
(a) carbon disulfide (b) benzene
(c) chloroform (d) all

570. Riboflavin is more soluble in
(a) water
(b) saline solution
(c) alcohol
(d) none

571. Which of the following is permitted as filler material in color prepara-tions
(a) potable water (b) common salt
(c) sugar (d) all

572. Which of the following is permitted as filler material in color preparations
(a) liquid glucose (b) tartaric acid
(c) glycerine (d) all

573. Which of the following is permitted as filler material in color preparations
(a) propylene glycol
(b) dilute acetic acid
(c) sorbitol
(d) all

574. Which of the following is permitted as filler material in color preparations
(a) citric acid
(b) sodium carbonate
(c) lactose
(d) all

575. Which of the following is permitted as filler material in color prepar-ations
(a) sodium alginate (b) dextrins
(c) ethylacetate (d) all

576. Which of the following is permitted as filler material in color prepara-tions
(a) starch (b) ethanol
(c) bee wax (d) all

577. Which of the following is permitted as filler material in color prepara-tions
(a) ammonium hydroxide
(b) lactic acid
(c) carrageenan
(d) all

578. Which of the following is permitted as filler material in color preparations
(a) gum arabic (b) gelatin
(c) pectin (d) all

579. Maximum limit (ppm) of acetic acid in bread is about
(a) 1500 (b) 2500
(c) 5000 (d) GMP

580. Maximum limit (ppm) of acetic acid in biscuits is about
(a) 1500 (b) 2500
(c) 5000 (d) GMP

581. Maximum limit (ppm) of sucroglycerides in biscuits is about
(a) 1000 (b) 2500
(c) 5000 (d) GMP

582. Maximum limit (ppm) of guar gum in bread is about
(a) 1000 (b) 2500
(c) 5000 (d) GMP

583. Maximum limit (ppm) of lecithin in bread is about
(a) 1000 (b) 2500
(c) 5000 (d) GMP

584. Maximum limit (ppm) of GMS in bread is about
(a) 1000 (b) 2500
(c) 5000 (d) GMP

585. Maximum limit (ppm) of polyglycerol/esters of fatty acids in bread is about
(a) 1000 (b) 2000
(c) 5000 (d) GMP

586. Maximum limit (ppm) of bacterial amylase in bread is about
(a) 1000 (b) 2500
(c) 5000 (d) GMP

587. Maximum limit (ppm) of bacterial amylase in biscuits is about
(a) 1000 (b) 2500
(c) 5000 (d) GMP

588. Maximum limit (ppm) of ammonium persulfate in bread (on flour weight basis) is about
(a) 1000 (b) 2500
(c) 5000 (d) GMP

589. Maximum limit (ppm) of calcium carbonate in bread (on flour weight basis) is about
(a) 1000 (b) 2500
(c) 5000 (d) GMP

590. Maximum limit (ppm) of potassium bromate in bread (on flour weight basis) is about
(a) 50 (b) 2500
(c) 5000 (d) GMP

591. Which of the following is a bread improver
(a) fungal amylase
(b) bacterial amylase
(c) both (d) none

592. Which of the following is a bread improver
(a) ammonium persulfate
(b) calcium phosphate
(c) calcium carbonate
(d) all

593. Which of the following is a bread improver
(a) potassium bromated
(b) potassium iodate
(c) both (d) none

594. Which of the following is a flour treatment agent
(a) ammonium chloride
(b) L-cysteine monohydrochloride
(c) both
(d) none

595. Which of the following is a flour treatment agent
(a) ammonium phosphate
(b) benzoyl peroxide
(c) both
(d) none

596. Maximum limit (ppm) of benzoyl peroxide in biscuits is
(a) 40 (b) 90
(c) 150 (d) 200

597. Maximum limit (ppm) of aspartame in bread is about
(a) 750 (b) 1000
(c) 2200 (d) GMP

598. Maximum limit (ppm) of acesulfame K in bread is
(a) 750 (b) 1000
(c) 2200 (d) GMP

599. Maximum limit (ppm) of sucralose in bread is
(a) 750 (b) 1000
(c) 2200 (d) GMP

600. Which of the following is a thickening agent
(a) guar gum (b) tragacanth gum
(c) pectin (d) all

601. Which of the following is a thickening agent
(a) xanthan gum (b) alginate
(c) CMC (d) all

602. Which of the following is a modified starch
(a) acid treated starch
(b) alkali treated starch
(c) balanced starch
(d) all

603. Which of the following is a modified starch
(a) distarch glycerol
(b) distarch phosphate
(c) monostarch phosphate
(d) all

604. Which of the following is a modified starch
(a) oxidized starch
(b) starch acetate
(c) hydroxyl propyl starch
(d) all

605. Which of the following is anti-foaming agent
(a) ascorbic acid
(b) dimethyl polysiloxane
(c) methyl cellulose
(d) none

606. Which of the following is firming agent
(a) calcium chloride
(b) calcium gluconate
(c) calcium carbonate
(d) all

607. Maximum limit (%) of methyl cellulose in snacks foods is about
(a) 0.1 (b) 0.2
(c) 0.5 (d) 1.0

608. Which of the following is used as stabilizer in ice cream
(a) guar gum (b) alginate
(c) CMC (d) all

609. Which of the following is a plant gum
(a) guar gum (b) xanthan gum
(c) agar (d) all

610. Maximum limit (%) of gellan gum in sugar boiled confectionery is about
(a) 1.0 (b) 0.5
(c) 0.2 (d) 0.1

611. Maximum limit (%) of sorbic acid in instant idli mix is about
(a) 1.0 (b) 0.5
(c) 0.2 (d) 0.1

612. Maximum limit (%) of sorbic acid in gulabjamun mix is about
(a) 1.0 (b) 0.5
(c) 0.2 (d) 0.1

613. Maximum limit (%) of sorbic acid in rice based papads is about
(a) 1.0 (b) 0.5
(c) 0.2 (d) 0.1

614. Maximum limit (%) of calcium carbonate as anticaking agent in idli mix is about
(a) 1.0 (b) 0.5
(c) 0.2 (d) 2.0

615. Maximum limit (ppm) of TiO_2 in powdered soft drink concentrate mix is about
(a) 100 (b) 200
(c) 500 (d) 1000

616. Which of the following can act as stabilizer or emulsifier
(a) gum arabic (b) gum ghatti
(c) ester gum (d) all

617. Maximum limit (%) of sodium aluminium silicate in powdered soft drink concentrate is
(a) 1.0 (b) 0.5
(c) 0.2 (d) 0.1

618. Maximum limit (%) of TBHQ in flavor emulsion is
(a) 1.0 (b) 0.01
(c) 0.2 (d) 0.1

619. Maximum limit (ppm) of propyl gallate in lard is
(a) 100 (b) 200
(c) 500 (d) 10

620. Maximum limit (ppm) of propyl gallate in table maragrine is
(a) 100 (b) 200
(c) 500 (d) 10

621. Maximum limit (ppm) of BHA in lard is
(a) 100 (b) 200
(c) 500 (d) 10

622. Which of the following is not a stabilizer
(a) BHA (b) CMC
(c) both (d) none

623. Maximum limit (ppm) of ascorbyl palmitate in lard is
(a) 100 (b) 200
(c) 500 (d) 10

624. Maximum limit (ppm) of propyl gallate in table margarine is
(a) 100 (b) 200
(c) 500 (d) 10

625. Maximum limit (ppm) of resin guaicae (as antioxidant) in lard is
(a) 100 (b) 200
(c) 500 (d) 10

626. Maximum limit (ppm) of resin guaicae (as antioxidant) in fat spread is
(a) 100 (b) 200
(c) 500 (d) 10

627. Maximum limit (ppm) of TBHQ in lard is
(a) 100 (b) 200
(c) 500 (d) 10

628. Maximum limit (ppm) of TBHQ in fat spread is
(a) 100 (b) 200
(c) 500 (d) 10

629. Maximum limit (ppm) of sodium citrate (as antioxidant synergist) in lard is
(a) 100 (b) 200
(c) 500 (d) GMP

630. Maximum limit (ppm) of phosphoric acid (as antioxidant synergist) in lard is
(a) 100 (b) 200
(c) 500 (d) GMP

631. Maximum limit (ppm) of phosphoric acid (as antioxidant synergist) in fat spread is
(a) 100 (b) 200
(c) 500 (d) GMP

632. Maximum limit (ppm) of dimethyl polysiloxane in lard is
(a) 100 (b) 200
(c) 500 (d) 10

633. Maximum limit (ppm) of dimethyl polysiloxane in fat spread is
(a) 100 (b) 200
(c) 500 (d) 10

634. Maximum limit (ppm) of dimethyl polysiloxane in edible vegetable oils is
(a) 100 (b) 200
(c) 500 (d) 10

635. Maximum limit (ppm) of lecithin in table margarine is
(a) 10 (b) 50
(c) 100 (d) GMP

636. Maximum limit (ppm) of sorbic acid in table margarine is
(a) 100 (b) 1000
(c) 500 (d) 200

637. Maximum limit (ppm) of sorbic acid in fat spread is
(a) 100 (b) 1000
(c) 500 (d) 200

638. Maximum limit (ppm) of beta-carotene in table margarine is
(a) 25 (b) 50
(c) 100 (d) 500

639. Maximum limit (ppm) of annatto extract in table margarine is
(a) 10 (b) 20
(c) 5 (d) 25

640. Maximum limit (ppm) of curcumin in fat spread is
(a) 10 (b) 20
(c) 5 (d) 25

641. Maximum limit (ppm) of beta-apo-8'-carotenal in table margarine is
(a) 10 (b) 20
(c) 5 (d) 25

642. Maximum limit (ppm) of beta-apo-8'-carotenal in fat spread is
(a) 10 (b) 20
(c) 5 (d) 25

643. Maximum limit (ppm) of calciumdisodium EDTA in table margarine is
(a) 10 (b) 20
(c) 50 (d) 100

644. Maximum limit (ppm) of calciumdisodium EDTA in fat spread is
(a) 50 (b) 100
(c) 150 (d) 200

645. Maximum limit (ppm) of diacetyl in table margarine is
(a) 4.0 (b) 8.0
(c) 15 (d) 40

646. Maximum limit (ppm) of diacetyl in fat spread is
(a) 4.0 (b) 8.0
(c) 15 (d) 40

647. Maximum limit (g/kg) of sodium ascorbate in frozen lobster is
(a) 1.0 (b) 10
(c) 50 (d) 100

648. Maximum limit (g/kg) of sodium ascorbate in frozen fish fillets is
(a) 1.0 (b) 10
(c) 50 (d) 100

649. Maximum limit (g/kg) of citric acid in minced meat flesh is
(a) 1.0 (b) 10
(c) 50 (d) 100

650. Maximum limit (g/kg) of potassium polyphosphate in frozen shrimp is
(a) 1.0 (b) 10
(c) 50 (d) 100

651. Maximum limit (ppm) of orthophosphoric acid in canned shrimp is
(a) 100 (b) 500
(c) 850 (d) 30

652. Maximum limit (ppm) of potassium bisulfite in frozen shrimps is
(a) 30 (b) 500
(c) 850 (d) 100

653. Maximum limit (ppm) of potassium bisulfite in cooked frozen shrimps is
(a) 30 (b) 500
(c) 850 (d) 100

654. Maximum limit (ppm) of calcium sorbate in salted fish is
(a) 100 (b) 200
(c) 500 (d) 50

655. Maximum limit (ppm) of potassium sorbate in salted fish is
(a) 100 (b) 200
(c) 500 (d) 50

656. Maximum limit (ppm) of sunset yellow in canned shrimp is
(a) 30 (b) 500
(c) 850 (d) 100

657. Maximum limit (g/kg) of pectin in canned finfish is
(a) 1.5 (b) 2.5
(c) 5 (d) 20

658. Maximum limit (g/kg) of xanthan gum in canned fish is
(a) 1.5 (b) 2.5
(c) 5 (d) 20

659. Maximum limit (g/kg) of sodium alginate in canned fish is
(a) 1.5 (b) 2.5
(c) 5 (d) 20

660. Maximum limit (g/kg) of CMC in canned shrimp is
(a) 1.5 (b) 2.5
(c) 5 (d) 20

661. Maximum limit (g/kg) of distarch glycerol in canned finfish is
(a) 60 (b) 50
(c) 40 (d) 20

662. Maximum limit (g/kg) of distarch phosphate in canned finfish is
(a) 20 (b) 50
(c) 40 (d) 60

663. Maximum limit (g/kg) of natural flavors in canned finfish is
(a) 10 (b) 5
(c) 20 (d) GMP

664. Maximum limit (ppm) of calciumdisodium EDTA in canned shrimps is
(a) 250 (b) 350
(c) 540 (d) 720

665. Maximum limit (ppm) of calciumdisodium EDTA in canned crab meat is
(a) 250 (b) 350
(c) 540 (d) 720

666. Maximum limit (ppm) of BHA in thermally processed chestnuts is
(a) 100 (b) 200
(c) 300 (d) 400

667. Maximum limit (ppm) of BHA in thermally processed groundnut paste is
(a) 100 (b) 200
(c) 300 (d) 400

668. Maximum limit (ppm) of annatto extract in icecream is
(a) 20 (b) 100
(c) 200 (d) 500

669. Maximum limit (ppm) of annatto extract in butter is
(a) 20 (b) 100
(c) 200 (d) 500

670. Maximum limit (ppm) of annatto extract in sweetened condensed milk is
(a) 20 (b) 100
(c) 200 (d) 500

671. Maximum limit (ppm) of beta-apo-8-carotenal in butter is
(a) 20 (b) 35
(c) 100 (d) 250

672. Maximum limit (ppm) of beta-apo-8-carotenal in yoghurt is
(a) 20 (b) 35
(c) 100 (d) 250

673. Maximum limit (ppm) of beta-apo-8-carotenal in cheese is
(a) 20 (b) 35
(c) 100 (d) 250

674. Maximum limit (ppm) of beta-apo-8-carotenal in icecream is
(a) 20 (b) 35
(c) 100 (d) 250

675. Maximum limit (ppm) of canthaxanthin in icecream is
(a) 20 (b) 35
(c) 100 (d) 250

676. Maximum limit (ppm) of riboflavin in icecream is
(a) 50 (b) 100
(c) 150 (d) 200

677. Maximum limit (ppm) of riboflavin in processed cheese is
(a) 50 (b) 100
(c) 150 (d) 200

678. Maximum limit (ppm) of riboflavin in ice candies is
(a) 50 (b) 100
(c) 150 (d) 200

679. Maximum limit (ppm) of curcumin in icecream is
(a) 50 (b) 100
(c) 150 (d) 200

680. Maximum limit (ppm) of beta-carotene in butter is
(a) 100 (b) 200
(c) 350 (d) 400

681. Maximum limit (ppm) of beta-carotene in icecream is
(a) 100 (b) 200
(c) 350 (d) 400

682. Maximum limit (ppm) of beta-carotene in cheese is
(a) 25 (b) 50
(c) 75 (d) 100

683. Maximum limit (ppm) of nisin in processed cheese is
(a) 12.5 (b) 25.5
(c) 35.5 (d) 40

684. Maximum limit (ppm) of sorbic acid (as preservative) in whey powder is
(a) 1000 (b) 1500
(c) 2000 (d) 3000

685. Maximum limit (ppm) of sorbic acid (as preservative) in processed cheese is
(a) 1000 (b) 1500
(c) 2000 (d) 3000

686. Maximum limit (ppm) of sorbic acid (as preservative) in processed cheese spread is
(a) 1000 (b) 1500
(c) 2000 (d) 3000

687. Maximum limit (ppm) of calcium chloride in cheddar cheese is
(a) 200 (b) 250
(c) 300 (d) 600

688. Maximum limit (ppm) of calcium chloride in cottage cheese is
(a) 200 (b) 250
(c) 300 (d) 600

689. Maximum limit (ppm) of annatto in cheddar cheese is
(a) 200 (b) 250
(c) 300 (d) 600

690. Maximum limit (ppm) of annatto in camembert cheese is
(a) 200 (b) 250
(c) 300 (d) 600

691. E number for curcumins is
(a) 100 (b) 101
(c) 102 (d) 103

692. E number for riboflavins is
(a) 100 (b) 101
(c) 102 (d) 103

693. E number of tartrazine is
(a) 100 (b) 101
(c) 102 (d) 103

694. E number of alkanet is
(a) 100 (b) 101
(c) 102 (d) 103

695. E number of sunset yellow FCF is
(a) 110 (b) 120
(c) 121 (d) 122

696. E number of carmine is
(a) 110 (b) 120
(c) 121 (d) 122

697. E number of amaranth is
(a) 120 (b) 121
(c) 123 (d) 124

698. E number of ponceau 4R is
(a) 120 (b) 121
(c) 123 (d) 124

699. E number of ponceau SX is
(a) 122 (b) 123
(c) 124 (d) 125

700. E number of erythrosine is
(a) 121 (b) 123
(c) 127 (d) 129

701. E number of allurared AC is
(a) 121 (b) 123
(c) 127 (d) 129

702. E number of patent blue V is
(a) 131 (b) 132
(c) 133 (d) 140

703. E number of indigotine is
(a) 131 (b) 132
(c) 133 (d) 140

704. E number of brilliant blue FCF is
(a) 131 (b) 132
(c) 133 (d) 140

705. E number of chlorophyll is
(a) 131 (b) 132
(c) 133 (d) 140

706. E number of copper chlorophyll is
(a) 140 (b) 141
(c) 142 (d) 143

707. E number of fastgreen FCF is
(a) 140 (b) 141
(c) 142 (d) 143

708. E number of caramel-I plain is
(a) 150a (b) 150b
(c) 150c (d) 150d

709. E number of caramel-II caustic sulphite process is
(a) 150a (b) 150b
(c) 150c (d) 150d

710. E number of caramel-III ammonia process is
(a) 150a (b) 150b
(c) 150c (d) 150d

711. E number of caramel-I ammonia sulphite process is
(a) 150a (b) 150b
(c) 150c (d) 150d

712. E number of brillian black PN is
(a) 151 (b) 154
(c) 155 (d) 160a

713. E number of brown FK is
(a) 151 (b) 154
(c) 155 (d) 160a

714. E number of brown HT is
(a) 151 (b) 154
(c) 155 (d) 160a

715. E number of carotenes is
(a) 151 (b) 154
(c) 155 (d) 160a

716. E number of annatto extracts is
(a) 160b (b) 160c
(c) 160d (d) 160e

717. E number of paprika oleoresins is
(a) 160b (b) 160c
(c) 160d (d) 160e

718. E number of lycopene is
(a) 160b (b) 160c
(c) 160d (d) 160e

719. E number of beta-apo-carotenal is
(a) 160b (b) 160c
(c) 160d (d) 160e

720. E number of flavoxanthin is
(a) 161a (b) 161b
(c) 161g (d) 162

721. E number of lutein is
(a) 161a (b) 161b
(c) 161g (d) 162

722. E number of canthaxanthin is
(a) 161a (b) 161b
(c) 161g (d) 162

723. E number of beet red is
(a) 161a (b) 161b
(c) 161g (d) 162

724. E number of anthocyanin is
(a) 163 (b) 170
(c) 171 (d) 172

725. E number of calcium carbonates is
(a) 163 (b) 170
(c) 171 (d) 172

726. E number of TiO_2 is
(a) 163 (b) 170
(c) 171 (d) 172

727. E number of iron oxide is
(a) 163 (b) 170
(c) 171 (d) 172

728. E number of aluminium oxide is
(a) 171 (b) 172
(c) 173 (d) 174

729. E number of food grade tannins is
(a) 180 (b) 181
(c) 182 (d) 200

730. E number of sorbic acid is
(a) 180 (b) 181
(c) 182 (d) 200

731. E number of sodium sorbate is
(a) 200 (b) 201
(c) 202 (d) 203

732. E number of potassium sorbate is
(a) 200 (b) 201
(c) 202 (d) 203

733. E number of benzoic acid is
(a) 210 (b) 209
(c) 211 (d) 180

734. E number of sodium benzoate is
(a) 210 (b) 209
(c) 211 (d) 180

735. E number of potassium benzoate is
(a) 212 (b) 213
(c) 210 (d) 180

736. E number of calcium benzoate is
(a) 212 (b) 213
(c) 210 (d) 180

737. E number of sulphurdioxide is
(a) 216 (b) 217
(c) 218 (d) 220

738. E number of sodium sulphite is
(a) 221 (b) 222
(c) 223 (d) 224

739. E number of sodium hydrogensulphite is
(a) 221 (b) 222
(c) 223 (d) 224

740. E number of sodium metabisulphite is
(a) 221 (b) 222
(c) 223 (d) 224

741. E number of potassium metabisulphite is
(a) 221 (b) 222
(c) 223 (d) 224

742. E number of potassium sulphite is
(a) 225 (b) 226
(c) 227 (d) 228

743. E number of potassium bisulphite is
(a) 225 (b) 226
(c) 227 (d) 228

744. E number of thiabendazole is
(a) 226 (b) 233
(c) 227 (d) 234

745. E number of nisin is
(a) 236 (b) 233
(c) 227 (d) 234

746. E number of natamycin is
(a) 236 (b) 233
(c) 235 (d) 234

747. E number of formic acid is
(a) 236 (b) 233
(c) 227 (d) 234

748. E number of sodium formate is
(a) 234 (b) 235
(c) 236 (d) 237

749. E number of formaldehyde is
(a) 240 (b) 241
(c) 242 (d) 243

750. E number of potassium nitrite is
(a) 240 (b) 241
(c) 242 (d) 243

751. E number of sodium nitrite is
(a) 249 (b) 250
(c) 251 (d) 252

752. E number of sodium nitrate is
(a) 249 (b) 250
(c) 251 (d) 252

753. E number of potassium nitrate is
(a) 249 (b) 250
(c) 251 (d) 252

754. E number of acetic acid (glacial) is
(a) 260 (b) 261
(c) 262 (d) 263

755. E number of potassium acetate is
(a) 260 (b) 261
(c) 262 (d) 263

756. E number of sodium acetate is
(a) 260 (b) 261
(c) 262 (d) 263

757. E number of calcium acetate is
(a) 260 (b) 261
(c) 262 (d) 263

758. E number of ammonium acetate is
(a) 264 (b) 265
(c) 270 (d) 280

759. E number of dehydroacetic acid is
(a) 264 (b) 265
(c) 270 (d) 280

760. E number of lactic acid is
(a) 264 (b) 265
(c) 270 (d) 280

761. E number of propionic acid is
(a) 264 (b) 265
(c) 270 (d) 280

762. E number of sodium propionate is
(a) 281 (b) 282
(c) 290 (d) 296

763. E number of calcium propionate is
(a) 281 (b) 282
(c) 290 (d) 296

764. E number of carbondioxide is
(a) 281 (b) 282
(c) 290 (d) 296

765. E number of malic acid is
(a) 281 (b) 282
(c) 290 (d) 296

766. E number of fumaric acid is
(a) 297 (b) 300
(c) 301 (d) 307

767. E number of ascorbic acid is
(a) 297 (b) 300
(c) 301 (d) 307

768. E number of sodium ascorbate is
(a) 297 (b) 300
(c) 301 (d) 307

769. E number of alpha-tocopherol is
(a) 297 (b) 300
(c) 301 (d) 307

770. E number of propyl gallate is
(a) 311 (b) 310
(c) 313 (d) 305

771. E number of octyl gallate is
(a) 311 (b) 310
(c) 313 (d) 305

772. E number of ethyl gallate is
(a) 311 (b) 310
(c) 313 (d) 305

773. E number of TBHQ is
(a) 319 (b) 320
(c) 321 (d) 322

774. E number of BHA is
(a) 319 (b) 320
(c) 321 (d) 322

775. E number of BHT is
(a) 319 (b) 320
(c) 321 (d) 322

776. E number of lecithins is
(a) 319 (b) 320
(c) 321 (d) 322

777. E number of ethoxyquin is
(a) 326 (b) 324
(c) 321 (d) 325

778. E number of sodium lactate is
(a) 326 (b) 324
(c) 321 (d) 325

779. E number of potassium lactate is
(a) 326 (b) 324
(c) 321 (d) 325

780. E number of calcium lactate is
(a) 325 (b) 326
(c) 327 (d) 328

781. E number of ammonium lactate is
(a) 325 (b) 326
(c) 327 (d) 328

782. E number of citric acid is
(a) 330 (b) 331
(c) 332 (d) 333

783. E number of sodium citrate is
(a) 330 (b) 331
(c) 332 (d) 333

784. E number of potassium citrate is
(a) 330 (b) 331
(c) 332 (d) 333

785. E number of calcium citrate is
(a) 330 (b) 331
(c) 332 (d) 333

786. E number of tartaric acid is
(a) 334 (b) 335
(c) 338 (d) 339

787. E number of sodium tartarate is
(a) 334 (b) 335
(c) 338 (d) 339

788. E number of orthophosphoric acid is
(a) 334 (b) 335
(c) 338 (d) 339

789. E number of sodium phosphate is
(a) 334 (b) 335
(c) 338 (d) 339

790. E number of potassium phosphates is
(a) 330 (b) 339
(c) 340 (d) 341

791. E number of calcium phosphates is
(a) 330 (b) 339
(c) 340 (d) 341

792. E number of ammonium phosphates is
(a) 330 (b) 342
(c) 344 (d) 355

793. E number of lecithin citrate is
(a) 330 (b) 342
(c) 344 (d) 355

794. E number of adipic acid is
(a) 330 (b) 342
(c) 344 (d) 355

795. E number of succinic acid is
(a) 363 (b) 362
(c) 364 (d) 365

796. E number of ammonium citrate is
(a) 380 (b) 385
(c) 386 (d) 391

797. E number of calcium disodium EDTA is
(a) 380 (b) 385
(c) 386 (d) 391

798. E number of disodium EDTA is
(a) 380 (b) 385
(c) 386 (d) 391

799. E number of phytic acid is
(a) 380 (b) 385
(c) 386 (d) 391

800. E number of alginic acid is
(a) 401 (b) 400
(c) 402 (d) 391

801. E number of sodium alginate is
(a) 401 (b) 400
(c) 402 (d) 391

802. E number of potassium alginate is
(a) 401 (b) 400
(c) 402 (d) 391

803. E number of carrageenan is
(a) 401 (b) 407
(c) 402 (d) 391

804. E number of arabinogalactan is
(a) 401 (b) 400
(c) 402 (d) 409

805. E number of carob bean gum is
(a) 410 (b) 400
(c) 402 (d) 391

806. E number of agar is
(a) 406 (b) 400
(c) 402 (d) 391

807. E number of oat gum is
(a) 401 (b) 411
(c) 402 (d) 391

808. E number of guar gum is
(a) 401 (b) 400
(c) 412 (d) 391

809. E number of tragacanth gum is
(a) 401 (b) 400
(c) 402 (d) 413

810. E number of acacia gum is
(a) 414 (b) 400
(c) 402 (d) 391

811. E number of xanthan gum is
(a) 401 (b) 415
(c) 402 (d) 391

812. E number of karaya gum is
(a) 401 (b) 415
(c) 402 (d) 391

813. E number of gellan gum is
(a) 401 (b) 400
(c) 402 (d) 418

814. E number of sorbitol is
(a) 420 (b) 400
(c) 402 (d) 391

815. E number of mannitol is
(a) 401 (b) 421
(c) 402 (d) 391

816. E number of glycerol is
(a) 401 (b) 400
(c) 422 (d) 391

817. E number of konjac flour is
(a) 425 (b) 400
(c) 402 (d) 391

818. E number of polyoxyethylene stearate is
(a) 401 (b) 400
(c) 402 (d) 430

819. E number of pectins is
(a) 440 (b) 400
(c) 402 (d) 391

820. E number of gamma cyclodextrin is
(a) 458 (b) 400
(c) 402 (d) 391

821. E number of beta cyclodextrin is
(a) 401 (b) 459
(c) 402 (d) 391

822. E number of cellulose is
(a) 401 (b) 400
(c) 460 (d) 391

823. E number of methyl cellulose is
(a) 401 (b) 400
(c) 402 (d) 461

824. E number of sodium CMC is
(a) 401 (b) 466
(c) 402 (d) 391

825. E number of mono & diglycerides of fatty acids is
(a) 471 (b) 400
(c) 402 (d) 391

826. E number of sodium lactylate is
(a) 401 (b) 481
(c) 402 (d) 391

827. E number of calcium lactylate is
(a) 482 (b) 400
(c) 402 (d) 391

828. E number of sorbitan monostearate is
(a) 491 (b) 400
(c) 402 (d) 391

829. E number of sorbitan tristearate is
(a) 401 (b) 492
(c) 402 (d) 391

830. E number of sorbitan monolaurate is
(a) 401 (b) 400
(c) 493 (d) 391

831. E number of sodium carbonates is
(a) 401 (b) 400
(c) 402 (d) 500

832. E number of ammonium carbonate is
(a) 401 (b) 503
(c) 402 (d) 391

833. E number of calcium chloride is
(a) 509 (b) 400
(c) 402 (d) 391

834. E number of ammonium chloride is
(a) 401 (b) 510
(c) 402 (d) 391

835. E number of magnesium oxide is
(a) 530 (b) 400
(c) 402 (d) 391

836. E number of sodium silicate is
(a) 550 (b) 400
(c) 402 (d) 391

837. E number of calcium silicate is
(a) 401 (b) 400
(c) 552 (d) 391

838. E number of magnesium silicate is
(a) 401 (b) 400
(c) 402 (d) 553

839. E number of aluminium silicate is
(a) 559 (b) 400
(c) 402 (d) 391

840. E number of potassium silicate is
(a) 401 (b) 560
(c) 402 (d) 391

841. E number of gluconic acid is
(a) 620 (b) 400
(c) 574 (d) 391

842. E number of glutamic acid is
(a) 620 (b) 400
(c) 402 (d) 391

843. E number of monosodium glutamate is
(a) 401 (b) 621
(c) 402 (d) 391

844. E number of inosinic acid is
(a) 630 (b) 400
(c) 402 (d) 391

845. E number of maltol is
(a) 401 (b) 400
(c) 402 (d) 636

846. E number of sodium L-aspartate is
(a) 638 (b) 400
(c) 402 (d) 391

847. E number of alanine is
(a) 401 (b) 639
(c) 402 (d) 391

848. E number of glycine is
(a) 401 (b) 400
(c) 640 (d) 391

849. E number of L-leucine is
(a) 401 (b) 400
(c) 402 (d) 641

850. E number of carnauba wax is
(a) 901 (b) 902
(c) 903 (d) 904

851. E number of bee wax is
(a) 901 (b) 902
(c) 903 (d) 904

852. E number of rice bran wax is
(a) 901 (b) 902
(c) 903 (d) 908

853. E number of calcium iodate is
(a) 916 (b) 902
(c) 903 (d) 904

854. E number of potassium iodate is
(a) 901 (b) 917
(c) 903 (d) 904

855. E number of nitrogen oxide is
(a) 901 (b) 902
(c) 918 (d) 904

856. E number of potassium bromate is
(a) 924a (b) 902
(c) 903 (d) 904

857. E number of calcium bromate is
(a) 901 (b) 924b
(c) 903 (d) 904

858. E number of chlorine is
(a) 901 (b) 902
(c) 925 (d) 904

859. E number of chlorine dioxide is
(a) 901 (b) 902
(c) 903 (d) 926

860. E number of benzoyl peroxide is
(a) 928 (b) 902
(c) 903 (d) 904

861. E number of dichloro difluro methane is
(a) 901 (b) 902
(c) 903 (d) 940

862. E number of nitrous oxide is
(a) 901 (b) 942
(c) 903 (d) 904

863. E number of acesulfame K is
(a) 950 (b) 951
(c) 952 (d) 954

864. E number of aspartame is
(a) 950 (b) 951
(c) 952 (d) 954

865. E number of saccharin is
(a) 950 (b) 951
(c) 952 (d) 954

866. E number of alitame is
(a) 950 (b) 951
(c) 956 (d) 954

867. E number of thaumatin is
(a) 950 (b) 951
(c) 952 (d) 957

868. E number of steviocide is
(a) 950 (b) 951
(c) 960 (d) 954

869. E number of maltitol is
(a) 966 (b) 967
(c) 968 (d) 965

870. E number of lactitol is
(a) 966 (b) 967
(c) 968 (d) 965

871. E number of xylitol is
(a) 966 (b) 967
(c) 968 (d) 965

872. E number of amylases is
(a) 1100 (b) 1101
(c) 1102 (d) 1104

873. E number of proteases is
(a) 1100 (b) 1101
(c) 1102 (d) 1104

874. E number of lipases is
(a) 1100 (b) 1101
(c) 1102 (d) 1104

875. E number of glucose oxidase is
(a) 1100 (b) 1101
(c) 1102 (d) 1104

876. E number of propylene glycol is
(a) 1503 (b) 1505
(c) 1518 (d) 1520

877. E number of oxidized starch is
(a) 1403 (b) 1404
(c) 1418 (d) 1420

878. E number of bleached starch is
(a) 1403 (b) 1404
(c) 1418 (d) 1420

879. E number of enzyme treated starch is
(a) 1403 (b) 1405
(c) 1418 (d) 1420

880. E number of acid treated starch is
(a) 1403 (b) 1404
(c) 1418 (d) 1401

881. E number of castor oil is
(a) 1503 (b) 1504
(c) 1518 (d) 1520

882. Which of the following is coloring agent
(a) E200 (b) E203
(c) E110 (d) E210

883. Which of the following is a preservative
(a) E200 (b) E203
(c) E110 (d) E210

884. Which of the following is acidity regulator
(a) E200 (b) E203
(c) E330 (d) E210

885. Which of the following is an antioxidant
(a) E310 (b) E203
(c) E110 (d) E210

886. Which of the following is sequestrant
(a) E200 (b) E203
(c) E110 (d) E339

887. Which of the following is anti caking agent
(a) E170 (b) E203
(c) E110 (d) E210

888. Which of the following is water retention agent
(a) E200 (b) E203
(c) E339 (d) E210

889. Which of the following is thickening agent
(a) E400 (b) E203
(c) E110 (d) E210

890. Which of the following is a stabilizer
(a) E200 (b) E415
(c) E110 (d) E210

891. Which of the following is an emulsifier
(a) E200 (b) E203
(c) E110 (d) E405

892. Which of the following is a sweetener
(a) E421 (b) E203
(c) E110 (d) E210

893. Which of the following is a humectant
(a) E500 (b) E603
(c) E422 (d) E110

894. Which of the following is flour treatment agent
(a) E517 (b) E203
(c) E110 (d) E210

895. Which of the following is firming agent
(a) E509 (b) E511
(c) E518 (d) all

896. Which of the following is anticaking agent
(a) E530 (b) E203
(c) E110 (d) E210

897. Which of the following is a dough conditioner
(a) E200 (b) E203
(c) E516 (d) E210

898. Which of the following is flavor enhancer
(a) E621 (b) E203
(c) E110 (d) E210

899. Which of the following is a glazing agent
(a) E901 (b) E910
(c) both (d) none

900. Which of the following is a sweetener
(a) E950 (b) E901
(c) E906 (d) E913

10.3 ANSWER KEY

Q.	A.	Q.	A.	Q.	A.	Q.	A.	Q.	A.	Q.	A.
1.	(b)	2.	(d)	3.	(d)	4.	(d)	5.	(d)	6.	(d)
7.	(b)	8.	(d)	9.	(d)	10.	(d)	11.	(d)	12.	(b)
13.	(a)	14.	(b)	15.	(b)	16.	(b)	17.	(d)	18.	(b)
19.	(a)	20.	(b)	21.	(a)	22.	(b)	23.	(d)	24.	(b)
25.	(a)	26.	(d)	27.	(d)	28.	(d)	29.	(a)	30.	(b)
31.	(d)	32.	(a)	33.	(b)	34.	(c)	35.	(d)	36.	(c)
37.	(d)	38.	(d)	39.	(d)	40.	(d)	41.	(c)	42.	(b)
43.	(a)	44.	(b)	45.	(d)	46.	(d)	47.	(c)	48.	(d)
49.	(d)	50.	(a)	51.	(c)	52.	(d)	53.	(a)	54.	(c)
55.	(d)	56.	(d)	57.	(d)	58.	(d)	59.	(d)	60.	(a)
61.	(b)	62.	(b)	63.	(c)	64.	(c)	65.	(d)	66.	(b)
67.	(b)	68.	(c)	69.	(a)	70.	(b)	71.	(d)	72.	(a)
73.	(b)	74.	(c)	75.	(d)	76.	(b)	77.	(b)	78.	(c)
79.	(a)	80.	(b)	81.	(d)	82.	(b)	83.	(a)	84.	(d)

Q. A.	Q. A.	Q. A.	Q. A.	Q. A.	Q. A.
85. (d)	86. (a)	87. (a)	88. (b)	89. (a)	90. (b)
91. (c)	92. (d)	93. (c)	94. (a)	95. (c)	96. (a)
97. (c)	98. (c)	99. (d)	100. (d)	101. (b)	102. (a)
103. (c)	104. (c)	105. (d)	106. (b)	107. (a)	108. (c)
109. (d)	110. (c)	111. (d)	112. (b)	113. (d)	114. (a)
115. (b)	116. (c)	117. (a)	118. (b)	119. (b)	120. (c)
121. (d)	122. (d)	123. (a)	124. (c)	125. (a)	126. (c)
127. (a)	128. (a)	129. (b)	130. (b)	131. (d)	132. (b)
133. (b)	134. (d)	135. (d)	136. (b)	137. (c)	138. (a)
139. (b)	140. (d)	141. (b)	142. (c)	143. (c)	144. (c)
145. (d)	146. (d)	147. (d)	148. (b)	149. (c)	150. (a)
151. (c)	152. (c)	153. (d)	154. (a)	155. (b)	156. (c)
157. (c)	158. (d)	159. (d)	160. (d)	161. (d)	162. (d)
163. (d)	164. (d)	165. (d)	166. (d)	167. (b)	168. (d)
169. (c)	170. (a)	171. (b)	172. (a)	173. (b)	174. (b)
175. (a)	176. (d)	177. (c)	178. (b)	179. (b)	180. (b)
181. (a)	182. (a)	183. (b)	184. (c)	185. (b)	186. (a)
187. (a)	188. (a)	189. (b)	190. (c)	191. (b)	192. (a)
193. (c)	194. (d)	195. (d)	196. (c)	197. (d)	198. (a)
199. (b)	200. (d)	201. (c)	202. (b)	203. (c)	204. (d)
205. (a)	206. (d)	207. (d)	208. (d)	209. (d)	210. (d)
211. (d)	212. (d)	213. (d)	214. (c)	215. (d)	216. (b)
217. (d)	218. (c)	219. (a)	220. (c)	221. (d)	222. (d)
223. (d)	224. (d)	225. (c)	226. (d)	227. (d)	228. (a)
229. (b)	230. (a)	231. (d)	232. (a)	233. (a)	234. (b)
235. (d)	236. (d)	237. (a)	238. (d)	239. (d)	240. (a)
241. (c)	242. (a)	243. (b)	244. (b)	245. (d)	246. (d)
247. (b)	248. (b)	249. (a)	250. (b)	251. (a)	252. (d)
253. (d)	254. (d)	255. (d)	256. (d)	257. (d)	258. (d)
259. (c)	260. (b)	261. (d)	262. (d)	263. (c)	264. (c)
265. (a)	266. (b)	267. (a)	268. (c)	269. (a)	270. (c)
271. (b)	272. (a)	273. (c)	274. (b)	275. (a)	276. (b)
277. (c)	278. (d)	279. (a)	280. (b)	281. (c)	282. (d)
283. (a)	284. (c)	285. (b)	286. (d)	287. (a)	288. (b)
289. (c)	290. (a)	291. (b)	292. (a)	293. (d)	294. (d)

Q. A.	Q. A.	Q. A.	Q. A.	Q. A.	Q. A.
295. (c)	296. (a)	297. (b)	298. (c)	299. (c)	300. (a)
301. (a)	302. (b)	303. (d)	304. (d)	305. (a)	306. (d)
307. (a)	308. (c)	309. (b)	310. (a)	311. (c)	312. (d)
313. (c)	314. (c)	315. (c)	316. (d)	317. (d)	318. (d)
319. (a)	320. (b)	321. (d)	322. (a)	323. (b)	324. (d)
325. (d)	326. (d)	327. (d)	328. (a)	329. (b)	330. (c)
331. (d)	332. (a)	333. (b)	334. (c)	335. (c)	336. (d)
337. (a)	338. (b)	339. (c)	340. (a)	341. (b)	342. (c)
343. (c)	344. (d)	345. (c)	346. (c)	347. (a)	348. (d)
349. (c)	350. (b)	351. (c)	352. (d)	353. (d)	354. (b)
355. (a)	356. (d)	357. (a)	358. (d)	359. (d)	360. (a)
361. (b)	362. (c)	363. (a)	364. (b)	365. (a)	366. (b)
367. (c)	368. (c)	369. (b)	370. (a)	371. (a)	372. (d)
373. (a)	374. (c)	375. (b)	376. (d)	377. (c)	378. (c)
379. (a)	380. (d)	381. (d)	382. (c)	383. (b)	384. (d)
385. (b)	386. (c)	387. (b)	388. (c)	389. (d)	390. (b)
391. (a)	392. (d)	393. (d)	394. (a)	395. (d)	396. (d)
397. (b)	398. (d)	399. (d)	400. (d)	401. (d)	402. (d)
403. (d)	404. (d)	405. (d)	406. (d)	407. (b)	408. (a)
409. (c)	410. (a)	411. (b)	412. (a)	413. (a)	414. (b)
415. (d)	416. (d)	417. (b)	418. (d)	419. (a)	420. (c)
421. (b)	422. (b)	423. (d)	424. (b)	425. (d)	426. (d)
427. (d)	428. (a)	429. (a)	430. (a)	431. (b)	432. (d)
433. (c)	434. (d)	435. (d)	436. (a)	437. (a)	438. (a)
439. (c)	440. (a)	441. (d)	442. (a)	443. (d)	444. (a)
445. (a)	446. (c)	447. (c)	448. (c)	449. (c)	450. (a)
451. (c)	452. (a)	453. (a)	454. (b)	455. (b)	456. (c)
457. (c)	458. (c)	459. (d)	460. (c)	461. (a)	462. (b)
463. (d)	464. (d)	465. (c)	466. (d)	467. (d)	468. (d)
469. (a)	470. (d)	471. (c)	472. (a)	473. (a)	474. (a)
475. (a)	476. (a)	477. (a)	478. (c)	479. (a)	480. (b)
481. (c)	482. (b)	483. (d)	484. (b)	485. (c)	486. (d)
487. (b)	488. (d)	489. (a)	490. (d)	491. (b)	492. (b)
493. (c)	494. (d)	495. (b)	496. (a)	497. (a)	498. (d)
499. (a)	500. (a)	501. (b)	502. (c)	503. (b)	504. (b)

Q. A.	Q. A.	Q. A.	Q. A.	Q. A.	Q. A.
505. (b)	506. (b)	507. (a)	508. (c)	509. (b)	510. (c)
511. (b)	512. (a)	513. (a)	514. (a)	515. (b)	516. (a)
517. (a)	518. (b)	519. (d)	520. (d)	521. (d)	522. (d)
523. (d)	524. (d)	525. (d)	526. (d)	527. (d)	528. (d)
529. (a)	530. (c)	531. (a)	532. (b)	533. (d)	534. (d)
535. (c)	536. (d)	537. (c)	538. (b)	539. (a)	540. (a)
541. (a)	542. (c)	543. (c)	544. (d)	545. (d)	546. (c)
547. (c)	548. (c)	549. (a)	550. (a)	551. (b)	552. (a)
553. (c)	554. (c)	555. (d)	556. (a)	557. (b)	558. (c)
559. (c)	560. (b)	561. (a)	562. (d)	563. (d)	564. (a)
565. (a)	566. (b)	567. (c)	568. (d)	569. (d)	570. (b)
571. (d)	572. (d)	573. (d)	574. (d)	575. (d)	576. (d)
577. (d)	578. (d)	579. (b)	580. (d)	581. (a)	582. (c)
583. (d)	584. (d)	585. (b)	586. (d)	587. (d)	588. (b)
589. (b)	590. (a)	591. (c)	592. (d)	593. (c)	594. (c)
595. (c)	596. (a)	597. (c)	598. (b)	599. (a)	600. (d)
601. (d)	602. (d)	603. (d)	604. (d)	605. (b)	606. (d)
607. (c)	608. (d)	609. (a)	610. (a)	611. (b)	612. (b)
613. (d)	614. (d)	615. (a)	616. (d)	617. (b)	618. (b)
619. (a)	620. (b)	621. (b)	622. (a)	623. (c)	624. (c)
625. (a)	626. (c)	627. (b)	628. (b)	629. (d)	630. (a)
631. (a)	632. (d)	633. (d)	634. (d)	635. (d)	636. (b)
637. (b)	638. (a)	639. (b)	640. (c)	641. (d)	642. (d)
643. (c)	644. (a)	645. (a)	646. (a)	647. (a)	648. (a)
649. (a)	650. (b)	651. (c)	652. (d)	653. (a)	654. (b)
655. (b)	656. (a)	657. (b)	658. (d)	659. (d)	660. (b)
661. (a)	662. (d)	663. (d)	664. (a)	665. (a)	666. (b)
667. (b)	668. (b)	669. (a)	670. (b)	671. (b)	672. (c)
673. (b)	674. (c)	675. (c)	676. (a)	677. (b)	678. (a)
679. (b)	680. (a)	681. (a)	682. (d)	683. (a)	684. (c)
685. (d)	686. (d)	687. (a)	688. (a)	689. (d)	690. (d)
691. (a)	692. (b)	693. (c)	694. (d)	695. (a)	696. (b)
697. (c)	698. (d)	699. (d)	700. (c)	701. (d)	702. (a)
703. (b)	704. (c)	705. (d)	706. (b)	707. (d)	708. (a)
709. (b)	710. (c)	711. (d)	712. (a)	713. (b)	714. (c)

Q.	A.	Q.	A.	Q.	A.	Q.	A.	Q.	A.	Q.	A.
715.	(d)	716.	(a)	717.	(b)	718.	(c)	719.	(d)	720.	(a)
721.	(b)	722.	(c)	723.	(d)	724.	(a)	725.	(b)	726.	(c)
727.	(d)	728.	(c)	729.	(b)	730.	(d)	731.	(b)	732.	(c)
733.	(a)	734.	(c)	735.	(a)	736.	(b)	737.	(d)	738.	(a)
739.	(b)	740.	(c)	741.	(d)	742.	(a)	743.	(d)	744.	(b)
745.	(d)	746.	(c)	747.	(a)	748.	(d)	749.	(a)	750.	(d)
751.	(b)	752.	(c)	753.	(d)	754.	(a)	755.	(b)	756.	(c)
757.	(d)	758.	(a)	759.	(b)	760.	(c)	761.	(d)	762.	(a)
763.	(b)	764.	(c)	765.	(d)	766.	(a)	767.	(b)	768.	(c)
769.	(d)	770.	(b)	771.	(a)	772.	(c)	773.	(a)	774.	(b)
775.	(c)	776.	(d)	777.	(b)	778.	(d)	779.	(a)	780.	(c)
781.	(d)	782.	(a)	783.	(b)	784.	(c)	785.	(d)	786.	(a)
787.	(b)	788.	(c)	789.	(d)	790.	(c)	791.	(d)	792.	(b)
793.	(c)	794.	(d)	795.	(a)	796.	(a)	797.	(b)	798.	(c)
799.	(d)	800.	(b)	801.	(a)	802.	(c)	803.	(b)	804.	(d)
805.	(a)	806.	(a)	807.	(b)	808.	(c)	809.	(d)	810.	(a)
811.	(b)	812.	(b)	813.	(d)	814.	(a)	815.	(b)	816.	(c)
817.	(a)	818.	(d)	819.	(a)	820.	(a)	821.	(b)	822.	(c)
823.	(d)	824.	(b)	825.	(a)	826.	(b)	827.	(a)	828.	(a)
829.	(b)	830.	(c)	831.	(c)	832.	(b)	833.	(a)	834.	(b)
835.	(a)	836.	(a)	837.	(c)	838.	(d)	839.	(a)	840.	(b)
841.	(c)	842.	(a)	843.	(b)	844.	(a)	845.	(d)	846.	(a)
847.	(b)	848.	(c)	849.	(d)	850.	(c)	851.	(a)	852.	(d)
853.	(a)	854.	(b)	855.	(c)	856.	(a)	857.	(b)	858.	(c)
859.	(d)	860.	(a)	861.	(d)	862.	(b)	863.	(a)	864.	(b)
865.	(d)	866.	(c)	867.	(d)	868.	(c)	869.	(d)	870.	(a)
871.	(b)	872.	(a)	873.	(b)	874.	(d)	875.	(c)	876.	(d)
877.	(b)	878.	(a)	879.	(b)	880.	(d)	881.	(a)	882.	(c)
883.	(b)	884.	(c)	885.	(a)	886.	(d)	887.	(a)	888.	(c)
889.	(a)	890.	(b)	891.	(d)	892.	(a)	893.	(c)	894.	(a)
895.	(d)	896.	(a)	897.	(c)	898.	(a)	899.	(c)	900.	(a)

CHAPTER 11

FOOD PACKAGING

11.1 SUBJECTIVE KEYNOTES

1. The package is the physical entity that contains the product.
2. A primary package is the one which is in direct contact with the contained product.
3. A secondary package contains a number of primary packages. It is outside the primary packaging perhaps used to group primary packages together.
4. A tertiary package is made up of a number of secondary packages. It is used for bulk handling.
5. A quaternary package is frequently used to facilitate the handling of tertiary packages.
6. Packaging has four primary functions i.e. containment, protection, convenience and communication.
7. A package functions as a "silent salesman".
8. Shelf life is the length of time that foods, beverages, and many other perishable items are given before they are considered unsuitable for sale, use, or consumption.
9. Shelf life is a function of the product, package and the environment through which the product is transported, stored and sold.
10. Controlled atmosphere packaging (CAP) is the enclosure of food in a gas impermeable package inside which the gaseous environment with respect to CO2, O2, N2, water vapor and trace gases has been changed, and is selectively controlled to increase shelf life.
11. The storage life of various fruits and vegetables can be increased by 2 to 4 times the normal life by employing Controlled Atmosphere storage technology.
12. Modified Atmosphere Packaging (MAP) can be defined as the enclosure of food in a package in which the atmosphere inside the package is modified or altered to provide an optimum atmosphere for increasing shelf life and maintaining quality of the food.
13. CAS is capital-intensive and expensive to operate thus it is more appropriate, for those foods that are amenable to long-term storage such as apples, kiwifruit, pears and meat.
14. The vegetables should be packed in new loosely woven gunny bags or wooden/plastic crates or in lined or unlined corrugated fiberboard boxes.

15. Tomatoes should be packed in baskets or wooden boxes or lined or unlined corrugated fiberboard boxes.
16. Fresh chillies should be packed in gunny bags or in bamboo baskets, or in corrugated boxes, lined or unlined.
17. Jams, jellies and marmalades should be packed in glass bottles or open top cans.
18. Meat should be wrapped in polyethylene sheets or bags and delivered in clean, rust free and closed containers.
19. Cooked meat product should be packed in butter paper or any grease proof clean wrapping material.
20. The fresh product should be packed in polyethylene lined insulated containers, made of plywood, country wood or plastic.
21. The bread should be wrapped in slice form in LDPE coated poster paper or clean waxed paper, grease –proof paper or any other suitable non-toxic wrapper.
22. The cakes should be wrapped or packed in clean waxed paper, grease-proof polyethylene, glassine/LDPE or any other suitable wrapper or tins.
23. The peanut butter should be packed in wide-mouthed glass jars or polystyrene tubs or any other suitable container or the required size and shape.
24. The flour should be packed in either LDPE coated jute bags or LDPE coated raffia bags.
25. Oil should be packed in tinplate containers, glass bottles, rigid plastic containers of HDPE, food grade PVC, PET and flexible pouches made of plastic film/foil/laminate.
26. Vanaspati should be packed in suitable sealed packages, such as, flexible packs and well closed tinplate containers.
27. Tea can be packed in flexible packaging materials or laminates such as LDPE, paper coated LDPE, PET/LDPE and BOPP/LDPE.
28. Carbonated beverages should be filled in glass containers or in PET bottles or in cans.
29. Packages are classified as 1) traditional or natural and 2) fabricated or modern packaging materials based on the availability of the materials.
30. Paper derives its name from the reedy plant "papyrus".
31. The first authentic papermaking - the formation of a cohesive sheet from the rebonding of separated fibers - has been attributed to Tsai-Lun of China in 105 AD, who used bamboo, mulberry bark and rags.
32. The cell wall of softwoods, which are preferred for most pulp products, typically contain 40 to 44% cellulose, 25 to 29% hemicelluloses and 25 to 31% lignin by weight.
33. Softwoods fibers are generally up to 2.5 times longer than hardwoods fibers.
34. Pulps that retain most of the wood lignin consist of stiff fibers that do not produce strong papers.
35. Brightness is an index of whiteness, measured as the reflectivity of a paper sample using light at 457 nm.
36. Unbleached pulps exhibit a range of brightness values from 15 to 60.

37. Cellulose and hemicellulose are inherently white and do not contribute to color; it is the chromophoric groups on the lignin that are largely responsible for the color of the pulp.
38. The most effective bleaching agent for most ground woods is hydrogen peroxide.
39. Calendering, a process that reorients the surface fibers in the base sheet of paper (or the coating applied to the surface) through the use of pressure.
40. Glass has the least resistance to temperature changes.
41. Glass is optically isotropic.
42. The principal ingredient of glass is silica derived from sand, flint or quartz.
43. Bottles are normally produced by a two-step Blow & Blow process; while glass jars are produced by two step Press and Blow process.
44. Four metals are commonly used for the packaging of foods: steel, aluminium, tin and chromium.
45. Two piece cans are manufactured by Drawn and Ironed process and Drawn and Redrawn process.
46. Plastics are organic polymers which are long chain molecules obtained by addition or condensation of one or more monomers.
47. The most common ones are polyolefins, polyvinyls and polyesters and these may be divided into thermosetting and thermoplastic resins.
48. Thermo plastic can be considered chemically as derivatives of ethylene ($CH2 = CH2$) in which one hydrogenation is replaced by CH3, chlorine, fluorine or phenyl groups.
49. The density of LDPE film is 910 to 940 kg/m3 while the density of HDPE ranges from 941 to 965 kg/m3.
50. Aseptic packaging is the filling of sterile containers with a commercially sterile product under aseptic conditions, and then sealing them so that re infection is prevented; that is, so that are hermetically sealed.
51. Paper is called board when it is heavier than 224 g/m^2.
52. The first machine to manufacture paper continuously was invented by the Frenchman Louis-Nicolas Robert in 1799.
53. Calendering is a process which reorients the surface fibres in the base sheet of paper (or the coating applied to the surface) by the use of pressure. This serves to smooth the surface, control surface texture and develop a glossy finish. Such papers are known as machine finished.
54. Kraft paper is a coarse paper with exceptional strength, often made on a fourdrinier machine and then either machine-glazed on a Yankee dryer or machine-finished on a calender.
55. Greaseproof paper is a translucent, machine-finished paper which has been hydrated to give oil and grease resistance. Prolonged beating or mechanical refining is used to fibrillate and break the cellulose fibres which absorb so much water that they become superficially gelatinized and sticky.
56. Glassine paper derives its name from its glassy, smooth surface, high density and transparency. It is produced by further treating greaseproof paper in a supercalender

where is it carefully dampened with water and run through a set of steam-heated rollers.

57. Colourless glass is derived from soda, lime and silica. It is also known as white flint.
58. Amber glass is usually obtained by melting a composition containing iron oxide under strongly reduced conditions. Carbon is also added. Amber glass has UV protection properties and could well be suited for use with light-sensitive products.
59. Blue glass is usually obtained by the addition of cobalt to a low-iron glass.
60. Annealing is a process which reduces the residual strain in the container that has been introduced in the forming process.
61. The beer can should be designed to contain pressures up to 7 kg/cm^2.
62. The reaction of metals in aqueous solutions or under moist conditions is electrochemical in nature, including the transfer of electric charge across the boundary of metal surface and its environment.
63. Plastics can meet the needs of a wide temperature range, from frozen food processing (−40°C) and storage (−20°C) to the retort sterilization (121°C), and reheating of packaged food products by microwave(100°C) and radiant heat (200°C).
64. The technique of pigmenting plastics has been developed using white compounds such as calcium carbonate or, more usually, titanium dioxide, to give a white appearance.
65. Screw cap and pressure fit closures with precise profiles are made by injection moulding.
66. LLDPE film has a density range similar to that of LDPE. It has short side chain branching and is superior to LDPE in most properties such as tensile and impact strength and also in puncture resistance. It is used for pillow packing of milk and other liquid foods.
67. PP is an addition polymer of propylene formed under heat and pressure using Zieger-Natta type catalysts to produce a linear polymer. PP is a harder and denser resin than PE and more transparent in its natural form.
68. Oriented PP (OPP) or biaxially oriented PP (BOPP) film was the first plastic film to successfully replace regenerated cellulose film (RCF) in major packaging applications such as biscuit packing.
69. The range of food products packed in PP films include biscuits, crisps (chips) and snack foods, chocolate and sugar confectionery, ice cream and frozen food, tea and coffee.
70. When terephthalic acid reacts with ethylene glycol and polymerises, the result is PET. PET can be made into film by blowing or casting. It can be blow moulded, injection moulded, foamed, extrusion coated on paperboard and extruded as sheet for thermoforming. PET melts at 260°C.
71. Polycarbonate is a polyester containing carbonate groups in its structure. It is formed by the polymerisation of the sodium salt of bisphenolic acid with phosgene. It is glass clear, heat resistant and very tough and durable. PC is mainly used as a glass replacement in processing equipment and for glazing applications.
72. Ionomers are polymers formed from metallic salts of acid copolymers and possess interchange ionic cross-links which provide the characteristic properties of the family of plastics.

73. Polyvinylidene chloride is a copolymer of vinyl chloride and vinylidene chloride.
74. Ethylene Vinyl alcohol is a copolymer of ethylene and vinyl alcohol. It is related to polyvinyl alcohol (PVOH), which is a water-soluble synthetic polymer with excellent film-forming, emulsifying and adhesive properties. It is a high-barrier material with respect to oil, grease, organic solvents and oxygen. It is moisture sensitive and, in film form, is water soluble.
75. Polytetrafluoroethylene (PTFE) is a high melting point, inert and waxy polymer. It is also known as Teflon. It is used in the form of tape and coatings on packaging machines to reduce adhesion.
76. Orientation of thermoplastic film is a process of stretching the material in such a manner as to line up the molecular chains in a predetermined direction.
77. Lamination is a process of bringing two or more webs together and bonding them with an adhesive or by heating. A laminate is defined as any combination of distinctly different plastic film materials or plastic plus non-plastic materials like paper, aluminium foil and regenerated cellulose, where each major web is generally thicker than six microns, regardless of the method of manufacture.
78. There are two main techniques employed in the fabrication of laminates: adhesive lamination and extrusion coating.
79. Bi-axially oriented polypropylene (BOPP), low density polyethylene (LDPE), and polyvinyl chloride (PVC) are synthetic polymers used as trays, pouches, or overwraps for packing fresh fruits and vegetables.
80. Packaging material in fresh bread include waxed paper, nitrocellulose coated cellophane, LDPE, polypropylene.
81. For designing packages, the primary requisite is the knowledge of relationship between moisture content and equilibrium relative humidity (ERH) and hence water activity denoted through moisture sorption isotherm.
82. The films used for milk pouch packing are of 45-75µ thickness.
83. The terpene containing aroma of fruit yoghurt has an effect on polystyrene.
84. Injection moulded polystyrene and Polypropylene/ HIPS cups introduced with aluminium foil based peelable lids are used for dahi packing.
85. Vegetable parchment paper has good wet strength, generally paper of 45 gsm and above and aluminum foil above 0.009 mm thickness are used for butter packing.
86. Cream cheese is packed in foil lined card board boxes of heat stable plastic packs. Saran is used as wrapping material for Neufchatel cheese.
87. Air evacuation and gas flushing is used for cottage cheese, green cheese is packed by waxing and paraffining or alternatively vacuum packed in polyethylene.
88. Ripened cheese is packed in laminated cellophane film.
89. Whole milk powder is generally packed in Lacquered tins under N2 gas packaging. For retail, milk powder is packed in laminate bags made of PET /Al / PE.
90. A high oxygen barrier is important in the application of films for packaging of meat and meat products. Films made of poly vinyl chloride (PVC), polyethylene (PE) or polypropylene (PP) have a relatively high oxygen permeability, whereas films of

polyvinyliedene chloride (PVDC), polyester (PET), polyamide (PA) or nylon and cellulose films are less or non-permeable to oxygen.

91. Vacuum packaging of meat is meant to retard or completely stop the oxidative reactions. But it frequently causes discoloration as meat becomes darker and purplish.

92. Beef having pH more than 5.8 is not suitable for vacuum packaging, due to chances of the growth of H_2S producing bacteria and H_2S reacts with myoglobin and forms a green coloured compound 'Sulphmyoglobin'.

93. MAP greatly extends not only the shelf life of muscle foods but also maintains colour, texture and flavour of the product.

94. The first successful canning of beer took place in 1933 in Newark, New Jersey when 2000 cans were produced for a test market. In January, 1935 the first beer cans went on sale.

95. Aluminum was used for beer can ends by Schlitz in 1960. A pull-tab aluminum top was developed in 1962 by Ermal Fraze. It was test marketed and emerged in 1965 as the now familiar ring-pull tab.

96. Microprocessors & microcomputers were first used in packaging machines in 1973. Microprocessor-based equipment controls were first used in 1977. Microprocessor-controlled packaging machinery was first commercially used in 1978.

97. Packaging that has an external or internal indicator to provide information about characteristics of the history of the package and/or the quality of the food packed is known as intelligent packaging.

98. Thermoset polyester plates were the first commercial application of microwaveable materials when used for frozen meals in the mid-1980s.

99. The gravure printing process, sometimes called rotogravure, uses a recessed-image plate cylinder to transfer the image to the substrate.

100. Flexography is a variation of letterpress printing used mainly for packaging applications. It is characterized by the use of an elastomeric printing plate, fast-drying inks, and an ink-metering roll system.

11.2 MULTIPLE CHOICE QUESTIONS

1. The insulating effect of a package is determined by its
(a) thermal conductivity
(b) reflectivity
(c) both
(d) none

2. Paperboard and polystyrene have
(a) low thermal conductivity
(b) high thermal conductivity
(c) no
(d) none

3. _______ are the most prone to thermal shock
(a) glass (b) plastics
(c) paper (d) metals

4. Fresh foods that are respiring require _______ degree of permeability in the material to allow exchange of oxygen and carbon dioxide
(a) high (b) low
(c) no (d) none

5. The package which is in direct contact with the contained product is called _____ package
(a) primary (b) secondary
(c) tertiary (d) quaternary

6. The package containing number of primary packages is called _____ package
(a) primary (b) secondary
(c) tertiary (d) quaternary

7. The package containing number of secondary packages is called _____ package
(a) primary (b) secondary
(c) tertiary (d) quaternary

8. Package used to facilitate the handling of tertiary package is called
(a) primary (b) secondary
(c) tertiary (d) quaternary

9. A metal can, paperboard carton and glass bottle are examples of ______ packaging
(a) primary (b) secondary
(c) tertiary (d) quaternary

10. A corrugated case and boxes are examples of _______ packaging
(a) primary (b) secondary
(c) tertiary (d) quaternary

11. Stretch wrapped pallet of corrugated cases are examples of __________ packaging
(a) primary (b) secondary
(c) tertiary (d) quaternary

12. __________ functions as a silent salesman
(a) package (b) product
(c) consumer (d) distributer

13. UPC stands for
(a) Universal Product Code
(b) United Packaging Code
(c) Universal Packaging Code
(d) none

14. RFID stands for
(a) Radio Frequency Identification
(b) Rapid Frequency Identification
(c) Radio Frequent Identification
(d) none

15. The time between the production and packaging of the product and the point at which the product first becomes unacceptable under defined environmental conditions is called
(a) self life
(b) shelf life
(c) product life
(d) market life

16. The degree of moisture protection of a package is measured in terms of
(a) water vapor transpiring rate
(b) water vapor turning rate
(c) water vapor transmission rate
(d) none

17. The degree of protection of packaged foods for oxygen sensitive foods is measured in terms of
(a) water vapor transmission rate
(b) oxygen transmission rate
(c) oxygen transfer rate
(d) none

18. Vertical, vibration, compression, racking and piercing are examples of _____ hazard
(a) impact (b) climatic
(c) biological (d) all

19. High and low temperature, low pressure and light are examples of _____ hazard
(a) impact (b) climatic
(c) biological (d) all

20. Rodents, micro organisms, moths and termites are examples of ________ hazard
(a) impact (b) climatic
(c) biological (d) all

21. To have greater shelf life, fruits and vegetables should be harvested at _______ stage
(a) optimum maturity (b) ripened
(c) over ripened (d) none

22. _______ is used as a flush gas in controlled atmospheric storage
(a) oxygen
(b) carbon dioxide
(c) nitrogen
(d) none

23. CAS stands for
(a) Controlled Atmospheric Storage
(b) Cautious Atmospheric Storage
(c) Cold Atmospheric Storage
(d) none

24. CAS can increase the storage life of fruits and vegetables by _____ times
(a) 0 (b) 4
(c) 10 (d) both (b) & (c)

25. MAP stands for
(a) Modified Atmospheric Storage
(b) Mango Atmospheric Storage
(c) Manmade Atmospheric Storage
(d) none

26. __________ is capital intensive and expensive to operate
(a) CAS (b) MAP
(c) MAS (d) none

27. The main gas used in MAP is
(a) oxygen
(b) carbon dioxide
(c) nitrogen
(d) all

28. Carbondioxide in MAP has ________ property
(a) bacteriostatic (b) fungistatic
(c) both (d) none

29. The solubility of carbon dioxide _______ with decrease in temperature
(a) increase (b) decrease
(c) no change (d) none

30. The antimicrobial activity of carbon dioxide is _______ at low temperature
(a) greater (b) less
(c) no change (d) none

31. Oxygen ________ fat oxidation, browning reaction and pigment oxidation
(a) promotes (b) retards
(c) no effect (d) none

32. ________ is an inert gas used in MAP
(a) nitrogen (b) oxygen
(c) carbon dioxide (d) all

33. Nitrogen is __________ gas
(a) tasteless (b) odorless
(c) both (d) none

34. ________ retards the growth of aerobic microbes but does not prevent the growth of anaerobic bacteria
(a) nitrogen (b) oxygen
(c) carbon dioxide (d) all

35. EOPO stands for
(a) Edible Oil Packaging Order
(b) Edible Oil Pack Order
(c) Edible Oil Patenting Order
(d) none

36. EOPO came in _________
(a) 1948 (b) 1968
(c) 1992 (d) 1998

37. SWWA stands for
(a) Standards of Weights & Measures Act
(b) Stationary Weights & Measures Act
(c) Stainless steel Weight & Measures Act
(d) none

38. Every commodity in packaged form should bear upon it
(a) net quantity
(b) number of commodity contained
(c) sale price
(d) all

39. Minimum package of bread is
(a) 25g (b) 50g
(c) 100g (d) 500g

40. Minimum package of biscuit is
(a) 25g (b) 50g
(c) 100g (d) 500g

41. Minimum package of baby food is
(a) 25g (b) 50g
(c) 100g (d) 200g

42. Minimum package of weaning food is
(a) 25g (b) 50g
(c) 100g (d) 200g

43. Minimum package of butter is
(a) 25g (b) 50g
(c) 100g (d) 500g

44. Minimum package of cereals and pulses is
(a) 25g (b) 50g
(c) 100g (d) 500g

45. Minimum package of coffee is
(a) 25g (b) 50g
(c) 100g (d) 500g

46. Minimum package of tea is
(a) 25g (b) 50g
(c) 100g (d) 500g

47. Minimum package of milk powder is
(a) 25g (b) 50g
(c) 100g (d) 500g

48. Minimum package of soft drinks is
(a) 25ml (b) 50ml
(c) 100ml (d) 500ml

49. Minimum package of salt is
(a) 10g (b) 50g
(c) 100g (d) 500g

50. The laws governing the packaging of commodity is governed by
(a) Packaged Consumers Rules
(b) Packaged Commodity Rules
(c) Packed Consumers Rules
(d) none

51. IS stands for
(a) Indian Statistics
(b) Indian Standards
(c) Indian Socialites
(d) none

52. PE means
(a) polymethylene
(b) polyethylene
(c) polyoxymethylene
(d) none

53. PVC means
(a) poly vinyl chlorine
(b) poly vinyl chloride
(c) polyester vinyl chloride
(d) none

54. PAT means
(a) poly alkylene terephthalate
(b) poly alkylene teraphthalate
(c) poly alkylene tetraphthalate
(d) none

55. Any packaged food which does not conform to the requirements under the PFA is deemed __________
(a) misleading (b) misbranded
(c) mismatching (d) none

56. A green colored circle and square on the packaged food indicates ____ food
(a) vegetarian
(b) non vegetarian
(c) both
(d) none

57. A red colored circle and square on the packaged food indicates _______ food
(a) vegetarian
(b) non vegetarian
(c) both
(d) none

58. FPO stands for
(a) Food Products Order
(b) Fruits Product Order
(c) Fun Products Order
(d) none

59. MFPO stands for
(a) Meat Food Products Order
(b) Meat Fruits Products Order
(c) Meat & Fowl Products Order
(d) none

60. EOPO comes under
(a) Essential Commodities Act
(b) Meat Food Products Order
(c) Fruits Product Order
(d) none

61. AGMARK stands for
(a) Agricultural Grading and Marking Act
(b) Agro Industry Grading and Marking Act
(c) Agricultural Grading and Marketing Act
(d) none

62. BIS stands for
(a) Bureau of Indian Standards
(b) Bureau of International Standards
(c) Bureau of India Standard
(d) none

63. For the packaging of pasteurized flavored milk ________________ are recommended
(a) glass bottles
(b) LDPE lined cartons
(c) both
(d) none

64. Condensed milk and sterilized cream are packed in
(a) hermetically sealed containers
(b) LDPE lined cartons
(c) both
(d) none

65. Jams, Jellies and Marmalades are packaged in
(a) glass bottles
(b) open top cans
(c) both
(d) none

66. Meat should be packed in
(a) polyethylene sheets
(b) glass bottles
(c) plastics
(d) none

67. Cooked meat products should be packed in
(a) butter paper
(b) grease proof paper
(c) wrapping paper
(d) all

68. Bread should be packed in
(a) LDPE
(b) waxed paper
(c) grease proof paper
(d) all

69. Biscuits should be packed in
(a) tinplate containers
(b) cardboard paper
(c) cellol LDPE/PET/LDPE
(d) all

70. Peanut butter should be packed in
(a) wide mouthed glass jar
(b) polystyrene tubs
(c) both (d) none

71. Flours and starches should be packed in
(a) LDPE coated jute bags
(b) LDPE coated raffia bags
(c) both (d) none

72. Edible spray dried potato flour should be packed in
(a) HDPE
(b) metalized polyester bags
(c) BOPP/LDPE
(d) all

73. Oil should be packed in
(a) tinplate containers
(b) glass bottles
(c) food grade PVC and PET
(d) all

74. Carbonated beverage should be packed in
(a) PET bottles (b) cans
(c) both (d) none

75. Chillies should be packed in
(a) PET/EVA
(b) BOPP/EVA
(c) LDPE coated raffia bags
(d) all

76. Paper derives its name form
(a) papyrus (b) paparus
(c) popar (d) none

77. The first authentic papermaking has been attributed to
(a) Tsai Lun (b) Robert Koch
(c) both (d) none

78. The property of paper packaging material is
(a) good stiffness
(b) good absorbent
(c) good printability
(d) all

79. Raw material for production of paper is
(a) pulp (b) wood
(c) bark (d) none

80. The main constituents of wood cell wall is
(a) cellulose (b) hemicelluloses
(c) lignin (d) all

81. The cell wall of softwood contain _____% cellulose
(a) 40-44 (b) 25-29
(c) 25-31 (d) none

82. The cell wall of softwood contain _____% hemicellulose
(a) 40-44 (b) 25-29
(c) 25-31 (d) none

83. The cell wall of softwood contain _____% lignin
(a) 40-44 (b) 25-29
(c) 25-31 (d) none

84. The fibers of _________ wood are 2.5 times longer
(a) soft (b) hard
(c) both are same (d) none

85. The pulps that retain most of the wood lignin produce _______ papers
(a) strong (b) weak
(c) no (d) none

86. The pulp is cooked with solutions of various chemicals to remove
(a) cellulose
(b) hemicelluloses
(c) lignin
(d) none

87. The bleaching agents to bleach the mechanical pulps are
(a) hydrogen peroxide
(b) sodium hypochlorite
(c) sodium hydrosulfite
(d) all

88. The pulps for paper making are
(a) chemical (b) mechanical
(c) both (d) none

89. The chemical commonly used in cooking of pulp in chemical pulping is
(a) sodium hydroxide
(b) sodium chloride
(c) magnesium chloride
(d) all

90. Unbleached pulps exhibit a range of brightness values from
(a) 0-15 (b) 15-60
(c) 60-90 (d) none

91. The color of the pulp is mainly due to
(a) lignin (b) cellulose
(c) hemicelluloses (d) all

92. The process of bleaching of pulp _____ the strength of pulp
(a) improve (b) reduce
(c) no effect (d) none

93. The most effective bleaching agent is
(a) hydrogen peroxide
(b) sodium hypochlorite
(c) sodium hydrosulfite
(d) all

94. Beating and refining operations ______ the strength of finished sheet
(a) improve (b) reduce
(c) no effect (d) none

95. The objective of beating of the fibers is to ______ of fibers
(a) increase the surface area
(b) increase water imbibing capacity
(c) both
(d) none

96. Sizing process renders the paper sheet more resistant to penetration by
(a) sunlight (b) water
(c) oxygen (d) none

97. The most common sizing agent is
(a) rusin
(b) hydrogen peroxide
(c) sodium hydroxide
(d) none

98. The weight of paper per unit area is called
(a) grammage (b) hammrage
(c) tamprage (d) none

99. According to ISO standards, the paperboard has a grammage of ______ g/square meter
(a) 224 (b) 324
(c) 424 (d) 524

100. The final moisture content of the paper is
(a) 1-4 (b) 4-10
(c) 10-15 (d) 15-20

101. ______ is the process used to smoothen the surface of paper and give it glossy finish
(a) candle (b) calendaring
(c) cylindering (d) none

102. Machine finished (MF) are those which have undergone
(a) calendaring
(b) chemical pulping
(c) bleaching (d) none

103. Sizing and coating are applied to ______ of paper
(a) improve appearance
(b) strength
(c) prevent excess water penetration
(d) all

104. The surface sizing agents include CMC which is
(a) cabon methyl cellulose
(b) carboxy methyl cellulose
(c) carbon methylene cellulose
(d) none

105. The commonly used adhesives used for paperboard is
(a) acrylic based emulsions
(b) starch
(c) CMC (d) PVC

106. ______ is used in the form of film for paperboard to produce a water vapor barrier
(a) paraffin wax
(b) guar gum
(c) acrylic based emulsion
(d) all

107. The papers used for writing, books and cover papers are
(a) fine (b) coarse
(c) both (d) none

108. Fine papers are made of
(a) bleached pulp
(b) unbleached pulp
(c) both
(d) none

109. Coarse papers are made of
(a) bleached pulp
(b) unbleached pulp
(c) both (d) none

110. Kraft paper is ________ paper
(a) fine (b) coarse
(c) both (d) none

111. Bleached paper is ____________ than unbleached paper
(a) weaker (b) stronger
(c) no strength (d) none

112. Greaseproof paper is resistant to
(a) oil (b) grease
(c) both (d) none

113. Glassine paper has
(a) smooth surface (b) high density
(c) transparency (d) all

114. Vegetable parchment paper is strongest when
(a) wet (b) dry
(c) both (d) none

115. The major types of waxed paper are
(a) wet waxed (b) dry waxed
(c) wax laminated (d) all

116. Paper is termed board when its grammage exceeds _____ g/m^2
(a) 224 (b) 324
(c) 424 (d) 524

117. Linerboard has atleast _______ plies
(a) 1 (b) 2
(c) 3 (d) 4

118. Food board is used for food packaging having a single or multiply construction and made for 100% ______
(a) bleached virgin
(b) unbleached virgin
(c) virgin
(d) none

119. ASTM stands for
(a) American Society for Testing and Materials
(b) American Society for Testing and Marketing
(c) American Socialites for Testing and Materials
(d) none

120. The main types of glass container used in food packaging are
(a) bottles (b) jars
(c) both (d) none

121. Bottles are those with ______ neck
(a) narrow (b) wide
(c) both (d) none

122. Jars are those with _______ neck
(a) narrow (b) wide
(c) both (d) none

123. The principal ingredient of glass is
(a) alumina (b) silica
(c) manganese (d) iron ore

124. Silica for glass making is derived from
(a) sand (b) flint
(c) quartz (d) all

125. Glass is in _____ state
(a) solid (b) liquid
(c) glassy (d) none

126. In the basic structural unit, silica is co-ordinated to __________ surrounding oxygen atoms
(a) 1 (b) 2
(c) 4 (d) 4

127. The closures of glass bottle or jar consists of
(a) cap (b) cork
(c) both (d) none

128. The actual sealing component of the closure of glas bottle is
(a) gasket (b) liner
(c) lug (d) both (a) & (b)

129. Gasket is made form
(a) rubbers (b) plastisols
(c) both (d) none

130. The crown cork helps to
(a) retain internal pressure
(b) protects contents
(c) maintain vacuum inside container
(d) secure contents inside container

131. The bark cork helps to
(a) retain internal pressure
(b) protects contents
(c) maintain vacuum inside container
(d) secure contents inside container

132. The pry off or press on twist cork helps to
(a) retain internal pressure
(b) protects contents
(c) maintain vacuum inside container
(d) secure contents inside container

133. Crown corks are used in
(a) carbonated drinks
(b) wine
(c) heat processed food
(d) peanut butter

134. Bark corks are used in
(a) carbonated drinks
(b) wine
(c) heat processed food
(d) peanut butter

135. Pry off corks are used in
(a) carbonated drinks
(b) wine
(c) heat processed food
(d) peanut butter

136. Which of the packaging material is most susceptible to heat shock
(a) paper (b) glass
(c) metal (d) plastic

137. Glass is optically
(a) isotropic (b) anisotropic
(c) both (d) none

138. The largest constituent of glass manufacturing is
(a) silica (b) cullet
(c) both (d) none

139. Which of the following gas is produced during refining process in the manufacture of glass
(a) CO_2 (b) SO_2
(c) both (d) none

140. The mass or lump of molten glass is called
(a) gob (b) mob
(c) tob (d) fob

141. The process to make glass containers from molten glass is
(a) blow and blow
(b) press and blow
(c) both
(d) none

142. Bottles are normally produced by
(a) B&B
(b) P&B
(c) neck press & blow
(d) all

143. Jars are mostly produced by
(a) B&B (b) P&B
(c) neck press & blow (d) all

144. The process of removal of stress from the glass is
(a) smelting (b) annealing
(c) fusion (d) blowing

145. The annealing temperature of the glass is _____°C
(a) 140 (b) 240
(c) 440 (d) 540

146. The annealing involves
(a) heating (b) cooling
(c) both (d) none

147. Which of the following is property of glass
(a) inert (b) impermeable
(c) odorless (d) all

148. The disadvantage of glass is
(a) fragile (b) heavy weight
(c) expensive (d) all

149. Metal commonly used for food packaging is
(a) steel (b) aluminium
(c) tin (d) all

150. ECCS stands for
(a) electron chromium coated steel
(b) electrolytically chromium coated steel
(c) electrically chromium coated steel
(d) none

151. TFS stands for
(a) Tin Free Steel
(b) Tin Filmed Steel
(c) Tin Featured Steel
(d) none

152. Metal used during soldering of three piece metal can is
(a) lead (b) copper
(c) both (d) none

153. Property of metal as a packaging material
(a) high mechanical strength
(b) low toxicity
(c) superior barrier to gases
(d) all

154. Property of metal as a packaging material
(a) barrier to moisture
(b) barrier to light
(c) ideal surface for lacquering
(d) all

155. Tinplate refers to ______ steel sheet
(a) low carbon (b) higher carbon
(c) no carbon (d) none

156. Method of tinplating is
(a) dipping (b) electroplating
(c) immersion (d) all

157. Method of electroplating of tin is
(a) ferrostan process
(b) halogen process
(c) both (d) none

158. The oil used in food packaging to be used on the metal plates is
(a) cotton seed oil
(b) dioctyl sebacate
(c) acetylbutyl citrate (d) all

159. ECCS has ________ color
(a) dull bluish (b) bright bluish
(c) dull yellow (d) bright yellow

160. Property of aluminium is
(a) weak (b) light weight
(c) ductile (d) all

161. Which of the following is a type of can manufacture
(a) 2 piece (b) 3 piece
(c) 4 piece (d) both (a) & (b)

162. Two piece can is manufactured by ________ process
(a) drawn and ironed
(b) drawn and redrawn
(c) both
(d) none

163. The function of protective coating on metal cans is to
(a) avoid contamination
(b) barrier to corrosion
(c) decoration
(d) all

164. The internal enamels for metal containers is
(a) vinyl
(b) vinyl organosol
(c) acrylic
(d) all

165. The internal enamels for metal containers is
(a) polybutadiene (b) phenolic
(c) epoxyphenolic (d) all

166. The thickness of aluminium foil is _______ μm
(a) 1-4 (b) 4-150
(c) 150-250 (d) 250-300

167. Homopolymers consist of
(a) single repeating unit
(b) 2 repeating units
(c) 2-10 repeating units
(d) none

168. Copolymers consist of
(a) single repeating unit
(b) more than 1 repeating units
(c) both
(d) none

169. Plastic is type of ________ polymer
(a) thermosetting (b) thermoplastic
(c) both (d) none

170. The common plastic used for food packaging is
(a) polyolefins (b) polyvinyls
(c) polyesters (d) all

171. Most of the plastic used in food packaging is
(a) thermosetting (b) thermoplastic
(c) both (d) none

172. Those plastics which soften upon heating and harden upon cololing are called
(a) thermosetting (b) thermoplastic
(c) both (d) none

173. Which of the following fall in thermoplastic group
(a) vinyl plastics (b) polyolefins
(c) both (d) none

174. The advantage of plastic is
(a) light weight
(b) good strength
(c) design flexibility
(d) all

175. The disadvantage of plastic is
(a) low cost (b) coloring
(c) disposability (d) all

176. Polyethylene is a polymer of
(a) ethylene (b) methylene
(c) styrene (d) propylene

177. The density of LDPE film is ______ kg/m^3
(a) 110-140 (b) 410-440
(c) 910-940 (d) 1040-1140

178. LLDPE stands for
(a) Linear Low Density Polyethylene
(b) Long Low Density Polyethylene
(c) Laminar Low Density Polyethylene
(d) none

179. LLDPE is ________ than LDPE
(a) stronger (b) weaker
(c) similar (d) none

180. LLDPE is ____________ transparent than LDPE
(a) more (b) less
(c) opaque (d) none

181. LLDPE has ___________ resistance to environmental stress cracking than LDPE
(a) more (b) less
(c) opaque (d) none

182. HDPE is _______ than LDPE
(a) stiffer (b) harder
(c) weaker (d) both (a) & (b)

183. The density of polypropylene is ________ kg/m^3
(a) 500 (b) 600
(c) 900 (d) 1200

184. The density of HDPE is ________ kg/m^3
(a) 900-940 (b) 941-965
(c) 966-1000 (d) 1001-1025

185. Polystyrene is made by polymerization if
(a) styrene (b) ethylene
(c) methylene (d) propylene

186. EVOH means
(a) ethylene vinyl alcohol
(b) ethanol vinyl alcohol
(c) ethyl vinyl alkene
(d) none

187. PET is ______ polymer
(a) thermoplastic (b) thermosetting
(c) both (d) none

188. The bonding of two or more webs is called
(a) coating (b) laminating
(c) both (d) none

189. The laminating is bonded by _____ process
(a) wet bond (b) dry bond
(c) extrusion (d) all

190. The filling of sterile containers with commercially sterile product under aseptic conditions is
(a) aseptic packaging (b) MAP
(c) CAS (d) all

191. The sterilization of packaging material for aseptic packaging material is
(a) irradiation (b) heat
(c) chemical (d) all

192. Irradiation of packaging material uses gamma rays from ____ as source
(a) cesium (b) cobalt
(c) both (d) none

193. A radiation dose of _____ kGy is sufficient to ensure sterility
(a) 25 (b) 50
(c) 75 (d) 100

194. Wavelength of UV radiation used for sterilization of aseptic packaging material is
(a) 100-200nm (b) 200-315nm
(c) 400-515nm (d) 600-615nm

195. Most effective wavelength of UV radiation used for sterilization of aseptic packaging material is
(a) 200-225nm (b) 226-247nm
(c) 248-280nm (d) 281-300nm

196. Steam is _______ efficient than dry heat for sterilization of packaging material
(a) more (b) less
(c) no (d) none

197. Steam sterilization for packaging material is done at ___°C/20 minutes
(a) 121 (b) 151
(c) 171 (d) 191

198. Dry heat sterilization for packaging material is done at ___°C/20 minutes
(a) 121 (b) 151
(c) 171 (d) 191

199. Chemical sterilants used in packaging material
(a) hydrogen peroxide
(b) peracetic acid
(c) both (d) none

200. The systems of cans was pioneered by
(a) WM Martin (b) KM Smith
(c) RL Gayle (d) none

11.3 ANSWER KEY

Q. A.	Q. A.	Q. A.	Q. A.	Q. A.	Q. A.
1. (c)	2. (a)	3. (a)	4. (a)	5. (a)	6. (b)
7. (c)	8. (d)	9. (a)	10. (b)	11. (c)	12. (a)
13. (a)	14. (a)	15. (b)	16. (c)	17. (b)	18. (a)
19. (b)	20. (c)	21. (a)	22. (c)	23. (a)	24. (d)
25. (a)	26. (a)	27. (d)	28. (c)	29. (a)	30. (a)
31. (a)	32. (a)	33. (c)	34. (a)	35. (a)	36. (d)
37. (a)	38. (d)	39. (c)	40. (a)	41. (d)	42. (d)

Q.	A.	Q.	A.	Q.	A.	Q.	A.	Q.	A.	Q.	A.
43.	(a)	44.	(c)	45.	(a)	46.	(a)	47.	(b)	48.	(c)
49.	(a)	50.	(b)	51.	(b)	52.	(b)	53.	(b)	54.	(a)
55.	(b)	56.	(a)	57.	(b)	58.	(b)	59.	(a)	60.	(a)
61.	(a)	62.	(a)	63.	(c)	64.	(c)	65.	(c)	66.	(a)
67.	(d)	68.	(d)	69.	(d)	70.	(c)	71.	(c)	72.	(d)
73.	(d)	74.	(c)	75.	(d)	76.	(a)	77.	(a)	78.	(d)
79.	(a)	80.	(d)	81.	(a)	82.	(b)	83.	(c)	84.	(a)
85.	(a)	86.	(c)	87.	(d)	88.	(c)	89.	(a)	90.	(b)
91.	(a)	92.	(a)	93.	(a)	94.	(a)	95.	(c)	96.	(b)
97.	(a)	98.	(a)	99.	(a)	100.	(b)	101.	(b)	102.	(a)
103.	(d)	104.	(b)	105.	(a)	106.	(a)	107.	(a)	108.	(a)
109.	(b)	110.	(b)	111.	(a)	112.	(c)	113.	(d)	114.	(a)
115.	(d)	116.	(a)	117.	(b)	118.	(a)	119.	(a)	120.	(c)
121.	(a)	122.	(b)	123.	(b)	124.	(d)	125.	(c)	126.	(d)
127.	(c)	128.	(d)	129.	(c)	130.	(a)	131.	(b)	132.	(c)
133.	(a)	134.	(b)	135.	(c)	136.	(b)	137.	(a)	138.	(a)
139.	(c)	140.	(a)	141.	(c)	142.	(a)	143.	(a)	144.	(b)
145.	(d)	146.	(c)	147.	(d)	148.	(d)	149.	(d)	150.	(b)
151.	(a)	152.	(c)	153.	(d)	154.	(d)	155.	(a)	156.	(d)
157.	(c)	158.	(d)	159.	(a)	160.	(d)	161.	(d)	162.	(c)
163.	(d)	164.	(d)	165.	(d)	166.	(b)	167.	(a)	168.	(b)
169.	(c)	170.	(d)	171.	(b)	172.	(b)	173.	(c)	174.	(d)
175.	(c)	176.	(a)	177.	(c)	178.	(a)	179.	(a)	180.	(b)
181.	(a)	182.	(d)	183.	(c)	184.	(b)	185.	(a)	186.	(a)
187.	(a)	188.	(b)	189.	(d)	190.	(a)	191.	(d)	192.	(c)
193.	(a)	194.	(b)	195.	(c)	196.	(a)	197.	(a)	198.	(c)
199.	(c)	200.	(a)								

CHAPTER 12

FOOD ANALYSIS

12.1 SUBJECTIVE KEYNOTES

1. Chemical analysis of foods is an important part of a quality assurance program in food processing from ingredients or raw materials to finished products.
2. Chemical analysis is also important in formulating and developing new products and evaluating new processes for making food products.
3. Process control samples are usually analyzed by rapid methods, whereas nutritive information for nutrition labelling generally requires the use of more time consuming methods of analysis.
4. Sampling is the initial point for sample identification.
5. Performing the assay is unique for each component or characteristic be analyzed and may be unique to a specific type of food product.
6. Selection of method depends largely on the objective of the measurement.
7. Proximate analysis of foods refers to determining the major components of moisture, ash, lipids, protein and carbohydrates.
8. The performance of many analytical methods is affected by the food matrix.
9. A major consideration for determining method validity is the analysis of materials used as controls, often referred to as standard reference materials or check samples.
10. AOAC international is an organization begun in 1884 to serve the analytical methods needs of government regulatory and research agencies.
11. The goal of AOAC International is to provide methods that will be fit for their intended purpose.
12. The AACC publishes a set of approved laboratory methods, applicable mostly to cereal products.
13. The AOCS publishes a set of official methods and recommended practices, applicable mostly to fat and oil analysis.
14. Accuracy refers to how close a particular measure is to the true or correct value.
15. Precision is a measure of how reproducible or how close replicate measurements become.
16. The range is simply the difference between the largest and smallest observation.

17. The standard deviation measures the spread of the experimental values and gives a good indication of how close the values are to each other.
18. Absolute error is the difference between the experimental value and true value.
19. Systematic error produces results that consistently deviate from expected value in one direction or the other.
20. Sensitivity relates to the magnitude of change of a measuring device with changes in compound concentration.
21. Ash refers to the inorganic residue remaining after either ignition or complete oxidation of organic matter in a foodstuff.
22. The ash content of foods can be expressed on either a wet basis or on a dry weight basis.
23. Dry ashing refers to the use of muffle furnace capable of maintaining temperatures of 500-600°C.
24. Wet ashing is a procedure for oxidising organic substances by using acids and oxidizing agents or their combination.
25. Non-protein nitrogen could come from free amino acids, small peptides, nucleic acids, phospholipids, amino sugars, porphyrins, and some vitamins, alkaloids, uric acid, urea and ammonium ions.
26. Kjeldahl, Dumas, and infrared spectroscopy methods are generally used for protein analysis.
27. Qualitative analysis ensures that ingredient labels present accurate compositional information.
28. Quantitative analysis ensures that added components are listed in the proper orderon ingredient labels.
29. Titratable acidity deals with measurement of the total acid concentration contained within a food.
30. Molarity is a concentration unit representing the number of moles of the solute per liter of solution.
31. Normality is a concentration unit representing the number of equivalents of a solute per liter of solution.
32. Percentage concentrations are the mass amount of solute or analyte per 100ml or 100 g of material.
33. The isoelectric point is defined as the pH at which a protein has no net charge in solution.
34. Proteins can be separated on the basis of their isoelectric points, in a technique called capillary isoelectric focusing.
35. SDS capillary gel electrophoresis techniques can be used to separate proteins by size to determine molecular masses.
36. The primary difference between capillary electrophoresis and conventional electrophoresis is that capillary tubing is used in place of acrylamide gels cast in tubes or slabs.
37. An antigen is any molecule that induces the formation of antibodies.

38. Antibodies are proteins produced by animals in response to an antigen.
39. The immobilized antibody is called a capture antibody.
40. The process of structural modification by chemical reaction is referred to as derivatization.
41. Analysis of extraneous matter is an important element both in the selection of raw materials for food manufacturing and for monitoring the quality of processed foods.
42. X-ray radiography is used for selecting grains for processing.
43. NIR spectroscopy used with single kernel characterization system is able to detect later stages of internal insect infestation in wheat with a 95% confidence.
44. Microscopy techniques including light microscopy, fluorescence microscopy and scanning electron microscopy are used to study the structure and functional relationship of foods.
45. The three main methods used to measure the oxygen demand of water and waste water are BOD, COD and TOC.
46. The oxygen demand of water and waste water is proportional to the amount of organic matter present in it.
47. UV range runs from 200-350 nm and the visible range from 350-700 nm.
48. The relationship between absorbance of solution and the concentration of the absorbing species is known as Beer's law.
49. A basic spectrophotometer is composed of light source, monochromator, sample holder, radiation detector and readout device.
50. The component that functions to isolate the specific, narrow, continuous group of wavelengths to be used in spectroscopic assay is the monochromator.
51. Adulteration is defined as the addition or subtraction of any substance to or from food, so that the natural composition and quality of food substance is affected. Food analysis has an important role in detection of adulteration.
52. Analysis of foods generally involves three main steps namely selection and preparation of samples, carrying out analysis and calculation & interpretation of results.
53. A predetermined procedure for the selection, withdrawal, preservation, transportation, and preparation of the portions to be removed from a lot as samples is known as sampling plan. A sampling plan should be selected on the basis of the sampling objective, nature of the population, the sample selection criteria, and the analysis procedures.
54. In attribute sampling plan, sampling is done to decide the acceptability of a population based on the number of unacceptable items in the sample. The acceptability of an item depends on the presence or absence of a characteristic.
55. Variable sampling plans are those for which a quality characteristic is measured quantitatively on each item inspected. The average measurement is compared with the standard and used for the acceptability decision. For e.g. salt content, moisture.
56. Acceptance sampling plans are commonly used to decide whether to accept or reject a lot. They may be used for evaluation of attributes or variables or both. These

acceptance sampling plans may be single, double, multiple, sequential or skip plans. The decision of accepting or rejecting a lot is based on a single random sample in single sampling plan.

57. Probability sampling deals with the selection of a representative sample from a population based on chance with the elimination of human bias. The probability of including any item in the sample is known and sampling error can be calculated.

58. Stratified sampling involves dividing the population (size *N*) into a certain number of mutually exclusive homogeneous subgroups (size *N*1, *N*2, *N*3, etc.). Random samples are obtained from each subgroup. This can be used only when subpopulations of similar characteristics can be observed within the whole population.

59. In cluster sampling, the population is divided into subgroups or clusters followed by randomly selecting only a certain number of clusters for analysis. The main difference between cluster sampling and stratified sampling is that in the latter samples are taken from every single subgroup, while in cluster sampling only randomly selected clusters are sampled.

60. Composite sampling is used to obtain samples from bagged products such as flour, seeds, and larger items in bulk. Small aliquots are taken from different bags, or containers, and combined into a single sample (the composite sample) that is used for analysis.

61. Judgment sampling is solely at the discretion of the sampler and therefore is highly dependent on the person taking the sample.

62. Convenience sampling is performed when ease of sampling is the key factor as the most accessible sample from a lot is selected. This is also called 'chunk sampling' or 'grab sampling'.

63. Physical methods of food analysis are based on the measurement of physical properties of food such as dielectric constant, specific gravity, conductivity, refractive index freezing point or absorption of energy in or near infrared region and comparison to known properties of standards. Physical methods yield rapid results and thus provide ease in handling large number of samples.

64. Methods for determination of moisture are divided as: drying methods, distillation methods, chemical methods and physical methods.

65. Direct methods for moisture determination involve removal of water from the food matrix using extraction, drying and other methods and then quantifying this water generally by weighing or titration. Indirect methods for moisture determination measure a property such as dielectric onstant conductivity, specific gravity etc. exhibited by the water present in the food sample in order to quantify its moisture content.

66. The dielectric constant of water contained in foods bears a linear relationship with moisture content of such foods till a moisture content of 30 %.

67. Conductivity methods for moisture determination are based on the principle that as the moisture content of the sample increases, its resistance decreases and conductivity increases.

68. Methods for protein analysis involve chemical methods (Kjeldahl's method, Dumas method) and instrumental methods.

69. Tryptophan and tyrosine absorb ultraviolet light strongly at 280 nm. The tryptophan and tyrosine content of many proteins remains fairly constant, and so the absorbance of protein solutions at 280nm can be used to determine their concentration.

70. In Biuret method, a violet-purplish color is produced when cupric ions (Cu^{2+}) of biuret reagent interact with peptide bonds under alkaline conditions. The reagent is mixed with a protein solution and then allowed to stand for 15-30 minutes before the absorbance is read at 540 nm.

71. The Lowry method combines the biuret reagent with another reagent (the Folin-Ciocalteau phenol reagent) which reacts with tyrosine and tryptophan residues in proteins. This gives a bluish color which can be read somewhere between 500 - 750 nm depending upon the required sensitivity. This method is more sensitive to low concentrations of proteins than the biuret method.

72. The Babcock, Gerber and Detergent methods are examples of non-solvent liquid extraction methods for determining the lipid content of milk and some other dairy products.

73. Automated Turbidimetric Method is used for determination of fat in milk samples. Turbidity is related to fat content in the sample. Initially EDTA is added to the sample. This dilutes the sample and removes turbidity resulting from casein protein. After this the sample is homogenized to form uniform fat globules. This if followed by turbidity measurement.

74. Fat content can be determined using methods that measure properties of fats like density, nuclear magnetic resonance, x-ray absorption, infrared spectra and dielectric constant. These methods are rapid and do not require extraction of fat from the sample. They are based on the difference between these properties possessed by fats and other components present in the food matrix.

75. Color measurement of food products serves as a subjective and consistent tool for color quality control. In the electromagnetic spectrum, visible region is represented by 380 to 780 nm wavelength region.

76. Color scales are standard schemes used in the food industry to describe color. The Commission International de l'Eclariage (CIE) system, the Hunter L, a, b system, and the Munsell color solid are some of the most often used color scales.

77. The CIE system is a trichromatic system. It is based on the fact that any color can be obtained by suitably mixing the three primary colors, red, green and blue. These primaries are represented by the imaginary primaries, X, Y and Z.

78. The HunterLab L*, a*, b* is widely used in food colorimetry. It measures three color dimensions: the degree of lightness (L), the degree of redness or greenness (+/-a), and the degree of yellowness or blueness (+/-b).

79. In the Munsell color classification, every color is expressed by a combination of three attributes namely: hue, value and chroma. The color of any surface can be identified by comparing it to the chips under proper illumination and viewing conditions. The color is then identified by its hue, value, and chroma.

80. Colorimeters are the instrument used for color measurement usually consist of two main parts: the optical sensor, which contains the light source and the micro-processor, which automatically converts the colors to numeric equations.

81. Spectrophotometers measure a ratio of the light reflected or transmitted from a food product to that from a known reference standard. Spectrophotometers are more accurate and more expensive than colorimeters.
82. In bench top water activity meters, an infrared beam is focused on a tiny mirror which in turn determines the precise dewpoint temperature of the sample. This dewpoint temperature is then converted into water activity.
83. Portable water activity meters require the sample to be placed in a standard cup and the lid be covered. A push of a button presents an accurate water activity reading within five minutes.
84. Water activity of a food sample may also be obtained using hygrometer. Hygrometers may be of three types: resistive electrolytic, capacitance or a dew point hygrometer.
85. Sensory evaluation requires three major components i.e. proper laboratory facilities, sensory panels and rigorous training programmes.
86. A temperature of 20 ˚C and RH of 62 % in the testing room is ideal for sensory evaluation.
87. Institute of food technologists (USA) recommends 3 to 10 panelists for trained; 8-25 panelists for semi-trained and 80 panelists for untrained for sensory evaluation procedures.
88. Discriminative testing is the one of the most useful analytical tools available to the sensory professionals. It is on the basis of a perceived difference between two products that one can justify proceeding to a descriptive test . Within this general class are a variety of specific methods e.g. Paired comparison test, Duo trio test, Triangle test, Multiple sample test, dual and multiple standard tests etc.
89. In quantitative descriptive analysis (QDA), the judges evaluate all of the products on an attribute on more than a single occasion. Descriptive Analysis is used for storage testing, product development, quality control an instrument sensory relationship.
90. Refractometry is a method of testing the physicochemical properties of a substance (commonly liquids) by measuring its refractive index. Refractive index of a substance is the ratio of speed of light in vacuum to the speed of light in that substance. Refractive index is reported at 20°C.
91. The sodium D line at 589 nm is the most frequently used wavelength of light for a refractometer.
92. Refractometer is the device used to find the refractive index of a liquid sample by measuring the extent to which light is refracted when it moves from a medium into the sample.
93. Polarimetry is a non-destructive method used to measure the concentration and purity of a substance in a solution based on its rotation of the plane polarised light.Polarimetry can be used only for optically active substances.
94. A substance is said to be optically active if it rotates the plane polarised light,either left or right. Optical activity of a substance depends on its chirality. A chiral substance is one that cannot be superposed on its mirror image.
95. Depending on the nature of support medium used the electrophoresis can be divided into paper, starch, polyacrylamide, and gel electrophoresis.

96. In capillary electrophoresis the capillary tubing is used in place of acrylamide gels and a high voltage of 10-50KV is applied. Capillary electrophoresis has a very high resolving power and speed and is versatile and automated. Capillary electrophoresis technique can be used to separate a wide spectrum of biological materials including amino acids, peptides, proteins, nucleic acids etc.

97. Antinutrients are the natural or synthetic compounds that interfere with the absorption of nutrients present in the food.

98. Food toxins are substances which are ingested through food and cause cellular or tissue injury by an actual or potentially harmful alteration of cellular or tissue constituents and processes which may or may not be manifested by clear clinical symptoms.

99. Chromatography is a non-destructive procedure for separating a multi component mixture of trace, minor, or major constituents into its individual fractions. Different variations may be applied to liquids, solids and gases.

100. In gas chromatography a gas is used as the mobile phase instead of a liquid. Gas-solid chromatography (GSC) is based upon selective adsorption on a solid, whereas in gas-liquid chromatography the separation based on selective adsorption between the gas and an immobile liquid phase.

12.2 MULTIPLE CHOICE QUESTIONS

1. Which of the following belongs to color measurement
(a) CIE system
(b) Hunter system
(c) Munsell system
(d) all

2. Which of the following is a type of colorimeter
(a) Hunter (b) Gardner
(c) both (d) none

3. CIE system of color measurement first came in
(a) 1931 (b) 1941
(c) 1951 (d) 1971

4. CIELAB system of color measurement first came in
(a) 1931 (b) 1941
(c) 1951 (d) 1971

5. Commonly used test for starch detection is
(a) Amylasc (b) iodine
(c) phosphate (d) iron

6. PCA stands for
(a) Protein composition analysis
(b) principle
(c) purification
(d) none

7. TFA stands for
(a) Target Factor Analysis
(b) Tertiary Factor Analysis
(c) both
(d) none

8. PSD stands for
(a) Particle Size Distribution
(b) Powder Size Distribution
(c) both (d) none

9. The principle food reserve polysaccharide of plants is
(a) Starch (b) cellulose
(c) glycogen (d)hemicelluloses

10. Which of the starch granules are larger in size
(a) A (b) B
(c) both are equal (d) none

11. Acoustic spectroscopy for principle size measurement in suspensions uses ultrasound waves of frequency
(a) 1-200 MHz
(b) 500-1000 MHz
(c) 2000-5000 MHz
(d) none

12. ELS stands for
(a) Electrophoretic Light Scattering
(b) Electron Light Scattering
(c) both
(d) none

13. Technique used to test the quality and purity of material is DSC which stands for
(a) Differential Scanning Calorimetry
(b) Double scanning calorimetry
(c) Differential Spectral Calorimetry
(d) none

14. An endothermic reaction is one in which energy is
(a) Absorbed (b) released
(c) both (d) none

15. An exothermic reaction is one in which energy is
(a) Absorbed (b) released
(c) both (d) none

16. The melting of ice is an example of
(a) Endothermic (b) exothermic
(c) both (d) none

17. Boiling of water is an example of
(a) Endothermic (b) exothermic
(c) both (d) none

18. Freezing of water is an example of
(a) Endothermic (b) exothermic
(c) both (d) none

19. Undamaged starch granules are ________ in cold water
(a) Soluble
(b) insoluble
(c) sparingly soluble
(d) none

20. Which of the following process occurs at a higher temperature
(a) Gelatinization
(b) retrogradation
(c) both
(d) none

21. Solid fat index is the ratio of
(a) Solid to liquid fat
(b) liquid to solid fat
(c) starch to fat
(d) none

22. Maillard reaction is due to ________ browning
(a) Enzymatic (b) non-enzymatic
(c) both (d) none

23. The final product of Maillard reaction is
(a) Melanin (b) melanoidins
(c) peptides (d) COOH

24. Furosine is widely used for the evaluation of ________ damage in food material
(a) Lysine (b) leucine
(c) phenylalanine (d) tyrosine

25. Lactulose is found in foods containing
(a) Lactose (b) glucose
(c) sucrose (d) maltose

26. Hydrometer is used to measure
(a) Water activity
(b) specific gravity
(c) temperature
(d) water vapours

27. Lactulose levels (mg/litre) in raw milk is
(a) 20 mg/litre (b) 0 mg/litre
(c) 30 mg/litre (d) 40 mg/litre

28. The determination of latctulose is used to assess undergone by milk
(a) Thermal treatment
(b) freezing treatment
(c) both (d) none

29. Carotenoids are ________ soluble
(a) Water (b) fat
(c) both (d) none

30. Which of the following is animal source of carotenoids
(a) Salmon (b) shrimps
(c) egg yolk (d) all

31. Which of the following is a column chromatography method
(a) Zone chromatography
(b) stepwise elution chromatography
(c) gradient elution chromatography
(d) all

32. Which of the following is true about TLC method
(a) Reliable (b) inexpensive
(c) portable (d) all

33. Suberin is a derivative of
(a) Carbohydrate (b) amino acid
(c) fatty acid (d) none

34. Which of the following is not used in dietary fiber analysis
(a) Amylase
(b) lipase
(c) amylogluycosidase
(d) protease

35. Which of the following is not used in dietary fiber analysis
(a) TDF (b) SDF
(c) IDF (d) none

36. Which of the following method is related to dietary fiber analysis
(a) Faulks & Timms
(b) Englyst & Hudson
(c) Monro
(d) all

37. Which of the following method is related to dietary fiber analysis
(a) Uppsala
(b) Mongeau & Brassard
(c) Prosky
(d) all

38. Which of the following method is related to dietary fiber analysis
(a) Aspetal (b) Li & Andrews
(c) Li & Cardozo (d) all

39. Which of the following method is related to dietary fiber analysis
(a) Englyst et al (b) Muir & O'Dea
(c) Goni et al (d) all

40. Which of the following method is related to dietary fiber analysis
(a) Akerberg et al (b) Champ et al
(c) McLeary et al (d) all

41. Which of the following is not a dietary fiber
(a) Glucan
(b) fructooligosaccharide
(c) lactulose
(d) all

42. Which of the following is a dietary fiber
(a) Polydextrose
(b) resistant maltodextrin
(c) FOS
(d) all

43. Dry ashing is related to
(a) Carbohydrates (b) minerals
(c) proteins (d) fat

44. Wet ashing is related to
(a) Carbohydrates (b) minerals
(c) proteins (d) fat

45. Minerals analysis involves
(a) Dry ashing (b) wet ashing
(c) UV photolysis (d) all

46. Which of the following technique can be used in minerals analysis
(a) XRF (b) AAS
(c) NIR (d) all

47. Which of the following technique is used in mycotoxin analysis
(a) TLC (b) HPLC
(c) ELISA (d) all

48. Which of the following is screening method for mycotoxin analysis
(a) ELISA
(b) TLC
(c) HPLC
(d) capillary electrophoresis

49. Which of the following is confirmatory method for analysis of mycotoxins
(a) TLC
(b) HPLC
(c) capillary electrophoresis
(d) all

50. Which of the following step is involved in mycotoxins analysis
(a) Extraction
(b) cleanup
(c) determination
(d) all

51. Zearalenone is a
(a) Non steroidal toxin
(b) estrogenic toxin
(c) both
(d) none

52. Which of the following is a visualization technique
(a) Principle Component Analysis
(b) Orthogonal Rotation
(c) Target Factor Analysis
(d) Evolving Factor Analysis

53. Which of the following is visualization technique
(a) Non-linear Mapping
(b) Orthogonal Rotation
(c) Target Factor Analysis
(d) Evolving Factor Analysis

54. Which of the following is visualization technique
(a) Projection Pursuit
(b) Orthogonal Rotation
(c) Target Factor Analysis
(d) Evolving Factor Analysis

55. Which of the following is visualization technique
(a) Three-way PCA
(b) Orthogonal Rotation
(c) Target Factor Analysis
(d) Evolving Factor Analysis

56. Which of the following is clustering technique
(a) Hierarchical method
(b) PCA
(c) Projection Pursuit
(d) all

57. Which of the following is clustering technique
(a) K-means
(b) minimum spanning tree
(c) Kohonen self-organising maps
(d) all

58. Which of the following is an optimization technique
(a) ANN (b) PCA
(c) both (d) none

59. Which of the following is an optimization technique
(a) RSM (b) ANN
(c) both (d) none

60. Property related to powdered food material
(a) PSD (b) segregation
(c) agglomeration (d) all

61. Which of the following is related to agglomeration in powdered food material
(a) Size enlargement
(b) enhanced dispersibility
(c) enhanced flowability
(d) all

62. Which one is related to particle size analysis
(a) Equivalent sphere diameter
(b) particle size distribution
(c) both (d) none

63. Property that can be used for particle size analysis
(a) Sedimentation velocity
(b) intensity of scattered light
(c) conductivity
(d) all

64. Which of the following can be used in particle fractionation
(a) Sieve analysis
(b) gravitation
(c) chromatographic technique
(d) all

65. Which of the following can be used in particle size analysis
(a) Laser light scattering
(b) NIR reflectance spectroscopy
(c) both
(d) none

66. Which of the following can be used in particle size analysis
(a) Dynamic light scattering
(b) Refractometer
(c) both
(d) none

67. Which of the following can be used in particle size analysis
(a) Laser light scattering
(b) rheometer
(c) both
(d) none

68. Which of the following can be used in particle size analysis
(a) LALLS (b) NIR
(c) DLS (d) all

69. Which of the following is related to surface charge analysis
(a) Zeta potential (b) water activity
(c) ferementation (d) all

70. Surface charge analysis helps in
(a) Creaming (b) flocculation
(c) thickening (d) all

71. Surface charge determination can be done by
(a) Colorimetric titration
(b) particle charge detector
(c) both
(d) none

72. DSC is a ________ technique
(a) Microscopic (b) thermal
(c) spectroscopic (d) irradiation

73. Which of the following is not a thermal treatment
(a) Steaming (b) fermentation
(c) refrigeration (d) roasting

74. Which of the following is not a thermal treatment
(a) Freezing
(b) pasteurization
(c) thermization
(d) none

75. Which of the following is a part of differential scanning calorimeter
(a) Spindle (b) amplifier
(c) probe (d) all

76. Which of the following is a part of differential scanning calorimeter
(a) Transducer (b) transformer
(c) needle (d) all

77. Which of the following is a part of differential scanning calorimeter
(a) Lens (b) eye-piece
(c) furnace (d) all

78. Which of the following is a part of differential scanning calorimeter
(a) Cooling unit (b) z-blades
(c) both (d) none

79. Which of the following is a part of differential scanning calorimeter
(a) Reference pan
(b) aluminium pan
(c) both
(d) none

80. In DSC analysis, the graph is plotted between
(a) Temperature & pressure
(b) heat flow & temperature
(c) heat flow & pressure
(d) temperature & volume

81. Which of the following is related to DSC analysis
(a) On-set temperature
(b) enthalpy
(c) heat capacity
(d) all

82. DSC can be useful in analyzing
(a) Refractive index
(b) ripening index
(c) denaturation
(d) all

83. DSC can be useful in analyzing
(a) Gelatinization (b) carmelization
(c) both (d) none

84. DSC can be useful in analyzing
(a) Microbiology (b) Retrogradation
(c) Oxidation (d) all

85. DSC can be useful in analyzing
(a) Glass transition
(b) phase transition
(c) both
(d) none

86. DSC analysis of fat helps in analysing
(a) Polymorphism (b) rancidity
(c) both (d) none

87. DSC analysis of fat helps in analysing
(a) Oxidation (b) solid fat index
(c) both (d) none

88. Reaction that give brown color to the food products
(a) Non-enzymatic browning
(b) Maillard reaction
(c) Carmelization
(d) all

89. Furfural derivatives can be analyzed using
(a) Colorimetric method
(b) flow injection analysis
(c) both (d) none

90. Furfural derivatives can be analyzed using _______ method
(a) Spectrophotometric
(b) chromatographic
(c) both (d) none

91. Furfural derivatives can be analyzed using
(a) HPLC (b) MEKC
(c) gas chromatography
(d) all

92. _______ is used for preparation of sample for analysis of volatile compounds
(a) Distillation
(b) solvent extraction
(c) simultaneous distillation-extraction
(d) all

93. _______ is used for preparation of sample for analysis of volatile compounds
(a) Supercritical fluid extraction
(b) static headspace
(c) gas purge-and-trap
(d) all

94. _________ is used for preparation of sample for analysis of volatile compounds
(a) Thermal desorption
(b) solid phase microextraction
(c) both
(d) none

95. Hydroxy methyl furfural can be determined using
(a) Gas chromatography
(b) Mass spectrometry
(c) both
(d) none

96. Hydroxy methyl furfural can be determined using
(a) HPLC (b) electronic noses
(c) both (d) none

97. Which of the following is a product from Maillard reaction
(a) Pentosidine (b) pyrraline
(c) pyrralyne (d) all

98. Which of the following is related to Maillard reaction
(a) Cyclization (b) dehydration
(c) both (d) none

99. Which of the following is related to Maillard reaction
(a) Adolization (b) condensation
(c) both (d) none

100. Data from electronic noses can be analyzed using
(a) Multivariate analysis
(b) neural network
(c) fuzzy logic
(d) all

101. Flow injection analysis is promising technique due to
(a) Versatility (b) economy
(c) reproducibility (d) all

102. Electron spin resonance is a _____ __________ method
(a) Thermal (b) spectroscopic
(c) rheometric (d) all

103. Identification of origin of meat can be done by
(a) PAGE
(b) SDS-PAGE
(c) isoelectric focusing
(d) all

104. Identification of origin of meat can be done by
(a) Immunodiffusion
(b) ELISA
(c) both
(d) none

105. Identification of origin of meat can be done by
(a) Counter-immunoelectrophoresis
(b) DNA hybridization
(c) polymerase chain reaction
(d) all

106. Which of the following is a physical parameter related to meat quality
(a) Color (b) texture
(c) both (d) none

107. Which of the following is a physical parameter related to meat quality
(a) Drip loss
(b) water holding capacity
(c) both
(d) none

108. Which one is used for analyzing protein quality of meat
(a) SDS-PAGE (b) PCR
(c) both (d) none

109. Which of the following can analyzed using NIR
(a) Beef tenderness
(b) meat composition
(c) beef quality
(d) all

110. Which meat quality parameter can be analyzed using ultrasound
(a) Beef texture
(b) intra-mascular fat content
(c) beef carcass composition
(d) all

111. Which meat quality parameter can be analyzed using NMR
(a) Meat characterization
(b) water content
(c) both
(d) none

112. Degradation of thiamine in meat leads to formation of
(a) Thiazoles (b) thiophenes
(c) thiofurans (d) all

113. ________ are derived from oxidation of unsaturated fatty acids in cooked meat
(a) Aldehydes (b) ketones
(c) both (d) none

114. ________ are derived from oxidation of saturated fatty acids in cooked meat
(a) Aldehydes (b) ketones
(c) both (d) none

115. Dilution technique used in meat flavor
(a) AEDA (b) CHARM
(c) DHDA (d) all

116. Method used for nitrogen estimation
(a) Kjeldahl's method
(b) Soxlet method
(c) both (d) none

117. Method used for nitrogen estimation
(a) Chemiluminescence method
(b) Dumas method
(c) both (d) none

118. DSC stands for
(a) Differential Scanning Calorimetry
(b) Direct Scanning Calorimetry
(c) Duplicate Scanning Calorimetry
(d) Derived Simulated Calorimetry

119. AAS stands for
(a) Atomic Adsorption Spectra
(b) Atomic Absorption Spectra
(c) Atomic Adsorption Stimulation
(d) Atomic Absorption Stimulation

120. MS stands for
(a) Mass spectroscopy
(b) Mechanical
(c) Motion spectroscopy
(d) all

121. GC stands for
(a) Gas Calorimetry
(b) Gas Chromatography
(c) both
(d) none

122. FTIR stands for
(a) Fourier Transform IR
(b) Fourier Transmission IR
(c) Far Transform IR
(d) Far Transmission

123. SEM stands for
(a) Spectroscopic Electron Microscope
(b) Static Electron Microscope
(c) Scanning Electron Microscope
(d) Standard Electron Microscope

124. TEM stands for
(a) Tunneling Electron Microscope
(b) Transmission Electron Microscope
(c) Transform Electron Microscope
(d) none

125. Which one is essential in viscosity measurement using viscsometer
(a) Spindle (b) rpm
(c) temperature (d) all

126. Cup & Bob is related to
(a) Refractometer
(b) rheometer
(c) water activity meter
(d) all

127. Plate and plate geometry probes are related to
(a) Refractometer
(b) rheometer
(c) water activity meter
(d) all

128. Cone and plate geometry probes are related to
(a) Refractometer
(b) rheometer
(c) water activity meter
(d) all

129. Which of the following contains prism
(a) Refractometer
(b) rheometer
(c) water activity meter
(d) all

130. Texture analyzer can analyze ______ ______ property of food
(a) Hardness (b) brightness
(c) sweetness (d) all

131. Texture analyzer can analyze ______ ______ property of food
(a) Gumminess (b) Springiness
(c) both (d) none

132. Texture analyzer can analyze ______ ______ property of food
(a) Adhesiveness (b) cohesiveness
(c) both (d) none

133. Mixolab is related to
(a) Cereals lab
(b) fruits & vegetables lab
(c) dairy lab (d) all

134. Falling number apparatus is related to ______ analysis
(a) Protein (b) carbohydrate
(c) lipid (d) all

135. Falling number apparatus is related to ______ activity
(a) Amylase (b) protease
(c) lipase (d) all

136. Farinograph is related to
(a) Dairy
(b) fruits & vegetables
(c) beverage (d) cereals

137. Extensograph is related to
(a) Dairy
(b) fruits & vegetables
(c) beverage (d) cereals

138. Loaf volume is related to
(a) Biscuits (b) bread
(c) both (d) none

139. Specific volume is related to
(a) Biscuits (b) bread
(c) both (d) none

140. Spread ratio is related to
(a) Cookies (b) bread
(c) pizza base (d) none

141. Cookies hardness ______ with increase in spread ratio
(a) Increases (b) decreases
(c) do not change (d) none

142. TPA is related to
(a) Color (b) flavor
(c) taste (d) texture

143. ______ is used for bread dough fermentation
(a) Proving cabinet (b) farinograph
(c) moulder (d) none

144. ______ seeds are used to determine bread loaf volume
(a) Poppy seeds
(b) tamarind seeds
(c) rye seeds
(d) all

145. In texture analyzer, ______ is used to determine cookies hardness
(a) Plunger (b) blade
(c) spindle (d) all

146. In texture analyzer, ______ is used to determine bread loaf hardness
(a) Plunger (b) blade
(c) spindle (d) all

147. Eye piece is related to
(a) Viscometer
(b) microscope
(c) water activity meter
(d) all

148. Which one is related to polarimetry
(a) Plane polarised light
(b) specific rotation
(c) both
(d) none

149. Which one is related to polarimetry
(a) Optical activity (b) concentration
(c) path length (d) all

150. Which one is related to AAS
(a) Monochromator (b) Lamp
(c) electrode (d) all

151. Which one is related to inductively coupled plasma
(a) ICP-OES (b) ICP-MS
(c) both (d) none

152. Which is related to mass spectroscopy analysis
(a) Chaneltron (b) cap electrodes
(c) ring electrodes (d) all

153. Which is related to mass spectroscopy analysis
(a) Electron gate
(b) emitting filament
(c) capillary tip
(d) all

154. Which of the following is a coupled mass spectroscopic technique
(a) GC-MS (b) LC-MS
(c) both (d) none

155. Which of the following is a coupled mass spectroscopic technique
(a) SFC-MS (b) CE-MS
(c) both (d) none

156. Which one is a component of chromatography
(a) Injector (b) column
(c) detector (d) all

157. Which one is a component of liquid chromatography instrument
(a) Solvent reservoirs
(b) injection device
(c) column (d) all

158. Which one is a component of liquid chromatography instrument
(a) Detector
(b) solvent delivery system
(c) data acquisition
(d) all

159. Which one is used in sediment test of raw milk
(a) Sediment tester
(b) sampling dipper
(c) sieves (d) all

160. COB test in milk is used to determine
(a) Developed acidity (b) fat
(c) T.S. (d) all

161. Alcohol test in milk is useful in indication of
(a) Fat content
(b) protein content
(c) mineral balance
(d) water content

162. MBRT is related to ______ analysis
(a) Milk
(b) fruits and vegetables
(c) cereals
(d) none

163. Butyrometer is used for ______ analysis
(a) Carbohydrate (b) fat
(c) protein (d) minerals

164. Hehner test is used for detection of ______ in milk
(a) Formaldehyde (b) boric acid
(c) water (d) none

165. Rosalic acid test is done for detection of
(a) Neutralizer (b) starch
(c) water (d) all

166. Bromothymol blue test is used for detection of
(a) Neutralizer (b) mastitis
(c) benzoic acid (d) all

167. Rose-Gottlieb method is related to analysis of
(a) Carbohydrate (b) protein
(c) fat (d) minerals

168. Kjeldahl method is related to analysis of
(a) Carbohydrate (b) protein
(c) fat (d) minerals

169. Polarimetric method is used for determination of
(a) Protein (b) fat
(c) lactose (d) all

170. Munson & Walker volumetric method is used for determination of
(a) Protein (b) fat
(c) lactose (d) all

171. Lane-Eynon method is used for determination of
(a) Protein (b) fat
(c) lactose (d) all

172. Which one is used in determination of sucrose
(a) HCl (b) NaOH
(c) ethyl alcohol (d) all

173. Which one is used in determination of copper
(a) Sulphuric acid
(b) nitric acid
(c) potassium iodide
(d) all

174. Which is not used in determination of phosphorus
(a) Ammonium hydroxide
(b) KI
(c) nitric acid
(d) phenolphthalein

175. Which one is used in determination of citric acid
(a) Tartaric acid (b) KBr
(c) sulphuric acid (d) all

176. Which test is used for boric acid detection
(a) Turmeric paper test
(b) glycerol test
(c) both
(d) none

177. Leach test is used for detection of ______
(a) Formaldehyde (b) boric acid
(c) benzoic acid (d) all

178. Hehner test is used for detection of ______
(a) Formaldehyde (b) boric acid
(c) benzoic acid (d) all

179. Ferric chloride test is used for detection of
(a) Salicyclic acid (b) benzoic acid
(c) formaldehyde (d) all

180. Hortvet method is used for determination of
(a) SNF content (b) fat content
(c) freezing point (d) none

181. Gerber method is used for the determination of
(a) Carbohydrate (b) protein
(c) fat (d) minerals

182. Which of the following is used in Gerber method
(a) Sulphuric acid (b) amyl alcohol
(c) butyrometer (d) all

183. Which of the following is a type of sampling
(a) Simple random
(b) stratified random
(c) systematic random
(d) all

184. Which of the following term is related to sampling
(a) Composite (b) gross sample
(c) increment (d) all

185. Which of the following term is related to sampling
(a) Bulk sampling (b) homogeneity
(c) lot (d) all

186. Which of the following term is related to sampling
(a) Population (b) laboratory sample
(c) both (d) none

187. Which of the following term is related to sampling
(a) Sample (b) segment
(c) strata (d) all

188. Which of the following term is related to sampling
(a) Subsample (b) test portion
(c) both (d) none

189. Which of the following is used in preparation of sample
(a) Tissue grinder
(b) ball mill
(c) ultracentrifugal mill
(d) all

190. Siemens is the unit of
Electric conductance
(b) electric capacitance
(c) electric potential
(d) electric resistance

191. Farad is the unit of
(a) Electric conductance
(b) electric capacitance
(c) electric potential
(d) electric resistance

192. Volt is the unit of
(a) Electric conductance
(b) electric capacitance
(c) electric potential
(d) electric resistance

193. Ohm is the unit of
(a) Electric conductance
(b) electric capacitance
(c) electric potential
(d) electric resistance

194. Joule is the unit of
(a) Energy (b) work
(c) quantity of heat (d) all

195. Newton is the unit of
(a) Force (b) frequency
(c) illuminance (d) inductance

196. Hertz is the unit of
(a) Force (b) frequency
(c) illuminance (d) inductance

197. Lux is the unit of
(a) Force (b) frequency
(c) illuminance (d) inductance

198. Henry is the unit of
(a) Force (b) frequency
(c) illuminance (d) inductance

199. Lumen is the unit of
(a) Luminous flux
(b) magnetic flux
(c) magnetic flux density
(d) power

200. Weber is the unit of
(a) Luminous flux
(b) magnetic flux
(c) magnetic flux density
(d) power

201. Tesla is the unit of
(a) Luminous flux
(b) magnetic flux
(c) magnetic flux density
(d) power

202. Watt is the unit of
(a) Luminous flux
(b) magnetic flux
(c) magnetic flux density
(d) power

203. Pascal is the unit of
(a) Pressure (b) stress
(c) both (d) none

204. Coulomb is the unit of
(a) Quantity of electricity
(b) electric charge
(c) both
(d) none

205. Deka refers to
(a) 10 (b) 100
(c) 1000 (d) 0.1

206. Hecto refers to
(a) 10 (b) 100
(c) 1000 (d) 0.1

207. Kilo refers to
(a) 10 (b) 100
(c) 1000 (d) 0.1

208. Deci refers to
(a) 10 (b) 100
(c) 1000 (d) 0.1

209. Centi refers to
(a) 0.01 (b) 0.001
(c) 10 (d) 100

210. Milli refers to
(a) 0.01 (b) 0.001
(c) 10 (d) 100

211. Mega refers to
(a) 10^6 (b) 10^9
(c) 10^{-6} (d) 10^{-9}

212. Giga refers to
(a) 10^6 (b) 10^9
(c) 10^{-6} (d) 10^{-9}

213. Micro refers to
(a) 10^6 (b) 10^9
(c) 10^{-6} (d) 10^{-9}

214. Nano refers to
(a) 10^6 (b) 10^9
(c) 10^{-6} (d) 10^{-9}

215. Pico refers to
(a) 10^{-9} (b) 10^{-12}
(c) 10^{-15} (d) 10^{-18}

216. Femto refers to
(a) 10^{-9} (b) 10^{-12}
(c) 10^{-15} (d) 10^{-18}

217. Atto refers to
(a) 10^{-9} (b) 10^{-12}
(c) 10^{-15} (d) 10^{-18}

218. Reliability of analytical methods depend on
(a) Specificity (b) accuracy
(c) both (d) none

219. Reliability of analytical methods depend on
(a) Precision (b) sensitivity
(c) both (d) none

220. C.V. refers to
(a) Coefficient of variation
(b) Coefficient of variable
(c) Coefficient of vectors
(d) all

221. Which is related to spectrophotometer
(a) UV-Vis (b) IR
(c) both (d) none

222. MRI stands for
(a) Magnetic Resonance Imaging
(b) Magnetic Resonance Imaging
(c) Magnetic Radioisotopic Imaging
(d) Magnetic Radioisotopic Induction

223. Which of the following is related to spectrophotometer
(a) Light source
(b) monochromator
(c) cuvette
(d) all

224. Which of the following is related to spectrophotometer
(a) Photoresistor (b) amplifier
(c) both (d) none

225. Spectrophotometric methods are based on
(a) Beer's law (b) Fick's law
(c) Power law (d) all

226. Ultrasonic wave have frequency (kHz) greater than
(a) 5 kHz (b) 10 kHz
(c) 20 kHz (d) 100 kHz

227. Which of the following is a testing geometry for dynamic rheometer
(a) Parallel plates
(b) cone & plate
(c) concentric cylinder
(d) all

228. Which one is related to large strain methodology
(a) Compression (b) torsion
(c) tension (d) all

229. Which of the following is an apparent elastic property
(a) Poison's ratio
(b) Young's modulus
(c) both (d) none

230. TPA refers to
(a) Texture Profile Analysis
(b) Total Protein Analysis
(c) Taste Profile Analysis
(d) none

231. Which of the following is related to TPA
(a) Cohesiveness (b) springiness
(c) hardness (d) all

232. Which of the following is related to texture analysis
(a) Uniaxial compression
(b) needle puncture
(c) cone penetration
(d) all

233. Which of the following is related to non-Newtonian flow
(a) Shear thinning
(b) shear thickening
(c) plastic
(d) all

234. Shear thinning behaviour is also known as ______ behaviour
(a) Pseudoplastic
(b) dilatant
(c) both (d) none

235. Shear thickening behaviour is also known as ______ behaviour
(a) Pseudoplastic (b) dilatant
(c) both (d) none

236. Which is not a type of viscometer
(a) Rotational (b) vibrating
(c) tube (d) all

237. Ultrasonic Doppler velocimetry is used for determining
(a) Viscosity (b) velocity
(c) current (d) all

238. Viscosity can be measured using
(a) Tomographic velocity profiling
(b) Ultrasonic Doppler velocimetry
(c) Ultrasonic reflectance sensor
(d) all

239. Crispness can be measured using ______ analysis
(a) Mechanical (b) acoustic
(c) ultrasonic (d) all

240. Image analysis can be used for analysis of
(a) Texture (b) appearance
(c) both (d) none

241. NMR refers to
(a) Nuclear Magnetic Resonance
(b) Neutral Magnetic Resonance
(c) Nuclear Molecular Resonance
(d) Neutral Molecular Resonance

242. Mealiness is related to lack of
(a) Juiciness (b) viscosity
(c) both (d) none

243. Wooliness is related to lack of
(a) Juiciness (b) viscosity
(c) both (d) none

244. Refractive index is ______ property of material
(a) Rheological (b) optical
(c) mechanical (d) none

245. Translucency is ______ property of material
(a) Rheological (b) optical
(c) mechanical (d) none

246. Gloss is ______ property of material
(a) Rheological (b) optical
(c) mechanical (d) none

247. Flourescence is ______ property of material
(a) Rheological (b) optical
(c) mechanical (d) none

248. Which of the following is wavelength dependent process
(a) Reflectance (b) absorption
(c) transmittance (d) all

249. Kubelka-Munk theory is related to
(a) Translucency (b) viscsosity
(c) water activity (d) all

250. Which of the following is related to surface plasmon resonance (SPR)
(a) Gold layer (b) Plasmon
(c) detector (d) all

251. Which of the following is an electrical property
(a) Conductance (b) capacitance
(c) loss tangent (d) all

252. Drying is related to
(a) Heat transfer (b) mass transfer
(c) both (d) none

253. Which is required to determine terminal velocity of solid foods
(a) Fan (b) flow variator
(c) floating zone (d) all

254. Drag coefficient of corn is
(a) 0.56 (b) 0.45
(c) 0.50 (d) 0.52

255. Drag coefficient of soybeans is
(a) 0.56 (b) 0.45
(c) 0.50 (d) 0.52

256. Drag coefficient of Alfaalfa seeds is
(a) 0.56 (b) 0.45
(c) 0.50 (d) 0.52

257. Drag coefficient of cotton seeds is
(a) 0.56 (b) 0.45
(c) 0.50 (d) 0.52

258. Drag coefficient of barley seeds is
(a) 0.56 (b) 0.45
(c) 0.50 (d) 0.52

259. Electrical conductivity (mS/cm) of pear is
(a) 0.41 (b) 0.23
(c) 0.38 (d) 0.26

260. Electrical conductivity (mS/cm) of apple is
(a) 0.41 (b) 0.23
(c) 0.38 (d) 0.26

261. Electrical conductivity (mS/cm) of potato is
(a) 0.41 (b) 0.23
(c) 0.38 (d) 0.26

262. Electrical conductivity (mS/cm) of turnip is
(a) 0.41 (b) 0.23
(c) 0.38 (d) 0.26

263. Electrical conductivity (mS/cm) of onion is
(a) 0.22 (b) 0.45
(c) 0.48 (d) 0.80

264. Electrical conductivity (mS/cm) of tomato is
(a) 0.22 (b) 0.45
(c) 0.48 (d) 0.80

265. Electrical conductivity (mS/cm) of pepper is
(a) 0.22 (b) 0.45
(c) 0.48 (d) 0.80

266. Electrical conductivity (mS/cm) of mushrooms is
(a) 0.22 (b) 0.45
(c) 0.48 (d) 0.80

267. Electrical conductivity (mS/cm) of chicken is
(a) 0.22 (b) 0.45
(c) 0.48 (d) 0.80

268. Electrical conductivity (mS/cm) of egg white is
(a) 0.21 (b) 0.37
(c) 0.41 (d) 0.71

269. Electrical conductivity (mS/cm) of egg yolk is
(a) 0.21 (b) 0.37
(c) 0.41 (d) 0.71

270. Image analysis can be used to analyze
(a) Shape
(b) volume
(c) mass estimation
(d) all

271. Image analysis can be used to analyze
(a) Color (b) texture
(c) mass (d) all

272. TDRS is
(a) Time-resolved diffuse reflectance spectroscopy
(b) Time diffuse reflectance spectroscopy
(c) Temperature diffuse reflectance spectroscopy
(d) none

273. In grains analysis, ratio of maximum length & maximum width is known as
(a) Perimeter (b) area
(c) elongation (d) none

274. Which of the following is related to grains analysis
(a) Compactness (b) elongation
(c) both (d) none

275. Which of the following is related to grains analysis
(a) Particle mass
(b) polishing percentage
(c) maximum length
(d) all

276. Which of the following is a hygroscopic property
(a) Equilibrium moisture content
(b) water diffusivity
(c) grain specific weight
(d) all

277. Which of the following is used in wheat analysis
(a) Farinograph (b) mixograph
(c) consistograph (d) all

278. Which of the following is used in wheat analysis
(a) Extensograph
(b) alveograph
(c) amylograph
(d) all

279. AOAC international begun in _______ year
(a) 1884 (b) 1895
(c) 1906 (d) 1917

280. Federal Food, Drug and Cosmetic (FD&C) Act came into existence in
(a) 1906 (b) 1928
(c) 1936 (d) 1948

281. Food & Drug Act came into existence in
(a) 1906 (b) 1928
(c) 1936 (d) 1948

282. Food Quality Protection Act came into existence in
(a) 1966 (b) 1976
(c) 1986 (d) 1996

283. Nutrition Labelling & Education Act came into existence in
(a) 1970 (b) 1980
(c) 1990 (d) 1860

284. Dietary Supplement Health & Education Act came into existence in
(a) 1991 (b) 1994
(c) 1998 (d) 2001

285. Codex Alimentarius Commission was established in
(a) 1962 (b) 1967
(c) 1972 (d) 1977

286. Daily value is a generic term used to describe
(a) Reference daily intake
(b) daily reference value
(c) both
(d) none

287. Which of the following is a non-probability sampling
(a) Composite sampling
(b) cluster sampling
(c) stratified sampling
(d) judgement sampling

288. Which of the following is a non-probability sampling
(a) Simple random
(b) systematic
(c) stratified
(d) convenience sampling

289. Which of the following is a non-probability sampling
(a) Restricted (b) quarter
(c) both (d) none

290. Which of the following is probability sampling
(a) Stratified (b) cluster
(c) composite (d) all

291. Which of the following is a state of water in food product
(a) Free water
(b) adsorbed water
(c) water of hydration
(d) all

292. Approximate % moisture in raw watermelon
(a) 63.2 (b) 91.5
(c) 83.9 (d) 75.3

293. Approximate % moisture in beef
(a) 63.2 (b) 91.5
(c) 83.9 (d) 75.3

294. Approximate % moisture in egg
(a) 63.2 (b) 91.5
(c) 83.9 (d) 75.3

295. Approximate % moisture in raw apples
(a) 63.2 (b) 91.5
(c) 83.9 (d) 75.3

296. Approximate % moisture in walnuts
(a) 1.6 (b) 4.4
(c) 15.4 (d) 17.1

297. Approximate % moisture in brown sugar
(a) 1.6 (b) 4.4
(c) 15.4 (d) 17.1

298. Approximate % moisture in raisins
(a) 1.6 (b) 4.4
(c) 15.4 (d) 17.1

299. Approximate % moisture in honey
(a) 1.6 (b) 4.4
(c) 15.4 (d) 17.1

300. Hydrometry is used for measuring
(a) Specific gravity (b) density
(c) both (d) none

301. Pycnometer is used for measuring
(a) Specific gravity
(b) water activity
(c) refractive index
(d) none

302. Which of the following is a type of hydrometer
(a) Lactometer
(b) alcoholometer
(c) Baume hydrometer
(d) all

303. Westphal Balance function on ________ principle/law
(a) Beer law
(b) Lamberts law
(c) Archimedes
(d) all

304. Hydrometer function on ________ principle/law
(a) Beer law (b) Lamberts law
(c) Archimedes (d) all

305. Which of the following is a low resolution NMR
(a) Frequency domain NMR
(b) continuous wave NMR
(c) Pulsed NMR
(d) all

306. Which of the following is related to protein analysis method
(a) Kjeldahl
(b) Dumas
(c) Infrared spectroscopy
(d) all

307. Which of the following is related to protein analysis method
(a) Biuret method
(b) Lowry method
(c) Bradford Dye-binding method
(d) all

308. Which of the following is related to protein analysis method
(a) Bicinchoninic acid method
(b) UV 280nm absorption method
(c) both
(d) none

309. BCA method is used for ________ analysis
(a) Carbohydrate (b) protein
(c) fat (d) minerals

310. Somogyi-Nelson method is used for determining
(a) Carbohydrate (b) protein
(c) fat (d) minerals

311. Iodine value is related to______ analysis
(a) Carbohydrate (b) protein
(c) fat (d) minerals

312. Cloud point is related to ________ analysis
(a) Carbohydrate (b) protein
(c) fat (d) minerals

313. Which of the following is related to fat analysis
(a) Saponification value
(b) FFA
(c) cold test
(d) all

314. Peroxide value is related to ________ analysis
(a) Carbohydrate (b) protein
(c) fat (d) minerals

315. Which of the following is related to fat oxidation analysis
(a) Peroxide value (b) anisidine value
(c) totox value (d) all

316. Which of the following is related to fat oxidation analysis
(a) Thiobarbituric acid
(b) Oil stability index
(c) both
(d) none

317. Which of the following is type of electrophoresis
(a) Zone (b) paper
(c) disc gel (d) all

318. esu is a unit of
(a) Charge (b) density
(c) volume (d) none

319. Electrophoresis is used for analysis
(a) Carbohydrate (b) protein
(c) fat (d) none

320. Which one is objective method of consistency measurement
(a) Consistometer (b) flowmeter
(c) spread meter (d) all

321. Which of the following is a lipid
(a) Sphingomyelin (b) cerebroside
(c) ganglioside (d) all

322. Which solvent is used in fat extraction
(a) Petroleum ether (b) butanol
(c) chloroform (d) all

323. Which solvent is not used in fat extraction
(a) Methanol (b) chloroform
(c) water (d) none

324. Cerimetric methods are used for determination of
(a) Sugar (b) protein
(c) fat (d) all

325. Iodometric methods are used for determination of
(a) Sugar (b) protein
(c) fat (d) all

326. Alkaline ferricyanide method is used for determination of
(a) Sugar (b) protein
(c) fat (d) all

327. Emission spectroscopy is used for analysis of
(a) Carbohydrate (b) protein
(c) fat (d) mineral

328. Flame photometry is used for analysis of
(a) Carbohydrate (b) protein
(c) fat (d) mineral

329. Atomic absorption spectroscopy is used for analysis of
(a) Carbohydrate (b) protein
(c) fat (d) mineral

330. Which of the following is used in moisture determination
(a) Karl Fisher titration
(b) Infrared determination
(c) NMR method
(d) all

12.3 ANSWER KEY

Q.	A.	Q.	A.	Q.	A.	Q.	A.	Q.	A.	Q.	A.
1.	(d)	2.	(c)	3.	(a)	4.	(d)	5.	(b)	6.	(b)
7.	(a)	8.	(a)	9.	(a)	10.	(a)	11.	(a)	12.	(a)
13.	(a)	14.	(a)	15.	(b)	16.	(a)	17.	(a)	18.	(b)
19.	(b)	20.	(a)	21.	(a)	22.	(b)	23.	(b)	24.	(a)
25.	(a)	26.	(b)	27.	(b)	28.	(a)	29.	(b)	30.	(d)
31.	(d)	32.	(d)	33.	(c)	34.	(b)	35.	(d)	36.	(d)
37.	(d)	38.	(d)	39.	(d)	40.	(d)	41.	(c)	42.	(d)
43.	(b)	44.	(b)	45.	(d)	46.	(d)	47.	(d)	48.	(a)
49.	(d)	50.	(d)	51.	(c)	52.	(a)	53.	(a)	54.	(a)
55.	(a)	56.	(d)	57.	(d)	58.	(a)	59.	(c)	60.	(d)
61.	(d)	62.	(c)	63.	(d)	64.	(d)	65.	(c)	66.	(a)
67.	(c)	68.	(d)	69.	(a)	70.	(d)	71.	(c)	72.	(b)
73.	(b)	74.	(d)	75.	(b)	76.	(a)	77.	(c)	78.	(a)
79.	(c)	80.	(b)	81.	(d)	82.	(c)	83.	(a)	84.	(b)
85.	(c)	86.	(a)	87.	(b)	88.	(d)	89.	(c)	90.	(c)
91.	(d)	92.	(d)	93.	(d)	94.	(c)	95.	(c)	96.	(c)
97.	(d)	98.	(c)	99.	(c)	100.	(d)	101.	(d)	102.	(b)
103.	(d)	104.	(c)	105.	(d)	106.	(c)	107.	(c)	108.	(a)
109.	(d)	110.	(d)	111.	(c)	112.	(d)	113.	(a)	114.	(a)
115.	(d)	116.	(a)	117.	(c)	118.	(a)	119.	(b)	120.	(a)
121.	(b)	122.	(a)	123.	(c)	124.	(a)	125.	(d)	126.	(b)
127.	(b)	128.	(b)	129.	(a)	130.	(a)	131.	(c)	132.	(c)
133.	(a)	134.	(b)	135.	(a)	136.	(d)	137.	(d)	138.	(b)
139.	(b)	140.	(a)	141.	(b)	142.	(d)	143.	(a)	144.	(c)
145.	(b)	146.	(a)	147.	(b)	148.	(c)	149.	(d)	150.	(d)

Q. A.	Q. A.	Q. A.	Q. A.	Q. A.	Q. A.
151. (c)	152. (d)	153. (d)	154. (c)	155. (c)	156. (d)
157. (d)	158. (d)	159. (d)	160. (a)	161. (c)	162. (a)
163. (b)	164. (a)	165. (a)	166. (b)	167. (c)	168. (b)
169. (c)	170. (c)	171. (c)	172. (d)	173. (d)	174. (b)
175. (d)	176. (c)	177. (a)	178. (a)	179. (d)	180. (c)
181. (c)	182. (d)	183. (d)	184. (d)	185. (d)	186. (c)
187. (d)	188. (c)	189. (d)	190. (a)	191. (b)	192. (c)
193. (d)	194. (d)	195. (a)	196. (b)	197. (c)	198. (d)
199. (a)	200. (b)	201. (c)	202. (d)	203. (c)	204. (c)
205. (a)	206. (b)	207. (c)	208. (d)	209. (a)	210. (b)
211. (a)	212. (b)	213. (c)	214. (d)	215. (b)	216. (c)
217. (d)	218. (c)	219. (c)	220. (a)	221. (c)	222. (a)
223. (d)	224. (c)	225. (a)	226. (c)	227. (d)	228. (d)
229. (c)	230. (a)	231. (d)	232. (d)	233. (d)	234. (a)
235. (b)	236. (d)	237. (a)	238. (d)	239. (d)	240. (c)
241. (a)	242. (a)	243. (a)	244. (b)	245. (b)	246. (b)
247. (b)	248. (d)	249. (a)	250. (d)	251. (d)	252. (c)
253. (d)	254. (a)	255. (b)	256. (c)	257. (d)	258. (c)
259. (a)	260. (b)	261. (c)	262. (d)	263. (a)	264. (b)
265. (c)	266. (a)	267. (d)	268. (d)	269. (b)	270. (d)
271. (d)	272. (a)	273. (c)	274. (c)	275. (d)	276. (d)
277. (d)	278. (d)	279. (a)	280. (c)	281. (a)	282. (d)
283. (c)	284. (b)	285. (a)	286. (c)	287. (d)	288. (d)
289. (c)	290. (d)	291. (d)	292. (b)	293. (a)	294. (d)
295. (c)	296. (b)	297. (a)	298. (c)	299. (d)	300. (c)
301. (a)	302. (d)	303. (c)	304. (c)	305. (d)	306. (d)
307. (d)	308. (c)	309. (b)	310. (a)	311. (c)	312. (c)
313. (d)	314. (c)	315. (d)	316. (c)	317. (d)	318. (a)
319. (b)	320. (d)	321. (d)	322. (d)	323. (c)	324. (a)
325. (a)	326. (a)	327. (d)	328. (d)	329. (d)	330. (d)

CHAPTER 13

FUNCTIONAL FOODS

13.1 SUBJECTIVE KEYNOTES

1. Hippocrates is the father of medicine.
2. The term 'functional foods' comprises some bacterial strains and products of plant and animal origin containing physiologically active compounds beneficial for human health and reducing the risk of chronic disease.
3. The term functional food was first introduced in Japan in mid 1980's.
4. The functional food comprises of conventional foods containing naturally occurring bioactive substances, food enriched with bioactive substances and synthesized food ingredients introduced to traditional foods.
5. Functional components include probiotics, prebiotics, soluble fiber, omega-3-poly unsaturated fatty acids, conjugated linoleic acid, plant antioxidants, vitamins and minerals, some proteins, peptides, amino acids as well as phospholipids are frequently mentioned.
6. Functional foods are not pills or capsules but are consumed as part of a normal everyday diet.
7. Most intensively investigated class of physiologically active components derived from animal products are the (n-3) fatty acids, predominantly found in fatty fish such as salmon, tuna etc.
8. The two primary (n-3) fatty acids are EPA and DHA.
9. DHA is an essential component of the phospholipids of cellular membranes, especially in the brain and retina of the eye and is necessary for their proper functioning.
10. The use of (n-3) fatty acid supplements is safe provided daily intake of EPA and DHA from supplements do not exceeds 2 g per day.
11. Probiotics are defined as selected, viable microbial dietary supplements that when introduced in sufficient quantities beneficially affect human organism through their effects in the intestinal.
12. Therapeutic and preventive use of these functional ingredients for various health concerns including cancer, intestinal tract function, immune function, allergy, stomach health, urogenital health, cholesterol lowering hypertension.
13. Prebiotics are non digestible or low digestible food ingredients that benefit the host organism by selectively stimulating the growth or activity of one or a limited number of probiotic bacteria in the colon.

14. Lactulose, GOS, FOS, inulin and its hydrolysates, malto-oligosaccharides and resistant starch are prebiotics commonly used in human nutrition.
15. The main end products of carbohydrate metabolism are SCFA namely acetate, butyrate and propionate which have positive effects on colonic cell growth and stability.
16. The term polyphenol includes simple phenols and flavonoids, which are found in fruits and vegetables and nuts and their products and possess important antioxidant properties.
17. Flavonoids include proanthrocyanidins, quercetin and epicatechin found mainly in chocolate, tea and wine.
18. Red wine contains resveratrol, a non polyphenol anti-oxidant product of grape skins.
19. In addition to antioxidant effects on LDL, other potentially cardio protective effects of polyphenols include inhibition of platelet aggregation and vascular relaxation through the production of nitric oxide.
20. Green tea is particularly abundant in specific polyphenolic components known as catechins.
21. The major catechins in green tea are epicatechin, epicatechin-3-gallats, epigallo-catechin and epigallocatechin-3-gallate (EGCG).
22. Soy isoflavones are phytoesterogens that are derived from the protein fraction of the soybean and its food products (e.g., soy milk, soy flour, tofu, miso) include genistein and daidzein, and possess estrogenic properties because of the similarities of their chemical structures to estrogenic compounds.
23. Soy isoflavones helps in the prevention of heart disease, osteoporosis, breast and prostate cancer.
24. There are severed plant derived carotenoids in the human diet, of which ß-carotene, α-carotene, leutin, zeaxanthin and lycopene appears to have most significance for health.
25. Beta-carotene is a limited precursor of vitamin A, and excessive amounts of ß carotene lead to reversible carotenemia but not to vitamin A toxicity.
26. Free radicals are a major cause of many degenerative diseases, such as atherosclerosis, cancer, cardiovascular diseases, inflammatory bowel disease, skin aging, old age dementia and arthritis.
27. Free radicals can damage DNA and cause mutagenicity and cytotoxicity.
28. Fibre fermentation products also induce death of intestinal oncogenic cells, decreasing the risk of gastro intestinal tumours.
29. Phytochemicals are naturally occurring, biologically active chemical compounds in plants.
30. Phytochemicals are protective and disease-preventing particularly for some forms of cancer and heart diseases.
31. Phytochemicals include the widely studied, flavanoids, phyto-oestrogens, phytosterols and carotenoids.
32. Flavanoids enhance the effects of vitamin C and function as antioxidants.

33. Flavonoids are also known to be biologically active against liver toxins, tumours, viruses and other microbes, allergies and inflammation.
34. Hesperidin raises blood levels of the "good" cholesterol and lowers the "bad" cholesterol.
35. Querecitin reduces inflammation associated with allergies.
36. Tangeretin induces programmed cell death in cancer cells (leukemia) but does not affect normal healthy cells.
37. Resveratol may reduce the risk of blood clots, cancer, heart disease and stroke.
38. Anthocyanins have been shown to improve balance coordination and short term me mory in the elderly.
39. Isoflavone, a phyto-oestrogen, is associated with cancer prevention, improved cardiovascular health and improved bone health.
40. Isoflavones contained in soy are believed to be helpful in the treatment of menopausal symptoms.
41. Phytosterols reduce the amount of dietary cholesterol absorbed by the body by blocking uptake in the intestine.
42. Phytosterols also facilitate cholesterol excretion from the body.
43. Carotenoids are plant pigments found in bright yellow, orange and red fruits and vegetables.
44. Carotenoids are generally well known as vitamin A precursors.
45. In the carotene group only alpha, beta and epsilon carotene function as vitamin A precursors.
46. Xanthophylls protect vitamin A, other carotenoids and vitamin E from oxidation.
47. Zeaxanthin and lutein may help to prevent loss of sight in persons over fifty years.
48. Xanthophyll protects the skin from sunlight; it was once popular as a tanning product.
49. Cryptoxanthin is thought to have a protective effect on the female reproductive tissues, namely vaginal, uterine and cervical tissues.
50. Phytochemicals are also found to boost cancer-fighting enzymes, promote growth of beneficial bacteria and help in blood clotting.
51. Supercritical extraction is a novel separation technique that utilizes the solvent properties of fluids near their thermodynamic critical points.
52. Supercritical fluids such as carbon dioxide, ethane, propane, butane, pentane, ethylene, ammonia, sulfur dioxide, water, chlorodifluoromethane, etc. are used in supercritical extraction processes.
53. Supercritical fluid extraction technology have applications in decaffeinating coffee, tea, and extraction of essential oils from spices and bioactive components from various food sources which are used as ingredients in foods.
54. Supercritical fluid extraction technology can also be used in development of low or cholesterol free food products.
55. Pressurized low polarity water (PLPW) extraction is a technology that modifies the properties of water by increasing the temperature up to 374°C and keeping the

pressure high enough to maintain the water in the liquid state to improve its extraction ability. It is also known as subcritical water extraction technology.

56. Distillation is one of the main methods of extracting essential oils from plants. Distillation is a process of heating a substance until the most volatile constituents change into the vapour phase, and then cooling the vapours to recover the constituents in liquid form by condensation.

57. Vacuum distillation is a method of distilling substances at temperatures below their normal boiling point. High vacuum distillation may be used for certain classes of chemical compounds that decompose, polymerize, react, or are destroyed by conventional distillation methods.

58. Dietary fiber reduces the caloric density of foods by physical dilution and by entrapment of nutrients, which reduces the rate of digestion and absorption (bioavailability) of nutrients, especially glucose and lipids.

59. Soluble dietary fibers that form viscous hydrated masses suppress appetite better than the less viscous fibers. This is due to the large amount within the fiber that increases stomach distension and triggers the feeling of fullness.

60. Short-chain fatty acids produced during fermentation of fibers in the colon have appetite-suppressing effect.

61. Dietary fiber promotes increased fecal bulk, which helps to trap and remove bile acids and cholesterols from the intestinal tract. Dietary fibers also helps in reducing rate of reabsorption of cholesterol which leads to lower plasma levels of LDL cholesterol and triglycerides, which reduces the risk of atherosclerosis and associated cardiovascular diseases.

62. There is an inverse relationship between dietary fiber consumption and breast cancer. Fiber within the intestinal lumen retards reabsorption of estrogens, which increases fecal estrogen level but leads to lower levels of blood and urine estrogen.

63. Soluble dietary fibers mix very well with water to form highly hydrated masses that are almost completely fermented in the colon by microorganisms.

64. Pectin is largely undigested in the upper GIT but is easily broken down through fermentation by colon microorganisms.

65. The basic mechanism behind the hypocholesterolemic effects of guar gum is the ability to reduce concentration of free cholesterol in the liver.

66. The mechanism for the physiological effects of beta-glucan is related to its ability to form a hydrated viscous mass in the GIT. The increased viscosity of GIT contents leads to trapping and reduced absorption of glucose and bile acids, which reduces their plasma levels and enhances excretion in the feces.

67. Resistant starch refers to a starch fraction and its degradation products that are resistant to enzyme digestion and will pass unchanged and unabsorbed from the stomach to the small intestine of healthy individuals. In the large intestine, the resistant starch particles are fermented to variable extents by colonic micro flora.

68. Potential health benefits of prebiotics include increased bioavailability of minerals, and reduced risks of various diseases such as cancer, intestinal infections, cardiovascular disorders, obesity, and diabetes.

69. Some prebiotics such as arabinoxylans and arabinoxylan-oligosaccharides contain covalently bound hydroxycinnamic acids (e.g. ferulic acid) and exhibits anti-oxidant properties.
70. Prebiotic components also have immune stimulating effects. Inulin and other fructooligosaccharides have been shown to stimulate activity of natural killer cells and enhance SCFA-mediated anti-inflammation responses.
71. Prebiotic components also shown to have anticarcinogenic effects.
72. Inulin, also known as raftiline, is found in several foods such as wheat, onion, garlic, bananas, fruits, and vegetables, but industrial production utilizes chicory roots.
73. Oligofructose, also known as raftilose, is obtained by enzymatic hydrolysis of inulin to give a product composed of linear G-Fn and Fn chains with DP ranging from 2 to 8.
74. Consumption of oligofructose-enriched inulin promotes improved dietary calcium absorption and could be used in the formulation of functional foods for the prevention of osteoporosis.
75. Increased solubility of minerals, enlargement of absorption surface, and increased expression of calcium binding proteins are some of the mechanisms that have been proposed to explain the enhanced mineral absorption associated with dietary prebiotics.
76. Lactulose is a synthetic disaccharide analog of lactose consisting of fructose chemically bonded to galactose. Lactulose does not occur naturally but is produced by isomerization of lactose in basic media and in the presence of different types of catalyst. Lactulose can be used as prebiotic.
77. One of the key aspects of the beneficial effects of dietary fibers, prebiotics, and indigestible polyphenolic compounds is their ability to induce production of large quantities of SCFAs in the colon.
78. Butyrate is commonly found as part of the lipid component of dairy milk but is also one of the main by-products of fiber fermentation in the colon and has been shown to induce various beneficial metabolic effects.
79. Medium chain triglycerides diets also reduced blood levels of several types of LDL as well as LDL-cholesterol to greater extent than traditional oil that contained long-chain triglycerides.
80. The simplest omega-6 fatty acid is linoleic acid (C18:2), while linolenic acid (C18:3) is the simplest omega-3 fatty acid. Both fatty acids have been reported to protect against cardiovascular and inflammatory diseases, though linolenic acid has greater health benefits.
81. Combinations of docosahexaenoic acid (DHA) and eicosapentaenoic acid (EPA) with alpha -linolenic acid can lower the risk of fatal ischemic heart disease in older adults.
82. Fish oil contains a high n-3:n-6 ratio and has been shown to decrease serum triglyceride and cholesterol levels.
83. Omega-3 fatty acids are also reported to alter the metabolic pathways in skeletal muscle and promote protein synthesis to maintain lean muscle mass.
84. Conjugated Linoleic Acid is a mixture of positional and geometrical isomers of linoleic acid formed by rumen microorganisms. CLA occurs naturally and can be found at low levels in ruminant fats such as beef tallow and milk fat.

85. CLA can also be synthesized from linoleic acid or vegetable oils that have high levels of linoleic acid such as corn, canola, soybean, safflower, and sunflower.
86. Conjugated Eicosapentaenoic Acid (CEPA) can be prepared by alkaline treatment of eicosapentaenoic acid (EPA) and has been shown to induce strong and selective in vitro apoptosis of tumor cells through a lipid peroxidation mechanism.
87. Bioactive peptides are short-chain protein molecules (usually <20 amino acid residues) that when ingested can provide physiological benefits such as reduction in the risk of cardiovascular diseases, negative modulation of tumor growth, and reduction in blood sugar level.
88. Low molecular weight peptides are desirable because they can be resistant to enzymatic hydrolysis during passage through the gastrointestinal tract.
89. Commercially, bioactive peptides are produced in vitro from customized enzymatic protein hydrolysis, and they provide specialized therapeutic support to patients with particular physiological and nutritional needs.
90. Opioid peptides are the peptides that possess some form of affinity for opiate receptors and can exert effects on the nervous system. Naturally occurring opioid peptides in the body are called endorphins and enkephalins while opioid peptides obtained from enzymatic digests of food proteins are called exorphins.
91. Caseinophosphopeptides are phosphorylated peptide fragments obtained from enzymatic hydrolysis of casein (milk protein), which is known to contain several phosphoserine residues.
92. Calmodulin is a calcium-activated protein that plays important roles in maintaining physiological functions of cells and body organs including cell division, cell proliferation, and neurotransmission.
93. Antithrombotic peptides are peptides that inhibit blood clotting and may be used to treat coagulation of blood in the circulatory system e.g. casoplatelin, casopiastrin etc.
94. Resveratrol is a non-flavonoid polyphenolic compound found in the skin of dark-colored grapes and products made from such grapes e.g. wine and grape juice.
95. Grape seeds contain several polyphenolic components, especially the proanthocyanidins or condensed tannins that are considered to possess health promoting properties.
96. Glycomacropeptide which is released from kappa–casein by rennet or chymosin during cheese making, is believed to play a role in satiety by stimulating increased production of cholecystokinin in the intestine.
97. Whole-grain flaxseed contains bioactive fatty acids and lignan (secoisolariciresinol diglucoside).
98. Tea consists of mostly the polyphenols called flavonoids that can be as high as 150 mg in a 150-mL cup of brewed black tea.
99. Epigallo-catechin gallate is a potential agent that can reduce the development and severity of Alzheimer's disease because of its ability to inhibit aggregation of amyloid-β, which is a major pathogenic factor in the disease.
100. Nuts contain many phytochemicals (flavonoids, phenolic compounds, isoflavones, sterols, etc.), which have been shown to be inversely related to coronary heart disease.

13.2 MULTIPLE CHOICE QUESTIONS

1. Which of the following belongs to functional food
(a) Carbohydrate (b) protein
(c) prebiotic (d) none

2. Which of the following belongs to functional food
(a) Carbohydrate (b) protein
(c) probiotic (d) none

3. In functional foods, probiotics are
(a) Carbohydrate (b) microorganism
(c) protein (d) none

4. In functional foods, prebiotics are
(a) Carbohydrate (b) microorganism
(c) protein (d) none

5. Which of the following is a prebiotic
(a) LAB (b) Bifidobacteria
(c) both (d) none

6. Which of the following is a probiotic
(a) LAB (b) Bifidobacteria
(c) both (d) none

7. Which of the following is a function of probiotic
(a) Improve intestinal microflora
(b) reduce diarrhea
(c) both
(d) none

8. Which of the following is a function of probiotic
(a) Reduce constipation
(b) reduce cholesterol
(c) both (d) none

9. Which of the following is a prebiotic
(a) Oligosaccharides
(b) resistant starch
(c) pectin (d) all

10. Which of the following is a function of probiotic
(a) Improve intestinal microflora
(b) reduce osteoporosis
(c) both (d) none

11. Which of the following can act as functional food ingredient
(a) Probiotics (b) prebiotic
(c) vitamins (d) all

12. Which of the following can act as functional food ingredient
(a) Proteins (b) antioxidants
(c) minerals (d) all

13. Which of the following can act as functional food ingredient
(a) Fatty acids (b) phytochemicals
(c) both (d) none

14. Folic acid intake reduces risk of
(a) Cardiovascular diseases
(b) osteoporosis
(c) both
(d) none

15. Vitamin K & vitamin B_6 intake reduces risk of
(a) Cardiovascular diseases
(b) osteoporosis
(c) both
(d) none

16. Minerals such as Ca & Mg perform following function
(a) Reduce osteoporosis
(b) strengthen immune system
(c) both
(d) none

17. Minerals such as Zinc & Selenium perform following function
(a) Reduce osteoporosis
(b) strengthen immune system
(c) both
(d) none

18. Which of the following is an antioxidant
(a) Vitamin E (b) vitamin C
(c) carotenoids (d) all

19. Which of the following is an antioxidant
(a) Flavonoids (b) tea polyphenols
(c) both (d) all

20. Antioxidant intake reduces risk of
(a) Atherosclerosis (b) cancer
(c) both (d) none

21. Antioxidant intake reduces risk of
(a) Oxidative DNA damage
(b) ageing
(c) both
(d) none

22. Which of the following is a function of antioxidant
(a) Reduce atherosclerosis
(b) anti-inflammatory
(c) anti-ageing
(d) all

23. Which of the following can act as functional food ingredient
(a) Tocopherol (b) tocotrienol
(c) ascorbic acid (d) all

24. Which of the following can act as functional food ingredient
(a) Alpha-carotene (b) beta-carotene
(c) both (d) none

25. Which of the following can act as functional food ingredient
(a) Lycopene (b) lutein
(c) zeaxanthin (d) all

26. Which of the following can act as functional food ingredient
(a) Cryptoxanthin (b) zeaxanthin
(c) both (d) none

27. Which of the following can act as functional food ingredient
(a) Anthocyanin (b) cyanidin
(c) quercetin (d) all

28. Which of the following can act as functional food ingredient
(a) Catechin (b) resveratrol
(c) both (d) none

29. Which of the following can act as functional food ingredient
(a) Isoflavonoids (b) gernistein
(c) daidzein (d) all

30. Which of the following can act as functional food ingredient
(a) Glycitein (b) coumestrol
(c) both (d) none

31. Tri-peptides (milk) intake can perform following function
(a) Reduce B.P.
(b) influence cognitive function
(c) both
(d) none

32. Which of the following fatty acid can act as functional food ingredient
(a) Omega-3 fatty acid
(b) GLA
(c) CLA
(d) all

33. Fatty acid like GLA & CLA intake reduces risk of
(a) Cardiovascular disease
(b) arthritis
(c) cancer
(d) all

34. Omega-3 fatty acid intake reduces risk of
(a) Cardiovascular disease
(b) arthritis
(c) cancer
(d) all

35. Which of the following is a phyto-chemical
(a) Phytosterol (b) beta-glucan
(c) isoflavones (d) all

36. Which of the following is a phyto-chemical
(a) Lignans
(b) green tea polyphenols
(c) both
(d) none

37. Phytosterols intake reduces
(a) Serum cholesterol
(b) cancer development
(c) both
(d) none

38. Green tea polyphenol intake reduces
(a) Serum cholesterol
(b) cancer development
(c) both
(d) none

39. Which of the following is a source of tocopherols
(a) Oilseeds (b) cereal oil
(c) nuts (d) all

40. Which of the following is a source of tocotrienols
(a) Oilseeds (b) cereal oil
(c) nuts (d) all

41. Which of the following is a source of ascorbic acid
(a) Acerola cherry (b) citrus
(c) both (d) none

42. Which of the following is not a carotenoid
(a) Lycopene (b) lutein
(c) glycitein (d) none

43. Which of the following is not a carotenoid
(a) Beta-carotene (b) zeaxanthin
(c) daidzein (d) none

44. Which of the following is not a carotenoid
(a) Alpha-carotene (b) gernistein
(c) cryptoxanthin (d) none

45. Which of the following is a carotenoid
(a) Lycopene (b) lutein
(c) beta-carotene (d) all

46. Which of the following is a carotenoid
(a) Zeaxanthin
(b) alpha-carotene
(c) cryptoxanthin
(d) all

47. Which of the following is a isoflavonoid
(a) Gernistein (b) anthocyanin
(c) cyanidins (d) all

48. Which of the following is a isoflavonoid
(a) Daidzein (b) catechin
(c) quercetin (d) all

49. Which of the following is a isoflavonoid
(a) Resveratrol (b) glycetein
(c) cyanidin (d) all

50. Which of the following is a isoflavonoid
(a) Gernistein (b) daidzein
(c) both (d) none

51. Which of the following is a isoflavonoids
(a) Glycetein (b) coumestrol
(c) both (d) none

52. Which of the following is a flavonoid
(a) Lycopene (b) cyanidin
(c) lutein (d) all

53. Which of the following is a flavonoid
(a) Zeaxanthin (b) quercitin
(c) lutein (d) all

54. Which of the following is a flavonoid
(a) Cryptoxanthin (b) catechin
(c) gernistein (d) all

55. Which of the following is a flavonoid
(a) Resveratrol (b) carotene
(c) daidzein (d) all

56. Which of the following is a flavonoid
(a) Anthocyanin (b) cyanidin
(c) both (d) none

57. Which of the following is a flavonoid
(a) Quercitin
(b) catechin
(c) resveratrol
(d) all

58. Which of the following is a good source of alpha-carotene
(a) Carrot (b) tomatoes
(c) spinach (d) none

59. Which of the following is a good source of beta-carotene
(a) Carrot (b) tomatoes
(c) spinach (d) none

60. Which of the following is a good source of lycopene
(a) Carrot (b) tomatoes
(c) spinach (d) none

61. Which of the following is a good source of lutein
(a) Carrot (b) tomatoes
(c) spinach (d) none

62. Which of the following is a good source of zeaxanthin
(a) Maize (b) citrus
(c) spinach (d) none

63. Which of the following is a good source of cryptoxanthin
(a) Maize (b) citrus
(c) spinach (d) none

64. Which of the following is a good source of resveratrol
(a) Grapes (b) maize
(c) spinach (d) none

65. Which of the following is a good source of resveratrol
(a) wheat (b) maize
(c) wine (d) none

66. Which of the following is a good source of catechin
(a) Green tea (b) maize
(c) spinach (d) none

67. Which of the following is a good source of anthocyanins
(a) berries (b) oilseed
(c) nuts (d) none

68. Which of the following is a good source of isoflavonoids
(a) Sugarcane (b) soy
(c) spinach (d) none

69. Which of the following is a good source of lignans
(a) Flaxseed (b) soy
(c) rye (d) all

70. Which of the following is a good source of secoisolariciresinol
(a) Flaxseed (b) soy
(c) both (d) none

71. Which of the following is a good source of matairesinol
(a) Flaxseed (b) soy
(c) both (d) none

72. Which of the following is a type of lignan
(a) secoisolariciresinol
(b) matairesinol
(c) both
(d) none

73. Which of the following is a good source of polyphenols
(a) Fruits (b) berries
(c) vegetables (d) all

74. Which of the following is a polyphenol
(a) Hydroxytyrosol (b) caffeic acid
(c) both (d) none

75. Which of the following is a polyphenol
(a) Oleuropein (b) ferulic acid
(c) phytic acid (d) all

76. Which of the following is related to herb extract
(a) Rosemarinic acid
(b) carnosic acid
(c) both
(d) none

77. Which of the following is related to herb extract
(a) Carnosol (b) carvacrol
(c) both (d) none

78. Which one is related to dietary fiber
(a) Carcinogen inactivation
(b) fermentation in colon
(c) both
(d) none

79. Which one is present in green tea
(a) Catechin (b) epicatechin
(c) epigallocatechin (d) all

80. Which one is present in green tea
(a) Epicatechin gallate
(b) epigallocatechin gallate
(c) both
(d) none

81. Which one is a good source of theaflavins
(a) Black tea (b) watermelon
(c) tomato (d) milk

82. Which is a function of polyphenols
(a) Antioxidant
(b) anti-inflammation
(c) deodorizing
(d) all

83. Dietary source of phytic acid is
(a) Whole legume seed
(b) cereal grains
(c) both (d) none

84. Dietary source of fiber is
(a) Whole wheat (b) fruits
(c) vegetables (d) all

85. Dietary source of trypsin inhibitor is
(a) Soy (b) cereal
(c) potatoes (d) all

86. Dietary source of polyphenols is
(a) Tea (b) coffee
(c) beans (d) all

87. Dietary source of tanins is
(a) Tea (b) coffee
(c) beans (d) all

88. Dietary source of cyanogens is
(a) Cassava (b) linseed
(c) both (d) none

89. Dietary source of cyanogens is
(a) Peas (b) beans
(c) both (d) none

90. Dietary source of glucosinolates is
(a) cabbage (b) broccoli
(c) both (d) none

91. Dietary source of saponins is
(a) soyabeans (b) peanuts
(c) sugar beet (d) all

92. Dietary source of goitrogens is
(a) broccoli (b) garlic
(c) both (d) none

93. Dietary source of oxalic acid is
(a) Spinach (b) rhubarb
(c) both (d) none

94. Which of the following company deals in probiotic products
(a) Danone (b) Valio
(c) Yakult (d) all

95. Which of the following company deals in probiotic products
(a) Muller (b) Bioferme
(c) Olma (d) all

96. Which of the following company deals in prebiotic products
(a) Ingmann foods (b) Campina
(c) Lifeway (d) all

97. Which cereal is used for preparation of functional food
(a) Oat (b) barley
(c) both (d) none

98. Which of the following is a functional food component from cereals
(a) Glucan (b) arabinoxylan
(c) both (d) none

99. Which of the following is a functional food component from cereals
(a) Galactooligosaccharide
(b) FOS
(c) resistant starch
(d) all

100. Which of the following can act as prebiotics
(a) FOS (b) IMO
(c) polydextrose (d) all

101. Which of the following can act as prebiotics
(a) SOS (b) GOS
(c) XOS (d) all

102. Which of the following can act as prebiotics
(a) Inulin (b) oligofructose
(c) both (d) none

103. Which of the following helps in reducing the risk of coronary heart disease
(a) Soluble oat fiber (b) soy protein
(c) phytosterol (d) all

104. Olive oil is a significant dietary source of
(a) Squalene (b) resveratrol
(c) both (d) none

105. Squalene content (mg/kg) in olive oil is
(a) 10-20 (b) 100-200
(c) 500-700 (d) none

106. Resveratrol is present in
(a) Peanut (b) grape
(c) both (d) none

107. Peanut is a source of __________ phytochemical
(a) Ethyl protocatechuate
(b) resveratrol
(c) both
(d) none

108. USFA content is ________ than SFA content in peanut oil USFA
(a) Higher (b) lower
(c) equal (d) none

109. Peanut oil contains
(a) Tocopherols (b) phytosterols
(c) both (d) none

110. Functional ingredient present in peanut is
(a) Resveratrol (b) flavonoids
(c) tocopherols (d) all

111. Functional ingredient present in peanut is
(a) ethyl protocatechuate
(b) phytosterol
(c) both (d) none

112. Which of the following phytochemical is beneficial to human health
(a) Carotenoids (b) terpenoids
(c) phytosterols (d) all

113. Which of the following phytochemical is beneficial to human health
(a) Flavonoids (b) indoles
(c) isoflavones (d) all

114. Which of the following phytochemical is beneficial to human health
(a) Isothiocyanates (b) lignin
(c) both (d) none

115. Which of the following phytochemical is beneficial to human health
(a) Phytate (b) soluble fiber
(c) insoluble fiber (d) all

116. Mechanism of biological action of phytochemicals
(a) Oxidative stress
(b) carcinogen metabolism
(c) both (d) none

117. Lycopene is present in
(a) Tomatoes (b) watermelon
(c) pink guava (d) all

118. Lycopene is present in
(a) Pink grapefruit (b) papaya
(c) both (d) none

119. Which of the following has highest lycopene content
(a) Fresh tomatoes
(b) cooked tomatoes
(c) Tomato sauce
(d) none

120. Which of the following has highest lycopene content
(a) Tomato soup (b) tomato paste
(c) tomato juice (d) none

121. Psyllium husk is a good source of
(a) Soluble fiber (b)insoluble fiber
(c) both (d) none

122. Flavonoids have ________ property
(a) Anticarcinogenic
(b) anti-inflammatory
(c) anti-allergic
(d) all

123. Which of the following is a whole grain phytochemical
(a) Phenolic acids
(b) alkylresorcinol
(c) flavonoids
(d) all

124. Which of the following is a whole grain phytochemical
(a) Carotenoids
(b) vitamin E
(c) gamma-oryzanol
(d) all

125. Which of the following is a whole grain phytochemical
(a) Dietary fiber (b) beta-glucan
(c) both (d) none

126. Phenolic acids are________ compounds
(a) Hydroxylated (b) oxygenated
(c) dehydrogenated (d) none

127. Phenolic acids can be derived from
(a) Benzoic acid (b) cinnamic acid
(c) both (d) none

128. Ferulic acid is a derivative of
(a) Cinnamic acid (b) acetic acid
(c) citric acid (d) none

129. Ferulic acid is present in
(a) Corn (b) wheat
(c) oats (d) all

130. Which of the following contains highest ferulic acid content
(a) Corn (b) wheat
(c) oats (d) rice

131. Alkylresorcinols are generally found in ________ fraction of grain
(a) Bran (b) endosperm
(c) germ (d) none

132. Which of the following is a class of dietary flavonoids
(a) Flavonols (b) Flavones
(c) Flavanones (d) all

133. Vitamin E is a collective name for eight lipid-soluble compounds consisting of
(a) Chromanol ring (b) phytyl tail
(c) both (d) none

134. Gamma-oryzanol consists of
(a) Phenolic acid (b) sterol
(c) both (d) none

135. Gamma-oryzanol content (mg/kg) in rice is around
(a) 10 (b) 30
(c) 250 (d) 3000

136. Gamma-oryzanol is present in
(a) Rice (b) wheat
(c) rye (d) all

137. Which one is related to gamma-oryzanol
(a) Antioxidant activity
(b) serum cholesterol lowering
(c) both (d) none

138. Which one is a dietary fiber
(a) Lignin (b) hemicelluloses
(c) cellulose (d) all

139. Which of the following has probiotic effects
(a) Bifidobacterium
(b) Streptococcus
(c) Lactococcus
(d) all

140. Prebiotics are _____food ingrerdients.
(a) Non-digestible (b) low digestible
(c) both (d) none

141. Probiotics are helpful in
(a) Stomach health
(b) allergy
(c) urogenital health
(d) all

142. CLA content is highest in _____
(a) Cereals (b) fruits
(c) milk (d) none

143. Probiotics are helpful in
(a) Cholesterol lowering
(b) hypertension
(c) gut function
(d) all

144. Flavonoids include
(a) Proanthrocyanidin
(b) quercetin
(c) epicatechin
(d) all

145. Green tea is a source of
(a) Catechin (b) epicatechin
(c) both (d) none

146. Which of the following is a flavonoids
(a) Hesperidin (b) quercetin
(c) both (d) none

147. Which of the following is a flavonoids
(a) Tangeretin (b) reveratol
(c) both (d) none

148. Which of the following is a flavonoids
(a) Anthocyanin (b) flavonols
(c) both (d) none

149. Xanthophylls protect vitamin_____ from oxidation
(a) A (b) E
(c) both (d) none

150. Cryptoxanthin is
(a) Xanthophyll (b) chlorophyll
(c) both (d) none

13.3 ANSWER KEY

Q.	A.	Q.	A.	Q.	A.	Q.	A.	Q.	A.	Q.	A.
1.	(c)	2.	(c)	3.	(b)	4.	(a)	5.	(d)	6.	(c)
7.	(c)	8.	(c)	9.	(d)	10.	(c)	11.	(d)	12.	(d)
13.	(c)	14.	(c)	15.	(c)	16.	(c)	17.	(c)	18.	(d)
19.	(c)	20.	(c)	21.	(c)	22.	(d)	23.	(d)	24.	(c)
25.	(d)	26.	(c)	27.	(d)	28.	(c)	29.	(d)	30.	(c)
31.	(c)	32.	(d)	33.	(d)	34.	(d)	35.	(d)	36.	(c)
37.	(c)	38.	(c)	39.	(d)	40.	(d)	41.	(c)	42.	(c)
43.	(c)	44.	(b)	45.	(d)	46.	(d)	47.	(a)	48.	(a)
49.	(b)	50.	(c)	51.	(c)	52.	(b)	53.	(b)	54.	(b)
55.	(a)	56.	(c)	57.	(d)	58.	(a)	59.	(a)	60.	(b)
61.	(c)	62.	(a)	63.	(b)	64.	(a)	65.	(c)	66.	(a)
67.	(a)	68.	(b)	69.	(a)	70.	(c)	71.	(c)	72.	(c)
73.	(d)	74.	(c)	75.	(d)	76.	(c)	77.	(c)	78.	(c)
79.	(d)	80.	(c)	81.	(a)	82.	(d)	83.	(c)	84.	(d)

Q.	A.	Q.	A.	Q.	A.	Q.	A.	Q.	A.	Q.	A.
85.	(d)	86.	(d)	87.	(d)	88.	(c)	89.	(c)	90.	(c)
91.	(d)	92.	(c)	93.	(c)	94.	(d)	95.	(d)	96.	(d)
97.	(c)	98.	(c)	99.	(d)	100.	(d)	101.	(d)	102.	(c)
103.	(d)	104.	(a)	105.	(c)	106.	(c)	107.	(c)	108.	(a)
109.	(c)	110.	(d)	111.	(c)	112.	(d)	113.	(d)	114.	(c)
115.	(d)	116.	(c)	117.	(d)	118.	(c)	119.	(c)	120.	(b)
121.	(a)	122.	(d)	123.	(d)	124.	(d)	125.	(c)	126.	(a)
127.	(c)	128.	(a)	129.	(d)	130.	(a)	131.	(a)	132.	(d)
133.	(c)	134.	(c)	135.	(d)	136.	(d)	137.	(c)	138.	(d)
139.	(d)	140.	(c)	141.	(d)	142.	(c)	143.	(d)	144.	(d)
145.	(c)	146.	(c)	147.	(c)	148.	(c)	149.	(c)	150.	(a)

CHAPTER 14

BIOCHEMISTRY

14.1 SUBJECTIVE KEYNOTES

1. Carbohydrates, proteins and lipids are three major components of any food system.
2. Carbohydrates are the poly hydroxyl aldehydes or ketones, or compounds which produce them on hydrolysis.
3. D-Glucose is the most predominant naturally occurring aldose monosaccharide.
4. Polysaccharides are the polymers of monosaccharides or their derivatives, held together by glycosidic bonds. Homopolysaccharides are composed of a single monosaccharide (e.g., starch, glycogen, cellulose, inulin). Heteropolysaccharides contain a mixture of few monosaccharides or their derivatives (e.g., mucopolysacaharides).
5. Lipids are important to the body as constituents of membranes, source of fat soluble (A, D, E & K) vitamins and metabolic regulators (steroid hormones and prostaglandins.
6. The unsaturated fatty acids-linoleic and linolenic acid are essential to humans, the deficiency of which causes phrynoderma or toad skin.
7. Cholesterol is essential for the synthesis of bile acids, hormones(sex and cortical) and vitamin D.
8. Steroids contain the ring cyclopentanophenanthrene.
9. Collagen is the most abundant protein in mammals. It is rich in hydroxyproline and hydroxylysine. Heating of collagen converts it to gelatin.
10. Proteins are polymers composed of L-α-amino acids.
11. The secondary structure of protein mainly consists of α-helix and/or β pleated sheet. α-helix is stabilized by extensive hydrogen bonding. Β-pleated sheet is composed of two or more segments of fully extended polypeptide chains.
12. Enzymes are the protein biocatalysts synthesized by living cells.
13. All enzymes are proteins but all proteins are not enzymes.
14. Enzymes can be classified into six major classes- oxidoreductases, transferases, hydrolases, lyases, isomerases and ligases.
15. Vitamins are accessory food factors required in the diet. They are classified as fat soluble (A, D, E and K) and water soluble (B-complex and C).
16. The active form of vitamin D is calcitriol which functions like a steroid hormone and regulate plasma levels of calcium and phosphate.

17. Digestion is a process that converts complex food stuffs into simpler ones which can be readily absorbed by the gastrointestinal tract.

18. The digestion of carbohydrates is initiated in the mouth by salivary amylase and is completed in the small intestine by pancreatic amylase, oligosaccharidases and disaccharidases.

19. Lactose intolerance due to a defect in the enzyme lactase (β-galactosidase) resulting in the inability to hydrolyse lactose is the common abnormality of carbohydrate digestion.

20. Hemoglobin is primarily responsible for the delivery of O_2 from lungs to tissue and the transport of CO_2 from tissues to lungs.

21. Thalassemias are hemolytic disorders caused by impairment/ imbalance in the synthesis of globin chains of Hb. These include α-thalassemia trait, hydrops fetalis and β-thalassemias.

22. Heme is the most important porphyrin compound,primarily synthesized in the liver from the pre-cursors-glycine and succinyl CoA.

23. Heme is degraded mainly to bilirubin, a yellow colour bile pigment.

24. Gluconeogenesis is the synthesis of glucose from noncarbohydrate precursors like amino acids (except leucine and lysine), lactate, glycerol, propionate etc.

25. In glycogenolysis, glycogen stored in the liver and muscles, is converted first to glucose-1- phosphate and then into glucose-6-phosphate. Two hormones which control glycogenolysis are a peptide, glucagon from the pancreas and epinephrine from the adrenal glands.

26. End products of glycolysis are two pyruvate, two NADH and two ATP.

27. Hexose monophosphate shunt (HMP shunt) is the direct oxidative pathway of glucose.

28. Triacylglycerols (TG) are the highly concentrated form of energy, stored in adipose tissue.

29. Atherosclerosis is o complex disease characterized by thickening of arteries due to the, accumulation of lipids.

30. The amino acids undergo transamination and deamination to liberate ammonia for the synthesis of urea, the end product of protein metabolism.

31. The carbon skeleton of amino acids is involved either in the synthesis of glucose (glycogenic) or fat (ketogenic), or both-glucose and fat.

32. Calcium is required for the development of bones and teeth, muscle contraction, blood coagulation, nerve transmission etc.

33. Sodium, potassium and chlorine are involved in the regulation of acid-base equilibrium, fluid balance and osmotic pressure in the body.

34. Fluorine in trace amounts prevents dental caries while its higher intake leads to fluorosis.

35. Anterior pituitary gland is the master endocrine organ.

36. The calorific values of carbohydrates, fats and proteins respectively are 4, 9 and 4 Cal/g.

37. Specific dynamic action (SDA) is the extra heat produced by the body over and above the calculated calorific value of foodstuff. It is higher for proteins (30%), lower for carbohydrates (5%) and for a mixed diet, it is around 10%.
38. The RDA for protein is 1g/kg body weight/day.
39. Protein-energy malnutrition (PEM) is the most common nutritional disorder in the developing countries. Kwashiorkar is primarily due to inadequate protein intake while marasmus is mainly caused by calorie deficiency.
40. Transcription is the process in which RNA is synthesized from DNA, which is carried out in 3 stages-initiation, elongation and termination.
41. Ribosomes are the factories of protein biosynthesis.
42. Insulin is a peptide hormone while inulin is a polysaccharide composed of fructose.
43. Van den Bergh reaction is used for the detection of serum bilirubin.
44. The RQ is essentially the ratio of carbon dioxide eliminated to oxygen consumed.
45. A RQ closer to 1 typically means more carbohydrate is being burned and a value closer to 0.7 typically means fat is being burned. A value of 0.7 indicates that lipids are being metabolized, 0.8 for proteins, and 1.0 for carbohydrates.
46. Innate immunity is present right from birth, while acquired immunity consists of antibody mediated and T cell mediated responses.
47. The cells responsible for innate immune system are phagocytes, natural killer cells (NKC), mast cells and dendritic cells.
48. Chitin is made up of the a mino sugar, glucosamine and hyaluronic acid is composed of glucuronic acid and glucosamine.
49. Heparin is blood anticoagulant. It is present in liver, lung thymus spleen and blood. It is polymer of D-glucuronic acid and D-glucosamine.
50. Glycolysis is the splitting of glucose. The first step of the process occurs in the cytoplasm of the cell and yields a small amount of energy (measured in units of ATP) and two molecules of pyruvate. The other steps occur in the mitochondria.
51. Citric acid cycle is also called Tricarboxylic acid (TCA) cycle or Krebs cycle.
52. Phenylketonuria (PKU) is a disorder that causes a buildup of the amino acid phenylalanine, which is an essential amino acid that cannot be synthesized in the body. Excess phenylalanine is normally converted to tyrosine, another amino acid, and eliminated from the body. Without the enzyme phenylalanine hydroxylase that converts it to tyrosine, phenylalanine builds up in the blood and is toxic to the brain, causing intellectual disability. To prevent it, people must restrict phenylalanine intake.
53. Homocystinuria is caused by lack of the enzyme needed to metabolize homocysteine. This disorder can cause a number of symptoms, including decreased vision and skeletal abnormalities.
54. Tyrosinemia is caused by lack of the enzyme needed to metabolize tyrosine. The most common form of this disorder mostly affects the liver and the kidneys.
55. Albinism is a condition in which there is total absence of tyrosinase inside the melanocytes in the skin, as a result black pigment melanin is not formed in the skin, eyes and hair.

56. Many legumes have root nodules that provide a home for symbiotic nitrogen-fixing bacteria called rhizobia. The Rhizobia convert nitrogen gas from the atmosphere into ammonia, which is then used in the formation of amino acids and nucleotides .
57. The color of the root nodules is pink due to the presence of leghaemoglobin.
58. Galactosemia is a high blood level of galactose.
59. Tangier disease or hypoalphalipoproteinemia is a rare inherited disorder characterized by a severe reduction in the amount of high density lipoprotein (HDL), often referred to as "good cholesterol", in the bloodstream.
60. Hypolipoproteinemia refers to unusually low levels of fats (lipids) in the blood.

14.2 MULTIPLE CHOICE QUESTIONS

1. The most abundant substance in living beings is
(a) Water (b) protein
(c) fat (d) minerals

2. Which of the following is a weakest bond
(a) Hydrogen bond
(b) Ionic
(c) hydrophobic
(d) Vander Waals

3. The HOH bond angle is
(a) 104.5° (b) 110.5°
(c) 115.5° (d) 100.5°

4. Melting point (°C) of water is
(a) 0 (b) 10
(c) 100 (d) none

5. The dielectric constant for water at 25°C is
(a) 78.5 (b) 89.5
(c) 98.5 (d) 15.5

6. Amphipathic compounds are
(a) Polar (b) non-polar
(c) both (d) none

7. The pH of 1 M NaOH is
(a) 7 (b) 10
(c) 12 (d) 14

8. The pH of 1 M HCl is
(a) 7 (b) 10
(c) 0 (d) 14

9. pH of blood plasma of people with reverse diabetes is
(a) 7.4 (b) < 7.4
(c) > 7.4 (d) 0

10. Human blood pH is normally
(a) 7.35-7.45 (b) 8.15-8.25
(c) 9.11-35 (d) 6.5

11. Which of the following is not a aromatic amino acid
(a) Phenyl alanine
(b) tyrosine
(c) tryptophan
(d) serine

12. The simplest amino acid is
(a) Glycine (b) alanine
(c) valine (d) serine

13. Which is a sulphur containing amino acid
(a) Methionine (b) valine
(c) tyrosine (d) lysine

14. Isoelectric point is the point at which net electric charge is
(a) 0 (b) 10
(c) 100 (d) 1000

15. Two amino acids are joined by_______ linkage/bond
(a) Ester (b) peptide
(c) glycosidic (d) none

16. The non-amino acid part of conjugated protein is
(a) Conjugated protein
(b) prosthetic group
(c) lipoproteins
(d) glycolipids

17. The technique for the separation of proteins bases on the migration of charged proteins in an electric field is called
(a) Electrophoresis
(b) chromatography
(c) both
(d) none

18. SDS stands for
(a) Sodium dodecyl sulphate
(b) Sodium diethyl sulphate
(c) Sodium diphenyl sulphate
(d) none

19. ______ is used to determine the isoelectric point of a protein
(a) Isoelectric focusing
(b) isoelectric gelling
(c) ion migration
(d) none

20. The number of enzyme units/mg of total protein is
(a) Specific activity (b) activity
(c) both (d) none

21. PDB stands for
(a) Protein data bank
(b) protein digestion base
(c) protein delivery bank
(d) none

22. Structural asymmetry in a molecule can be measured by
(a) Chromatography
(b) electrophoresis
(c) circular dichorism
(d) centrifugation

23. Alpha-keratin is a form of ________ protein
(a) Globular (b) fibrous
(c) both (d) none

24. If a carbonyl group is at end of carbon chain, the monosaccharide is
(a) Aldose (b) ketose
(c) both (d) none

25. If a carbonyl group is at any position (non-terminal) of carbon chain, the monosaccharide is
(a) Aldose (b) ketose
(c) both (d) none

26. Simplest monosaccharide is
(a) Glyceraldehyde
(b) dihydroxyacetone
(c) both
(d) none

27. Disaccharide in milk is
(a) Lactose (b) galactose
(c) glucose (d) arabinose

28. Which of the following is table sugar
(a) Lactose (b) sucrose
(c) fructose (d) invert sugar

29. Glycogen is found in abundance in
(a) Liver (b) gall bladder
(c) spleen (d) lungs

30. The net yield is ______ molecules of ATP per molecule of glucose in glycolysis
(a) 2 (b) 3
(c) 4 (d) 6

31. The net yield is ______ molecules of NADP per molecule of glucose in glycolysis
(a) 2 (b) 3
(c) 4 (d) 6

32. The enzyme in saliva for carbohydrates digestion is
(a) Alpha-amylase (b) beta-amylase
(c) lipase (d) ligase

33. Glycogen in animals is stored in
(a) Liver (b) spleen
(c) lungs (d) gall bladder

34. In galactosemia, there is a enhance concentration of _______ in blood and urine
(a) Proteins (b) galactose
(c) sucrose (d) fructose

35. Brain requires _______ g of glucose each day for its activity
(a) 120 g (b) 140g
(c) 160g (d) 180g

36. Synthesizing glucose from non-carbohydrates precursors is called
(a) Gluconeogenesis (b) glycolysis
(c) both (d) none

37. Gluconeogenesis takes place in
(a) Liver (b) spleen
(c) lungs (d) gall bladder

38. How many pyruvate molecules are required for synthesizing 1 molecule of glucose
(a) 2 (b) 4
(c) 1 (d) 3

39. Which of the following is not an intermediate of citric acid cycle
(a) Pyruvate (b) succinate
(c) fumarate (d) malate

40. The conversion of glucose to pyruvate is
(a) Gluconeogenesis (b) glycolysis
(c) both (d) none

41. The conversion of pyruvate to glucose is
(a) Gluconeogenesis (b) glycolysis
(c) both (d) none

42. The hormone glucagon _______ the glucose
(a) Release (b) inhibit
(c) no effect (d) none

43. Excessive production of urine occurs in diabetes
(a) Insipidus (b) Mellitus
(c) both (d) none

44. Insulin is produced by _______ cells of pancreas
(a) Alpha (b) beta
(c) gamma (d) none

45. Glycogen in liver represents_____% of the weight of liver
(a) 10 (b) 20
(c) 30 (d) 40

46. Niacin exists in co-enzyme forms as
(a) NAD (b) FMN
(c) TPP (d) all

47. Citric acid cycle is also known as
(a) TCA cycle (b) Kreb cycle
(c) both (d) none

48. In citric acid cycle in aerobes, glucose is oxidised to ______ ultimately
(a) Pyruvate (b) succinate
(c) CO_2 & H_2O (d) O_2

49. PDH complex stands for
(a) Pyruvate dehydrogenase
(b) pyruvate deoxygenase
(c) both
(d) none

50. The starting material of citric acid cycle is
(a) Acetyl-CoA (b) pyruvate
(c) malate (d) succinate

51. One gene one enzyme hypothesis was given by
(a) Louis Pasteur
(b) Beadle & Tatum
(c) Tathom
(d) Peter Shewry

52. Enzymes that catalyze cleavages are
(a) Lyases (b) synthases
(c) both (d) none

53. Enzymes that catalyze condensation reaction in which two atoms are joined
(a) Ligases (b) synthases
(c) both (d) none

54. Enzymes that transfer phosphoryl group from nucleoside triphosphate to an acceptor molecule is
(a) Lyases (b) kinases
(c) both (d) none

55. The last reaction of citric acid cycle is conversion of ________ to oxaloacetaete
(a) Succinate (b) fumarate
(c) malate (d) pyruvate

56. CA cycle generates ______ ATP per cycle
(a) 1 (b) 2
(c) 3 (d) 4

57. CA cycle takes place in
(a) Mitochondria (b) kidney
(c) liver (d) spleen

58. Triacylglycerols are delivered to tissues by
(a) Chylomicrons (b) FA
(c) NADH (d) ATP

59. Vitamin B_{12} can be synthesized by
(a) Plants (b) animals
(c) microorganisms (d) all

60. Zell wager syndrome individuals cannot make
(a) Peroxisome (b) mitochondria
(c) both (d) none

61. The enzyme for oxidation of fat are present in
(a) Liver (b) kidney
(c) both (d) none

62. The enzyme for oxidation of fat are present in
(a) Glyoxysome
(b) mitochondrion
(c) Endoplasmic reticulum
(d) all

63. The ketone bodies are formed in
(a) Liver (b) spleen
(c) gall bladder (d) none

64. Over production of ketone bodies lead to
(a) Acidosis (b) ketosis
(c) both (d) none

65. Apoptosis of cell is
(a) Natural death of cell
(b) birth of cell
(c) regeneration of cell
(d) all

66. Glucose has 4 asymmetric carbons, and thus has ______ isomers
(a) Two (b) eight
(c) sixteen (d) twenty four

67. If D and L isomers are present in equal concentration, it is known as ________ mixture
(a) Ideal (b) racemic
(c) tautomeric (d) optical

68. A racemic mixture has ______ optical activity
(a) 0% (b) 50%
(c) 75% (d) 100%

69. Which of the following sugar cannot form osazone
(a) Fructose (b) maltose
(c) sucrose (d) glucose

70. Native sucrose is
(a) Dextrorotatory (b) Levorotatory
(c) Both (d) none

71. The major fraction of starch is that of
(a) Amylase (b) amylopectin
(c) both equal (d) none

72. Dextrins are the breakdown products of
(a) Proteins (b) starch
(c) TGA (d) nucleic acids

73. Inulin is a polymer of
(a) Fructose (b) glucose
(c) galactose (d) maltose

74. Which of the following is used for assessing the kidney function through measurement of glomerular filtration rate
(a) Glucose (b) starch
(c) inulin (d) dextrins

75. Which of the following vitamin cannot be synthesized by human body
(a) B (b) C
(c) D (d) A

76. Urea synthesis occurs in
(a) Liver (b) kidney
(c) gall bladder (d) spleen

77. One urea cycle requires ______ ATP molecules
(a) 1 (b) 2
(c) 3 (d) 4

78. Sulphur containing amino acid is
(a) Methionine (b) valine
(c) leucine (d) asparagines

79. The biosynthesis of urea begins in the liver ________
(a) Cytosol (b) microsomes
(c) nucleus (d) mitochondria

80. High iodine value of lipid indicates
(a) Polymerization (b) unsaturation
(c) saturation (d) oxidation

81. Most fatty acids are degraded by the process
(a) β- oxidation (b) α-oxidation
(c) Lipolysis (d) All

82. β-oxidation occurs in
(a) Mitochondria (b) cytosol
(c) glyoxisomes (d) all

83. The glycogen degradation is stimulated in the liver by
(a) Epinephrine (b) glucagon
(c) both (d) none

84. The synthesis of fatty acid begin with
(a) Acetyl CoA (b) succinate
(c) malate (d) pyruvate

85. Which of the following is an anti-coagulant
(a) Ceruplasmin (b) fibronectin
(c) heparin (d) all

86. The imino acid found in protein structure
(a) Arginine (b) proline
(c) histidine (d) lysine

87. The following is a non-protein amino acid
(a) Ornithine (b) homocysteine
(c) histamine (d) all

88. The bonds in protein structure that are not broken on denaturation
(a) Hydrogen (b) peptide
(c) ionic (d) disulphide

89. The reaction given by two or more peptide linkages is
(a) Biuret test
(b) ninhydrin test
(c) xanthoproetic test
(d) pauley's test

90. Which of the following has the highest saponification value
(a) Human fat (b) butter
(c) coconut oil (d) milk

91. The nitrogenous base present in lecithin
(a) Choline (b) ethanolamine
(c) inositol (d) serine

92. The number of double bonds in arachidonic acid is
(a) 1 (b) 2
(c) 3 (d) 4

93. One of the following is an amphipathilc lipid
(a) Phospholipids (b) fatty acids
(c) bile salts (d) all the above

94. Esterification of cholesterol occurs at carbon position
(a) 1 (b) 2
(c) 3 (d) 4

95. Name the test employed t o check the purity of butter through the estimation o f volatile fatty acids
(a) Iodine number
(b) saponification no.
(c) acid no.
(d) Reichert Meissl no.

96. Which of the following amino acid is ketogenic
(a) Leucine (b) lysine
(c) both a & b (d) isoleucine

97. Which of the following amino acid is glycogenic
(a) Alanine (b) aspartate
(c) glycine (d) all

98. Which of the following amino acid is ketogenic as well as glycogenic
(a) Isoleucine (b) phenylalanine
(c) tyrosine (d) all

99. The amino acids which are sweet
(a) Glycine (b) alanine
(c) valine (d) all

100. Which of the following amino acid is tasteless
(a) Isoleucine (b) phenylalanine
(c) tyrosine (d) leucine

101. Which of the following amino acid is bitter
(a) Arginine (b) isoleucine
(c) both (d) methionine

102. Which of the following amino acid does not have isomer
(a) Leucine (b) glycine
(c) valine (d) alanine

103. The number of polypeptide chains in a protein can be identified by
(a) Ninhydrin test
(b) dansyl chloride
(c) cynogen bromide
(d) all

104. The nitrogenous base not present in DNA structure is
(a) Adenine (b) thymine
(c) cytosine (d) uracil

105. The number of base pairs present in each turn (pitch) of B-form of DNA helix
(a) 9 (b) 10
(c) 11 (d) 12

106. The backbone of nucleic acid structure is constructed by
(a) Peptide bonds
(b) glycosidic bonds
(c) phosphodiester bonds
(d) all

107. The following coenzyme is a nucleotide
(a) FAD (b) NAD+
(c) CoASH (d) all

108. The nucleotide that serves as an intermediate for biosynthetic reaction
(a) UDP-glucose
(b) CDP-acylglycerol
(c) S-adenosylmethionine
(d) all

109. Which of the following base pair is more strong
(a) GC base pair (b) AT base pair
(c) both equal (d) none

110. The building block of nucleic acids is
(a) Nucleotides (b) nucleosides
(c) pentose sugar (d) nitrogen base

111. Which of the following is not a purine base
(a) Adenine (b) guanine
(c) thymine (d) both a & b

112. Which of the following is not a pyrimidine base
(a) Adenine (b) thymine
(c) cytosine (d) uracil

113. The number of hydrogen bonds in AT pair is
(a) 1 (b) 2
(c) 3 (d) 4

114. The number of hydrogen bonds in GC pair is
(a) 1 (b) 2
(c) 3 (d) 4

115. Pepsin is an example of the class of enzymes namely
(a) Oxidoreductases
(b) transferases
(c) hydrolases
(d) ligases

116. The coenzyme not involved in hydrogen transfer
(a) FMN (b) FAD
(c) NAD+ (d) FH_4

117. In the feedback regulation the end product binds at
(a) Active site (b) allosteric site
(c) E-S complex (d) none

118. γ-glutamyl transpeptidase activity in serum is elevated in
(a) Pancreatis
(b) muscular dystrophy
(c) myocardial infarction
(d) alcoholism

119. In recent years a non-protein compound has been identified to bring about catalysis in biological system. The name of the compound is
(a) DNA (b) RNA
(c) lipids (d) carbohydrates

120. The elements with same atomic number but different mass number is called
(a) Isomer (b) isotope
(c) both (d) none

121. The functionally active form of vitamin D is
(a) Cholecalciferol
(b) ergocalciferol
(c) dehydrocholesterol
(d) calcitriol

122. The metabolite excreted in urine in thiamine deficiency
(a) Pyruvate
(b) xanthurenic acid
(c) glucose
(d) FIGLU

123. Folic acid antagonist used in the treatment of cancer
(a) Methotrexate (b) trimethoprim
(c) sulphonamide (d) all

124. Macrocytic anaemia is due to deficiency of
(a) Folic acid (b) thiamine
(c) cobalamin (d) niacin

125. The A in CoA stands for
(a) Acyl (b) acetylation
(c) antagonistic (d) aryl

126. The least toxic among the fat soluble vitamins is
(a) A (b) D
(c) E (d) K

127. The key enzyme that converts trypsinogen to trypsin is
(a) Secretin (b) chymotrypsin
(c) elastase (d) enteropeptidase

128. Salivary-amylase becomes inactive in the stomach primarily due to
(a) Inactivation by low pH
(b) degradation by gastric pepsin
(c) inhibition by Cl^-
(d) inhibition by peptides

129. Cellulose is not digested in humans due to lack of the enzyme that hydrolyses______bond
(a) Alpha (b) beta
(c) gamma (d) omega

130. Lactose intolerance is caused by the deficiency o f the enzyme
(a) Amylase
(b) lactase
(c) beta- galactosidase
(d) both (b) and (c)

131. Which of the following enzyme is absent in adults
(a) Lactase (b) rennin
(c) amylase (d) lipase

132. Hemophilia A is due to the deficiency of clotting factor
(a) X (b) V
(c) VIII (d) II

133. The immunoglobulin present in most abundant quantity is
(a) IgG (b) IgA
(c) IgM (d) IgE

134. Name the immunoglobulin involved in body allergic reactions
(a) IgE (b) IgA
(c) IgM (d) IgD

135. The following anticoagulant binds with Ca^{2+} and prevents blood clotting
(a) Heparin (b) Oxalate
(c) protein C (d) all

136. The difference between plasma and serum is the presence or absence of
(a) Albumin (b) globulin
(c) fibrinogen (d) minerals

137. The characteristic red colour of haemoglobin is due to
(a) Heme (b) α-globin
(c) β-globin (d) all

138. The number of heme groups present in myoglobin is
(a) 1 (b) 2
(c) 3 (d) 4

139. The patients of sickle-cell anemia are resistant to
(a) Filaria (b) malaria
(c) diabetes (d) trypanosoma

140. Name the amino acid that directly participates in the synthesis of heme is
(a) Methionine (b) aspartate
(c) glycine (d) tryptophan

141. 1 g of haemoglobin on degradation finally yields about ______ mg of bilirubin
(a) 15 (b) 25
(c) 35 (d) 45

142. Name the compound with the greatest standard free energy
(a) ATP
(b) phosphocreatine
(c) cyclic AMP
(d) phosphoenolpyruvate

143. The P:O ratio for the oxidation of $FADH_2$ is
(a) 1 (b) 2
(c) 3 (d) 4

144. ATP synthase activity is associated with the mitochondrial enzyme complex
(a) V (b) III
(c) IV (d) I

145. One of the following enzymes in glycolysis catalyses an irreversible reaction
(a) Hexokinase
(b) phosphofructokinase
(c) pyruvate kinase
(d) all

146. Synthesis of 2, 3-bisphosphoglycerate occurs in the tissue namely
(a) Liver (b) kidney
(c) erythrocytes (d) brain

147. The number of ATP produced when a molecule of acetyl CoA is oxidized through citric acid cycle
(a) 12 (b) 24
(c) 36 (d) 15

148. The connecting link between HMP shunt and lipid synthesis is
(a) Ribose
(b) NADPH
(c) sedo-heptulose 7-phosphate
(d) NADH

149. The following substance(is) (are) ketogenic
(a) Fatty acids (b) leucine
(c) lysine (d) all

150. The lipoprotein possessing the highest quantity of phospholipid
(a) HDL (b) LDL
(c) VLDL (d) chylomicrons

151. Hypercholesterolemia is observed in the disorder
(a) Hypothyroidism
(b) diabetes mellitus
(c) nephritic syndrome
(d) all

152. Van den bergh reaction is used for the detection of serum ________
(a) Creatinine (b) sugar
(c) bilirubin (d) protein

153. Hormone sensitive lipase activity is inhibited by the hormone
(a) Epinephrine (b) insulin
(c) thyroxine (d) glucocorticoids

154. The synthesis of urea occurs in
(a) Kidney (b) liver
(c) muscle (d) brain

155. The amino acid required for the formation of glutathione
(a) Glycine (b) cysteine
(c) glutamate (d) all

156. In the synthesis of cysteine the carbon skeleton is provided by
(a) Serine (b) methionine
(c) glutamate (d) alanine

157. The amino acids are said to be ketogenic when the carbon skeleton is finally degraded to
(a) Succinyl CoA (b) fumarate
(c) acetyl CoA (d) pyruvate

158. The amino acid that does not participate in transamination
(a) Lysine (b) glutamate
(c) alanine (d) tryptophan

159. The energy supplied by TGA is
(a) 9 Cal/g (b) 18 Cal/g
(c) 24 Cal/g (d) 30 Cal/g

160. TGA are stored in the body in ________ form
(a) Hydrous
(b) anhydrous
(c) both
(d) not stored at all

161. Glycogen is stored in the body in ________ form
(a) Hydrous
(b) anhydrous
(c) both
(d) not stored at all

162. Which of the following comprises the highest percentage of total cholesterol in human body
(a) LDL (b) HDL
(c) VLDL (d) none

163. The precursor for bile acids and steroid hormones is
(a) Phospholipids (b) glycolipids
(c) cholesterol (d) both b & c

164. Which of the following is/are ketone bodies
(a) Acetone (b) acetoacetate
(c) cholesterol (d) both a & b

165. Cholesterol is ___________ in nature
(a) Hydrophobic (b) hydrophilic
(c) amphipathic (d) none

166. The normal blood urea concentration is _____ mg/dl
(a) 10-40 (b) 40-80
(c) 80-120 (d) 120-160

167. The following substance(s) are involved in the regulation of plasma calcium level
(a) Calcitrol
(b) parathyroid hormone
(c) calcitonin
(d) all

168. The following is a sulphur containing essential amino acid
(a) Methionine (b) cysteine
(c) cystine (d) all

169. The following element is involved in wound healing
(a) Calcium (b) sodium
(c) zinc (d) magnesium

170. The element that prevents the development of dental caries
(a) Fluorine (b) calcium
(c) phosphorous (d) sodium

171. The most abundant mineral in the body is
(a) Phosphorous (b) iron
(c) calcium (d) sodium

172. Tetany occurs when body calcium levels
(a) Decrease
(b) increase
(c) not affected by calcium
(d) none

173. The recommended ration of Ca:P for an adult is
(a) 1:1 (b) 1:2
(c) 2:1 (d) 3:2

174. The chief cation of the extracellular fluid is
(a) Calcium (b) magnesium
(c) sodium (d) potassium

175. Addison's disease occurs due to low levels of ________ in body
(a) Calcium (b) magnesium
(c) sodium (d) potassium

176. The chief cation of the intracellular fluid is
(a) Calcium (b) magnesium
(c) sodium (d) potassium

177. The daily requirement of chloride as NaCl is
(a) 1-5 g (b) 5-10 g
(c) 10-50 g (d) 50-100 g

178. The protein containing non-heme iron is
(a) Transferrin (b) ferritin
(c) hemosiderin (d) all

179. The storage form of iron is
(a) Transferrin
(b) ferritin
(c) hemosiderin
(d) both (b) & (c)

180. The disease due to excessive accumulation of iron in the body is
(a) Anaemia
(b) hemosiderosis
(c) addison disease
(d) cushing's disease

181. Which of the following is protein hormone
(a) Insulin (b) glucagon
(c) oxytocin (d) all

182. Which of the following hormone is secreted by posterior pituitary gland
(a) ACTH (b) FSH
(c) ADH (d) TSH

183. Master endocrine gland is
(a) Anterior pituitary
(b) middle pituitary
(c) posterior pituitary
(d) none

184. ADH is also known as
(a) Insulin (b) prolactin
(c) vasopressin (d) oxytocin

185. Diabetes insipidus occurs due to ________ hormone
(a) Insulin
(b) prolactin
(c) vasopressin
(d) oxytocin

186. Gastrin is a hormone containing ________ amino acids
(a) 15 (b) 17
(c) 20 (d) 23

187. Impairment in the synthesis of dopamine by the brain is a major causative factor for the disorder
(a) Parkinson's disease
(b) Addison's disease
(c) Cushing's syndrome
(d) Goiter

188. Name the hormone, predominantly produced in response to fight, fright and flight
(a) Thyroxine (b) aldosterone
(c) epinephrine (d) ADH

189. A diet rich in animal proteins would result in _______ acid production
(a) Less (b) more
(c) neutral (d) none

190. A vegetarian diet has a ___________ effect on the body
(a) Acidifying (b) alkalizing
(c) neutralizing (d) none

191. The most abundant protein in mammals is
(a) Collagen (b) fibrillin
(c) laminin (d) fibronectin

192. One of the following minerals is lacking in milk
(a) Calcium (b) sodium
(c) iron (d) potassium

193. Glucose has a glycaemic index of
(a) 1 (b) 20
(c) 50 (d) 100

194. Phrynoderma occurs due to deficiency of ________ in human diet
(a) Calcium
(b) sodium
(c) essential fatty acids
(d) all

195. The daily normal requirement of protein in an adult is
(a) 1g/kg body wt (b) 2g/kg body wt
(c) 3g/kg body wt (d) 4g/kg body wt

196. The specific dynamic action (SDA) is the greatest for the following foodstuff
(a) Protein (b) carbohydrates
(c) fats (d) vitamins

197. The reference protein for the calculation of chemical score
(a) Meat protein (b) fish protein
(c) milk protein (d) egg protein

198. One of the following is the most important essential fatty acid in the diet
(a) Linoleic acid
(b) arachidonic acid
(c) oleic acid
(d) palmitic acid

199. The essential amino acid limiting in rice
(a) Methionine (b) tryptophan
(c) lysine (d) histidine

200. When the particle size is < 1 nm, it is called ________ solution
(a) True (b) colloidal
(c) suspension (d) all

14.3 ANSWER KEY

Q.	A.	Q.	A.	Q.	A.	Q.	A.	Q.	A.	Q.	A.
1.	(a)	2.	(d)	3.	(a)	4.	(a)	5.	(a)	6.	(c)
7.	(d)	8.	(c)	9.	(b)	10.	(a)	11.	(d)	12.	(a)
13.	(a)	14.	(a)	15.	(b)	16.	(b)	17.	(a)	18.	(a)
19.	(a)	20.	(a)	21.	(a)	22.	(c)	23.	(b)	24.	(a)
25.	(b)	26.	(c)	27.	(a)	28.	(b)	29.	(a)	30.	(a)
31.	(a)	32.	(a)	33.	(a)	34.	(b)	35.	(a)	36.	(a)
37.	(a)	38.	(a)	39.	(a)	40.	(b)	41.	(b)	42.	(a)
43.	(a)	44.	(b)	45.	(a)	46.	(a)	47.	(c)	48.	(c)
49.	(a)	50.	(a)	51.	(b)	52.	(a)	53.	(a)	54.	(b)
55.	(c)	56.	(a)	57.	(a)	58.	(a)	59.	(c)	60.	(a)
61.	(c)	62.	(c)	63.	(a)	64.	(c)	65.	(a)	66.	(c)
67.	(b)	68.	(a)	69.	(c)	70.	(a)	71.	(b)	72.	(b)
73.	(a)	74.	(c)	75.	(b)	76.	(a)	77.	(c)	78.	(a)
79.	(d)	80.	(b)	81.	(b)	82.	(b)	83.	(c)	84.	(a)
85.	(c)	86.	(b)	87.	(a)	88.	(b)	89.	(a)	90.	(c)
91.	(a)	92.	(d)	93.	(a)	94.	(c)	95.	(d)	96.	(c)
97.	(d)	98.	(d)	99.	(d)	100.	(d)	101.	(c)	102.	(b)
103.	(b)	104.	(d)	105.	(b)	106.	(c)	107.	(d)	108.	(d)
109.	(a)	110.	(a)	111.	(c)	112.	(a)	113.	(b)	114.	(c)
115.	(c)	116.	(d)	117.	(b)	118.	(d)	119.	(b)	120.	(b)
121.	(d)	122.	(a)	123.	(a)	124.	(a)	125.	(b)	126.	(c)
127.	(d)	128.	(a)	129.	(b)	130.	(d)	131.	(b)	132.	(c)
133.	(a)	134.	(a)	135.	(b)	136.	(c)	137.	(a)	138.	(a)
139.	(b)	140.	(c)	141.	(c)	142.	(d)	143.	(b)	144.	(a)
145.	(d)	146.	(c)	147.	(a)	148.	(b)	149.	(c)	150.	(a)
151.	(d)	152.	(c)	153.	(b)	154.	(b)	155.	(a)	156.	(b)
157.	(c)	158.	(a)	159.	(a)	160.	(b)	161.	(a)	162.	(a)
163.	(c)	164.	(d)	165.	(c)	166.	(a)	167.	(d)	168.	(d)
169.	(c)	170.	(a)	171.	(c)	172.	(a)	173.	(a)	174.	(c)
175.	(c)	176.	(d)	177.	(b)	178.	(d)	179.	(d)	180.	(b)
181.	(d)	182.	(c)	183.	(a)	184.	(c)	185.	(c)	186.	(b)
187.	(a)	188.	(c)	189.	(b)	190.	(b)	191.	(a)	192.	(c)
193.	(d)	194.	(c)	195.	(a)	196.	(a)	197.	(d)	198.	(a)
199.	(c)	200.	(a)								

CHAPTER 15

EMERGING FOOD PROCESSING TECHNOLOGIES

15.1 SUBJECTIVE KEYNOTES

1. Novel food processing technologies are being explored and implemented to provide safe, fresher tasting, nutritive foods without the use of heat or chemical preservatives.
2. Novel food processing techniques include high pressure, pulsed electric fields, radio-frequency, high intensity pulsed light, ultrasound, irradiation, microwave, ohmic, dielectric heating, dehydration and drying techniques.
3. High Pressure Processing (HPP) is a method of food processing where food is subjected to elevated pressures (400-600 MPa), with or without the addition of heat, to achieve microbial inactivation or to alter the food attributes in order to achieve consumer-desired qualities.
4. High Pressure Processing is governed by three principles; Le Chatelier's principle, Principle of microscopic ordering and isostatic principle.
5. High pressure affects the permeability of the cellular membrane, which is responsible for the nutrient and respiration transport mechanisms of the cell.
6. HPP can be applied to a wide range of different foods including meat based products, fish, pre-cooked dishes and fruits, vegetables and juices.
7. High intensity pulsed electric field (PEF) processing involves the application of pulses of high voltage (typically 20–80 kV/cm) for short time periods (less than 1 second) to fluid foods placed between 2 electrodes
8. The PEF technology is applicable to the treatment of microorganisms in liquids such as *Escherichia coli, Salmonella and Listeria monocytogenes, Zygosaccharomyces bailii, Obesumbacterium proteus* and *Lactobacillus brevis*.
9. When exposed to high electrical field pulses (usually for a couple of microseconds), cell membranes of microbes develop pores either by enlargement of existing pores or by creation of new ones (electroporation).
10. Pulsed light technology involves treatment of food, equipments or package material surfaces with intense, short duration (<1 up to 2 millisecond) pulses of broad spectrum "white light" that includes wavelengths in the UV to near-infrared region (200-1000 nm).

11. Main mechanism of microbial inactivation by PL is explained through the photochemical effect, which consists in the formation of pyrimidines dimers in the DNA of bacteria, viruses, and other pathogens, thus preventing the cell from replicating .
12. PL treatment is more effective for the sterilization of surfaces than liquid media.
13. Ultraviolet (UV) processing involves use of radiation of wavelengths ranging from 315-400nm (UVA), 280-315nm (UVB), 200-280nm (UVC), and 100-200nm (vacuum UV).
14. The antimicrobial effects of UV processing are mainly due to DNA damage, which prevents cross linking between neighbouring pyrimidines and ultimately DNA transcription.
15. Ultrasound, utilizes the energy generated by sound waves of 20,000 or more vibrations per second.
16. The ultrasonic shock waves disrupt cellular structural and functional components to the point of cell lysis.
17. High voltage arc discharge is a technique to pasteurize liquid foods by applying rapid discharge voltages through an electrode gap below the surface of aqueous suspensions of microorganisms.
18. High voltage arc discharge method inactivates microorganisms and enzymes.
19. Oscillating magnetic fields is applied in the form of constant amplitude or decaying amplitude sinusoidal waves. This technology is still being researched as inconsistent results of microbial inactivation studies make it impossible to clearly state the effects of magnetic fields on microbial populations.
20. Covalent bonds within deoxyribose nucleic acid (DNA) may be disrupted by oscillating magnetic fields leading to loss of growth ability.
21. The irradiation sources that are permitted for use in food processing are gamma rays (from radioisotopes cobalt 60 and cesium 137), electron beams, and x-rays
22. One gray is one joule of energy absorbed per kilogram of irradiated product.
23. The amount of energy absorbed by the food depends on the mass, bulk density and thickness of the food.
24. Irradiation creates damage in the genetic material of the cell causing a lesion of the DNA or breaking both the strands of DNA.
25. Food irradiation may be used to reduce losses caused by premature ripening of fruits, sprouting of vegetables and pests.
26. Ohmic heating is direct resistance heating for sterilization of static liquid foods or in-can foods.
27. The basic principle of ohmic heating, or Joule effect, is the dissipation of electrical energy in the form of heat using electrical conductor.
28. Electrical conductivity is the main parameter in heating rate and in ohmic heating treatment.
29. The electrical conductivity is measured by the quantity of electricity transferred across a unit area, per unit potential gradient and per unit time.
30. The electrical conductivity of food products usually increases with water content and temperature.

31. Microwave vacuum-drying is a dehydration process that uses microwave radiation for heat generation in the absolute pressure range from above the triple point of water to less than atmospheric pressure.
32. Dielectric heating lies in the electromagnetic spectrum in the range of frequencies from 300 kHz to 300GHz.
33. Combination of microwaves with halogen lamp heating is a recent developmentin microwave baking.
34. Tempering is defined as bringing the temperature of food upto a few degrees below complete thawing.
35. Irradiated foods are those that have been treated with ionizing radiations.
36. Primary or reference dosimeters are used to calculate the amount of energy absorbed or to measure the dose directly in irradiation.
37. Power ultrasound is a promising and innovative method to assist the extraction of valuable compounds from fruits, vegetables and food products.
38. Oscillatory particle motion produced by high-intensity ultrasonic waves can also induce secondary flows, known as acoustic streaming.
39. Power ultrasound is an attractive means of drying heat sensitive foods because they can be dried more rapidly and at a lower temperature than in the conventional hot-air driers.
40. Radio frequency electric fields (RFEF) processing include a radio frequency power supply and a treatment chamber that is capable of applying high electric fields to liquid foods.
41. In RFEF, power supply is continuous whereas in PEF, it is pulsed.
42. In RFEF, increasing the electric field strength, number of treatment steps and temperature enhance the microbiological inactivation.
43. Pulsed light technology (PLT) can be used to sterilize, sanitize or reduce microbial load in foods, food packaging materials, as well as surfaces, environments, plants, devices and media involved in food processes.
44. The electromagnetic radiations are emitted and propagated by means of waves which differ in wavelength, frequency and energy.
45. PLT does not involve toxic chemicals or photolytic by-products.
46. PLT allows achievement of an effective microbial inactivation requiring much lower processing time and energy.
47. The major limitation of PLT is due to the intrinsically poor penetrating power of light and to the requirements of transparency and surface smoothness for the product to be treated.
48. Integrated membrane processes involve microfiltration, Ultrafiltration, reverse osmosis, osmotic membrane distillation and direct osmosis.
49. The integrated membrane processes allow improvements in the process efficiency and economy.
50. The extent of concentration achievable in the direct osmosis process depends on the osmotic pressure of osmotic agent.

51. Plasma, a quasi-neutral gas, is considered to be the fourth state of matter, following the more familiar states of solid, liquid & gas and constitutes more than 99% matter of the universe. It is more or less an electrified gas with a chemically reactive media that consists of a large number of different species such as electrons, positive and negative ions, free radicals, gas atoms and molecules in the ground or any higher state of any form of excited species.
52. Plasma can be produced at low-pressure or atmospheric pressure by coupling energy to a gaseous medium by several means such as mechanical, thermal, chemical, radiant, nuclear, or by applying a voltage, or by injecting electromagnetic waves and also by a combination of these to dissociate the gaseous component molecules into a collection of ions, electrons, charge-neutral gas molecules, and other species.
53. Cold plasmas refer to the plasmas where most of the coupled electrical energy is primarily channelled to the electron component of the plasma, thereby producing energetic electrons instead of heating the entire gas stream; while the plasma ions and neutral components remain at or near room temperature.
54. Cold plasma processing is useful for microbial inactivation while maintaining quality of fresh produce. This process is not effective for in vitro model food systems for inactivation of microbes or enzymes which are present in intact tissues, as it is a surface phenomenon.
55. Plasma technology is considered as modern non conventional technique which is used for the preparation of modified starches, altering its physical and chemical properties. It also offers many potential applications in food packaging.
56. A capacity coupled plasma (CCP) sources is one of the most common types of technological plasma sources. It essentially consists of two metal electrodes separated by a small distance, placed in a chamber. The gas pressure in the chamber can be lower or equal to atmospheric. A typical CCP system is driven by a single RF power supply, typically at 13.56 MHz. One of two electrodes is connected to the power supply, and the other one is grounded.
57. An inductive coupled plasma (ICP) is a type of plasma source in which the energy is supplied by electrical currents which are produced by electromagnetic induction, that is, by time-varying magnetic fields.
58. Plasma technology can be used for the modification of food packaging polymers.
59. Plasma technology can be used for food packaging surface sterilization.
60. Treatment of food packaging materials with cold plasma can enhance food packaging barrier properties.

15.2 MULTIPLE CHOICE QUESTIONS

1. Radioisotopes are also known as
(a) Radionucliotides
(b) radioactive isotpes
(c) both
(d) none

2. The Becquerel (Bq) is the unit of radioactivity and equals________ disintegration per second.
(a) One (b) two
(c) five (d) hundred

3. Which of the following is high frequency (low amplitude) wave
   (a) AC power (b) radiowaves
   (c) cosmic rays (d) microwaves

4. Which is true about high pressure processing
   (a) Desirable texture changes possible
   (b) in package processing possible
   (c) positive consumer appeal
   (d) all

5. Which of the following is high frequency (low amplitude) wave
   (a) Cosmic rays (b) gamma rays
   (c) x-rays (d) all

6. Which of the following is low frequency (high amplitude) wave
   (a) Infrared (b) radiowaves
   (c) microwaves (d) all

7. Low dose irradiation upto 1kGy can be done for
   (a) Sprout inhibition
   (b) delay of ripening
   (c) insect disinfestation
   (d) all

8. Medium dose irradiation can reduce numbers of
   (a) Spoilage microorganisms
   (b) non-spore forming pathogens
   (c) both
   (d) none

9. High dose irradiation is generally done for
   (a) Sterility
   (b) delay of ripening
   (c) insect disinfestations
   (d) none

10. HPP processing technique of foods involve
    (a) Hazard (b) power
    (c) pressure (d) none

11. HHP refers to ________ processing
    (a) High Hazard Pressure
    (b) High Hazard Pasteurization
    (c) High Hydrostatic Pasteurization
    (d) High Hydrostatic Pressure

12. UHP refers to ________ processing
    (a) Ultra High Pressure
    (b) Upstream High Pressure
    (c) Under High Pressure
    (d) all

13. HPP processing of food is ________ technique
    (a) Thermal (b) non-thermal
    (c) both (d) none

14. HPP ____________ can be used for
    (a) Liquid foods (b) solid foods
    (c) both (d) none

15. HPP is done between pressure of
    (a) 50-100mPa (b) 50-1000MPa
    (c) 50-100Pa (d) 50-1000Pa

16. HPP technique contains
    (a) Pressure vessel
    (b) pressure transmitting medium
    (c) both
    (d) none

17. Which of the following principle is used in HPP processing
    (a) Archimedes principle
    (b) Le Chatelier's principle
    (c) Beer's law
    (d) none

18. Which of the following principle is used in HPP processing
    (a) Lambert's law (b) Isostatic rule
    (c) both (d) none

19. Microbial inactivation in HPP processing occurs due to
    (a) Enzyme denaturation
    (b) ribosomes disruption
    (c) membrane permeability change
    (d) all

20. Cell wall of gram negative bacteria are ________ than gram negative bacteria
(a) weaker (b) stronger
(c) same (d) none

21. Bacterial spores can be killed at pressure higher than
(a) 10MPa (b) 80MPa
(c) 400MPa (d) 800MPa

22. Which of the following can influence microbial sensitivity to higher pressure
(a) pH (b) water activity
(c) temperature (d) all

23. PSF stands for
(a) Pressure Shift Freezing
(b) Pressure Sterilization Freezing
(c) Primary Sterilization Freezing
(d) Positive Shift Freezing

24. Exudation of moisture refers to
(a) Ice crystal (b) drip loss
(c) thawing (d) none

25. HPT stands for
(a) High Pressure Thawing
(b) High Pasteurization Thawing
(c) High Pressure technique
(d) none

26. PEF stands for
(a) Pressure Emission Force
(b) Pressurized Electric Field
(c) Pulsed Electric Field
(d) none

27. Electroporation occurs in
(a) HPP (b) PEF
(c) UHT (d) none

28. PEF treatment systems involves
(a) Treatment chamber
(b) storage capacitor
(c) resistor
(d) all

29. PEF treatment systems involves
(a) HV power supply
(b) inductors
(c) treatment chamber
(d) all

30. Which of the following is a PEF processing parameter
(a) Electric field strength
(b) treatment time
(c) specific energy
(d) all

31. Which of the following is a PEF processing parameter
(a) Pulse geometry
(b) treatment temperature
(c) conductivity
(d) all

32. PEF processing is a ____________ technique
(a) Thermal (b) non-thermal
(c) both (d) none

33. Which of the following is a component of PEF system
(a) Transformer (b) rectifier
(c) resistor (d) all

34. Which of the following is a component of PEF system
(a) Capacitor (b) food cell
(c) oscilloscope (d) all

35. Which of the following is a component of PEF system
(a) Thyratron (b) inductor
(c) trigger circuit (d) all

36. Wavelength of UV light ranges from ________ to ______ nm
(a) 10-100 (b) 100-400
(c) 500-800 (d) 1200-1400

37. Wavelength of UVA radiations ranges from ________ to ______ nm
(a) 100-200 (b) 200-280
(c) 280-315 (d) 315-400

38. Wavelength of UVB radiations ranges from ________ to ______ nm
(a) 100-200 (b) 200-280
(c) 280-315 (d) 315-400

39. Wavelength of UVC radiations ranges from ________ to ______ nm
(a) 100-200 (b) 200-280
(c) 280-315 (d) 315-400

40. Wavelength of vacuum UV radiations ranges from ________ to ______ nm
(a) 100-200 (b) 200-280
(c) 280-315 (d) 315-400

41. X rays have penetration depth in food
(a) 1cm (b) 5cm
(c) 60-400cm (d) 1000-2500cm

42. Which of the following is a size reduction technique
(a) Cutting (b) milling
(c) pulping (d) all

43. Which of the following is a size reduction technique
(a) Slicing (b) dicing
(c) mincing (d) all

44. Which of the following is a size reduction technique
(a) Shredding (b) flaking
(c) both (d) none

45. Which of the following is a size reduction technique
(a) Emulsification
(b) homogenization
(c) chopping
(d) all

46. Which of the following law is related to size reduction of food material
(a) Kick's law (b) Rittinger's law
(c) Bond's law (d) all

47. Bond's law is used to calculate ________ for size reduction
(a) Surface area (b) volume
(c) energy (d) none

48. Which mill is used for size reduction of food material
(a) pin and disc (b) wing-beater
(c) disc beater (d) all

49. Which mill is used for size reduction of food material
(a) Vertical toothed disc mill
(b) hammer mill
(c) ball mill
(d) all

50. Which operation uses slurry in size reduction
(a) Choke feeding
(b) closed circuit grinding
(c) wet milling
(d) all

51. Osmotic dehydration reduces
(a) Water loss
(b) loss of water soluble solutes
(c) both
(d) none

52. In sonication, cavitation refers to formation of ________ in liquid
(a) Bubbles (b) crystals
(c) precipitates (d) all

53. Osmotic dehydration can be used in combination with
(a) Sonication
(b) gamma-irradiation
(c) vacuum
(d) all

54. Osmotic dehydration can be used in combination with
(a) Centrifugal force
(b) HIPEF
(c) HHP
(d) all

55. Osmotic dehydration can be used for
(a) Partial concentration
(b) concentration
(c) dehydration
(d) all

56. Osmotic dehydration can be used for
(a) Dehydration
(b) concentration
(c) partial dehydration
(d) all

57. Osmotic dehydration can be used with air-drying to improve ________ property of food.
(a) nutrition (b) sensory
(c) functional (d) all

58. Osmotic dehydration ________ the sugar to acid ratio of food
(a) increases (b) decreases
(c) do not affect (d) none

59. Which is true about osmotic dehydration of food
(a) improve texture
(b) stability of pigment
(c) both
(d) none

60. Osmotic dehydration of fruits enhances________ behaviour
(a) viscous (b) elastic
(c) viscoelastic (d) none

61. Infusion of sugar in fruit during osmotic dehydration causes
(a) Elasticity (b) plasticity
(c) both (d) none

62. Which of the following is most eligible energy saving method
(a) Osmotic dehydration
(b) sun drying
(c) vacuum dehydration
(d) all

63. Which of the following is most eligible energy saving method
(a) Spray drying
(b) osmotic dehydration
(c) drum drying
(d) all

64. Which one is non thermal processing method
(a) Osmotic dehydration
(b) membrane technology
(c) both (d) none

65. Which is a limitation of membrane processing
(a) Concentration polarization
(b) fouling
(c) shear damage
(d) all

66. Evaporative concentration can be achieved using ________ evaporator
(a) Open pan (b) plate
(c) rising film (d) all

67. Evaporative concentration can be achieved using ________ evaporator
(a) Falling film
(b) agitated thin-film
(c) both
(d) none

68. Ultrafiltration is used for
(a) Clarification (b) concentration
(c) purification (d) all

69. Ultrafiltration membranes are made up of
(a) Polysuphone
(b) polyvinyldene fluoride
(c) cellulose acetate
(d) all

70. In ultrafiltration membrane, the pore size ranges from________ to ________ nm.
(a) 1-100 (b) 100-200
(c) 200-500 (d) 500-700

71. In ultrafiltration technique, applied pressure ranges from ______ to ______ bar
(a) 1-10 (b) 10-100
(c) 80-100 (d) 100-500

72. In reverse osmosis, excess pressure is applied ranges from______ to ______ bar
(a) 1-10 (b) 10-100
(c) 80-100 (d) 100-500

73. Reverse osmosis removes ________ % of ions

(a) 10 (b) 50
(c) 99 (d) 99.9

74. Reverse osmosis removes ________ % of viruses and bacteria

(a) 10 (b) 50
(c) 99 (d) 99.9

75. Reverse osmosis removes ________ % of pyrogens

(a) 10 (b) 50
(c) 99 (d) 99.9

76. Cook-chill products are stored above their

(a) Boiling point
(b) freezing point
(c) atmospheric pressure
(d) all

77. Cook-chill is a type of __________ processing

(a) non-thermal (b) minimal
(c) irradiation (d) none

78. Cook-chill system can be used by

(a) hospital (b) school
(c) airline catering (d) all

79. In cook-chill system, initial cooking is done above ______°C temperature

(a) 70 (b) 80
(c) 90 (d) 100

80. In cook-chill system, initial cooking is done above 70°C for not less than

(a) Two (b) five
(c) ten (d) fifteen

81. In cook-chill system, time for chilling to begin after initial cooking cycle

(a) 10 min (b) 20 min
(c) 30 min (d) 40 min

82. In cook-chill system, reheating temperature is above _________°C.

(a) 70 (b) 80
(c) 90 (d) 100

83. In cook-chill system, reheating is done at above 70°C for not less than ______min

(a) Two (b) five
(c) ten (d) fifteen

84. Cook freeze is a type of _________ processing

(a) minimal (b) irradiation
(c) pressure (d) PEF

85. Storage temperature in cook freeze system is

(a) -10°C (b) -18°C
(c) -27°C (d) -40°C

86. In cook freeze system time for freezing to begin after initial cooking cycle

(a) 10 min (b) 20 min
(c) 30 min (d) 40 min

87. In cook freeze system, reheating temperature is above _______°C

(a) 70 (b) 80
(c) 90 (d) 100

88. Sous-vide is a type of __________ processing

(a) Minimal (b) irradiation
(c) pressure (d) PEF

89. Sous-vide is a specific type of _________ processing

(a) Cook-freeze (b) cook-chill
(c) irradiation (d) microwave

90. In sous-vide system, initial cooking is done above_________°C

(a) 60 (b) 70
(c) 80 (d) 90

91. In sous-vide system, initial cooking is done above 70°C for________min

(a) 10 (b) 20
(c) 30 (d) time not specified

92. In sous-vide system, portioning is done at less than____________°C

(a) 10 (b) 20
(c) 30 (d) 50

93. In sous-vide system, portioning is done at less than 10∘C within _____ minutes
(a) 10 (b) 20
(c) 30 (d) time not specified

94. In sous-vide system, chilling time is _________ hours
(a) 1.5 (b) 2.5
(c) 4 (d) 6

95. In sous-vide system, storage temperature is ________∘C
(a) 10 (b) -10
(c) -18 (d) 15

96. Which is related to sous-vide sytem of processing
(a) Par-cooked
(b) vacuum packed
(c) in-pack pasteurization
(d) all

97. Which is true about sous-vide sytem of processing
(a) prevent evaporative loss of volatile flavor
(b) reduce oxidation
(c) reduce leaching of nutrient
(d) all

98. Sous-vide system reduces risk of
(a) Listeria monocytogenes
(b) Salmonella typhirium
(c) Clostridium botulinum
(d) all

99. Which pasteurization system takes least time
(a) Forced air convection
(b) Pressurized steam
(c) microwave
(d) none

100. Ohmic heating uses
(a) Electrodes (b) pressure
(c) both (d) none

101. Which one is related to minimal processing of fresh fruits and vegetables
(a) Trimming (b) peeling
(c) cutting (d) all

102. Which one is related to minimal processing of fresh fruits and vegetables
(a) Washing (b) drying
(c) packaging (d) all

103. Spoilage change affecting minimally fresh processed fruits and vegetables
(a) Off-flavors (b) discoloration
(c) softening (d) all

104. Spoilage change affecting minimally fresh processed fruits and vegetables
(a) Loss of crispness (b) water loss
(c) both (d) none

105. Which one affects quality of minimally fresh processed fruits & vegetables
(a) Mechanical damage
(b) biochemical damage
(c) physiological ageing
(d) all

106. Trimming is generally done at ____∘C (max) in fresh processed fruits and vegetables
(a) 5 (b) 10
(c) 15 (d) 20

107. Active or passive MAP is done at ____∘C (max) in fresh processed fruits and vegetables
(a) 5 (b) 10
(c) 15 (d) 20

108. Which one is related to clean area in minimally fresh processed vegetable processing plant
(a) Product washing (b) pre-cooling
(c) trimming (d) dewatering

109. Which one is related to clean area in minimally fresh processed vegetable processing plant
(a) Manual selection
(b) classification
(c) trimming (d) rinsing

110. Concentration of ozone used as disinfection agent in fresh processed vegetables
(a) 0.5-2 ppm (b) 3-5 ppm
(c) 25-200 ppm (d) none

111. Concentration of chlorine dioxide used as disinfection agent in minimally fresh processed vegetables
(a) 0.5-2 ppm (b) 3-5 ppm
(c) 25-200 ppm (d) none

112. Concentration of sodium hypochlorite used as disinfection agent in minimally fresh processed vegetables
(a) 0.5-2 ppm (b) 3-5 ppm
(c) 25-200 ppm (d) none

113. In minimally processing of fresh fruits and vegetables, which disinfe-ction agent causes higher population reduction on produce surface
(a) sodium hypochlorite
(b) calcium hypochlorite
(c) chlorine dioxide
(d) ozone

114. Which disinfection agent used in minimal processing has low irritancy
(a) sodium hypochlorite
(b) chlorine dioxide
(c) ozone
(d) all

115. Which disinfection agent is highly unstable in water
(a) sodium hypochlorite
(b) calcium hypochlorite
(c) chlorine dioxide
(d) ozone

116. Biocontrol is related to
(a) Chemistry (b) microbiology
(c) food physics (d) none

117. AFP stands for
(a) Anti freeze protein
(b) anti fungal protein
(c) both (d) none

118. Which is true about high pressure freezing
(a) Reduced duration of phase transition
(b) less mechanical stress during formation of ice crystals
(c) both
(d) none

119. Which of the following is related to magnetostrictive transducers
(a) Electrical coil
(b) oscillating magnetic field
(c) mechanical vibration
(d) all

120. Which of the following is related to piezoelectric transducers
(a) Electrode
(b) electrical insulator
(c) piezoelectric element
(d) all

121. Which is a type of ultrasound transducer
(a) Magnetostrictive
(b) piezoelectric
(c) both
(d) none

122. Power generator (in ultrasound) convert a standard electrical frequency (5-60 Hz) into high alternative frequency above
(a) 10kHz (b) 20kHz
(c) 50kHz (d) 100kHz

123. Which is a type of power ultrasonic system
(a) Ultrasonic bath
(b) ultrasonic probe
(c) air-borne power ultrasonic system
(d) all

124. Transducers in ultrasonic bath is bonded at
(a) Base (b) side walls
(c) both (d) none

125. Which is a component of ultrasonic bath
(a) Transducer (b) process medium
(c) both (d) none

126. Which is a component of ultrasonic bath
(a) Jacketed vessel (b) refrigerant
(c) timer (d) all

127. Which is a component of ultrasonic probe system
(a) Transducer housing
(b) ttransducer
(c) horn
(d) all

128. Which is a component of air-borne power ultrasonic system
(a) Power generator
(b) stepped plate transducer
(c) both
(d) none

129. Which is a component of air-borne power ultrasonic system
(a) Piezoelectric transducer
(b) mechanical amplifier
(c) radiating surface
(d) all

130. CFD stands for
(a) Computational Fluid Dynamics
(b) Central Food Dynamics
(c) Computational Food Design
(d) Central Food Design

131. Food freezing is a ________ process
(a) Heat transfer (b) mass transfer
(c) both (d) none

132. Which is a function of power ultrasound in assisting food freezing
(a) Initiation of ice nucleation
(b) acceleration of freezing process
(c) control of crystal size distribution in product
(d) all

133. Which affects the power ultrasound efficiency
(a) Acoustic power level
(b) acoustic duration
(c) both
(d) none

134. Which chemical change occur during food dehydration
(a) Color loss (b) solubility
(c) aroma loss (d) all

135. Which chemical change occur during food dehydration
(a) Rehydration (b) lipid oxidation
(c) shrinkage (d) all

136. Which chemical change occur during food dehydration
(a) Browning reaction
(b) gelatinization
(c) color loss
(d) all

137. Which physical change occur during food dehydration
(a) Rehydration (b) solubility
(c) both (d) none

138. Which physical change occur during food dehydration
(a) Texture and shrinkage
(b) aroma loss
(c) both
(d) none

139. Which nutritional change occur during food dehydration
(a) Vitamin loss
(b) protein loss
(c) microbial survival
(d) all

140. Which is related to hybrid system of heat pump drier
(a) Modified atmosphere heat pump
(b) solar assisted
(c) heat pump assisted heating drier
(d) all

141. Which is a component of RF-assisted heat pump drier
(a) Evaporator (b) compressor
(c) expansion valve (d) all

142. Which is a component of RF-assisted heat pump drier
(a) Condenser (b) RF generator
(c) axial fan (d) all

143. RFD process can be used for
(a) pre-heating (b) pre-cooking
(c) both (d) none

144. RFD process can be used for
(a) Sterilization (b) tempering
(c) post baking (d) all

145. Thicker materials can be heated in microwave at __________ frequency
(a) Higher (b) lower
(c) zero (d) none

146. Penetration depth increases with _________ in microwave frequency
(a) Decrease (b) increase
(c) do not change (d) none

147. Which one has good electrical conductivity
(a) Condiments (b) yoghurt
(c) egg (d) all

148. Which one has low electrical conductivity
(a) Milk desserts (b) powders
(c) fruit juices (d) all

149. Which one has low electrical conductivity
(a) Marmalade (b) hydrocolloids
(c) gelatine (d) all

150. Which one has low electrical conductivity
(a) Wine (b) margarine
(c) egg (d) all

151. Which one has good electrical conductivity
(a) Wine (b) gelatine
(c) hydrocolloids (d) all

152. Which one has poor electrical conductivity
(a) Frozen food (b) yoghurt
(c) fruit juices (d) all

153. Which one has poor electrical conductivity
(a) Foam (b) syrup
(c) liquor (d) all

154. Which one has poor electrical conductivity
(a) Fat (b) condiment
(c) egg (d) wine

155. Which one is configuration of ohmic heating
(a) Batch configuration
(b) transverse ohmic heating
(c) collinear ohmic heating
(d) all

156. Which one affect in ohmic heating
(a) Slip velocity
(b) volume fraction
(c) particle diameter
(d) all

157. Which of the following is a medium dielectric heating frequency range
(a) 300-3000 kHz
(b) 3-30 MHz
(c) 30-300 MHz
(d) 300-3000 MHz

158. Which of the following is a high dielectric heating frequency range
(a) 300-3000 kHz
(b) 3-30 MHz
(c) 30-300 MHz
(d) 300-3000 MHz

159. Which of the following is a very high dielectric heating frequency range
(a) 300-3000 kHz
(b) 3-30 MHz
(c) 30-300 MHz
(d) 300-3000 MHz

160. Which of the following is a ultra high dielectric heating frequency range
(a) 300-3000 kHz
(b) 3-30 MHz
(c) 30-300 MHz
(d) 300-3000 MHz

161. Which of the following is a super high dielectric heating frequency range
(a) 30-300 MHz
(b) 300-3000 MHz
(c) 3-30 GHz
(d) 30-300 GHz

162. Which of the following is a extremely high dielectric heating frequency range
(a) 30-300 MHz (b)300-3000 MHz
(c) 3-30 GHz (d) 30-300 GHz

163. Microwave can be used for
(a) Heating (b) drying
(c) cooking (d) all

164. Microwave can be used for
(a) Thawing (b) tempering
(c) pasteurization (d) all

165. Microwave can be used for
(a) Roasting (b) blanching
(c) sterilization (d) all

166. Infrared heating can be applied for
(a) Drying (b) baking
(c) roasting (d) all

167. Infrared heating can be applied for
(a) Blanching (b) pasteurization
(c) both (d) none

168. Near-infrared (NIR) radiations correspond to spectral range of ________ μm
(a) 0.75-1.4 (b) 1.4-3.0
(c) 3.0-1000 (d) none

169. Mid-infrared (MIR) radiations correspond to spectral range of ________ μm
(a) 0.75-1.4 (b) 1.4-3.0
(c) 3.0-1000 (d) none

170. Far-infrared radiations correspond to spectral range of ____ μm
(a) 0.75-1.4 (b) 1.4-3.0
(c) 3.0-1000 (d) none

171. Which of the following law is related to infrared radiations
(a) Planck's law
(b) Wien's displacement law
(c) both
(d) none

172. Which of the following law is related to infrared radiations
(a) Stefan-Boltzmann's law
(b) Modified Beer's law
(c) both
(d) none

173. As food is exposed to infrared radiations, it ________ the radiations
(a) Absorbed (b) reflected
(c) scattered (d) all

174. Hydroxyl group shows infrared absorption bands at wavelength (μm)
(a) 2.7-3.3 (b) 3.25-3.7
(c) 5.71-5.76 (d) 5.92

175. Aliphatic carbon-hydrogen bond shows infrared absorption bands at wavelength (μm)
(a) 2.7-3.3 (b) 3.25-3.7
(c) 5.71-5.76 (d) 5.92

176. Carbonyl group (ester) shows infrared absorption bands at wavelength (μm)
(a) 2.7-3.3 (b) 3.25-3.7
(c) 5.71-5.76 (d) 5.92

177. Carbonyl group (amide) shows infrared absorption bands at wavelength (μm)
(a) 2.7-3.3 (b) 3.25-3.7
(c) 5.71-5.76 (d) 5.92

178. Nitrogen-hydrogen group shows infrared absorption bands at wavelength (μm)

(a) 2.83-3.33 (b) 4.44-4.76
(c) 5.95-6.05 (d) 7

179. Carbon-carbon double bond shows infrared absorption bands at wavelength (μm)
(a) 2.83-3.33 (b) 4.44-4.76
(c) 5.95-6.05 (d) 7

180. Electric field strength ($kVcm^{-1}$) in PEF technique ranges from ________ to ________
(a) 1.2-3.5 (b) 12-35
(c) 112-350 (d) none

181. Mechanism by which microorganisms are destroyed by electric field
(a) Electroporation (b) electrolysis
(c) oxidation (d) all

182. Degree of inactivation of micro-organisms by HIPEF increases at
(a) High electric field intensity
(b) higher number of pulses
(c) larger duration of pulses
(d) all

183. Which of the following affect degree of inactivation of microorganisms in HIPEF
(a) Temperature of food
(b) pH
(c) both
(d) none

184. Which of the following affect degree of inactivation of microorganisms in HIPEF
(a) Ionic strength
(b) electrical conductivity
(c) both
(d) none

185. Which is a component of HIPEF plant
(a) Pulse generator
(b) capacitors
(c) discharge chamber
(d) all

186. Inductors in HIPEF equipment are used to modify ______ of electric field pulse
(a) Shape (b) width
(c) both (d) none

187. Which of the following is present in HIPEF equipment
(a) Refrigeration coil
(b) discharge switch
(c) inductors
(d) all

188. Component of high pressure system is
(a) Pressure generation system
(b) temperature control device
(c) material handling system
(d) all

189. 'Monobloc' is related to ________ processing
(a) HPP (b) PEF
(c) irradiation (d) none

190. Pulsed light technology inactivate microorganisms by ______ effect
(a) Photochemical (b) photothermal
(c) both (d) none

191. If the force is perpendicular to surface of material generated by ultr-asonic wave after heating it produces
(a) Compression wave
(b) shearing wave
(c) both (d) none

192. If the force is parallel to surface of material generated by ultrasonic wave after heating it produces
(a) Compression wave
(b) shearing wave
(c) both (d) none

193. Cavitation is related to ________ processing
(a) HPP
(b) ultrasonication
(c) pulse light technology
(d) none

194. Ultrasound produces very rapid changes in pressure and temperature that causes
(a) Shear disruption
(b) cavitation
(c) thinning of cell membrane
(d) all

195. Mano-thermo-sonication involves
(a) Pressure (b) temperature
(c) sonication (d) all

196. Which is related to dielectric heating
(a) Polarization (b) loss factor
(c) both (d) none

197. OMD stands for
(a) Osmotic Membrane Distillation
(b) Osmotic Membrane Digestion
(c) Osmophillic Microbial Decay
(d) all

198. Osmotic membrane distillation is __________ process
(a) Athermal
(b) non-pressure driven
(c) both
(d) none

199. Driving force in osmotic membrane distillation is
(a) Vapor pressure difference
(b) pressure difference
(c) temperature difference
(d) concentration difference

200. Osmotic membrane distillation can be done for
(a) Pre-concentration
(b) concentration
(c) both
(d) none

201. Osmotic membrane distillation is a separation process in which micro-porous ________ membrane separates the two aqueous solution
(a) Hydrophillic (b) hydrophobic
(c) both (d) none

202. In osmotic membrane distillation, _______ can be used as osmotic agent
(a) $MgSO_4$ (b) NaCl
(c) $CaCl_2$ (d) all

203. In osmotic membrane distillation, ___ ______ can be used as osmotic agent
(a) Glycerol (b) brine
(c) $MgCl_2$ (d) all

204. Osmotic membrane distillation involves _______ transfer
(a) Heat (b) mass
(c) both (d) none

205. In osmotic membrane distillation, membranes are made up of
(a) PTFE (b) PP
(c) PVDF (d) all

206. In osmotic membrane distillation, __________ of membrane is decisive parameter to make the process viable
(a) Hydrophobicity (b) pore size
(c) porosity (d) none

207. Which of the following is related to osmotic membrane distillation
(a) Wettability
(b) Laplace equation
(c) both
(d) none

208. Which of the following affect osmotic membrane distillation flux
(a) Feed
(b) Osmotic agent concentration
(c) both
(d) none

209. Transmembrane flux increases with __________ in temperature
(a) increase (b) decrease
(c) do not change (d) none

210. Direct osmosis is ________ process
(a) Athermal
(b) non-pressure driven
(c) both
(d) none

211. Which one is related to radio frequency electric field equipment
(a) Amplifier (b) transformer
(c) function generator
(d) all

212. Which one is related to radio frequency electric field equipment
(a) Voltage divider (b) current probe
(c) oscilloscope (d) all

213. Comminution is also known as
(a) Size reduction (b) mixing
(c) irradiation (d) none

214. Ultrasonic homogenisers use ______ frequency sound waves
(a) Low (b) high
(c) both (d) none

215. Ultrasonic homogenisers use high frequency sound waves to cause ______ ______ in low viscosity liquids
(a) Compression (b) tension
(c) both (d) none

216. Ultrasonic homogenisers are used for production of
(a) Salad cream
(b) baby foods
(c) essential oil emulsion
(d) all

217. Which of the following has mixing action
(a) Extruders
(b) size reduction equipment
(c) both (d) none

218. Electroporation during PEF processing causes destabilization of ____ in cell membranes
(a) Lipid bilayer (b) protein
(c) both (d) none

219. In PEF processing, which one is economical
(a) Higher field strength
(b) shorter pulses
(c) both (d) none

220. Degree of inactivation of vegetative cells by PEF is greater at
(a) Higher electric field intensities
(b) higher number of pulses
(c) larger duration of pulses
(d) all

221. In PEF processing, electric field pulses are
(a) Monopolar (b) bipolar
(c) both (d) none

222. In PEF processing, wave form of electric pulses are
(a) Sinusoidal (b) square
(c) rectangular (d) all

223. Which of the following wave is least energy efficient for microbial inactivation in PEF processing
(a) Oscillatory
(b) square wave
(c) bipolar
(d) exponential decaying

224. Which of the following wave is more energy efficient for microbial inactivation in PEF processing
(a) square wave
(b) bipolar
(c) exponential decaying
(d) all

225. Which pulse is more deadly for microorganisms in PEF processing
(a) Monopolar (b) bipolar
(c) both are equal (d) none

226. Products having higher ionic strength have_______ electrical conductivity
(a) Higher (b) lower
(c) equal (d) none

227. In PEF processing, inactivation of microorganisms _______ with increasing conductivity of food material
(a) Increases (b) decreases
(c) do not change (d) none

228. In PEF processing, ______ arrangement is used for treatment chamber
(a) Coaxial (b) co-field
(c) both (d) none

229. In PEF processing, co-field chamber having two hollow cylindrical electrodes are separated by______
(a) Insulator (b) capacitor
(c) resistor (d) none

230. PEF treatment generally causes a rise in product temperature upto ____°C
(a) 30 (b) 70
(c) 80 (d) 90

231. PEF processing is restricted to
(a) Liquid foods
(b) low conductivity food
(c) both
(d) none

232. In PEF processing, presence of bubbles in food causes
(a) non-uniform treatment
(b) safety problems
(c) both (a) & (b)
(d) higher inactivation

233. Vegetative microbial cells are more sensitive to PEF in ______ phase of growth
(a) Logarithmic (b) stationary
(c) both (d) none

234. Which of the following is more resistant to PEF processing
(a) Bacterial spores
(b) yeast ascospores
(c) both (a) & (b)
(d) cells

235. In PEF processing, which is more effective at inactivating microorganisms
(a) Higher electric field strength
(b) longer treatment time
(c) both
(d) none

236. Which is more resistant to PEF treatment
(a) Enzymes
(b) microorganisms
(c) both are equal sensitive
(d) none

237. Which affects enzyme inactivation in PEF treatment
(a) Elelctric field strength
(b) active site
(c) suspension medium
(d) all

238. Which affects enzyme inactivation in PEF treatment
(a) Number of pulses
(b) pulse duration
(c) pulse width
(d) all

239. Which affects enzyme inactivation in PEF treatment
(a) Total treatment time
(b) secondary structure
(c) tertiary structure
(d) all

240. Which is true about oscillating magnetic field
(a) Kills vegetative cells
(b) preserve color and flavor
(c) low energy input
(d) all

241. Which is true about PEF technology
(a) Kills vegetative cells
(b) no toxicity
(c) shorter treatment time
(d) all

242. Which is true about pulse light technology
(a) Medium cost
(b) very rapid process
(c) low energy input
(d) all

243. Which is true about pulse light technology
(a) Suitable for dry foods
(b) little or no change to foods
(c) both
(d) none

244. Which is true about ultrasound processing
(a) Effective against vegetative cells
(b) effective against spores
(c) effective against enzymes
(d) all

245. Which is true about ultrasound processing
(a) Heat transfer increased
(b) reduction of processing time
(c) affect enzyme activity
(d) all

246. Which is true about photodynamic system
(a) low cost
(b) no additive requirement
(c) natural light used
(d) all

247. Which is true about gamma irradiation
(a) Excellent penetration to food
(b) suitable for sterilization
(c) low energy cost
(d) all

248. Which is true about gamma irradiation
(a) Little quality loss
(b) suitable for dry foods
(c) insecticidal
(d) all

249. Which is true about high pressure processing
(a) Kills vegetative bacteria
(b) no toxicity
(c) preserve color
(d) all

250. Which is true about high pressure processing
(a) Preserve flavor
(b) reduced processing time
(c) uniform treatment
(d) all

15.3 ANSWER KEY

Q. A.	Q. A.	Q. A.	Q. A.	Q. A.	Q. A.
1. (c)	2. (a)	3. (c)	4. (d)	5. (d)	6. (d)
7. (d)	8. (c)	9. (a)	10. (c)	11. (d)	12. (a)
13. (b)	14. (c)	15. (b)	16. (c)	17. (b)	18. (b)
19. (d)	20. (a)	21. (d)	22. (d)	23. (a)	24. (b)
25. (a)	26. (c)	27. (b)	28. (d)	29. (d)	30. (d)
31. (d)	32. (b)	33. (d)	34. (d)	35. (d)	36. (b)
37. (d)	38. (c)	39. (b)	40. (a)	41. (c)	42. (d)
43. (d)	44. (c)	45. (d)	46. (d)	47. (c)	48. (d)
49. (d)	50. (c)	51. (b)	52. (a)	53. (d)	54. (d)
55. (a)	56. (c)	57. (d)	58. (a)	59. (c)	60. (a)
61. (b)	62. (b)	63. (b)	64. (c)	65. (d)	66. (d)

Q.	A.	Q.	A.	Q.	A.	Q.	A.	Q.	A.	Q.	A.
67.	(c)	68.	(d)	69.	(d)	70.	(a)	71.	(a)	72.	(b)
73.	(c)	74.	(d)	75.	(d)	76.	(b)	77.	(b)	78.	(d)
79.	(a)	80.	(a)	81.	(c)	82.	(a)	83.	(a)	84.	(a)
85.	(b)	86.	(c)	87.	(a)	88.	(a)	89.	(b)	90.	(b)
91.	(d)	92.	(a)	93.	(c)	94.	(a)	95.	(c)	96.	(d)
97.	(d)	98.	(d)	99.	(c)	100.	(c)	101.	(d)	102.	(d)
103.	(d)	104.	(c)	105.	(d)	106.	(b)	107.	(a)	108.	(d)
109.	(d)	110.	(a)	111.	(b)	112.	(c)	113.	(d)	114.	(a)
115.	(d)	116.	(b)	117.	(a)	118.	(c)	119.	(d)	120.	(d)
121.	(c)	122.	(b)	123.	(d)	124.	(a)	125.	(c)	126.	(d)
127.	(d)	128.	(c)	129.	(d)	130.	(a)	131.	(c)	132.	(d)
133.	(c)	134.	(a)	135.	(b)	136.	(d)	137.	(c)	138.	(c)
139.	(d)	140.	(d)	141.	(d)	142.	(d)	143.	(c)	144.	(d)
145.	(b)	146.	(a)	147.	(d)	148.	(b)	149.	(a)	150.	(b)
151.	(d)	152.	(a)	153.	(d)	154.	(a)	155.	(d)	156.	(d)
157.	(a)	158.	(b)	159.	(c)	160.	(d)	161.	(c)	162.	(d)
163.	(d)	164.	(d)	165.	(d)	166.	(d)	167.	(c)	168.	(a)
169.	(b)	170.	(c)	171.	(c)	172.	(c)	173.	(d)	174.	(a)
175.	(b)	176.	(c)	177.	(d)	178.	(a)	179.	(b)	180.	(b)
181.	(d)	182.	(d)	183.	(c)	184.	(c)	185.	(d)	186.	(c)
187.	(d)	188.	(d)	189.	(a)	190.	(c)	191.	(a)	192.	(b)
193.	(b)	194.	(d)	195.	(d)	196.	(c)	197.	(a)	198.	(c)
199.	(a)	200.	(a)	201.	(b)	202.	(d)	203.	(d)	204.	(c)
205.	(d)	206.	(a)	207.	(c)	208.	(c)	209.	(a)	210.	(c)
211.	(d)	212.	(d)	213.	(a)	214.	(b)	215.	(c)	216.	(d)
217.	(c)	218.	(c)	219.	(c)	220.	(d)	221.	(c)	222.	(d)
223.	(a)	224.	(a)	225.	(b)	226.	(a)	227.	(b)	228.	(c)
229.	(a)	230.	(a)	231.	(c)	232.	(c)	233.	(a)	234.	(c)
235.	(c)	236.	(a)	237.	(d)	238.	(d)	239.	(d)	240.	(d)
241.	(d)	242.	(d)	243.	(c)	244.	(d)	245.	(d)	246.	(d)
247.	(d)	248.	(d)	249.	(d)	250.	(d)				

CHAPTER 16

MISCELLANEOUS

16.1 SUBJECTIVE KEYNOTES

1. The initial Good Manufacturing Practice Regulations for foods was published in 1969, and is often referred to as the Umbrella GMP.
2. Quality criteria for the consumers include appearance, texture, flavor, nutritive value and safety.
3. Cohesiveness is the degree to which the sample deforms before rupturing when biting with molars.
4. Gummines is the energy required disintegrating a semi-solid food to a state ready for swallowing.
5. Viscosity is the force required for drawing a liquid from a spoon over the tongue.
6. Flavor is a combination of aroma.
7. Factors affecting quality of foods include genetic, pre-harvest, harvest and post harvest factors.
8. Gibberlic acid increases size and weight of grapes, berries, apricot and strawberry and causes parthenocarpic fruits and reduces disorder like water spots and corky spots in citrus.
9. Pruning affects the size, colour, acidity and sugar content of the fruits such as grapes.
10. The Alveograph measures air pressure inside of a dough bubble.
11. Mixograph and Farinograph give information regarding optimum dough water absorption, strength, mixing time, and tolerance to over mixing.
12. Succulometer is used to measure the maturity of sweet corn and also used to determine the storage life of apples.
13. Tendrometer is used to measure the tenderness of pea.
14. Instruments used to measure texture of fruits are texture analyzer, puncture meter and fibrometer.
15. Asrigency is dry puckery sensation due to precipitation of the proteins in the saliva and in the mucous membrane lining of the mouth which deprives them of their lubricating character.
16. Water activity refers to amount of water available for microorganisms in food.
17. Fat soluble vitamins are vitamin A, D, E and K.

18. Water soluble vitamins are vitamin B and vitamin C.
19. Essential minerals include Fe, Cu, I, Co, Mn, Zn, Cr, Ni, Si, F, Mo and Se.
20. Freezing is effective treatment for food preservation.
21. Milling of wheat helps to separate bran and germ from endosperm.
22. Controlled atmospheric storage is a method in which selected atmospheric concentrations of gases are maintained throughout food storage in order to extend product shelf life.
23. Sodium chloride (NaCl) is a curing agent.
24. Sausages are the meat products usually made from ground meat with a skin around it.
25. Mango is a climacteric fruit.
26. Nitrates and nitrites are curing agents which help in color retention of meat.
27. Myoglobin is an iron- and oxygen-binding protein found in the muscle tissue of vertebrates in general and in almost all mammals.
28. Citrus peel contains high levels of pectin.
29. Folic acid deficiency causes pernicious anaemia.
30. Hedonic scale is used for sensory analysis.
31. Triangle test is related to sensory analysis of food products.
32. Blackening of the pickle occurs due to iron and formation of hydrogen sulfide.
33. Potassium metabisulphite is used to preserve the taste of dehydrated foods and to make these foods more palatable. Potassium metabisulphite cannot preserve certain kinds of foods, including raw fruits, vegetables and meats.
34. In glycogenolysis, glycogen stored in the liver and muscles, is converted first to glucose-1- phosphate and then into glucose-6-phosphate. Two hormones which control glycogenolysis are a peptide, glucagon from the pancreas and epinephrine from the adrenal glands.
35. *Diabetes mellitus* commonly referred to as diabetes, is a group of metabolic disorders in which there are high blood sugar levels over a prolonged period. Medication involves- insulin and metformin.
36. APEDA Agricultural and Processed Food Products Export Development Authority (APEDA) was established by the Government of India under the Agricultural and Processed Food Products Export Development Authority Act passed by the Parliament in December, 1985.
37. WHO- World Health Organization established on 7 April 1948, headquartered in Geneva, Switzerland.
38. Legumes are low in glycemic index.
39. The protein content of egg is approx. 13% by wt of egg, egg white and egg yolk comprises of 45% and 31% protein, by weight.
40. Barley malt is used for the preparation of beer.
41. Natural aging occurs when freshly milled "green" flour is exposed to air for several days or weeks. Flour is naturally aged when it comes into contact with the oxygen in

the air. The gluten network gets strengthened by oxidizing gluten-forming proteins, allowing them to create stronger gluten.

42. Bleaching of flour yields flour with better gluten elasticity characteristics.
43. In Golden Rice two genes have been inserted into the rice genome by genetic engineering, to restart the carotenoid biosynthetic pathway leading to the production and accumulation of β-carotene in the grains. Both genes are naturally involved in carotene biosynthesis.
44. ALA is found in certain vegetable oils, walnuts, flaxseeds and soy products. EPA and DHA are found in fish, seafood and fish oils.
45. Kimchi is a traditional Korean dish of fermented vegetables.
46. Wheat/ cereals are deficient in lysine amino acid, while legumes are limiting in cysteine and methionine.
47. Spread ratio or spread factor of cookie is calculated as ratio of diameter to thickness of cookie.
48. Sorption isotherm is plotted between moisture content and aw of food product.
49. Chloropicrin is still registered for grain fumigation as well as empty binfumigation to combat insect infestations. Methyl bromide is an effective grain fumigant for the control of stored grain insects at all stages of development.
50. Preparation of jelly requires pectin, acid, water and sugar.
51. The enzyme lipase should be inactivated, preferably prior to homogenization or immediately afterwards.
52. The term "aflatoxin" is derived from the name of one of the molds that produce it, Aspergillus flavus.
53. The specific heat is the amount of heat per unit mass required to raise the temperature by one degree Celsius.
54. Rennin is an enzyme used in preparation of cheese.
55. The ultraviolet (UV) region scanned is normally from 200 to 400 nm, and the visible portion is from 400 to 800 nm. The range of visible spectrophotometer is 400-800nm.
56. The temperature range should be high enough for the meat to cure properly and dry, but low enough so bad bacteria and mold doesn't grow. Ideal temperatures are between 50-60F (10C-15C). Humidity is equally as important as temperature. The humidity of your hanging environment needs to be between 65% and 80%.
57. Saffron is obtained from stigma of flower of Crocus sativus plant.
58. The Class-I Preservative includes salt, vinegar, salt, vegetable oil, honey, sugar and wood smoke. Sorbates, sulfites and benzoates are some of the Class II preservatives used.
59. Poly propylene (oriented and metalized) and PET metalized have lowest gas and moisture permeability.
60. Fat in commercial ice cream is 10%.
61. Oats are rich source of beta glucan.
62. End products of glycolysis are two pyruvate, two NADH, and two ATP.

63. Anti nutritional factor in egg is avidin
64. Enzymatic meat tenderizers are Papain and Bromelain.
65. Phosphatase test is done to check the adequacy of pasteurization.
66. Nisin is a natural preservative which is used in various food products for increasing shelf life and inhibiting growth of bacteria.
67. Sorbic acid and calcium propionate are used for preservation of cake and other baked products.
68. Solid carbon dioxide is called dry ice.
69. Hurdle technology deliberately combines existing and new preservation techniques to establish a series of preservative factors (hurdles) that the microorganisms in question are unable to overcome (jump over). These hurdles may be temperature, water activity, acidity, redox potential, preservatives, and others.
70. Sorbic acid is most active at pH 4.4.
71. Hard wheat is different from soft because of difference in its protein content, endosperm structure, strength and damaged starch content.
72. Avocados are about 77% fat, by calories, making them even higher in fat than most animal foods. The main fatty acid is a monounsaturated fat called oleic acid.
73. Pycnometer is used for measurement of specific gravity.
74. Refractometer is used to measure Total Soluble Solids.
75. Force of attraction between similar molecules is called cohesive forces.
76. Gibberellic acid is used to extend the shelf life of Fruits and vegetables.
77. Moisture content for paddy storage is less than 14% for grain storage and less than 9% for long term seed preservation.
78. Parboiling of paddy increases its vitamins and minerals content.
79. Tomato ketchup is a bingham fluid.
80. The quality of egg is checked by "candling".
81. Wax edible coating improves the shelf life, imparts glossiness and prevents dehydration of fresh fruits and vegetables.
82. Black bread mold (Rhizopus stolonifer) is a widely distributed thread-like mucoralean mold found on bread surfaces.
83. Fat bloom is caused by a thin layer of fat crystals on the surface of the chocolate. Fat bloom happens if the chocolate gets too warm. The cocoa butter melts and then re-solidifies, leaving those gray streaks.
84. Sugar bloom happens if the chocolate was stored in a damp area. Moisture collects on the surface of the chocolate and draws out the sugar. When the moisture evaporates, it leaves behind a grit of sugar crystals across the surface.
85. Leaching is the process of extracting a substance from a solid material that has come into contact with a liquid.
86. Packaging tests include bursting strength, compression test, WVTR.
87. Enrobing is a process that involves covering a confection or snack with chocolate or chocolate coatings.

88. Irradiation is the process of exposing a material to ionizing radiation whose source is photons (γ-rays, x-rays), or high energy electrons. Commonly, γ-rays are produced by radioactive isotopes such as cobalt-60 and cesium-137.

89. In heat transfer at a boundary (surface) within a fluid, the Nusselt number (Nu) is the ratio of convective to conductive heat transfer across (normal to) the boundary. In this context, convection includes both advection and diffusion.

90. The Prandtl Number is a dimensionless number approximating the ratio of momentum diffusivity and thermal diffusivity. Sponsored Links. The Prandtl Number is a dimensionless number approximating the ratio of momentum diffusivity (kinematic viscosity) and thermal diffusivity and can be expressed as. $Pr = v / \alpha$ (1).

91. Holstein Friesians are a breed of dairy cattle originating from the Dutch provinces of North Holland and Friesland, and what is now Schleswig-Holstein in Northern Germany and Jutland. They are known as the world's highest-production dairy animals.

92. The white leghorn lays pure white eggs, usually about 300 per year, the highest for poultry.

93. The RQ is essentially the ratio of carbon dioxide eliminated to oxygen consumed. Roughly: RQ = CO_2 eliminated / O_2 consumed. A RQ closer to 1 typically means more carbohydrate is being burned and a value closer to 0.7 typically means fat is being burned. A value of 0.7 indicates that lipids are being metabolized, 0.8 for proteins, and 1.0 for carbohydrates.

94. BMI stands for Basal metabolic rate and it quantifies the amount of tissue mass (muscle, fat, and bone) in an individual, and then categorize that person as underweight, normal weight, overweight, or obese based on that value. It is ratio of $(\text{Height})^2/\text{Wt}$ in Kg.

95. Hops are responsible for bitter flavor of beer.

96. Psychrophilic bacteria, otherwise known as psychrophiles, are bacteria that are able to grow at low-temperatures. These types of bacteria include Coliforms, Vibrio, Pseudomonas, and Listeria, which are all in fact quite harmful to the human health.

97. Acid produced in sauerkraut is lactic acid. Sauerkraut is finely cut cabbage that has been fermented by various lactic acid bacteria, including Leuconostoc, Lactobacillus, and Pediococcus.

98. Increase in volume of dough is due to CO2 gas produced as a result of growth of yeast.

99. Aflatoxins are highly toxic and carcinogenic chemical substances produced by Aspergillus spp. on variety of agricultural commodities including groundnut.

100. Class I preservatives are generally preservatives that are found in the common kitchen. The Class-I Preservative includes salt, vinegar, salt, vegetable oil, honey, sugar and wood smoke. Sorbates, sulfites and benzoates are some of the Class II preservatives used.

16.2 MULTIPLE CHOICE QUESTIONS

1. NIFTEM stands for
   (a) National Institute of Food Technology, Energy & Management
   (b) Novel Institute of Food Technology, Energy & Management
   (c) National Institute of Food Technology, Enterpreneurship & Management
   (d) none

2. NIFTEM is located at
   (a) Karnal (b) Sonepat
   (c) New Delhi (d) Panipat

3. NIFTEM is located at
   (a) Haryana (b) Karnataka
   (c) Punjab (d) New Delhi

4. NDRI stands for
   (a) National Draught Resistant Institute
   (b) North Delhi Research Institute
   (c) National Dairy Research Institute
   (d) none

5. NDRI is located at
   (a) Haryana (b) Karnataka
   (c) Punjab (d) New Delhi

6. NDRI is located at
   (a) Karnal (b) Sonepat
   (c) New Delhi (d) Panipat

7. IARI stands for
   (a) Institute of Agricultural Research, India
   (b) Indian Agricultural Research Institute
   (c) Internal Agricultural & Research Institute
   (d) none

8. IARI is located at
   (a) Haryana (b) Karnataka
   (c) Punjab (d) New Delhi

9. IARI is located at
   (a) Karnal (b) Sonepat
   (c) New Delhi (d) Panipat

10. NABI stands for
    (a) National Antibiotic Institute
    (b) National Agri - Food Biotechnology Institute
    (c) Non- Agricultural Biotechnology Institute
    (d) all

11. NABI is located at
    (a) Haryana (b) Karnataka
    (c) Punjab (d) New Delhi

12. NABI is located at
    (a) Karnal (b) Mohali
    (c) New Delhi (d) Ranchi

13. CIPHET stands for
    (a) Central Institute of Pharma Education and Technology
    (b) Central Institute of Post Harvest Engineering and Technology
    (c) Central Institute for Physical Handicapped Education and Training
    (d) none

14. CIPHET is located at
    (a) Karnal (b) Mohali
    (c) New Delhi (d) Ranchi

15. SLIET is located at
    (a) Karnal (b) Longowal
    (c) Kharagpur (d) Ranchi

16. CFTRI stands for
    (a) Central Food Technology and Research Institute
    (b) Central Fashion Technology and Research Institute
    (c) Centre for Training and Research In India
    (d) none

17. CFTRI is located at
(a) Mumbai (b) Anand
(c) Mysore (d) Ranchi

18. NDDB stands for
(a) National Dairy Development Board
(b) National Drumstick Development Board
(c) National Drum-drier Bussiness
(d) none

19. NDDB is located at
(a) Mumbai (b) Anand
(c) Mysore (d) Ranchi

20. IICPT stands for
(a) International Institute of Crop Processing & Technology
(b) Indian Institute of Crop Processing & Technology
(c) Indian Institute of Chemical Protection Technology
(d) none

21. IICPT is located at
(a) Haryana (b) West Bengal
(c) Tamil Nadu (d) Rajasthan

22. ICAR stands for
(a) Indian Council of Agricultural Research
(b) Indian Council of Antibiotics Research
(c) Indian Council of Aeronautics Research
(d) none

23. ICAR is located at
(a) Mumbai (b) Anand
(c) Mysore (d) New Delhi

24. CPRI stands for
(a) Council of Product Research in India
(b) Council of Potato Research In India
(c) Central Potato Research Institute
(d) none

25. CPRI is located at
(a) Shimla (b) Srinagar
(c) New Delhi (d) Mumbai

26. DWR stands for
(a) Directorate of Wheat Research
(b) Division of Wheat Research
(c) Drectorate of Weaning Food Research
(d) none

27. DWR is located at
(a) Karnal (b) Anand
(c) Mysore (d) Ranchi

28. DGR stands for
(a) Directorate of Grape Research
(b) Directorate of Groundnut Research
(c) Directorate of Guava Research
(d) none

29. DGR is located at
(a) Haryana (b) Rajasthan
(c) Gujarat (d) Uttar Pradesh

30. NIN stands for
(a) National Institute of Nutrition
(b) National Institute of Netherland
(c) International Institute of Neurology
(d) none

31. CIFE stands for
(a) Central Institute of Food Education
(b) Central Institute of Fisheries Education
(c) Central Institute of Fashion Education
(d) none

32. CIFE is located at
(a) Mumbai (b) Chennai
(c) New Delhi (d) Lucknow

33. Central Agricultural Research Institute is located at
(a) New Delhi (b) Port Blair
(c) Ranchi (d) Jhansi

34. Central Inland Fisheries Research Institute is located at
(a) Hisar (b) Jaipur
(c) Barrackpore (d) Mumbai

35. Central Institute for Research on Buffaloes is located at
(a) Hisar (b) Jaipur
(c) Barrackpore (d) Mumbai

36. Central Institute for Research on Goats is located at
(a) New Delhi (b) Makhdoom
(c) Barrackpore (d) Mumbai

37. Central Institute of Agricultural Engineering is located at
(a) Bhopal (b) Jaipur
(c) Barrackpore (d) Mumbai

38. Central Institute of Arid Horticulture is located at
(a) Bikaner (b) Nagpur
(c) Bhubneshwar (d) Mumbai

39. Central Institute of Cotton Research is located at
(a) Bikaner (b) Nagpur
(c) Bhubneshwar (d) Mumbai

40. Central Institute of Freshwater Aquaculture is located at
(a) Bikaner (b) Nagpur
(c) Bhubneshwar (d) Mumbai

41. Central Institute of Subtropical Horticulture is located at
(a) Bikaner (b) Nagpur
(c) Srinagar (d) Mumbai

42. Central Institute of Fisheries Technology is located at
(a) Bikaner (b) Nagpur
(c) Bhubneshwar (d) Cochin

43. Central Plantation Crops Research Institute is located at
(a) Kasargod (b) Cochin
(c) Chennai (d) Cuttack

44. Centra Rice Research Institute is located at
(a) Kasargod (b) Cochin
(c) Chennai (d) Cuttack

45. Central Institute of Temperate Horticulture is located at
(a) Kasargod (b) Lucknow
(c) Chennai (d) Cuttack

46. Central Soil and Water Conservation Research & Training is located at
(a) Dehradun (b) Delhi
(c) Karnal (d) Rajahmundry

47. Central Soil Salinity Research Institute is located at
(a) Dehradun (b) Delhi
(c) Karnal (d) Rajahmundry

48. Central Tobacco Research Institute is located at
(a) Dehradun (b) Delhi
(c) Karnal (d) Rajahmundry

49. Central Tuber Crops Research Institue is located at
(a) Trivandrum (b) Delhi
(c) Karnal (d) Rajahmundry

50. Indian Agricultural Statistical Research Institute is located at
(a) New Delhi (b) Mumbai
(c) Pune (d) Patna

51. Indian Grassland and Fodder Research Institute is located at
(a) New Delhi (b) Mumbai
(c) Jhansi (d) Patna

52. Indian Institute of Agricultural Biotechnology is located at
(a) Ranchi (b) Mumbai
(c) Pune (d) Patna

53. Indian Institute of Horticultural Ressearch is located at
(a) New Delhi (b) Mumbai
(c) Kolkatta (d) Patna

54. Indian Institute of Natural Resin & Gums is located at
(a) Ranchi (b) Mumbai
(c) Pune (d) Patna

55. Indian Institute of Pulses Research is located at
(a) Kanpur (b) Bhopal
(c) Calicut (d) Lucknow

56. Indian Institute of Soil Sciences is located at
(a) Kanpur (b) Bhopal
(c) Calicut (d) Lucknow

57. Indian Institute of Spices Research is located at
(a) Kanpur (b) Bhopal
(c) Calicut (d) Lucknow

58. Indian Institute of Sugarcane Research is located at
(a) Kanpur (b) Bhopal
(c) Calicut (d) Lucknow

59. Indian Institute of Vegetables Research is located at
(a) Varanasi (b) Hyderabad
(c) Bangalore (d) Coimbatore

60. National Academy of Agricultural Research and is located at
(a) Varanasi (b) Hyderabad
(c) Bangalore (d) Coimbatore

61. National Institute of Animal Nutrition andPhysiology is located at
(a) Varanasi (b) Hyderabad
(c) Bangalore (d) Coimbatore

62. Sugarcane Breeding Institute is located at
(a) Varanasi (b) Hyderabad
(c) Bangalore (d) Coimbatore

63. National Centre for Agricultureal Economics and Policy Research is located at
(a) New Delhi (b) Jhansi
(c) Trichi (d) Nagpur

64. National Centre for Integrated Pest Management is located at
(a) New Delhi (b) Jhansi
(c) Trichi (d) Nagpur

65. National Research Centre for Agroforestry is located at
(a) New Delhi (b) Jhansi
(c) Trichi (d) Nagpur

66. National Research Centre for Banana is located at
(a) New Delhi (b) Jhansi
(c) Trichi (d) Nagpur

67. National Research Centre for Citrus is located at
(a) New Delhi (b) Jhansi
(c) Trichi (d) Nagpur

68. National Research Centre for Grapes is located at
(a) Pune (b) Muzaffarpur
(c) Solapur (d) Bikaner

69. National Research Centre for Litchi is located at
(a) Pune (b) Muzaffarpur
(c) Solapur (d) Bikaner

70. National Research Centre for Pomegranate is located at
(a) Pune (b) Muzaffarpur
(c) Solapur (d) Bikaner

71. National Research Centre on Camel is located at
(a) Pune (b) Muzaffarpur
(c) Solapur (d) Bikaner

72. National Research Centre on Equines is located at
(a) Hisar (b) Hyderabad
(c) New Delhi (d) Ajmer

73. National Research Centre on Meat is located at
(a) Hisar (b) Hyderabad
(c) New Delhi (d) Ajmer

74. National Research Centre on Plant Biotechnology is located at
(a) Hisar (b) Hyderabad
(c) New Delhi (d) Ajmer

75. National Research Centre on Seed Spices is located at
(a) Hisar (b) Hyderabad
(c) New Delhi (d) Ajmer

76. National Bureau of Plant Genetics Resources is located at
(a) Delhi (b) Karnal
(c) Lucknow (d) Hyderabad

77. National Bureau of Animal Genetics Resources is located at
(a) Delhi (b) Karnal
(c) Lucknow (d) Hyderabad

78. National Bureau of Fish Genetics Resources is located at
(a) Delhi (b) Karnal
(c) Lucknow (d) Hyderabad

79. Directorate of Maize Research is located at
(a) Delhi (b) Hyderabad
(c) Mau (d) Junagarh

80. Directorate of Rice Research is located at
(a) Delhi (b) Hyderabad
(c) Mau (d) Junagarh

81. Directorate of Oilseed Research is located at
(a) Delhi (b) Hyderabad
(c) Mau (d) Junagarh

82. Directorate of Seed Research is located at
(a) Delhi (b) Hyderabad
(c) Mau (d) Junagarh

83. Directorate of Groundnut Research is located at
(a) Delhi (b) Hyderabad
(c) Mau (d) Junagarh

84. Directorate of Soybean Research is located at
(a) Indore (b) Bharatpur
(c) Solan (d) Pune

85. Directorate of Rapeseed Mustard Research is located at
(a) Indore (b) Bharatpur
(c) Solan (d) Pune

86. Directorate of Mushroom Research is located at
(a) Indore (b) Bharatpur
(c) Solan (d) Pune

87. Directorate of Cashew Research is located at
(a) Indore (b) Bharatpur
(c) Solan (d) Puttur

88. Directorate of Onion and Garlic Research is located at
(a) Indore (b) Bharatpur
(c) Solan (d) Pune

89. Directorate of Medicinal ans Aromatic Plant Research is located at
(a) Anand (b) Bharatpur
(c) Solan (d) Pune

90. Directorate of Research on Women in Agriculture is located at
(a) Bhubaneshwar (b) Kolkatta
(c) Solan (d) Pune

91. National Meat and Poultry Processing Board is located at
(a) Bhubaneshwar (b) Kolkatta
(c) Solan (d) Pune

92. National Meat and Poultry Processing Board was constituted in
(a) 1999 (b) 2003
(c) 2009 (d) 2011

93. Preservation of Food Adulteration Act came in
(a) 1944 (b) 1954
(c) 1957 (d) 1973

94. Fruit Products Order came in
(a) 1955 (b) 1973
(c) 1967 (d) 1998

95. Meat Food Products Order came in
(a) 1955 (b) 1973
(c) 1967 (d) 1998

96. Solvent extracted Oil Deoiled Meal and Edible Flour (control) Order came in
(a) 1955 (b) 1973
(c) 1967 (d) 1998

97. Edible oils Packaging Order came in
(a) 1955 (b) 1973
(c) 1967 (d) 1998

98. Vegetable Oil Products Order came in
(a) 1955 (b) 1973
(c) 1967 (d) 1998

99. Food Safety and Standards Act came in
(a) 1955 (b) 1973
(c) 2006 (d) 1998

100. FSSA rules and Regulations came in
(a) 2011 (b) 2009
(c) 1992 (d) 1998

101. Milk and Milk Products Amendment Regulations came in
(a) 2011 (b) 2009
(c) 1992 (d) 1998

102. Milk and Milk Products Order came in
(a) 2011 (b) 2009
(c) 1992 (d) 1998

103. Flavor consists of
(a) taste (b) aroma
(c) both (d) none

104. The persons are classified as taster's and non tasters depending on their ability to taste
(a) phenylthiocarbamide
(b) phenylthioglutamate
(c) both
(d) none

105. Chewing the leaves of Gymnemma sylvestre destroys the ability to taste ______ substances
(a) sweet (b) sour
(c) both (d) none

106. Berries of Synsepalum dulcificum abolishes ______ sensation
(a) sweet (b) sour
(c) both (d) none

107. Lemonene is a type of
(a) terpenoid (b) flavonoid
(c) sulfur compound (d) none

108. Hesperidin is a type of
(a) terpenoid
(b) flavonoid
(c) sulfur compound
(d) none

109. Naringin is a type of
(a) terpenoid
(b) flavonoid
(c) sulfur compound
(d) none

110. Which of the following compound is mainly responsible for aroma of banana
(a) isopentyl acetate
(b) citral
(c) benzaldehyde
(d) all

111. Which of the following compound is mainly responsible for aroma of lemon
(a) isopentyl acetate
(b) citral
(c) benzaldehyde
(d) all

112. Which of the following compound is mainly responsible for aroma of almonds
(a) isopentyl acetate
(b) citral
(c) benzaldehyde
(d) all

113. With an increase in pressure, boiling point of a fluid ______
(a) increases (b) decreases
(c) donot change (d) none

114. Diphasic colloidal dispersion of solid in liquid is
(a) emulsion (b) sol
(c) foam (d) solid foam

115. Diphasic colloidal dispersion of liquid in liquid is
(a) emulsion (b) sol
(c) foam (d) solid foam

116. Diphasic colloidal dispersion of gas in liquid is
(a) emulsion (b) sol
(c) foam (d) solid foam

117. Diphasic colloidal dispersion of gas in solid is
(a) emulsion (b) sol
(c) foam (d) solid foam

118. Diphasic colloidal dispersion of solid in gas is
(a) aerosol (b) sol
(c) foam (d) solid foam

119. Meringue is an example of
(a) aerosol (b) sol
(c) foam (d) solid foam

120. Skimmed milk is an example of
(a) aerosol (b) sol
(c) foam (d) solid foam

121. Smoke for flavoring food is an example of
(a) aerosol (b) sol
(c) foam (d) solid foam

122. Chlorine dioxide in flour is used as
(a) bleaching agent (b) improver
(c) both (d) none

123. Benzoyl peroxide in flour is used as
(a) bleaching agent (b) improver
(c) both (d) none

124. Nitrosyl Chloride in flour is used as
(a) bleaching agent (b) improver
(c) both (d) none

125. Potassium bromate in flour is used as
(a) bleaching agent
(b) improver
(c) both
(d) none

126. Potassium iodate in flour is used as
(a) bleaching agent (b) improver
(c) both (d) none

127. Calcium peroxide in flour is used as
(a) bleaching agent (b) improver
(c) both (d) none

128. GM in GM foods stands for
(a) germ mediated
(b) green modified
(c) genetically modified
(d) gregor mendel

129. Organic foods are similar to genetically modified foods
(a) true (b) false
(c) not determined (d) none

130. The commonly cultivated GM crop in India is
(a) *Bt.* Cotton (b) *Bt.* Brinjal
(c) *Bt.* Corn (d) *Bt.* Tomato

131. The advantage of GM foods over conventional food is
(a) pest resistance
(b) herbicide resistance
(c) cold tolerance
(d) all

132. Organic food involves the use of
(a) insecticides (b) herbicides
(c) pesticides (d) none

133. The cause of celiac disease is
(a) gluten (b) zein
(c) hodein (d) globulin

134. The body part affected by celiac disease is
(a) heart (b) intestine
(c) stomach (d) liver

135. The cure for celiac disease is
(a) irradiation of wheat
(b) gluten free diet
(c) cold storage of wheat
(d) none

136. HPP stands for
(a) High Pressure Processing
(b) High Power Processing
(c) Hot Pressure Processing
(d) none

137. The pressure commonly used in HPP is _____ MPa
(a) 100-200 (b) 400-700
(c) 300-400 (d) none

138. The adulteration of starch in milk can be detected by ________test
(a) alcohol (b) iodine
(c) DMAB (d) urease

139. Addition of cow to buffalo milk is adulteration of milk
(a) yes
(b) no
(c) cannot be determined
(d) none

140. Chikki, a nutritive snack in India has _____% protein
(a) 13-14 (b) 1-2
(c) 5-7 (d) 20-30

141. Which of the following are prepared from khoa
(a) peda (b) burfi
(c) gulab jamun (d) all

142. Rosogolla and rasmalai are prepared from
(a) khoa (b) chhana
(c) flour (d) none

143. Sonpapdi and Mysore pak are prepared from
(a) khoa (b) chhana
(c) flour (d) none

144. Shrikhand is prepared from
(a) dahi (b) khoa
(c) chhana (d) flour

145. Khoa made by the slow evaporation of milk with constant stirring has a moisture content of ____%
(a) 1-2 (b) 9-11
(c) 30-35 (d) 60-65

146. The average yield of chhana from cow milk is _____%
(a) 1-2 (b) 9-11
(c) 16-18 (d) 60-65

147. The average yield of chhana from buffalo milk is _____%
(a) 1-2 (b) 9-11
(c) 16-18 (d) 22-24

148. The average moisture content of chnna is _____%
(a) 1-2 (b) 9-11
(c) 16-18 (d) 50

149. Which of the following sugar is also known as icing sugar
(a) granulated (b) cube
(c) powdered (d) brown

150. Which of the following is a type of sugar
(a) rock (b) diamond
(c) powdered (d) all

151. Molasses is obtained from
(a) sugarcane (b) barley
(c) sorghum (d) wheat

152. Molasses has an average _______% sucrose
(a) 1-2 (b) 9-11
(c) 16-18 (d) 35

153. Commercially sugar is obtained from
(a) sugarcane (b) maple
(c) barley (d) palm

154. Gur contains _______ % sucrose
(a) 50-55 (b) 96-97
(c) 67-87 (d) 35-46

155. Gur contains
(a) sucrose (b) invert sugar
(c) both (d) none

156. Which of the following is present in butter fat
(a) butyric acid (b) caproic acid
(c) both (d) none

157. Which of the following is present in coconut oil
(a) caprylic acid (b) capric acid
(c) lauric (d) all

158. Which of the following is present in fish oil
(a) gadoleic acid (b) EPA
(c) DHA (d) all

159. Approximate acid dissociation constant of arginine is
(a) 4.4 (b) 8.5
(c) 6.5 (d) 12

160. Approximate acid dissociation constant of glutamate is
(a) 4.4 (b) 8.5
(c) 6.5 (d) 12

161. Approximate acid dissociation constant of cysteine is
(a) 4.4 (b) 8.5
(c) 6.5 (d) 12

162. Approximate acid dissociation constant of histidine is
(a) 4.4 (b) 8.5
(c) 6.5 (d) 12

163. Approximate acid dissociation constant of lysine is
(a) 4.4 (b) 8.5
(c) 6.5 (d) 10

164. Approximate acid dissociation constant of tyrosine is
(a) 4.4 (b) 8.5
(c) 6.5 (d) 10

165. Which of the following pigment is highly stable to light
(a) anthocyanin (b) bixin
(c) carotenes (d) all

166. Which of the following pigment is highly stable to light
(a) betalains (b) norbixin
(c) curcumin (d) all

167. Which of the following pigment is highly stable to light
(a) xanthophyll (b) cochineal
(c) chlorophyll (d) all

168. Which of the following pigment is highly stable to heat
(a) anthocyanin (b) cochineal
(c) caramel (d) all

169. ________ show low stability towards light
(a) curcumin (b) quinones
(c) caramel (d) all

170. Which of the following pigment is highly stable to pH change
(a) betalains (b) carotenes
(c) caramel (d) all

171. Which of the following pigment show low stability to pH change
(a) xanthophyll
(b) oxyhemoglobin
(c) anthocyanin
(d) all

172. Which of the following pigment is water soluble
(a) anthocyanin (b) bixin
(c) canthaxanthin (d) all

173. Which of the following pigment is water soluble
(a) betalains (b) carotenes
(c) cochineal (d) all

174. Which of the following pigment is water soluble
(a) curcumin (b) norbixin
(c) xanthophylls (d) all

175. Which of the following pigment is oil soluble
(a) bixin
(b) canthaxanthin
(c) carotenes
(d) all

176. Which of the following pigment shows high stability towards oxygen
(a) anthocyanin (b) betalains
(c) bixin (d) all

177. Which of the following pigment shows high stability towards oxygen
(a) caramel (b) chlorophyll
(c) norbixin (d) all

178. Which of the following pigment shows high stability towards oxygen
(a) oxyhemoglobin (b) xanthophyll
(c) both (d) none

179. Which of the following pigment shows low stability towards oxygen
(a) carotene (b) curcumin
(c) both (d) none

180. Continuous phase in smoke is
(a) gas (b) liquid
(c) solid (d) none

181. Dispersed phase in smoke is
(a) gas (b) liquid
(c) solid (d) none

182. Continuous phase in foam is
(a) gas (b) liquid
(c) solid (d) none

183. Dispersed phase in foam is
(a) gas (b) liquid
(c) solid (d) none

184. Continuous phase in emulsion is
(a) gas (b) liquid
(c) solid (d) none

185. Dispersed phase in emulsion is
(a) gas (b) liquid
(c) solid (d) none

186. Which of the following have basic pH
(a) Camembert cheese
(b) Cheddar cheese
(c) egg yolk
(d) none

187. Which of the following have basic pH
(a) cucumber (b) guava nectar
(c) egg white (d) none

188. Which of the following have acidic pH
(a) banana (b) blue berry
(c) honey (d) all

189. Which of the following have acidic pH
(a) grape fruit (b) fruit jam
(c) ketchup (d) all

190. Which of the following have acidic pH
(a) peanut butter (b) vinegar
(c) soy sauce (d) all

191. Which of the following have basic pH
(a) tofu (b) molasses
(c) milk (d) none

192. Pearson square method is related to
(a) fluid flow
(b) thermodynamic
(c) cream
(d) none

193. Which of the following technique can be used in foreign body food inspection
(a) magnetic (b) capacitance
(c) optical (d) all

194. Which of the following technique can be used in foreign body food inspection
(a) X-rays (b) ultrasonics
(c) ultraviolet (d) all

195. Which of the following is related to biscuit processing
(a) rotary moulder
(b) moulding roller
(c) wire cut machine
(d) all

196. Bt. in GM crops is
(a) Bacillus thuringiensis
(b) Bacillus thermophillus
(c) Bacillus thoreus
(c) none

197. Golden rice developed to synthesize the precursor of ______ in edible par
(a) Vitamin C (b) β-carotene
(c) minerals (d) all

198. Irradiation helps in
(a) control of ripening
(b) disinfestations
(c) inhibition of sprouting
(d) all

199. Which of the following is a novel technique used in food processing
(a) HPP (b) PEF
(c) both (d) none

200. PEF stands for
(a) Pulse Electric Field
(b) Poor Electric Field
(c) Pulse Electromagnetic Field
(d) none

201. The dietary fiber in kidney bean and green gram
(a) 25% (b) 2%
(c) 5% (d) 50%

202. The hypocholesterolemic effect of dietary fiber fraction of legumes is attributed to
(a) higher pectin content
(b) gums
(c) galactans
(d) all

203. Which of the following is anti-nutritional factor (ANF)
(a) lectins (b) saponins
(c) oestrogens (d) all

204. Most of the ANF are
(a) heat stable (b) heat labile
(c) no effect (d) none

205. Which of the following ANF is heat stable
(a) lysinoalanine (b) saponins
(c) phytates (d) all

206. The safe storage moisture for pulses is
(a) 10-12% (b) 20%
(c) <5% (d) <10%

207. TSP stands for
(a) Thermo set protein
(b) Texturized set protein
(c) texurized soy protein
(d) none

208. Legumes are deficient in
(a) methionine & tryptophan
(b) phenylalanine & threonine
(c) valine & leucine
(d) all

209. Recommended amout (g/day) of grain legumes for men
(a) 25 (b) 50
(c) 75 (d) 100

210. Recommended amout (g/day) of grain legumes for women
(a) 25 (b) 45
(c) 65 (d) 85

211. Recommended amout (g/day) of cereals for men
(a) 220 (b) 320
(c) 420 (d) 520

212. Recommended amout (g/day) of cereals for women
(a) 240 (b) 340
(c) 440 (d) 540

213. Major portion of yield after legumes milling
(a) dhal (b) grits
(c) husk (d) powder

214. % dhal obtained after milling
(a) 10-25 (b) 30-50
(c) 50-80 (d) 95

215. % husk obtained after milling
(a) 10-25 (b) 30-50
(c) 50-80 (d) 95

216. Which of the following is fermented black skin soybean
(a) Tao si (b) soyasauce
(c) sufu (d) Thua-nao

217. Which of the following is natural fermented soybean
(a) Tofu (b) Thua nao
(c) both (d) none

218. Which of the following is fermented curd
(a) Tao si (b) soyasauce
(c) sufu (d) Thua-nao

219. Which of the following is fermented soyasauce
(a) Tao si (b) Si-iu
(c) sufu (d) Thua-nao

220. Soya sauce is ____ stage fermentation
(a) single (b) two
(c) three (d) four

221. Production of Koji for preparation of soya sauce is ________ fermentation
(a) aerobic (b) anaerobic
(c) both (d) none

222. Which of the following is soya paste
(a) Tao-hu-yi (b) Tao-cheow
(c) both (d) none

223. Which of the following is fermented soya curd
(a) Tao-hu-yi (b) Tao-cheow
(c) both (d) none

224. Tao-hu-yi is made by organism
(a) Actinomucar elegans
(b) *Aspergillus niger*
(c) Bacillus cereus
(d) none

225. Which of the following is fermented whole soybean
(a) Tao-hu-yi (b) Tao-cheow
(c) Thuo-nao (d) none

226. The flatulence sugars escape digestion due to lack of
(a) Alpha-galactosidase
(b) amylase
(c) trypsin
(d) all

227. Major portion of phosphorus in pulses is present as
(a) phytate phosphorus
(b) sodium phosphate
(c) both
(d) none

228. The major constituent responsible for the unique flavor of aromatic long grains of rice is
(a) 2-acetyl-1-pyrroline
(b) 6-octyl-proline
(c) both
(d) none

229. The average length of matted rice is
(a) 5-7.5 mm (b) 6.2-8.5 mm
(c) 7.2-9.9 mm (d) 7.8-10.1 mm

230. The main ingredient of premium quality pasta is
(a) durum wheat flour
(b) durum wheat semolina
(c) Aestivum flour
(d) Compactum semolina

231. Pasta is __________ product
(a) fried (b) extruded
(c) fermented (d) baked

232. The pasta dough temperature during extrusion should not above __________ °C to prevent damage to the gluten network
(a) 40 (b) 50
(c) 60 (d) 70

233. The pasta is dried down to a moisture content of around ______% fro long term storage
(a) 12.5 (b) 15.5
(c) 18.5 (d) 20.5

234. The fiber content of bread can be increased by fortifying bread with
(a) bran (b) rice flour
(c) both (d) none

235. A high Gli: Glutenin ratio in wheat proteins results in
(a) more viscous dough
(b) more elastic dough
(c) no effect
(d) none

236. A low Gli: Glutenin ratio in wheat proteins results in
(a) more viscous dough
(b) more elastic dough
(c) no effect
(d) none

237. Active form of bakers yeast is
(a) compressed yeast
(b) active dry yeast
(c) cream yeast
(d) all

238. Cream yeast contain ____________% moisture
(a) 15 (b) 35
(c) 50 (d) 85

239. Which of the following is a variety of wheat grown in India
(a) Sonalika (b) HUW234
(c) WH542 (d) all

240. Which of the following wheat varieties is recommended for bread
(a) Sonalika (b) HUW234
(c) WH542 (d) all

241. Which of the following wheat varieties is recommended for bread
(a) Sonalika (b) VL892
(c) WH542 (d) both (a) & (b)

242. Which of the following wheat varieties is recommended for pasta
(a) MACS-1967 (b) HI-8498
(c) NIAW-917 (d) both (a) & (b)

243. Which of the following has strong gluten characteristics
(a) HI-977 (b) CBW-38
(c) WH-1025 (d) both (a) & (b)

244. Which of the following has weak gluten characteristics
(a) WH-1025 (b) VL-892
(c) WH-1021 (d) both (a) & (b)

245. Good source of β-amylase
(a) wheat (b) barley
(c) potato (d) none

246. Which of the following is acid protease
(a) pepsin (b) rennin
(c) chymosin (d) all

247. Protein hydrolysates can be obtained by action of
(a) acid (b) alkali
(c) enzyme (d) all

248. Which of the following is present in carrot
(a) α-carotene (b) β-carotene
(c) xanthophylls (d) all

249. Which of the following is present in apricot
(a) β-carotene (b) γ-carotene
(c) lycopene (d) all

250. Which of the following is present in peach
(a) cryptoxanthin (b) xanthophyll
(c) zeaxanthin (d) all

251. Which of the following is present in soybean
(a) α-carotene (b) β-carotene
(c) both (a) & (b) (d) none

252. Which of the following is present in orange
(a) lycopene (b) xanthophyll
(c) zeaxanthin (d) all

253. Which of the following is present in red pepper
(a) α-carotene (b) β-carotene
(c) capsanthain (d) all

254. Which of the following is present in corn

(a) zeaxanthin (b) cryptoxanthin
(c) α & β-carotene (d) all

255. Citric acid content (%) in banana is about

(a) 0.32 (b) 0.10
(c) 0.65 (d) 0.19

256. Citric acid content (%) in artichokes is about

(a) 0.32 (b) 0.10
(c) 0.65 (d) 0.19

257. Citric acid content (%) in broccoli is about

(a) 0.32 (b) 0.10
(c) 0.21 (d) 0.65

258. Citric acid content (%) in cauliflower is about

(a) 0.32 (b) 0.10
(c) 0.21 (d) 0.65

259. Citric acid content (%) in blueberries is about

(a) 0.32 (b) 1.10
(c) 0.65 (d) 1.56

260. Citric acid content (%) in cranberries is about

(a) 0.32 (b) 1.10
(c) 0.65 (d) 1.56

261. Citric acid content (%) in grapefruit is about

(a) 0.32 (b) 0.65
(c) 1.33 (d) 2.65

262. Citric acid content (%) in lemon is about

(a) 0.65 (b) 1.33
(c) 3.84 (d) 6.08

263. Citric acid content (%) in lemon juice is about

(a) 0.65 (b) 1.33
(c) 3.84 (d) 6.08

264. Citric acid content (%) in oranges is about

(a) 0.32 (b) 0.65
(c) 0.98 (d) 1.33

265. Citric acid content (%) in strawberries is about

(a) 0.65 (b) 1.08
(c) 1.33 (d) 1.45

266. Citric acid content (%) in tomatoes is about

(a) 0.30 (b) 0.60
(c) 0.90 (d) 1.20

267. Malic acid content (%) in dried apricot is about

(a) 0.33 (b) 0.81
(c) 0.07 (d) 0.17

268. Malic acid content (%) in banana is about

(a) 0.37 (b) 0.87
(c) 0.07 (d) 1.17

269. Malic acid content (%) in blue berries is about

(a) 0.02 (b) 0.10
(c) 0.33 (d) 0.65

270. Malic acid content (%) in broccoli is about

(a) 0.02 (b) 0.5
(c) 0.12 (d) 0.33

271. Malic acid content (%) in canned peaches is about

(a) 0.05 (b) 0.39
(c) 0.50 (d) 0.69

272. Malic acid content (%) in plum is about

(a) 0.92 (b) 1.92
(c) 1.77 (d) 1.42

273. Malic acid content (%) in rhubarb is about

(a) 0.92 (b) 1.92
(c) 1.77 (d) 1.42

274. Malic acid content (%) in watermelon is about
(a) 0.1 (b) 0.2
(c) 0.3 (d) 0.4

275. Milk fat content of Jersey cow breed is ______%
(a) 4.0 (b) 3.4
(c) 3.94 (d) 5.37

276. Milk fat content of Brown Swiss cow breed is ______ %
(a) 4.1 (b) 3.4
(c) 3.94 (d) 5.37

277. Milk fat content of Holstein cow breed is ______ %
(a) 4.0 (b) 3.4
(c) 3.94 (d) 5.37

278. Protein content of Jersey cow milk is ______%
(a) 3.58 (b) 3.32
(c) 3.61 (d) 3.92

279. Protein content of Holstein cow milk is ______%
(a) 3.58 (b) 3.32
(c) 3.61 (d) 3.92

280. Protein content of Brown Swiss cow milk is ______%
(a) 3.78 (b) 3.32
(c) 3.61 (d) 3.92

281. Lactose content of Jersey cow milk is ______%
(a) 4.93 (b) 4.87
(c) 5.04 (d) 4.67

282. Lactose content of Holstein cow milk is ______%
(a) 4.93 (b) 4.87
(c) 5.04 (d) 4.67

283. Lactose content of Brown Swiss cow milk is ______%
(a) 4.93 (b) 4.87
(c) 5.04 (d) 4.67

284. Ash content of Jersey cow milk is ______%
(a) 0.71 (b) 0.68
(c) 0.73 (d) 1.03

285. Ash content of Holstein cow milk is ______%
(a) 0.71 (b) 0.68
(c) 0.73 (d) 1.03

286. Ash content of Brown Swiss cow milk is ______%
(a) 0.71 (b) 0.68
(c) 0.73 (d) 1.03

287. Total solids in Jersey cow milk is ______%
(a) 12.9 (b) 13.41
(c) 12.26 (d) 14.91

288. Total solids in Holstein cow milk is ______%
(a) 12.9 (b) 13.41
(c) 12.26 (d) 14.91

289. Total solids in Brown Swiss cow milk is ______%
(a) 12.9 (b) 13.41
(c) 12.26 (d) 14.91

290. Water content in Jersey cow milk is ______%
(a) 85.09 (b) 86.59
(c) 87.74 (d) 89.39

291. Water content in Holstein cow milk is ______%
(a) 85.09 (b) 86.59
(c) 87.74 (d) 89.39

292. Water content in Brown Swiss cow milk is ______%
(a) 85.09 (b) 86.59
(c) 87.74 (d) 89.39

293. SNF content in Jersey cow milk is ______%
(a) 9.4 (b) 9.54
(c) 8.86 (d) 8.99

294. SNF content in Holstein cow milk is ______%
(a) 9.4 (b) 9.54
(c) 8.86 (d) 8.99

295. SNF content in Brown Swiss cow milk is ______%
(a) 9.4 (b) 9.54
(c) 8.86 (d) 8.99

296. Fat content in Jersey cow milk is ______%
(a) 1.0 (b) 2.1
(c) 4.1 (d) 11.9

297. Fat content of wheat bran is ______%
(a) 1.0 (b) 2.1
(c) 4.1 (d) 11.9

298. Fat content wheat germ is ______%
(a) 1.0 (b) 2.1
(c) 4.1 (d) 11.9

299. Pentosans content of wheat is about _____%
(a) 5.1 (b) 6.2
(c) 25.1 (d) 3.6

300. Pentosans content of wheat bran is about _____%
(a) 5.1 (b) 6.2
(c) 25.1 (d) 3.6

301. Pentosans content of wheat germ is about _____%
(a) 5.1 (b) 6.2
(c) 25.1 (d) 3.6

302. Ash content of wheat is about _____%
(a) 0.5 (b) 6.38
(c) 4.8 (d) 1.73

303. Ash content of wheat bran is about _____%
(a) 0.5 (b) 6.38
(c) 4.8 (d) 1.73

304. Ash content of wheat germ is about _____%
(a) 0.5 (b) 6.38
(c) 4.8 (d) 1.73

305. Which of the following is used in SanFrancisco dough bread making
(a) Torulopsis
(b) Saccharomyces
(c) Lactobacillus
(d) all

306. Sarcino sickness in beer is characterized by
(a) sourness (b) turbidity
(c) ropiness (d) all

307. Which of the following mold causes spoilage in cheese
(a) Geotrichum
(b) Cladosporium
(c) Penicillium
(d) all

308. Food grade ethanol is produced by
(a) yeast (b) bacteria
(c) both (d) none

309. Amylase can be obtained from
(a) Aspergillus (b) Bacillus
(c) Rhizopus (d) all

310. Cellulase is chiefly obtained from
(a) Aspergillus b) Bacillus
(c) Rhizopus (d) none

311. Which of the following can produce cellulase
(a) Aspergillus (b) Trichoderma
(c) both (d) none

312. Which of the following can be obtained from *Aspergillus niger*
(a) Lipase
(b) Glucose oxidase
(c) Pectinase (d) all

313. Bacillus subtilis is the source of
(a) protease (b) pectinase
(c) cellulase (d) all

314. Food intoxication can be caused by
(a) Staphylococcus aureus
(b) Clostridium botulinum
(c) both (d) none

315. Food infection is generally caused by
(a) Salmonella
(b) Bacillus cereus
(c) *E. coli* (d) all

316. Energy flow rate (kJ/sec) in DSI preheater is about
(a) 1451 (b) 1680
(c) 1814 (d) 3500

317. Shrinkage is associated with
(a) cooking
(b) dehydration
(c) microwave heating
(d) none

318. Which of the following is present in glass
(a) SiO_2 (b) Na_2O
(c) CaO (d) all

319. Which of the following is present in highest amount in glass
(a) SiO_2 (b) Na_2O
(c) CaO (d) none

320. Which of the following is present in lowest amount in glass
(a) Na_2O (b) Al_2O_3
(c) CaO (d) none

321. Glass transistion temperature (°C) of high density polyethylene is
(a) -18 (b) -25
(c) -125 (d) 100

322. Glass transistion temperature (°C) of low density polyethylene is
(a) -18 (b) -25
(c) -125 (d) 100

323. Glass transistion temperature (°C) of polypropylene is
(a) -18 (b) -25
(c) -125 (d) 100

324. Glass transistion temperature (°C) of polystyrene (packaging material) is
(a) -18 (b) -25
(c) -125 (d) 100

325. Glass transistion temperature (°C) of polyvinylchloride is
(a) -33 (b) -5
(c) 48 (d) 87

326. Melting temperature (°C) of high density polyethylene is
(a) 98 (b) 137
(c) 176 (d) 240

327. Melting temperature (°C) of low density polyethylene is
(a) 98 (b) 137
(c) 176 (d) 240

328. Melting temperature (°C) of polypropylene is
(a) 98 (b) 137
(c) 176 (d) 240

329. Melting temperature (°C) of polystyrene is
(a) 98 (b) 137
(c) 176 (d) 240

330. Melting temperature (°C) of polyvinylchloride is
(a) 99 (b) 112
(c) 172 (d) 212

331. Which of the following is a property of PET
(a) good moisture, O_2 and aroma barrier
(b) glossy and clear
(c) excellent heat resistance
(d) all

332. Which of the following is not a property of polystyrene
(a) excellent moisture, O_2 and aroma barrier
(b) strong and stiff
(c) glossy & clear
(d) none

333. Myoglobin content of more active muscles is _______ than passive ones
(a) more (b) less
(c) equal (d) none

334. Color of mutton is
(a) red (b) pink
(c) purple (d) brown

335. The color of pork is
(a) red (b) greyish pink
(c) purple (d) brown

336. The color of poultry is
(a) grey white to dull red
(b) greyish pink
(c) purple
(d) brown

337. The color of chevon is
(a) red (b) grayish pink
(c) purple (d) brown

338. The color of buffalo meat is
(a) cherry red (b) grayish pink
(c) purple (d) brown

339. The globin part of myoglobin is
(a) protein (b) fat
(c) mineral (d) none

340. In intact meat, iron in heme ring of myoglobin exists in ________ state
(a) reduced (b) excited
(c) ground (d) none

341. The water in meat is avalaible as
(a) free water
(b) immobilized water
(c) bound water
(d) all

342. Marbling in meat is
(a) intermuscular fat
(b) intramuscular fat
(c) both
(d) none

343. The shortening of pre rigor meat at low temperature is called
(a) cold shortening
(b) rigor mortis
(c) chiiling injury
(d) none

344. The shortening of rigor meat when it si thawed after freezing is called
(a) rigor meat
(b) thaw rigor
(c) cold shortening
(d) none

345. Juiciness of meat depends on
(a) intermediate moisture
(b) intramuscular fat
(c) water holding capacity
(d) all

346. Meat flavor depends on
(a) intramuscular fat
(b) myofibrillar proteins
(c) intermuscular fat
(d) none

347. Electrical stimulation ____________ tender-ness in beef from older animals
(a) increase (b) decrease
(c) no effect (d) none

348. CASF stands for
(a) calcium activated sarcoplamic factor
(b) cesium activated sarcoplamic factor
(c) chlorine activated sarcoplamic factor
(d) none

349. Desmin is a
(a) M line protein (b) Z line protein
(c) both (d) none

350. Cathepsin is activated at
(a) higher pH (b) lower pH
(c) neutral pH (d) none

351. Fresh meat has a water activity of
(a) 0.99 (b) 0.90
(c) 0.88 (d) 0.80

352. Putrid meat has a pH ________ than ground meat
(a) higher (b) lower
(c) similar (d) none

353. In case of spoiled meat and products, TBA values are
(a) higher (b) lower
(c) similar (d) none

354. Permitted level of nitrates as preservative by inhibiting the growth of bacteria in meat is ______ ppm
(a) 100 (b) 200
(c) 300 (d) 500

355. Permitted level of nitrites as preservative by inhibiting the growth of bacteria in meat is ______ ppm
(a) 100 (b) 200
(c) 300 (d) 500

356. Pasteurization of meat is done at __°C
(a) 20-35 (b) 58-75
(c) 85-90 (d) 95-99

357. ________ is processed into bacon
(a) pork bellies (b) buffalo meat
(c) veal meat (d) none

358. Sausage is derived from latin word 'salsus' which means
(a) sugar (b) salt
(c) dry (d) meat

359. Bologna is __________ type sausage
(a) emulsion (b) spicy
(c) dry (d) none

360. IMM stands for
(a) Intermediate moisture meat
(b) Intermediate meat moisture
(c) inrermuscular moisture meat
(d) intramuscular moisture meat

361. Mortadella is ________ type sausage
(a) emulsion (b) spicy
(c) dry (d) none

362. IMM have moisture content in range
(a) 10-20% (b) 20-50%
(c) 50-70% (d) >80%

363. IMM have water activity in range
(a) 0.20-0.40 (b) 0.40-0.60
(c) 0.60-0.85 (d) none

364. The highest proportion of albumen is
(a) ovalbumin (b) conalbumin
(c) ovomucoid (d) ovomucin

365. The lowest proportion of albumen is
(a) ovalbumin (b) avidin
(c) ovomucoid (d) ovomucin

366. Biological value of egg is
(a) 1 (b) 3
(c) 5 (d) 10

367. Raw chicken meat has ________ moisture
(a) 10-24% (b) 25-50%
(c) 50-70% (d) 70-75%

368. The fat content of meat is ________ proportional to moisture content
(a) directly (b) inversely
(c) no relation (d) none

369. Poultry meat has ________ protein content than most of red meat
(a) higher (b) lower
(c) similar (d) none

370. Male poultry has _____ protein content as compared to female poultry
(a) higher (b) lower
(c) similar (d) none

371. Cholesterol content of poultry meat/ 100g is ______mg
(a) 20 (b) 40
(c) 60 (d) 80

372. Ranikhet is a disease found in
(a) buffalo (b) birds
(c) sheep (d) goat

373. Singeing is done to
(a) remove feather (b) slaughter
(c) washing (d) none

374. Offal is ________ part of slaughtered animal
(a) edible (b) inedible
(c) both (d) none

375. Vicera means organs of
(a) thoracic cavity
(b) abdominal cavity
(c) both
(d) none

376. National Meat and Poultry Processing Board is situated at
(a) New Delhi (b) Bhopal
(c) Karnal (d) Bangalore

377. Chairman of NMPPB is
(a) Sh. K.D. Singh
(b) Sh. Ajit Kumar
(c) Sh. Sharad Pawar
(d) none

378. The euillibrium between water intake and output is called
(a) water balance
(b) water consumption
(c) water usage
(d) none

379. Cholesterol and phospholipids are transported via vehicles called
(a) fatty acids (b) chylomicrons
(c) thoracic duct (d) all

380. The energy yielding foods are
(a) carbohydrates, proteins and fat
(b) water and vitamins
(c) vitamins and minerals
(d) none

381. The body building foods are
(a) carbohydrates (b) protein
(c) fat (d) minerals

382. The body protective food are
(a) carbohydrates (b) proteins
(c) vitamins & minerals
(d) fat

383. The base of the food pyramd is formed of
(a) cereals
(b) fruits and vegetabes
(c) pulses (d) meat

384. The food guide pyramid os based on the principles of
(a) balance (b) variety
(c) moderation (d) all

385. Sauteeing is a ______ cooking method
(a) moist heat (b) dry heat
(c) combination (d) none

386. Braising is a ______ cooking method
(a) moist heat (b) dry heat
(c) combination (d) none

387. Poaching is a ______ cooking method
(a) moist heat (b) dry heat
(c) combination (d) none

388. Poaching involves the cooking in a minimum amount of liquid at temperature of ______°C
(a) 80-85 (b) >100
(c) <60 (d) <80

389. ________ is a method used to prepare cakes
(a) baking (b) frying
(c) braising (d) poaching

390. When food is totally immersed in hot oil, it is called
(a) baking (b) frying
(c) braising (d) poaching

391. Liver is an excellent source of iron and vitamin
(a) A (b) C
(c) K (d) D

392. ________ is a good substitute for butter
(a) ghee (b) margarine
(c) hydrogenated fat (d) none

393. The development of disagreeable flavors and odors in fat is
(a) rancidity (b) putrification
(c) glycolysis (d) all

394. Fats with ________ smoking pont are suitable for frying
(a) low (b) high
(c) very low (d) none

395. _______ is used as a flavoring agent
(a) turmeric (b) saffron
(c) cardamom (d) tamarind

396. Sugar has _________ g carbohydrates/100g
(a) 90 (b) 94
(c) 99.4 (d) 104

397. Jaggery has ________ g carbohydrates/100g
(a) 90 (b) 95
(c) 99.4 (d) 104

398. Honey has ________ g carbohydrates/100g
(a) 79.5 (b) 94
(c) 99.4 (d) 104

399. The good source of iron among the following is
(a) sugar (b) jaggery
(c) honey (d) none

400. Jaggery has ______ mg iron/100g
(a) 0.15 (b) 0.30
(c) 0.45 (d) 0.60

401. Sugar has ______ mg iron/100g
(a) 2.65 (b) 3.65
(c) 5.65 (d) 7.65

402. Honey has ______ mg iron/100g
(a) 0.69 (b) 0.99
(c) 1.99 (d) 2.99

403. The nutrients most stable to baking are:
(a) proteins (b) vitamins
(c) carbohydrates (d) minerals

404. One of the following is not used in bakery and confectionary items:
(a) wheat flour (b) glycogen
(c) diamond sugar (d) brown sugar

405. Wheat flour contains a yellow pigment which may require bleaching and that is:
(a) lycopene (b) metanil
(c) xanthophyll (d) none

406. The steeping and steaming treatment given to paddy before milling is called
(a) bleaching (b) parboiling
(c) ageing (d) debranning

407. When the bread leaves the oven crumb temp is around 100°C & the moisture content at the centre is around:
(a) 45 % (b) 20 %
(c) 30% (d) 05%

408. Chromatography is based on
(a) adsorption (b) absorption
(c) reflectance (d) transmittance

409. Which is not a part of HPLC
(a) injector (b) transmitter
(c) detector (d) recorder

410. Kjeldahl method is used for estimating
(a) protein (b) fat
(c) carbohydrate (d) vitamins

411. Which of the following acids is not used in wet ashing
(a) HCl
(b) sulphuric acid
(c) perchloric acid
(d) nitric acid

412. Which of the solvent is not used for lipid extraction
(a) acetic acid (b) ethyl ether
(c) petroleum ether (d) n-butanol

413. Moisture content on dry basis is determined by
(a) Ww-Wd/Wd (b) Ww+Wd/Wd
(c) Ww-Wd (d) Wd-Ww/Wd

Where Ww-moisture content on wet basis, Wd-m.c. on dry basis

414. Which of the following is not colorimeter
(a) tintometer
(b) munsell disc
(c) nephelometer
(d) electrophoresis

415. pH is defined as
a) $-\log(H^-)$ (b) $-\log(H^+)$
(c) $\log(OH^+)$ d) $\log(-H^+)$

416. Which of the following is not used for pH measurement
(a) glass electrodes
(b) refractrometer
(c) isfet sensor
(d) colomel

417. Lyophilizer is based on principle of
(a) evaporation (b) sublimation
(c) freezing d) diffusion

418. A satellite in space exchange heat with the surrounding essentially by
(a) conduction (b) convection
(c) radiation (d) both (a) & (b)

419. Indicate the metal with highest value of thermal conductivity
(a) steel (b) silver
(c) copper (d) aluminum

420. Reynolds's number dimension is
(a) MLT^{-2} (b) ML^2T
(c) MLT^2 (d) none

421. For perfect black body absorptivity is equal to
(a) 1 (b) 2
(c) 3 (d) 4

422. Stephan – Boltzman law states
(a) $E_b = \sigma T^4$ (b) $E_b = \sigma T^3$
(c) $E_b = \sigma T^2$ (d) $E_b = \sigma T$

423. In leaching there is separation of solid from ________ by a solvent
(a) solid (b) liquid
(c) gas (d) all above

424. The units of diffusivity are
(a) m/s (b) m^2/s
(c) $Kmol/m^2.s$ (d) none

425. Process of mixing of 2 fluids is
(a) reversible (b) irreversible
(c) isothermal (d) none

426. Which law in mass transfer is similar to Fourier's law of heat transport
(a) coulombs' law (b) fick's law
(c) stephan's law (d) curie's law

427. In case of _______ boiling vaporization takes place directly from the surface
(a) cavitation (b) nucleate
(c) central (d) core

428. In boiling point diagram the saturated vapor curve is called
(a) triple point (b) boiling point
(c) dew point (d) saturation point

429. In distillation __________ takes place in both the direction
(a) heat transfer (b) mass transfer
(c) evaporation (d) none

430. The reciprocity theorem states that:-
(a) $F_{12} = F_{21}$ (b) $A_1F_{12} = A_2F_{21}$
(c) $\alpha F_{12} = \alpha_2F_{21}$ (d) $A_2F_{12} = A_1F_{21}$

431. Which of the following Fungi is responsible for the production of aflatoxin?
(a) Aspergillus flavus
(b) A. paraciticus
(c) A. normius
(d) all

432. Cheddar cheese is
(a) hard cheese
(b) soft cheese
(c) semi hard
(d) external mould cheese

433. Which of the following is an acid-alcohol fermented milk?
(a) bulgarican buttermilk
(b) kefir
(c) yoghurt
(d) kumis

434. The sparkling clear sweetened fruit juice is
(a) RTS (b) squash
(c) puree (d) cordial

435. Which of the following fruit component of fruit acts as antioxidant?
(a) malic acid (b) pectin
(c) vitamin C
(d) none of the above

436. Lactic acid fermentation does not produce
(a) ATP (b) CO_2 & NADH
(c) CO_2 (d) NADH

437. Rum is
(a) distilled liquor
(b) undistilled liquor
(c) fortified wine
(d) by-product of brewing

438. Which acid is found in carbonated non alcoholic beverages?
(a) propionic acid (b) acetic acid
(c) citric acid
(d) phosphoric acid

439. Cider is fermented
(a) apple (b) banana
(c) orange (d) cherries

440. Sarcina sickness is the defect of
(a) wine (b) sauerkraut
(c) beer (d) bread

441. Alcohol content of champagne is
(a) 4-8% (b) 10-14%
(c) 15-25% (d) 35-45%

442. Microorganism responsible for discoloration in fish during spoilage
(a) Alcaligenes (b) Flavobacterium
(c) Pseudomonas (d) Rhodotorula

443. The fat content in a low-fat yogurt is
(a) $< 0.5\%$ (b) 0.5 to 2.0 %
(c) 2.5 to 3% (d) more than 3%

444. Common microbial contaminant in poultry (freshly dressed, eviscerated) food product is
(a) Coliforms
(b) *Clostridium* sp.
(c) Bacillus subtilis
(d) Pseudomonas

445. Ripening of cheese in presence of fungus Penicillium roquefortii results change in
(a) texture (b) flavor
(c) colour (d) both (a) & (b)

446. Bottom fermenting yeast is
(a) S. cerevisiae (b) S. uvarum
(c) both (d) none

447. For beer production the fermentation is carried out at temperature
(a) 10 to 20 ºC (b) 20-25 ºC
(c) 30-35 ºC (d) 35-40 ºC

448. Best method to show the association between height and weight of sample in a lot is by
(a) bar chart
(b) line diagram
(c) scatter diagram
(d) histogram

449. Square root of deviation is also known as
(a) standard deviation
(b) standard error
(c) mean deviation
(d) range

450. In a 3 × 3 table, the number of degrees of freedom is
(a) 4 (b) 9
(c) 3 (d) 81

451. The best sampling procedure that can be followed is
(a) simple random
(b) stratified random
(c) convenient
(d) systematic

452. As the sample size increases standard deviation
(a) decreases
(b) increases
(c) remains same
(d) may increase or decrease

453. In food, carbohydrates supply _______ Kcal. per gram.
(a) 4 (b) 5
(c) 6 (d) 7

454. In food, proteins supply ______ Kcal. per gram.
(a) 4 (b) 5
(c) 6 (d) 7

455. Protein is required for
(a) production of antibodies
(b) bacteria inhibition
(c) proper bowel function
(d) absorption of water

456. Which of the following is a macromineral needed by our bodies to maintain health
(a) copper (b) tin
(c) magnesium (d) iron

457. In which of the following foods is solanine considered a toxin
(a) potato (b) tomato
(c) coffee (d) tea

458. If the legal maximum of nitrite (NO_2) is 156 ppm, how much sodium nitrite can you legally add to 1 kg. of meat?
(a) 156 mg (b) 31.2 oz
(c) 78 mg (d) 15.6 ounces

459. Glucose, a simple sugar, melts at 150°C. This is equivalent to
(a) 278.1°F (b) 238°F
(c) 65.5°F (d) 302°F

460. ________________ is an alternative name for baking soda.
(a) carbon dioxide
(b) potassium bitartrate
(c) sodium bicarbonate
(d) calcium carbonate

461. If a food product contains 10,000,000 (10~) microbes per gram, and experiences a
(a) §§§§ (b) §§§§
(c) §§§§ (d) §§§§

462. .9999 per cent kill rate, then _____ microbes per gram will survive.
(a) 1 (b) 10
(c) 100 (d) 1,000

463. The technical name for freeze drying is ________________
(a) lyophilization (b) sublimation
(c) condensation (d) evaporation

464. First commercial aseptic filling machines was commissioned by
(a) James Dole Corporation
(b) Monsanto corporation
(c) Jill corporation
(d) none

465. Cans used for aseptic
(a) tin plate (b) ECCS
(c) both (d) none

466. Glass bottles are sterilized by
(a) saturated steam (b) dry heat
(c) both (d) none

467. Twist wrapping is done for
(a) toffees (b) butter
(c) condensed milk (d) all

468. Mostly biscuits are wrapped in
(a) BOPP (b) PET
(c) glass (d) all

469. GTR stands for
(a) Gas Transmission Rate
(b) Glass Transfer Rate
(c) Gas Transfer Rate
(d) none

470. Thickness of plastic film is measured by
(a) dial gauge (b) vernier caliper
(c) micrometer (d) all

471. Paper density is called
(a) basis weight (b) grammage
(c) both (d) none

16.3 ANSWER KEY

Q.	A.	Q.	A.	Q.	A.	Q.	A.	Q.	A.	Q.	A.
1.	(c)	2.	(b)	3.	(a)	4.	(c)	5.	(a)	6.	(a)
7.	(b)	8.	(d)	9.	(c)	10.	(b)	11.	(c)	12.	(b)
13.	(b)	14.	(b)	15.	(b)	16.	(a)	17.	(c)	18.	(a)
19.	(c)	20.	(b)	21.	(c)	22.	(a)	23.	(d)	24.	(c)
25.	(a)	26.	(a)	27.	(a)	28.	(b)	29.	(c)	30.	(a)
31.	(b)	32.	(a)	33.	(b)	34.	(c)	35.	(a)	36.	(b)
37.	(a)	38.	(a)	39.	(b)	40.	(c)	41.	(c)	42.	(d)
43.	(a)	44.	(d)	45.	(b)	46.	(a)	47.	(c)	48.	(d)
49.	(a)	50.	(a)	51.	(c)	52.	(a)	53.	(c)	54.	(a)
55.	(a)	56.	(b)	57.	(c)	58.	(d)	59.	(a)	60.	(b)
61.	(c)	62.	(d)	63.	(a)	64.	(a)	65.	(b)	66.	(c)
67.	(d)	68.	(a)	69.	(b)	70.	(c)	71.	(d)	72.	(a)
73.	(b)	74.	(c)	75.	(d)	76.	(a)	77.	(b)	78.	(c)
79.	(a)	80.	(b)	81.	(b)	82.	(c)	83.	(d)	84.	(a)
85.	(b)	86.	(c)	87.	(d)	88.	(d)	89.	(a)	90.	(a)
91.	(b)	92.	(c)	93.	(b)	94.	(a)	95.	(b)	96.	(c)
97.	(d)	98.	(d)	99.	(c)	100.	(a)	101.	(b)	102.	(c)
103.	(c)	104.	(a)	105.	(a)	106.	(b)	107.	(a)	108.	(b)
109.	(b)	110.	(a)	111.	(b)	112.	(c)	113.	(a)	114.	(b)
115.	(a)	116.	(c)	117.	(d)	118.	(a)	119.	(c)	120.	(b)
121.	(a)	122.	(c)	123.	(a)	124.	(c)	125.	(b)	126.	(b)
127.	(b)	128.	(c)	129.	(b)	130.	(a)	131.	(d)	132.	(d)
133.	(a)	134.	(b)	135.	(b)	136.	(a)	137.	(b)	138.	(b)
139.	(b)	140.	(a)	141.	(d)	142.	(b)	143.	(c)	144.	(a)
145.	(c)	146.	(c)	147.	(d)	148.	(d)	149.	(c)	150.	(d)
151.	(a)	152.	(d)	153.	(a)	154.	(c)	155.	(c)	156.	(c)
157.	(d)	158.	(d)	159.	(d)	160.	(a)	161.	(b)	162.	(c)
163.	(d)	164.	(d)	165.	(a)	166.	(a)	167.	(d)	168.	(a)
169.	(a)	170.	(d)	171.	(d)	172.	(a)	173.	(d)	174.	(d)
175.	(d)	176.	(d)	177.	(d)	178.	(c)	179.	(c)	180.	(a)
181.	(c)	182.	(b)	183.	(a)	184.	(b)	185.	(b)	186.	(a)
187.	(c)	188.	(d)	189.	(d)	190.	(d)	191.	(a)	192.	(c)

Q. A.	Q. A.	Q. A.	Q. A.	Q. A.	Q. A.
193. (d)	194. (d)	195. (d)	196. (a)	197. (b)	198. (d)
199. (c)	200. (a)	201. (a)	202. (d)	203. (d)	204. (b)
205. (c)	206. (a)	207. (c)	208. (a)	209. (b)	210. (b)
211. (d)	212. (c)	213. (a)	214. (c)	215. (a)	216. (a)
217. (b)	218. (c)	219. (b)	220. (b)	221. (a)	222. (b)
223. (a)	224. (a)	225. (c)	226. (a)	227. (a)	228. (a)
229. (a)	230. (b)	231. (b)	232. (b)	233. (a)	234. (a)
235. (a)	236. (b)	237. (d)	238. (d)	239. (d)	240. (c)
241. (d)	242. (d)	243. (d)	244. (d)	245. (b)	246. (d)
247. (d)	248. (d)	249. (d)	250. (d)	251. (c)	252. (d)
253. (d)	254. (d)	255. (a)	256. (b)	257. (c)	258. (c)
259. (d)	260. (b)	261. (c)	262. (c)	263. (d)	264. (c)
265. (b)	266. (a)	267. (b)	268. (a)	269. (b)	270. (c)
271. (d)	272. (a)	273. (c)	274. (b)	275. (d)	276. (a)
277. (b)	278. (d)	279. (b)	280. (c)	281. (a)	282. (b)
283. (c)	284. (a)	285. (b)	286. (c)	287. (d)	288. (c)
289. (b)	290. (a)	291. (c)	292. (b)	293. (b)	294. (c)
295. (a)	296. (b)	297. (c)	298. (d)	299. (a)	300. (c)
301. (b)	302. (d)	303. (b)	304. (c)	305. (d)	306. (d)
307. (d)	308. (c)	309. (d)	310. (a)	311. (c)	312. (d)
313. (a)	314. (c)	315. (d)	316. (b)	317. (b)	318. (d)
319. (a)	320. (b)	321. (c)	322. (b)	323. (a)	324. (d)
325. (d)	326. (b)	327. (a)	328. (c)	329. (d)	330. (d)
331. (d)	332. (a)	333. (a)	334. (a)	335. (b)	336. (a)
337. (a)	338. (a)	339. (a)	340. (a)	341. (d)	342. (b)
343. (a)	344. (b)	345. (d)	346. (b)	347. (a)	348. (a)
349. (b)	350. (b)	351. (a)	352. (a)	353. (a)	354. (d)
355. (b)	356. (b)	357. (a)	358. (b)	359. (a)	360. (a)
361. (c)	362. (b)	363. (c)	364. (a)	365. (b)	366. (d)
367. (d)	368. (b)	369. (a)	370. (a)	371. (c)	372. (b)
373. (a)	374. (c)	375. (c)	376. (a)	377. (a)	378. (a)
379. (b)	380. (a)	381. (b)	382. (c)	383. (a)	384. (d)
385. (b)	386. (c)	387. (a)	388. (a)	389. (a)	390. (b)
391. (a)	392. (b)	393. (a)	394. (b)	395. (c)	396. (c)

Q.	A.	Q.	A.	Q.	A.	Q.	A.	Q.	A.	Q.	A.
397.	(c)	398.	(a)	399.	(b)	400.	(a)	401.	(a)	402.	(a)
403.	(d)	404.	(b)	405.	(c)	406.	(b)	407.	(a)	408.	(a)
409.	(b)	410.	(a)	411.	(a)	412.	(a)	413.	(a)	414.	(d)
415.	(b)	416.	(b)	417.	(b)	418.	(c)	419.	(b)	420.	(d)
421.	(a)	422.	(a)	423.	(a)	424.	(b)	425.	(b)	426.	(b)
427.	(b)	428.	(c)	429.	(b)	430.	(b)	431.	(d)	432.	(a)
433.	(a)	434.	(d)	435.	(c)	436.	(b)	437.	(a)	438.	(a)
439.	(a)	440.	(c)	441.	(b)	442.	(c)	443.	(b)	444.	(d)
445.	(b)	446.	(b)	447.	(b)	448.	(c)	449.	(a)	450.	(a)
451.	(b)	452.	(a)	453.	(a)	454.	(a)	455.	(a)	456.	(c)
457.	(a)	458.	(a)	459.	(d)	460.	(c)	461.	(b)	462.	(a)
463.	(a)	464.	(c)	465.	(c)	466.	(a)	467.	(a)	468.	(a)
469.	(d)	470.	(c)								

ASRB NET QUESTION PAPER 2013 (MEMORY BASED)

1. The microwave frquency used in food industry is/are
   (a) 915 MHz (b) 2450 MHz
   (c) both (d) none
2. The reaction between amino acids and reducing sugars that gives browning of food is
   (a) maillard reaction
   (b) osazone formation
   (c) sugar inversion
   (d) none
3. The microwaves in microwave oven is emitted by
   (a) magnetron (b) capacitor
   (c) convection coil (d) none
4. Economy ice creams have fat content of
   (a) 5% (b) 10%
   (c) 15% (d) 20%
5. The elasticity of dough is due to
   (a) gliadin (b) glutenin
   (c) pentosan (d) none
6. The Milk Production in India in 2011-12
   (a) 140 MT (b) 132MT
   (c) 128MT (d) 150MT
7. A gluten free diet is recommended to a person suffering from ____________ disease
   (a) goitre (b) celiac
   (c) diabetes (d) piles
8. The antinutritional factor present in legumes is
   (a) Trypsin inhibitors
   (b) Avidin
   (c) Saponin
   (d) both (a) and (c)
9. Cereals are deficient in which amino acid
   (a) methionine (b) lysine
   (c) arginine (d) cysteine
10. Banana is an example of
    (a) climacteric fruit
    (b) non-climacteric fruit
    (c) both
    (d) none
11. Hard wheat is tempered to a moisture content of
    (a) 14.5-15.5% (b) 15.5-16.5%
    (c) 16.5-17.5% (d) 17.5-18.5%
12. Ideally, chapatti flour should have protein content of
    (a) 5-10% (b) 10-15%
    (c) 15-20% (d) None
13. Egg shell comprises of 11% of the whole egg
    (a) 6% (b) 11%
    (c) 16% (d) 21%
14. For tenderization of meat, the enzyme used is
    (a) papain (b) rennin
    (c) lipase (d) all
15. Curcumin is present in
    (a) beetroot (b) turmeric
    (c) onion (d) capsicum

16. The sweetest sugar is
(a) sucrose (b) fructose
(c) dextrose (d) glucose

17. TSS of tomato puree should be
(a) 5% (b) 9%
(c) 21% (d) 26%

18. Which of the following food items raise the glycemic index
(a) potatoes (b) white bread
(c) white rice (d) all

19. Which of the following has highest protein content
(a) soy isolate
(b) soy concentrate
(c) soy grits
(d) soy chunks

20. Beta glucan, is an example of
(a) resistant starch
(b) soluble fiber
(c) insoluble fiber
(d) none

21. The alcohol is used for the precipitation of
(a) soluble fiber (b) pectin
(c) both (d) none

22. Heating of collagen results in formation of
(a) agar (b) gelatin
(c) lecithin (d) avidin

23. Iodine test is used for the detection of
(a) protein (b) starch
(c) vegetable fat (d) sugar

24. The ANF in pulses can be reduced by
(a) soaking (b) fermentation
(c) germination (d) all

25. Which of the following is used as preservative in juice
(a) sulfur dioxide
(b) potassium metabisulfite
(c) citric acid
(d) hydrogen peroxide

26. Moisture content in whole wheat flour is
(a) 9% (b) 12%
(c) 14% (d) 16%

27. The optimum harvest moisture content for corn is
(a) 12-15% (b) 23-25%
(c) 27-32% (d) 5-9%

28. The optimum pH for benzoic acid is
(a) 2.5 (b) 3.5
(c) 4.5 (d) 5.5

29. The most important sensory parameter is
(a) appearance (b) taste
(c) flavor (d) mouthfeel

30. Which among the following requires lower sugar levels to gel
(a) high-methoxyl pectin
(b) low-methoxyl pectin
(c) medium-methoxyl pectin
(d) none

31. Gluten and starch can be separated by
(a) centrifugation (b) filtration
(c) washing (d) both a and c

32. The proportion of absorbed protein from a food which becomes incorporated into the proteins of the organism's body is called
(a) NPU
(b) biological value
(c) PER
(d) none

33. The world consumer day is celebrated on
(a) 15 Jan (b) 15 Feb
(c) 15 Mar (d) 15 Apr

34. The astringency is caused by
(a) proteins (b) fatty acids
(c) polyphenols (d) minerals

35. Oryzanol, an anti-oxidant is present in
(a) wheat (b) maize
(c) rice (d) sorghum

36. Gingerol is present in
(a) ginger (b) garlic
(c) onion (d) none

37. Allicin is an organosulfur compound obtained from
(a) ginger (b) garlic
(c) onion (d) turmeric

38. Platform test for milk does not include
(a) COB test
(b) alcohol test
(c) sediment test
(d) protein evaluation

39. Which chemical is added as preservative for milk during conducting platform tests
(a) mercuric chloride
(b) silver nitrate
(c) hydrogen peroxide
(d) none

40. Casein is present in milk in which state
(a) true solution (b) colloidal
(c) micelle (d) none

41. Fishes excrete nitrogen in the form of
(a) urea (b) ammonia
(c) nitrogen dioxide (d) all

42. Double toned milk has fat content of
(a) 3.5% (b) 1.5%
(c) 0.5% (d) 4.5%

43. Whole egg powder can be obtained by
(a) spray drying (b) drum drying
(c) sun drying (d) tray drying

44. The most common causative organism for food poisoning is
(a) Clostridium (b) Salmonella
(c) Penicillium (d) Candida

45. Index organism for pasteurization of milk is
(a) Coxiella burnetii
(b) Mycobacterium tuberculosis
(c) Clostridium botulinum
(d) none

46. Egg is added to cake batter as
(a) protein supplement
(b) emulsifier
(c) stabilizing agent
(d) thickener

47. Parboiling of rice
(a) Improves its processing
(b) Boosts its nutritional value
(c) Increases the milling yield
(d) all

48. The color changes during parboiling of rice is due to
(a) diffusion of pigments from outer layers to centre of rice kernel
(b) non-enzymatic browning
(c) both
(d) no change in color of parboiled rice

49. Polishing of rice results in loss of
(a) protein
(b) carbohydrates
(c) vitamins
(d) none

50. Red wine involves the fermentation of
(a) red grapes without skin
(b) red grapes with skin
(c) green grapes without skin
(d) green grapes with skin

51. Milling of wheat results in separation of
(a) bran and germ
(b) endosperm from bran and germ
(c) kernel from husk
(d) germ from endosperm and bran

52. Bananas are best stored at a temperature of
(a) 7°C (b) 14°C
(c) 21°C (d) 28°C

53. MAP involves control of
(a) temperature of package
(b) gas in package
(c) the level of gas in package
(d) both (b) and (c)

54. Bitter pit is a disorder in
(a) banana (b) kiwi
(c) apple (d) grapes

55. FSSA came is
(a) 2006 (b) 2009
(c) 2011 (d) 2014

56. The strong lemon odour is due to
(a) Citral (b) malic acid
(c) tartaric acid (d) acetic acid

57. Limonoids are phytochemicals present in
(a) wheat (b) tea
(c) orange (d) lettuce

58. Which of the following functions as an antibiotic
(a) saponin (b) avidin
(c) lecithin (d) heparin

59. The bitterness of beer is due to
(a) hops (b) bran of barley
(c) molasses (d) none

60. The flavonoid present in green tea is
(a) acetaldehyde (b) catechin
(c) betalain (d) none

61. Which of the following is size reduction law
(a) Rittinger's law (b) Kick' law
(c) Bond Law (d) all

62. Stoke's law is fricional force exerted on spherical objects with
(a) large Reynold's no.
(b) small Reynold's no.
(c) both
(d) none

63. The richest plant source of omega 3 fatty acid is
(a) sesame (b) linseed
(c) flaxseed (d) sunflower

64. The preservative used in meat which also helps in retention of color of meat is
(a) sulfur dioxide (b) sodium nitrite
(c) sodium chloride (d) none

65. The damaged starch levels for bread flour is
(a) 2-3% (b) 5-7%
(c) 7-9% (d) 10-14%

66. The stiffening of the body of animal after death, called rigor mortis occurs due to
(a) build up of ATP
(b) loss of ATP
(c) loss of glucose
(d) none

67. Kgray is the unit of
(a) energy
(b) irradiaion dose
(c) pressure
(d) heat

68. Diacetyl flavor in butter should not exceed
(a) 2 ppm (b) 4 ppm
(c) 6 ppm (d) 8 ppm

69. Which product is regarded as Indian cheese
(a) paneer (b) khoa
(c) channa (d) chakka

70. According to FSSA, the sugar in Sweetened Condensed Milk should be
(a) NMT 40% (b) NLT 40%
(c) NMT 50% (d) NLT 50%

71. Which type of cheese is produced from whey
(a) Cottage (b) Ricotta
(c) Cheddar (d) Mozarella

72. Expresso coffee originated in
(a) USA (b) Italy
(c) Korea (d) India

73. Sterilization of plastic films is done by
(a) heat (b) H_2O_2
(c) both (d) none

74. Which of the following is an antimicrobial enzyme
(a) plasmin (b) lysozyme
(c) peroxidase (d) zymase

75. Which of the following vitamins act as anti-oxidants
(a) A&B (b) B&C
(c) C&E (d) D&E

76. The flavor of mint is due to
(a) limonoids (b) catechin
(c) menthol (d) caffeine

77. Enzymatic browning occurs due to enzyme
(a) lipase
(b) peroxidase
(c) polyphenol oxidase
(d) proteinase

78. Aflatoxins are produced by
(a) Penicillium notatum
(b) Aspergillus flavus
(c) Lactobacillus plantarum
(d) Rhizopus stolonifer

79. The Intermediate Moisture Foods have a moisture content of
(a) 10-30% (b) 20-50%
(c) 50-70% (d) < 5%

80. Which of the following is a type of rancidity that can occur in food
(a) oxidative (b) hydrolytic
(c) ketonic (d) all

81. Which ISO standard integrates the seven principles of HACCP system
(a) 14000 (b) 22000
(c) 31000 (d) 26000

82. Vinegar is
(a) 10% acetic acid
(b) 4% acetic acid
(c) 4% formaldehyde
(d) none

83. The Codex Alimentarius is
(a) food code (b) food law
(c) food principle (d) food fact

84. The roasting of pulses is done at a temperature of
(a) 80°C (b) 110°C
(c) 130°C (d) 200°C

85. Which of the following is used as food glazing agent
(a) waxes (b) GMS
(c) carrageenan (d) lecithin

86. The ERH of a food is 67%, what is its water activity
(a) 67 (b) 6.7
(c) 0.67 (d) 0.067

87. The syneresis of jelly is due to
(a) excess of acid
(b) high storage temperature
(c) sealing with paraffin wax
(d) all

88. Which of the following is an inert gas
(a) O_2 (b) CO
(c) CO_2 (d) N_2

89. Which of the following is an enzyme used for tenderization of meat to make it softer
(a) papain (b) bromelain
(c) actinidin (d) all

90. The antinuritional factor in cereals is
(a) saponin (b) raffinose
(c) hemagglutenin (d) phytic acid

91. The cholesterol levels in body can be reduced by
(a) soluble fiber (b) beta glucan
(c) isoflavones (d) all

92. Aspartame, an artificial sweetener, is formed of
(a) aspartic acid and valine
(b) aspartic acid and alanine
(c) aspartic acid and phenylalanine
(d) aspartic acid and glycine

93. Acid whey has lactic acid content of
(a) 1-4% (b) 4-6%
(c) 6-10% (d) 10-14%

94. One IU of beta-carotene is equivalent to
(a) 0.6mcg beta-carotene
(b) 6 mcg beta-carotene
(c) 0.6mg beta-carotene
(d) 6 mg beta-carotene

95. Most common type of food borne disease is by micro-organism
(a) Vibrio cholera (b) Salmonella
(c) Mycobacterium (d) none

96. Canning is performed in which type of packaging material
(a) paper
(b) plastic
(c) foil
(d) tin plated steel cans

97. Which micro-organisms are involved in sauerkraut preparation
(a) Leuconostoc mesenteroides and Lactobacillus plantarum
(b) Leuconostoc mesenteroides and Lactobacillus bulgaricus
(c) Lactobacillus plantarum and Leuconostoc pseudomesenteroides
(d) Lactobacillus plantarum and Leuconostoc lactis

98. Common problem occurring in intermediate moisture foods is
(a) enzymatic browning
(b) non-enzymatic browning
(c) oxidation
(d) both (b) and (c)

99. Six sigma is a set of techniques for
(a) food fortification
(b) food safety
(c) food packaging
(d) none

100. Which of the following technique is used in flavor extraction
(a) supercritical fluid extraction
(b) molecular distillation
(c) liq-liq counter current extraction
(d) all

101. F-value holds importance in case of
(a) pasteurization (b) canning
(c) homogenization (d) drying

ANSWER KEY ASRB NET-2013

Q. A.	Q. A.	Q. A.	Q. A.	Q. A.	Q. A.
1. (c)	2. (a)	3. (a)	4. (b)	5. (b)	6. (c)
7. (b)	8. (d)	9. (b)	10. (a)	11. (b)	12. (b)
13. (b)	14. (a)	15. (b)	16. (b)	17. (b)	18. (d)
19. (a)	20. (b)	21. (c)	22. (b)	23. (b)	24. (d)
25. (b)	26. (a)	27. (b)	28. (c)	29. (c)	30. (b)
31. (d)	32. (b)	33. (c)	34. (c)	35. (c)	36. (a)
37. (b)	38. (d)	39. (a)	40. (b)	41. (b)	42. (b)
43. (a)	44. (b)	45. (a)	46. (b)	47. (d)	48. (c)
49. (c)	50. (b)	51. (b)	52. (b)	53. (d)	54. (c)
55. (a)	56. (a)	57. (c)	58. (b)	59. (a)	60. (b)
61. (d)	62. (b)	63. (c)	64. (b)	65. (c)	66. (b)
67. (b)	68. (b)	69. (a)	70. (b)	71. (b)	72. (b)
73. (c)	74. (b)	75. (c)	76. (c)	77. (c)	78. (b)
79. (b)	80. (d)	81. (b)	82. (b)	83. (a)	84. (c)
85. (a)	86. (c)	87. (d)	88. (d)	89. (d)	90. (d)
91. (d)	92. (c)	93. (c)	94. (a)	95. (b)	96. (d)
97. (a)	98. (d)	99. (b)	100. (d)	101. (b)	

ASRB NET QUESTION PAPER 2014 (MEMORY BASED)

1. Fat and SNF standards for buffalo milk are
(a) 1.5% and 8.5% (b) 4.5% and 8.5%
(c) 6.0% and 8.5% (d) 6.0% and 9.0%

2. Base material used for beer making is
(a) wheat (b) rice
(c) barley (d) sugarcane

3. Dole process is related to
(a) aseptic canning (b) blanching
(c) comminution (d) ohmic heating

4. Egg whites are rich source of
(a) Vitamin D (b) zinc
(c) selenium (d) all

5. Yellow color of egg yolk is due to
(a) lutein (b) zeaxanthin
(c) both (d) none

6. Zein is a principal protein of which crop
(a) wheat (b) mustard
(c) corn (d) sesame

7. Cans treated with acid resistant lacquer cans are also known as
(a) A-enamel (b) C-enamel
(c) R-enamel (d) none

8. Minimum expansion of water is occurs in
(a) slow freezing (b) blast freezing
(c) IQF (d) all

9. Celiac disease is related to
(a) oryzin (b) gliadin
(c) zein (d) hordein

10. Salometer is used to measure the strength of
(a) brine solution (b) sugar solution
(c) acid solution (d) alkali solution

11. Who is considered as "father of canning"
(a) Robert Hooke (b) Louis Pasteur
(c) Nicholas Appert (d) V. Kurien

12. Compound responsible for anti-microbial activity of clove is
(a) eugenol (b) cinnamldehyde
(c) camphene (d) terpineol

13. Characteristic of bakery fat is
(a) low melting (b) high melting
(c) sharp melting (d) all

14. Function performed by nitrates in cured meat
(a) pink color
(b) prevent spoilage
(c) both
(d) none

15. Time-temperature combination used for milk pasteurization
(a) 63°C/30 min (b) 63°C/15 min
(c) 72°C/30 min (d) 72°C/15 min

16. Which one is present in highest amount in finger millet (ragi)
(a) thiamin (b) niacin
(c) iron (d) calcium

17. Cider is prepared from
(a) wheat (b) barley
(c) apple (d) grapes

18. Meat fat is rich in
(a) saturated fat (b) unsaturated fat
(c) both (d) none

19. Common adulterant in black pepper is
(a) brick sand
(b) papaya seed
(c) urea
(d) pomegranate seeds

20. TSS in jam as per FSSAI
(a) NLT 60 (b) NMT 60
(c) NLT 65 (d) NMT 65

21. Which one is not available for fermentation by yeast during bread making
(a) glucose (b) amylose
(c) starch (d) none

22. In food industry, rotameter is used to measure
(a) viscosity (b) flow rate
(c) rpm (d) all

23. TQM stand for
(a) total quality management
(b) temperature quotient measurement
(c) tertiary quality management
(d) tertiary quotient measurement

24. Fat standards for double toned milk is
(a) 0.5% (b) 1.5%
(c) 3.0% (d) 6.0%

25. Diet containing __________ is helpful in controlling pellagra
(a) cauliflower (b) cabbage
(c) both (d) none

26. Mold growth in bread is due to
(a) Rhizopus stolonifer
(b) Rhizopus oryzae
(c) Aspergilus niger
(d) Aspergilus oryzae

27. Saffron is obtained from which portion of the plant
(a) seed (b) stigma
(c) sepal (d) petal

28. Food poisoning is caused by
(a) Salmonella (b) Lactobaccillus
(c) Pencillium (d) Candida

29. Marasmus is caused due to deficiency of
(a) Vitamin (b) minerals
(c) protein (d) fat

30. Freezing has most pronounced effect on which of the following attribute of fruits and vegetables
(a) color (b) flavor
(c) texture (d) taste

31. Rotavane is used in tea processing for
(a) withering (b) rolling
(c) fermentation (d) drying

32. Each degree of salinometer scale corresponds to
(a) 0.265% NaCl (b) 2.65% NaCl
(c) 12.65% NaCl (d) 26.5% NaCl

33. Consumer Protection Act came in year
(a) 1973 (b) 1986
(c) 1993 (d) 2006

34. Base material used for rum making is
(a) wheat (b) rice
(c) barley (d) sugarcane

35. Which of the following component is present in highest amount in whole egg
(a) water (b) protein
(c) fat (d) carbohydrate

36. Smoking treatment is used for the preservation of
(a) Fruits (b) milk
(c) meat (d) vegetables

37. Veal is the meat of
(a) Pig (b) calf
(c) goat (d) buffalo

38. Barges is
a) A cocoa variety
(b) processing method of cocoa beans
(c) cocoa based drink
(d) cocoa shipment method

39. Total Plate Count is expressed as
(a) Cfu/ml (b) cells/ml
(c) viable cells/ml (d) none

40. HACCP stands for
(a) Hazard Analysis Central Critical Parameters
(b) Hazard Analysis Critical Control Point
(c) Hazard Analysis Combined Critical Point
(d) Health Analysis Critical Control Point

41. The scientific name of the liquerice plant is
(a) Glycyrrhiza glabra
(b) *Cinnamomum tamala*
(c) Ocimum basilicum
(d) Syzygium aromaticum

42. Which one is used for the adulteration of dried coriander powder
(a) horse dung powder
(b) cow dung powder
(c) buffalo dung powder
(d) all

43. The most preferred type of yeast used in baking is
(a) Instant yeast
(b) compressed yeast
(c) dry yeast
(d) Brewer's yeast

44. Intrinsic factor for microbial growth is
(a) temperature (b) pressure
(c) humidity (d) water activity

45. Potassium metabisulphite should not be used for the foods containing
(a) meat (b) vitamin B_1
(c) raw fruits (d) all

46. Gluten proteins are
(a) Water soluble
(b) water insoluble
(c) fat soluble
(d) none

47. The bioactive compound in ginger is
(a) Ginseng (b) gingerol
(c) sinigrin (d) allicin

48. Which of the following is regarded as king of spices
(a) Chilly (b) pepper
(c) turmeric (d) aniseed

49. Which of the following prevents the spoilage of fat
(a) Sequestrants (b) emulsifier
(c) anti-oxidants (d) lipase

50. Central Agmark Laboratory is situated at
(a) Chennai (b) New Delhi
(c) Nagpur (d) Kolkatta

51. Most commonly used fumigant in stored grains
(a) methyl bromide (b) phosphine
(c) carbondioxide (d) both (a) & (b)

52. Compound responsible for butter flavor is
(a) diacetyl (b) vanillin
(c) both (a) & (b) (d) diethyl

53. Test used for the detection of vanaspati adulteration in ghee
(a) furfural test (b) baudouin test
(c) both (d) none

54. Which one is used for iron fortification of food products
(a) ferrous sulphate
(b) ferrous gluconate
(c) ferrous lactate
(d) all

55. Minor ingredient used in bread making
(a) sugar (b) fat
(c) maida (d) water

56. The gluten formation starts at which stage in bread making
(a) hydration (b) development
(c) pick up (d) clean

57. Spoilage of fruit juice is due to
(a) lactic acid fermentation
(b) organic acid fermentation
(c) slime production
(d) all options

58. Anti-nutritional factor present in egg
(a) avidin (b) phytic acid
(c) saponin (d) tannin

59. Red color of tomatoes is due to
(a) anthocyanin (b) carotene
(c) riboflavin (d) lycopene

60. Which of the following share highest portion of meat export in india
(a) cattle meat (b) goat meat
(c) pig meat (d) poultry meat

61. Temperature used in UHT processing of milk
(a) 63°C (b) 72°C
(c) 115°C (d) 135°C

62. Which of the following is a component of myofibril
(a) actin (b) myosin
(c) both (d) none

63. Flavour components in food products can be analyzed by
(a) calorimetry (b) chromatography
(c) rheology (d) psychrometry

64. Bioactive component of garlic having health benefits
(a) gallicin (b) allicin
(c) mellicin (d) allyl sulphate

65. Food sterilization via irradiation is also known as
(a) gas sterilization
(b) cold sterilization
(c) neutral sterilization
(d) all

66. Identify the correct statement
(a) Cereals are rich in methionine and lysine
(b) Cereals are deficient in methionine and lysine
(c) Cereals are rich in methionine and deficient in lysine
(d) Cereals are deficient in methionine and rich in lysine

67. Defense Food Research Laboratory is located at
(a) Delhi (b) Hyderabad
(c) Nagpur (d) Mysore

68. Common salt affects __________of meat during curing
(a) color (b) flavor
(c) appearance (d) all

69. According to BIS standards, minimum fat content in khoa should be
(a) 15% (b) 23%
(c) 37% (d) 50%

70. Glazing of fish is practiced to protect it from
a) oxidation (b) freezer burn
(c) both (d) none

71. Poultry eggs are rich source of nutrients except
(a) malic acid (b) ascorbic acid
(c) adipic acid (d) all

72. Which of the following is true about wax coating of fruits
(a) increases surface gloss
(b) decreases rate of transpiration
(c) increases shelf life
(d) all

73. ISO 14000 is related to management of
(a) food safety (b) environment
(c) energy (d) food quality

74. Which of the following compound is characterized by the presence of $-NH_2$ and carboxyl -COOH group
(a) carboxylic acid (b) amino acid
(c) ascorbic acid (d) retinol

75. Which of the following is specifically related to fruits and vegetables
(a) FPO (b) FSSAI
(c) AGMARK (d) all

76. Which one is not dependent on food product size
(a) pasteurization (b) ohmic heating
(c) both (d) none

77. The deficiency of calcium and vitamin D in children leads to
(a) Osteomalacia (b) rickets
(c) scurvy (d) pellagra

ANSWER KEY ASRB NET-2014

Q. A.	Q. A.	Q. A.	Q. A.	Q. A.	Q. A.
1. (d)	2. (c)	3. (a)	4. (d)	5. (c)	6. (c)
7. (c)	8. (c)	9. (b)	10. (a)	11. (c)	12. (a)
13. (c)	14. (c)	15. (a)	16. (d)	17. (c)	18. (b)
19. (b)	20. (c)	21. (a)	22. (b)	23. (a)	24. (b)
25. (c)	26. (a)	27. (b)	28. (a)	29. (c)	30. (c)
31. (b)	32. (a)	33. (b)	34. (d)	35. (a)	36. (c)
37. (b)	38. (d)	39. (a)	40. (b)	41. (a)	42. (a)
43. (b)	44. (d)	45. (d)	46. (b)	47. (b)	48. (b)
49. (c)	50. (c)	51. (d)	52. (a)	53. (c)	54. (d)
55. (a)	56. (a)	57. (d)	58. (a)	59. (d)	60. (a)
61. (d)	62. (c)	63. (b)	64. (b)	65. (b)	66. (c)
67. (d)	68. (b)	69. (c)	70. (c)	71. (d)	72. (d)
73. (b)	74. (b)	75. (a)	76. (b)	77. (b)	

ASRB NET QUESTION PAPER 2015 (MEMORY BASED)

1. Yoghurt is a fermented product which contains
   (a) virus (b) bacteria
   (c) yeast (d) all

2. Freezing of food is a good method of preservation as
   (a) it does not allow growth of pathogens
   (b) it does not allow growth of psychrophiles
   (c) it retards the enzymic reactions
   (d) all

3. Pulse electric field works by
   (a) electric breakdown of cellwall
   (b) application of high pressure
   (c) high heat treatment
   (d) none

4. Oil content in sesame seeds is
   (a) 30% (b) 40%
   (c) 50% (d) 60%

5. The pectin content is citrus peel is
   (a) 20-30% (b) 10-20%
   (c) 30-40% (d) 40-50%

6. Milling of wheat is done to
   (a) separate endosperm from germ and bran
   (b) separate oil from germ
   (c) separate bran from germ
   (d) none

7. Controlled Atmospheric Packaging
   (a) monitors the gases inside the package
   (b) monitors the external gases
   (c) monitors and controls the gases inside the package
   (d) none

8. Which of the following is a climacteric fruit
   (a) mango (b) orange
   (c) grapes (d) strawberry

9. The fat content in double toned milk is
   (a) 0.5% (b) 1.5%
   (c) 1% (d) 3.5%

10. For optimum jelly preparation which one is correct
    (a) pH-3.1; Acid-1%; sugar-67.5%
    (b) pH-5.5; Acid-1%; sugar-60.5%
    (c) pH-3.1; Acid-2%; sugar-55.5%
    (d) pH-3.1; Acid-1%; sugar-75.5%

11. Which of the following helps to retain color of meat
    (a) NaCl (b) nitrates
    (c) bromates (d) none

12. Myoglobins binds to which gas in muscles
    (a) CO_2 (b) O_2
    (c) N_2 (d) CO

13. Heating of collagen leads to formation of
    (a) gelatin (b) actin
    (c) myosin (d) none

14. Ground meat with skin around it is called
    (a) sausage (b) beef
    (c) mutton (d) none

15. Folic acid deficiency leads to
(a) pellagra (b) anaemia
(c) scurvy (d) beri-beri

16. Pascal is the unit of meaurement of
(a) temperature (b) pressure
(c) heat (d) none

17. Hedonic scale is used for
(a) sensory analysis
(b) protein analysis
(c) fat analysis
(d) none

18. Which of the following test involves two similar and one dissimilar sample
(a) triangle (b) duo-trio
(c) paired comparison (d) none

19. Which of the following is responsible for sour taste
(a) citric acid (b) quinine
(c) caffeine (d) glutamates

20. Most common type of spoilage in cans is
(a) TA spoilage (b) hydrogn swell
(c) soft swell (d) all

21. The radiation dose for inhibiting sprouting in potatoes is
(a) 0.05-0.15kGy (b) 5-10 kGy
(c) 0.5-1.0 kGy (d) 0.01-0.02 kGy

22. The hormone involved in the conversion of glycogen to glucose in the liver is
(a) glucagon (b) insulin
(c) vasopressin (d) none

23. Disease which arises due to inefficient glucose metabolism is
(a) diabetes mellitis
(b) diabetes insipidus
(c) celiac disease
(d) goitre

24. Scientific name of rice
(a) Oryza sativa (b) Allium cepa
(c) Hordeum vulgare
(d) Triticum aestivum

25. White revolution/ Operation flood started in year
(a) 1950 (b) 1960
(c) 1970 (d) 1980

26. ISO 9000- Quality systems came in year
(a) 1977 (b) 1987
(c) 1997 (d) 1999

27. NDDB was established in
(a) 1965 (b) 1970
(c) 1975 (d) 1980

28. Which of the following organization was set up jointly by FAO and WHO
(a) Codex Alimentarius (b) FSSAI
(c) EU (d) USDA

29. The letter P in HACCP stands for
(a) point (b) protection
(c) preservation (d) none

30. The principal anti-nutritional factor in soyabean is
(a) saponin (b) avidin
(c) lecithin (d) none

31. PFA came in year
(a) 1954 (b) 1955
(c) 1964 (d) 1965

32. The protein content of lean meat is
(a) 20-22% (b) 30-32%
(c) 40-42% (d) 10-12%

33. Barley malt is used for the preparation of
(a) wine (b) beer
(c) brandy (d) rum

34. Bleaching of flour results in flour with
(a) improved dough strength
(b) decreased dough strength
(c) improved protein content
(d) decreased protein content

35. Genetically modified rice with carotene is
(a) golden rice (b) silver rice
(c) red rice (d) yellow rice

36. The insertion of BT gene causes a change in
(a) DNA (b) RNA
(c) both (d) none

37. Which of the following has the highest protein content
(a) soya chunks
(b) soya grits
(c) soya protein concentrate
(d) soya protein isolate

38. Kimchi is a product of which country
(a) Japan (b) China
(c) India (d) Korea

39. Noodles originated in which country
(a) India (b) Japan
(c) SriLanka (d) China

40. Dry ice is
(a) solid N_2 (b) solid O_2
(c) solid CO_2 (d) solid H_2

41. Wheat is lacking in which amino acid
(a) cysteine
(b) methionine
(c) lysine
(d) tryptophan

42. What is the ratio of diameter to thickness of cookie called
(a) spread ratio (b) cookie ratio
(c) spread factor (d) both (a) & (c)

43. Sorption isotherm is plotted between moisture content and
(a) aw of food product
(b) temp of product
(c) RH of product
(d) none

44. The production of milk in India in 2015-16 has been
(a) 145 MT (b) 155 MT
(c) 165MT (d) 175MT

45. The preparation of jelly does not require
(a) acid (b) sugar
(c) pectin (d) salt

46. An effective fumigant in grain storage is
(a) Methyl bromide
(b) Calcium Carbide
(c) acetylene
(d) none

47. The detection of ghee with vanaspati can be done by
(a) *Baudouin test*
(b) *Bromo-Thymol test*
(c) *Starch test*
(d) *Resorcinol test*

48. Which enzyme should be inactivated immediately after homogenization of milk
(a) phosphatase (b) lipase
(c) peroxidase (d) proteinase

49. Mycotoxin is produced by
(a) Aspergillus flavus
(b) Penicillium
(c) Coxiella burnetii
(d) All

50. Rennin is a
(a) protein (b) enzyme
(c) fatty acid (d) antibiotic

51. The color of anthocyanin at low pH is
(a) red (b) blue
(c) green (d) brown

52. Angle of repose of wheat grain is
(a) 27° (b) 37°
(c) 47° (d) 57°

53. Father of canning is
(a) Nicholas Appert (b) Pebbles
(c) Huan Chang (d) M. Dirchoff

54. The visible range of spectrophotometer is
(a) 400-800nm (b) 200-400nm
(c) 800-1200nm (d) None

55. When milk is coagulated, the part that remains is called
(a) casein (b) whey
(c) both (d) none

56. Pectin is precipitated by
(a) alcohol
(b) ammonium hydroxide
(c) Petroleum Ether
(d) all

57. The biological value of egg is
(a) 93 (b) 63
(c) 73 (d) 43

58. Saffron is obtained from ____ of flower
(a) petals (b) stigma
(c) stamen (d) leaflets

59. The taste buds have a life cycle of ____ days
(a) 15 (b) 30
(d) 60 (d) 90

60. Rhodopsin is a mixture of
(a) opsin and 11-cis-retinal
(b) rhodsin and opsin
(c) it is not a mixture
(d) none

61. Banana like flavor
(a) malic acid (b) limonin
(c) iso amyl alcohol (d) tartaric acid

62. Which of the following is not central tendency
(a) mean (b) median
(c) mode (d) standard deviation

63. Which among the following has the lowest penetration power
(a) gamma rays (b) x rays
(c) microwaves (d) infra red rays

64. The immunity with which a person is born is
(a) innate (b) acquired
(c) adaptive (d) none

65. TSS of fruits is measured by
(a) refractometer (b) penetrometer
(c) hydrometer (d) lactometer

66. Which enzyme is responsible for browning of cut fruits and vegetables
(a) PPO (b) amylase
(c) lipase (d) proteinase

67. Which of following is not a preservation technique
(a) fermentation (b) packaging
(c) pickling (d) micro filtration

68. Which one is class 1 preservative
(a) sorbates (b) carbonates
(c) salt (d) benzoates

69. Carbonated drinks and water is packaged in
(a) PET (b) PS
(c) polyamide (d) none

70. Kernel weight of rice
(a) 10-20mg (b) 20-30mg
(c) 30-40mg (d) 40-50mg

71. Which of the following is used as an adhesive in laminates
(a) silicates (b) starch
(c) casein (d) whey

72. Which of the following cannot be prevented by consumption of probiotics
(a) urinary tract infections
(b) digestive disorders
(c) respiratory disorders
(d) all

73. Which of the following is amphoteric in nature
(a) carbohydrates (b) proteins
(c) fats (d) fibers

74. Deficiency of niacin can cause
(a) Beri-beri (b) Pellagra
(c) Scurvy (d) Rickets

75. Which of the following is an example of cryogenic agent
(a) liquid H_2 (b) liquid N_2
(c) liquid CO (d) None

76. Which of the following contains antimicrobial and anti-inflammatory factors
(a) spices (b) condiments
(c) both (d) none

77. Which of the following differentiates a sample based on its intensity
(a) Duo-trio
(b) Triangle test
(c) Ranking Test
(d) Paired comparison

78. Heat required to raise the temperature of 1g of substance by 1°C is called
(a) specific heat (b) enthalpy
(c) latent heat (d) none

79. The enzymatic reactions are most affected by
(a) Temperature (b) pH
(c) Both (d) None

80. 10°C is equivalent to ______ °F
(a) 20 (b) 30
(c) 40 (d) 50

81. Blackening of pickle occurs due to formation of
(a) hydrogen sulfide (b) bromates
(c) oil (d) vinegar

82. Which of the following mill uses compressive and shearing forces
(a) ball mill (b) hammer mill
(c) roller mill (d) pin mill

83. Which of the following gas is important for climacteric fruits
(a) CO_2 (b) O_2
(c) C_2H_4 (d) CO

84. Which of the following is the cooking technique that will allow the retention of most vitamins
(a) pickling (b) roasting
(c) frying (d) boiling

85. Dry curing method of cooking meat is
(a) broiling (b) roasting
(c) frying (d) all

86. Laminates are prepared from
(a) paper (b) cellophane
(c) aluminium (d) all

87. Essential oil of spices is obtained by
(a) distillation (b) evaporation
(c) expression (d) both a and c

88. Resins are addded during manufacturing of paper to
(a) Improve water repellancy
(b) improve tear resistance of paper
(c) Improve the gas barrier property
(d) all

89. Which of the following is responsible for spoilage of eggs
(a) Penicillium (b) Calosporium
(c) Thamnidium (d) all

90. A disadvantage of freezing is
(a) Enzymes inactivation
(b) Freezer burn
(c) Growth of mo's
(d) Shelf life

ANSWER KEY ASRB NET-2015

Q.	A.	Q.	A.	Q.	A.	Q.	A.	Q.	A.	Q.	A.
1.	(b)	2.	(d)	3.	(a)	4.	(c)	5.	(a)	6.	(a)
7.	(c)	8.	(a)	9.	(b)	10.	(a)	11.	(b)	12.	(b)
13.	(a)	14.	(a)	15.	(b)	16.	(b)	17.	(a)	18.	(a)
19.	(a)	20.	(a)	21.	(a)	22.	(a)	23.	(a)	24.	(a)
25.	(c)	26.	(b)	27.	(a)	28.	(a)	29.	(a)	30.	(a)
31.	(a)	32.	(a)	33.	(b)	34.	(a)	35.	(a)	36.	(a)
37.	(a)	38.	(d)	39.	(d)	40.	(c)	41.	(c)	42.	(d)
43.	(a)	44.	(b)	45.	(d)	46.	(a)	47.	(a)	48.	(b)
49.	(a)	50.	(b)	51.	(a)	52.	(a)	53.	(a)	54.	(a)
55.	(b)	56.	(a)	57.	(a)	58.	(b)	59.	(a)	60.	(a)
61.	(c)	62.	(d)	63.	(c)	64.	(a)	65.	(a)	66.	(a)
67.	(d)	68.	(c)	69.	(a)	70.	(d)	71.	(c)	72.	(c)
73.	(b)	74.	(b)	75.	(b)	76.	(c)	77.	(c)	78.	(a)
79.	(c)	80.	(d)	81.	(a)	82.	(c)	83.	(c)	84.	(a)
85.	(d)	86.	(d)	87.	(d)	88.	(a)	89.	(d)	90.	(b)

ASRB NET QUESTION PAPER 2016 (MEMORY BASED)

1. Fat and SNF standards for double toned milk are
   (a) 1.5% and 8.5% (b) 1.5% and 9.0%
   (c) 3.0% and 8.5% (d) 3.0% and 9.0%
2. Consumer protection bill came in
   (a) 1985 (b) 1993
   (c) 2006 (d) 2015
3. How many net ATP generated in glycolysis
   (a) one (b) two
   (c) four (d) six
4. Red rot disease is found in which of the following crop
   (a) wheat (b) pearl millet
   (c) mustard (d) sugarcane
5. Study of air-water vapour environment is called
   (a) psychrometrics (b) psychometric
   (c) psycometry
   (d) gas-hydrometry
6. Full form of OHSAS-
   (a) Operational Hurdle and Sanitation Analysis System
   (b) Operational Health and Sanitation Assessment Series
   (c) Occupational Hurdle and Sanitation Analysis System
   (d) Occupational Health and Safety Assessment Series
7. Which of the following is not an end product of glycolysis
   (a) pyruvate (b) NADH
   (c) ATP (d) glucose
8. The *amount* of energy *required to raise* the *temperature* of one mole or gram of a substance by one degree celsius without any change of phase
   (a) specific heat (b) latent heat
   (c) enthalpy (d) none
9. Force during cutting is
   (a) frictional force (b) shear force
   (c) impact (d) none
10. Anti nutritional factor in egg is
    (a) avidin (b) saponin
    (c) phytate (d) tannin
11. Fat and SNF standards for toned milk are
    (a) 1.5% and 8.5%
    (b) 1.5% and 9.0%
    (c) 3.0% and 8.5%
    (d) 3.0% and 9.0%
12. Flour of whole cereals contains
    (a) soluble fiber
    (b) insoluble fiber
    (c) both
    (d) none
13. An antioxidant found in milk is
    (a) lactoferrin (b) lecithin
    (c) casein (d) lactose
14. *A ton of refrigeration is defined as the quantity of heat required to be removed from one tonne of ice within ________ hour when the initial condition of water is 0°C.*
    (a) 1 hour (b) 6 hours
    (c) 12 hours (d) 24 hours

15. *Thermal death time* is defined as the *time* required to kill a population of the target microorganism in a water-based solution at a given temperature is also known as
(a) D-value (b) F-value
(c) z-value (d) T-value

16. Ohmic heating is also known as
(a) Joule Heating
(b) Electrical Resistance Heating
(c) both (d)
none

17. Hormone controlling blood sugar levels is
(a) Ptylin (b) insulin
(c) epinephrine (d) glucagon

18. Electromagnetic radiation in the range of wavelengths 400-700 nm is called
(a) visible light (b) UV light
(c) IR light (d) NIR radiations

19. Frequency used in microwave heating is
(a) 915MHz (b) 2450 MHz
(c) both (d) none

20. Which of the following is meat tenderizer enzyme
(a) Papain (b) Bromelain
(c) both (d) none

21. Which of the following is most important for the success of a new food product in the market?
(a) Product's composition
(b) packaging
(c) sensory
(d) marketing

22. Boiling point of milk is
(a) 93.5 °C (b) 97.8 °C
(c) 100.07°C (d) 102 °C

23. Vitamin having anti-oxidant properties
(a) vitamin A (b) vitamin B
(c) vitamin C (d) vitamin D

24. Which of the following is fat soluble vitamin
(a) retinol (b) thiamine
(c) riboflavin (d) ascorbic acid

25. Which of the following is not a tetrose
(a) erythrose (b) threose
(c) erythulose (d) ribose

26. Which of the following is a measure of central tendency
(a) mean (b) mode
(c) median (d) all options

27. Highest cereal producing countries are
(a) China (b) United States
(c) India (d) France

28. Major crop of India in terms of production is
(a) rice (b) wheat
(c) barley (d) pearl millet

29. *Which of the following natural preservative* is used in various food products for increasing shelf life and inhibiting growth of bacteria
(a) ascorbic acid (b) sorbic acid
(c) nisin (d) propionate

30. Which of the following is used for preservation of cake and other baked products
(a) sorbic acid
(b) calcium propionate
(c) both
(d) none

31. Cheese prepared using skim milk is
(a) cottage cheese (b) ricotta cheese
(c) parmesan (d) all options

32. HTST pasteurization of milk is carried out at
(a) 71.7°C for 15 sec
(b) 71.7°C for 30 min
(c) 62.8°C for 15 sec
(d) 62.8°C for 30 min

33. Phosphatase test is done to check the adequacy of
(a) blanching (b) pasteurization
(c) parboiling (d) sterilization

34. Cryovac is related to
(a) food packaging (b) food processing
(c) pasteurization (d) all options

35. Dry ice is also known as
(a) liquid nitrogen (b) solid nitrogen
(c) solid CO_2 (d) liquid CO_2

36. Cheese is the dairy product prepared with the help of
(a) rennet
(b) recombinant chymosin
(c) both
(d) none

37. *The hurdle included in hurdle technology* is
(a) temperature (b) water activity
(c) redox potential (d) all options

38. Leuconostoc mesenteroides produce _______ acid in Sauerkraut
(a) lactic acid (b) citric acid
(c) acetic acid (d) fumaric acid

39. *Which of the following* is a *fumigant causes ozone depletion.*
(a) methyl bromide (b) chloropicrin
(c) formaldehyde (d) iodoform

40. Aflatoxin is a poison produced by
(a) Aspergillus niger
(b) Aspergillus flavus
(c) Aspergillus affinis
(d) Aspergillus fumigatus

41. T.A. spoilage is caused by **thermophilic anaerobes** that produce acid and gas in low-acid goods. Clostridium thermosaccharolyticum, an obligate thermophile, causes spoilage. The can swells due to
(a) CO_2 (b) H_2
(c) both (d) none

42. Sorbic acid is most active at *pH*
(a) 4.4 (b) 5.5
(c) 6.5 (d) 7.0

43. pH of moderate acid foods
(a) 4.5-5.0 (b) 5.0-5.5
(c) 5.5-6.0 (d) 6.0-6.5

44. *Which of the following fatty acids is essential* for humans
(a) linolenic acid (b) linoleic acid
(c) both (d) none

45. EPA and DHA, both are present in
(a) cotton seed oil (b) corn oil
(c) fish oil (d) mustard oil

46. Which of the following contains highest amount of vitamin C
(a) Barbados cherry (b) guava
(c) amla (d) pomegranate

47. Eggshell is made almost entirely of _________ crystals.
(a) calcium carbonate
(b) calcium hydroxide
(c) calcium chloride
(d) calcium phosphate

48. Hard wheat is different from soft because of difference in its
(a) protein content
(b) endosperm structure
(c) damaged starch content
(d) all options

49. Which of the following contains highest fat content
(a) cookies (b) avocado
(c) whipping cream (d) milk powder

50. Pycnometer is used for measurement of
(a) specific gravity (b) pH
(c) humidity (d) fat content

51. Refractometer is used to measure
(a) acid/brix ratio (b) acidity
(c) TSS (d) water activity

52. Which of the following is the strongest bond
(a) ionic (b) covalent
(c) hydrogen (d) vanderwall

53. Force of attraction between similar molecules is called ______forces.
(a) cohesive (b) adhesive
(c) vanderwall (d) shear

54. __________is used to extend the shelf life of Fruits and vegetables.
(a) ethylene
(b) calcium carbonate
(c) gibberellic acid
(d) all

55. Moisture content for paddy storage is less than ________ % for grain storage and less than __________ % for long term seed preservation.
(a) 14% & 9% (b) 9% & 14%
(c) 14% & 12% (d) 12% & 14%

56. Parboiling of paddy increases its
(a) vitamins (b) minerals
(c) both (d) none

57. Which of the following is a non-newtonian fluid
(a) fruit juice (b) tomato ketchup
(c) water (d) all

58. Tomato ketchup is a
(a) bingham fluid (b) pseudoplastic
(c) rheopectic (d) all

59. The quality of egg is checked by
(a) water activity meter
(b) candling
(c) Senn's process
(d) all

60. Which of the following function in fruits and vegetables
(a) shelf life
(b) imparts glossiness
(c) prevents dehydration of fresh fruits and vegetables
(d) all

61. Black *bread* mold (Rhizopus stolonifer) is a widely distributed ________ mucoralean mold found on *bread* surfaces.
(a) thread-like (b) dotted
(c) both (d) none

62. Instant coffee is packaged in
(a) Aluminum foil (b) PET
(c) AL+PET (d) PP

63. Which of the following is ethylene scavenger
(a) $CaCO_3$ (b) $KMnO_4$
(c) KCl (d) none

64. Mineral water is packaged in
(a) PE (b) PET
(c) AL+PET (d) PP

65. Carrageenan is extracted from
(a) red seaweeds
(b) blue seaweeds
(c) green seaweeds
(d) algae

66. The hardening of the surface during drying is regarded as
(a) dry surface
(b) hard surface
(c) case hardening
(d) none

67. Which one is the sweetest sugar
(a) sucrose (b) glucose
(c) fructose (d) dextrin

68. The thickness of starch paste is due to
(a) amylose (b) amylopectin
(c) dextrin (d) amylase

69. Chemical that reduces the surface tension
(a) preservatives (b) emulsifiers
(c) stabilizers (d) humectant

70. Which is responsible for the umami taste
(a) aspartic acid (b) citric acid
(c) butyrate (d) glutamate

71. The category of food which help in the prevention and cure of diseases
(a) pharmaceuticals (b) nutraceuticals
(c) medical food (d) all

72. Fat bloom in chocolates is caused by
(a) dehydration of surface
(b) high temperature
(c) both
(d) none

73. Respiratory quotient of protein is
(a) 0.5 (b) 0.7
(c) 0.8 (d) 1.0

74. Sugar bloom in chocolates is caused by
(a) high sugar content
(b) moisture on surface
(c) both
(d) none

75. The process of extracting a substance from a solid material that has come into contact with a liquid.
(a) sublimation (b) expression
(c) comminution (d) leaching

76. Packaging tests include
(a) bursting strength
(b) compression test
(c) WVTR
(d) all

77. Spray drying involves removal of moisture from food material by
(a) atomization (b) evaporation
(c) both (d) none

78. Liquid is forced to flow in upward direction in
(a) rising film evaporator
(b) falling film evaporator
(c) multiple film evaporator
(d) all

79. Which one has low density
(a) VDL (b) HDL
(c) LDL (d) all

80. *Process* that involves covering a confection or snack with chocolate or other materials.
(a) coating (b) enrobing
(c) conching (d) none

81. Irradiation is the process of exposing a material to ionizing radiation whose source is photons (γ-rays, x rays), or high energy electrons. Commonly, γ-rays are produced by radioactive isotope
(a) cobalt-60 (b) cesium-137
(c) both (a) & (b) (d) none

82. In heat transfer at a boundary (surface) within a fluid, the ratio of convective to conductive heat transfer across (normal to) the boundary is known as
(a) Nusselt number
(b) Prndtl number
(c) Sherwood number
(d) Reynolds number

83. Dimensionless number approximating the ratio of momentum diffusivity (kinematic viscosity) and thermal diffusivity is known as
(a) Nusselt number
(b) Prndtl number
(c) Sherwood number
(d) Reynolds number

84. The ratio of the total rate of mass transfer to the rate of diffusive mass transport alone.
(a) Nusselt number
(b) Prndtl number
(c) Sherwood number
(d) Reynolds number

85. Incorporation of air into cream is known as
(a) whipping
(b) feathering
(c) aeration
(d) enrobing

86. Dimensionless number that gives a measure of the ratio of inertial forces to viscous forces for given flow conditions.
(a) Nusselt number
(b) Prndtl number
(c) Sherwood number
(d) Reynolds number

87. 100 g of spirulina has ________g of protein
(a) 27 (b) 37
(c) 47 (d) 57

88. Which of the following cow breed give highest milk
(a) Holstein (b) Jersey
(c) Brangus (d) Shorthorn

89. Highest egg laying chicken breed
(a) white leghorn (b) Plymouth rock
(c) Ancona (d) Barnevelder

90. Respiratory quotient of carbohydrate is
(a) 0.5 (b) 0.7
(c) 0.8 (d) 1.0

91. World food day is celebrated on
(a) 15th Sept (b) 16th Oct
(c) 5th March (d) 1st November

92. *BMI* stands for
(a) bone muscles index
(b) bone mass index
(c) body muscles index
(d) body mass index

93. Which is responsible for bitter flavor of beer
(a) yeast
(b) alcohol
(c) hops
(d) flavouring agent

94. Microorganisms that can grow at refrigerated temperature
(a) thermophiles
(b) mesophiles
(c) psychrophiles
(d) zerophiles

95. Heat stability of milk can be assessed by
(a) acidity test (b) alcohol test
(c) MBR test (d) phosphatase test

96. Maillard reaction in milk upon heating is responsible for flavor in milk.
(a) bitter (b) salty
(c) caramelized (d) sweet

97. Good qulity cakes can be prepared using wheat varieties.
(a) soft wheat (b) hard wheat
(c) durum wheat (d) all

98. Which of the following is most economical evaporator
(a) single effect (b) double effect
(c) triple effect (d) 4-effect

99. Acid produced in sauerkraut is
(a) acetic acid (b) citric acid
(c) lactic acid (d) fumaric acid

100. Increase in volume of dough is due to
(a) gluten formation (b) CO_2 gas
(c) baking (d) all

101. Aflatoxins are highly toxic and carcinogenic chemical substances produced by Aspergillus spp. on variety of agricultural commodities and found at highest level in
(a) groundnut *(b) wheat*
(c) barley *(d) pearl millet*

102. Oleoresins are prepared from
(a) basil (b) cardamom
(c) celery (d) all

103. Which of the following is a liquefying enzyme.
(a) Alpha amylase (b) pectinase
(c) protease (d) cellulase

104. The major portion in cholesterol is that of
(a) LDL (b) HDL
(c) CDL (d) ferritin

105. Which one is not a Class I preservatives
(a) salt (b) sugar
(c) acetic acid (d) benzoic acid

106. Yellow color of cow milk is due to
(a) carotene (b) riboflavin
(c) annatto (d) all

107. An efficient method of drying milk is
(a) spray drying
(b) drum drying
(c) freeze drying
(c) cabinet drying

108. The pungency of chilli is due to the presence of
(a) capsaicin (b) capsicin
(c) capcisin (d) capcaisin

109. Staling of bread is due to
(a) retrogradation (b) gelatinization
(c) fermentation (d) none

110. The grain size can be enlarged by
(a) gibberellic acid (b) auxin
(c) fumaric acid (d) all

111. The fruits affected by ethylene gas
(a) climacteric
(b) non-climacteric
(c) both
(d) none

112. __________are a class of carbohydrate derivatives formed when sugars are reacted with excess of phenylhydrazine
(a) osazones
(b) formaldehydes
(c) ozone
(d) organic acids

113. Zero energy cool chambers are related to
(a) storage
(b) 1° processing
(c) 2° processing
(d) 3° processing

114. Site of protein synthesis in cells is
(a) ribosomes
(b) mitochondria
(c) golgi body
(d) all

115. Penetration of EM waves will be high at
(a) high frequency and high loss value
(b) high frequency and low loss value
(c) low frequency and high loss value
(d) low frequency and low loss value

116. Energy required to heat milk in comparison to water is
(a) 90 % (b) 96 %
(c) 100% (d) 107%

ANSWER KEY ASRB NET-2016

Q.	A.	Q.	A.	Q.	A.	Q.	A.	Q.	A.	Q.	A.
1.	(b)	2.	(d)	3.	(b)	4.	(d)	5.	(a)	6.	(d)
7.	(d)	8.	(a)	9.	(b)	10.	(a)	11.	(c)	12.	(c)
13.	(b)	14.	(d)	15.	(b)	16.	(c)	17.	(b)	18.	(a)
19.	(c)	20.	(c)	21.	(c)	22.	(c)	23.	(c)	24.	(a)
25.	(d)	26.	(d)	27.	(a)	28.	(a)	29.	(c)	30.	(c)
31.	(d)	32.	(a)	33.	(b)	34.	(a)	35.	(c)	36.	(c)
37.	(d)	38.	(a)	39.	(a)	40.	(b)	41.	(c)	42.	(a)
43.	(d)	44.	(c)	45.	(c)	46.	(a)	47.	(a)	48.	(d)
49.	(b)	50.	(a)	51.	(c)	52.	(b)	53.	(a)	54.	(c)
55.	(a)	56.	(c)	57.	(b)	58.	(a)	59.	(b)	60.	(d)
61.	(a)	62.	(c)	63.	(b)	64.	(b)	65.	(a)	66.	(c)
67.	(c)	68.	(b)	69.	(b)	70.	(d)	71.	(b)	72.	(b)
73.	(c)	74.	(b)	75.	(d)	76.	(d)	77.	(c)	78.	(a)
79.	(a)	80.	(b)	81.	(c)	82.	(a)	83.	(b)	84.	(c)
85.	(a)	86.	(d)	87.	(d)	88.	(a)	89.	(a)	90.	(d)
91.	(b)	92.	(d)	93.	(c)	94.	(c)	95.	(b)	96.	(c)
97.	(a)	98.	(d)	99.	(c)	100.	(b)	101.	(a)	102.	(d)
103.	(a)	104.	(a)	105.	(d)	106.	(a)	107.	(a)	108.	(a)
109.	(a)	110.	(a)	111.	(a)	112.	(a)	113.	(a)	114.	(a)
115.	(d)	116.	(d)								

ASRB NET QUESTION PAPER 2017 (MEMORY BASED)

1. The indicator organism for pasteurized milk is
(a) Mycobacterium tuberculosis
(b) Coxiella burnetii
(c) Clostridium butulinum
(d) Salmonella typhi

2. Asepsis means
(a) Absence of micro-organisms
(b) Absence of air
(c) Absence of moisture
(d) Absence of light

3. Which among the following is a source of thickening agent
(a) Lecithin
(b) Guar gum
(c) GMS
(d) Sodium nitrite

4. The acid present in carbonated drink is
(a) Citric acid
(b) Carbonic acid
(c) Phosphoric acid
(d) all

5. Carageenan is obtained from
(a) Brown algae
(b) Green algae
(c) Red algae
(d) Blue green algae

6. Which gas is responsible for the ripening of fruits
(a) O_2 (b) CO_2
(c) C_2H_4 (d) N_2

7. Tea is the national drink of
(a) China (b) Sri Lanka
(c) India (d) Nepal

8. Extensibility of dough is attributed to
(a) Globulin (b) Gliadin
(c) Glutenin (d) Albumin

9. Elasticity of dough is due to
(a) Glutenin (b) Globulin
(c) Albumin (d) Gliadin

10. The allergic disease due to consumption of gluten containing food products is
(a) Parkinson's disease
(b) Celiac disease
(c) Phenylketonuria
(d) None

11. A clear drink having 30% TSS is
(a) Nector (b) Juice
(c) Cordial (d) Squash

12. Scientific name of rice is
(a) Triticum aestivum
(b) Zea mays
(c) Oryza sativa
(d) Hordeum vulgare

13. Salt balance theory was given by
(a) Peebles (b) Somer & Hart
(c) Fischer (d) None

14. Flavor which is not easily detectable is
(a) Salty (b) Bitter
(c) Umami (d) Sweet

15. The microwave frequency used in food processing are
(a) 215 & 915 MHz
(b) 915 & 2450 MHz
(c) 215 & 2450 MHz
(d) 415 & 2450 MHz

16. Canning is also known as
(a) Appertization
(b) Pasteurization
(c) Irradiation
(d) Vacreation

17. Isoelectric point of milk protein is
(a) 2.6 (b) 3.6
(c) 4.6 (d) 5.6

18. The platform test done to check the milk quality is
(a) COB test (b) Alcohol test
(c) Acidity (d) all

19. Which of the following is a curing agent
(a) Sodium Nitrite (b) Sodim nitrate
(c) Salt (d) all

20. Indicator of faecal contamination of water is
(a) Salmonella (b) E.Coli
(c) both (d) none

21. Yoghurt contains
(a) Yeast (b) Mold
(c) Bacteria (d) All

22. Black spots in meat is due to
(a) Penicillium (b) Candida
(c) Cladosporium (d) Rhizopus

23. Semolina is obtained from
(a) Triticum compactum
(b) Triticum aestivum
(c) Triticum durum
(d) none

24. The deficiency of niacin causes
(a) Scurvy (b) Beri-beri
(c) Pellagra (d) Anaemia

25. Which of the following is a proteolytic enzyme
(a) Ficin (b) Papain
(c) Actinindin (d) all

26. Enzyme used in cheese manufacturing is
(a) Rennin (b) Zymase
(c) Peroxidase (d) none

27. Fat content in economy ice cream is
(a) 5% (b) 10%
(c) 20% (d) 30%

28. Beta glucan, a soluble fiber is present in
(a) Oats (b) rice
(c) corn (d) wheat

29. Yeast in bread making is responsible for
(a) Flavor of bread
(b) Loaf volume of bread
(c) both
(d) none

30. The triple point of water is
(a) 0.01°C and 4.58mm Hg
(b) 1°C and 4.58mm Hg
(c) 10°C and 4.58mm Hg
(d) 0.1°C and 4.58mm Hg

31. LTLT process for pasteurization of milk is at 63°C for 30 mins, HTST process
(a) 72°C/15 sec (b) 72°C/15 min
(c) 75°C/15 sec (d) 75°C/15 min

32. Aspartame, an artificial sweetener is made of
(a) Aspartic acid & valine
(b) Aspartic acid & phenylalanine
(c) Aspartic acid & histidine
(d) Aspartic acid & alanine

33. Waxy wheat has a low content of
(a) Amylose
(b) amylopectin
(c) both
(d) none

34. Ropiness in bread is due to
(a) Bacillus mesentaricus
(b) Rhizopus stolonifer
(c) Candida torulopsis
(d) Lactobacillus

35. Antimicrobial component of milk is
(a) lysozyme (b) lactose
(c) casein (d) phospholipids

36. One cup of coffee has a caffeine content of
(a) 95 mg (b) 55 mg
(c) 45 mg (d) 75 mg

37. Caramelized flavor in milk is due to
(a) Enzymatic browning
(b) Non-enzymatic browning
(c) Both
(d) none

38. The scum formation in pickles is due to the growth of
(a) Bacteria (b) Virus
(c) Molds (d) all

39. Which of the following maturing agent is no longer used in wheat flour maturation
(a) Potassium bromate
(b) Ascorbic acid
(c) Azodicarbonamide
(d) all

40. Flavor reversion in food is due to
(a) Linolenic acid (b) Linoleic acid
(c) Butyric acid (d) Stearic acid

41. The flavor of wine is not be affected by
(a) Ageing of wine
(b) Maturation of wine
(c) Type of container
(d) none

42. Colostrum has lower content of _____ than normal bovine milk
(a) Lactose (b) Protein
(c) Minerals (d) Vitamins

43. According to ICMR, 1 mcg of retinol is equivalent to
(a) 1 mcg of beta carotene
(b) 4 mcg of beta carotene
(c) 6 mcg of beta carotene
(d) 8 mcg of beta carotene

44. Symba process utilizes __________ wastes in SCP production
(a) Protein (b) Starch
(c) Fat (d) Mineral

45. Stoke's law is used to find out
(a) Terminal velocity
(b) Drag coefficient
(c) Surface tension
(d) Specific gravity

46. Carcinogenic compound formed in cured meat is
(a) Nitroso compound
(b) Ketonic compound
(c) carbonyl compound
(d) none

47. Which container should not be used for wine maturation
(a) New oak (b) old oak
(c) white oak (d) all

48. Which of the following is not a universal taste
(a) Salt (b) Sweet
(c) Sour (d) None

49. The deterioration of fat due to oxidation and hydrolysis is called
(a) Denaturation
(b) Decomposition
(c) Rancidity
(d) Saponification

50. Which of the following has a higher protein content among the following
(a) Sorghum
(b) Pearl Millet
(c) Maize
(d) Oats

51. Milk gets coagulated upon
(a) Increase in acidity
(b) Decrease in pH
(c) High heat treatment
(d) all

52. Which of the following is regarded as saccharifying enzyme
(a) Alpha amylase (b) beta amylase
(c) both (d) none

53. Which one is similar to functional foods
(a) Nutraceutical
(b) therapeutic foods
(c) both (a) & (b)
(d) GM foods

54. Which of the following is an example of natural emulsifier
(a) Honey
(b) Lecithin
(c) carrageenan
(d) trisodium citrate

55. Which of the following is a Lab fermented product
(a) Yoghurt (b) Kefir
(c) Sauerkraut (d) all

56. Maximum microwaves absorbed by
(a) Water
(b) salt solution
(c) sugar solution
(d) salt & sugar solution

57. Pickles commonly contains
(a) Lactobacillus (b) Leuconostoc
(c) Pediococcus (d) all

58. Which of the following is mostly non-pathogenic
(a) Bacteria (b) yeast
(c) mold (d) virus

59. Concept used for killing of most resistant spores of Clostridium botulinum in vegetables
(a) 5D (b) 7D
(c) 12D (d) 0.5D

60. Compound having flavour similar to banana
(a) Isoamyl acetate
(b) ethyl butyrate
(c) diacetyl
(d) methyl anthranilate

61. Essential amino acid present in low amount in legumes
(a) Lysine (b) methionine
(c) leucine (d) valine

62. Which of the following starch have lowest gelatinization temperature
(a) potato (b) rice
(c) maize (d) sorghum

63. Micro-organism responsible for production of vinegar from alcohol
(a) Aspergillus (b) Acetobacter
(c) Bacillus (d) Penicillium

64. Codex Alimentarius means
(a) Food code (b) food law
(c) food standard (d) none

65. Which of the following has the highest calcium content
(a) Milk (b) sorghum
(c) pearl millet (d) rice

66. Parboiling is done in case of
(a) Wheat (b) rice
(c) corn (d) sorghum

67. Which of the following is intrinsic factor for growth of micro-organism
(a) Water activity (b) pH
(c) redox potential (d) all

68. The toughening of meat after slaughter is due to
(a) Drying (b) rigor mortis
(c) cold shortening (d) none

69. Indicator enzyme for pasteurization of milk is
(a) lipase
(b) peroxidase
(c) alkaline phosphatase
(d) lactase

70. Which of the following is not a fat soluble vitamin
(a) A (b) B
(c) D (d) K

71. Annatto is obtained from
(a) Seeds (b) leaf
(c) stem (d) roots

72. Sal marina is obtained from
(a) animal skin (b) sea water
(c) salmon (d) plant

73. P in HACCP stands for
(a) Prevention (b) pollutants
(c) points (d) peaks

74. SO_2 is used in the milling of
(a) Rice (b) wheat
(c) corn (d) legumes

75. The process in which both dehusking and polishing are involved is
(a) Husking (b) milling
(c) parboiling (d) none

76. After drying, the final moisture content of parboiled rice is
(a) 20% (b) 14%
(c) 18% (d) 25%

77. Which of the following fatty acid is most susceptible to rancidity
(a) lauric acid (b) oleic acid
(c) linoleic acid (d) stearic acid

78. FSSAI standards for total dissolved solids in packaged drinking water
(a) NLT 500mg/litre
(b) NMT 500mg/litre
(c) NLT 100mg/litre
(d) NMT 100mg/litre

79. Alternative for citric acid and tartaric acid in food flavor enhancer in beverages
(a) malic acid
(b) MSG
(c) potassium glutamate
(d) disodium guanylate

80. Rubber rollers in paddy husking works on the principle of
(a) friction (b) pressure
(c) impact (d) crushing

81. Emery coated rollers are used for
(a) legume milling (b) rice milling
(c) oil extraction (d) none

82. Which is not measurable
(a) pressure (b) temperature
(c) momentum (d) none

83. What happened to fruit, if it is stored at lower temperature than its optimum temperature
(a) chilling injury
(b) freezing injury
(c) frost injury
(d) none

84. Non-climacteric fruits are harvested at what % of ripening
(a) 75% (b) 85%
(c) 95% (d) 100%

85. Phytochemical absent in cereals (except oats)
(a) phytosterol (b) carotenoids
(c) saponins (d) lignans

86. Protein not present in Cereals
(a) gliadin (b) glutenin
(c) oryzenin (d) casein

87. Best method for extraction of essential oils
(a) super critical fluid extraction
(b) steam distillation
(c) solvent extraction
(d) expression

88. Instrument used for measuring apparent density of cereal grains
(a) single kernel characterization system
(b) extensiograph
(c) farinograph
(d) butyrometer

89. Latest milk packaging technique
(a) PET bottles (b) glass bottles
(c) pouch (d) cartons

90. Nisin is related to
(a) broad spectrum antibiotic
(b) narrow spectrum antibiotic
(c) broad spectrum bacteriocin
(d) narrow spectrum bacteriocin

91. Which of the following is used as indicator of starch conversion
(a) alpha-amylase
(b) beta-amylase
(c) diastase
(d) iodine

ANSWER KEY ASRB NET-2017

Q.	A.	Q.	A.	Q.	A.	Q.	A.	Q.	A.	Q.	A.
1.	(b)	2.	(a)	3.	(b)	4.	(d)	5.	(c)	6.	(c)
7.	(c)	8.	(b)	9.	(a)	10.	(b)	11.	(c)	12.	(c)
13.	(b)	14.	(c)	15.	(b)	16.	(a)	17.	(c)	18.	(d)
19.	(d)	20.	(b)	21.	(c)	22.	(c)	23.	(c)	24.	(c)
25.	(d)	26.	(a)	27.	(b)	28.	(a)	29.	(c)	30.	(a)
31.	(a)	32.	(b)	33.	(a)	34.	(a)	35.	(a)	36.	(a)
37.	(b)	38.	(c)	39.	(a)	40.	(a)	41.	(d)	42.	(a)
43.	(d)	44.	(b)	45.	(a)	46.	(a)	47.	(a)	48.	(d)
49.	(c)	50.	(d)	51.	(d)	52.	(b)	53.	(c)	54.	(b)
55.	(d)	56.	(a)	57.	(d)	58.	(b)	59.	(c)	60.	(a)
61.	(b)	62.	(a)	63.	(b)	64.	(a)	65.	(a)	66.	(b)
67.	(d)	68.	(b)	69.	(c)	70.	(b)	71.	(a)	72.	(b)
73.	(c)	74.	(c)	75.	(b)	76.	(b)	77.	(c)	78.	(b)
79.	(a)	80.	(a)	81.	(a)	82.	(d)	83.	(a)	84.	(d)
85.	(c)	86.	(d)	87.	(a)	88.	(a)	89.	(c)	90.	(c)
91.	(d)										